A Specialist Periodical Report

Spectroscopic Properties of Inorganic and Organometallic Compounds

Volume 10

A Review of the Literature Published during 1976

Senior Reporter

E. A. V. Ebsworth FRSE, *Department of Chemistry, University of Edinburgh*

Reporters

D. M. Adams, *University of Leicester*
A. P. Cox, *University of Bristol*
G. Davidson, *University of Nottingham*
J. D. Donaldson, *Chelsea College, University of London*
P. Gans, *University of Leeds*
P. N. Gates, *Royal Holloway College, University of London*
M. Goldstein, *Sheffield Polytechnic*
R. J. Lynch, *University of Durham*
B. E. Mann, *University of Sheffield*
M. J. Tricker, *Heriot-Watt University, Edinburgh*

The Chemical Society
Burlington House, London, W1V 0BN

ISBN: 0 85186 093 1
ISSN: 0584-8555

Library of Congress Catalog Card No. 74-6662

Organic formulae composed by Wright's Symbolset method

PRINTED IN GREAT BRITAIN BY JOHN WRIGHT AND SONS LTD., AT THE STONEBRIDGE PRESS, BRISTOL BS4 5NU

Foreword

It is a privilege for me to take over from Professor N. N. Greenwood as Senior Reporter for *Spectroscopic Properties of Inorganic and Organometallic Compounds.* Under Professor Greenwood's guidance this series of volumes has become an essential source-book for anyone interested in the spectroscopy of inorganic species; the chemical community owe him and his colleagues an immense debt for the way in which they have so perfectly done what the Chemical Society had in mind when it launched the series of *Specialist Periodical Reports.*

Although the Senior Reporter for Volume 10 is new, the general format of the volume is the same as that of Volume 9, and indeed I am delighted that many of the same reporters have been willing to continue to contribute. The only significant change derives from the difficulties caused on the one hand by the continuing increase in relevant publications and on the other by the equally continuing increase in the cost of producing a book like this. As a result of these pressures, the length of Volume 10 is less than that of Volume 9, and the proportion of relevant published material that is reported is substantially less. Reporters have had to be more selective than in the past, and this of course makes their task more difficult, for the coverage of published material is no less comprehensive. I am most grateful to all the reporters for this Volume for the way in which they have accepted these additional constraints; this is but one of the many ways in which they have made my work as Senior Reporter both easy and pleasant.

E. A. V. EBSWORTH

August 1977

Contents

Conversion Factors

1 kJ mol^{-1}

2.3901×10^{-1} kcal mol^{-1}
1.0364×10^{-2} eV atom^{-1}
8.3593×10 cm^{-1}
2.5061×10^{6} MHz

1 kcal mol^{-1}

4.1840 kJ mol^{-1}
4.3364×10^{-2} eV atom^{-1}
3.4976×10^{2} cm^{-1}
1.0486×10^{7} MHz

1 cm^{-1}

1.1963×10^{-2} kJ mol^{-1}
2.8592×10^{-3} kcal mol^{-1}
1.2399×10^{-4} eV atom^{-1}
2.9979×10^{4} MHz

1 MHz

3.9903×10^{-7} kJ mol^{-1}
9.5370×10^{-8} kcal mol^{-1}
4.1357×10^{-9} eV atom^{-1}
3.3356×10^{-5} cm^{-1}

1 eV atom^{-1}

9.6485×10 kJ mol^{-1}
2.3060×10 kcal mol^{-1}
8.0655×10^{3} cm^{-1}
2.4180×10^{8} MHz

Mössbauer Spectra: $E_\gamma(^{57}Fe) = 14.413$ keV

1 mm s^{-1}

4.639×10^{-6} kJ mol^{-1}
1.109×10^{-6} kcal mol^{-1}
4.808×10^{-8} eV atom^{-1}
3.878×10^{-4} cm^{-1}
1.162×10 MHz

For other Mössbauer nuclides, multiply the above conversion factors by E_γ(keV)/14.413

1
Nuclear Magnetic Resonance Spectroscopy

BY B. E. MANN

1 Introduction

The growth in papers containing n.m.r. data on inorganic and organometallic compounds has continued with the result that only papers containing information on nuclei other than ^{1}H, or papers wherein ^{1}H n.m.r. spectroscopy makes a significant contribution, are included. A more selective approach has also been applied within the area of solid-state n.m.r. spectroscopy, and papers which are considered to be principally of interest to physicists have been omitted. As usual, no attempt is made to cover n.m.r. spectroscopy in depth, and any reader who requires such an approach is referred to the excellent *Specialist Periodical Report* on 'Nuclear Magnetic Resonance'.[1]

Volume 8 of 'Advances in Magnetic Resonance' has appeared and contains chapters on 'The Theory of Chemically Induced Dynamic Spin Polarization', by H. H. Freed and J. B. Pedersen, and 'N.M.R. in Flowing Systems', by D. W. Jones and T. F. Child.[2] In addition, the first supplement to this series on 'High Resolution N.M.R. in Solids, Selective Averaging', by U. Haeberlen has appeared.[3] Volume 6B of 'Annual Reports on N.M.R. Spectroscopy' contains chapters on 'N.M.R. and Conformations of Amino Acids, Peptides, and Proteins', by W. A. Thomas, and 'Fluorine-19 N.M.R. Spectroscopy', by L. Cavalli.[4] The series 'N.M.R., Basic Principles and Progress' has proved to be prolific with books on 'Van der Waals Forces and Shielding Effects', by F. H. A. Rummens,[5] 'High Resolution N.M.R. Spectroscopy in Solids', by M. Mehring,[6] 'Chlorine, Bromine, and Iodine N.M.R. Physico-chemical and Biological Applications', by B. Lindman and S. Forsén,[7] and 'Introductory Essays', edited by M. M. Pintar, containing 11 essays on various n.m.r. topics, none of which is of direct relevance to this chapter.[8] 'Progress in Nuclear Magnetic Resonance

1 'Nuclear Magnetic Resonance', ed. R. K. Harris (Specialist Periodical Reports), The Chemical Society, London, 1976, Vol. 5, 1977, Vol. 6.
2 'Advances in Magnetic Resonance', ed. J. S. Waugh, Academic Press, New York, 1976, Vol. 8.
3 'Advances in Magnetic Resonance', ed. J. S. Waugh, Supplement 1, 'High Resolution N.M.R. in Solids, Selective Averaging', by U. Haeberlen, Academic Press, New York, 1976.
4 'Annual Reports on N.M.R. Spectroscopy', ed. E. F. Mooney, Vol. 6B, Academic Press, New York and London, 1976.
5 'N.M.R., Basic Principles and Progress', ed. P. Diehl, E. Fluck, and R. Kosfeld, Vol. 10; 'Van der Waal Forces and Shielding Effects', by F. H. A. Rummens, Springer-Verlag, Berlin, 1975.
6 'N.M.R., Basic Principles and Progress', ed. P. Diehl, E. Fluck, and R. Kosfeld, Vol. 11; 'High Resolution N.M.R. Spectroscopy in Solids', by M. Mehring, Springer-Verlag, Berlin, 1976.
7 'N.M.R., Basic Principles and Progress', ed. P. Diehl, E. Fluck, and R. Kosfeld, Vol. 12; 'Chlorine, Bromine, and Iodine N.M.R. Physico-chemical and Biological Applications', by B. Lindman and S. Forsén, Springer-Verlag, Berlin, 1976.
8 'N.M.R., Basic Principles and Progress', ed. P. Diehl, E. Fluck, and R. Kosfeld, Vol. 13; 'Introductory Essays', ed. M. M. Pintar, Springer-Verlag, Berlin, 1976.

Spectroscopy' has produced a review on 'Spin–Spin Coupling and the Conformational States of Peptide Systems', by V. F. Bystrov.[9] The second volume of 'Topics in Carbon-13 N.M.R. Spectroscopy' has now appeared and contains a number of chapters on various aspects of ^{13}C n.m.r. spectroscopy, and in particular 'Carbon-13 N.M.R. Studies of Organometallic and Transition Metal Complex Compounds', by O. A. Gansow and W. D. Vernon.[10]

A number of books devoted to n.m.r. spectroscopy have been published, including 'Proton Nuclear Magnetic Resonance', by P. Joseph-Nathan;[11] 'The Dynamics of Spectroscopic Transition: Illustrated by Magnetic Resonance and Laser Effects', by J. D. Macomber;[12] 'Spectroscopic Methods for Determining the Structure of Organic Compounds. Section "Nuclear Magnetic Resonance Spectroscopy"', by J. Prochorow;[13] 'Magnetic Resonance', edited by C. A. McDowell,[14] which contains chapters on 'High Resolution N.M.R. Studies on Solids', by E. R. Andrew, 'The Studies of Ionic Processes at Membranes by N.M.R. Methods', by L. W. Reeves, and 'Tunnelling Processes as Studied by Magnetic Resonance Techniques', by R. Srinivasan; 'Fourier Transform N.M.R. Spectroscopy', by D. Shaw;[15] 'Use of Nuclear Magnetic Relaxation in the Analysis of Inorganic Compounds', by A. A. Popel;[16] 'Interpretation of Carbon-13 Nuclear Magnetic Resonance Spectra', by F. Wehrli and T. Wirthlin;[17] 'Spectroscopic Methods in Pharmacy, Vol. 2: Nuclear Magnetic Resonance Spectroscopy and Mass Spectrometry', by G. Ruecker;[18] 'Magnetic Resonance of Biomolecules', by P. F. Knowles, D. Marsh, and H. W. E. Rattle;[19] 'Nuclear Magnetic Resonance in Biological Research: Peptides and Proteins', by K. Wuethrich;[20] 'Nuclear Magnetic Resonance in Biochemistry', by T. L. James;[21] a collection of lectures in 'Magnetic Resonance in Chemistry and Biology';[22] and 'Corso Teorico-Pratico di Resonanza Magnetica Nucleare', edited by A. Frigerio.[23]

[9] V. F. Bystrov, *Progr. N.M.R. Spectroscopy*, 1976, **10**, 41.

[10] 'Topics in Carbon 13 N.M.R. Spectroscopy', ed. G. C. Levy, Wiley, London, 1976, Vol. 2.

[11] P. Joseph-Nathan, *Ser. Quim.-Programa Reg. Desarrollo Cient. Tecnol.*, 1973, Vol. 9 (*Chem. Abs.*, 1976, **84**, 81 782).

[12] J. D. Macomber, 'Monographs in Chemical Physics: The Dynamics of Spectroscopic Transition: Illustrated by Magnetic Resonance and Laser Effects', Wiley–Interscience, New York, 1975.

[13] J. Prochorow, 'Spectroscopic Methods for Determining the Structure of Organic Compounds. Section "Nuclear Magnetic Resonance Spectroscopy"', Panst. Wyd. Nauk., Warsaw, 1974 (*Chem. Abs.*, 1976, **84**, 135 025).

[14] 'International Review of Science', Physical Chemistry Series 2, Vol. 4, ed. C. A. McDowell, Butterworth, London, 1975.

[15] D. Shaw, 'Fourier Transform N.M.R. Spectroscopy', Elsevier, Amsterdam, 1976.

[16] A. A. Popel, 'Use of Nuclear Magnetic Relaxation in the Analysis of Inorganic Compounds', Izd. Kazan, Univ., Kazan, U.S.S.R., 1975 (*Chem. Abs.*, 1976, **84**, 83 759).

[17] F. Wehrli and T. Wirthlin, 'Interpretation of Carbon-13 Nuclear Magnetic Resonance Spectra', Heyden, London, 1976.

[18] G. Ruecker, 'Spectroscopic Methods in Pharmacy, Vol. 2: Nuclear Magnetic Resonance Spectroscopy and Mass Spectrometry', Wiss. Verlagages, Stuttgart, 1976.

[19] P. F. Knowles, D. Marsh, and H. W. E. Rattle, 'Magnetic Resonance of Biomolecules', Wiley, Chichester, England, 1976.

[20] K. Wuethrich, 'Nuclear Magnetic Resonance in Biological Research: Peptides and Proteins', North-Holland, London, 1976.

[21] T. L. James, 'Nuclear Magnetic Resonance in Biochemistry', Academic Press, New York, 1975.

[22] 'Magnetic Resonance in Chemistry and Biology', ed. J. N. Herak and K. J. Adamić, Marcel Dekker, New York, 1973.

[23] 'Corso Teorico-pratico di Resonanza Magnetica Nucleare', ed. A. Frigerio, Tamburini Editore, Milan, 1973.

Several other books have appeared which contain sections or chapters devoted to n.m.r. spectroscopy. These books have articles on 'Physical Methods and Techniques; II. Nuclear Magnetic Resonance', by R. B. Jones and L. Phillips;[24] 'Nuclear Magnetic Resonance Spectroscopy', by I. O. Sutherland;[25] 'N.M.R. of Nuclei other than Proton and Fluorine-19', by G. A. Webb;[26] 'Correlations in Nuclear Magnetic Shielding, Part 1', by J. Mason;[27] '^{29}Si Nuclear Magnetic Resonance', by J. Schraml and J. M. Bellama;[28] 'The Nuclear Overhauser Effect', by J. K. Saunders and J. W. Easton;[28] 'Molecular Structure by N.M.R. in Liquid Crystals', by L. Lunazzi;[28] and 'N.m.r. Parameters of the Proton Directly Bonded to Phosphorus'.[29]

Once again numerous reviews have appeared. Some of the reviews are of n.m.r. spectroscopy in general.[30–35] Other reviews cover 'High-resolution N.M.R. Method',[36] 'N.M.R. as an Analytical Technique',[37] 'Nuclear Magnetic Resonance as the Stereochemists' Tool',[38] 'Determination of Molecular Conformation in Solution',[39] 'Conformational Analysis using High-resolution N.M.R. Spectroscopy',[40] 'Nuclear Magnetic Resonance Techniques for the Determination of the Structure of Molecules',[41] 'Recent N.M.R. Applications in Organic Chemistry',[42] 'Nuclear Magnetic Resonance at High Pressures',[43] 'Basics of N.M.R. Spectroscopy',[44] 'Practical Interpretations of N.M.R. Spectra',[45] 'Pulsed N.M.R. Spectrometry',[46] 'Proton N.M.R. Spectrometry',[47] 'Analysis of N.M.R. Spectra',[48] 'Allylic Interproton Spin–Spin Coupling',[49] 'Influence of

[24] R. B. Jones and L. Phillips, *Ann. Reports* (*B*) 1975, **71**, 17.
[25] I. O. Sutherland, in ref. 14, 'Organic Chemistry', Vol. 1, p. 55.
[26] G. A. Webb, *Magn. Resonance Rev.*, 1974, **3**, 327.
[27] J. Mason, *Adv. Inorg. Chem. Radiochem.*, 1976, **18**, 197 (*Chem. Abs.*, 1976, **85**, 84 794).
[28] 'Determination of Organic Structures by Physical Methods', ed. F. C. Nachod, J. J. Zuckerman, and E. W. Randall, Academic Press, New York, 1976.
[29] 'Topics in Phosphorus Chemistry', ed. E. J. Griffith and M. Grayson, Wiley–Interscience, New York, 1976, Vol. 8.
[30] J. R. Wasson and D. R. Lorenz, *Analyt. Chem.*, 1976, **48**, 246R.
[31] J. N. Shoolery, *Ind. Res.*, 1975, **17**, 67 (*Chem. Abs.*, 1976, **85**, 28 293).
[32] P. T. Narasimhan, *J. Sci. Ind. Res.*, 1976, **35**, 11 (*Chem. Abs.*, 1976, **85**, 53 815).
[33] R. T. Parfitt, *Pract. Pharm. Chem.*, 3rd edn., 1976, **2**, 361 (*Chem. Abs.*, 1976, **85**, 10 462).
[34] A. V. Kessenykh, *Khim. Zhizn*, 1973, **4**, 18 (*Chem. Abs.*, 1976, **84**, 36 755).
[35] I. Morishima and T. Inubushi, *Kagaku* (*Kyoto*), 1975, **30**, 442 (*Chem. Abs.*, 1976, **84**, 14 233).
[36] I. Ando, *Sen'i Gakkaishi*, 1975, **31**, P400 (*Chem. Abs.*, 1976, **84**, 81 776).
[37] R. Kaplan, *Ind. Res.*, 1976, **18**, 73 (*Chem. Abs.*, 1976, **84**, 184 774).
[38] J. C. Jochims, *Konstanzer Universitaetsreden*, 1973, 56, 'Nuclear Magnetic Resonance as the Stereochemists' Tool' (*Chem. Abs.*, 1976, **85**, 11 662).
[39] M. R. Willcott and R. E. Davis, *Science*, 1975, **190**, 850.
[40] A. Zschunke, *Wiss. Z. Martin-Luther-Univ., Halle-Wittenberg, Math.-Naturwiss. Reihe*, 1975, **24**, 29 (*Chem. Abs.*, 1976, **84**, 104 553).
[41] A. V. Xavier, *Phys. Chem. Sci. Res. Rep.* 1 *Nat. Seawater, Rep. Dahlem Workshop*, 1975, 69 (*Chem. Abs.*, 1976, **84**, 97 177).
[42] T. Takeuchi, *Kagaku To Yakugaku No Kyoshitsu*, 1973, **41**, 19 (*Chem. Abs.*, 1976, **84**, 104 552).
[43] J. Jonas, *Ann. Rev. Phys. Chem.*, 1975, **26**, 167 (*Chem. Abs.*, 1976, **84**, 23 803).
[44] J. Murto, *Taeydennyskoulutuskurssi – Suom. Kem. Seura*, 1973, **18**, 1 (*Chem. Abs.*, 1976, **84**, 29 915).
[45] J. Paasivirta, *Taeydennyskoulutuskurssi – Suom. Kem. Seura*, 1973, **18**, 31 (*Chem. Abs.*, 1976, **84**, 29 916).
[46] E. Rahkamaa, *Taeydennyskoulutuskurssi – Suom. Kem. Seura*, 1973, **18**, 48 (*Chem. Abs.*, 1976, **84**, 29 917).
[47] E. Rahkamaa, *Taeydennyskoulutuskurssi – Suom. Kem. Seura*, 1973, **18**, 85 (*Chem. Abs.*, 1976, **84**, 29 919).
[48] J. Paasivirta, *Taeydennyskoulutuskurssi – Suom. Kem. Seura*, 1973, **18**, 115 (*Chem. Abs.*, 1976, **84**, 29 920).
[49] M. Barfield, R. J. Spear, and S. Sternhell, *Chem. Rev.*, 1976, **76**, 593.

Solvents on Spectroscopy',[50] 'Solute–Solute and Solute–Solvent Interactions: N.M.R. Studies',[51] 'Forbidden Transition in Magnetic Resonance',[52] 'Spin Relaxation in Fluids',[53] 'Magnetic Relaxation',[54] 'Reorientation and Dipole Relaxation',[55] 'N.M.R. and the Periodic Table',[56] 'The Application of Carbon-13 N.M.R. to Oil Chemistry. Part 2. Review of Recent Topics',[57] 'Carbon-13 N.M.R. Spectroscopy',[58, 59] 'Carbon-13 Fourier Transform N.M.R. An Important New Analysis Tool',[60] 'Applications of ^{13}C and ^{1}H N.M.R. and I.R. Spectroscopies in Structural and Quantitative Analysis',[61] '^{13}C N.M.R. Use in Organic Chemistry, Biochemistry, and Technology',[62] 'The Application of Carbon-13 N.M.R. Spectroscopic Techniques to Biological Problems',[63] '^{13}C N.M.R. Spectroscopy in Organic Chemistry',[64] 'The Use of Alkali Metal N.M.R. in the Study of Solvation and Complexation of Alkali Metal Ions',[65] 'Dynamic N.M.R. of Carbon',[66] 'Unusually High Barriers to Rotation Involving the Tetrahedral Carbon Atom',[67] 'Nuclear Magnetic Resonance Studies of Coordination Compounds',[68] 'Nuclear Magnetic Resonance Spectroscopy of Porphyrins and Metalloporphyrins',[69] 'Investigations of Hydrides of the Group IVa and Va Metals by Nuclear Magnetic Resonance',[70] 'N.M.R. Spectroscopy of Iron–Sulphur Proteins',[71] 'Apicophilicity and Ring Strain in Five-coordinate Phosphoranes',[72] 'Organic and Inorganic Reaction Mechanisms Using Oxygen-17',[73] 'Chemically Induced Dynamic Nuclear Polarization and Chemically Induced Dynamic Electron Polarization',[74] 'N.M.R. in Drug Metabolism',[75] 'N.M.R. in the Field of Biochemistry',[76] and 'N.M.R. Spectroscopy in Biological Sciences',[77] 'N.M.R. Spectroscopy in Biological Research',[78] 'Use of N.M.R.

[50] M. Jauquet and P. Laszlo, *Tech. Chem. (N.Y.) 8. Solutions and Solubilities*, Pt. 1, 1975, 195 (*Chem. Abs.*, 1976, **84**, 186 720).
[51] M. Rabinovitz, *Jerusalem Symp. Quantum Chem. Biochem. 8. Environ. Eff. Mol. Struct. Prop.*, 1976, 229 (*Chem. Abs.*, 1976, **85**, 20 211).
[52] B. F. Alekseev, *Izvest. V. U. Z., Radiofiz.*, 1975, **18**, 1727 (*Chem. Abs.*, 1976, **84**, 128 085).
[53] H. Gabriel, *Hyperfine Interact.*, 1976, **2**, 91 (*Chem. Abs.*, 1976, **85**, 151 034).
[54] K. Kuwata and G. Soda, *Kanwa Gensho No Kagaku*, 1973, 74 (*Chem. Abs.*, 1976, **84**, 186 697).
[55] N. Nakamura, *Bunseki Kiki*, 1976, **14**, 73 (*Chem. Abs.*, 1976, **84**, 171 522).
[56] R. K. Harris, *Chem. Soc. Rev.*, 1976, **5**, 1.
[57] O. Suzuki, *Yukagaku*, 1976, **25**, 2 (*Chem. Abs.*, 1976, **84**, 137 567).
[58] E. Breitmaier and G. Bauer, *Pharm. Unserer Z.*, 1976, **5**, 97 (*Chem. Abs.*, 1976, **85**, 176 394).
[59] Y.-W. Wang, *Hua Hsueh Tung Pao*, 1976, **3**, 180 (*Chem. Abs.*, 1976, **85**, 151 021).
[60] W. B. Moniz, C. F. Poranski, jun., and S. A. Sojka, *Rep. NRL Prog. Aug.*, 1975, 1 (*Chem. Abs.*, 1976, **84**, 122 361).
[61] I. K. O'Neill, *Proc. Analyt. Div. Chem. Soc.*, 1975, **12**, 243.
[62] Z. Majerski, B. Goricnik, and D. Skare, *Kem. Ind.*, 1975, **24**, 443 (*Chem. Abs.*, 1976, **84**, 171 520).
[63] J. Feeney, *New Tech. Biophys. Cell Biol.*, 1975, **2**, 287 (*Chem. Abs.*, 1976, **84**, 176 000).
[64] M. J. Gasic, *Glas. Hem. Drus., Beograd*, 1974, **39**, 693 (*Chem. Abs.*, 1976, **84**, 157 300).
[65] A. I. Popov, *Solute–Solvent Interact.*, 1976, **2**, 271 (*Chem. Abs.*, 1976, **85**, 149 666).
[66] G. J. Martin, *Inf. Chim.*, 1975, **150**, 127, 137 (*Chem. Abs.*, 1976, **84**, 142 657).
[67] M. Oki, *Angew. Chem. Internat. Edn.*, 1976, **15**, 87.
[68] L. M. Venanzi, *Chimia*, 1976, **30**, 227 (*Chem. Abs.*, 1976, **84**, 171 531).
[69] H. Scheer and J. J. Katz, *Porphyrins Metalloporphyrins*, 1975, 399 (*Chem. Abs.*, 1976, **85**, 38 751).
[70] K. Nakamura, *Trans. Nat. Res. Inst. Met.*, 1975, **17**, 305 (*Chem. Abs.*, 1976, **84**, 142 653).
[71] W. D. Phillips and M. Poe, *Iron–Sulfur Proteins*, 1973, **2**, 255 (*Chem. Abs.*, 1976, **84**, 175 310).
[72] S. Trippett, *Phosphorus and Sulphur*, 1976, **1**, 89.
[73] B. L. Silver, *C. R. Colloq. Internat. Isot. Oxygene*, 1975, 88 (*Chem. Abs.*, 1976, **84**, 22 619).
[74] K. Y. Chu, *Hwahak Kwa Kongop Ui Chinbo*, 1975, **15**, 300 (*Chem. Abs.*, 1976, **84**, 171 521).
[75] D. E. Case, *Chem. and Ind.*, 1976, **9**, 391 (*Chem. Abs.*, 1976, **85**, 71 835).
[76] M. Tsuda, *Kagaku To Yakugaku No Kyoshitsu*, 1975, **47**, 23 (*Chem. Abs.*, 1976, **84**, 70 403).
[77] P. J. Sadler, *Progr. Med. Chem.*, 1975, **12**, 159 (*Chem. Abs.*, 1976, **84**, 86 220).
[78] E. Rahkamaa, P. Holmberg, K. Kiviniitty, A. Koivula, and K. Raisanen, *Finnish Arkhimedes*, 1976, **28**, 21 (*Chem. Abs.*, 1976, **85**, 74 340).

for the Study of the Structure and Function of Enzymes',[79] and 'The Calculation of N.M.R. Lineshapes of Glasses'.[80]

I.U.P.A.C. have published revised recommendations for the 'Presentation of N.M.R. data for publication in chemical journals – B. Conventions relating to spectra from nuclei other than protons'.[81] All workers in this area are strongly urged to read these recommendations. Difficulties are commonly encountered as a result of the use of the wrong chemical shift sign convention and not defining the sign convention. These difficulties are particularly common for ^{31}P n.m.r. spectroscopy where a ^{31}P chemical shift of say δ 10 can mean a shift to either high or low frequency (high frequency would be assumed) of an unknown reference, normally assumed to be 85% H_3PO_4. Similar difficulties can be encountered with other nuclei, especially ^{1}H, ^{13}C, and ^{19}F. Throughout this chapter, authors permitting, the sign convention that high frequency is positive is adopted.

In addition to these books and reviews, papers have appeared which are too broadly based to fit into any of the later sections of this chapter and are included here. The valence bond treatment of the Fermi contact term for nucleus–nucleus coupling constants gave simplified expressions in which a constant term for each nucleus was introduced. The order of the bond separating the coupled nuclei was estimated. This approach was applied to long-range hydrogen–metal and carbon–metal coupling constants in organometallic compounds.[82] ^{17}O, ^{35}Cl, ^{51}V, ^{53}Cr, ^{55}Mn, and ^{95}Mo n.m.r. spectra have been determined for aqueous solutions of $[ClO_4]^-$, $[VO_4]^{3-}$, $[CrO_4]^{2-}$, $[MnO_4]^-$, and $[MoO_4]^{2-}$, partly enriched in ^{17}O. The following coupling constants were determined: $^{1}J(^{17}O, ^{35}Cl) = 81.4 \pm 4.0$ Hz; $^{1}J(^{17}O, ^{51}V) = 61.2 \pm 2.5$ Hz; $^{1}J(^{17}O, ^{53}Cr) = 10 \pm 2$ Hz; $^{1}J(^{17}O, ^{55}Mn) = 28.9 \pm 2.8$ Hz; and $^{1}J(^{17}O, ^{95}Mo) = 40.5 \pm 0.8$ Hz. A linear dependence of the reduced coupling constant on the atomic number of the central nucleus in the isoelectric series $[VO_4]^{3-}$, $[CrO_4]^{2-}$, and $[MnO_4]^-$ was found.[83] The mutual influence of ligands in main-group element co-ordination compounds AL_n has been analysed on the basis of the orbitally deficient scheme of the structure of these compounds. It was shown that a decrease of the A–^{19}F bond strength can be accompanied by an increase of $^{1}J(A, ^{19}F)$.[84] A statistical method has been suggested to analyse the temperature-dependent n.m.r. relaxation in order to determine structural and kinetic parameters of metal ion complexes in solution.[85] ^{19}F Solvate isotope shifts of several inorganic fluoro-complexes of Group II, III, IV, V, and VI metals have been carried out. Isotope shifts were compared with the effective charge and their distribution in fluoride-containing anions. The possibility of using ^{19}F solvate isotope shifts to establish unfamiliar complex anions and their hydration was discussed.[86] The ^{19}F n.m.r.

[79] S. M. Dudkin and V. G. Sakharovskii, *Itogi Nauki Tekh., Mol. Biol.*, 1975, **6**, 105 (*Chem. Abs.*, 1976, **85**, 58 514).
[80] G. E. Jellison, P. J. Bray, and P. C. Taylor, *Phys. Chem. Glasses*, 1976, **17**, 35 (*Chem. Abs.*, 1976, **85**, 70 384).
[81] *Pure Applied Chem.*, 1976, **45**, 219.
[82] G. Barbieri, R. Benassi, and F. Taddei, *Gazzetta*, 1975, **105**, 807.
[83] O. Lutz, W. Nepple, and A. Nolle, *Z. Naturforsch.*, 1976, **31a**, 1046.
[84] E. M. Shustorovich and Yu. A. Buslaev, *Inorg. Chem.*, 1976, **15**, 1142.
[85] V. A. Glebov, Yu. D. Knyazev, and I. M. Nikitina, *Koord. Khim.*, 1976, **2**, 700.
[86] B. N. Chernyshov and V. A. Shcherbakov, *Inorg. Nuclear Chem.* (Herbert H. Hyman Memorial Volume), 1976, p. 127.

spectra of *m*- and *p*-$FC_6H_4N{=}N$ derivatives show that ^{19}F shielding is a useful probe of $M \rightarrow N_2Ar$ π-bonding effects in ArN_2 complexes as long as the compound is confined to complexes of the same structural type (singly-bent $[ArN_2]^+$ in the present case). For the complexes studied a total range of 10 p.p.m. was recorded for $(\delta_p - \delta_m)$, the shielding parameters becoming progressively more positive as conjugative transfer of electron density along the $M \rightarrow N_2Ar$ chain decreased. Arranging complexes in order of increasing $\delta_p - \delta_m$ gives a sequence which is in excellent agreement with the ranking of $M \rightarrow N_2Ar$ π-bonding anticipated on the basis of relative donor and acceptor powers of co-ligands and the formal oxidation state of the metal atom. The ^{19}F n.m.r. technique was extended to provide the first unambiguous evidence that the $[ArN_2]^+$ ligand is a poorer π-acceptor than $[NO]^+$.[87] The ^{13}C n.m.r. spectra of several metal enolates have been examined in various aprotic solvents and compared with those of enol acetates and Me_3SiO ethers. The chemical shift difference between enolates and enolacetates appears to be related both to the π-electron density and reactivity at the enolate π-carbon atom.[88] Diamagnetic metal complexes of propane 1,3-diamine-*NNN'N'*-tetra-acetate($[pyta]^{4-}$) have been studied by ^{13}C n.m.r. spectroscopy. The results confirm previous studies in showing that complexes in which the ligand is quinquedentate are not formed (except possibly with Hg^{II}) but also show that this is due not to decreased ring strain in the six-co-ordinate complex but rather to steric compression of the glycinate chelate arms arising from the bunched backbone methylene groups. At low pH the complex formed between $[pyta]^{4-}$ and Al^{III} is more inert than the analogous $[edta]^{4-}$ complex despite the lower equilibrium formation constant, and this is in line with the reduced tendency of $[pyta]^{4-}$ to form complexes in which the ligand is quinquedentate.[89]

2 Stereochemistry

This section is subdivided into ten parts which contain n.m.r. information about hydrogen, lithium, sodium, beryllium, magnesium, and transition-metal complexes, presented by Groups, according to the Periodic Table. Within each Group classification is by ligand type. As far as possible, cross-references are given at the beginning of each sub-group to compounds discussed elsewhere in this chapter. In this cross-referencing, it has not proved possible within the space available to include the many compounds that occur within the sections on dynamic systems, paramagnetic systems, and solid-state n.m.r. spectroscopy. Thus many more compounds of relevance to this section appear elsewhere.

Complexes of Group I and II Elements.—Information on $B_5H_{10}BeBH_4$ is given later (ref. 1136). A review entitled 'Nuclear Magnetic Resonance and its Use in the Study of Chlorophylls' has appeared.[90]

87 W. E. Carroll, M. Deane, and F. J. Lalor, 'Proceedings of the 16th International Conference on Coordination Chemistry, 1974, 2.24B (*Chem. Abs.*, 1976, **85**, 54 247).
88 H. O. House, A. V. Prabbu, and W. V. Phillips, *J. Org. Chem.*, 1976, **41**, 1209.
89 M. J. Farmer, O. W. Howarth, and P. Moore, *J.C.S. Dalton*, 1976, 1445.
90 F. Sh. Isangalin, L. A. Sibel'dina, L. P. Kayushin, B. R. Prokhorenko, and V. M. Kutyurin, *Metody Issled. Fotokhim. Reakts. Fotosint. In Vitro In Vivo, Mater. Vses. Semin., 6th* (1974), 1975, 76 (*Chem. Abs.*, 1976, **85**, 174 200).

The contact contribution to $^1J(^1H, ^2H)$ in HD has been calculated to be 39 Hz, *cf.* 42.94 Hz (experimental).[91] ^{13}C N.m.r. spectroscopy and $^1J(^1H, ^{13}C)$ have been used to investigate α-lithio-sulphoxides and sulphones.[92] CNDO/2 Calculations have been performed on $(Me_2O)_2Li$(pentadienyl), and the calculated charge densities when compared with ^{13}C n.m.r. data gave qualitative agreement.[93] The optical pumping method of alkali molecules by atom–molecule exchange collisions has been applied to obtain the magnetic shielding difference $\sigma(^{23}Na) - \sigma(^{23}Na)_2 = (29 \pm 16) \times 10^{-6}$ between sodium atoms and Na_2 molecules, and the scalar nuclear spin–spin coupling constant $d_s = (306 \pm 30)\ s^{-1}$ for $^{23}Na^{39}K$.[94] Magnetic quantum transitions for spin $\frac{3}{2}$ nuclei in n.m.r. spectra in lyotropic liquid crystals have been observed for $^{23}Na^+$.[95] The magnetic shielding of ^{133}Cs in Cs^+, Cs_2, and free atoms has been determined.[96] $\delta(^{13}C)$, $^1J(^1H, ^{13}C)$, and $^{2,3}J(^1H, ^{13}C)$ have been reported for $M(C_5H_5)$ (M = Li, Na, or K), $PhBe(C_5H_5)$, $BrBe(C_5H_5)$, $(C_5H_5)_2Mg$, and related molecules. The results indicate that neither the ^{13}C chemical shift nor $^1J(^1H, ^{13}C)$ may be used as a criterion for assigning a π- or σ-bonded structure. For $(C_5H_5)BeBr$, $^1J(^9Be, ^{13}C) = 1.1$ Hz.[97] ^1H Spin–lattice relaxation measurements have been used to assign signals of metalloporphyrin and chlorophyll n.m.r. spectra.[97] ^1H and ^{13}C n.m.r. spectroscopies have been used to demonstrate that aggregation of Mg^{II} porphyrins involves the magnesium in one molecule and the propionate carbonyl in another, whereas in zinc porphyrins, aggregation is independent of the side-chains.[99] A similar ^{13}C n.m.r. study of the relative donor properties of the three carbonyl groups of chlorophyll *a* has provided evidence that the donor–acceptor interactions that bind the chlorophyll dimer together involve a substantial participation by the ring V keto-carbonyl and minimal participation by the *Z* ester carbonyl groups.[100]

Complexes of Sc, Y, U, Ti, and Zr.—Information concerning complexes of these elements can be found at the following sources: $[M(C_2Me_2B_{10}H_{10})_2]^{2-}$ (M = Ti, Zr, or Hf),[1188] [(η⁵-cycloheptadienyl)(η⁷-cycloheptatrienyl)Ti],[154] and [Ti(tetraphenylporphyrin)O].[306]

The ^1H and ^{13}C n.m.r. spectra of (1; M = Sc or Y) have been measured. For (1; M = Y), $\delta(^{13}C_5H_5) = 112.2$, $\delta(\mu\text{-}^{13}CH_3) = 7.86$, and $\delta(^{13}CH_3) =$

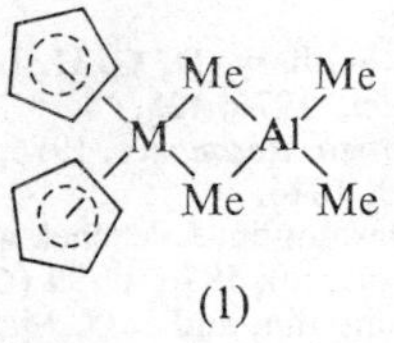

(1)

[91] W. Saenger and J. Voitlaender, *Z. Naturforsch.*, 1975, **30a**, 1491.
[92] R. Lett, G. Chassaing, and A. Marquet, *J. Organometallic Chem.*, 1976, **111**, C17.
[93] J. F. Sebastian, B. Hsu, and J. R. Grunwell, *J. Organometallic Chem.*, 1976, **105**, 1.
[94] F. Konig and H. G. Weber, *Chem. Phys. Letters*, 1976, **44**, 293.
[95] G. Lindblom, H. Wennerström, and B. Lindman, *J. Magn. Resonance*, 1976, **23**, 177.
[96] R. Huber, H. G. Weber, H. Nienstaedt, and G. Z. Putlitz, *Phys. Letters* (*A*), 1975, **55**, 141.
[97] P. Fischer, J. Stadelhofer, and J. Weidlein, *J. Organometallic Chem.*, 1976, **116**, 65.
[98] I. S. Denniss, J. K. M. Sanders, and J. C. Waterton, *J.C.S. Chem. Comm.*, 1976, 1049.
[99] R. J. Abrahams, F. Eivazi, H. Pearson, and K. M. Smith, *J.C.S. Chem. Comm.*, 1976, 698.
[100] L. L. Shipman and T. R. Janson, *Proc. Nat. Acad. Sci. U.S.A.*, 1975, **72**, 2873.

-7.9. For the μ-$^{13}CH_3$ group, the first $^1J(^{13}C, ^{89}Y) = 12.2$ Hz was observed.[101] The 100 MHz n.m.r. spectra of Sc, Tl, and Pb octaethylporphyrin and of a μ-oxo dimer of Sc octaethylporphyrin show the methylene protons of the ethyl side-chains to be anisochronous. The 100 and 300 MHz n.m.r. spectra of Tl aetioporphyrin show that the methylene protons of the four ethyl side-groups of this porphyrin are also anisochronous. It was suggested that these phenomena result from the inherent asymmetry of metalloporphyrins rather than from hindered rotation of the side-chains.[102] ^{19}F N.m.r. spectra have been reported for $U(OTeF_5)_{6-n}F_n$,[103] $[UF_7]^-$, and $[UF_8]^{2-}$.[104, 105]

The ^1H and ^{13}C n.m.r. spectra of $[\{(\eta^5\text{-}C_5H_5)_2ZrH\}_2(O^{13}CH{=}^{13}CHO)]$ show $^1J(^{13}C, ^{13}C) = 99$ Hz.[106] The Lewis acidity of $RTi(OCMe_3)_3$ has been related to the ^1H shifts of the methyl group and attributed to π-bonding.[107] The ^{13}C chemical shifts of a large series of cyclopentadienyl derivatives of titanium and iron have been measured. Substituent- and *trans*-effects were investigated.[108] The ^{13}C n.m.r. spectra of $[(\eta^5\text{-}C_5H_5)M(\eta^7\text{-}C_7H_7)]$ (M = Ti, Zr, Cr, or Mo) have yielded information on the charge on the carbon atoms. For M = Cr, the carbon atoms of the five-membered ring are more shielded than those of the seven-membered ring. For M = Ti, the reverse is true. Chemical shift considerations make it clear that for M = Cr, the carbon atoms of the five-membered rings are more negatively charged than those of the seven-membered ring. For M = Ti, the highest negative charge is on the carbon atoms of the seven-membered ring. The replacement of the 3*d* metal by the corresponding 4*d* metal causes a high-frequency shift for the signals of the carbon atoms of the five-membered ring and a low-frequency shift for the carbon atoms of the seven-membered ring. This results in an inversion of signals for M = Mo with respect to M = Cr. However, it was not clear in this conference abstract how the charge effect on chemical shifts was separated from all the other contributions.[109] The solution ^{15}N n.m.r. spectrum of $[\{(\eta^5\text{-}C_5Me_5)_2Zr(^{15}N_2)\}_2{}^{15}N_2]$ supports the solid-state *X*-ray structure with two signals with $^1J(^{15}N, ^{15}N) = 6.2$ Hz at $\delta = 89.8$ and at $\delta = 160.4$ a low-frequency singlet due to μ-$^{15}N_2$.[110]

Complexes of V, Nb, and Ta.—The ^{13}C n.m.r. spectra of $[(Me_3SiCH_2)_4M(CSiMe_3)_2]$ (M = Nb, Ta, or W) and $[(Me_3SiCH_2)_6M_2]$ (M = Mo or W) have been reported. $\delta(M^{13}CSi)$ is *ca.* 400; for $[(Me_3SiCH_2)_4W(CSiMe_3)_2]$,

101 J. Holton, M. F. Lappert, G. R. Scollary, D. G. H. Ballard, R. Pearce, J. L. Atwood, and W. E. Hunter, *J.C.S. Chem. Comm.*, 1976, 425.
102 C. A. Busby and D. Dolphin, *J. Magn. Resonance*, 1976, **23**, 211.
103 K. Seppelt, *Chem. Ber.*, 1976, **109**, 1046.
104 R. Bougon, R. M. Costes, J. P. Desmoulin, J. Michel, and J. L. Person, *Inorg. Nuclear Chem.* (Herbert H. Hyman Memorial Volume), 1976, p. 99 (*Chem. Abs.*, 1976, **85**, 114 145).
105 R. Bougon, P. Charpin, J. P. Desmoulin, and J. G. Malm, *Inorg. Chem.*, 1976, **15**, 2532.
106 J. M. Manriquez, D. R. McAlister, R. D. Sanner, and J. E. Bercaw, *J. Amer. Chem. Soc.*, 1976, **98**, 6734.
107 R. Choukroun, M. Basso-Bert, C. Blandy, R. Guerreiro, and D. Gervais, 'Proceedings of the 16th International Conference on Coordination Chemistry', 1974, R77 (*Chem. Abs.*, 1976, **85**, 176 632).
108 A. N. Nesmeyanov, E. I. Fedin, P. V. Petrovskii, V. A. Dubovitskii, O. V. Nogina, and N. S. Kochetkova, *Zhur. strukt. Khim.*, 1975, **16**, 759.
109 C. J. Groenenboom, H. J. De Liefde Meijer, and F. Jellinek, 'Proceedings of the 16th International Conference on Coordination Chemistry', 1974, R2 (*Chem. Abs.*, 1976, **85**, 27 142).
110 J. M. Manriquez, R. D. Sanner, R. E. Marsh, and J. E. Bercaw, *J. Amer. Chem. Soc.*, 1976, **98**, 3042.

$^1J(^{13}C, ^{183}W) = 74$ Hz and $\delta(^{13}CH_2) = 70$.[111] For $[(\eta^5\text{-}C_5H_5)V(CO)_2\{Ph_2P(CH_2)_nPPh_2\}]$ there is a relationship between $\delta(^{51}V)$ and the chelate ring size; $\delta(^{31}P)$ was also reported.[112] Qualitative MO treatments of the complexes $[(\eta^5\text{-}C_5H_5)V(CO)_3L]$ suggest that ^{51}V chemical shifts should correlate with the π-acceptor abilities of L, and consequently a sequence of π-acceptor abilities $P(OR)_3 > CN^- > PMe_3 \sim PEt_3 \sim SbPh_3 \sim PPhF_2 > PBu^i{}_3 \sim P(NR^1{}_2)_3 > PPh_3 > AsPh_3$ was derived. $^1J(^{31}P, ^{51}V)$ and ^{51}V linewidths were also reported.[113] $\delta(^{51}V)$ and $^1J(^{31}P, ^{51}V)$ have been reported for 49 mono- and di-substituted tertiary phosphine derivatives of $[(\eta^5\text{-}C_5H_5)V(CO)_4]$ and $[V(CO)_6]^-$. $^1J(^{31}P, ^{51}V)$ decreases in the order $PBu^i{}_3 > PF_3 > P(OR)_3 > P(NR^1{}_2)_3 > PR_3 \sim PH_3$ and is larger for $[V(CO)_5PR_3]^-$, *cis*-$[V(CO)_4(PR_3)_2]^-$, and *trans*-$[(\eta^5\text{-}C_5H_5)V(CO)_2(PR_3)_2]$ as compared with *cis*-$[(\eta^5\text{-}C_5H_5)V(CO)_2(PR_3)_2]$ and $[(\eta^5\text{-}C_5H_5)V(CO)_3PR_3]$. Variations in $^1J(^{31}P, ^{51}V)$ were shown to result from (i) indirect vanadium–phosphorus π-interaction, (ii) inductive effects of the substituent attached to phosphorus, and (iii) the average energy separation between the highest occupied and lowest unoccupied MO.[114] ^{31}P N.m.r. data have also been reported for $[NbH_5(dppe)_2]$,[115] $[TaMe_5(Me_2PCH_2CH_2PMe_2)_2]$,[116] and $[(\eta^3\text{-allyl})(dppe)(OC)_3V]$.[117]

^{51}V N.m.r. spectra of a number if iso- and hetero-polyvanadate anions have been reported. Spectra of solutions containing $[V_{10}O_{28}]^{6-}$ (pH 6.5—3.5) are consistent with the known solid-state structure of this anion. The $[VO_2]^+$ cation in aqueous solution almost certainly has an octahedral *cis*-dioxo-structure as its n.m.r. linewidth is comparable with those for the *cis*-compounds $[VO_2(O_4C_2)_2]^{3-}$ and $[VO_2(edta)]^{3-}$ which are broader than those for pseudo-tetrahedral $[VCl_2O_2]^-$ and $[VF_2O_2]^-$. Chemical shifts of the pseudo-octahedral vanadium atoms in $[V_xW_{6-x}O_{19}]^{n-}$, $[PV_xW_{12-x}O_{40}]^{n-}$, and related anions range from 506 to 545 p.p.m. (w.r.t. VCl_3O) and linewidths from 60 to 200 Hz. Most of the possible geometric isomers were found. The ^{31}P and ^{51}V n.m.r. chemical shifts progressively decrease with the introduction of vanadium atoms into the heteropolyanion structures.[118–120] The polymeric $[M(OEt)_nF_{5-n}]$ complexes (M = Nb or Ta) have been investigated by ^1H and ^{19}F n.m.r. spectroscopies.[121] In MeCN, the ^{19}F n.m.r. spectrum shows that $MePOCl_2$ and TaF_5 react to give $[TaF_nCl_{6-n}]^-$, $[TaF_xCl_y(NCMe)_{6-x-y}]^{5-x-y}$, POF_nCl_{3-n}, and $[TaF_4(MePOF_2)_2]^+$. The effect of HF was examined and MeCN is replaced.[122] $[Ta(S_2CNMe_2)_4]^+$ is the first

[111] R. A. Andersen, A. L. Galyer, and G. Wilkinson, *Angew. Chem. Internat. Edn.*, 1976, **15**, 609.
[112] D. Rehder, L. Dahlenburg, and I. Müller, *J. Organometallic Chem.*, 1976, **122**, 53.
[113] D. Rehder, W. L. Dorn, and J. Schmidt, *Transition Metal Chem.*, 1976, **1**, 74.
[114] D. Rehder, W. L. Dorn, and J. Schmidt, *Transition Metal Chem.*, 1976, **1**, 233.
[115] R. R. Schrock, *J. Organometallic Chem.*, 1976, **121**, 373.
[116] R. R. Schrock, *J. Organometallic Chem.*, 1976, **122**, 209.
[117] U. Franke and E. Weiss, *J. Organometallic Chem.*, 1976, **121**, 355.
[118] S. E. O'Donnell and M. T. Pope, *J.C.S. Dalton*, 1976, 2290.
[119] M. T. Pope, S. E. O'Donnell, and R. A. Prados, in 'Inorganic Compounds with Unusual Properties' (Advances in Chemistry Series No. 150), The American Chemical Society, Washington, 1976, p. 85 (*Chem. Abs.*, 1976, **85**, 85 110).
[120] L. P. Kazanskii, M. A. Fedotov, M. N. Ptushkina, and V. I. Spitsyn, *Doklady Akad. Nauk S.S.S.R.*, 1975, **224**, 866.
[121] Yu. V. Kokunov, V. D. Kopanev, M. P. Gustyakova, and Yu. A. Buslaev, *Koord. Khim.*, 1976, **2**, 605.
[122] E. G. Il'in, M. N. Shcherbakova, and Yu. A. Buslaev, *Koord. Khim.*, 1976, **2**, 790.

eight-co-ordinate tetrakis chelate which is stereochemically rigid on the n.m.r. time-scale.[123]

Complexes of Cr, Mo, and W.—Information concerning complexes of these elements can be found at the following sources: $[(Me_3SiCH_2)_4W(CSiMe_3)_2]$, $[(Me_3SiCH_2)_6M_2]$ (M = Mo or W),[111] $[(\eta^5\text{-}C_5H_5)M(\eta^7\text{-}C_7H_7)]$ (M = Cr or Mo),[109] $[V_xM_{12-x}O_{40}]^{n-}$ (M = Mo or W),[118] $[PV_2M_{10}O_{50}]^{5-}$ (M = Mo or W),[119] and $[\{(Ph_2PMe)_2M\}_2WS_4]$ (M = Cu, Ag, or Au).[482]

The ^{13}C n.m.r. spectrum of Me_6W has been remeasured and $^1J(^{13}C, ^{183}W)$ has now been found to be 43.2 Hz rather than 1000 Hz as previously reported. ^{13}C N.m.r. data were also given for $[WMe_8]^{2-}$.[124] $^1J(^{13}C, ^{183}W)$, $^2J(^{13}C, ^{183}W)$, and $^3J(^{13}C, ^{183}W)$ have been determined for 10 tungsten carbene and nine tungsten carbyne complexes. $^1J(^{13}C, ^{183}W)$ is of analytical importance, being characteristically greater for carbyne than for carbene bonds because of different hybridization at the carbon atom and thus providing information about bond angles and polarities of W—C and W—C—R units. The substituents R^1 and R^2 in $[(OC)_5W{=}CR^1R^2]$ and $[X(OC)_4W{\equiv}CR^1]$ as well as the halogens (X) lead to minor changes in $^1J(^{13}C, ^{183}W)$. These changes are comparable to those of $^1J(^1H, ^{13}C)$ in corresponding substituted methanes. Unexpectedly $^1J(^{13}C, ^{183}W)$ increases with X = Cl, Br, or I. $^2J(^{13}C, ^{183}W)$, though being much smaller than $^1J(^{13}C, ^{183}W)$, reflects different hybridization at the β-carbon atom.[125] For $[(OC)_5CrC(NR_2)(naphthyl)]$, the ^1H n.m.r. spectrum shows that the 1-naphthyl group is strongly restricted even at temperatures up to +80 °C, whereas the 2-naphthyl group rotates rapidly down to −90 °C.[126] $^1J(^{15}N, ^{13}C)$ has been determined for (2; M = Cr or W). For (2; M = Cr), $^1J[^{13}C(1), ^{15}N]$ = 4.0 Hz,

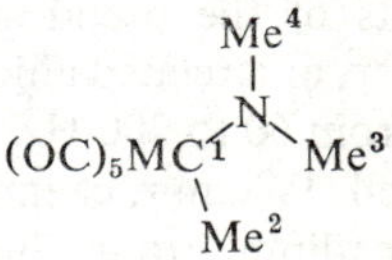

(2)

$^1J[^{13}C(2), ^{15}N]$ = 0.5 Hz, $^1J[^{13}C(3), ^{15}N]$ = 8.3 Hz, and $^1J(^{13}C, ^{15}N)$ = 12.4 Hz. It was concluded that if the nitrogen atom is sp^2 hybridized then the carbon–nitrogen bond has only 10% s_C character and hence the metal requires a great deal of carbon 2*s* orbital in bond formation.[127] ^{13}C N.m.r. data have also been reported for $[(OC)_5CrC(SiPh_3)(NMe_2)]$,[128] $[(\eta^6\text{-}C_6H_6)Cr(CO)_2CPh]^+$, $[(\eta^6\text{-}C_6H_6)\text{-}Cr(CO)_2CPh{=}PMe_3]^+$,[129] $[(OC)_5CrCNEt_2]^+$, and $[(OC)_5CrC(NEt_2)F]$, including ^{19}F n.m.r. data,[130] $[(\eta^6\text{-arene})Cr(CO)_2CPh]^+$ and $[(\eta^6\text{-arene})Cr(CO)_2CPhOMe]$,[131]

123 D. F. Lewis and R. C. Fay, *Inorg. Chem.*, 1976, **15**, 2219.
124 A. L. Galyer and G. Wilkinson, *J.C.S. Dalton*, 1976, 2235.
125 F. H. Köhler, H. J. Kalder, and E. O. Fischer, *J. Organometallic Chem.*, 1976, **113**, 11.
126 H. Brunner, J. Doppelberger, E. O. Fischer, and M. Lappus, *J. Organometallic Chem.*, 1976, **112**, 65.
127 F. R. Kreissl, E. W. Meineke, E. O. Fischer, and J. D. Roberts, *J. Organometallic Chem.*, 1976, **108**, C29.
128 E. O. Fischer, H. Hollfelder, F. R. Kreissl, and W. Uedelhoven, *J. Organometallic Chem.*, 1976, **113**, C31.
129 F. R. Kreissl and P. Stückler, *J. Organometallic Chem.*, 1976, **110**, C9.
130 E. O. Fischer, W. Kleine, and F. R. Kreissl, *Angew. Chem. Internat. Edn.*, 1976, **15**, 616.
131 E. O. Fischer, P. Stückler, H.-J. Beck, and F. R. Kreissl, *Chem. Ber.*, 1976, **109**, 3089.

$[XCr(CO)_3(EMe_3)CMe]$ (including ^{31}P n.m.r. data; E = P, As, or Sb),[132,133] *mer*-$[(BF_4)Cr(CO)_3(PMe_3)CMe]$ (including ^{31}P n.m.r. data),[134] $[X(OC)_4MCR]$ (M = Cr or W),[135] $[(OC)_4Cr(ER_3)C(OMe)Me]$ (E = P, As, or Sb),[136] $[(OC)_5$-$WCROW(CO)_4CR]$,[137] $[(\eta^5$-$C_5H_5)M(CO)_2OCMe{=}CHCH]$ (M = Mo or W),[138] $[(OC)_5CrCR^1NHR^2]$,[139] and (3; M = Mo or W),[140] $[(OC)_5CrC(NEt_2)X]$ (X = Cl or OEt),[141] (4),[142] $[Br(OC)_3(Me_3P)WC(p$-tolyl$){=}PMe_3]$,[143] $[(\eta^5$-$C_5H_5)(OC)_2$-$W{\equiv}CR]$,[144] $[(\eta^5$-$C_5H_5)\{\eta^5$-$C_5H_4C{\equiv}W(CO)_4Br\}Fe]$,[145] $[I(OC)_4WCSC(O)Me]$,[146] $[(OC)_5W{=}C{=}C{=}CPhNMe_2]$, including $J(^{13}C, ^{183}W)$,[147] (5),[148] $[(\eta^5$-$C_5H_5)$-$W(CO)_2(PR_3)X]$, including ^{31}P n.m.r. data (X = $PbMe_3$, $GeMe_3$, or Me),[149] and $[(OC)_5MSbF\{M'(CO)_3(\eta^5$-$C_5H_5)\}_2]$, only ^{19}F n.m.r. data (M = Cr or W; M′ = Mo or W).[150]

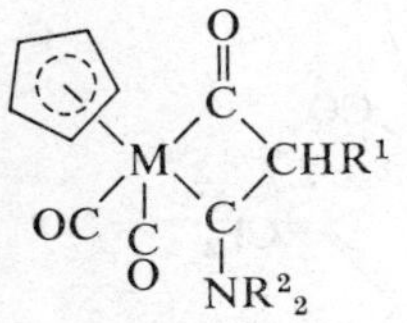
(3)

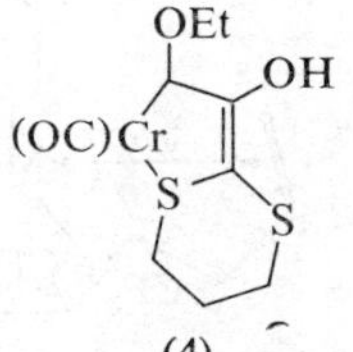
(4)

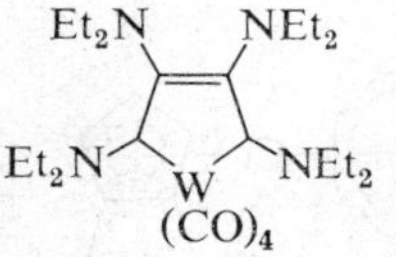
(5)

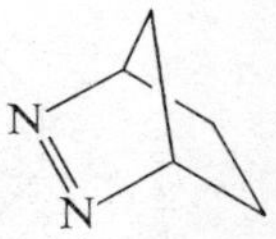
(6)

The ^1H n.m.r. spectrum of $[(\eta^5$-$C_5H_5)CrH\{P(OMe)_3\}_3]$ is a 1 : 3 : 3 : 1 quartet with $^2J(^1H, ^{31}P)$ = 75 Hz, while the $^{31}P\{^1H\}$ spectrum is a singlet, implying either an improbable trigonal-bipyramidal structure or fluxionality.[151] The ^{13}C n.m.r. spectrum of 2,3-diazobicyclo[2,2,1]hept-2-ene (6) in $[(\eta^5$-$C_5H_5)Cr(CO)(NO)L]$, L = (6), shows an apparent plane of symmetry for L, implying that L is either η^2-bonded to the chromium or is *N*-bonded and fluxional.[152] $^1H\{^{183}W\}$, $^1H\{^{31}P\}$, and $^1H\{^{183}W^{31}P\}$ nuclear magnetic double- and triple-resonance experiments have been used to determine ^{183}W chemical shifts and other parameters in 35 tungsten(0) complexes with tertiary phosphines, η^5-C_5H_5, and other ligands. The observed chemical shifts are several thousand p.p.m. to high field of those for derivatives of WF_6, and this was attributed to the occupancy of the d_{xy}, d_{yz}, and

132 E. O. Fischer and K. Richter, *Chem. Ber.*, 1976, **109**, 2547.
133 E. O. Fischer and K. Richter, *Chem. Ber.*, 1976, **109**, 3079.
134 K. Richter, E. O. Fischer, and C. G. Kreiter, *J. Organometallic Chem.*, 1976, **122**, 187.
135 E. O. Fischer and G. Kreis, *Chem. Ber.*, 1976, **109**, 1673.
136 E. O. Fischer and K. Richter, *Chem. Ber.*, 1976, **109**, 1140.
137 E. O. Fischer and K. Weiss, *Chem. Ber.*, 1976, **109**, 1128.
138 H. G. Alt, *Angew. Chem. Internat. Edn.*, 1976, **15**, 759.
139 K. Weiss and E. O. Fischer, *Chem. Ber.*, 1976, **109**, 1868.
140 W. Beck, H. Brix, and F. H. Köhler, *J. Organometallic Chem.*, 1976, **121**, 211.
141 E. O. Fischer, W. Kleine, and F. R. Kreissl, *J. Organometallic Chem.*, 1976, **107**, C23.
142 H. G. Raubenheimer, S. Lotz, and J. Coetzer, *J.C.S. Chem. Comm.*, 1976, 732.
143 F. R. Kreissl, W. Uedelhoven, and A. Ruhs, *J. Organometallic Chem.*, 1976, **113**, C55.
144 E. O. Fischer, T. L. Lindner, and F. R. Kreissl, *J. Organometallic Chem.*, 1976, **112**, C27.
145 E. O. Fischer, M. Schluge, and J. O. Besenhard, *Angew. Chem. Internat. Edn.*, 1976, **15**, 683.
146 B. Duane Dombek and R. J. Angelici, *Inorg. Chem.*, 1976, **15**, 2397.
147 E. O. Fischer, H.-J. Kalder, A. Frank, F. H. Köhler, and G. Huttner, *Angew. Chem., Internat. Edn.*, 1976, **15**, 683.
148 R. B. King and C. A. Harmon, *Inorg. Chem.*, 1976, **15**, 879.
149 T. A. George and C. D. Sterner, *Inorg. Chem.*, 1976, **15**, 165.
150 P. Panster and W. Malisch, *Chem. Ber.*, 1976, **109**, 2112.
151 J. D. Koola and H. H. Brintzinger, *J.C.S. Chem. Comm.*, 1976, 388.
152 M. Herberhold and H. Alt, *Annalen*, 1976, 292.

d_{xz} valence orbitals in addition to the $d_{x^2-y^2}$ and d_{z^2}, which reduces the *d*-electron imbalance and gives a smaller paramagnetic term. π-Back-bonding with ligands produces a high-field shift, and the marked sensitivity of the shielding to temperature and isotopic substitution was attributed to changes in electronic excitation energies.[153] The ^{13}C, ^{19}F, and ^{31}P n.m.r. spectra of [(η^6-cycloheptatriene)-Cr(PF_3)$_3$] show only one type of PF_3 group at room temperature and two at −90 °C.[154] N.m.r. data have also been reported for [Mo(CO)$_3$(MeO_2CCH=$CHCO_2Me$)$_2$]$_2$ (^{13}C n.m.r. data),[155] [(η^4-cyclohexadiene)$_2$Cr(PF_3)$_2$] (^{19}F n.m.r. data),[156] (7) (^{13}C n.m.r. data), [(η^5-cyclo-octadienyl)$_2$Cr],[157] [(η^5-C_5H_5)(OC)$_3$-MP(E)Cl_2] (^{31}P n.m.r. data; E = S or Se),[158] (8) (^{19}F n.m.r. data),[159] and [(η-C_7H_7)Mo(CO)$_2${$Ph_2P(CH_2)_nPPh_2$}]$^+$ (^{31}P n.m.r. data).[160]

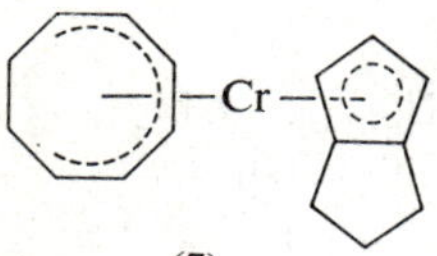

(7)

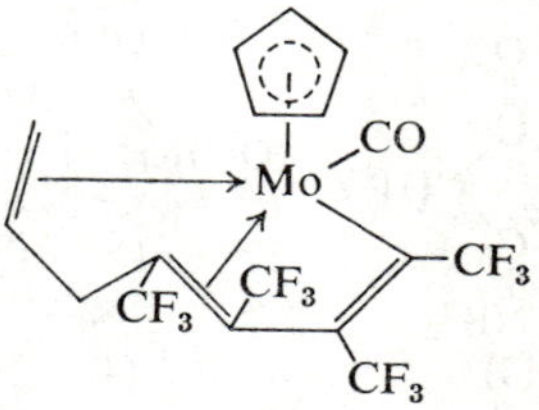

(8)

The ^{1}H n.m.r. spectra of a series of [(η^6-arene)$_2$Cr] complexes have been investigated. Simple first-order spectra were obtained in most cases. A significant reduction in J(^{1}H, ^{1}H) values relative to the free ligands were observed for the complexed arenes. In FC_6H_5 complexes, J(^{1}H, ^{19}F) is so diminished as not to be observed in the majority of the spectra. *Meta*-disubstituted arenes in which the substituents are not identical yield two distinguishable [(η^6-arene)$_2$Cr] diastereomers.[161] This work was subsequently extended to ^{13}C n.m.r. spectra. No transmission of substituent effects across the ring was found and this was interpreted to mean that a significant reduction in ring aromaticity occurs upon complexation to chromium, an effect attributed to the donation of arene π-electron density into vacant metal orbitals.[162] In the ^{1}H and ^{13}C n.m.r. spectra of [(η^6-C_6H_6)$_2$M] and [(η^6-EtC_6H_5)$_2$M] (M = Cr or Mo), there is a large shift attributable to an increase in electron density on the ring by the metal.[163] The ^{1}H n.m.r. spectra of 30 [(η^6-arene)Cr(CO)$_3$] complexes have been measured. The chemical shift increments of aromatic protons on complex formation of $PhCHBu^t_2$ have been used as a reference for the eclipsed conformation with the substituent staggered with respect to the CO ligands. Then the contribution of

[153] H. C. E. McFarlane, W. McFarlane, and D. S. Rycroft, *J.C.S. Dalton*, 1976, 1616.
[154] P. L. Timms and T. W. Turney, *J.C.S. Dalton*, 1976, 2021.
[155] H. Schaper and H. Behrens, *J. Organometallic Chem.*, 1976, **113**, 361.
[156] J. R. Blackborow, R. H. Grubbs, A. Miyashita, A. Scrivanti, and E. A. Koerner von Gustorf, *J. Organometallic Chem.*, 1976, **122**, C6.
[157] J. Muller, W. Holyinger, and F. H. Köhler, *Chem. Ber.*, 1976, **109**, 1222.
[158] W. Malisch and R. Alsmann, *Angew. Chem. Internat. Edn.*, 1976, **15**, 769.
[159] J. L. Davidson, M. Green, F. G. A. Stone, and A. J. Welch, *J.C.S. Dalton*, 1976, 2044.
[160] E. E. Isaacs and W. A. G. Graham, *J. Organometallic Chem.*, 1976, **120**, 407.
[161] V. Graves and J. J. Lagowski, *J. Organometallic Chem.*, 1976, **120**, 397.
[162] V. Graves and J. J. Lagowski, *Inorg. Chem.*, 1976, **15**, 577.
[163] A. N. Nesmeyanov, L. P. Yur'eva, B. I. Kozyrkin, N. N. Zaitseva, V. A. Svoren, V. I. Robas, E. B. Zavelovich, and P. V. Petrovskii, *Izvest, Akad. Nauk S.S.S.R., Ser. khim.*, 1975, **9**, 2006.

this conformer to the conformational equilibrium of the other monoalkyl substituted complexes was estimated.[164] The influence of substituents on the ^{1}H and ^{13}C n.m.r. spectra of mono- and di-substituted arenes complexed to $Cr(CO)_3$ has been discussed. Special attention was paid to diastereotopy resulting from different types of chirality.[165] For $[(\eta^6\text{-}PhC_6H_5)Cr(CO)_3]$ and related compounds the complexed ring ^{13}C chemical shifts are *ca.* 35 p.p.m. to low frequency from the free ligand while the uncomplexed change by no more than 1 to 2 p.p.m.[166] ^{13}C N.m.r. spectroscopy has been used to investigate the protonation of $[(\eta^6\text{-}MeOC_6H_5)Cr(CO)_3]$. The arene signals move to low field but there is no sign of C—H bond formation. It was therefore concluded that the protonation occurred at the metal.[167] The ^{13}C chemical shifts of $[R_3ECr(CO)_2(\eta^6\text{-arene})]$ have been discussed in terms of the deshielding effect on the arene ring carbon atom *trans* to the Group V atom in the preferred conformation of these molecules.[168] The unusual stability of $[(\eta^7\text{-cumyl})Cr(CO)_3]^+$ has been investigated using ^{13}C n.m.r. spectroscopy. The σ^+ substituent constant for the $Cr(CO)_3$ group and the fraction of unit positive charge transmitted into the $M(CO)_3$ groups were qualitatively estimated.[169] N.m.r. data have also been reported for $[(\eta^6\text{-}C_6H_6)(\eta^6\text{-}C_6F_6)Cr]$ (^{19}F n.m.r. data),[170] $[(\eta^6\text{-}C_6H_6)(\eta^6\text{-}C_6F_5H)Cr]$ (^{19}F n.m.r. data),[171] (9) (^{13}C n.m.r. data),[172] $[(\eta^6\text{-}L)Cr(CO)_3]$ [^{11}B n.m.r. data, L = (10)],[173] and $[(\eta^6\text{-}L)M(CO)_3]$ [^{19}F and ^{31}P n.m.r. data; L = (11); X = OMe or F; M = Cr, Mo, or W].[174]

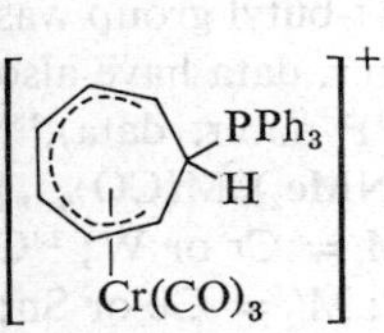
(9)

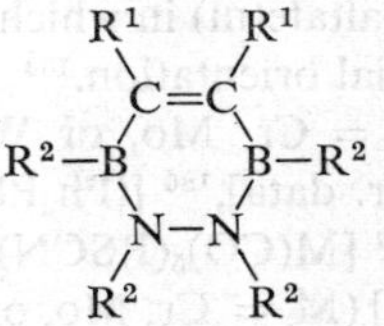
(10)

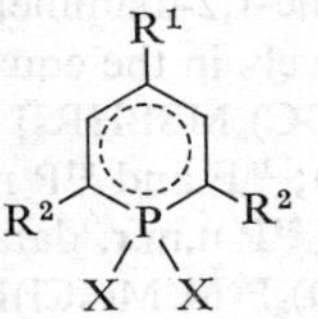
(11)

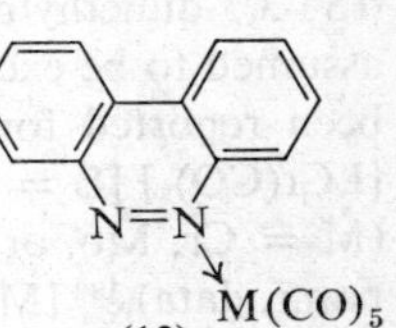
(12)

For (12; M = Cr or Mo) there is only one set of phenyl resonances in the ^{13}C n.m.r. spectrum, whereas for (12; M = W) two sets were observed. It was therefore suggested that (12; M = Cr or Mo) is fluxional.[175] ^{13}C N.m.r. spectroscopy has been used to demonstrate specific labelling with ^{13}CO in $[W(CO)_5L]$ species, and $^{1}J(^{13}C, ^{183}W)$ has been reported for $[W(CO)_5P(4\text{-}ClC_6H_4)_3]$, $[W(CO)_5(CNC_6H_{11})]$, and $[ClW(CO)_5]^-$.[176] ^{13}C N.m.r. spectroscopy has been used to

164 F. van Meurs, J. M. van der Toorn, and H. van Bekkum, *J. Organometallic Chem.*, 1976, **113**, 341.
165 J. Tirouflet, J. Besancon, F. Mabon, and M. L. Martin, *Org. Magn. Resonance*, 1976, **8**, 444.
166 D. A. Brown, N. J. Fitzpatrick, I. J. King, and N. J. Mathews, *J. Organometallic Chem.*, 1976, **104**, C9.
167 G. A. Olah and S. H. Yu, *J. Org. Chem.*, 1976, **41**, 717.
168 W. R. Jackson, C. F. Pincombe, I. D. Rae, D. Rash, and B. Wilkinson, *Austral. J. Chem.*, 1976, **29**, 2431.
169 G. A. Olah and S. H. Yu, *J. Org. Chem.*, 1976, **41**, 1695.
170 M. J. McGlinchey and T.-S. Tan, *J. Amer. Chem. Soc.*, 1976, **98**, 2271.
171 T.-S. Tan and M. J. McGlinchey, *J.C.S. Chem. Comm.*, 1976, 155.
172 A. Salzer, *Inorg. Chim. Acta.*, 1976, **18**, L31.
173 W. Siebert and R. Full, *Angew. Chem. Internat. Edn.*, 1976, **15**, 45.
174 M. Lückoff and K. Dimroth, *Angew. Chem. Internat. Edn.*, 1976, **15**, 503.
175 M. Kooti and J. F. Nixon, *J. Organometallic Chem.*, 1976, **105**, 217.
176 B. D. Dombek and R. J. Angelici, *J. Amer. Chem. Soc.*, 1976, **98**, 4110.

show that ^{13}CO incorporates purely *cis* into *fac*-[Cr(CO)$_3$(PPh$_3$)(phen)] to give [Cr(CO)$_3$(^{13}CO)(phen)].[177] $^1J(^{31}P, ^{95,97}Mo)$ has been measured in [(OC)$_5$MoL] complexes and is between 133 and 284 Hz.[178] For [(OC)$_5$CrPMe$_2$SMe], [(OC)$_5$-CrPMe$_2$S]$^-$, and [(OC)$_5$CrPMe$_2$SCr(CO)$_5$]$^-$, there is a correlation between the ^{31}P chemical shift and ν(PS).[179] A correlation between ν(CO), $^1J(^{31}P, ^{183}W)$ and ionization energies for [R$_3$PW(CO)$_5$] has been noted. It was concluded that $^1J(^{31}P, ^{183}W)$ values increase because the *s*-character contribution to the phosphorus lone pair in the tungsten–phosphorus σ-bond increases.[180] For [(OC)$_5$MP$_4$O$_6$] (M = Cr, Mo, or W), the ^{31}P chemical shifts of the unco-ordinated P^{III} atoms have been found to be a measure of the extent of interaction between the co-ordinated phosphorus atom and the metal. The P_4O_6 cage appears to be a strong π-acceptor ligand.[181] The ^{31}P n.m.r. spectrum of [Cr(NO)(CO)(Cl)-{P(OMe)$_3$}$_3$] is AB$_2$ with $^2J(^{31}P, ^{31}P) = 89$ Hz.[182] The ^{13}C chemical shift of the isopropyl CH carbon at δ 88.4 in [{HB(3,5-Me$_2$C$_3$HN$_2$)$_3$}Mo(NO)(CO)$_2$Br(OPr^i)] was attributed to very strong electron removal by the electron-deficient molybdenum.[183] The ^1H n.m.r. spectra of a number of Cr^0, Mo^0, Co^{III}, Pd^{II}, and Pt^{II} complexes of (*R*)-propane-1,2-diamine, (*S*)-3,3-dimethylbutane-1,2-diamine, (*RS*)-*NNN'N'*-tetramethylpropane-1,2-diamine, and ethane-1,2-diamine have been analysed and the coupling constant data utilized to calculate the conformational populations for the various complexes. Necessary parameters for the conformational analysis were obtained from a study of tetracyano-{(*S*)-3,3-dimethylbutane-1,2-diamine}cobaltate(III) in which the t-butyl group was assumed to be exclusively in the equatorial orientation.[184] N.m.r. data have also been reported for [(OC)$_5$MSPHR$_2$] (M = Cr, Mo, or W; ^{31}P n.m.r. data),[185] [LCr(CO)$_5$] [L = (13); ^{19}F and ^{31}P n.m.r. data],[186] [(Ph$_2$PNRNMe$_2$)$_n$M(CO)$_{6-n}$] (M = Cr, Mo, or W; ^{31}P n.m.r. data),[187] [M(CO)$_5$(RSCN)] (M = Cr or W; ^{13}C n.m.r. data),[188] [M(CO)$_5$P(M'Me$_2$Cl)Bu$^t_2$] (M = Cr, Mo, or W; M' = Ge or Sn; ^{31}P n.m.r. data),[189] [(OC)$_5$MoL] [^{31}P n.m.r. data; L = (14)],[190] [Mo(CO)$_2$-

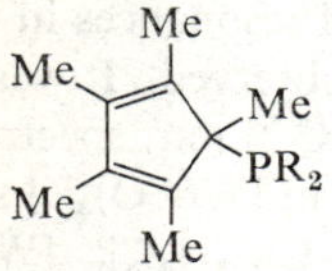

(13)

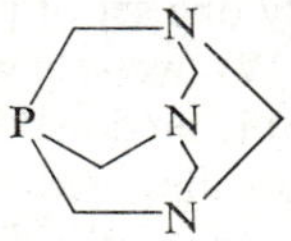

(14)

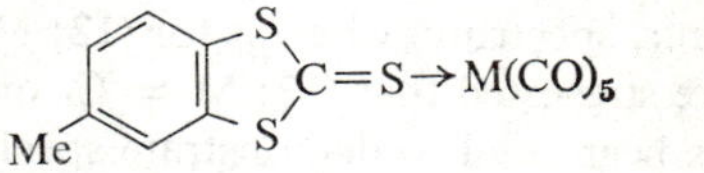

(15)

[177] M. A. Cohen and T. L. Brown, *Inorg. Chem.*, 1976, **15**, 1417.
[178] D. S. Milbrath, J. G. Verkade, and R. J. Clark, *Inorg. Nuclear Chem. Letters*, 1976, **12**, 921.
[179] E. Lindner and W. P. Meier, *Chem. Ber.*, 1976, **109**, 3323.
[180] M. S. A. Abd El-Mottaleb, *J. Mol. Structure*, 1976, **32**, 203.
[181] M. L. Walker and J. L. Mills, *J. Organometallic Chem.*, 1976, **120**, 355.
[182] M. Herberhold and W. Bernhagen, *Angew. Chem. Internat. Edn.*, 1976, **15**, 617.
[183] J. McCleverty, D. Seddon, N. A. Bailey, and N. W. J. Walker, *J.C.S. Dalton*, 1976, 898.
[184] C. J. Hawkins and R. M. Peachey, *Austral. J. Chem.*, 1976, **29**, 33.
[185] E. Lindner and W. P. Meier, *J. Organometallic Chem.*, 1976, **114**, 67.
[186] P. Jutzi, H. Saleske, and D. Nadler, *J. Organometallic Chem.*, 1976, **118**, C8.
[187] G. E. Graves and L. W. Houk, *Inorg. Chem.*, 1976, **15**, 7.
[188] M. H. Quick and R. J. Angelici, *Inorg. Chem.*, 1976, **15**, 160.
[189] H. Schumann, J. Held, H.-J. Kroth, and W.-W. du Mont, *Chem. Ber.*, 1976, **109**, 246.
[190] J. R. Delerno, L. M. Trefonas, M. Y. Darensbourg, and R. J. Majeste, *Inorg. Chem.*, 1976, **15**, 816.

$(dppe)_2X]^+$ (X = F, Br, or I; ^{19}F and ^{31}P n.m.r. data),[191] (15; M = Cr or Mo) (^{13}C n.m.r. data),[192] $[M(CO)_5(CS)]$ and *trans*-$[W(CO)_4(CS)I]^-$ [M = Cr or W; ^{13}C n.m.r. data including $^1J(^{13}C, ^{183}W)$],[193] $[LM(CO)_4]$ [M = Cr, Mo, or W; L = $R_2PCH_2CH_2PR_2$, $RP(CH_2CH_2PR_2)_2$, or $P(CH_2CH_2PR_2)_3$; R = neopentyl; ^{31}P n.m.r. data],[194] $[W(CO)_5$(substituted pyridine)] (^{13}C n.m.r. data),[195] and $[(OC)_5WSCH$(piperidine)] (^{13}C n.m.r. data).[196]

In the ^{19}F n.m.r. spectrum of $[WF(NMe)(NCMe)_4]$, $^2J(^{14}N, ^{19}F)$ was observed to the *trans*-NCMe and INDOR was used to measure $\delta(^{14}N)$. ^{19}F N.m.r. data were also given for $[WF_5NMe]^-$ and $[W_2F_9(NMe)_2]^-$.[197] The ^{13}C n.m.r. spectrum of $[W(dppe)_2Br\{NN{=}CH(CH_2)_3OH\}]$ shows N=CH at δ 171.4.[198] ^1H, ^{19}F, and ^{31}P n.m.r. data have been published for $[Mo(N_2H_4)F(dppe)_2]^+$.[199]

In a variety of alkali chromate solutions in H_2O and D_2O, the dependence of the Larmor frequency of ^{53}Cr on the concentration of the solution was measured and a large solvent isotope effect was found. In a 2.2 molar solution of K_2CrO_4 in D_2O, a ratio of the Larmor frequency of ^{53}Cr to ^{39}K was measured to high accuracy and then extrapolated to infinite dilution. $\delta(^{53}Cr)$ for $Cr(CO)_6$ was also measured.[200] $\delta(^{95}Mo)$ and $\delta(^{97}Mo)$ have been reported for several molybdenum compounds in different oxidation states, and the chemical shifts related to $[MoO_4]^{2-}$ ranged from +500 to −1900 p.p.m.[201,202] Fourier transform ^{17}O and ^{33}S n.m.r. spectra have been reported for several aqueous sulphate, thiosulphate, molybdate, and thiomolybdate solutions. Using information from ^{17}O signals, the ^{33}S n.m.r. lines were assigned. Experimental parameters, chemical shifts and linewidths were given.[203] After enrichment to 2% with ^{17}O, the ^{17}O n.m.r. spectrum of $[As_2Mo_6O_{26}]^{6-}$ shows separate signals for OMo at δ 819, OMo_2 at δ 386, and OAs at δ 102. The ^{17}O n.m.r. spectrum of $[(PhAs)_2Mo_6O_{24}]^{4-}$ was also reported.[204] Similar work has been reported for polyoxomolybdates.[205] In the ^{19}F n.m.r. spectrum of $[(FC_6H_4O)_nWF_{6-n}]$, the signals move to low frequency with increasing n with approximate sensitivities of $\delta(^{19}FW)$: $\delta(p\text{-}^{19}F)$: $\delta(m\text{-}^{19}F)$ of $1 : \frac{1}{20} : \frac{1}{30}$, respectively. All the chemical shifts were sensitive to molecular stereochemistry with the substituents *trans* to oxygen shifted to lower frequency than those *trans* to fluorine.[206] N.m.r. data have also

191 M. R. Snow and F. L. Wimmer, *Austral. J. Chem.*, 1976, **29**, 2349.
192 A. R. Siedle, *J. Organometallic Chem.*, 1976, **120**, 369.
193 B. D. Dombek and R. J. Angelici, *Inorg. Chem.*, 1976, **15**, 1089.
194 R. B. King, J. C. Cloyd, jun., and R. H. Reimann, *Inorg. Chem.*, 1976, **15**, 449.
195 M. A. M. Meester, R. C. J. Vriends, D. J. Stufkens, and K. Vrieze, *Inorg. Chim. Acta*, 1976, **19**, 95.
196 B. D. Dombek and R. J. Angelici, *Inorg. Chem.*, 1976, **15**, 2403.
197 O. R. Chambers, D. S. Rycroft, D. W. A. Sharp, and J. M. Winfield, *Inorg. Nuclear Chem. Letters*, 1976, **12**, 559.
198 P. C. Bevan, J. Chatt, R. A. Head, P. B. Hitchcock, and G. J. Leigh, *J.C.S. Chem. Comm.*, 1976, 509.
199 M. Hidai, T. Kodama, M. Sato, M. Harakawa, and Y. Uchida, *Inorg. Chem.*, 1976, **15**, 2694.
200 B. W. Epperlein, H. Krueger, O. Lutz, A. Nolle, and W. Mayr, *Z. Naturforsch.*, 1975, **30a**, 1237.
201 O. Lutz, A. Nolle, and P. Kroneck, *Z. Naturforsch.*, 1976, **31a**, 454.
202 W. D. Kautt, H. Krueger, O. Lutz, H. Maier, and A. Nolle, *Z. Naturforsch.*, 1976, **31a**, 351.
203 O. Lutz, W. Nepple, and A. Nolle, *Z. Naturforsch.*, 1976, **31a**, 978.
204 M. Filowitz and W. G. Klemperer, *J.C.S. Chem. Comm.*, 1976, 233.
205 M. Filowitz, W. G. Klemperer, L. Messerle, and W. Shum, *J. Amer. Chem. Soc.*, 1976, **98**, 2345.
206 L. B. Handy, C. Benham, F. E. Brinckman, and R. B. Johannesen, *J. Fluorine Chem.*, 1976, **8**, 55.

been reported for $[MoO_2Al_2(OPr^i)_4]$ (^{13}C n.m.r. data)[207] and $[(CF_3CH_2O)_n$-$MoF_{6-n}]$ (^{19}F n.m.r. data).[208]

Complexes of Mn and Re.—Information concerning complexes of these elements can be found at the following sources: $[(OC)_4Mn(Et_2NC{=}C{=}NEt_2)Mn(CO)_4]$,[148] $[RP(CH_2CH_2PR_2)_2M(CO)_3Br]$ (R = neopentyl; M = Mn or Re),[194] $[(\eta^5$-$C_5H_4BRX)Mn(CO)_3]$,[296] $[(OC)_3Mn(PhBCH{=}CHCH{=}CEt)Mn(CO)_3]$,[1198] $[(1,2$-$C_2B_9H_{11})Re(CO)_3]^-$,[1178] and cymantrene.[294]

Newly measured values and literature values of ^{55}Mn n.m.r. linewidths of ^{55}Mn containing organometallic compounds in THF solution have been compared with measured quadrupole coupling constants. A linear relationship between $(e^2qQ/h)^2$ and linewidth was observed; the slope corresponding to a rotational correlation time of 1.3 ps. Using scaled partial quadrupole splitting parameters (pqs) for X derived from Mössbauer data on Fe^{II} compounds, the pqs parameter for CO in $XMn(CO)_5$ complexes was determined to be -0.50 mm s^{-1}. On this basis the sign and magnitude of the quadrupolar coupling constants in $XMn(CO)_5$ compounds can be readily predicted from the use of pqs parameters determined for Fe^{II} compounds and tested by the observation of ^{55}Mn n.m.r. linewidths in THF solution.[209]

The ^{31}P n.m.r. chemical shifts of some metallated phosphorus ligands including $[(OC)_4MnP(OPh)_2OC_6H_4]$ have been examined and it was concluded that ^{31}P chemical shifts of *ortho*-metallated $(PhO)_3P$ occur at unusually high frequency compared with an unmetallated ligand having an equivalent stereochemical environment.[210] The ^{13}C n.m.r. spectrum of (16) and the corresponding carbinols have been reported and the structure and bonding discussed.[211, 212] On co-ordination of diphenylketen in (17) C^1 moves from δ 47.0 to δ 4.1 and C^2 moves from δ 201.3 to δ 237.5. Although two sets of phenyl signals should be observed, only one set is observed even at -55 °C.[213] The ^{13}C n.m.r. spectrum of (18) shows that the positive charge is delocalized on both metallocene moieties.[214] N.m.r. data have also been reported for two isomers of $[(\eta^5$-$C_5H_5)Re(CO)_2$-$(PPh_3)H]^+$ (^{13}C n.m.r. data),[215] $[\{Mn(CO)_4H\}P(CF_3)_2Fe(\eta^5$-$C_5H_5)(CO)_2]$ (^{19}F n.m.r. data),[216] (19) (^{19}F n.m.r. data),[217] $[Re(CO)_3(dppe)(COSiPh_3)]$ (^{13}C n.m.r. data),[218] $[(OC)_4Re(\mu$-$Br)(\mu$-$CPh)Re(CO)_4]$ (^{13}C n.m.r. data),[219] $[(\eta^5$-$C_5H_5)Mn$-

[207] T. Ouhadi, J. P. Bioul, C. Stevens, R. Warin, L. Hocks, and Ph. Teyssié, *Inorg. Chim. Acta*, 1976, **19**, 203.
[208] L. B. Handy, *J. Fluorine Chem.*, 1976, **7**, 641 (*Chem. Abs.*, 1976, **85**, 103 123).
[209] P. S. Ireland, C. A. Deckert, and T. L. Brown, *J. Magn. Resonance*, 1976, **23**, 485.
[210] R. P. Stewart, jun., L. R. Isbrandt, and J. J. Benedict, *Inorg. Chem.*, 1976, **15**, 2011.
[211] N. M. Loim, P. V. Petrovskii, V. I. Robas, Z. N. Parnes, and D. N. Kursanov, *J. Organometallic Chem.*, 1976, **117**, 265.
[212] A. G. Ginzburg, V. N. Setkina, P. V. Petrovskii, Sh. G. Kasumov, G. A. Panosyan, and D. N. Kursanov, *J. Organometallic Chem.*, 1976, **121**, 381.
[213] A. D. Redhouse and W. A. Herrmann, *Angew. Chem. Internat. Edn.*, 1976, **15**, 615.
[214] A. G. Ginzburg, V. N. Setkina, Sh. G. Kasumov, G. A. Panosyan, P. V. Petrovskii, and D. N. Kursanov, *Doklady Akad. Nauk S.S.S.R.*, 1976, **228**, 1368.
[215] N. I. Pyshnograyeva, V. N. Setkina, G. A. Panosyan, P. V. Petrovskii, Yu. V. Makarov, N. E. Kolobova, and D. N. Kursanov, *J. Organometallic Chem.*, 1976, **108**, 85.
[216] R. C. Dobbie and P. R. Mason, *J.C.S. Dalton*, 1976, 189.
[217] A. N. Nesmeyanov, N. E. Kolobova, I. B. Zlotina, B. V. Lokshin, I. F. Leshcheva, G. K. Znobina, and K. N. Anisimov, *J. Organometallic Chem.*, 1976, **110**, 339.
[218] J. R. Anglin and W. A. G. Graham, *J. Amer. Chem. Soc.*, 1976, **98**, 4679.
[219] E. O. Fischer, G. Huttner, T. L. Lindner, A. Frank, and F. R. Kreissl, *Angew. Chem. Internat. Edn.*, 1976, **15**, 231.

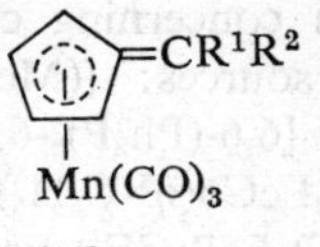

(16)

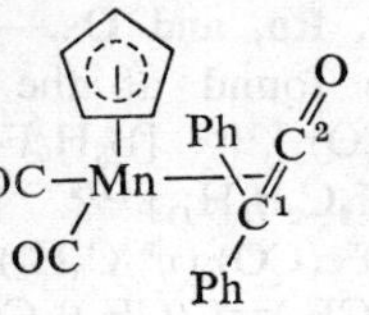

(17)

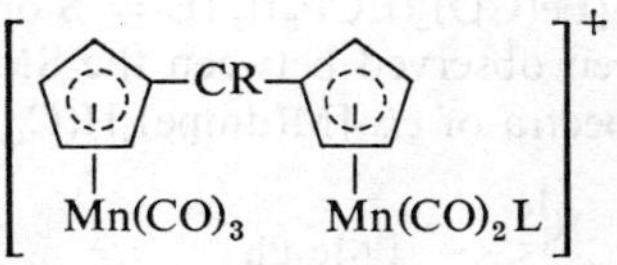

(18)

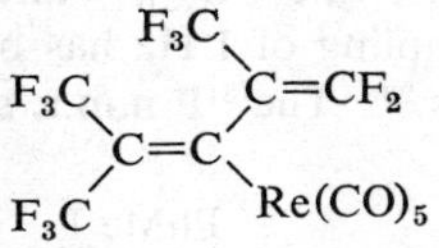

(19)

$(CO)_2CMe_2]$ (^{13}C n.m.r. data),[220] $[(\eta^5\text{-}C_5H_5)(OC)_2MnC{=}C{=}CBu^t{}_2]$ (^{13}C n.m.r. data),[221] $[(\eta^5\text{-}C_5H_5)(OC)_2ReCPh]^+$, and $[(\eta^5\text{-}C_5H_5)(OC)_2ReCMePh]^+$ (^{13}C n.m.r. data).[222]

^{13}C N.m.r. spectra of several carbonyl complexes of manganese and cobalt show that line broadening due to the spin–spin coupling of carbon nuclei by metals having $I > \frac{1}{2}$ is eliminated by thermal decoupling or by using viscous solutions. The latter technique is useful for variable-temperature n.m.r. studies.[223] The two isomers $[(OC)_4XMnSPHMe_2]$ and $[(OC)_4XMnP(SH)Me_2]$ were easily proved by ^{31}P n.m.r. spectroscopy. In the former case $^1J(^1H, ^{31}P)$ is *ca.* 500 Hz, and in the latter $^2J(^1H, ^{31}P)$ is *ca.* 4 Hz.[224] The ^{13}C n.m.r. signal of the CH_2 group in *cis*-$[ReOCl_3(Et_3P)]$ is complex and thought to be AA′X.[225] N.m.r. data have also been reported for $[Re_2(CO)_8(ECF_3)_2]$ (^{19}F n.m.r. data; E = P or As),[226] $[Mn(CO)_4(ECF_3)]_2$ (^{19}F n.m.r. data; E = S or Se),[227] $[Mn(CO)_4\text{-}(SCHAr_2)]_2$ (^{13}C n.m.r. data),[228] $[Mn_2(CO)_5(Ph_2PCH_2PPh_2)_2]$ (^{31}P n.m.r. data; $[AX]_2$ at −80 °C),[229] $[(OC)_4BrMSeHPPh_2]$ (^{31}P and ^{77}Se n.m.r. data; M = Mn or Re),[230] and (20) (^{31}P n.m.r. data).[231]

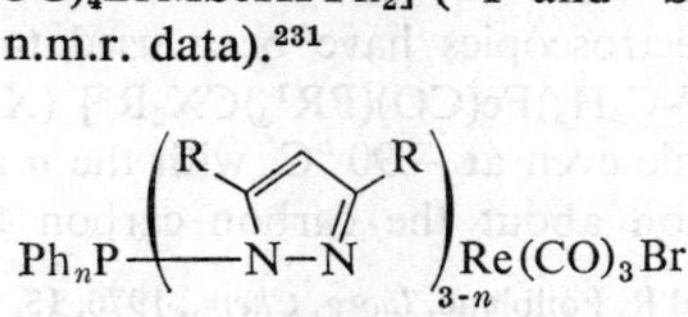

(20)

[220] E. O. Fischer, R. L. Clough, G. Besl, and F. R. Kreissl, *Angew. Chem. Internat. Edn.*, 1976, **15**, 543.
[221] H. Berke, *Angew. Chem. Internat. Edn.*, 1976, **15**, 624.
[222] E. O. Fischer, R. L. Clough, and P. Stückler, *J. Organometallic Chem.*, 1976, **120**, C6.
[223] S. Aime, G. Gervasio, L. Milone, and E. Rosenberg, *Transition Metal Chem.*, 1976, **1**, 177.
[224] E. Lindner and H. Dreher, *J. Organometallic Chem.*, 1976, **104**, 331.
[225] J. E. Fergusson and P. F. Heveldt, *J. Inorg. Nuclear Chem.*, 1976, **38**, 1976.
[226] J. Grobe and W. Mohr, *J. Fluorine Chem.*, 1976, **8**, 145.
[227] M. K. Chaudhuri, A. Haas, and A. Wensky, *J. Organometallic Chem.*, 1976, **116**, 323.
[228] H. Alper and H.-N. Paik, *J. Organometallic Chem.*, 1976, **121**, 225.
[229] K. G. Caulton and P. Adair, *J. Organometallic Chem.*, 1976, **114**, C11.
[230] E. Lindner and B. Schilling, *J. Organometallic Chem.*, 1976, **114**, C23.
[231] J. Hoyano and L. K. Peterson, *Canad. J. Chem.*, 1976, **54**, 2697.

Complexes of Fe, Ru, and Os.—Information concerning complexes of these elements can be found at the following sources: $[(Me_2C_2B_4H_4)FeH_2]$,[1150] $[(B_5H_3)(CO)_2Fe(CO)_3]$,[1154] $[B_5H_9Fe(CO)_3]$,[1155] $[6,6\text{-}(Ph_3P)_2\text{-}6,2,3\text{-}RuC_2B_7H_9]$,[1166] $3,1,2\text{-}[(Ph_3P)_2RuH_2C_2B_9H_{11}]$,[1179] $[Me_4C_4B_8H_8FeCo(\eta^5\text{-}C_5H_5)]$,[1189] $[(CF_3)_2C{=}C\text{-}\{C(CF_3){=}CF_2\}Fe(CO)_2(\eta^5\text{-}C_5H_5)]$,[217] $[(OC)_4FeP(CF_3)_2Fe(\eta^5\text{-}C_5H_5)(CO)_2]$,[216] $[(\eta^5\text{-}C_5H_5)_2Fe_2\{C(CF_3){=}C(CF_3)\}_2CO]$,[159] $[(\eta^5\text{-}C_7H_9)(\eta^5\text{-}C_7H_7)Fe]$,[154] substituted ferrocenes,[108] $[(OC)_3Mn\{\eta^5\text{-}C_5H_4CR(OH)C_5H_4\text{-}\eta^5\}Fe(\eta^5\text{-}C_5H_5)]$,[214] $[(\eta^5\text{-}Br\text{-}(OC)_4W{\equiv}CC_5H_4)(\eta^5\text{-}C_5H_5)Fe]$,[145] $[(\eta^5\text{-}C_5H_5)Fe(CO)_2\{PPh_2(CH_2)_nPPh_2\}]^+$,[160] $[(OC)_3Fe(\mu\text{-}NCEt_2)_2Fe(CO)_3]$,[148] and $[Fe(CO)_3(ECF_3)]_2$ (E = S or Se).[227]

In (21), a coupling of 1 Hz has been observed between the SMe and hydride hydrogen atoms.[232] The ^{31}P n.m.r. spectra of *cis*-$[M(dmpe)_2H(C_{10}H_7)]$ (M = Fe

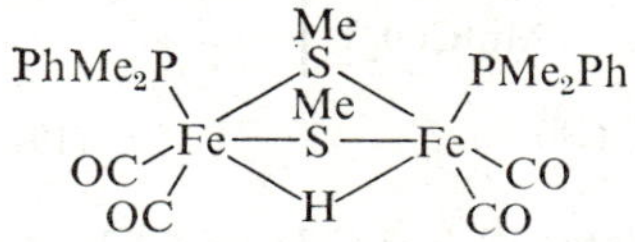

(21)

or Ru) have been fully analysed as ABCD.[233] It was claimed that the ^{31}P n.m.r. spectrum of $HRu(diop)_2^+$ shows the presence of two inequivalent phosphorus atoms with $^2J(^{31}P, ^{31}P) = 40$ Hz. This is probably an oversimplification as it should be $[AX]_2$.[234] The hydride signal of $[HRu_6(CO)_{18}]^-$ is at δ −16.41 with T_1 of 20 ± 3.5 s, yet for $[HOs_6(CO)_{18}]^-$ the hydride signal is at δ −11.26. The ^{13}C n.m.r. spectrum of $[HRu_6(CO)_{18}]^-$ is a singlet even at −104 °C,[235] whereas $[H_2Os_3(CO)_9PEt_3]$ shows six signals for the ^{13}CO ligands.[236] At −85 °C, the ^{13}C n.m.r. spectrum of $[H_2Os_6(CO)_{18}]$ shows 10 signals in the ratio 1 : 1 : 1 : 1 : 2 : 2 : 2 : 2 : 2 : 4, in agreement with the solid-state structure.[237] ^{31}P N.m.r. data have also been reported for $[(OC)(\eta^5\text{-}C_5H_5)FeHPPh_3]$ (including ^{13}C n.m.r. data),[238] $[RuHX(PPh_3)_3]$,[239] $[RuL(CO_2Me)_2(PPh_3)_3]$ (L = H, CO, or H_2O),[240] $[RuH(\eta^6\text{-}C_6H_5PPh_2)(PPh_3)_2]^+$,[241] $[HOs_3(CO)_9C_9H_{13}]$ and (22) (only ^{13}C n.m.r. data).[242]

^1H and ^{19}F n.m.r. spectroscopies have been used to demonstrate that the preferred rotamer of $[(\eta^5\text{-}C_5H_5)Fe(CO)(PR^1_3)CX_2R^2]$ (X = H or F) is (23).[243] Compound (24) is dynamic even at −90 °C, with the σ and π roles of the iron interchanging and rotation about the carbon–carbon bonds.[244] The $^{13}C\{^1H\}$

[232] K. Fauvel, R. Mathieu, and R. Poilblanc, *Inorg. Chem.*, 1976, **15**, 976.
[233] S. D. Ittel, C. A. Tolman, A. D. English, and J. P. Jesson, *J. Amer. Chem. Soc.*, 1976, **98**, 6073.
[234] B. R. James and D. K. W. Wang, *Inorg. Chim. Acta*, 1976, **19**, L17.
[235] C. R. Eady, B. F. G. Johnson, J. Lewis, M. C. Malatesta, P. Machin, and M. McPartlin, *J.C.S. Chem. Comm.*, 1976, 945.
[236] E. G. Bryan, W. G. Jackson, B. F. G. Johnson, J. W. Kelland, J. Lewis, and K. T. Schorpp, *J. Organometallic Chem.*, 1976, **108**, 385.
[237] M. McPartlin, C. R. Eady, B. F. G. Johnson, and J. Lewis, *J.C.S. Chem. Comm.*, 1976, 883.
[238] D. L. Reger and E. C. Culbertson, *J. Amer. Chem. Soc.*, 1976, **98**, 2789.
[239] A. Spencer, *Inorg. Nuclear Chem. Letters*, 1976, **12**, 661.
[240] R. J. Young and G. Wilkinson, *J.C.S. Dalton*, 1976, 719.
[241] D. J. Cole-Hamilton, R. J. Young, and G. Wilkinson, *J.C.S. Dalton*, 1976, 1995.
[242] E. G. Bryan, B. F. G. Johnson, and J. Lewis, *J. Organometallic Chem.*, 1976, **122**, 249.
[243] K. Stanley, R. A. Zelonka, J. Thompson, and M. C. Baird, 'Proceedings of the 16th International Conference on Coordination Chemistry', 1974, 4.19 (*Chem. Abs.*, 1976, **85**, 77 490).
[244] R. C. Kerber, W. P. Giering, T. Bauch, P. Waterman, and E.-Hua Chou, *J. Organometallic Chem.*, 1976, **120**, C31.

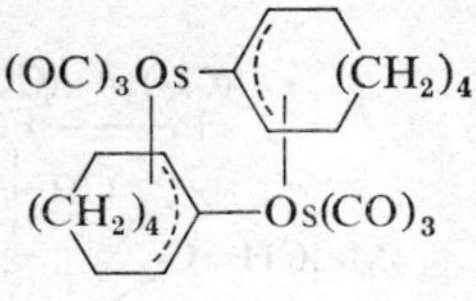

(22)

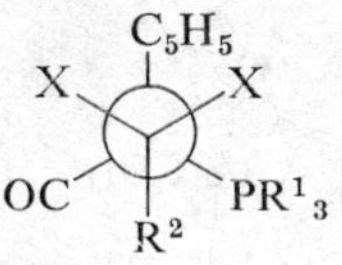

(23)

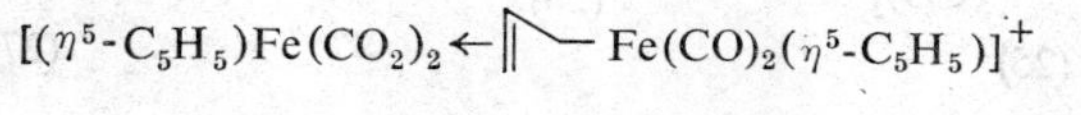

(24)

n.m.r. spectra of $[(\eta^5\text{-}C_5H_5)FeC_6H_4OP(OPh)_2(CO)]$ and $[(\eta^5\text{-}C_5H_5)FeC_6H_4OP\text{-}(OPh)_2\{P(OPh)_3\}]$ have been determined in order to demonstrate the use of ^{13}C n.m.r. in confirming *ortho*-metallation of co-ordinated $P(OPh)_3$. The metallated phenyl ring signals are spread over a range of more than 50 p.p.m., whereas the other phenyl ring signals are about the same as free $P(OPh)_3$.[245] The carbene signal of $[(MeNC)_4FeC(NHMe)NMeCNHMe]^{2+}$ is at δ 210.5, whereas for $[(Bu^tNC)_2MeIRhC(NHBu^t)NBu^nCNHBu^t]$ it is at δ 187, with $^1J(^{13}C, ^{103}Rh)$ = 35 Hz.[246] N.m.r. data have also been reported for *cis*- and *trans*-[Fe(*trans*-σ-CF=CFCF$_3$)(CO)$_4$I] (^{19}F n.m.r. data),[247] $[C_3F_7Fe(CO)_2L]^+$ (^{19}F n.m.r. data),[248] $[(\eta^5\text{-}C_5H_5)Fe(CO)_2(CH_2)_nFe(CO)_2(\eta^5\text{-}C_5H_5)]$ (^{13}C n.m.r. data),[249] $[(\eta^5\text{-}C_5H_5)\text{-}Fe(CO)(PPh_3)(CH_2CHMe_2)]$ (^{13}C n.m.r. data),[250] $[(\eta^5\text{-}C_5H_5)(Ph_3P)\text{-}RuC_6F_4N{=}NAr]$ (^{19}F n.m.r. data),[251] (25) (^{13}C n.m.r. data),[252] $[(OC)_3\text{-}Fe(\eta^3\text{-}C_3H_4CF_2CF_2)]$ (^{19}F n.m.r. data),[253] (26) (^{13}C n.m.r. data),[254] $[(OC)_4\text{-}FeCO_2Me]^-$ (^{13}C n.m.r. data),[255] (27) (^{13}C n.m.r. data),[256] *trans*-[Ru(CO)Cl-(CNMeCH$_2$CH$_2$NMe)]$^+$ (^{13}C n.m.r. data),[257] and (28) (^{13}C n.m.r. data).[258]

Eu(CF$_3$COcamphorate)$_3$ has been used to demonstrate optical activity in $[(MeO_2CCH{=}CHCO_2Me)Fe(CO)_3]$.[259] The ^{13}C n.m.r. spectrum of (29) shows three Si—Me signals.[260] Completely coupled and ^1H decoupled ^{13}C n.m.r. spectra

[245] R. P. Stewart, jun., L. R. Isbrandt, J. J. Benedict, and J. G. Palmer, *J. Amer. Chem. Soc.*, 1976, **98**, 3215.
[246] P. R. Branson, R. A. Cable, M. Green, and M. K. Lloyd, *J.C.S. Dalton*, 1976, 12.
[247] K. Stanley and D. W. McBride, *Canad. J. Chem.*, 1976, **54**, 1700.
[248] D. W. Hensley and R. P. Stewart, jun., *Inorg. Chem.*, 1976, **15**, 2418.
[249] L. Pope, P. Sommerville, M. Laing, K. J. Hindson, and J. R. Moss, *J. Organometallic Chem.*, 1976, **112**, 309.
[250] D. L. Reger and E. C. Culbertson, *Syn. React. Inorg. Metal.-Org. Chem.*, 1976, **6**, 1.
[251] M. I. Bruce, R. C. F. Gardner, and F. G. A. Stone, *J.C.S. Dalton*, 1976, 81.
[252] F. R. Young, D. H. O'Brien, R. C. Pettersen, R. A. Levenson, and D. L. von Minden, *J. Organometallic Chem.*, 1976, **114**, 157.
[253] R. C. Kerber and E. A. Koerner von Gustorf, *J. Organometallic Chem.*, 1976, **110**, 345.
[254] R. Aumann and J. Knecht, *Chem. Ber.*, 1976, **109**, 174.
[255] W. Petz, *J. Organometallic Chem.*, 1976, **105**, C19.
[256] E. Sappa, L. Milone, and A. Tiripicchio, *J.C.S. Dalton*, 1976, 1843.
[257] P. B. Hitchcock, M. F. Lappert, and P. L. Pye, *J.C.S. Chem. Comm.*, 1976, 644.
[258] H. Alper, *Org. Magn. Resonance*, 1976, **8**, 58 7.
[259] V. Schurig, *Tetrahedron Letters*, 1976, 1269.
[260] H. Sakurai, Y. Kamiyama, and Y. Nakadaira, *J. Amer. Chem. Soc.*, 1976, **98**, 7453.

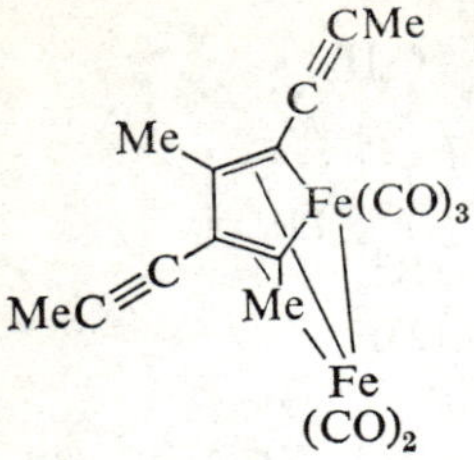

(25)

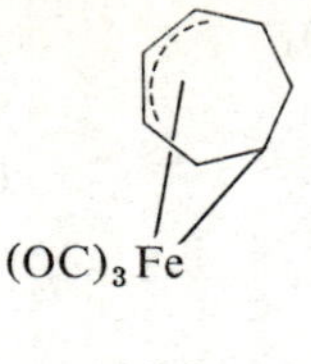

(26)

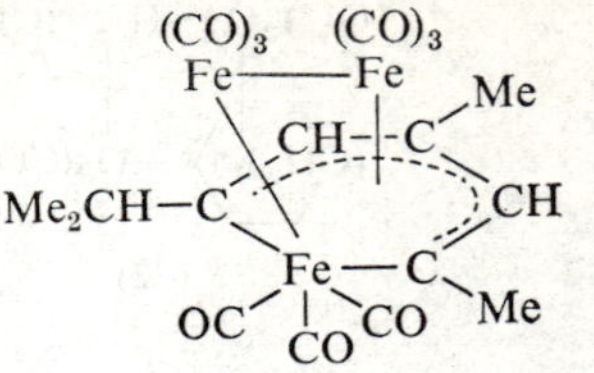

(27)

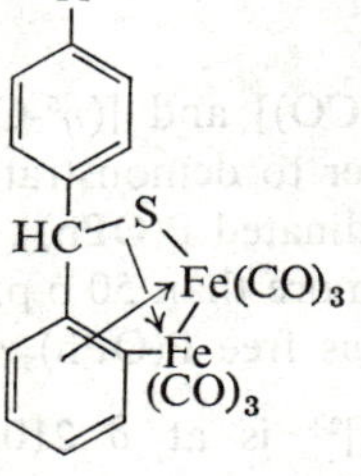

(28)

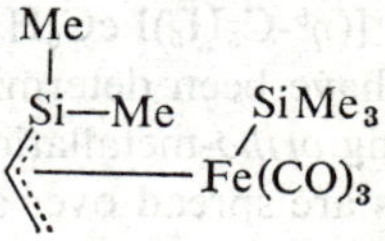

(29)

of [(η^4-butadiene)Fe(CO)$_3$] have been presented. The structure of the ligand was interpreted, on the basis of coupling data, in terms of a non-planar CH skeleton in which the carbon–carbon bond lengths are nearly equal and terminal carbon atoms exhibit some rehybridization towards sp^3. The results agree with *X*-ray crystallographic data.[261] The ^{19}F n.m.r. spectra of [(η^4-FC$_6$H$_4$CH=CHCH=CH$_2$)-Fe(CO)$_3$] and [{η^4-CH$_2$=C(C$_6$H$_4$F)CH=CH$_2$}Fe(CO)$_3$], where the fluorine atom is in the *meta* or *para* position, indicate that position 1 of the butadienyl ligand has a higher π-electron density than position 2. The σ-electron density is distributed equally. The inductive and resonance σ-constants for the [(η^4-1-butadienyl)Fe(CO)$_3$] substituent are 0.10 and -0.07 and for the [(η^4-2-butadienyl)Fe(CO)$_3$] substituent are 0.11 and 0.00, respectively.[262] From the ^{13}C n.m.r. spectrum of (30), $^1J(^1\text{H}, ^{13}\text{C})$ for the cyclobutadiene ring is 182 Hz.[263]

(30)

For (31, X = [C^1=O]$^-$), $\delta(^{13}\text{C}^1\text{=O})$ = 240, which is consistent with an acetyl, but when X = C^{1}OMe, $\delta(^{13}\text{C}^1\text{O})$ = 270 which is consistent with carbene bonding.[264] On the basis of ^{13}C n.m.r. evidence on [(η^4-C$_4$H$_3$CHR)Fe(CO)$_3$]$^+$ it was suggested that there is extensive donation of electron density from the iron *via* the cyclobutadiene to the exocyclic electron-deficient carbon atom, thus

[261] K. Bachmann and W. von Philipsborn, *Org. Magn. Resonance*, 1976, **8**, 648.
[262] N. S. Nametkin, A. I. Nekhaev, S. P. Gubin, and V. D. Tyurin, *Izvest. Akad. Nauk S.S.S.R., Ser. khim.*, 1976, **7**, 1636.
[263] A. Sanders and W. P. Giering, *J. Organometallic Chem.*, 1976, **104**, 49.
[264] T. Mitsudo, H. Nakanishi, T. Inubushi, I. Morishima, Y. Watanabe, and Y. Takegami, *J.C.S. Chem. Comm.*, 1976, 416.

giving highly stabilized carbonium ions. Thus when R = H, $\delta(^{13}CH_2)$ = 104.5.[265] From a comparison of β-effects in the ^{13}C n.m.r. spectrum given by Me substituents in $[(\eta^4\text{-diene})Fe(CO)_3]$ complexes with those of the corresponding uncomplexed 1,3-dienes, it was concluded that bonding to iron results in a change in the π-bond orders of the ligand. This change is consistent with (32) making an important contribution to the bonding in these compounds.[266] In contrast, on the basis of the effect of $[(\eta^4\text{-butadiene})Fe(CO)_3]$ substituents on ^{19}F n.m.r. chemical shifts in *m*- and *p*-substituted fluorobenzenes, it was concluded that all the carbon atoms are sp^2 hybridized.[267] The ^1H n.m.r. spectrum of (33) has been fully analysed as $[AX]_2$.[268] At −90 °C, the ^{13}C n.m.r. spectrum

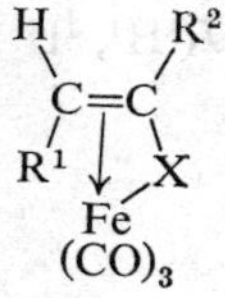

(31)

(32)

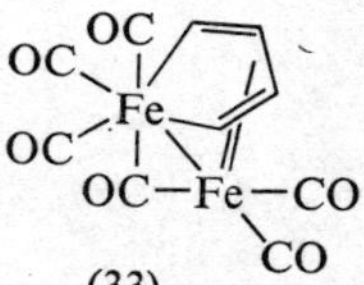

(33)

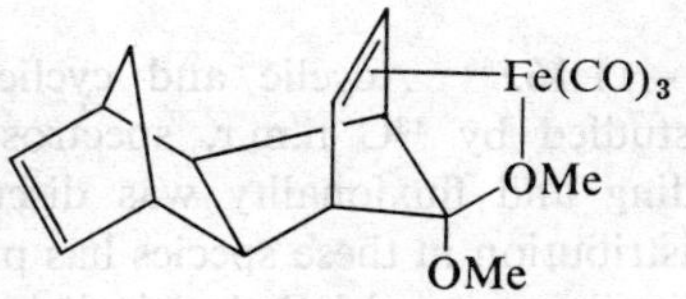

(34)

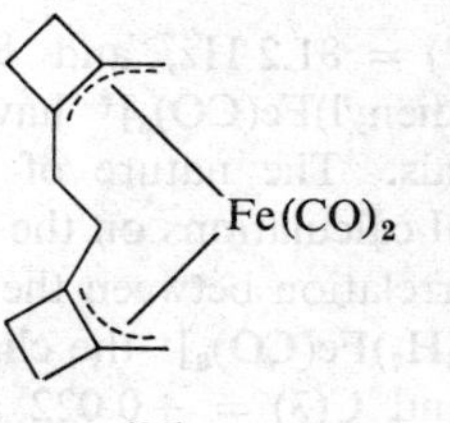

(35)

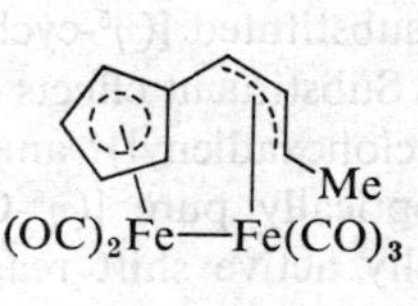

(36)

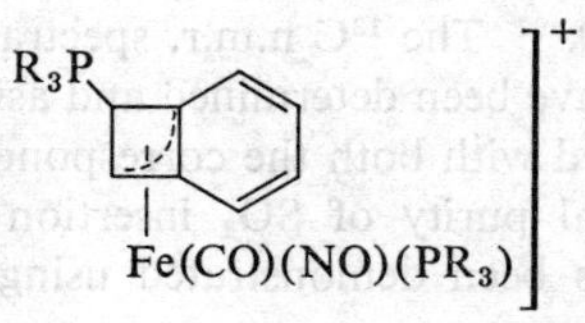

(37)

of $[(\eta^4\text{-butadiene})Os_3(CO)_{10}]$ shows nine ^{13}CO n.m.r. signals.[269] N.m.r. data have also been reported for $[\{\eta^2\text{-}CH{=}CH_2P(O)Ph_2\}Fe(CO)_3]$ (^{13}C n.m.r. data),[270] (34) (^{13}C n.m.r. data),[271] (35) (^{13}C n.m.r. data),[272] (36) (^{13}C n.m.r. data),[273] (37) (^{31}P

[265] C. S. Scott, D. Seyferth, and P. C. Reeves, *J. Organometallic Chem.*, 1976, **104**, 363.
[266] A. J. Pearson, *Austral. J. Chem.*, 1976, **29**, 1679.
[267] N. S. Nametkin, S. P. Gubin, A. I. Nekhaev, and V. D. Tyurin, *J. Organometallic Chem.*, 1976, **114**, 299.
[268] G. Dettlaf and E. Weiss, *J. Organometallic Chem.*, 1976, **108**, 213.
[269] E. G. Bryan, B. F. G. Johnson, J. W. Kelland, J. Lewis, and M. McPartlin, *J.C.S. Chem. Comm.*, 1976, 254.
[270] B. F. G. Johnson, J. Lewis, and S. R. Postle, *J. Organometallic Chem.*, 1976, **121**, C7.
[271] P. Laszlo and A. Stockis, *J. Organometallic Chem.*, 1976, **117**, C41.
[272] R. B. King and C. A. Harmon, *J. Amer. Chem. Soc.*, 1976, **98**, 2409.
[273] J. D. Edwards, S. A. R. Knox, and F. G. A. Stone, *J.C.S. Dalton*, 1976, 1813.

n.m.r. data),[274] various η^4-butadiene iron carbonyl derivatives (^{13}C [275–281] and ^{31}P [277] n.m.r. data), and $[RuCl_2(\eta^4\text{-diene})(amine)_2]$ (^{13}C n.m.r. data).[282]

The species generated from protonation of $[(\eta^4\text{-butadiene})Fe(CO)_3]$ and $[(\eta^4\text{-cyclohexadiene})Fe(CO)_3]$ in strongly acidic media have been studied by i.r. and ^{13}C n.m.r. spectroscopies. Protonation occurs at the iron in these species to yield monoprotonated metal hydride species. Based on ^{13}C n.m.r. data, the best description of the bonding is where the iron is σ-bonded to a terminal carbon atom and π-bonded to the remaining allyl moiety. Analysis of $^1J(^1H, ^{13}C)$ suggests that the hydride may be bridging between the iron and the carbon in the protonated species. When FSO_3D is used as the acid deuteriation occurs at both ends of the butadiene.[283, 284] When $[(\eta^4\text{-cyclobutadiene})Fe(CO)_3]^+$ is protonated it is similarly thought that (38) is formed with $J(^1H^1, ^1H^5) = 29$ Hz,

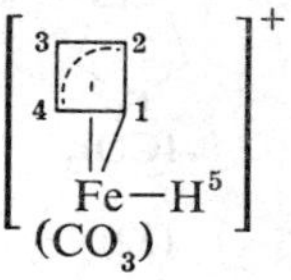

(38)

$J(^1H^5, ^{13}C^1) = 81.2$ Hz, and $\delta(^1H^5) = -11.16$.[285] Acyclic and cyclic *cisoid*-$[(\eta^5\text{-pentadienyl})Fe(CO)_3]^+$ have been studied by ^{13}C n.m.r. spectroscopy in strong acids. The nature of the bonding and fluxionality was discussed.[286] Theoretical calculations on the charge distribution in these species has produced a good correlation between the charge distribution and ^{13}C chemical shift. Thus for $[(\eta^5\text{-}C_5H_7)Fe(CO)_3]^+$ the charge distribution is C(1, 5) = -0.204, C(2, 4) = $+0.137$, and C(3) = $+0.022$ and the ^{13}C chemical shifts of $[(\eta^5\text{-1-MeC}_5H_6)\text{-}Fe(CO)_3]^+$ are $\delta(C^1) = 91.0$, $\delta(C^2) = 103.2$, $\delta(C^3) = 95.3$, $\delta(C^4) = 103.7$, and $\delta(C^5) = 62.0$.[287] The ^{13}C n.m.r. spectra of 11 substituted $[(\eta^5\text{-cyclohexadienyl})\text{-}Fe(CO)_3]^+$ have been determined and assigned. Substituent effects were deduced and compared with both the corresponding [cyclohexadienyl]$^+$ and reactivity.[288]

The optical purity of SO_2 insertion into optically pure $[(\eta^5\text{-}C_5H_5)Fe(CO)\text{-}(PPh_3)R]$ has been demonstrated using optically active shift reagents.[289] The

[274] A. Efraty, D. Liebman, J. Sikora, and D. Z. Denney, *Inorg. Chem.*, 1976, **15**, 886.
[275] H. Lumbroso and D. M. Bertin, *J. Organometallic Chem.*, 1976, **108**, 111.
[276] P. McArdle and H. Sherlock, *J.C.S. Chem. Comm.*, 1976, 537.
[277] R. A. Cable, M. Green, R. E. Mackenzie, P. L. Timms, and T. W. Turney, *J.C.S. Chem. Comm.*, 1976, 270.
[278] P. McArdle and H. Sherlock, *J. Organometallic Chem.*, 1976, **116**, C23.
[279] J. H. Eekhof, H. Hogeveen, and R. M. Kellogg, *J.C.S. Chem. Comm.*, 1976, 657.
[280] A. J. Birch and A. J. Pearson, *J.C.S. Chem. Comm.*, 1976, 601.
[281] M. S. Brockhart, G. W. Koszalka, G. O. Nelson, G. Scholes, and R. A. Watson, *J. Amer. Chem. Soc.*, 1976, **98**, 8155.
[282] C. Potvin, J. M. Manoli, G. Pannetier, R. Chevalier, and N. Platzer, *J. Organometallic Chem.*, 1976, **113**, 273.
[283] M. Brockhart, T. H. Whitesides, and J. M. Crockett, *Inorg. Chem.*, 1976, **15**, 1550.
[284] G. A. Olah, G. Liang, and S. H. Yu, *J. Org. Chem.*, 1976, **41**, 2227.
[285] G. A. Olah and G. Liang, *J. Org. Chem.*, 1976, **41**, 2659.
[286] G. A. Olah, S. H. Yu, and G. Liang, *J. Org. Chem.*, 1976, **41**, 2383.
[287] P. A. Dobosh, D. G. Gresham, C. P. Lillya, and E. S. Magyar, *Inorg. Chem.*, 1976, 2311.
[288] A. J. Birch, P. W. Westerman, and A. J. Pearson, *Austral. J. Chem.*, 1976, **29**, 1671.
[289] T. C. Flood, F. J. DiSanti, and D. L. Miles, *Inorg. Chem.*, 1976, **15**, 1910.

Ln(fod)$_3$ lanthanide shift reagents will complex to [(η^5-C_5H_5)Fe(CO)$_2$CN].[290] In [(η^5-C_5H_5)Fe(CO)(CS)L]$^+$, δ(^{13}CS) = *ca.* 320.[291] Although in the crystal structure the two rings in [Fe(η^5-C_7H_7)(η^5-C_7H_9)] are staggered by *ca.* 50°, there is partial rotation to generate an apparent plane of symmetry on the ^{1}H and ^{13}C n.m.r. time-scales at room temperature.[292] Natural abundance ^{13}C–^{57}Fe and ^{13}C–^{13}C satellite spectra have been obtained from ^{13}C n.m.r. spectra of ferrocene, 1,1′-dimethylferrocene, and [(η^4-cyclobutadiene)Fe(CO)$_3$]. Spectral analysis yielded ^1J(^{13}C, ^{57}Fe) and J(^{13}C, ^{13}C). J(^{13}C, ^{57}Fe) were calculated using two MO descriptions and compared with experimental values. The J(^{13}C, ^{13}C) are outside the range found for aromatic compounds.[293] A Hammett correlation of the n.m.r. chemical shifts of the α, β, and 1′ protons of 29 substituted phenylferrocenes has shown that the resonance effect is transmitted approximately equally to the α- and β-positions, that the inductive effect is transmitted more strongly to the α- than to the β-position, and that transmission across the iron atom resembles transmission through a *p*-phenylene group. Similar correlations were discussed for substituted ferrocenes, [(η^6-C_6H_6)(η^5-C_5H_5)Fe]$^+$, [(η^5-C_5H_5)$_2$-Co]$^+$, and cymantrenes.[294] In the ^{1}H n.m.r. spectrum of [(η^5-XC_5H_4)Fe-(η^5-C_5H_4CHMe)]$^+$ in strong acid the hydrogen atoms are inequivalent, demonstrating hindered rotation of the CHMe group relative to the C_5H_4 group.[295] ^{1}H and ^{13}C n.m.r. spectroscopies indicate that in contrast to [(η^5-C_5H_4BRX)-Mn(CO)$_3$], in [(η^5-C_5H_5)(η^5-C_5H_4BRX)Fe] the 3,4 protons are more deshielded than the 2,5 protons.[296] The ^{1}H n.m.r. spectra of the *E*- and *Z*-isomers of [39; Fc = (η^5-C_5H_5)(η^5-C_5H_4)Fe] showed that unlike other cyclopropane derivatives, signals due to the H^1 protons in the *E*-isomer occur at lower field relative to those in the *Z*-isomers.[297] N.m.r. data have also been reported for [M(C_5H_5BR)$_2$] (^{11}B n.m.r. data; M = Ru or Os),[298] (40) (^{11}B n.m.r. data),[299] (41) (^{13}C n.m.r. data),[300] [(η^5-C_5H_5)(η^5-C_5H_4CHPr^i)Fe]$^+$ (^{13}C n.m.r. data),[301] [(η^5-C_5H_5)(η^5-C_5H_4SiPh$_2$F)Fe] (^{19}F n.m.r. data),[302] and [(η^5-C_5H_5)(η^6-halonaphthalene)Fe]$^+$ (^{13}C n.m.r. data).[303]

290 T. J. Marks, R. Porter, and D. F. Shriver, *Proc. Rare Earth Res. Conf., 10th, 1 Conf-730402-P1*, 1973, 372 (*Chem. Abs.*, 1976, **85**, 37 817).
291 L. Busetto and A. Palazzi, *Inorg. Chim. Acta*, 1976, **19**, 233.
292 J. R. Blackborow, K. Hildenbrand, E. A. Koerner von Gustorf, A. Scrivanti, C. R. Eady, and D. Ehntolt, *J.C.S. Chem. Comm.*, 1976, 16.
293 P. S. Nielsen, R. S. Hansen, and H. J. Jakobsen, *J. Organometallic Chem.*, 1976, **114**, 145.
294 S. Toma, E. Solcaniova, and A. Maholanyiova, *Chem. Zvesti*, 1975, **29**, 526 (*Chem. Abs.*, 1976, **84**, 120 876).
295 L. I. Kazakova, I. F. Leshcheva, and L. A. Kazitsyna, *Zhur. org. Khim.*, 1976, **12**, 131.
296 T. Renk, W. Ruf, and W. Siebert, *J. Organometallic Chem.*, 1976, **120**, 1.
297 A. Ratajczak, B. Czech, and H. Zyzik, *Roczniki Chem.*, 1975, **49**, 953 (*Chem. Abs.*, 1976, **84**, 3921).
298 G. E. Herberich, H. J. Becker, K. Carsten, C. Engelke, and W. Koch, *Chem. Ber.*, 1976, **109**, 2382.
299 W. Seibert, T. Renk, K. Kinberger, M. Bochmann, and C. Krüger, *Angew. Chem. Internat. Edn.*, 1976, **15**, 779.
300 D. Astruc, R. Dabard, M. Martin, P. Batail, and D. Grandjean, *Tetrahedron Letters*, 1976, 829.
301 C. C. Lee, S. C. Chen, W. J. Pannekoek, and R. G. Sutherland, *J. Organometallic Chem.*, 1976, **118**, C47.
302 P. Bickart, F. M. Llort, and K. Mislow, *J. Organometallic Chem.*, 1976, **116**, C1.
303 R. G. Sutherland, S. C. Chen, W. J. Pannekoek, and C. C. Lee, *J. Organometallic Chem.*, 1976, **117**, 61.

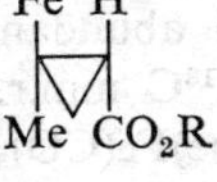

(39)

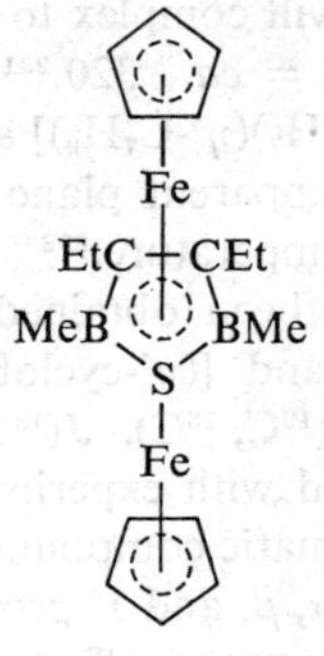

(40)

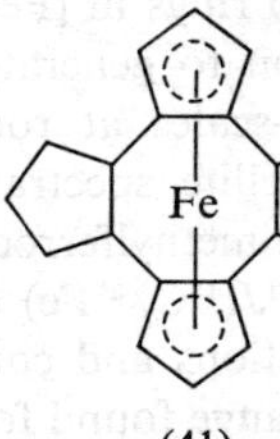

(41)

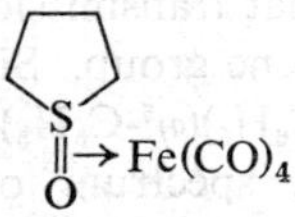

(42)

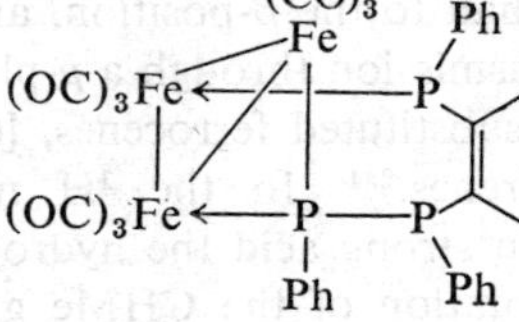

(43)

The ^{13}C n.m.r. spectrum of (42) shows only one ^{13}CO resonance even at -90 °C.[304] Separate signals are observed for the *cis* and *trans* ^{13}CN carbon atoms in $[Fe^{II}(CN)_5X]^{n-}$.[305] For a series of complexes [M(porphyrin)XY] (M = Ru, In, or Ti) the ^{13}C chemical shift is significantly dependent on the metal.[306] ^{31}P N.m.r. spectroscopy has been used to show that when $[RhCl_3L_3]$ is mixed with $[RuCl_2(PPh_3)_3]$, the equilibrium lies strongly in favour of $[(Ph_3P)$-$LClRuCl_3RhClL_2]$ ($L = PR_2Ph$).[307] ^{31}P N.m.r. data have also been reported for $[Fe_3(CO)_9(\mu_3\text{-}PR)_2]$, *e.g.* R = NEt$_2$, $\delta(^{31}P) = 401.6$,[308] (43),[309] $[(Ph_3P)_2ClRuCl_3Ru$-$(CO)(PPh_3)_2]$,[310] $[Ru(CO)_3(PPh_3)_2]$,[311] $[Fe(NO)(dppe)_2]^+$,[312] and $[Ru\{Ph_2P(CH_2)_n$-$PPh_2\}_2Cl_2]$.[313] N.m.r. data have also been reported for $[OsO_2$(pyridine)-$OCMe_2CMe_2O]$ (^{13}C n.m.r. data)[314] and $[MF_6]^{2-}$ (M = Ru or Os; ^{19}F n.m.r. data).[315]

Complexes of Co, Rh, and Ir.—Information concerning complexes of these elements can be found at the following sources: $[HRh(CO)(PPh_3)_3]$,[312] (HCl-

[304] J. H. Eekhof, H. Hogeveen, R. M. Kellogg, and E. P. Schudde, *J. Organometallic Chem.*, 1976, **105**, C35.
[305] A. D. James and R. S. Murray, *Inorg. Nuclear Chem. Letters*, 1976, **12**, 739.
[306] S. S. Eaton and G. R. Eaton, *Inorg. Chem.*, 1976, **15**, 134.
[307] R. A. Head and J. F. Nixon, *J.C.S. Chem. Comm.*, 1976, 62, 368.
[308] J. P. Lampin and F. Mathey, *Compt. rend.*, 1976, **282**, *C*, 979.
[309] R. B. King and R. H. Reimann, *Inorg. Chem.*, 1976, **15**, 184.
[310] P. W. Armit, W. J. Sime, and T. A. Stephenson, *J.C.S. Dalton*, 1976, 2121.
[311] H. B. Kuhnhen, *J. Organometallic Chem.*, 1976, **105**, 357.
[312] C. B. Ungermann and K. G. Caulton, *J. Amer. Chem. Soc.*, 1976, **98**, 3862.
[313] R. Mason, D. W. Meek, and G. R. Scollary, *Inorg. Chim. Acta*, 1976, **16**, L11.
[314] L. G. Marzilli, B. E. Hanson, T. J. Kistenmacher, L. A. Epps, and R. C. Stewart, *Inorg. Chem.*, 1976, **15**, 1661.
[315] S. V. Zemskov, S. P. Gabuda, V. N. Mit'kin, L. V. Lavrova, V. G. Isakova, B. I. Obmoin, Yu. I. Nikonorov, and A. M. Panich, *Zhur. strukt. Khim.*, 1976, **17**, 483.

$(R^1{}_3P)_2Rh(\mu\text{-}Cl)_2PtCl(PR^2{}_3)]$,[458] $[(\eta^5\text{-}C_5H_5)Co(Me_2C_2B_3H_3)CoH(Me_2C_2B_3H_3)Co(\eta^5\text{-}C_5H_5)]$,[1156] [6,6-$(R_3P)_2$-6-H-6,2,3-$MC_2B_7H_9$] (M = Rh or Ir), 2-[$(Ph_3P)_2$IrHCl-1-$SB_9H_8$],[1168] $[(\eta^5\text{-}C_5H_5)_4Co_4B_4H_4]$,[1152] 1,7,2,3-$[(\eta^5\text{-}C_5H_5)_2Co_2C_2B_3H_5]$,[1151] $[(\eta^5\text{-}C_5H_5)Co(Me_2C_2B_3H_3)Fe(CO)_3]$,[1150] $[(\eta^5\text{-}C_5H_5)CoCB_9H_{10}]^-$,[1167] $[(B_{10}H_{12})Ir(CO)\{S_2C_2(CN)_2\}(PPh_3)_2]^-$,[1169] $[(\eta^5\text{-}C_5H_5)Co(7\text{-}B_{10}H_{10}PR)]$,[1172] $[(\eta^5\text{-}C_5H_5)_2Co_2(CB_9H_{10})]^-$,[1157] [8,8′-$RE(C_2B_9H_{10})_2Co$] (E = O, Se, or Te),[1180] $[(Me_4C_4B_8H_8)FeCo(\eta^5\text{-}C_5H_5)]$,[1189] $[(\eta^3\text{-}C_7H_7)Co(PF_3)_3]$,[154] [$(\eta^4$-cyclo-octadiene)$Rh(\eta^6\text{-}C_5H_5BR)$],[298] $[(\eta^5\text{-}C_5H_5)Rh\{\eta^4\text{-}(EtNC{=}CNEt)_2CO\}]$,[148] $[(\eta^5\text{-}C_5H_5)_2Co]^+$,[294] Co carbonyl complexes,[223] $[Co(CO)_2(NO)P(CF_3)_2Fe(\eta^5\text{-}C_5H_5)(CO)_2]$,[216] Co^{III} diamine complexes,[184] $[(PPh_3)LClRuCl_3RhClL_2]$,[307] and $[\{R_nP(CH_2CH_2PR_2)_{3-n}\}_2Rh]^+$ (R = neopentyl).[194]

Reviews on cobalt(III) compounds entitled 'N.M.R. of Metal Complexes. Interpretation of Chemical Shifts' [316] and '^{13}C of Cobalt(III) Complexes Containing the Polyaminocarboxylic Acid Ion' [317] have appeared.

$[Ir(BH_4)H_2(PRBu^t{}_2)_2]$ has been found by ^{1}H, ^{11}B, and ^{31}P n.m.r. to be non-fluxional.[318] At −95 °C, (44) has three types of hydride signal and two types of

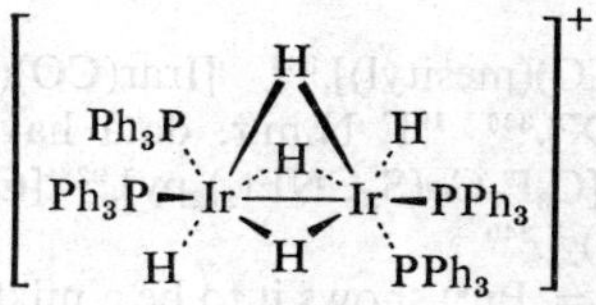

(44)

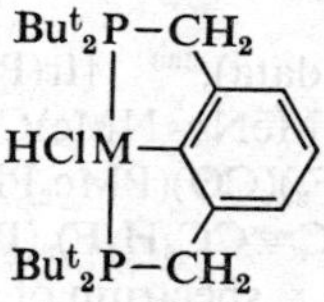

(45)

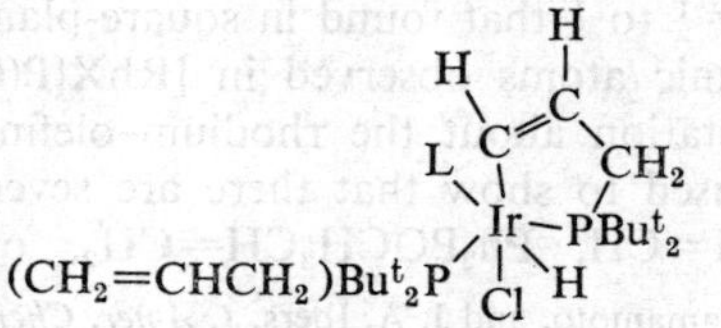

(46)

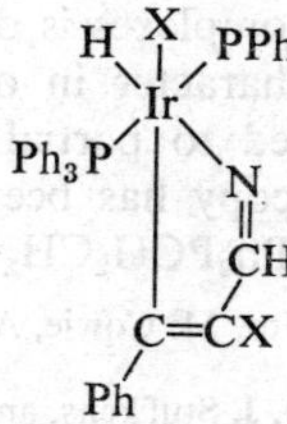

(47)

^{31}P n.m.r. signal with $^2J(^{31}P,^{31}P)$ = 95 Hz.[319] The ABC, ^{31}P n.m.r., spectrum of *fac*-$[\{(RO)_3P\}_3IrHCH_2C_6H_2\text{-}4,6\text{-}Me_2]$ has been analysed.[320] ^{31}P N.m.r. data have also been reported for $[CoH(CO)(Ph_2PCH_2)_3CMe]$,[321] $[\{(RO)_3P\}_3CoHX]$,[322] $[RhHCl\{(Ph_2PCH_2CH_2CH_2)_2PPh\}L]^+$,[323] (45; M = Rh or Ir),[324] [RhH{CH=

[316] H. Yoneda, *Kagaku (Kyoto)*, 1976, 31, 224 (*Chem. Abs.*, 1976, **85**, 132 921).
[317] K. Igi and J. Hidaka, *Kagaku (Kyoto)*, 1976, 31, 302 (*Chem. Abs.*, 1976, **85**, 132 920).
[318] H. D. Empsall, E. M. Hyde, B. L. Shaw, and M. F. Uttley, *J.C.S. Dalton*, 1976, 2069.
[319] R. H. Crabtree, H. Felkin, G. E. Morris, T. J. King, and J. A. Richards, *J. Organometallic Chem.*, 1976, **113**, C7.
[320] L. Dahlenburg, D. Rehder, W. Reith, W. Stränz, and D. Thoennes, *Transition Metal Chem.*, 1976, **1**, 206.
[321] J. Ellermann, J. F. Schindler, H. Behrens, and H. Schlenker, *J. Organometallic Chem.*, 1976, **108**, 239.
[322] L. W. Gosser, *Inorg. Chem.*, 1976, **15**, 1348.
[323] J. A. Tiethof, J. L. Peterson, and D. W. Meek, *Inorg. Chem.*, 1976, **15**, 1365.
[324] C. J. Moulton and B. L. Shaw, *J.C.S. Dalton*, 1976, 1020.

CMeCO(OR)}$(PPh_3)_2$],[325] and (46).[326] ^{13}C N.m.r. data have been reported for (47).[327]

^1H N.m.r. spectroscopy has been used to show that protonation of [(η^5-C_5H_5)-Rh(η^2-olefin)$_2$] occurs to give complexes such as [(η^5-C_5H_5)RhClEt(η^2-CH_2=CHOAc)].[328] The ^{59}Co chemical shifts of [$R_nX_{m-n}Sn\{Co(CO)_4\}_{4-m}$] are mainly sensitive to π-effects in the metal–metal bond.[329] ^{31}P N.m.r. data have also been published for [$R^1_2(R^2_3P)_2Co(acac)$],[330] [MeCo(dppe)$_2$],[331] (48),[332] (49) (including

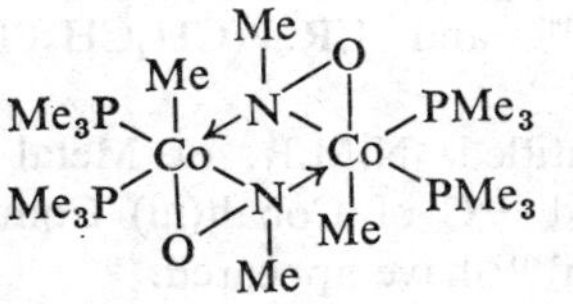

(48)

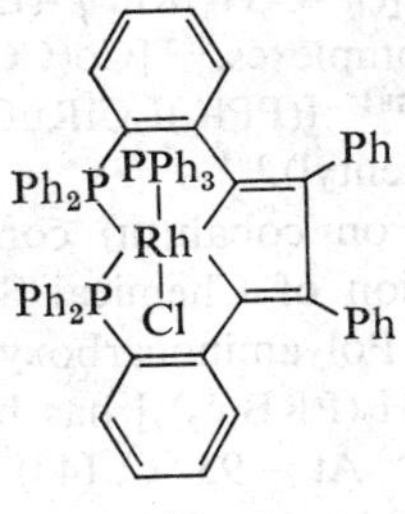

(49)

^{13}C n.m.r. data),[333] [Ir$(PPh_3)_2$(CO)(mesityl)],[334] [Irar(CO)$(PPh_3)_2$],[335] and [$(Ph_3P)_2$(CONMeN=NMe)ClRhHgX].[336] ^{19}F N.m.r. data have been reported for [RhBrI(CF_3)(CO)$(PMe_2Ph)_2$],[337] [$C_3F_7Co(S_2CNEt_2)_2$py],[338] [C_6F_5Co(BAE)],[339] and [$Rh_2Cu_4(C{\equiv}CC_6H_4F)_8(PMePh_2)_2$].[340]

The ^{31}P n.m.r. spectrum of (50; R = Bu^t) shows it to be a mixture, presumably *cis*- and *trans*-isomers.[341] The ^{13}C n.m.r. spectra of [RhX{P(CH_2CH_2CH=CH_2)$_3$}] and [RhX{P($CH_2CH_2CH_2CH$=CH_2)$_3$}] have revealed that $^1J(^{13}C, ^{103}Rh)$ of the olefin in these complexes is only $\frac{1}{2}$ to $\frac{1}{3}$ that found in square-planar complexes. The fluxional character in olefinic atoms observed in [RhX{P(CH_2CH_2CH=CH_2)$_3$}] is related to partial rotation about the rhodium–olefin bond.[342] ^{31}P N.m.r. spectroscopy has been used to show that there are several isomers of [RhXL] [L = $Ph_2PCH_2CH_2CH$=CH, Ph_2POCH_2CH=CH_2, or P(CH_2CH_2-

325 S. Komiya, T. Ito, M. Cowie, A. Yamamoto, and J. A. Ibers, *J. Amer. Chem. Soc.*, 1976, **98**, 3874.
326 S. Hietkamp, D. J. Stufkens, and K. Vrieze, *J. Organometallic Chem.*, 1976, **122**, 419.
327 J. F. van Baar, J. M. Klerks, P. Overbosch, D. J. Stufkens, and K. Vrieze, *J. Organometallic Chem.*, 1976, **112**, 95.
328 L. P. Seiwell, *Inorg. Chem.*, 1976, **15**, 2560.
329 L. F. Wuyts and G. P. Van der Kelen, 'Proceedings of the 16th International Conference on Coordination Chemistry', 1974, 2.38A, (*Chem. Abs.*, 1976, **85**, 54 244).
330 T. Ikariya and A. Yamamoto, *J. Organometallic Chem.*, 1976, **116**, 239.
331 T. Ikariya and A. Yamamoto, *J. Organometallic Chem.*, 1976, **116**, 231.
332 H.-F. Klein and H. H. Karsch, *Chem. Ber.*, 1976, **109**, 1453.
333 W. Winter, *Angew. Chem. Internat. Edn.*, 1976, **15**, 241.
334 L. Dahlenburg and R. Nast, *Angew. Chem. Internat. Edn.*, 1976, **15**, 110.
335 L. Dahlenburg and R. Nast, *J. Organometallic Chem.*, 1976, **110**, 395.
336 P. I. van Vliet, J. Kuyper, and K. Vrieze, *J. Organometallic Chem.*, 1976, **122**, 99.
337 H. C. Clark and K. J. Reimer, *Canad. J. Chem.*, 1976, **54**, 2077.
338 A. L. Marchese, M. Scudder, A. M. van den Bergen, and B. O. West, *J. Organometallic Chem.*, 1976, **121**, 63.
339 M. F. Corrigan, G. B. Deacon, B. O. West, and D. G. Vince, *J. Organometallic Chem.*, 1976, **105**, 119.
340 O. M. A. Salah and M. I. Bruce, *Austral. J. Chem.*, 1976, **29**, 531.
341 R. Mason, G. Scollary, B. Moyle, K. I. Hardcastle, B. L. Shaw, and C. J. Moulton, *J. Organometallic Chem.*, 1976, **113**, C49.
342 P. W. Clark and A. J. Jones, *J. Organometallic Chem.*, 1976, **122**, C41.

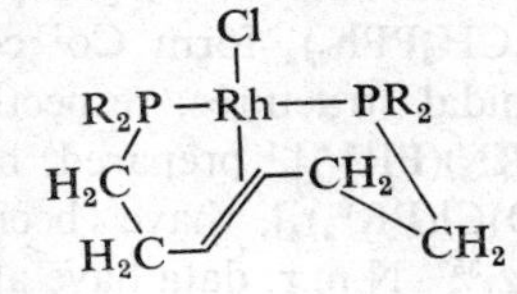

(50)

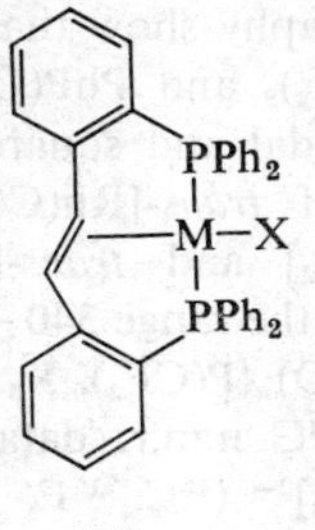

(51)

$CH{=}CH_2)_3]$.[343] The ^{13}C n.m.r. spectra of iridium complexes of cyclo-octa-1,5-diene have been reported and discussed. For compounds containing β-diketonato ligands, a correlation was found between the olefin ^{13}C chemical shifts and the Swain and Lupton *F* and *R* values of the substituent of the β-diketonato ligand. The relationship was explained by the Dewar–Chatt–Duncanson model of bonding. A tentative interpretation of the observed chemical shifts of the other carbon atoms was presented.[344] The ^{13}C chemical shifts of $[(\eta^5\text{-}C_5H_5)Co(\eta^5\text{-}C_5H_4CPh_2)]^+$ have been interpreted as showing delocalization of the positive charge onto the phenyl group, comparable to that of $[Ph_2CH]^+$ and $[Ph_3C]^+$ and delocalization to a much lesser extent onto the cobalticinium group.[345] Deprotonation of $[(\eta^5\text{-}C_5Me_5)Rh(\eta^6\text{-indene})]^{2+}$ to give $[(\eta^5\text{-}C_5Me_5)Rh(\eta^5\text{-indenyl})]^+$ has been shown by ^{13}C n.m.r. spectroscopy to be accompanied by the migration of the rhodium from the six- to the five-membered indene ring.[346] N.m.r. data have also been reported for (51; M = Rh or Ir) (^{31}P n.m.r. data),[347] (50; R = Ph) (^{13}C n.m.r. data),[348] $[(\eta^5\text{-}C_5H_5)(\eta^4\text{-MeHC}{=}\text{CEtCH}{=}S)Co]$ (^{13}C n.m.r. data),[349] $[(\eta^4\text{-butadiene})Co\{P(OMe)_3\}_3]^+$ (^{31}P n.m.r. data),[350] $[(\eta^4\text{-tetraphenylcyclobutadiene})Rh(hfac)]^+$ (^{19}F n.m.r. data),[351] $[(\eta^5\text{-}C_5H_5\text{-}Co\{\eta^4\text{-}(EtC_2Et)_2CO\}]$ (^{13}C n.m.r. data),[352] and $[(\eta^5\text{-}C_5Me_5)_2M_2(\mu\text{-}O_2PF_2)_3]^+$ (^{19}F and ^{31}P n.m.r. data; M = Rh or Ir).[353]

The ^{13}C n.m.r. apical resonances for $[Co_3(CO)_9CY]$ are at δ 230—310. Variable-temperature spectra have shown 'thermal decoupling' of ^{13}C from ^{59}Co and scrambling of the carbonyls.[354] The ^{13}C chemical shifts of the badly broadened resonances of $[Co_6(CO)_{15}C]^{2-}$ are δ(carbide) = 330.5, δ(^{13}Co-edge-bridging) = 234.9 and 252, and δ(^{13}CO-terminal) = 224.9.[355] ^{31}P N.m.r. spectroscopy and

343 P. E. Garrou, J. L. S. Curtis, and G. E. Hartwell, *Inorg. Chem.*, 1976, **15**, 3094.
344 R. Bonnaire and N. Platzer, *J. Organometallic Chem.*, 1976, **104**, 107.
345 J. E. Sheats, E. J. Sabol, jun., D. Z. Denney, and N. El Murr, *J. Organometallic Chem.*, 1976, **121**, 73.
346 C. White, S. J. Thompson, and P. M. Maitlis, *J.C.S. Chem. Comm.*, 1976, 409.
347 M. A. Bennett, R. N. Johnson, and I. B. Tomkins, *J. Organometallic Chem.*, 1976, **118**, 205.
348 P. W. Clark, *J. Organometallic Chem.*, 1976, **110**, C13.
349 D. C. Dittmer, K. Takahashi, M. Iwanami, A. I. Tsai, P. L. Chang, B. B. Blidner, and I. K. Stamos, *J. Amer. Chem. Soc.*, 1976, **98**, 2795.
350 E. L. Muetterties and P. L. Watson, *J. Amer. Chem. Soc.*, 1976, **98**, 4665.
351 J. F. Nixon and M. Kooti, *J. Organometallic Chem.*, 1976, **104**, 231.
352 R. S. Dickson and S. H. Johnson, *Austral. J. Chem.*, 1976, **29**, 2189.
353 S. J. Thompson, P. M. Bailey, C. White, and P. M. Maitlis, *Angew. Chem. Internat. Edn.*, 1976, **15**, 490.
354 S. Aime, L. Milone, and M. Valle, *Inorg. Chim. Acta*, 1976, **18**, 9.
355 V. G. Albano, P. Chini, G. Ciani, M. Sansoni, D. Strumolo, B. T. Heaton, and S. Martinengo, *J. Amer. Chem. Soc.*, 1976, **98**, 5027.

X-ray crystallography show that the two similar chelating triphosphine ligands $PhP(CH_2CH_2PPh_2)_2$ and $PhP(CH_2CH_2CH_2PPh_2)_2$ form Co^I complexes having trigonal-bipyramidal and square-pyramidal structures, respectively.[356] The ^{31}P n.m.r. spectra of *trans*-$[Rh(CO)Cl(PR^1_3)(PR^2_3)]$, prepared by mixing *trans*-$[Rh(CO)Cl(PR^1_3)_2]$ and *trans*-$[Rh(CO)Cl(PR^2_3)_2]$, have been recorded and $^2J(^{31}P, ^{31}P)$ are in the range 340—540 Hz.[357] N.m.r. data have also been reported for $[Co(NO)(CO)_n\{P(CF_3)_mX_{3-m}\}_{3-n}]$ (^{19}F n.m.r. data),[358] $[Rh(CO)_2Cl]_2$, $[Rh(CO)_2ClL]$ (^{13}C n.m.r. data),[359] $[RhCl(CO)\{P(OR)_3\}_2]$ (^{31}P n.m.r. data),[360] $[Ir(CO)Cl(PF_2O)_2]^{2-}$ (^{13}C, ^{19}F, and ^{31}P n.m.r. data),[361] and $[IrCl\{ON(CF_3)_2\}_2(CO)L_2]$ (^{19}F n.m.r. data).[362]

^{13}C N.m.r. spectroscopy has been used to show that the thermal reaction between *trans*-$[Co(CN)_4(SO_3)_2]^{5-}$ and $^{13}CN^-$ yields exclusively *trans*-$[Co(CN)_4(^{13}CN)SO_3]^{4-}$, whereas the photochemical reaction is random. At 14 °C the spectra were broad, but sharpened by −16 °C owing to the faster relaxation of ^{59}Co nuclei.[363] The ABX spectrum of L-phenylalanine in the $Co(NH_3)_4$ complex has been analysed. It was found that the rotamer population had been changed by complexation.[364] In $[(H_3N)_5Co(\text{L-histidine})]^{2+}$, the imidazole C-5 proton is coupled to only one of the methylene protons with $^4J(^1H, ^1H) = 0.9$ Hz and internal rotation of the imidazole ring around the C_β—C_γ bond is hindered, presumably due to intramolecular hydrogen-bonding.[365] The ^{13}C n.m.r. spectra of Co^{III} complexes with L-aspartic acid and L-glutamic acid as unidentate, bidentate, or bridging ligands have been measured. The relationship between structure and shifts was discussed.[366] ^1H N.m.r. spectroscopy has been shown to be a reliable way of determining the stereochemistry of amminenitrocobalt(III) complexes.[367] Amine ^1H chemical shifts have been measured for some *cis*-bis(ethylenediamine)cobalt(III) complexes in CF_3CO_2H. The chemical shifts were explained quantitatively by the magnetic anisotropy of the central metal ion and by assuming fast conformational inversion of the chelate rings. A discrepancy between the observed and computed shift values is apparent for the *trans*-hydrogens of the dicyanide complex which may due to the ground-state *trans*-effect of the $[CN]^-$ ion.[368] ^{13}C N.m.r. spectroscopy has been used to determine the structure of some cobalt(III) complexes of $HN(CH_2CH_2NH_2)_2$,[369]

356 R. Mason, G. R. Scollary, D. L. Dubois, and D. W. Meek, *J. Organometallic Chem.*, 1976, **114**, C30.
357 P. E. Garrou and G. E. Hartwell, *Inorg. Chem.*, 1976, **15**, 646.
358 R. C. Dobbie and S. Morton, *J.C.S. Dalton*, 1976, 1421.
359 A. J. Pribula and R. S. Drago, *J. Amer. Chem. Soc.*, 1976, **95**, 2784.
360 E. N. Yurchenko, A. D. Troitskaya, L. S. Gracheva, and L. Ya. Al't, *Tezisy Dokl.-Vses. Chugaevskoe Soveshch. Khim. Kompleksn. Soedin., 12th*, 1975, **2**, 319 (*Chem. Abs.*, 1976, **85**, 186 072).
361 J. Grosse and R. Schmutzler, *J.C.S. Dalton*, 1976, 412.
362 B. L. Booth, R. N. Haszeldine, and R. G. G. Holmes, *J.C.S. Chem. Comm.*, 1976, 489.
363 K. F. Miller and R. A. D. Wentworth, *Inorg. Chem.*, 1976, **15**, 1467.
364 U. Sakaguchi and H. Yoneda, *Chem. Letters*, 1976, **6**, 613 (*Chem. Abs.*, 1976, **85**, 70 386).
365 U. Sakaguchi, T. Taura, and H. Yoneda, *Chem. Letters*, 1976, **9**, 979 (*Chem. Abs.*, 1976, **85**, 151 485).
366 T. Yasui and T. Ama, *Bull. Chem. Soc. Japan*, 1976, **48**, 3171.
367 S. Balt, M. W. G. de Bolster, and C. Dekker, *Inorg. Chim. Acta*, 1976, **17**, 154.
368 U. Sakaguchi, S. Yamazaki, and H. Yoneda, *Bull. Chem. Soc. Japan*, 1976, **49**, 402.
369 F. D. Sancilio, L. F. Druding, and D. M. Lukaszewski, *Inorg. Chem.*, 1976, **15**, 1626.

and $H_2NCHMeCH_2CHMeNH_2$.[370, 371] The chemical shifts of the methyl group protons in *trans*-$[Co(dmg)_2(H_2O)_2]^+$, *trans*-$[CoX(dmg)_2H_2O]$, and $[CoCl(dmg)_2]^-$ depend on X. The methyl resonance frequency shifts in the order $H_2O < Cl^- < Br^- < I^-$ indicate an increase in electron density on the dimethylglyoximato. The size of the chemical shifts of intramolecular hydrogen bonds in *trans*-$[CoX(dmg)_2(OH_2)]$ follows the series $H_2O > Cl^- > Br^- > I^-$.[372] ^1H N.m.r. spectroscopy has been used to investigate the pH dependence of configurational changes in Co^{III} *cis*-dioximes.[373] The ^1H chemical shifts of $[Co(salen)(PR_3)_2]^+$ and related compounds appear to be dominated by anisotropic rather than inductive effects whereas the relative tertiary phosphine ligand field strengths appear to correlate with steric requirements of the tertiary phosphine as well as their expected σ-donor strengths.[374] Lanthanides have been used to show that the conformation of cobalamins in solution is principally the same as in the solid,[375] The $^{13}C\{^2H\}$ n.m.r. spectrum of dicyanocobalamin synthesized from $[^2H_3{}^{13}C]$methionine shows the intact incorporation of seven methyl groups into the macrocycle of vitamin B_{12} without exchange of ^2H with ^1H of the medium.[376] Similarly $^1H\{^{13}C\}$ difference spectra have been used to demonstrate intact incorporation of $[^2H_3{}^{13}C]$methionine into the C-1 and C-12 methyl groups of vitamin B_{12}.[377] The ^{13}C n.m.r. spectra of ^{13}C-enriched cyanide in some cyanocobalamins have been reported. As the *trans*-effect increases, the cyanide shift moves towards the position of free cyanide.[378] The ^{13}C n.m.r. spectra of cobalt(III) compounds containing various amino-acids as uni- or bi-dentate ligands have been measured. The relationship between the co-ordination types and changes in chemical shifts was clarified. The changes of chemical shift on co-ordination of the α-amino-acid to cobalt(III) were *ca.* 7 and 1 p.p.m. for the carboxy and α-carbon atoms, respectively. On the other hand, those on chelation were *ca.* 15 and 5 p.p.m. for each carbon atom. β-Alanine proved to be an exception.[379] N.m.r. data have been reported for $[Co(1,2\text{-dimethylethylenediamine})_3]^{3+}$ (^{13}C n.m.r. data),[380] *cis*- and *trans*-$[Co(dmg)_2(OH_2)_2]^+[NO_3]^-$ (^{59}Co n.m.r. data),[381] $[Co\{(O_2CCH_2CH_2)(O_2CCH_2)NCH_2CH_2N(CH_2CO_2)(CH_2CH_2CO_2)\}]^-$ (^{13}C n.m.r. data),[382] $[Rh(NO_3)(NO)_2(OPPh_3)_2]$ (^{31}P n.m.r. data),[383] $[(Ph_3P)_3RhOCO_2Rh(PPh_3)_2]$ (^{31}P n.m.r. data),[384] $[M(NO)(MeCN)_2(PPh_3)_2]^{2+}$ (^{31}P n.m.r. data;

370 H. Boucher and B. Bosnich, *Inorg. Chem.*, 1976, **15**, 1471.
371 H. Boucher and B. Bosnich, *Inorg. Chem.*, 1976, **15**, 2364.
372 A. P. Gulya, A. V. Ablov, O. A. Bologa, and V. A. Shcherbakov, *Koord. Khim.*, 1976, **2**, 14.
373 A. V. Ablov, A. P. Gulya, and O. A. Bologa, *Koord. Khim.*, 1976, **2**, 545.
374 C. W. Smith, G. W. Van Loon, and M. C. Baird, *Canad. J. Chem.*, 1976, **54**, 1875.
375 O. D. Hensens, H. A. O. Hill, J. Thornton, A. M. Turner, and R. J. P. Williams, *Phil. Trans. Roy. Soc.*, 1976, **B273**, 353.
376 A. Battersby, R. Hollenstein, E. McDonald, and D. C. Williams, *J.C.S. Chem. Comm.*, 1976, 543.
377 A. I. Scott, M. Kajiwara, T, Takahashi, I. M. Armitage, P. Demou, and D. Petrocine, *J.C.S. Chem. Comm.*, 1976, 544.
378 P. L. Gaus and A. L. Crumbliss, *Inorg. Chem.*, 1976, **15**, 739.
379 T. Ama and T. Yasui, *Bull. Chem. Soc. Japan*, 1976, **49**, 472.
380 S. Bagger and O. Bang, *Acta Chem. Scand.*, 1976, **A30**, 765.
381 A. P. Gulya and V. A. Shcherbakov, *Tezisy Dokl.-Vses. Chugaevskoe Soveshch., Khim. Kompleksn. Soedin., 12th*, 1975, **3**, 541 (*Chem. Abs.*, 1976, **85**, 186 084).
382 K. D. Gailey and B. E. Bodie, *J. Coordination Chem.*, 1975, **5**, 23.
383 P. B. Critchlow and S. D. Robinson, *J. Organometallic Chem.*, 1976, **114**, C46.
384 S. Krogsrud, S. Komiya, T. Ito, J. A. Ibers, and A. Yamamoto, *Inorg. Chem.*, 1976, **15**, 2798.

M = Rh or Ir),[385] $[RhCl_2(PMe_2Ph)_3SFO_3]$ (^{31}P n.m.r. data),[386] and $[\{\overline{SCNH(CH_2)_5}\}_4RhCl_2]^+$ (^{13}C n.m.r. data).[387]

Complexes of Ni, Pd, and Pt.—Information concerning complexes of these elements can be found at the following sources: $[(\eta^5\text{-}C_5H_5)_3Ni_3CB_5H_6]$,[1157] $[(B_{10}H_{12})_2Ni]^{2-}$, $[(B_{10}H_{12}Pd)_2C_{10}H_8N_2]^{2-}$,[1169] $[(1,2\text{-}C_2B_9H_{11})Pt(PPh_3)_2]$,[1178] *closo*-3,8-$[(Ph_3P)_2\text{-}3H\text{-}3,1,2\text{-}NiC_2B_9H_{10}]$,[1181] $[(Ph_3P)_2PtBMe_2Br_3]$,[1202] $[(Bu^t{}_2PCH_2C_6H_3\text{-}CH_2PBu^t{}_2)MCl]$ (M = Ni, Pd, or Pt),[324] $[(\eta^5\text{-}C_5H_5)Pd\{C_4(CF_3)_4\}Pd(\eta^3\text{-}C_3H_5)]$,[159] $[PtMe_3(C_5H_5BR)]$,[298] $[PhN_2C_6H_4(C{=}NCy)_2Pd(CNCy)_2Cl]$ (Cy = cyclohexyl),[327] Pd^{II} and Pt^{II} complexes of various diamine,[184] $[R_nP(CH_2CH_2PR_2)_{3-n}MCl_2]$ (R = neopentyl; M = Ni, Pd, or Pt),[194] $[M(PF_2O)_4]^{2-}$ (M = Ni, Pd, or Pt),[361] and $[\{PPh_nEt_{3-n})_2Ag\}Ni\{S_2C_2(CN)_2\}_2]$.[475]

For the compound $[o\text{-}C_6H_4(CH_2PBu^t{}_2)_2PtH_2]$, $^1J(^1H, ^{195}Pt)$ = 1008 Hz and $^1J(^{31}P, ^{195}Pt)$ = 2112 Hz,[388] while for (52; R = H), $^1J(^{31}P, ^{195}Pt)$ = 3468 Hz and $^3J(^{31}P, ^{195}Pt)$ = 150 Hz and for (52; R = $SiEt_3$), $^1J(^{31}P, ^{195}Pt)$ = 3228 Hz

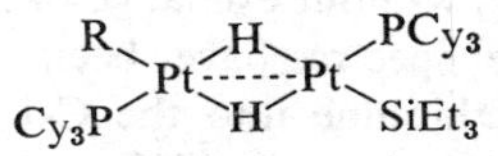

(52)

and $^3J(^{31}P, ^{195}Pt)$ = 60 Hz.[389] Variations in $^1J(^1H, ^{195}Pt)$ for *trans*-$[PtH(PEt_3)_2CN]$ (L = PEt_3) with added Lewis acid have been used to establish the strength of the Lewis acid to co-ordinate to cyanide. They decrease in the order $AlCl_3 > BAr_3 > BR_3 \sim CoCl_2 > AlAr_3 > ZnCl_2 > B(OAr)_3 > AlR_3 > AlR_2(OR)$.[390] N.m.r. data have also been published for $[Ni(\eta^5\text{-}C_5H_5)(CF_3C_2CF_3)_2H]$ (^{19}F n.m.r. data),[391] *trans*-$[PtH(CH_2CN)(PPh_3)_2]$ (^{31}P n.m.r. data),[392] and *trans*-$[PtH(PEt_3)_2(ESiH_3)]$ (^{29}Si, ^{31}P, ^{77}Se, ^{125}Te, and ^{195}Pt n.m.r. data; E = O, S, Se, or Te).[393]

The signs and magnitudes of $^3J(^1H, ^{31}P)$, $^3J(^{19}F, ^{31}P)$, *etc.*, have been determined by multiple resonance methods for CH_3 and CF_3 derivatives of Pt^{II}, Pt^{IV}, and Au^{III}. Whilst $^3J(^1H, ^{31}P)$ is more negative when the coupling path is *trans* than when it is *cis*, the converse is true of $^3J(^{19}F, ^{31}P)$. The coupling constants fall into ranges which are sufficiently distinct to permit their diagnostic use, and their origin was discussed in terms of the bonding in the complexes. ^{19}F, ^{31}P, and ^{195}Pt chemical shifts were given.[394] ^{195}Pt Chemical shifts and coupling constants have been reported for 78 platinum compounds which have platinum–carbon bonds. Experimental trends were noted and, where possible,

385 M. Ghedini, G. Dolcetti, O. Gandolfi, and B. Giovannitti, *Inorg. Chem.*, 1976, **15**, 2385.
386 C. Eaborn, N. Farrell, J. L. Murphy, and A. Pidcock, *J.C.S. Dalton*, 1976, 58.
387 P. Dini, J. C. J. Bart, E. Santoro, G. Cum, and N. Giordano, *Inorg. Chim. Acta*, 1976, **17**, 97.
388 C. J. Moulton and B. L. Shaw, *J.C.S. Chem. Comm.*, 1976, 365.
389 M. Green, J. A. K. Howard, J. Proud, J. L. Spencer, F. G. A. Stone, and C. A. Tsipis, *J.C.S. Chem. Comm.*, 1976, 671.
390 L. E. Manzer and G. W. Parshall, *Inorg. Chem.*, 1976, **15**, 3114.
391 J. L. Davidson and D. W. A. Sharp, *J.C.S. Dalton*, 1976, 1123.
392 R. Ros, R. Bataillard, and R. Roulet, *J. Organometallic Chem.*, 1976, **118**, C53.
393 E. A. V. Ebsworth, J. M. Edward, and D. W. H. Rankin, *J.C.S. Dalton*, 1976, 1667.
394 J. D. Kennedy, W. McFarlane, and R. J. Puddephatt, *J.C.S. Dalton*, 1976, 745.

assessed in terms of current theories. Electronic excitation energy, interbond angle, and substituent bulk and/or polarizability appear to make significant contributions to ^{195}Pt nuclear shielding.[395] In the ^{31}P n.m.r. spectrum of [{$Ph_2P(CH_2)_nPPh_2$}PtClMe], two ^{31}P signals are observed but $^2J(^{31}P, ^{31}P)$ was zero when $n = 1$ or 2 but 21 Hz when $n = 3$. $^1J(^{31}P, ^{195}Pt)$ increases markedly from 1434 Hz when $n = 1$ to 1794 Hz when $n = 2$. The low value was attributed to ring strain.[396] The ^{13}C n.m.r. spectra of the five-co-ordinate platinum complexes [PtMe{pz_3BH}olefin] have been reported. The olefinic carbon resonances are substantially shifted upfield from those of the free olefin and the magnitude of the shift related to metal–olefin π-back-bonding.[397] [$(\eta^1$-$C_5H_5)_2$Pt(η^4-cyclo-octa-1,5-diene)] shows only one η^1-C_5H_5 resonance even at -90 °C. The ^{13}C n.m.r. spectrum of [$(\eta^5$-$C_5H_5)$PtMe(CO)] has $^1J(^{13}CO, ^{195}Pt) = 2459$ Hz and $\delta(^{13}CO) = 163.4$.[398] The ^{13}C n.m.r. spectrum of (53) shows a complex behaviour which was attributed to a large $^1J(^{195}Pt, ^{195}Pt)$, although the species containing only ^{195}Pt is 11.5% abundant.[399]

The doubling methyl and methylene groups in (54), attributed to phosphorus coupling, vanish on warming.[400] The imino and isocyanide groups in (55) have a higher *trans*-effect than alkyl groups, whereas their *trans*-influence as indicated by $^1J(^{31}P, ^{195}Pt)$ is smaller.[401] For (55; L = PPh_3, X = NR^1R^2) there is no

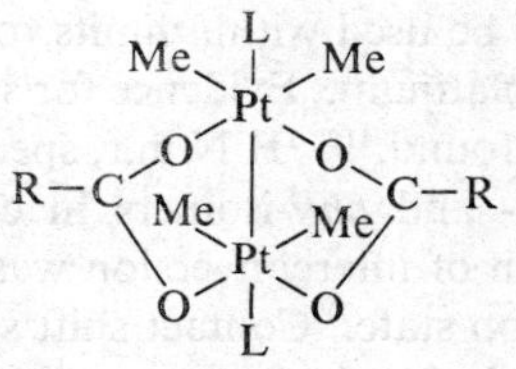

(53)

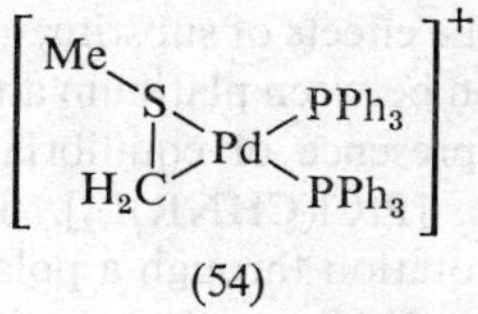

(54)

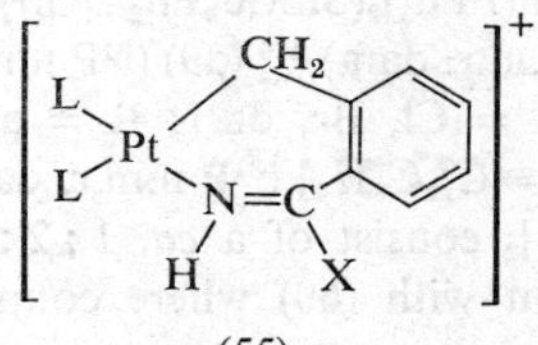

(55)

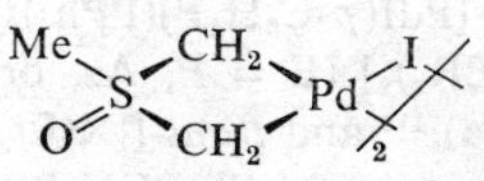

(56)

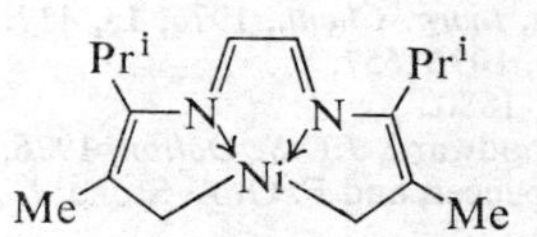

(57)

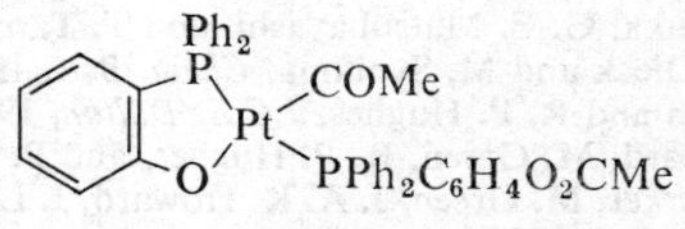

(58)

395 J. D. Kennedy, W. McFarlane, R. J. Puddephatt, and P. J. Thompson, *J.C.S. Dalton*, 1976, 874.
396 T. G. Appleton, M. A. Bennett, and I. B. Tomkins, *J.C.S. Dalton*, 1976, 439.
397 L. E. Manzer, *Inorg. Chem.*, 1976, **15**, 2354.
398 H. C. Clark and A. Shaver, *Canad. J. Chem.*, 1976, **54**, 2068.
399 J. Kuyper and K. Vrieze, *Transition Metal Chem.*, 1976, **1**, 208.
400 G. Yoshida, H. Kurosawa, and R. Okawara, *J. Organometallic Chem.*, 1976, **113**, 85.
401 R. Ros, J. Renaud, and R. Roulet, *J. Organometallic Chem.*, 1976, **104**, 271.

correlation between $^1J(^{31}P, ^{195}Pt)$ and $\delta(^{31}P)$ or $^2J(^1H, C, ^{195}Pt)$.[402] The ^1H n.m.r. spectrum of (56) shows an $[AB]_2$ spectrum for the CH_2 group with $^2J(A, B) =$ 9.15 Hz.[403] $^2J(^1H, C, ^{31}P)$ in *trans*-$[PtCl_2LY]$ [Y = *p*-$MeC_5H_4NCH(COPh)$, $MePh_2PCH(COPh)$, $Ph_3AsPCH(COPh)$, or $Me_2SCH(COPh)$] increases with increasing basicity of the ylide.[404] ^{13}C N.m.r. data have been reported for (57),[405] $[R^6HC{=}CR^5CR^4R^3CR^2R^1PdL_2]$,[406] $[CH_2{=}CHCH(CO_2Me)CH(CO_2Me)PdL_2$-$Cl]$,[407] $[L_2PtCH_2CMe{=}CMeCH_2]$, [1–3,8-η^4-$CH_2CHCH(CH_2)_2CH{=}CHCH_2$-$PtPMe_3]$,[408] [1,2,5-$\eta^3$-cyclo-octenylPdCl$]_2$,[409] and (58) (including ^{31}P n.m.r. data).[410]

The ^1H n.m.r. spectra of the *o*-tolyl methyl groups in [MX{*o*-(*o*-tolyl)$_2PC_6H_4C{=}$$CHC_6H_4P$(*o*-tolyl)$_2$-*o*}] (M = Ni, Pd, or Pt) vary with temperature. This behaviour was considered to be a consequence of interconversion of enantiomers arising from restricted rotation about the metal–phosphorus and metal–carbon bonds.[411] ^{13}C N.m.r. spectra have been measured for a series of compounds *trans*-$[(Et_3P)_2PtPhX]$. Correlations of the chemical shifts and coupling constants with sets of both Taft and modified Swain–Lupton substituent constants were examined and compared with previous work on ^{19}F chemical shifts for *trans*-$[(Et_3P)_2Pt(C_6H_4F)X]$. The observed correlations provide evidence that substituent parameters obtained from organic systems may be used within limits to describe accurately the effects of substituents bonded to platinum. Evidence for significant π-interaction between platinum and phenyl was found.[412] ^1H N.m.r. spectroscopy shows the presence of equilibria between *syn*- and *anti*-isomers in complexes such as *trans*-$[PtCl(CHNR)L_2]$. The mechanism of interconversion was thought to involve rotation through a polarized transition state. Contact shift studies on the complexes indicate that the basicity of the imidoyl nitrogen varies between complexes and between *syn*- and *anti*-isomers of the same complex.[413] N.m.r. data have also been reported for *cis*-$[PtPh(PPh_3)_2(SnMe_nPh_{2-n}Cl)]$ (^{31}P n.m.r. data),[414] *trans*-$[PdI(p\text{-}C_6H_4F)(PPh_3)_2]$ (^{19}F n.m.r. data),[415] (59) (^{31}P n.m.r. data),[416] *trans*-$[MXR(ER_3)_2]$ (E = P, As, or Sb; X = Cl, Br, or I; R = aryl or acyl; ^{31}P n.m.r. data),[417] and *trans*-$[\{X(Et_3P)_2PtC{\equiv}C\}_2C_6H_4]$ (^{31}P n.m.r. data).[418]

The CH_2 group of $[(Ph_2PCH_2PPh_2)PtCl]_2$ consist of a *ca.* 1 : 2 : 1 triplet in the ^1H n.m.r. spectrum. This is consistent with (60) where coupling to two

402 R. Ros, J. Renaud, and R. Roulet, *J. Organometallic Chem.*, 1976, **104**, 393.
403 P. Bravo, G. Fronza, and C. Ticozzi, *J. Organometallic Chem.*, 1976, **118**, C78.
404 H. Koezuka, G.-E. Matsubayashi, and T. Tanaka, *Inorg. Chem.*, 1976, **15**, 417.
405 H. tom Dieck and M. Svoboda, *Chem. Ber.*, 1976, **109**, 1657.
406 M. Green and R. P. Hughes, *J.C.S. Dalton*, 1976, 1880.
407 R. Goddard, M. Green, R. P. Hughes, and P. Woodward, *J.C.S. Dalton*, 1976, 1890.
408 G. K. Barker, M. Green, J. A. K. Howard, J. L. Spencer, and F. G. A. Stone, *J. Amer. Chem. Soc.*, 1976, **98**, 3373.
409 J. K. Stille and D. E. James, *J. Organometallic Chem.*, 1976, **108**, 401.
410 H. D. Empsall, B. L. Shaw, and B. L. Turtle, *J.C.S. Dalton*, 1976, 1500.
411 M. A. Bennett and P. W. Clark, *J. Organometallic Chem.*, 1976, **110**, 367.
412 D. R. Coulson, *J. Amer. Chem. Soc.*, 1976, **98**, 3111.
413 D. F. Christian, H. C. Clark, and R. F. Stepaniak, *J. Organometallic Chem.*, 1976, **112**, 209.
414 C. Eaborn, A. Pidcock, and B. R. Steele, *J.C.S. Dalton*, 1976, 767.
415 A. Sekiya and N. Ishikawa, *J. Organometallic Chem.*, 1976, **118**, 349.
416 L. C. Sawkins, B. L. Shaw, and B. L. Turtle, *J.C.S. Dalton*, 1976, 2053.
417 P. E. Garrou and R. F. Heck, *J. Amer. Chem. Soc.*, 1976, **98**, 4115.
418 R. Bast and J. Moritz, *J. Organometallic Chem.*, 1976, **117**, 81.

equivalent ^{195}Pt atoms produces a 1 : 8 : 18 : 8 : 1 quintet which could appear as a 1 : 2 : 1 triplet.[419] The ^{1}H n.m.r. spectrum of (61) shows an $[A_6X]_2M$ pattern for the methyl groups and the ^{31}P n.m.r. spectrum shows $^{1}J(^{31}P, ^{195}Pt)$.[420] N.m.r. data have also been reported for (62) (^{13}C and ^{31}P n.m.r. data),[421] and *trans*-$[PtI(PEt_3)_2(H_2SiNHSiH_3)]$ (^{15}N, ^{29}Si, ^{31}P, and ^{195}Pt n.m.r. data).[422]

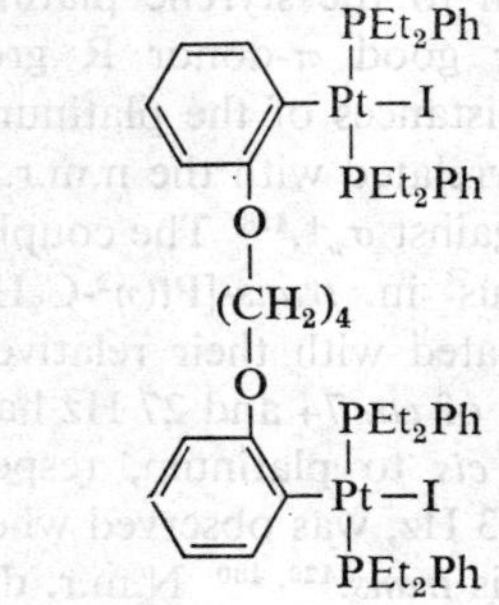

(59)

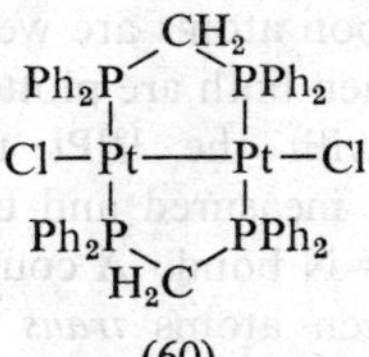

(60)

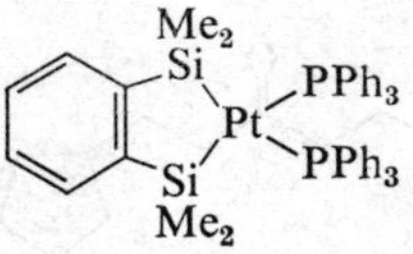

(61)

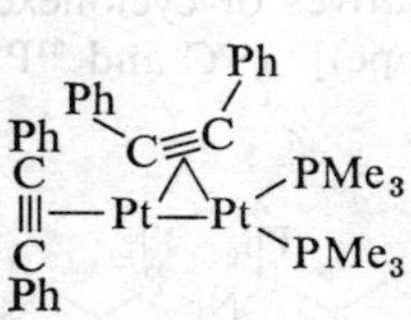

(62)

The ^{1}H n.m.r. spectrum of a freshly prepared $CDCl_3$ solution of *trans*-$[PtCl_2(\eta^2\text{-olefin})(pyridine)]$ shows no ^{195}Pt coupling to the α-protons of the pyridine as there is rapid exchange, but on standing the free pyridine is removed as $[PtCl_2(pyridine)_2]$ and coupling is observed.[423] The ^{13}C chemical shift and $^{1}J(^{13}C, ^{195}Pt)$ of *trans*-$[PtCl_2(\eta^2\text{-}C_2H_4)L]$ show a small dependence on L. A lowering of the carbon–carbon bond order in ethylene as shown by a decrease in $\nu(C{=}C)$ is linearly related to a higher upfield shift of the ^{13}C carbon resonance of this ligand.[424] The ^{13}C n.m.r. spectra of *trans*-$[PtCl_2(pyridine)(\eta^2\text{-olefin})]$ have been compared with i.r., Raman, and photoelectron spectroscopic results.[425] ^{13}C N.m.r. data have been reported for *trans*-$[(\eta^2\text{-olefin})PtCl_2(NC_5H_4Me)]$, $[(\eta^2\text{-olefin})PtCl_3]^-$, and $[(\eta^2\text{-olefin})PtCl(NH_3)_2]^+$. The observed trends in $^{1}J(^{13}C, ^{195}Pt)$ and $\delta(^{13}C)$ values of the olefinic carbon atoms have been discussed in terms of the nature of the olefin–platinum(II) bonding. The data can be rationalized qualitatively using a valence bond description of the bonding. In

419 M. P. Brown, R. J. Puddephatt, and M. Rashidi, *Inorg. Chim. Acta*, 1976, **19**, L33.
420 W. Fink, *Helv. Chim. Acta*, 1976, **59**, 606.
421 M. Green, D. M. Grove, J. A. K. Howard, J. L. Spencer, and F. G. A. Stone, *J.C.S. Chem. Comm.*, 1976, 759.
422 E. A. V. Ebsworth, J. M. Edwards, and D. W. H. Rankin, *J.C.S. Dalton*, 1976, 1673.
423 F. Pesa, L. Spaulding, and M. Orchin, *J. Coordination Chem.*, 1975, **4**, 225.
424 M. A. M. Meester, D. J. Stufkens, and K. Vrieze, *Inorg. Chim. Acta*, 1976, **16**, 191.
425 M. A. M. Meester, H. van Dam, D. J. Stufkens, and A. Oskam, *Inorg. Chim. Acta*, 1976, **20**, 155.

terms of the Chatt–Dewar model of the bonding, the data support the view that donation from olefin π- to a platinum σ-orbital is the predominant component of the olefin–platinum bond.[426] The effect of varying R^1 and R^2 for *trans*-$[(p\text{-}R^1C_6H_4CH{=}CH_2)PtCl_2L]$ (L = Cl or $4\text{-}R^2C_5H_4N$) on the nature of the styrene–platinum bond has been studied by ^{13}C n.m.r. and i.r. spectroscopies. The data showed a significant contribution to the styrene–platinum bond by the ionic valence-bond representation for good π-donor R groups.[427] For $[(p\text{-}YC_6H_4CH{=}CH_2)PtCl_2(NC_2H_4X)]$ the distances of the platinum atom from the two ethylenic carbon atoms are well correlated with the n.m.r. $^1J(^{13}C, ^{195}Pt)$ coupling constants when both are plotted against σ_p^+.[428] The coupling constants of ligand protons with the ^{195}Pt nucleus in *trans*-$[Pt(\eta^2\text{-}C_2H_4)Cl_2(R^1{}_2C{=}NNR^2R^3)]$ have been measured and correlated with their relative *cis* or *trans* position about the C=N bond. A coupling of *ca.* 74 and 27 Hz has been found for aldehydic hydrogen atoms *trans* and *cis* to platinum, respectively. For R = Me, a higher coupling constant, 11—13 Hz, was observed when the methyl group is *cis* to platinum and lower when it is *trans*.[429, 430] N.m.r. data have also been reported for $[(Ph_3P)_2Pt(\eta^2\text{-}HC{\equiv}CCMe_2OH)]$ (^{31}P n.m.r. data),[431] η^3-allylic palladium derivatives of cyclohexenes (^{13}C n.m.r. data),[432] $[(1\text{–}3\text{-}\eta^3\text{-cyclohepta-}2,4\text{-dienyl})Pd(dppe)]^+$ (^{13}C and ^{31}P n.m.r. data),[433] (63) (^{31}P n.m.r. data),[434] (64)

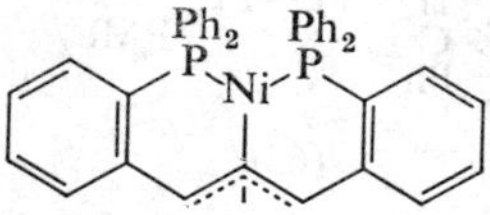

(63)

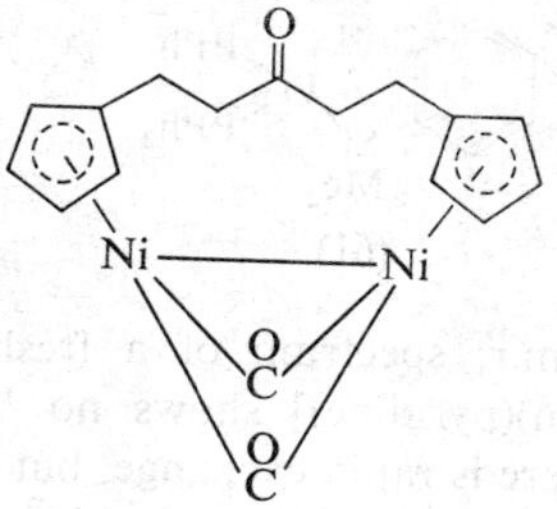

(64)

(^{13}C n.m.r. data),[435] and $[Bu^t{}_2(Me_2MCl)PNi(CO)_3]$ (^{31}P n.m.r. data; M = Ge or Sn).[436]

^{195}Pt Fourier transform n.m.r. spectra of some salts of platinum have been measured and solvent effects noted.[437] The ^1H and ^{13}C n.m.r. spectra have been recorded for a series of diamagnetic Ni^{II} complexes of open-chain quadridentate

[426] D. G. Cooper and J. Powell, *Inorg. Chem.*, 1976, **15**, 1959.
[427] D. G. Cooper, S. C. Nyburg, J. Powell, and K. A. Simpson, 'Proceedings of the 16th International Conference on Coordination Chemistry', 1974, R81 (*Chem. Abs.*, 1976, **85**, 191 638).
[428] S. C. Nyburg, K. Simpson, and W. Wong-Ng, *J.C.S. Dalton*, 1976, 1865.
[429] G. Natile, L. Maresca, and L. Cattalini, *J.C.S. Chem. Comm.*, 1976, 24.
[430] L. Maresca, G. Natile, L. Cattalini, and F. Gasparrini, *J.C.S. Dalton*, 1976, 1090.
[431] H. D. Empsall, B. L. Shaw, and A. J. Stringer, *J.C.S. Dalton*, 1976, 185.
[432] Y. Senda, H. Suda, J. Ishiyama, and S. Imaizumi, *Tetrahedron Letters*, 1976, 1983.
[433] D. J. Mabbott and P. M. Maitlis, *J.C.S. Dalton*, 1976, 2156.
[434] M. A. Bennett, R. N. Johnson, G. B. Robertson, I. B. Tomkins, and P. O. Whimp, *J. Amer. Chem. Soc.*, 1976, **98**, 3514.
[435] P. Eilbracht, *Chem. Ber.*, 1976, **109**, 3136.
[436] H. Schumann, J. Held, H.-J. Kroth, W.-W. du Mont, and B. Wöbke, *J. Organometallic Chem.*, 1976, **105**, 393.
[437] W. Freeman, P. S. Pregosin, S. N. Sze, and L. M. Venanzi, *J. Magn. Resonance*, 1976, **22**, 473.

ligands derived from a 2 : 1 ratio of ethylenediamine or propylenediamine with a β-diketone.[438] Substituent and solvent effects on the ^1H n.m.r. spectra of nickel(II) α-dioximates have been discussed. A correlation was found between the chemical shifts of the hydrogen-bonding hydrogen atoms and the stability of the complexes.[439] The ^1H n.m.r. spectrum of [M(BAE)] (M = Ni or Pt) with Eu(fod)$_3$ shows shifts and line broadenings. The line broadenings were attributed to exchange rather than paramagnetism.[440] The strength of aggregation in Ni^{II}, Pd^{II}, Zn^{II}, and Cd^{II} metalloporphyrins has been shown by n.m.r. spectroscopy to depend upon interaction of the metal ion with the porphyrin π-system within a selected series of 2,4-disubstituted deuterioporphyrin-IX dimethyl esters complexed with a constant metal ion. The aggregation is further enhanced by the presence of electron-withdrawing substituents on the porphyrin periphery.[441] On co-ordination of pyridine to Pt^{II} or Hg^{II} only small changes occur in the ^{13}C spectrum.[442] ^{15}N N.m.r. spectroscopy has been used to show that unsymmetric *OO'*-dihydroxydiarylazo ligands co-ordinate to Pt^{II} to yield two complexes (65).[443] The binding of complex nucleosides to [enPdCl$_2$] in neutral

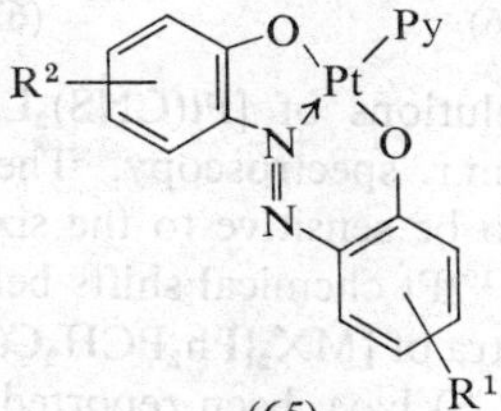

(65)

aqueous solution has been investigated by ^{13}C n.m.r. spectroscopy. Binding at N-3 in cytidine, thymidine, and uridine, at N-7 in 1-methylguanosine, and at N-1 in guanosine was demonstrated. For most carbon atoms adjacent to N-3 in pyrimidine nucleosides the complexation shifts of the basic ligand are *ca.* 30% of the corresponding upfield protonation shifts.[444] ^1H and ^{13}C n.m.r. spectroscopies have been used to show that uridine and thymidine undergo proton loss at N-3 on co-ordination to Pd^{II} and Pt^{II}.[445] The ^1H and ^{13}C n.m.r. spectra of [cytosine-PdCl$_2$] and [cytosine-PtCl$_2$] in $(CD_3)_2SO$ have been reported; N-3 complexation was found.[446] A large series of complexes of the type *cis*-[LL'PdCl$_2$], where L and L' are combinations of piperidine, Et$_3$N, phosphines, phosphites, phosphonites, and phosphinites, have been examined by ^1H, ^{13}C, and ^{31}P n.m.r. spectroscopies. The mixed ligand complexes are in equilibrium with their symmetric counterparts in solution. When both ligands are phosphorus donors, they are at least as thermodynamically stable and kinetically more inert

438 L. F. Lindoy, W. E. Moody, J. Lewis, and T. W. Matheson, *J.C.S. Dalton*, 1976, 1965.
439 I. N. Marov, A. T. Panfilov, E. K. Ivanova, and N. A. Plekhanov, *Koord. Khim.*, 1975, **1**, 1516.
440 J. K. Beattie, L. F. Lindoy, and W. E. Moody, *Inorg. Chem.*, 1976, **15**, 3171.
441 R. J. Abraham, F. Eivazi, H. Pearson, and K. M. Smith, *J.C.S. Chem. Comm.*, 1976, 699.
442 T. Ibusuki and Y. Saito, *Inorg. Chim. Acta*, 1976, **19**, 87.
443 P. S. Pregosin and E. Steiner, *Helv. Chim. Acta*, 1976, **59**, 376.
444 D. J. Nelson, P. L. Yeagle, T. L. Miller, and R. B. Martin, *Bioinorganic Chemistry*, 1976, **5**, 353.
445 M. C. Lim and R. B. Martin, *J. Inorg. Nuclear Chem.*, 1976, **38**, 1915.
446 F. Coletta, R. Ettorre, and A. Gambaro, *J. Magn. Resonance*, 1976, **22**, 453.

in solution than the symmetric complexes. The equilibrium position is dependent upon the nature of the ligands but seemingly solvent and temperature independent. $^2J(^{31}P, ^{31}P)$ are related to Tolman's electronic parameters in a parabolic way, consistent with current theories of spin–spin coupling. This allows the prediction of $^2J(^{31}P, ^{31}P)$ and discussion of lineshapes for $[X_nA]_2$ spin systems.[447] N.m.r. data have also been reported for $[Ni\{CH_2(CHMeNHCMeCH_2NHCMe_2)_2\text{-}CH_2\}]^{2+}$ (^{13}C n.m.r. data),[448] [1,2-bis{octaethyl-meso-porphyrinatonickel(II)}-ethane] (^{13}C n.m.r. data),[449] nickel derivatives of secocorrinoxide (^{13}C n.m.r. data),[450] (66) (^{13}C n.m.r. data),[451] and $[L_2M]$ [L = (67); M = Ni, Pd, or Pt; ^{31}P n.m.r. data].[452]

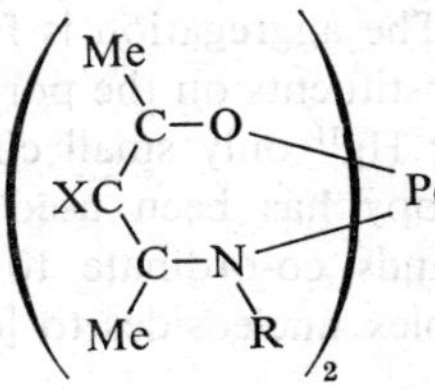

(66)

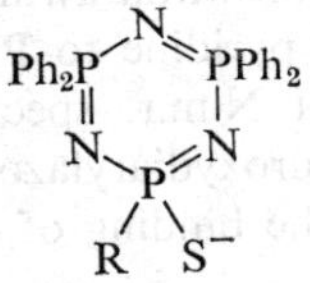

(67)

The isomers present in solutions of $[Pt(CNS)_2L_2]$ have been identified by $^1H\{^{195}Pt\}$ INDOR and ^{31}P n.m.r. spectroscopy. The mode of co-ordination of the SCN ligand was shown to be sensitive to the size of the *cis*-ligands. There is *ca.* 400 p.p.m. difference in ^{195}Pt chemical shifts between the N- and S-bonded species.[453] The ^{31}P n.m.r. spectra of $[MX_2\{Ph_2PCH_2C(CF_3){=}CHPPh_2\}]$ (M = Pd or Pt; X = Cl, NCS, or ^{15}NCS) have been reported. In spectra of isotopically substituted complexes, dithiocyanato and mixed thiocyanato–isothiocyanato isomers can be unambiguously identified owing to the occurrence of $^2J(^{31}P, ^{31}P)$ and $^2J(^{15}N, ^{31}P)$. $[Pt(NCS)(SCN)\{Ph_2PCH_2C(CF_3){=}CHPPh_2\}]$ and $[Pt(SCN)\text{-}(NCS)\{Ph_2PCH_2C(CF_3){=}CHPPh_3\}]$ also exhibit $^2J(^{14}N, ^{31}P)$ in *trans*-isothiocyanate groups.[454] The ^1H n.m.r. spectrum of the methyl groups in $[\overline{Pt(SCH_2CH_2S})(PMe_2Ph)_2]$ shows an intermediate pattern.[455] ^{13}C and ^{31}P n.m.r. spectra of *trans*-$[MX_2L_2]$ (M = Pd or Pt) have been reported. There is a small but marked dependence of values of $^1J(^{13}C, ^{31}P)$ on the nature of the group *trans* to phosphorus. The ^{31}P chemical shifts also reflect the nature of the *trans*-ligand.[456] Indeed, there is a correlation between the decrease in ^{31}P chemical shift and the increase in *trans*-effect in phosphorus-containing ligands in 12 palladium(II) complexes.[457] $^2J(^{31}P, ^{31}P)$ has been measured for $[MCl_2LL']$

447 A. W. Verstuyft, D. A. Redfield, L. W. Cary, and J. H. Nelson, *Inorg. Chem.*, 1976, **15**, 1128.
448 T. Ito, K. Soma, and M. Watanabe, *Chem. Letters*, 1976, **5**, 421 (*Chem. Abs.*, 1976, **85**, 27 133).
449 D. P. Arnold, A. W. Johnson, and M. Winter, *J.C.S. Chem. Comm.*, 1976, 797.
450 B. Kräutler, A. Pfaltz, R. Nordmann, K. O. Hodgson, J. D. Dunitz, and A. Eschenmoser, *Helv. Chim. Acta*, 1976, **59**, 924.
451 Y. Senda, A. Kasahara, A. Suzuki, and T. Izumi, *Bull. Chem. Soc. Japan*, 1976, **49**, 3337.
452 A. Schmidpeter, K. Blanck, and F. R. Ahmed, *Angew. Chem. Internat. Edn.*, 1976, **15**, 488.
453 S. J. Anderson, P. L. Goggin, and R. J. Goodfellow, *J.C.S. Dalton*, 1976, 1959.
454 A. J. Carty, *Inorg. Chem.*, 1976, **15**, 1956.
455 T. B. Rauchfuss, J. S. Shu, and D. M. Roundhill, *Inorg. Chem.*, 1976, **15**, 2096.
456 G. Balimann and P. S. Pregosin, *J. Magn. Resonance*, 1976, **22**, 235.
457 A. D. Troitskaya, G. A. Levshina, T. V. Zykova, L. F. Gogolyukhina, and R. A. Salakhutdinov, *Zhur. obshchei Khim.*, 1976, **46**, 1235.

(M = Pd or Pt; L, L′ = tertiary phosphine). ^{31}P N.m.r. spectra were measured on a number of dimeric complexes, *e.g.* $[Rh(PR^1_3)HCl(\mu\text{-}Cl)_2PtCl(PR^2_3)]$ and $[R_3PClPt(\mu\text{-}Cl)_2PtClPR_3]$. In the latter case $^2J(^{195}Pt, ^{195}Pt)$ = *ca.* 150 Hz.[458] ^{195}Pt Chemical shifts have been reported for a number of PMe_3, $AsMe_3$, $SbMe_3$, SMe_2, $SeMe_2$, and $TeMe_2$ complexes of Pt^0 and Pt^{II}. The chemical shifts for complexes involving more than one type of halide were used to analyse the effect of halide on $\delta(^{195}Pt)$ in different situations. $\delta(^{195}Pt)$ is relatively insensitive to the substituents on the donor atoms and useful in the determination of the nature of the platinum complexes. The results for $[PtX_3L]^-$ can be reconciled with the theoretical predictions of the covalency of the M—L bond.[459] Virtual coupling of $[\{(RO)_3P\}_2MCl_2]$ (M = Pd or Pt) has been examined by ^{13}C n.m.r. spectroscopy. In some cases a singlet was found owing to algebraic cancelling of $^2J(^{13}C, O, ^{31}P)$ and $^4J(^{13}C, O, P, M, ^{31}P)$.[460] ^{31}P Chemical shifts of a wide range of $[(R_3P)_2PdX_2]$ compounds have been measured, and it was shown that the linear relationship between the change in chemical shift on co-ordination and the chemical shift in the absence of the metal is generally applicable.[461] The ^{31}P n.m.r. spectrum of $[Pt\{Ph_2P(CH_2)_3PPh_2\}_2]$ is A_2B_2 at −115 °C, but there is rapid exchange at room temperature.[462] For $[PtCl_2(CF_3SCHMeCH_2SCF_3)]$, the ^{19}F n.m.r. spectrum shows the presence of four isomers which do not rapidly interconvert. It was concluded that ring fluxionality is fast as this would give rise to eight forms but the sulphur stereochemistry is rigid.[463] ^{31}P N.m.r. data have also been reported for *trans*-$[MCl(PEt_{3-n}Ph_n)(PF_2O)]$ (M = Pd or Pt; including ^{19}F n.m.r. data),[464] *trans*-$[MCl_2\{Bu^t_2P(CH_2)_mPBu^t_2\}]_n$ (M = Pd or Pt; m = 8—12; n = 1—3),[465] $[Pd(PEt_3)_4]$,[466] $[MCl_2\{Ph_2P(CH_2)_nPPh_2{=}CHC(O)R\}]$ (M = Pd or Pt),[467] $[L_2Pd(CN)_2]$ (including ^{13}C n.m.r. data),[468] $[PtX_3L]^-$ [including ^{19}F and ^{195}Pt n.m.r. data; L = PBr_nCl_{3-n}, PF_3, or $PX_n(OMe)_{3-n}$],[469] and MX_2L [L = (68)].[470]

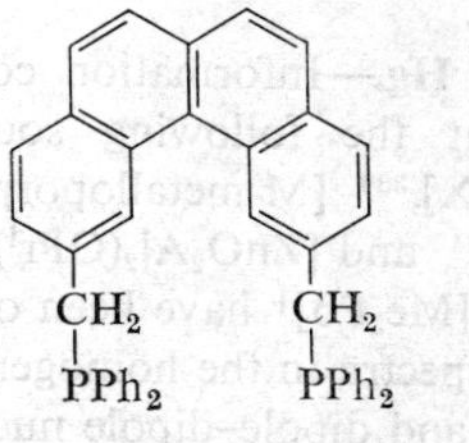

(68)

458 R. Huis and C. Masters, *J.C.S. Dalton*, 1976, 1796.
459 P. L. Goggin, R. J. Goodfellow, S. R. Haddock, B. F. Taylor, and I. R. H. Marshall, *J.C.S. Dalton*, 1976, 459.
460 A. W. Verstuyft, J. E. Nelson, and L. W. Cary, *Inorg. Chem.*, 1976, **15**, 732.
461 A. Verstuyft, J. H. Nelson, and L. W. Cary, *Inorg. Nuclear Chem. Letters*, 1976, **12**, 53.
462 F. C. March, R. Mason, D. W. Meek, and G. R. Scollary, *Inorg. Chim. Acta*, 1976, **19**, L25.
463 R. J. Cross, L. Manojlović-Muir, K. W. Muir, D. S. Rycroft, D. W. A. Sharp, T. Solomun, and H. T. Miguel, *J.C.S. Chem. Comm.*, 1976, 291.
464 J. Grosse and R. Schmutzler, *J.C.S. Dalton*, 1976, 405.
465 A. Pryde, B. L. Shaw, and B. Weeks, *J.C.S. Dalton*, 1976, 322.
466 R. A. Schunn, *Inorg. Chem.*, 1976, **15**, 208.
467 Y. Oosawa, H. Urabe, T. Saito, and Y. Sasaki, *J. Organometallic Chem.*, 1976, **122**, 113.
468 A. W. Verstuyft, L. W. Cary, and J. H. Nelson, *Inorg. Chem.*, 1976, **15**, 3161.
469 C. Crocker, P. L. Goggin, and R. J. Goodfellow, *J.C.S. Dalton*, 1976, 2494.
470 N. J. DeStefano, D. K. Johnson, and L. M. Venanzi, *Helv. Chim. Acta*, 1976, **59**, 2683.

Complexes of Cu, Ag, and Au.—Information concerning complexes of these elements can be found at the following sources: $[\{(Ph_3P)_2Cu\}_2B_{10}H_{10}]$,[1170] $[Rh_2Cu_4(C{\equiv}CC_6H_4F)_8(PMePh_2)_2]$,[340] Me, and CF_3 complexes of Au^{III}.[394]

E.p.r. and n.m.r. studies of the interaction of copper(II)bis(chelates) with electron-donor bases have been reviewed.[471] The ^1H and ^{13}C n.m.r. spectra show only one C_5H_5 signal in $[(\eta^1\text{-}C_5H_5)AgPPh_3]$ and the ^{31}P n.m.r. spectrum at -65 °C is a doublet due to $^{107,109}Ag$ coupling.[472] The ^{13}C n.m.r. spectra of $[Cu(CO)]^+$, $[Cu(CO)_3]^+$, $[Cu(CO)_4]^+$, and $[Ag(CO)_2]^+$ have been reported. No $^{107,109}Ag$ coupling was observed and it was suggested that exchange is fast.[473] For $[(RNC)_4Cu]^+$, the $^{13}C{\equiv}N$ resonance is broad and $^1J(^{13}C, ^{63,65}Cu)$ was not detected. The lineshape of the $^{13}C{\equiv}N$ resonance was discussed in terms of quadrupolar ^{14}N and $^{63,65}Cu$ relaxation. In most complexes exchange seems to be dominant, although in some cases quadrupolar ^{14}N relaxation is important. On co-ordination of CNR, the ^{13}CN resonance shifts upfield (15—20 p.p.m.) and $^1J(^{13}C, ^{14}N)$ increases three-fold. The upfield shift is associated with a larger excitation energy, and the increase of $^1J(^{13}C, ^{14}N)$ was ascribed to a larger effective nuclear charge on the carbon.[474] For $[\{(PPh_nEt_{3-n})_2Ag\}Ni\{S_2C_2(CN)_2\}_2]$, $^1J(^{31}P, ^{107}Ag)$ are 394, 413, 426, and 426 Hz for $n = 3, 2, 1$, and 0, respectively. These data, when correlated with the P—Ag—P angle values, were interpreted as showing the importance of steric effects in determining the *s*-character of the bond.[475] $[(R_3P)_2Ag(SSeCNR_2)]$ has $^1J(^{31}P, ^{107,109}Ag)$ = *ca.* 350 Hz, and restricted rotation of the carbon–nitrogen bond.[476] N.m.r. data have also been reported for $[Me_2Au(PPh_3)X]$ (^{31}P n.m.r. data),[477] $[AuMe_2(CF_3)(PR_3)]$ (^{19}F n.m.r. data),[478] $[Cu(O_2CCF_3)(C_{12}H_{18})]$ (^{13}C n.m.r. data),[479] $[\{(MeO)_3P\}_4Cu]^+$ (^{31}P and ^{63}Cu n.m.r. data),[480] [MXL] {^{31}P n.m.r. data; L = (68); M = Ag or Au},[481] and $[\{(Ph_2PMe)_2M\}_2WS_4]$ (^{31}P n.m.r. data; M = Cu, Ag, or Au).[482]

Complexes of Zn, Cd, and Hg.—Information concerning complexes of these elements can be found at the following sources: $[MeHg]^+$,[1289] $[(Ph_3P)_2(CONMeN{=}NMe)ClRhHgX]$,[336] [M-metalloporphyrins] (M = Zn or Cd),[441] pyridine complexes of Hg^{II},[442] and $[ZnO_2Al_2(OPr^i)_4]$.[207]

$[MeHg]^+$, $[Me_3Sn]^+$, and $[Me_3Pb]^+$ have been oriented in a lyotropic nematic media, and their ^1H n.m.r. spectra in the homogeneous oriented state have been analysed in terms of scalar and dipole–dipole nuclear couplings. These include coupling with ^{199}Hg, ^{117}Sn, ^{119}Sn, and ^{207}Pb. Metal–carbon bond lengths have been calculated from experimental data assuming a tetrahedral carbon atom and a carbon–hydrogen distance of 1.096 ± 0.005 Å. Such a procedure reveals a

471 A. A. Shklyaev and V. F. Anufrienko, *Zhur. strukt. Khim.*, 1975, **16**, 1082.
472 H. K. Hofstee, J. Boersma, and G. J. M. Van der Kerk, *J. Organometallic Chem.*, 1976, **120**, 313.
473 Y. Souma, J. Iyoda, and H. Sano, *Inorg. Chem.*, 1976, **15**, 968.
474 D. Knol, N. J. Koole, and M. J. A. de Bie, *Org. Magn. Resonance*, 1976, **8**, 213.
475 F. J. Hollander, Y. L. Ip, and D. Coucouvanis, *Inorg. Chem.*, 1976, **15**, 2230.
476 S. Araki, K. Matsumoto, K. Tanaka, and T. Tanaka, *J. Inorg. Nuclear Chem.*, 1976, **38**, 727.
477 S. Komiya and J. K. Kochi, *J. Amer. Chem. Soc.*, 1976, **98**, 7599.
478 A. Johnson and R. J. Puddephatt, *J.C.S. Dalton*, 1976, 1360.
479 D. L. Reger and M. D. Dukes, *J. Organometallic Chem.*, 1976, **113**, 173.
480 W. McFarlane and D. S. Rycroft, *J. Magn. Resonance*, 1976, **24**, 95.
481 D. K. Johnson, P. S. Pregosin, and L. M. Venanzi, *Helv. Chim. Acta*, 1976, **59**, 2691.
482 J. C. Huffman, R. S. Roth, and A. R. Siedle, *J. Amer. Chem. Soc.*, 1976, **98**, 4340.

systematic trend of shorter metal–carbon distances as determined by the n.m.r. method and when compared to the best recent electron diffraction results (*ca.* 3—5%). Relative comparison between the n.m.r. results for the ions reported here and the $[Me_2Sn]^{2+}$ ion published previously is much more reliable, and the magnitude for the decrease in metal–carbon bond length with charge on the metal atom and its co-ordination number was found.[483] ^{199}Hg Shielding anisotropies, $\sigma_{\parallel} - \sigma_{\perp}$, of 7475 and *ca.* 5500 p.p.m. in Me_2Hg and MeHgX have been obtained from studies of the molecules partially oriented by dissolution in the nematic phase. A survey of isotropic mercury shieldings shows that the total range is *ca.* 2500 p.p.m. and this, together with the values of the anisotropies, is consistent with the view that variations in the Ramsey paramagnetism dominate the changes in ^{199}Hg shielding. It also appears that the large chemical shift anisotropy will provide a significant mechanism for spin–lattice relaxation of the ^{199}Hg nucleus.[484] The ^{13}C chemical shifts of MeHg compounds and PhHg compounds have been measured. $J(^{13}C, ^{199}Hg)$ are strongly dependent on the solvent as well as on the other ligand. The observed variations of coupling constants were explicable either by a rehybridization scheme or by a departure from strict molecular linearity in nucleophilic solvents. There is a linear relationship between $^1J(^{13}C, ^{199}Hg)$ and $^2J(^1H, ^{199}Hg)$ through the origin for MeHgX.[485] The carbon–mercury force constant has been compared with $^1J(^{13}C, ^{199}Hg)$ and $^2J(^1H, ^{199}Hg)$ for MeHgX species and a crude linear relationship was found.[486] For some $[MeHgL]^+$ complexes, a linear correlation was found between $^2J(^1H, ^{199}Hg)$ and the pK_a of the ligand. ^{13}C N.m.r. data were also reported.[487] ^{1}H N.m.r. spectroscopy has been used to demonstrate that $[HSe(CH_2)_2NH_3]^+$ binds to $[MeHg]^+$.[488] ^{13}C Shieldings and $J(^{13}C, ^{199}Hg)$ for 14 phenyl and seven alkyl and alkenyl mercury compounds have been measured. Substituent effects on ^{13}C shieldings are similar to those in non-mercurated phenyl compounds with a similar relationship between ^{13}C shieldings and Hammett σ^+ values of *para* substituents. For phenylmercuric salts of carboxylic acids, the shieldings are insensitive to the particular anion. Trends in ^{13}C shieldings for a given compound type parallel ^{199}Hg shielding trends. $^1J(^{13}C, ^{199}Hg)$ range from 1186 to 3196 Hz, while $^2J(^{13}C, ^{199}Hg)$ range from 72 to 126 Hz and $^3J(^{13}C, ^{199}Hg)$ from 101 to 221 Hz in phenylmercury compounds. An alternation in the magnitude of $J(^{13}C, ^{199}Hg)$ with a number of intervening bonds is apparent. The dihedral angle dependence of $^3J(^{13}C, ^{199}Hg)$ in 2-norbornylmercury compounds was confirmed. No correlation was observed between directly bonded $^1J(^{13}C, ^{199}Hg)$ and the shielding of the mercury-bonding carbon, but this coupling increases significantly with increase in ^{199}Hg shielding.[489] ^{199}Hg Chemical shifts have been reported for 85 compounds and correlated with structure, and with resonance and inductive effects.[490, 491] On the basis of ^{31}P n.m.r. spectroscopy it was concluded that for R_3P and $(R_F)_2Hg$, the mercury–phosphorus bond is more stable

[483] L. W. Reeves, M. Suzuki, and J. A. Vanin, *Inorg. Chem.*, 1976, **15**, 1035.
[484] J. D. Kennedy and W. McFarlane, *J.C.S. Faraday II*, 1976, **72**, 1653.
[485] A. J. Brown, O. W. Haworth, and P. Moore, *J.C.S. Dalton*, 1976, 1589.
[486] N. Iwasaki, *Bull. Chem. Soc. Japan*, 1976, **49**, 2735.
[487] A. J. Canty and A. Marker, *Inorg. Chem.*, 1976, **15**, 425.
[488] Y. Sugiura, Y. Hojo, Y. Tamai, and H. Tanaka, *J. Amer. Chem. Soc.*, 1976, **98**, 2339.
[489] N. K. Wilson, R. D. Zehr, and P. D. Ellis, *J. Magn. Resonance*, 1976, **21**, 437.
[490] L. A. Fedorov, *Zhur. strukt. Khim.*, 1976, **17**, 247.
[491] L. A. Fedorov, *Zhur. strukt. Khim.*, 1976, **17**, 236.

than mercury–nitrogen or mercury–oxygen bonds.[492] ^{13}C N.m.r. spectroscopy has shown that (69) is fluxional.[493] Arene-mercurium ions have been shown by ^{1}H and ^{13}C n.m.r. spectroscopies to be in fast exchange.[494] N.m.r. data have also been reported for (70; M = Zn or Cd; ^{13}C and ^{31}P n.m.r. data),[495] $(R_F)_2Hg$

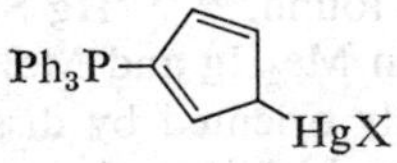

(69)

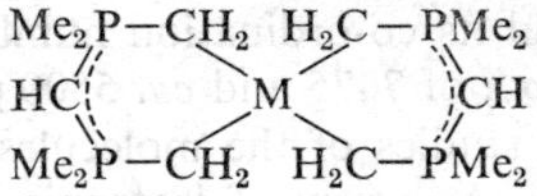

(70)

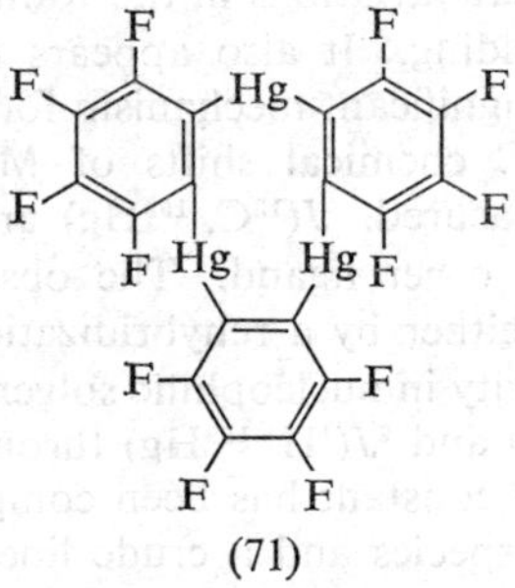

(71)

(^{19}F n.m.r. data),[496, 497] (71) (^{19}F n.m.r. data),[498] and $(Bu^t{}_3Si)_2Cd$ (^{13}C n.m.r. data).[499] ^{13}C N.m.r. studies have shown that $Hg_2{}^{2+}$-arene complexes exist in SO_2 solution, and allow the estimation of formation constants for Me_6C_6, *p*-xylene, and *p*-dichlorobenzene compounds, together with ^{13}C chemical shifts for bound substrates in these cases. ^{13}C Chemical shift changes that occur on the addition of $Hg_2{}^{2+}$ to an SO_2 solution of various methylated benzenes have been compared with those occurring on complexation of the same arenes by Ag^+ and the $Cr(CO)_3$ moiety. It was concluded that for the $Hg_2{}^{2+}$-arene complexes there are preferential binding sites.[500]

The ^{1}H n.m.r. spectra at 100 MHz have been described for some Zn complexes of the *E*- and *Z*-isomers of pyridine-2-carbaldehyde-2′-pyridylhydrazone in $(CD_3)_2SO$. Chemical shift data were discussed qualitatively in relation to factors such as the charge on the metal ion, the anisotropy of ligand N atoms, electric field effects, ring-current effects, and conformation.[501] A *modus operandi* has been published for the reproducible measurement of the ^{13}C n.m.r. spectra of metal porphyrins. Pyrrolidine is added to break up the aggregates, and the

492 Z. Stumbreviciute, L. A. Fedorov, B. L. Dyatkin, and B. I. Martynov, *Zhur. org. Khim.*, 1975, **11**, 2002.
493 N. L. Holy, N. C. Baenziger, R. M. Flynn, and D. C. Swenson, *J. Amer. Chem. Soc.*, 1976, **98**, 7823.
494 G. A. Olah, S. H. Yu, and D. G. Parker, *J. Org. Chem.*, 1976, **41**, 1983.
495 H. Schmidbaur and O. Gasser, *Angew. Chem. Internat. Edn.*, 1976, **15**, 502.
496 L. A. Fedorov, Z. Stumbreviciute, and E. I. Fedin, *Zhur. strukt. Khim.*, 1975, **16**, 976.
497 E. K. S. Liu and R. J. Lagow, *J. Amer. Chem. Soc.*, 1976, **98**, 8270.
498 C. M. Woodard, G. Hughes, and A. G. Massey, *J. Organometallic Chem.*, 1976, **112**, 9.
499 L. Rösch and H. Müller, *Angew. Chem. Internat. Edn.*, 1976, **15**, 620.
500 P. A. W. Dean, D. G. Ibbott, and J. B. Stothers, *Canad. J. Chem.*, 1976, **54**, 166.
501 C. F. Bell, G. R. Mortimore, and G. L. Read, *Org. Magn. Resonance*, 1976, **8**, 45.

method was applied to zinc coproporphyrins.[502] ^{113}Cd N.m.r. spectra of Cd^{II} carbonic anhydrase, Cd^{II} alkaline phosphatase, and model amino-acid complexes have been measured. The lines were broad and there are marked chemical shift changes.[503] Measurement of $^{2}J(^{1}H, ^{199}Hg)$ has shown that the strength of the MeHg—S bond is greater than that of the MeHg—N bond.[504] ^{113}Cd N.m.r. spectra of Cd^{II}–sulphur chelate complexes have been determined. The ^{113}Cd chemical shift of the chelate complexes exhibits no trend in terms of the extent of sulphur ligation, but the nuclear deshielding sequence $Cd(SR)_2 < Cd(SR)_3 < Cd(SR)_4$ has been indicated by measurements on the 1 : 2 cadmium–glutathione complex and the decanuclear cation which contains sites of the latter two types. The solution stability of $[Cd_{10}(SCH_2CH_2OH)_{16}]^{4+}$ was investigated, and it was concluded that small structural differences may exist between the crystalline and solution phases arising from partial rupture of $\overline{CdSCH_2CH_2OH}$ chelate rings known to be present in the crystal. From the data obtained in this and previous studies, a provisional ^{113}Cd chemical shift scale was presented in terms of the types of ligand atoms in the primary co-ordination sphere of Cd^{II}.[505] ^{13}C N.m.r. data have also been reported for zinc porphyrins[506] and $[Hg(PR_3)_2]^{2+}$ (including ^{19}F and ^{31}P n.m.r. data).[507]

3 Dynamic Systems

This section is in three parts: (i) 'Fluxional Molecules', dealing with rate processes involving no molecular change, (ii) 'Equilibria', dealing with the use of n.m.r. spectroscopy to measure the position of equilibria and ligand-exchange reactions, and (iii) 'Course of Reactions' dealing with the use of n.m.r. spectroscopy to monitor the course of reactions. Each section is ordered by the Periodic Table.

Reviews on 'Ligand Substitution Process. An N.M.R. Study. I. Solvation Shells of Diamagnetic Cations Al^{3+} and Be^{2+}',[508] 'The Errors in the Determination of Complex Formation by N.M.R.',[509] and 'Chemically Induced Dynamic Nuclear Polarization'[510] have appeared.

A permutational analysis of ^{1}H n.m.r. configurational rearrangements in *cis*-$[M(R^1COCHR^2COR^2)_2XY]$ compounds has been carried out.[511]

Fluxional Molecules.—*Lithium.* In the 300 MHz ^{1}H n.m.r. spectrum of the octahedral hexamer of $[EtMeCHCH_2Li]_6$ there are two magnetically equivalent sets of diastereomers in the ratio 3 : 2, whereas for $[(R)\text{-}EtMeCHCH_2Li]_6$ only

[502] R. J. Abraham, H. Pearson, and K. M. Smith, *J. Amer. Chem. Soc.*, 1976, **98**, 1604.
[503] I. M. Armitage, R. T. Pajer, A. J. M. S. Uiterkamp, J. F. Chlebowski, and J. E. Coleman, *J. Amer. Chem. Soc.*, 1976, **98**, 5710.
[504] Y. Hojo, Y. Sugiura, and H. Tanaka, *J. Inorg. Nuclear Chem.*, 1976, **38**, 641.
[505] R. A. Haberkorn, L. Que, jun., W. O. Gillum, R. H. Holm, C. S. Liu, and R. C. Lord, *Inorg. Chem.*, 1976, **15**, 2408.
[506] R. J. Abraham, H. Pearson, K. M. Smith, P. Loftus, and J. D. Roberts, *Tetrahedron Letters*, 1976, 877.
[507] P. A. W. Dean and D. G. Ibbott, *Canad. J. Chem.*, 1976, **54**, 177.
[508] J. J. Delpuech, M. R. Khaddar, A. A. Peguy, and P. R. Rubini, *Nato Adv. Study Inst. Ser., Ser.* C18, *Chem. Biol. Appl. Relaxation Spectrometry*, 1975, 311 (*Chem. Abs.*, 1976, **84**, 141 134).
[509] A. V. Bratchikov and G. L. Ryzhova, *Zhur. Fiz. Khim.*, 1976, **50**, 803.
[510] K. G. Siefert, *Chem. Unserer Z.*, 1976, **10**, 84.
[511] D. G. Bickley and N. Serpone, *Inorg. Chem.*, 1976, **15**, 948.

a single species exists. There are two exchange processes: inversion of the CH_2 group bonded to lithium and exchange of the RLi moieties between magnetically inequivalent sites. $\Delta H^{\ddagger}$ and $\Delta S^{\ddagger}$ for inversion are 15 kcal mol^{-1} and 0 e.u., respectively, and for exchange they are 3.3 kcal mol^{-1} and −34 e.u., respectively. It was concluded that the two processes are not correlated.[512] Variable-temperature ^{1}H and ^{13}C n.m.r. spectroscopies have been employed to establish the geometry of [(1,3,5-Ph_3pentadienyl)Li] in THF solution. The data show that the anion exists in both the *S* and *W* forms which are of a comparable energy. The interconversion between the two *S* forms is more rapid than the $W \rightleftharpoons S$ interconversion.[513]

Titanium, Niobium, and Tantalum. Rotation of the phenyl rings in titanyltetra-arylporphyrins have $\Delta G^{\ddagger}$ = 16 kcal mol^{-1}.[514] The $\eta^4 \leftrightarrow \eta^8$ ring interconversion for [M(η^8-C_8H_8)(η^4-C_8H_8)R] (M = Nb or Ta; R = Me or Ph), has $\Delta G_{337}{}^{\ddagger}$ = 15.7—17.5 kcal mol^{-1}.[515]

Chromium, Molybdenum, and Tungsten. ^{1}H and ^{13}C n.m.r. spectroscopies have been used to estimate the activation energy of *gauche* ⇌ *anti* interconversion in [Me(Et_2N)$_2$W≡W(NEt_2)$_2$Me] as *ca.* 21 kcal mol^{-1}. Internal rotations about the tungsten–nitrogen bonds allow interchange of proximal and distal ethyl groups. The activation energies for the two types in the *gauche*-rotamer differ by *ca.* 2.5 kcal mol^{-1}, while that in the *anti*-isomer is about halfway between.[516] For (72; M = Cr or Mo), there is NEt_2 coalescence with $\Delta G^{\ddagger}$ = 11.9 ± 0.2 kcal mol^{-1}.[517] The ^{1}H n.m.r. spectrum of *cis*-[M(CO)$_4$(CNMeCHCHNMe)$_2$]

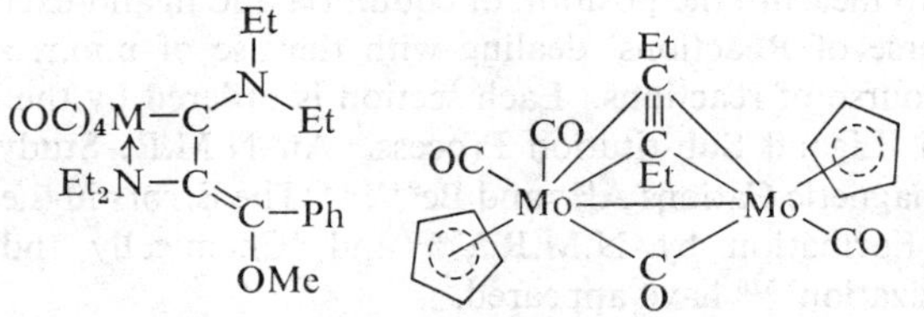

(72) (73)

(M = Cr, Mo, or W) shows that the molecule is dynamic owing to restricted rotation of the carbene about the metal–carbon bond which is due to steric hindrance with $\Delta G^{\ddagger}$ of *ca.* 9 kcal mol^{-1}.[518] $\Delta G^{\ddagger}$ = 11.6 ± 4 kcal mol^{-1} for restricted rotation of the carbon–oxygen bond in [(OC)$_5$CrC(OEt)$_2$].[519] Hindered rotation has been found for L in [(η^5-C_5H_5)Cr(CO)(NO)L] (L = η^2-olefin or η^2-acetylene).[520] The ^{13}C n.m.r. spectrum of (73) has six ^{13}CO signals due to two

512 G. Fraenkel, W. E. Beckenbaugh, and P. P. Yang, *J. Amer. Chem. Soc.*, 1976, **98**, 6878.
513 D. H. Hunter, R. E. Klinck, R. P. Steiner, and J. B. Stothers, *Canad. J. Chem.*, 1976, **54**, 1464.
514 P. Fournari, R. Guilard, M. Fontesse, J.-M. Latour, and J.-C. Marchon, *J. Organometallic Chem.*, 1976, **110**, 205.
515 R. R. Schrock, L. J. Guggenberger, and A. D. English, *J. Amer. Chem. Soc.*, 1976, **98**, 903.
516 M. H. Chisholm, F. A. Cotton, M. Extine, M. Millar, and B. R. Stults, *Inorg. Chem.*, 1976, **15**, 2244.
517 K. H. Dötz and C. G. Kreiter, *Chem. Ber.*, 1976, **109**, 2026.
518 C. G. Kreiter, K. Öfele, and G. W. Wieser, *Chem. Ber.*, 1976, **109**, 1749.
519 E. O. Fischer, K. Scherzer, and F. R. Kreissl, *J. Organometallic Chem.*, 1976, **118**, C33.
520 M. Herberhold, H. Alt, and C. G. Kreiter, *Annalen*, 1976, 300.

isomers at $-126\,°C$ but there is only one line at $-40\,°C$.[521] $[(\eta^5\text{-}C_5H_5)(\eta^2\text{-}MeC{\equiv}CMe)Mo(dppe)]^+$ shows two methyl signals at -50 and one at $110\,°C$ with $\Delta G_{Tc}^{\ddagger} = 14.3 \pm 0.5$ kcal mol^{-1}.[522] Ring inversion in $[(\eta^5\text{-}C_5H_5)Mo(CH_2CH_2CH_2AsMe_2)(CO)_2]$ has $\Delta G^{\ddagger} = 82.2$ kJ mol^{-1}.[523] Variable-temperature ^1H n.m.r. spectroscopy of $[(\eta^5\text{-}C_5H_5)MoCl(CO)(COC_8F_{12})]$ shows two cyclopentadienyl signals at room temperature and four in the ratio 1 : 1 : 3 : 3 at low temperatures.[524] The variable-temperature ^{13}C n.m.r. spectra of (74;

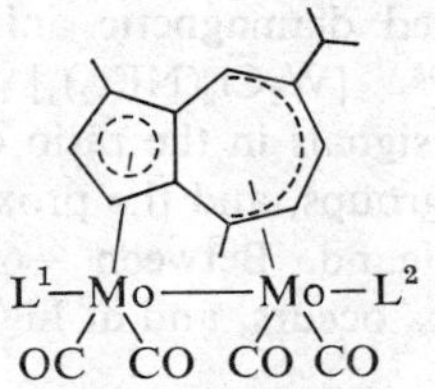

(74)

$L^1 = CO$, $L^2 = PEt_3$; $L^1 = PEt_3$, $L^2 = CO$) have been measured. It was shown that it is the $Mo(CO)_3$ group bonded to the cyclopentadienyl group that executes internal scrambling of its CO groups more rapidly.[525] The rate and pathway of CO scrambling in $[(azulene)Mo_2(CO)_6]$ has been measured by ^{13}C n.m.r. spectroscopy and compared with previously published results for analogous guaiazulene compounds. The results are qualitatively the same and show only slightly different rates, in accord with the idea that the presence of alkyl substituents on the azulene ligand has only a small effect on the fluxional processes. The dynamic properties of $[(guaiazulene)W_2(CO)_6]$ are virtually identical with the molybdenum analogue.[526] Variable-temperature ^1H n.m.r. spectroscopy of $[(\eta^6\text{-}C_6H_5CHBu^t_2)Cr(CO)_3]$ shows restricted rotation about the phenyl–carbon bond with $\Delta G^{\ddagger} = 16.9$ kcal mol^{-1}, which is substantially smaller than $\Delta G^{\ddagger}$ for the free ligand.[527] The $\delta \rightleftharpoons \lambda$ interconversion of the ring in $[(OC)_4MoNMe_2CH_2CH_2NMe_2]$ has $\Delta H^{\ddagger} = 41.5 \pm 1.9$ kJ mol^{-1} and $\Delta S^{\ddagger} = -8.9 \pm 9.0$ e.u.[528] $[M(CO)_5(RN{=}S{=}NR)]$ (M = Cr, Mo, or W) are fluxional with the metal atom gliding along the N—S—N system.[529] The variable-temperature spectra of the benzylic methylene protons in $[M(CO)_4(PhCH_2SCH_2CH_2SCH_2Ph)]$ (M = Cr, Mo, or W) have been attributed to the inversion of sulphur.[530] Activation parameters have been calculated for the non-rigid $[\{cis\text{-}Me_2AsC(CF_3){=}C(CF_3)AsMe_2\}M(CO)_3X_2]$ (M = Mo or W; X = Br or I).[531] Variable-temperature ^1H n.m.r.

521 W. I. Bailey, F. A. Cotton, J. D. Jamerson, and J. R. Kolb, *J. Organometallic Chem.*, 1976, **121**, C23.
522 J. A. Segal, M. L. H. Green, J.-C. Daran, and K. Prout, *J.C.S. Chem. Comm.*, 1976, 766.
523 M. Mickiewicz, K. P. Wainwright, and S. B. Wild, *J.C.S. Dalton*, 1976, 262.
524 J. L. Davidson, M. Green, F. G. A. Stone, and A. J. Welch, *J.C.S. Dalton*, 1976, 738.
525 F. A. Cotton, P. Lahuerta, and B. R. Stults, *Inorg. Chem.*, 1976, **15**, 1866.
526 F. A. Cotton and B. E. Hanson, *Inorg. Chem.*, 1976, **15**, 2806.
527 F. Van Meurs, J. M. A. Baas, J. M. Van der Toorn, and H. van Bekkum, *J. Organometallic Chem.*, 1976, **118**, 305.
528 C. J. Hawkins, R. M. Peachey, and C. L. Szoredi, *Inorg. Nuclear Chem. Letters*, 1976, **12**, 881.
529 R. Meij, J. Kuyper, D. J. Stufkens, and K. Vrieze, *J. Organometallic Chem.*, 1976, **110**, 219.
530 R. J. Cross, G. Hunter, and R. C. Massey, *J.C.S. Dalton*, 1976, 2015.
531 W. R. Cullen and L. M. Mihichuk, *Canad. J. Chem.*, 1976, **54**, 2548.

studies of the deuteriated analogue of 1,2-bis(isopropylseleno)ethane with $M(CO)_4$ (M = Cr, Mo, or W) and PdX_2 have been shown to result from inversion of configuration about the ligand atoms and not from reversal of the chelate ring.[532]

Variable-temperature n.m.r. spectra of $[Mo_2(NMe_2)_6]$ show that proximal and distal methyl groups are frozen out in solution below −30 °C. At higher temperatures they interconvert rapidly on the n.m.r. time-scale. A large chemical shift separation of *ca.* 2 p.p.m. between proximal and distal methyl proton resonances reflects the marked diamagnetic anisotropy of the molybdenum–molybdenum triple bond.[533] $[W_2Cl_2(NEt_2)_4]$ is similar.[534] At −60 °C, $[W_2(O_2CNMe_2)_6]$ shows four signals in the ratio 4 : 4 : 2 : 2 due to the bridging and non-bridging O_2CNMe_2 groups, and the proximal and distal methyl groups of the remaining O_2CNMe_2 ligand. Between −60 and −30 °C, interchange of the proximal and distal groups occurs, and at higher temperatures other groups become involved.[535]

Manganese and Rhenium. ^1H N.m.r. spectroscopy shows $[(\eta^5\text{-}C_5H_5)Re(CO)_2(PPh_3)H]^+$ to be a mixture of isomers which interconverts with E_a = 20 kcal mol^{-1}, $\Delta H^{\ddagger}$ = −20 kcal mol^{-1}, and $\Delta S^{\ddagger}$ = +10 e.u.[536] Hydrogen tautomerism in $[H_4Re_4(CO)_{15}]^{2-}$ has been studied by ^1H n.m.r. spectroscopy to determine $\Delta H^{\ddagger}$ = 11.73 ± 0.66 kcal mol^{-1} and $\Delta S^{\ddagger}$ = −1.35 ± 2.21 e.u.[537] The coalescence temperature for olefin rotation in $[(\eta^5\text{-}C_5H_5)(\eta^2\text{-olefin})Mn(CO)_2]$ has been determined.[538] ^1H and ^{13}C n.m.r. spectroscopies have been used to demonstrate that $[(\eta^5\text{-}C_7H_7)Mn(CO)_3]$ is fluxional *via* 1,2-shifts.[539]

Iron, Ruthenium, and Osmium. Variable-temperature ^{13}C n.m.r. spectra of $[HFe_3(CO)_{11}]^-$ have revealed that at least two dynamic processes are occurring over the range −107 to +30 °C. The bridging carbonyl unit is very basic and forms acid–base complexes with BF_3 as well as $[HNEt_3]^+$. These acid–base interactions apparently cause the fluxional process of the $[HFe_3(CO)_{11}]^-$ to stop at a higher temperature.[540] Variable-temperature ^{13}C n.m.r. studies have confirmed that the solid-state geometries of $[H_2M_3(CO)_9S]$ and $[H_3M_3(CO)_9CMe]$ (M = Ru or Os) persist in solution. A number of CO scrambling processes were considered.[541] ^{13}C N.m.r. spectra of hydrido-olefinic carbonyl clusters of ruthenium show that localized scrambling of CO occurs with different energy barriers which depend on the nature of the bonding between the ligand and metal cluster and on the nature of the substituent on the olefinic chain.

532 G. Hunter and R. C. Massey, *J.C.S. Dalton*, 1976, 2007.
533 M. H. Chisholm, F. A. Cotton, B. A. Frenz, W. W. Reichert, L. W. Shive, and B. R. Stults, *J. Amer. Chem. Soc.*, 1976, **98**, 4469.
534 M. H. Chisholm, F. A. Cotton, M. Extine, M. Millar, and B. R. Stults, *J. Amer. Chem. Soc.*, 1976, **98**, 4486.
535 M. Chisholm, M. Extine, F. A. Cotton, and B. R. Stults, *J. Amer. Chem. Soc.*, 1976, **98**, 4683.
536 G. A. Panosyan, P. V. Petrovskii, N. I. Pyshnograyeva, N. E. Kolobova, V. N. Setkina, and E. I. Fedin, *J. Organometallic Chem.*, 1976, **108**, 209.
537 P. Fantucci, *J. Organometallic Chem.*, 1976, **108**, 203.
538 M. Herberhold, G. O. Wiedersatz, and C. G. Kreiter, *J. Organometallic Chem.*, 1976, **104**, 209.
539 T. H. Whitesides and R. A. Budnik, *Inorg. Chem.*, 1976, **15**, 874.
540 J. R. Wilkinson and L. J. Todd, *J. Organometallic Chem.*, 1976, **118**, 199.
541 A. Forster, B. F. G. Johnson, J. Lewis, and T. W. Matheson, *J. Organometallic Chem.*, 1976, **104**, 225.

Resonances of carbon atoms σ-bonded to one ruthenium and π-bonded to the second one are in the range 162—198 p.p.m.[542] The ^{13}C n.m.r. spectrum of $[Os_5(CO)_{15}]^{2-}$ over the range from 0 to −110 °C shows only two signals (ratio 3 : 2), consistent with an idealized trigonal-bipyramidal geometry, whereas the ^{13}C n.m.r. spectrum of $[HOs_5(CO)_{15}]^-$ shows dynamic behaviour attributable to hydride migration around the three-fold axis.[543] Similar observations have been made for $[Os_6(CO)_{18}]^{2-}$ and $[HOs_6(CO)_{18}]^-$.[544]

$\Delta G^{\ddagger}$ for C—N bond rotation in $[(OC)_3Fe\{C(NMe_2)\}_2OSnMe_2]$ is 16.5 kcal mol^{-1}.[545] The ^{13}C n.m.r. spectra of 29 complexes of the type *cis*-$[M(CO)_4(M'Me_3)_2]$ (M = Ru or Os; M′ = Si, Ge, Sn, or Pb) have been assigned and barriers for CO scrambling determined.[546] The n.m.r. spectra of the (η^5-C_5H_5)- and (η^5-indenyl)–Fe–Sn–olefin compounds reveal orientational preferences and dynamic properties of the olefin ligand. The olefin rotates about the metal–olefin bond with a barrier of 11—14 kcal mol^{-1}. Evidence for restricted rotation about the tin–carbon bond was seen in the low-temperature n.m.r. spectra of the Ph_3Sn derivative. Chemical shift differences between the cyclopentadienyl and indenyl derivatives were due to a preferred orientation in the propene complex.[547] The compound $[(\mu\text{-}MePhSn)_2Fe_2(CO)_7]$ was synthesized to differentiate bridge deformation from Fe—Sn bond cleavage in R group interchange processes of stereochemically non-rigid $[(\mu\text{-}R_2Sn)_2Fe_2(CO)_7]$ complexes. It was found that both processes can occur, the former with $\Delta G_{298}^{\ddagger}$ = 11.8 ± 0.6 kcal mol^{-1} and the latter with $\Delta G_{298}^{\ddagger}$ = 19.2 ± 0.9 kcal mol^{-1}. The bridge deformation in $[(Bu^n{}_2Sn)_2Fe_2(CO)_7]$ occurs with simultaneous stereospecific interchange of bridge and terminal CO ligands. $[PhMeSnFe(CO)_4]_2$ exists in two isomeric forms which interconvert in solution, with $\Delta G_{298}^{\ddagger}$ = 25.5 ± 0.7 kcal mol^{-1}.[548]

The compounds $[Fe_2(CR^1{=}CR^2CR^3{=}CR^4)(CO)_6]$ and $[Fe_2(\mu\text{-}CR^1{=}CR^2COCR^3{=}CR^4)(CO)_6]$ are non-rigid and the $Fe(CO)_3$ fluxional.[549, 550] The ^{13}C n.m.r. spectra of $[Fe_3(CO)_8(C_2Ph_2)_2]$ and $[Co_4(CO)_{10}C_2Ph_2]$ show CO scrambling.[551] In (75), the CO groups undergo very different ^{13}CO exchange rates but inter-iron scrambling was not found.[552] The ^{13}C n.m.r. spectrum of (76) shows three ^{13}CO resonances at room temperature which coalesce on heating with $\Delta G^{\ddagger}$ = 17—18 kcal mol^{-1}.[553] Intramolecular CO exchange in several $[(\eta^4\text{-diene})Fe(CO)_3]$ complexes with heteroatoms in the 1 and 4 position have been studied by temperature-dependent ^{13}C n.m.r. spectroscopy. π,π-Bonded polar dienes show the highest barriers (10—14 kcal mol^{-1}), and

[542] S. Aime, L. Milone, D. Osella, M. Valle, and E. W. Randall, *Inorg. Chim. Acta*, 1976, **20**, 217.
[543] C. R. Eady, J. J. Guy, B. F. G. Johnson, J. Lewis, M. C. Malatesta, and G. M. Sheldrick, *J.C.S. Chem. Comm.*, 1976, 807.
[544] C. R. Eady, B. F. G. Johnson, and J. Lewis, *J.C.S. Chem. Comm.*, 1976, 302.
[545] W. Petz and A. Jonas, *J. Organometallic Chem.*, 1976, **120**, 423.
[546] L. Vancea, R. K. Pomeroy, and W. A. G. Graham, *J. Amer. Chem. Soc.*, 1976, **98**, 1407.
[547] J. W. Faller, B. V. Johnson, and C. D. Schaeffer, jun., *J. Amer. Chem. Soc.*, 1976, **98**, 1395.
[548] G. W. Grynkewich and T. J. Marks, *Inorg. Chem.*, 1976, **15**, 1307.
[549] S. Aime, L. Milone, and E. Sappa, *J.C.S. Dalton*, 1976, 838.
[550] F. A. Cotton, D. L. Hunter, and J. M. Troup, *Inorg. Chem.*, 1976, **15**, 63.
[551] S. Aime, L. Milone, and E. Sappa, *Inorg. Chim. Acta*, 1976, **16**, L7.
[552] D. G. Leppard, H.-J. Hansen, K. Bachmann, and W. von Philipsborn, *J. Organometallic Chem.*, 1976, **110**, 359.
[553] E. S. Magyar and C. P. Lillya, *J. Organometallic Chem.*, 1976, **116**, 99.

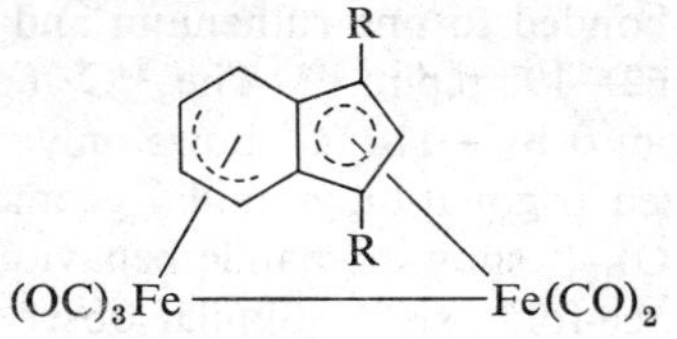

(75)

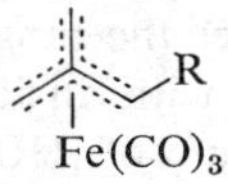

(76)

σ,π- and σ,σ-diazadiene chelates have barriers of activation <9 kcal mol^{-1}.[554] The fluxional behaviour of a series $[(\eta^4\text{-diene})M(CO)_3]$ has been studied by variable-temperature ^{13}C n.m.r. spectroscopy. A difference in rates of rearrangement between conjugated and non-conjugated dienes was noted. The increased activation barrier for CO scrambling in $[(\eta^4\text{-1,3-diene})Fe(CO)_3]$ complexes was attributed to the mixing of six-co-ordinate character into these formally five-co-ordinate molecules. The dependence of rearrangement energetics on the electronic structure of the conjugated diene is fully consistent with the above postulate. Possible mechanisms consistent with the permutational nature of the exchange were detailed and the apparent differences between conjugated and non-conjugated dienes were discussed. The variation in averaged ^{13}C chemical shifts of the CO groups can be satisfactorily accounted for by a metal–carbonyl π-back-bonding model. However, the individual CO resonances observed in low-temperature spectra do not agree with force constant calculations. $^1J(^{13}C, ^{57}Fe) = 27.9$ Hz for $[(\eta^4\text{-butadiene})Fe(CO)_3]$.[555] The fluxional behaviour of $[(\eta^4\text{-}C_8H_8)M(CO)_3]$ (M = Fe or Ru) *via* a 1,2-shift mechanism has been confirmed, with $E_a = 8.1$ kcal mol^{-1} for M = Fe and 8.6 kcal mol^{-1} for M = Ru. Averaging of the carbonyl resonances was also observed in the same general temperature range, but the mechanistic relation between the two processes is not unambiguously revealed by available data.[556] Compound (77) is fluxional at 75 °C, and at 120 °C it rearranges to the fluxional molecule (78).[557] ^1H and ^{13}C n.m.r. spectra have been used to determine $\Delta H^{\ddagger}$, $\Delta S^{\ddagger}$, and $\Delta G^{\ddagger}$ for $P(OR)_3$ scrambling in $[(\text{cyclo-octa-1,3-diene})Fe\{P(OMe)_3\}_3]$.[558] $[(\eta^4\text{-}C_7H_7GeMe_3)Fe(CO)_3]$ is fluxional.[559] The activation energy for CO scrambling in (79) is 12.8 ± 0.3 kcal mol^{-1}, which is 3 kcal mol^{-1} greater than for $[(\eta^4\text{-}C_4H_6)Fe(CO)_3]$; this was attributed to steric hindrance.[560] Compound (80) is also fluxional.[561] $[(\eta^6\text{-}C_8H_8)Ru(\eta^4\text{-norbornadiene})]$ is fluxional with the cyclo-octatetraene ligand rotating *via* the 'piano stool' or 1,3-shift mechanism with $E_a = 14.1 \pm 0.7$ kcal mol^{-1}.[562] ^{13}C N.m.r. spectroscopy has been used to demonstrate that (81) is fluxional.[563] $[(\eta^4\text{-}C_8H_8)Ru(\eta^6\text{-}C_6Me_6)]$ is also fluxional.[564]

[554] D. Leibfritz and H. tom Dieck, *J. Organometallic Chem.*, 1976, **105**, 255.
[555] L. Kruczynski and J. Takats, *Inorg. Chem.*, 1976, **15**, 3140.
[556] F. A. Cotton and D. L. Hunter, *J. Amer. Chem. Soc.*, 1976, **98**, 1413.
[557] R. Aumann, *Chem. Ber.*, 1976, **109**, 168.
[558] A. D. English, J. P. Jesson, and C. A. Tolman, *Inorg. Chem.*, 1976, **15**, 1730.
[559] L. K. K. Li Shing Man and J. Takats, *J. Organometallic Chem.*, 1976, **117**, C104.
[560] J. Elzinga and H. Hogeveen, *Tetrahedron Letters*, 1976, 2383.
[561] A. J. Carty, C. R. Jablonski, and V. Snieckus, *Inorg. Chem.*, 1976, **15**, 601.
[562] F. A. Cotton and J. R. Kolb, *J. Organometallic Chem.*, 1976, **107**, 113.
[563] L. J. Todd, J. P. Hickey, J. R. Wilkinson, J. C. Huffman, and K. Folting, *J. Organometallic Chem.*, 1976, **112**, 167.
[564] M. A. Bennett, T. W. Matheson, G. B. Robertson, A. K. Smith, and P. A. Tucker, *J. Organometallic Chem.*, 1976, **121**, C18.

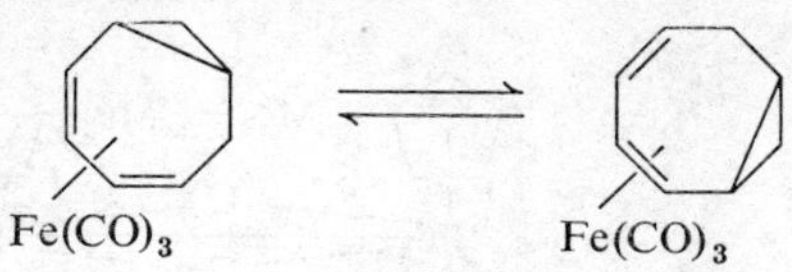

(77)

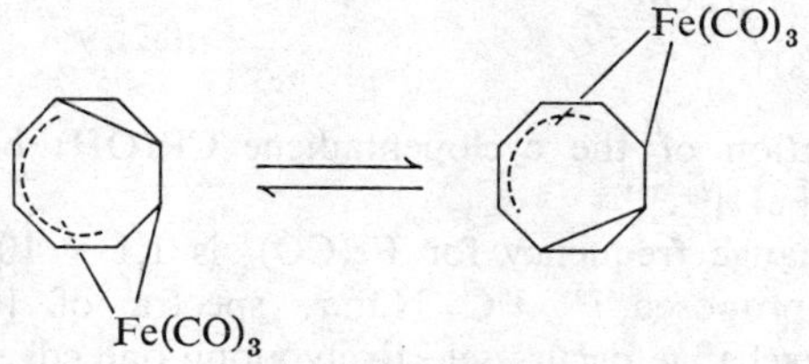

(78)

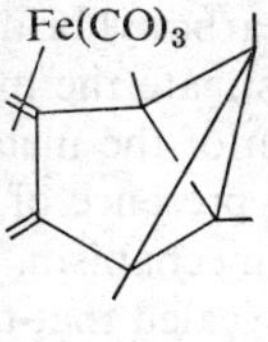

(79)

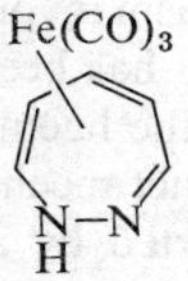

(80)

Low-temperature ^{13}C n.m.r. spectra of $[(C_7H_7)Fe(CO)_3M(CO)_n]$ (M = Mn or Re, $n = 3$; M = Rh, $n = 2$), show that while the local CO scrambling within the $M(CO)_3$ sets is indeed very ready, CO exchange between the two metals does not occur. The coupling between rearrangement of the ring and CO scrambling was offered as a rationale for the rapidity of the former process. The absence of CO exchange between the two ends of the molecule is consistent with the presence of the bridging C_7H_7 group and the associated long iron–rhodium bond length.[565] $\Delta G^{\ddagger}$ has been determined for $[Ru_3(CO)_8(C_8H_6)]$ and related molecules.[566]

The temperature dependence of the ^1H and ^{13}C n.m.r. spectra of $[(\eta^5\text{-dienyl})_2\text{-}M_2(CO)_4]$ (M = Fe or Ru) has been determined. Resonances characteristic of both the *cis*- and *trans*-bridged CO isomers were detected and both thermodynamic and kinetic parameters measured. For the ruthenium complexes there also appeared to be both bridged and unbridged species.[567] Compound (82) behaves similarly.[568] ^{13}C N.m.r. spectroscopy has been used to show that $[(\eta^5\text{-}C_5H_5)(OC)_2Fe(R^1N_3R^2)]$ is fluxional, involving CO exchange *via* a pseudo-Berry rotation.[569] The barriers to acyl group rotation in a series of acylferrocenes, including 1,1′-diacetylferrocene, have been determined by ^1H and ^{13}C n.m.r. spectroscopy; $\Delta G^{\ddagger}$, $\Delta H^{\ddagger}$, and $\Delta S^{\ddagger}$ were quoted.[570] $\Delta G^{\ddagger}$ has been determined

565 M. J. Bennett, J. L. Pratt, K. A. Simpson, L. K. K. Li Shing Man, and J. Takats, *J. Amer. Chem. Soc.*, 1976, **98**, 4810.
566 J. A. K. Howard, S. A. R. Knox, R. J. McKinney, R. F. D. Stansfield, F. G. A. Stone, and P. Woodward, *J.C.S. Chem. Comm.*, 1976, 557.
567 O. A. Gansow, A. R. Burke, and W. D. Vernon, *J. Amer. Chem. Soc.*, 1976, **98**, 5817.
568 F. A. Cotton, D. L. Hunter, P. Lahuerta, and A. J. White, *Inorg. Chem.*, 1976, **15**, 557.
569 E. Pfeiffer, J. Kuyper, and K. Vrieze, *J. Organometallic Chem.*, 1976, **105**, 371.
570 J. Sandström and J. Seita, *J. Organometallic Chem.*, 1976, **108**, 371.

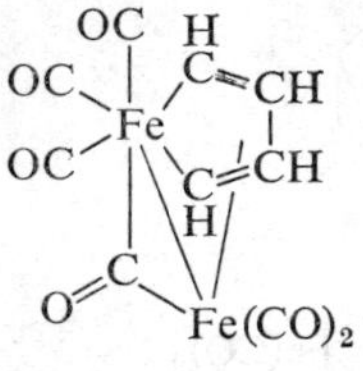

(81)

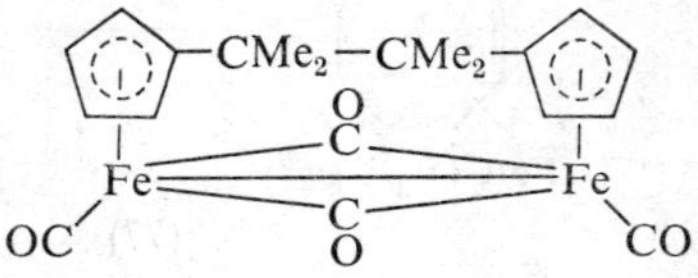

(82)

for restricted rotation of the cyclopentadiene–CR(OH) bond in $[(\eta^5\text{-}C_5H_5)\text{-}\{\eta^5\text{-}C_5H_4CR(OH)\}FeH]^{2+}$.[571]

The ligand exchange frequency for $Fe(CO)_5$ is $1.1 \times 10^{10}\ s^{-1}$, and a Berry mechanism was proposed.[572] ^{13}C N.m.r. spectra of $[Os_3(CO)_{12-n}(PEt_3)_n]$ indicate that CO exchange occurs selectively along one edge of the Os_3 triangle at low temperatures. This process involves only six of the CO groups when $n = 1$ or 2, and a mechanism involving a double carbonyl bridge was proposed.[573]

^1H N.m.r. spectroscopy has been used to investigate the mobility of the high-spin Fe^{III} ion relative to the haem plane. Motion of the metal through the hole was induced by two distinct mechanisms. In the presence of an excess of halide ion, the porphyrins inverted by an associative mechanism. Second-order rate constants for a variety of porphyrin complexes revealed that the rate of inversion decreased with decreasing porphyrin basicity, decreasing axial ligand field strength, and increasing solvent electric permittivity. Activation parameters obtained for four complexes indicate that the changes in rates were primarily enthalpy effects. The rapid rates of inversion obtained in this study reflected a high degree of mobility of the high-spin iron relative to the porphyrin plane, and suggest that both the Fe^{III} and Fe^{II} ions could pass readily through the porphyrin hole without being converted into the low-spin forms.[574] $[RuCl_3(NO)(ER_2)_2]$ (E = S or Se) shows sulphur inversion.[575] The racemization of $[Os(PMe_2Ph)_3\text{-}(S_2CNMe_2)Cl]$ has been followed, and activation parameters were determined.[576]

Cobalt, Rhodium, and Iridium. $[MH\{P(OCH_2)_3CPr^n\}_4]$ (M = Co, Rh, or Ir) are stereochemically non-rigid on the n.m.r. time-scale at 25 °C. However, the Rh and Ir analogues have given near-limiting and limiting spectra at low temperatures and activation parameters have been calculated.[577] A ^{19}F n.m.r. study has been carried out on $[R_FCo(PF_3)_n(CO)_{4-n}]$. In many cases, temperatures near −100 °C are required to freeze out the motion. Two types of dynamic process were found; one occurring between different isomers and the other between non-equivalent sites in a single isomer.[578] $\Delta G^{\ddagger}$ has been determined for olefin rotation coupled to Berry pseudorotation in $[Rh(S_2CNMe_2)(CNR)_2\text{-}(\eta^2\text{-olefin})]$.[579] The barriers to olefin rotation in $[(acac)Rh(\eta^2\text{-}C_2R_4)_2]$ increase

[571] P. Rimmelin, J. Sommer, J. Sandström, and J. Seita, *J. Organometallic Chem.*, 1976, **114**, 175.
[572] R. K. Sheline and H. Mahnke, *New Synth. Methods*, 1975, **3**, 219 (*Chem. Abs.*, 1976, **85**, 201 390).
[573] B. F. G. Johnson, J. Lewis, B. E. Reichert, and K. T. Schorpp, *J.C.S. Dalton*, 1976, 1403.
[574] R. V. Snyder and G. N. La Mar, *J. Amer. Chem. Soc.*, 1976, **98**, 4419.
[575] J. E. Fergusson and C. T. Page, *Austral. J. Chem.*, 1976, **29**, 2159.
[576] D. J. Cole-Hamilton and T. A. Stephenson, *J.C.S. Dalton*, 1976, 2396.
[577] E. M. Hyde, J. R. Swain, J. G. Verkade, and P. Meakin, *J.C.S. Dalton*, 1976, 1169.
[578] C. A. Udovich, M. A. Krevalis, and R. J. Clark, *Inorg. Chem.*, 1976, **15**, 900.
[579] T. Kaneshima, Y. Yumoto, K. Kawakami, and T. Tanaka, *Inorg. Chim. Acta*, 1976, **18**, 29.

with an increasing number of MeO substituents. When the olefin rotation is frozen out, the methoxy substituents of the olefins tend to be turned away from the acac ligand unless steric interaction occurs between the two π-co-ordinated ligands. An independent acac rotation was also found.[580] Variable-temperature ^{19}F n.m.r. spectroscopy shows that the PF_3 groups in $[Rh_2(PF_3)_6(R^1C_2R^2)]$ undergo intramolecular exchange; $\Delta G^{\ddagger} = 6$—10 kcal mol^{-1} is larger for electron-withdrawing and for sterically hindered acetylenes.[581–583] The ^{31}P n.m.r. spectrum of $[(\eta^3\text{-}C_3H_5)Rh\{P(OMe)_3\}_3]$ shows one resonance at 20 °C and two at -40 °C. At higher temperatures, first $P(OMe)_3$ exchange occurs, then *syn-anti* allyl exchange occurs.[584] The ^{13}C n.m.r. spectrum of (83) shows two cyclo-octadiene signals at room temperature and eight at -77 °C.[585] $[(\eta^4\text{-diene})\text{-}Rh(\eta^5\text{-cyclo-octatrienyl})]$ is fluxional.[586] ^{19}F N.m.r. spectroscopy shows that in

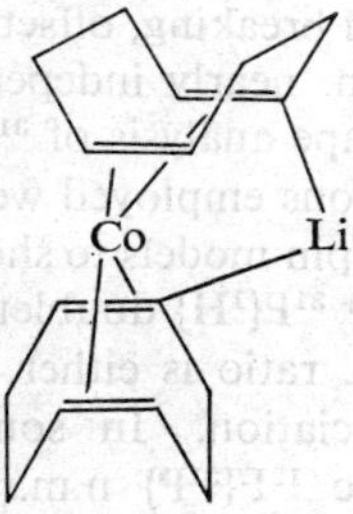

(83)

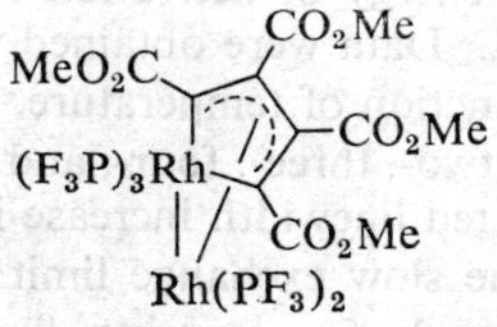

(84)

(84), the $Rh(PF_3)$ is fluxional with $\Delta G^{\ddagger} = 10.5$ kcal mol^{-1} while $Rh(PF_3)_2$ is not.[587] $[(\eta^5\text{-}C_5H_5)_3Rh_3(\mu_2\text{-}CO)_3]$ shows fluxional CO behaviour.[588] The ^{31}P n.m.r. spectra of $[CoL(CO)_2]^+$ and $[CoLL'(CO)]^+$ [L′ = phosphorus ligand, L = $(Ph_2PCH_2CH_2)_2PPh$, $(Ph_2PCH_2CH_2CH_2)_2PPh$, or $(Ph_2PCH_2)_3CH$] have been examined and dynamic behaviour was found. The stereochemistry depends on the ligand, *e.g.* $[Co\{(Ph_2PCH_2CH_2)_2PPh\}(CO)L]^+$ has a trigonal-bipyramidal structure with the equatorial position occupied by a carbonyl ligand and two terminal phosphorus ligands while the axial positions are occupied by the central phosphorus and L. $[Co\{(Ph_2PCH_2CH_2CH_2)_2PPh\}L(CO)]^+$ have square-pyramidal structures with L axial. $[Co\{(Ph_2PCH_2)_3CH\}L(CO)]$ is non-rigid whereas $[Co\{(Ph_2PCH_2CH_2)_2PPh\}L(CO)]$ is rigid. Many $^2J(^{31}P, ^{31}P)$ were reported.[589] $\Delta G^{\ddagger}$ for the dissociation

$$[M\{P(OEt)_3\}_5]^{n+} \rightleftharpoons [M\{P(OEt)_3\}_4]^{+} + P(OEt)_3$$

580 M. Herberhold, C. G. Kreiter, and G. O. Wiedersatz, *J. Organometallic Chem.*, 1976, **120**, 103.
581 M. A. Bennett, R. N. Johnson, and T. W. Turney, *Inorg. Chem.*, 1976, **15**, 90.
582 M. A. Bennett, R. N. Johnson, G. B. Robertson, T. W. Turney, and P. O. Whimp, *Inorg. Chem.*, 1976, **15**, 97.
583 M. A. Bennett, R. N. Johnson, and T. W. Turney, *Inorg. Chem.*, 1976, **15**, 111.
584 M. Bottrill and M. Green, *J. Organometallic Chem.*, 1976, **111**, C6.
585 K. Jonas, R. Mynott, C. Krüger, J. C. Sekutowski, and Y.-H. Tsay, *Angew. Chem. Internat. Edn.*, 1976, **15**, 767.
586 J. Müller, H.-O. Stühler, G. Huttner, and K. Scherzer, *Chem. Ber.*, 1976, **109**, 1211.
587 M. A. Bennett, R. N. Johnson, and T. W. Turney, *Inorg. Chem.*, 1976, **15**, 107.
588 R. J. Lawson and J. R. Shapley, *J. Amer. Chem. Soc.*, 1976, **98**, 7433.
589 D. L. DuBois and D. W. Meek, *Inorg. Chem.*, 1976, **15**, 3076.

for Co^I, Rh^I, Ir^I, Ni^{II}, Pd^{II}, and Pt^{II} and for

$$[RhL_5]^+ \rightleftharpoons [RhL_4]^+ + L$$

[L = $P(OR)_3$ or $P(OCH_2)_3CMe$; R = Me, Et, Pr^n, or Bu^t] have been determined. The relationship between these parameters and the corresponding $\Delta G^{\ddagger}$ for intramolecular exchange in ML_5 compounds was discussed and evidence was presented to suggest a similarity between the reaction co-ordinates for intramolecular rearrangement and ligand dissociation. In particular, the planar-association–trigonal-pyramidal reaction co-ordinate appears to involve tetragonal-pyramidal intermediates or transition states, implying that the axial ligands in the trigonal bipyramid do not remain co-linear with M throughout the reaction as has been assumed in the current picture for planar substitution. For M = Rh, L = $(RO)_3P$, it was suggested that two competing steric effects, crowding in the tetragonal pyramid and the steric assistance to bond breaking, offset one another to give a free energy of activation for dissociation, nearly independent of the steric size of L. Data were obtained from a lineshape analysis of $^{31}P\{^1H\}$ n.m.r. spectra as a function of temperature. The calculations employed were described in detail using two-, three-, four-, and five-site one-spin models to show the origin of the unexpected linewidth increase in the $[RhL_5]^+$ $^{31}P\{^1H\}$ doublet at temperatures above the slow exchange limit as the Rh : L ratio is either increased or decreased from 1 : 5, assuming first-order dissociation. In some instances equilibria were measured.[590] Variable-temperature $^{19}F\{^{31}P\}$ n.m.r. spectra of $[Rh_2(PF_3)_8]$ show exchange between two sites with $\Delta G^{\ddagger} = 13.21 \pm 0.05$ kcal mol^{-1}.[591] ^{31}P N.m.r. spectroscopy has been used to determine $\Delta G^{\ddagger}$ for ligand dissociation and five-co-ordinate rearrangements for $[ML_5]^{n+}$ (M = Co, Rh, Ir, Ni, Pd, or Pt). It was found that $\Delta G^{\ddagger}$ for intramolecular exchange increases with increasing steric bulk while $\Delta G^{\ddagger}$ for intermolecular exchange is almost constant.[592] Variable-temperature ^1H n.m.r. studies of *mer*-$[MCl_3\{S(CH_2SiMe_3)_2\}_3]$ (M = Rh or Ir) have yielded slightly different sulphur inversion barriers for the two different sulphur ligand environments in these complexes. The barrier is somewhat lower for the sulphur ligand *trans* to chlorine.[593]

Nickel, Palladium, and Platinum. $[M(PPh_3)_2Cl(ArNXNAr)]$ (M = Pd or Pt) and *trans*-$[Pt(PPh_3)_2H(ArN_3Ar)]$ are stereochemically non-rigid *via* a pseudo penta-co-ordinate intermediate.[594] Variable-temperature ^1H and ^{13}C n.m.r. studies of exchange in the five-co-ordinate complexes $[PtMe(HBpz_3)L]$ have been carried out and up to three different exchange processes found.[595] ^1H N.m.r. spectroscopy has been used to examine the dynamic behaviour of $[Pt(CH_2CH{=}CHMe)Cl(PPh_3)_2]$ and $[Pt(CH_2CH{=}CHMe)ClL]$ in various solvents.[596] $\Delta G^{\ddagger}$ have been determined for 1,3-shifts for a number of $[(\eta^3$-cycloheptadienyl)$PdL_2]$ complexes. A relationship was found between $\Delta G^{\ddagger}$ and the *trans*-effectiveness of L.[597] $[(MeNC)_5(Ph_3P)Pd_2]^{2+}$ is fluxional. At -8 °C three

[590] A. D. English, P. Meakin, and J. P. Jesson, *J. Amer. Chem. Soc.*, 1976, **98**, 7590.
[591] M. A. Bennett, R. N. Johnson, and T. W. Turney, *Inorg. Chem.*, 1976, **15**, 2938.
[592] A. D. English, P. Meakin, and J. P. Jesson, *Inorg. Chem.*, 1976, **15**, 1233.
[593] E. W. Abel, G. W. Farrow, and K. G. Orrell, *J.C.S. Dalton*, 1976, 1160.
[594] L. Toniolo, A. Immirzi, U. Croatto, and G. Bombieri, *Inorg. Chim. Acta*, 1976, **19**, 209.
[595] L. E. Manzer and P. Z. Meakin, *Inorg. Chem.*, 1976, **15**, 3117.
[596] H. Kurosawa and G. Yoshida, *J. Organometallic Chem.*, 1976, **120**, 297.
[597] B. E. Mann and P. M. Maitlis, *J.C.S. Chem. Comm.*, 1976, 1058.

methyl signals are observed, but at higher temperatures only two signals are observed.[598]

$[Ni\{S(CH_2)_3S(CH_2)_2S(CH_2)_3S(CH_2)_2\}]^{2+}$ shows fluxional behaviour due to ring conformations; activation parameters were determined.[599] Variable-temperature n.m.r. spectra of *trans*-$[MX_2(EEt_2)_2]$ (M = Pd, Pt = E = S, Se, or Te) indicate that inversion of configuration at the pyramidal chalcogen atoms proceeds readily. Retention of $^3J(^1H, ^{195}Pt)$ above the coalescence temperature for the sulphur and selenium compounds and for *cis*-$[Pt(C_6H_4Me\text{-}4)_2(TeEt_2)_2]$ confirms that the dynamic process is intramolecular for these complexes. The chelate complexes $[MX_2(EtSeCH_2CH_2SeEt)]$ show related spectral behaviour which was interpreted in terms of single inversions at selenium atoms. Barriers to inversion are in the order Te > Se > S and are sensitive to the *trans*-influence of the opposite ligand. These barriers are lower than in the related pyramidal systems such as sulphoxides and selenoxides and this may be due to stabilization of the transition state by the π-acceptor capacity of the metal atoms. The barriers to inversion are less in the palladium complexes than in the analogous platinum complexes.[600] The ^1H n.m.r. spectra of $[MX_2L_2]$ [M = Pd or Pt; L = $O(CH_2CH_2)_2E$; E = S or Se] are temperature dependent, and it was shown that the coalescence phenomena are caused by site inversion about the ligand atom rather than by ring reversal of the cyclic ligand. Thermodynamic parameters were reported.[601]

Copper and Zinc. The ^1H and ^{11}B n.m.r. spectra of $[(Ph_3P)_2CuB_3H_8]$ show dynamic scrambling of the copper around the ring; thermodynamic parameters were determined.[602] The ^1H n.m.r. spectra of N-substituted thioureas and their Zn^{II} complexes show two rate processes: (i) rotation of the carbon–nitrogen bonds and (ii) ligand exchange. Thermodynamic properties were determined.[603]

Boron and Aluminium. Variable-temperature ^1H and ^{11}B n.m.r. spectra of $[Mg(THF)_2(B_6H_9)_2]$, $[Zn(THF)_2(B_6H_9)_2]$, and $[Cd(B_6H_9)_2]$ indicate that they are stereochemically non-rigid with bridging hydrogen and metal being involved in the dynamic process. The low-temperature spectra are consistent with the metal being inserted into a basal boron–boron bond of the anion.[604] ^{13}C Spin–lattice relaxation times have been measured for $[Et_3Hex^nB]^-$ and related compounds. The spin–lattice relaxation times are dominated by internal rotation of the carbon–carbon bonds and the correlation times and E_a were estimated.[605] ^{13}C N.m.r. spectroscopy has been used to investigate hindered rotation about the B—N bond in $B(NR_2)(NPr^i_2)_2$. For R = Et, $\Delta G_{193}^{\ddagger}$ = 9.4 kcal mol^{-1} and for R = Me, $\Delta G_{193}^{\ddagger}$ = 7.9 kcal mol^{-1}.[606] $\Delta H^{\ddagger}$ for oxygen inversion in $BzEtOBF_3$ is 4.1 ± 0.3 kcal mol^{-1}.[607] The temperature dependence of the ^{13}C

598 J. R. Boehm and A. L. Balch, *J. Organometallic Chem.*, 1976, **112**, C20.
599 N. Herron, O. W. Howarth, and P. Moore, *Inorg. Chim. Acta*, 1976, **20**, L43.
600 R. J. Cross, T. H. Green, and R. Keat, *J.C.S. Dalton*, 1976, 1150.
601 J. C. Barnes, G. Hunter, and M. W. Lown, *J.C.S. Dalton*, 1976, 1227.
602 C. H. Bushweller, H. Beall, and W. J. Dewkett, *Inorg. Chem.*, 1976, **15**, 1739.
603 D. R. Eaton and K. Zaw, *J. Inorg. Nuclear Chem.*, 1976, **38**, 1007.
604 D. L. Denton, W. R. Clayton, M. Mangion, S. G. Shore, and E. A. Meyers, *Inorg. Chem.*, 1976, **15**, 541.
605 W. T. Ford, *J. Amer. Chem. Soc.*, 1976, **98**, 2727.
606 K. K. Curry and J. W. Gilje, *J. Amer. Chem. Soc.*, 1976, **98**, 8262.
607 S. Brownstein, *J. Amer. Chem. Soc.*, 1976, **98**, 2663.

n.m.r. lineshape of $[Et_nAlCl_{3-n}]_2$ has been determined. The exchange reaction of terminal ethyl groups for chloride ligands in $Et_3Al_2Cl_3$ was investigated to yield $E_a = 12.3 \pm 1.5$ kcal mol^{-1}.[608] Activation parameters for stereoisomerism of $[(Bu^tCOCHCOBu^t)_3Al]$ and $[Al(hfac)_2(dpm)]$ have been determined. The reactions were found to be first order and intramolecular. Permutational analysis of the combined results of this study and those of all other dynamic n.m.r. studies of [Al(β-diketonate)$_3$] has been reported and reveals that the steric course of stereoisomerism involves the effective simultaneous interchange of the terminal groups within two of the three bidentate ligands and enantiomerization at the metal centre. Consideration of the magnitudes of the activation parameters, the effective steric course, and a topological analysis leads to the following mechanistic conclusions. Chelates with alkyl or aryl substituents on the β-diketonate rings are most likely to stereoisomerize by a rhombic twist mechanism. Chelates with fluorocarbon substituents on the β-diketonate rings stereoisomerize by a bond-rupture mechanism which proceeds *via* an actual square-pyramidal-apical five-co-ordinate intermediate.[609]

Silicon, Germanium, and Tin. ^{13}C and ^{29}Si Spin–lattice relaxation and nuclear Overhauser enhancements have been measured for Ph_nSiH_{4-n}, Ph_2SiHMe, and Ph_2SiCl_2. Dipolar and other contributions to relaxation rates were separated. The dipolar rates were used to discuss internal rotation about the phenyl—silicon bond and the overall motional anisotropy.[610] ^1H N.m.r. relaxation data have been used to investigate molecular motion in Cl_3SiH and PhSeH.[611] Activation parameters for carbon–nitrogen bond rotation of $Me_2NCEE'MMe_3$ (E, E′ = S, O; M = Si, Ge, Sn, or Pb) have been determined using ^1H and ^{13}C n.m.r. spectroscopies.[612] Dynamic ^1H n.m.r. spectroscopic behaviour in $[(2\text{-}Me_2NCH_2C_6H_4)MePhSnBr]$ has been attributed to breaking of the tin–nitrogen bond.[613] For (85), the parameters of activation have been calculated by

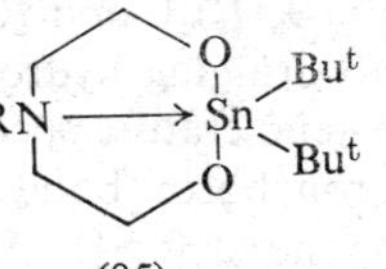

(85)

means of coalescence of both Bu^t signals. A mechanism of dissociation–inversion was discussed. The dependence of $J(^1H, C, C, ^{119}Sn)$ as well as the activation parameters on the donor strength of the solvents indicated the presence of a six-co-ordinate species.[614] $Bu^t{}_3SiX$ shows hindered rotation about the Bu^t—Si bond with $\Delta G^{\ddagger}$ around 40 kJ mol^{-1}.[615] ^1H N.m.r. spectroscopy has been used to study the influence of Ph_3C on the metallotropic rearrangement in $[C_5H_4CPh_3SiMe_3]$ and $[C_5H_4CPhSnMe_3]$. The most likely mechanism is a

608 R. Rottler, C. G. Kreiter, and G. Fink, *Z. Naturforsch.*, 1976, **31b**, 730.
609 M. Pickering, B. Jurado, and C. S. Springer, jun., *J. Amer. Chem. Soc.*, 1976, **98**, 4503.
610 R. K. Harris and B. J. Kimber, *Adv. Mol. Relaxation Processes*, 1976, **8**, 23.
611 A. Briguet, *Report FRNC-TH*-552, 1974, (*Chem. Abs.*, 1976, **85**, 122 847).
612 A. E. Lemire and J. C. Thompson, *Canad. J. Chem.*, 1975, **53**, 3732.
613 G. van Koten and J. G. Noltes, *J. Amer. Chem. Soc.*, 1976, **98**, 5393.
614 A. Zschunke, A. Tzschach, and K. Jurkschat, *J. Organometallic Chem.*, 1976, **112**, 273.
615 M. Weidenbruch, W. Peter, and C. Pierrard, *Angew. Chem. Internat. Edn.*, 1976, **15**, 43.

1,2-shift. The spectrum of the tin compound has been analysed and the free energy of activation derived.[616] ^1H N.m.r. spectroscopy has been used to follow 1,2-shifts in [$(C_5H_5)GeX_3$]; thermodynamic parameters were determined. There was a marked increase in $\Delta G^{\ddagger}$ from 15 kcal mol^{-1} to 21.4 kcal mol^{-1} when X_3 was changed from $(OMe)_3$ to the potentially four-co-ordinate ligand $(OCH_2CH_2)_3N$.[617] [$(MeOCH_2SiMe_2)_2NCOMe$][618] and [$Bu^n{}_3SnCH_2CONMe_2$][619] show restricted rotation of the nitrogen–carbon bond. Lineshape analysis of [$X_2Sn(S_2CMPr^i{}_2)_2$] and [$X_2Sn(Se_2CNPr^i)_2$] in $CHCl_3$ has been used to determine E_a for internal rotation of the N—Pr^i bond. E_a was shown to vary between 18.3 and 13.9 kcal mol^{-1} and was discussed in terms of differences in steric repulsion between the isopropyl group and the sulphur or selenium atom.[620]

Phosphorus and Arsenic. A perturbational MO model and *ab initio* MO calculations indicate that the nitrogen geometry of H_2NPH_2 changes from trigonal planar to approximately tetrahedral when the dihedral angle increases from 90° to 180°, thus implying that the topomerization of aminophosphines is a hybrid of nitrogen–phosphorus bond rotation and pyramidal inversion of nitrogen.[621] On the basis of ^1H n.m.r. spectra and the lack of changes in the $^{31}P\{^1H\}$ dynamic n.m.r. spectrum of $Bu^t{}_2PPBu^t{}_2$, there is slow rotation about the phosphorus–carbon bond with $\Delta H^{\ddagger} = 15.7 \pm 0.7$ kcal mol^{-1}, $\Delta S^{\ddagger} = 14 \pm 5$ e.u., and $\Delta G^{\ddagger} = 12.6 \pm 0.1$ kcal mol^{-1}.[622] ^{19}F Exchange in Me_2PF_3, Me_3PF_2, Et_2PF_3, and $(Me_2N)_2PF_3$ has been shown by use of ^1H, ^{19}F, and ^{31}P n.m.r. spectroscopies to be intramolecular in Teflon cells and not intermolecular as reported previously using Pyrex tubes.[623] ^1H and ^{31}P n.m.r. spectroscopies have been used to investigate the temperature-dependent n.m.r. spectra of $MeP(2,6\text{-}OMe_2C_6H_3)_4$. Hydrogen–deuterium exchange between the P—Me group and $CDCl_3$ was also found.[624] $\Delta G^{\ddagger}$ has been determined for the inversion of [$Me_2\overline{MPRCH_2CH_2CH_2}$] (M = Si, Ge, or Sn).[625] At low temperatures, ^{13}C, ^{19}F, and ^{31}P n.m.r. spectra of some phosphoranes, *e.g.* [$(CF_3)_3PCl(NMe_2)$], show the presence of different CF_3 environments consistent with substitution at axial or equatorial positions of the assumed trigonal-bipyramidal framework. The former are characterized by relatively small $^2J(^{19}F, ^{31}P)$ and the latter by relatively large $^2J(^{19}F, ^{31}P)$ values. New ^{31}P n.m.r. data for [$(CF_3)_3P(NMe_2)_2$] strongly support the location of both NMe_2 groups in the equatorial plane, as suggested earlier. The barrier to the averaging of CF_3 environments appears to decrease in the order NMe_2 > SMe > OMe. The presence of halogen substituents appears to lower the barrier to the averaging process compared to the doubly substituted molecules, and a chlorophosphorane appears to have a lower barrier than the corresponding

[616] W. Kläui and H. Werner, *Helv. Chim. Acta*, 1976, **59**, 844.
[617] V. S. Shriro, Yu. A. Strelenko, Yu. A. Ustynyuk, N. M. Zemlyansky, and K. A. Kocheshkov, *J. Organometallic Chem.*, 1976, **117**, 321.
[618] J. Kowalski and Z. Lasocki, *J. Organometallic Chem.*, 1976, **116**, 75.
[619] A. Roubineau and J. C. Pommier, *J. Organometallic Chem.*, 1976, **107**, 63.
[620] Y.-Ichi Takeda, N. Watanabe, and T. Tanaka, *Spectrochim. Acta*, 1976, **32A**, 1553.
[621] A. H. Cowley, M. W. Taylor, M.-H. Whangbo, and S. Wolfe, *J.C.S. Chem. Comm.*, 1976, 838.
[622] J. A. Brunelle, C. H. Bushweller, and A. D. English, *J. Phys. Chem.*, 1976, **80**, 2598.
[623] C. G. Moreland, G. O. Doak, L. B. Littlefield, N. S. Walker, J. W. Gilje, R. W. Braun, and A. H. Cowley, *J. Amer. Chem. Soc.*, 1976, **98**, 2161.
[624] I. Szele, S. J. Kubisen, jun., and F. H. Westheimer, *J. Amer. Chem. Soc.*, 1976, **98**, 3533.
[625] C. Couret, J. Escudié, J. Satgé, and G. Redoulès, *Angew. Chem. Internat. Edn.*, 1976, **15**, 429.

fluorophosphorane.[626] ^{13}C N.m.r. spectroscopy has been used to determine $\Delta G^{\ddagger}$ in (86).[630]* The ^{13}C n.m.r. spectra of fluorophosphoranes containing the ring system $-OC(CF_3)_2C(CF_3)_2O-$, *e.g.* (87), have been reported. The results give evidence for a two-stage intramolecular exchange.[631] Variable-temperature

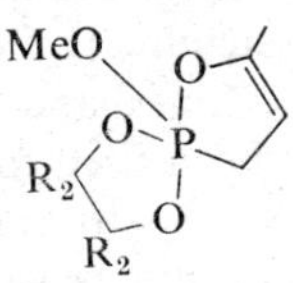

(86)

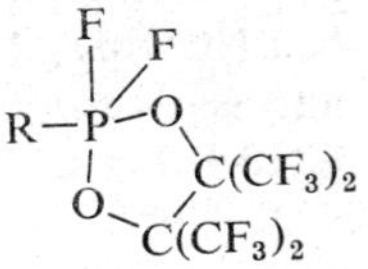

(87)

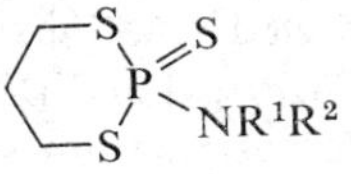

(88)

^{1}H and ^{19}F n.m.r. spectra for difluorophenylphosphoranes containing four-membered rings indicate that the favoured structure is one in which the four-membered ring spans an apical and equatorial position of the trigonal bipyramid. The fluorine atoms occupy equatorial and apical positions and the phenyl group is equatorial. There is rapid fluorine exchange, even at low temperatures.[632] The ^{19}F n.m.r. spectrum of $PhPF_3NHCHMeCPh_3$ shows fluxional behaviour.[633] The barrier to phosphorus–nitrogen rotation in (88) has been determined using ^{1}H and ^{31}P n.m.r. spectroscopies; $\Delta G^{\ddagger}$ was calculated and appears to depend on steric hindrance.[634]

Selenium and Chlorine. ^{1}H N.m.r. spectroscopy has been used to determine the barrier to rotation about the carbon–nitrogen bond in $R^1R^2C{=}NSeR$.[635] A study of chlorine and fluorine relaxation times of ClF_3 and ClF_5 has allowed the determination of reorientational and angular correlation times. The comparison between these two times shows that reorientation must occur by a large angle jump process rather than a diffusion process. Micellar solutions were also studied. The inside of a micelle must be regarded as a fluid. The determination of chlorine–fluorine scalar coupling and the chemical shift between inequivalent fluorine atoms in $ClOF_3$ was also reported.[636]

Equilibria.—*Solvation Studies of Ions.* The ^{1}H chemical shift of CF_3CO_2H has been measured as a function of added salt. The changes were attributed to breaking of the hydrogen-bonded network.[637] A method has been described for the determination of the contribution of single ions to the ^{1}H n.m.r. chemical shifts of water in which the metal ion is masked by the formation of a chelate using Na_2H_2edta. The chelate contribution is independent of the charge on the metal ions. The chemical shift of the water protons changes linearly with concentration. Within the complex group of different cations of the same charge

626 K. I. The and R. G. Cavell, *Inorg. Chem.*, 1976, **15**, 2518.
630 G. Buono and J. R. Llinas, *Tetrahedron Letters*, 1976, 749.
631 J. A. Gibson and G.-V. Röschenthaler, *J.C.S. Dalton*, 1976, 1440.
632 N. J. De'Ath, D. B. Denney, D. Z. Denney, and Y. F. Hsu, *J. Amer. Chem. Soc.*, 1976, **98**, 768.
633 M. Sanchez and A. H. Cowley, *J.C.S. Chem. Comm.*, 1976, 690.
634 J. Martin and J. B. Robert, *Tetrahedron Letters*, 1976, 2475.
635 C. O. Meese, W. Walter, and H. Schmidt, *Tetrahedron Letters*, 1976, 3133.
636 M. Alexandre, *Report CEA-R*-4557, 1975 (*Chem. Abs.*, 1976, **84**, 24 102).
637 A. Akisanya and F. A. M. Ayanbadejo, *Spectrochim Acta*, 1976, **32A**, 223.

* Here, and elsewhere in this chapter, reference numbers are missing, as several references were omitted in a revision of the original Report.

no difference in the curve was observed.[638] Nuclear relaxation of spin-$\frac{1}{2}$ nuclei of heavy atoms in electrolyte solutions may occur *via* transient anisotropic chemical shielding interactions and transient spin rotation interactions. Theoretical expressions were derived for spin-$\frac{1}{2}$ nuclei.[639] Pulsed n.m.r. spectroscopy has been used to study the structural characteristics of sea-water.[640] Variable-temperature ^1H n.m.r. spectra of water have been measured. With increasing temperature, the water resonance shifts to higher field. The partial breaking of hydrogen bonds corresponds to an increase of screening.[641]

Group IA. ^1H N.m.r. relaxation time measurements on 6M aqueous solutions of ^7LiI and ^6LiI have been recorded in D_2O containing small amounts of H_2O. The measurements were carried out at low temperatures and yielded maxima of a plot of relaxation rate against T^{-1}. From the maxima of relaxation rates, proton–iodide and proton–Li^+ distances in the first co-ordination sphere of the ions were determined. From a knowledge of ion–water oxygen distances it was shown that for I^- a somewhat broadened hydrogen-bonded configuration is valid and that for Li^+ the electric dipole orientation deviates from the radial direction. In order to test the reliability of the method, a proton–^{127}I interaction study was also performed in KI solution in glycerol. The iodine–hydrogen distance obtained was in satisfactory agreement with that found in aqueous solutions.[642] The spin–lattice relaxation time of ^6Li has been measured and was found to be as long as 1000 s.[643] Ion-association in aqueous $LiIO_3$ solution has been investigated by ^7Li spin–lattice relaxation times. In solutions containing less than 3M-$LiIO_3$ the $[IO_3]^-$ does not replace water in the Li^+ first co-ordination sphere. Addition of KIO_3 shortens the relaxation time but $RbIO_3$ and $CsIO_3$ have no effect.[644] The ion outer-sphere co-ordination number has been determined for Li^+, Na^+, and $[NH_4]^+$ in ethylenediamine and agrees well with values calculated from the apparent molar volume dependence on the square root molar concentration of electrolytes using the Masson equation.[645] ^{23}Na N.m.r. shift of $NaBPh_4$ has been measured in a variety of solvents and in the presence of various ligands. Shifts and linewidths were correlated with a number of electron-donor properties of the same material to see whether ^{23}Na n.m.r. shifts are a useful probe of ion–ligand interactions. Correlations were poor for ^{23}Na chemical shifts in pure ligand solvents and not much better for dilute acetone solutions of the ligand containing $NaBPh_4$. Normalized values in acetone solution gave a better correlation. Of all electron-donor properties, the solvent interaction with Na^+ as measured by ^{23}Na chemical shifts correlates best with the ability of a solvent to accept hydrogen bonds.[646] The ^{19}F and ^{23}Na n.m.r. spectra

[638] H. Geppert, G. Rudakoff, and R. Radegia, *Z. Chem.*, 1976, **16**, 201.
[639] R. N. Schwartz, *J. Mangn. Resonance*, 1976, **24**, 205.
[640] N. A. Mel'nichenko, *V SB., Ispol'z. Neorgan. Resursov Okeanich. Vody*, 1975, **1**, 43 (*Chem. Abs.*, 1976, **85**, 148 821).
[641] G. Widenlocher, E. Dayan, and B. Vodar, *Proc. Internat. Conf. Prop. Water Steam, 8th*, 1974, **2**, 1028 (*Chem. Abs.*, 1976, **85**, 197 837).
[642] A. Geiger and H. G. Hertz, *J. Solution Chem.*, 1976, **5**, 365.
[643] F. W. Wehrli, *J. Magn. Resonance*, 1976, **23**, 527.
[644] L. A. Azarova, N. V. Bryushkova, E. E. Vinogradov, I. M. Karataeva, and R. K. Mazitov, *Zhur. neorg. Khim.*, 1976, **21**, 3.
[645] E. F. Ivanova, A. I. Kruglyak, N. V. Bondarev, P. A. Grigorov, V. F. Lavrushin, and N. S. Pivnenko, *Zhur. fiz. Khim.*, 1976, **50**, 1052.
[646] E. M. Arnett, *U.S. NTIS, PB Rep. PB*-250773, 1974 (*Chem. Abs.*, 1976, **85**, 100 079).

of $NaPF_6$ have been measured in various solvents. The ^{19}F chemical shift moved upfield with increasing salt concentration indicating anion–cation interactions which may be solvent-separated or contact ion pairs. In acetic acid and THF the salt is essentially completely associated in the 0.1—1 mol l^{-1} concentration range, but in DMF, DMSO, and formamide there is little association. KPF_6 shows no concentration dependence whereas $[Bu^n{}_4N][PF_6]^-$ shows a little downfield shift with increasing concentration of the salt.[647] N.m.r. studies at 25 and 50 °C have established that the ions Na^+ and K^+ preserve their structure-forming and structure-destroying properties independently of whether KI or NaI is added to pure water or to a 1 or 4 mol l^{-1} solution of these salts. At −5 °C, Na^+ behaves as a structure destroyer if NaI is added to pure H_2O or to 1M-KI solution (a concentration below the boundary of complete hydration) but when added to 4M-KI at −5 °C, Na^+ behaves as a structure former.[648] ^1H N.m.r. relaxation times have been reported for the system of KI or RbI in D_2O–MeOH and extrapolated to zero proton concentration. Apart from the magnetic dipole–dipole interaction there seems to be an electronic interaction causing the relaxation mechanism. There is no preferential solvation for I^- but there is preferential hydration of Rb^+.[649] Solute–solvent and solute–solute interactions on ^{133}Cs chemical shifts have been investigated. The chemical shifts are strongly concentration-dependent, indicating ion-pair formation even with $[ClO_4]^-$ and $[NO_3]^-$. Solvent shifts of 130 p.p.m. were found.[650] Tetra-co-ordination of sodium is indicated by the ^{23}Na n.m.r. chemical shift for $NaClO_4$ in binary mixtures of ether and alcohol solvents. Solvent exchange occurs by a sequential process and involves one or several intermediates in competition with tetrahydrofurfuryl alcohol; whereas glyme gives monocyclic intermediates, diglyme and triglyme favour the formation of bicyclic intermediates. The conformational factors were analysed and also serve to explain the better co-ordination abilities of diglyme and triglyme as compared to glyme. The mechanism of solvent exchange was discussed.[651] The inference of a single bicyclic intermediate in the competition of diglyme or triglyme with tetrahydrofurfuryl alcohol for ^{23}Na was confirmed by analysis of linewidths. The dissymmetric bicyclic species are characterized by linewidths which increase markedly with respect to the pure solvents without an accompanying change in ^{23}Na chemical shift. Conversely the competition of glyme with the same alcohol involves three equi-probable intermediates.[652] An investigation of RCO_2H with CsF added has shown that the chemical shift of the carboxylic hydrogen is independent of the nature of the acid and varies with pK_a.[653] The chemical shifts induced in (89) by the Li^+, Na^+, K^+, and $[NH_4]^+$ ions have been investigated. The solvent lines shift as a function of concentration. The ions $[S_2O_3]^{2-}$, $[ClO_4]^-$, and I^- shift the signals downfield while $[BPh_4]^-$ shifts the signal upfield.[654]

[647] W. J. DeWitte and A. I. Popov, *J. Solution Chem.*, 1976, **5**, 231.
[648] V. V. Kushchenko, K. P. Mishchenko, and T. G. Soboleva, *Izvest. V.U.Z., Khim. i khim. Tekhnol.*, 1976, **19**, 1168 (*Chem. Abs.*, 1976, **85**, 184 432).
[649] D. S. Gill, H. G. Hertz, and R. Tutsch, *J.C.S. Faraday I*, 1976, **72**, 1559.
[650] W. J. DeWitte, R. C. Schoening, and A. I. Popov, *Inorg. Nuclear Chem. Letters*, 1976, **12**, 251.
[651] C. Detellier and P. Laszlo, *Helv. Chim. Acta*, 1976, **59**, 1333.
[652] C. Detellier and P. Laszlo, *Helv. Chim. Acta*, 1976, **59**, 1346.
[653] J. Emsley and O. P. A. Hoyte, *J.C.S. Dalton*, 1976, 2219.
[654] J. Rosenfarb and J. A. Caruso, *Canad. J. Chem.*, 1976, **54**, 3492.

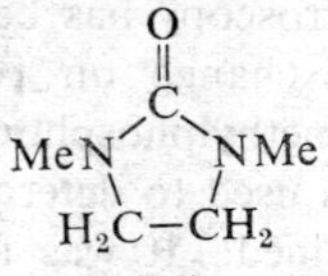

(89)

Beryllium and magnesium. ^1H N.m.r. spectra of $Be(NO_3)_2$ in acetone–water mixtures have been measured. Nitrate, hydroxy, and $[Be(H_2O)_4]^{2+}$ coexist in aqueous acetone with polymeric diaquahydroxyoxo complexes. The existence of hydrolysed species was established by analysis of ^{17}O n.m.r. relaxation in aqueous $Be(NO_3)_2$.[655] Spin–lattice relaxation mechanisms of ^9Be in aqueous $Be(NO_3)_2$ have been discussed and T_1 of between 2 and 10 s found. Contributions to T_1 due to spin–rotation, quadrupolar, and dipole–dipole mechanisms were also found. From ^9Be n.m.r. spectroscopy in solid $Be(NO_3)_2,4H_2O$, e^2qQ/h = 9.8 kHz.[656] ^{25}Mg N.m.r. spectra have been determined by the Fourier transform method as a function of concentration for aqueous solutions of $Mg(ClO_4)_2$, $Mg(NO_3)_2$, $MgBr_2$, $MgCl_2$, and $MgSO_4$. The linewidth data were discussed in terms of four models for quadrupolar nuclear spin relaxation in these systems. For $MgCl_2$ and $MgSO_4$, the linewidth variations cannot be accounted for in terms of variations of viscosity. It was found that the chemical shift is independent of concentration to ± 0.4 p.p.m.[657] Chemical shifts of the hydroxy-groups of water and methanol have been determined for mixed solvates of Mg^{II}. The anion, $[ClO_4]^-$, does not enter the first co-ordination sphere. Data at -70 °C were obtained by adding acetone to lower the freezing point.[658] The relative complexing abilities of various organic bases with Mg^{2+} have been determined as THF > dioxan > methanol from ^1H chemical shifts.[659]

Lanthanides and actinides. A matrix for the spin–spin relaxation time in spin-$\frac{7}{2}$ systems has been derived, solved numerically, and applied to ^{139}La relaxation times in the presence of bovine serum albumin to determine the correlation time, stability constant, and the rate of exchange.[660] Methanolic solutions of $LaCl_3$ have been studied by ^{35}Cl and ^{139}La n.m.r. spectra, and the species was found to be $LaCl_3(MeOH)_5$.[661] Rate constants and activation energies have been determined for $(Me_2N)_3PO$ exchange between solution and $[Eu(fod)_3\{OP(NMe_2)_3\}_2]$ from 0 to -54 °C with an activation energy of 5.5 kcal mol^{-1} and entropy of -28 e.u.[662] Aqueous and aqueous acetone solutions of $Lu(NO_3)_3$ have been studied by ^1H n.m.r. spectroscopy. It was deduced that $[Lu(OH_2)_6]^{3+}$ is formed in aqueous solution.[663] Low-temperature ^1H n.m.r. spectra of $UO_2(NO_3)_2$, $UO_2(ClO_4)_2$, and UO_2Cl_2 in aqueous acetone give a hydration number of six

655 V. A. Shcherbakov and O. G. Golubovskaya, *Zhur. neorg. Khim.*, 1976, **21**, 51.
656 F. W. Wehrli, *J. Magn. Resonance*, 1976, **23**, 181.
657 L. Simeral and G. E. Maciel, *J. Phys. Chem.*, 1976, **80**, 552.
658 B. N. Chernyshov and V. V. Kon'shin, *Koord. Khim.*, 1976, **2**, 365.
659 M. F. Mayahi and K. H. Al-Obaidi, *Bull. Coll. Sci., Univ. Baghdad*, 1975, **16**, 57 (*Chem. Abs.*, 1976, **84**, 187 198).
660 J. Reuben and Z. Luz, *J. Phys. Chem.*, 1976, **80**, 1357.
661 L. S. Smith, jun., D. C. McCain, and D. L. Wertz, *J. Amer. Chem. Soc.*, 1976, **98**, 5125.
662 V. A. Bidzilya, N. K. Davidenko, and L. P. Golovka, *Teor. i eksp. Khim.*, 1975, **11**, 687.
663 V. A. Shcherbakov and O. G. Golubovskaya, *Zhur. neorg. Khim.*, 1976, **21**, 314.

for $[UO_2]^{2+}$.[664] ^1H N.m.r. spectroscopy has been used to determine activation parameters for Me_2NCMeO exchange on $[UO_2(OCMeNMe_2)_5]^{2+}$.[665] When $[UO_2]^{2+}$ is added to dimethyl methylphosphonate (dmmp) $[UO_2(dmmp)_5]^{2+}$ is formed. Lineshape analysis was used to determine the rate of dmmp exchange and rate parameters were obtained. It was found that the exchange is predominantly dissociative in nature.[666] ^{17}O N.m.r. spectra have been investigated for aqueous $[UO_2]^{2+}$, $[NpO_2]^{2+}$, $[NpO_2]^{+}$, and $[PuO_2]^{2+}$ using 1—2% ^{17}O enrichment. The appearance of $5f$ electrons in $[NpO_2]^{2+}$ resulted in an increase of the shielding of oxygen in $[NpO_2]^{2+}$ as compared with $[UO_2]^{2+}$. This revealed a preferential localization of the Np electron in the equatorial plane. A minimum lifetime of water molecules in the $[NpO_2]^{2+}$ co-ordination sphere was 4×10^{-5} s. In the $[NpO_2]^{2+}$ aquo-complex, 6 ± 0.5 water molecules were bonded. The bond with the ligands is considerably weaker for $[NpO_2]^{2+}$ than $[PuO_2]^{2+}$, although the electronic configurations (for f-electrons) are similar for both ions. The ^{35}Cl and ^{14}N n.m.r. spectra revealed a strong association with anions in solutions of $[PuO_2]^{2+}$ and $[NpO_2]^{2+}$ salts.[667]

Molybdenum, tungsten, and manganese. The solvation of phosphomolybdic and phosphotungstic acid in aqueous acetone and cyclohexanone has been studied by n.m.r. spectroscopy. The solvent composition affects the degree of ionization of the acids in the same way as in HCl solution by changing the anion hydration. The ketones replace some of the water molecules bound to the heteropolyanion.[668] ^1H N.m.r. relaxation times of $Mn(ClO_4)_2$ in DMSO have been measured at 8 and 60 MHz, and salt concentrations from 10^{-2} to 10^{-1} mol l^{-1} at 20 to 165 °C. The contact displacement, hyperfine coupling constants, kinetic parameters for DMSO exchange, the correlation time, the hydrogen–Mn^{II} distance, and electronic relaxation time were determined.[669]

Iron, cobalt, nickel, and copper. A model for the solvation of $[Fe(phen)_3]^{2+}$ cation has been proposed based on the geometry of the complex. The solvation energy was considered to be the sum of two parts: that due to the ion–dipole interaction with the solvent and that due to the van der Waals interaction between the solvent and the aromatic ligands. The model was tested and verified by solubility and n.m.r. chemical shifts of the 5,6-protons.[670] Addition of organic solvents to aqueous cobalt(II) nitrate gives a shift of the ^1H n.m.r. signal to high field. The chemical shift increases linearly with increasing cobalt salt concentration, but if the $Co(NO_3)_2$ is replaced with $CoSO_4$ or $CoCl_2$ the shift does not change significantly. $Ni(NO_3)_2$ solutions exhibit no shift. The outer-sphere complexing of organic solvents to the oxygen of the water bound to Co^{2+} in the first co-ordination sphere was suggested.[671] The n.m.r. relaxation times T_1 and

664 V. A. Shcherbakov and L. L. Shcherbakova, *Radiokhimiya*, 1976, **18**, 207 (*Chem. Abs.*, 1976, **85**, 10 873).
665 R. P. Bowen, S. F. Lincoln, and E. H. Williams, *Inorg. Chem.*, 1976, **15**, 2126.
666 J. Crea, S. F. Lincoln, and E. H. Williams, *Austral. J. Chem.*, 1976, **29**, 2183.
667 L. G. Mashirov, D. N. Suglobov, and V. A. Shcherbakov, *Radiokhimiya*, 1975, **17**, 881 (*Chem. Abs.*, 1976, **84**, 97 280).
668 L. I. Lebedeva and E. V. Vanchikova, *Zhur. obshchei Khim.*, 1976, **46**, 201.
669 J. C. Boubel and J. J. Delpuech, *Adv. Mol. Relaxation Processes*, 1975, **7**, 209 (*Chem. Abs.*, 1976, **84**, 141 205).
670 F. M. Van Meter and H. M. Neumann, *J. Amer. Chem. Soc.*, 1976, **98**, 1382.
671 V. V. Matveev and A. S. Balashov, *Koord. Khim.*, 1976, **2**, 424.

T_2 have been determined for solutions of $Ni(ClO_4)_2$ and $Co(ClO_4)_2$ in DMSO and $(Me_2N)_3PO$. Kinetic parameters were deduced for solvent exchange.[672] The kinetics of ligand exchange of eight Co^{II} complexes of thiourea and substituted thioureas have been measured. In each case two mechanisms are involved. The first is a direct associative exchange and the second requires an initial displacement of the thiourea by the solvent (acetone). The rate constant and activation parameters were reported. The lability of the complexes arises from the low enthalpies of activation.[673] Cu^{2+}, $[Cu(en)_2]^{2+}$, Ni^{2+}, and $[Ni(en)_n]^{2+}$ have been examined in DMSO, and the rate of DMSO exchange was measured. At low temperatures the axial DMSO exchange with Cu^{2+} was determined but the rate for $[Cu(en)_2]^{2+}$ was too great for n.m.r. spectroscopy. At higher temperatures Cu^{2+} equatorial DMSO exchange was observed. For Ni^{2+} the rates are in the order $Ni^{2+} < [Ni(en)]^{2+} < [Ni(en)_2]^{2+}$. These results were explained in terms of the *trans*-effect.[674] ^{17}O N.m.r. studies have been carried out on $[Ni(dien)(H_2O)_3]^{2+}$, $[Ni(trien)(H_2O)_2]^{2+}$, and $[Ni(tetren)(H_2O)]^{2+}$. The scalar coupling constants (A/h) based on the above formulae are 2.27 ± 0.04, 2.25 ± 0.32, and $2.52 \pm 0.03 \times 10^7$ Hz, respectively. The nickel–dien system shows at least two water exchange rates which are not well resolved. Thermodynamic paramaters were determined.[675]

Silver, zinc, cadmium, and mercury. ^{109}Ag N.m.r. spectra have been measured for silver salts in solution after doping with Ni^{2+} to reduce the relaxation time. The chemical shift was concentration dependent, changing by up to 110 p.p.m. The effect was more marked in acetonitrile solution. There appears to be specific solvation by acetonitrile in acetonitrile–water mixtures.[676] The solvation of $AgClO_4$ in methanol–DMSO mixtures has been studied by ^1H n.m.r. spectroscopy. The ^1H chemical shifts depend strongly on the concentration of $AgClO_4$. It was deduced that Ag^+ is solvated by four molecules of DMSO.[677] Molal shifts induced in the resonance of protons of water and methanol by the d^{10} cations Ag^+, Zn^{2+}, Cd^{2+}, and Hg^{2+} are systematically downfield of the shifts induced by corresponding alkali and alkaline-earth cations. The cause of this effect is thought to be primarily an increased covalency in the solvent bonding. Since Zn^{2+} is known to be octahedrally co-ordinated in methanol a suggestion that the enhanced downfield shift is caused by an asymmetric solvation for Ag^+ was questioned.[678] Spin–lattice relaxation times have been measured for ^{113}Cd in aqueous $Cd(ClO_4)_2$ solution over the range from -17 to 84 °C with H_2O, D_2O, and $D_2{}^{17}O$ as solvents. In H_2O, dipolar interaction between ^{113}Cd and water protons is mainly responsible for relaxation, and from this a rotational correlation time of 3.9×10^{-11} s at 25 °C for the hydrated Cd^{2+} ion was deduced. Spin–rotation relaxation was observed in this solution as well as with solvent D_2O. In $D_2{}^{17}O$ solution, however, scalar relaxation of the first kind strongly dominates

672 J. C. Boubel, J. J. Delpuech, and A. A. Peguy, *Nato Adv. Study Inst. Ser., Ser. C*18, *Chem. Biol. Appl. Relaxation Spectrometry*, 1975, p. 323 (*Chem. Abs.*, 1976, **84**, 141 293).
673 D. R. Eaton and K. Zaw, *Inorg. Chim. Acta*, 1976, **16**, 61.
674 C. L. Watkins and G. S. Vigee, *J. Phys. Chem.*, 1976, **80**, 83.
675 D. P. Rablen, H. W. Dodgen, and J. P. Hunt, *Inorg. Chem.*, 1976, **15**, 931.
676 A. K. Rahimi and A. I. Popov, *Inorg. Nuclear Chem. Letters*, 1976, **12**, 703.
677 L. Rodehüser and H. Schneider, *Z. phys. Chem.*, 1976, **100**, 119.
678 R. D. Brown and M. C. R. Symons, *J.C.S. Dalton*, 1976, 426.

with a scalar coupling of 248 Hz between ^{113}Cd and ^{17}O.[679] ^1H and ^{31}P n.m.r. spectroscopies have been used to show that there is rapid exchange between the several ligands present in HgX_2 in $(Me_2N)_3PO$.[680]

Aluminium, gallium, and thallium. Solutions of $Al_8(OH)_{20}(OH_2)_nCl_4$ have been studied by ^1H n.m.r. spectroscopy as a function of concentration, temperature, and solvent. A ^1H n.m.r. signal of the non-equivalent hydroxy-group was observed at room temperature and high concentration, but only one signal was observed for co-ordinated and bulk water. It was deduced that $n = 10$ and $k = 60\ s^{-1}$ for exchange between bulk and co-ordinated water.[681] Separate n.m.r. signals from the solvation sphere and the bulk have been observed in methanolic solutions of $Al(ClO_4)_3$ for the methyl protons below +45 °C and in ethanolic solutions for the methyl and CH_2 protons below +5 °C. The OH protons in both cases gave separate signals below −30 °C. The lifetimes of the alcohol molecules and hydroxyl protons in the solvation sphere of Al^{3+} were determined.[682] The solvation of Al^{3+} in water–DMSO–actone has been investigated by ^1H n.m.r. spectroscopy, and no preference was found between water and DMSO.[683] The ^1H n.m.r. spectrum of $Al(ClO_4)_3$ in acetonitrile containing a little water shows 24 water absorptions, which were analysed on the basis of their concentration and temperature dependence. Separate signals were also observed for co-ordinated acetonitrile molecules. From integrated intensities as well as the number of lines it was concluded that at certain water contents, the total solvation number of Al^{3+} is smaller than six and the $[ClO_4]^-$ ion participates in the first co-ordination sphere. This is in accord with a positive deviation from Raoult's Law. Equilibrium constants were determined and thermodynamic parameters calculated.[684] The competition of halide ions, water, and acetonitrile for Ga^{3+} in a water–acetonitrile mixture has been studied by ^1H, ^{69}Ga, and ^{71}Ga n.m.r. spectroscopies. Equilibria such as

$$[Ga(NCMe)_6]^{3+} + nH_2O \rightleftharpoons [Ga(NCMe)_{6-n}(OH_2)_n]^{3+} + nMeCN$$

$$[GaX_4]^- + nH_2O \rightleftharpoons [GaX_{4-n}(OH_2)_n]^{n-1} + nX^-$$

have been observed. The donor capacity of ligands towards Ga^{3+} decrease in the order $Cl^- > Br^- > I^- > H_2O >$ MeCN.[685] The solvent dependence of $^{205}Tl^I$ n.m.r. chemical shifts has been found to correlate linearly with the solvent's relative solvating ability towards Tl^+. The relative solvating ability is quantitatively measured using the mole fraction dependence of chemical shift and analysed by a solvent exchange equilibrium constant theory. The analysis provides a method of obtaining chemical shifts in solvents that do not dissolve thallous salts. The correlation of solvating ability with the chemical shift was explained using MO theory. The relative solvating ability determined by ^{205}Tl

679 M. Holz, R. B. Jordan, and M. D. Zeidler, *J. Magn. Resonance*, 1976, **22**, 47.
680 J. L. Vidal and G. E. Ryschkewitsch, *J. Inorg. Nuclear Chem.*, 1976, **38**, 1933.
681 I. Zlateva, D. Trendafelov, and G. S. Nikolov, *Izvest. Khim.*, 1975, **8**, 433 (*Chem. Abs.*, 1976, **84**, 187 188).
682 I. Wawer and Z. Kecki, *Uniw. Adama Mickiewicza Poznaniu, Wydz. Mat., Fiz. Chem. (Pr.), Ser. Fiz.*, 1975, **19**, 317 (*Chem. Abs.*, 1976, **85**, 142 213).
683 H. Chi, C.-H. Ng, and N. C. Li, *J. Inorg. Nuclear Chem.*, 1976, **38**, 529.
684 Y. Ruben and J. Reuben, *J. Phys. Chem.*, 1976, **80**, 2394.
685 Yu. A. Buslaev, V. P. Tarasov, S. P. Petrosyants, and N. N. Mel'nikov, *Koord. Khim.*, 1975, **1**, 1435.

n.m.r. spectroscopy was compared to several current thermodynamic models of Lewis acid–base interactions.[686]

Nitrogen, sulphur, chlorine, and iodine. Analysis of the intramolecular contribution calculated from selective T_1 measurements on the OH and Me protons of methanol solutions of diamagnetic salts has provided evidence of negative solvation effects for $[CNS]^-$, $[ClO_4]^-$, and I^-. At low concentrations, the microdynamic behaviour of the hydroxy-group is strongly affected whereas that of the methyl group appears to be unchanged.[687] Medium effects on ^{35}Cl and ^{37}Cl n.m.r. spectra have been examined and the linewidth and shift were found to be solvent dependent.[688]

Ionic Equilibria. Group IA and IIA metals. ^7Li N.m.r. spectra of Li^+ complexes of 1,4-polymethylenetetrazoles and substituted glutarimides have been determined in $MeNO_2$. The concentration dependence of the values was explained by competitive ion pair formation. Glutarimide complexes were somewhat more stable than pentamethylene complexes.[689] Large variations in ^7Li and ^{35}Cl nuclear magnetic relaxation times have been observed when nucleosides, bases, and ribose are added to LiCl solution in DMSO. The molar shortening of T_1 is very sensitive to the presence of specific functional groups in the solutes studied and it was shown to reflect binding by Cl^- and Li^+ to specific sites on the solutes. Binding to guanosine, 1-methylguanosine, ionosine, adenosine, uridine, *d*-thymidine, cytidine, adenine, uracil, cytosine, and ribose was studied. Three binding sites were found: (i) Cl^- binds strongly to the (N^1—H, C^2—NH_2) region of guanosine, and more weakly to corresponding regions of its analogous 1-methylguanosine and inosine; (ii) Li^+ binds to the N^3 site of cytidine and cytosine, hindering rotation of the adjacent amino-group; (iii) LiCl binds, probably as an ion pair, to the furanose ring of all the ribosides studied. *d*-Thymidine shows no evidence of this binding in the ion relaxation data. Uracil, thymine, and adenine show no evidence of binding in the T_1 data although the proton resonances of the NH and NH_2 groups in these bases are chemically shifted. In all other cases proton chemical shifts generally confirm inferences drawn from relaxation data. Relations between equilibrium constants, relaxation rates, and total concentration were derived and used to estimate the association constant and relaxation rate for $^7Li^+$ in a ribose site.[690] The magnetic relaxation and quadrupole splitting of $^{23}Na^+$ counter-ions in the sodium octanoate–pentanol–water system have been studied as a function of temperature, composition, and the addition of other solvents. Amphiphiles cause an increase in micellar binding of Na^+, and the addition of NaCl also causes changes in both the fraction bound and the nature of ionic interaction. Solubilization of decanol and pentanol in the sodium octanoate micelle leads to enforced $^{23}Na^+$ relaxation.[691] The ^1H n.m.r. spectra of NaOH–H_2O–dioxan, $NaAl(OH)_4$–

[686] J. I. Dechter and J. I. Zink, *Inorg. Chem.*, 1976, **15**, 1690.
[687] L. Latanowicz and Z. Pajak, *Chem. Phys. Letters*, 1976, **38**, 166 (*Chem. Abs.*, 1976, **84**, 127 389).
[688] S. Forsén, H. Gustavsson, B. Lindman, and N.-O. Persson, *J. Magn. Resonance*, 1976, **23**, 515.
[689] Y. M. Cahen and R. F. Beisel, *Inorg. Nuclear Chem.* (Herbert H. Hyman Memorial Volume), 1976, p. 209 (*Chem. Abs.*, 1976, **85**, 192 020).
[690] A. C. Plaush and R. R. Sharp, *J. Amer. Chem. Soc.*, 1976, **98**, 7973.
[691] J. B. Rosenholm and B. Lindman, *J. Colloid Interface Sci.*, 1976, **57**, 362 (*Chem. Abs.*, 1976, **85**, 167 539).

H_2O–dioxan, and NaOH–$NaClO_4$–H_2O–dioxan have been interpreted in terms of the effect of adding dioxans and various ions on ice-like clusters of water. The $[OH]^-$ and $[Al(OH)_4]^-$ ions stabilize the water structure and reduce the dioxan solubility, but $[ClO_4]^-$ contributes to the structure-breaking effect of dioxan.[692] ^1H and ^{13}C n.m.r. spectroscopies have been used to investigate the complexes formed between benzo-18-crown-6, dibenzo-18-crown-6, and dibenzo-30-crown-10 and Na^+, K^+, Cs^+, and Ba^{2+}. The structure of the complexes was discussed.[693] ^{133}Cs N.m.r. spectra have been measured for crown and cryptate complexes of Cs^+. When a crown ether is added the signal shifts to a constant position which depends on solvent and on counter-ion.[694] ^{23}Na N.m.r. spectroscopy has been used to investigate the complexation of Na^+ to sugars. The spectra were used to determine the stability constant, the linewidth in the complex, and activation energy.[695] ^{13}C N.m.r. spectroscopy has been used to investigate the interaction of K^+ and Ca^{2+} with carriers,[696] and Group IA and IIA cations with macrotetrolide antibiotics.[697] ^1H N.m.r. spectroscopy has been used to investigate the binding of Ca^{2+} and La^{3+} to aldoses. Chiefly 1 : 1 complexes were found in a pyranose conformation with the axial–equatorial–axial arrangement for three consecutive *cis* hydroxy-groups.[698] ^7Li and ^{23}Na n.m.r. relaxation rates have been measured in aqueous polyphosphate solution. The Li^+–polyion interaction was shown to be qualitatively different from Na^+–polyion interactions.[699] Nematic lyotropic phases have been prepared using mixed detergents with the decyl sulphate anion and were examined by ^7Li, ^{23}Na, ^{133}Cs, and ^2H n.m.r. spectroscopies. All the nuclei show first-order quadrupole splittings. The influence of counter-ion and temperature variation were examined.[700] Alkali and halide ion n.m.r. spectra and shieldings have been studied in both aqueous and non-aqueous surfactant solutions. Changes in ^{23}Na n.m.r. properties were studied at the transition from normal micelles to micelles of the reversed type. The applicability of the H_2O isotope effect was studied as a probe into hydration properties.[701] The change in $\delta(^{19}F)$ for the CF_3 group in aqueous ammonium perfluoropropylphosphinate allows the calculation of the degree of aggregation in micelle formation reactions and critical micelle formation concentrations. The free energy, enthalpy, and entropy of micelle formation were given.[702] The work was extended to $[(C_nF_{2n+1})_2PO_2]^-Na^+$.[703] The ^{31}P chemical shifts of $(EtO)_3PO$, $[(EtO)_2PO_2]^-$, $[EtOPO_3]^{2-}$, $[HPO_4]^{2-}$, and $[EtPO_3]^{2-}$ depend on the

692 Yu. A. Volokhov, N. G. Dovbysh, V. B. Lebedev, and V. E. Mironov, *Zhur. strukt. Khim.*, 1975, **16**, 1013.
693 D. Live and S. I. Chan, *J. Amer. Chem. Soc.*, 1976, **98**, 3769.
694 E. Mei, J. L. Dye, and A. I. Popov, *J. Amer. Chem. Soc.*, 1976, **98**, 1619.
695 C. Detellier, J. Grandjean, and P. Laszlo, *J. Amer. Chem. Soc.*, 1976, **98**, 3375.
696 R. Büchi, E. Pretsch, W. E. Morf, and W. Simon, *Helv. Chim. Acta*, 1976, **59**, 2407.
697 R. Büchi, E. Pretsch, and W. Simon, *Helv. Chim. Acta*, 1976, **59**, 2327.
698 R. E. Lenkinski and J. Reuben, *J. Amer. Chem. Soc.*, 1976, **98**, 3089.
699 H. S. Kielman, J. M. A. M. Van Der Hoeven, and J. C. Leyte, *Biophys. Chem.*, 1976, **4**, 103 (*Chem. Abs.*, 1976, **84**, 128 457).
700 K. Radley, L. W. Reeves, and A. S. Tracey, *J. Phys. Chem.*, 1976, **80**, 174.
701 H. Gustavsson and B. Lindman, *Proc. Internat. Conf. Colloid Surf. Sci.*, 1975, **1**, 625 (*Chem. Abs.*, 1976, **85**, 54 286).
702 N. N. Kalibabchuk, L. K. D'yachek, D. I. Kurlyand, and V. Ya. Semenii, *Teor. i eksp. Khim.*, 1975, **11**, 838.
703 N. N. Kälibabchuk, L. K. D'yachek, and V. Ya. Semenii, *Kolloidn. Zhur.*, 1976, **38**, 155 (*Chem. Abs.*, 1976, **84**, 170 100).

concentration of added salt. Shifts of up to 2.5 p.p.m. were found; these depend on the added ion and the sign depends on the salt. Thus Et_4NCl produces the opposite shift to NaCl.[704] The behaviour of the alkyl proton resonances of symmetric $[R_4N]^+$ ions has been studied in aqueous and non-aqueous solvents. Information about ion-pairing has been obtained by investigating

$$\Delta\delta = \delta(CH_2) - \delta(CH_3)$$

of the cationic protons as a function of concentration, temperature, and anion. In organic solvents it was possible to determine an equilibrium constant, whereas aqueous solutions were more complex. At low and medium concentrations cation–cation interaction is dominant, whereas in concentrated solutions there are multiple ion aggregates.[705] A ternary complex between thiamine, Mg^{2+}, and pyruvate has been detected in aqueous solution at pH 6.2 by ^1H n.m.r. spectroscopy.[706]

The lanthanides. An investigation of the effect of bovine serum albumin on the relaxation of ^{139}La in $LaCl_3$ has been carried out.[707] ^1H N.m.r. spectroscopy has been used to investigate the complexation of La^{III}, Nd^{III}, and Er^{III} by edta. Equilibrium constants were calculated from the induced shifts and broadenings.[708] The complexation of $SmCl_3$ by edta has been investigated using ^1H n.m.r. spectroscopy as a function of concentration, pD, and temperature. Exchange was found between bound and co-ordinated edta. At temperatures greater than 50 °C the broadening is dominated by exchange but at lower temperatures relaxation is dominant. The kinetics were analysed at 80 °C and kinetic parameters derived.[709] Subsequently the work was extended.[710] ^1H n.m.r. spectroscopy has been used to investigate the pH dependence of the La^{III} and Lu^{III} complexes of cyta [711] and nta,[712] and stability constants were determined. The formation constants have been determined for Nd^{III} complexation with acetate, alanine, and histidine as a function of pH by ^1H n.m.r. spectroscopy.[713] ^1H N.m.r. spectroscopy has been used to determine the formation constants and structure of L-histidine complexes of $PrCl_3$ and $EuCl_3$.[714] The binding constants of $Eu(fod)_3$ to P=O and to C=O in $(EtO)_2P(O)CH_2CH_2C(O)OEt$ are 6.91 ± 0.29 and 3.48 ± 0.16 l mol^{-1}.[715]

The transition metals. ^{13}C N.m.r. spectroscopy has been used to investigate the structure, bonding, and kinetic labilities of some diamagnetic complexes of $[edta]^{4-}$, $[cydta]^{4-}$, and $[pyta]^{4-}$. The work confirmed much of the earlier literature work. Clearer evidence was found for different structural isomers of

704 A. J. R. Costello, T. Glonek, and J. R. Van Wazer, *Inorg. Chem.*, 1976, **15**, 972.
705 G. Kabisch, *Ber. Bunsengesellschaft phys. Chem.*, 1976, **80**, 602.
706 D. R. Petzold and G. Fischer, *Stud. Biophys.*, 1976, **54**, 53 (*Chem. Abs.*, 1976, **84**, 131 794).
707 J. Reuben, *Proc. Rare Earth Res. Conf.*, 12*th*, 1976, **1**, 260 (*Chem. Abs.*, 1976, **85**, 133 627).
708 N. A. Kostromina and T. V. Ternovaya, *Zhur. neorg. Khim.*, 1976, **21**, 366.
709 E. D. Romanenko, L. I. Budarin, and K. B. Yatsimirskii, *Teor. i eksp. Khim.*, 1976, **12**, 26.
710 E. D. Romanenko and L. I. Budarin, *Teor. i eksp. Khim.*, 1976, **12**, 379.
711 N. A. Kostromina and N. N. Tananaeva, *Zhur. neorg. Khim.*, 1975, **20**, 2343.
712 N. A. Kostromina and N. N. Tananaeva, *Koord. Khim.*, 1976, **2**, 910.
713 E. R. Birnbaum, A. D. Sherry, and D. W. Darnall, *Proc. Rare Earth Res. Conf.*, 10*th*, 2 *Conf*-730402-*P*2, 1973, 753 (*Chem. Abs.*, 1976, **85**, 37 833).
714 D. L. Howland and R. L. Flurry, jun., *J. Inorg. Nuclear Chem.*, 1976, **38**, 1568.
715 N. K. Davidenko, V. A. Bidzilya, and K. B. Yatsimirskii, 'Proceedings of the 16th International Conference on Coordination Chemistry', 1974, R1 (*Chem. Abs.*, 1976, **85**, 26 008).

$[Co(edta)X]^{2-}$ and for relatively slow nitrogen inversion in $[Hcydta]^{3-}$. $J(^{13}C, M)$ ($M = {}^{111,113}Cd$, ^{199}Hg, or ^{207}Pb) was observed.[716] A ^{19}F n.m.r. study of the reactions of $[TiF_5]^-$ with neutral Lewis bases has been reported. ^{19}F N.m.r. data for more than 30 complexes were presented and the effects of donor basicity, $(p{-}d)\pi$-bonding and solvent interactions on the chemical shifts were discussed.[717] ^{13}C N.m.r. spectroscopy has been used to study the kinetic processes associated with conformational interconversions of two isomeric forms of $[(MoO_2)_2edta]^{2-}$. Activation parameters $\Delta H^{\ddagger} = 13 \pm 2$ kcal mol^{-1} and $\Delta S^{\ddagger} = 0.2 \pm 0.04$ e.u. were determined. ^{13}C N.m.r. data were also reported for the rigid compounds $[(MoO_2)_2(pdta)]^{2-}$ and $[(MoO_2)_2(cydta)]^{2-}$.[718] Complex formation between (90) and p-$XC_6H_4CONHNH_2$ and MZ_2 (M = Mn, Co, or Ni;

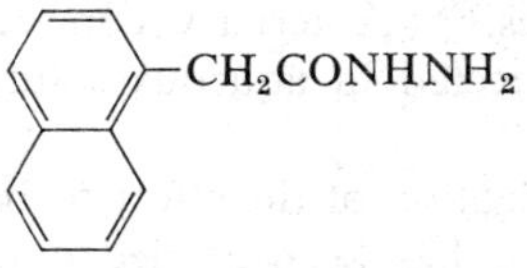

(90)

Z = Cl, Br, or I) has been studied using ^1H n.m.r. spectroscopy. A relationship was found between X and the donor capacity of the ligand.[719] ^1H N.m.r. studies of a 0.1 M citric acid solution (H_4L) and its interaction with $Fe(NO_3)_3$ indicates the formation of $FeHL(H_2O)_2$ and $[FeL(H_2O)_2]^-$ at pH 2.6 and 8 respectively.[720] ^1H N.m.r. spin–spin relaxation measurements have been made on the Fe^{III}–sulphosalicylic acid–edta Cu^{II}–ethylenediamine–$[S_2O_3]^{2-}$, Cu^{II}–$H_2NCH_2CH_2OH$–oxalate, and Cu^{II}–$H_2NCH_2CH_2OH$–lactate systems. Stability constants were determined for the complexes.[721] A concentration and temperature dependence study of $[(Me_3Si)_2PMe_2]^+[Co(CO)_4]^-$ has shown that the compound tends to dissociate to $Me_3SiCo(CO)_4$ and Me_3SiPMe_2. The extent of dissociation increases with the number of Me_3Si groups in the cation, increasing temperature, and decreasing concentration.[722] ^1H N.m.r. spectroscopy has been used to examine the structure and dynamics of Co^{II} and Ni^{II} complexes of pyridine in water. At high temperatures fast exchange prevails and the geometry of the complexes can be derived from spin–lattice relaxation times. At low temperatures slow exchange is dominant and the contact and dipolar shifts may be determined from the spectrum of the bound pyridine. In the intermediate range, exchange kinetics were studied. A value of *ca.* 3 Å for the metal–pyridine distance, an average number of 3.5 bound molecules per complex, and activation energies for chemical exchange of 16 kcal mol^{-1} for Ni^{II} and 22 kcal mol^{-1} for Co^{II} were calculated. The ligand molecules undergo free rotation about the metal–nitrogen bond.[723] Solutions of

[716] O. W. Howarth, P. Moore, and N. Winterton, 'Proceedings of the 16th International Conference on Coordination Chemistry', 1974, 3.32 (*Chem. Abs.*, 1976, **85**, 27 143).
[717] H. G. Lee, D. S. Dyer, and R. O. Ragsdale, *J.C.S. Dalton*, 1976, 1325.
[718] G. L. Blackmer, K. J. Johnson, and R. L. Roberts, *Inorg. Chem.*, 1976, **15**, 596.
[719] S. Yu. Chundak, P. I. Shman'ko, and A. P. Gulya, *Tezisy Dokl. Soobshch., Konf. Molodykh Uch. Mold., 9th*, 1974, 111 (*Chem. Abs.*, 1976, **84**, 173 148).
[720] N. A. Kostromina and V. V. Grigor'eva, *Zhur. neorg. Khim.*, 1975, **20**, 3029.
[721] A. A. Popel, A. V. Zakharov, Yu. I. Sal'nikov, and I. I. Kostrova, *Izvest. V.U.Z., Khim. i khim. Tekhnol.*, 1975, **18**, 1700 (*Chem. Abs.*, 1976, **84**, 96 196).
[722] H. Schäfer and A. G. MacDiarmid, *Inorg. Chem.*, 1976, **15**, 848.
[723] P. Mignot and J. Ph. Grivet, *Chem. Phys.*, 1976, **15**, 261.

$Co(NO_3)_2$ in aqueous glycols have been studied by ^1H n.m.r. spectroscopy and complex formation found.[724] The concentration association constant of $[NMe(C_8H_{17})_3]^+[CoBr_3PPh_3]^-$ in nitrobenzene has been determined at 34 °C using ^1H n.m.r. spectroscopy. On comparison of this result with those obtained earlier for $[NRBu^n_3]^+[CoBr_3PPh_3]^-$ (R = Me or Bu^n) it was concluded that the symmetry of the cation plays a very important role in determining the extent of ion association for this type of ion-pair complex. There is a linear correlation between the observed differences in isotropic shift of the *N*-methyl and *N*-methylene protons and the electric permittivity of the solvent, but only solvents with electric permittivities <10. These results were interpreted in terms of preferred orientation of the *N* methyl group towards the anion, the decrease in interionic distance with increasing electric permittivity, the formation of micelles, and 1 : 1 ion-pair formation.[725]

^1H N.m.r. spectroscopy has been used to investigate Ni^{II}–glutathione complexes in aqueous solution as a function of pH. There is an equilibrium between the octahedral and square-planar forms. At pH > 11 the Ni^{II} co-ordinates to sulphur and the deprotonated nitrogen of the peptide A linkage to form a square-planar complex.[726] The system

$$[HPd(PEt_3)_3]^+ + PEt_3 \underset{k_{-1}}{\overset{k_1}{\rightleftharpoons}} [HPd(PEt_3)_4]^+$$

has been investigated using ^{31}P n.m.r. spectroscopies. Four separate rate constants were derived, k_1, k_{-1}, k_m (the rate constant for intramolecular rearrangement of $[HPd(PEt_3)_4]^+$), and $k_m{}^1$ (the rate constant for intramolecular rearrangement of $[HPd(PEt_3)_3]^+$). In order to carry out this analysis, suitable computer programs had to be written.[727] Subsequently the work was extended to the nickel and platinum analogues. Evidence was found for the five-co-ordinate intermediate for both the nickel and platinum analogues. In the case of $[HPt(PEt_3)_4]^+$, there is preferential exchange of the PEt_3 which is *trans* to hydride.[728]

The n.m.r. method has been used to study complexes of Cu^{2+} with glycine in aqueous solution. In particular, the linewidth of the proton resonance of the solvent was examined. In acidic mixtures, water molecules exist between the Cu^{2+} ion and the glycine and outer-sphere complexes were formed. In neutral solutions or acidic solutions before protolysis of the glycine occurred, inner-sphere complexes were formed. The enhancement factor of the relaxation times and the correlation time, τ_c, were obtained.[729] The symmetric ligand exchange of $[edta]^{4-}$ with $[Cu(edta)]^{2-}$ in D_2O has been measured by n.m.r. line-broadening techniques from 7 to 94 °C. For ligand exchange, $E_a = 10.8$ kcal mol^{-1} and $\log A = 9.6$. In contrast to studies of other metal ions in analogous reactions,

724 V. V. Matveev, V. D. Guslyannikov, L. F. Malysheva, and E. E. Zaev, *Zhur. strukt. Khim.*, 1975, **16**, 924.
725 T. C. Tan, K. L. Chua, and Y. Y. Lim, *J.C.S. Dalton*, 1976, 863.
726 B. Jeżowska-Trzebiatowska, G. Formicka-Kozłowska, and H. Kozłowski, *Chem. Phys. Letters*, 1976, **42**, 242.
727 P. Meakin, A. D. English, and J. P. Jesson, *J. Amer. Chem. Soc.*, 1976, **98**, 414.
728 A. D. English, P. Meakin, and J. P. Jesson, *J. Amer. Chem. Soc.*, 1976, **98**, 422.
729 R. Freymann and A. Dezor, *Adv. Mol. Relaxation Processes*, 1976, **8**, 313 (*Chem. Abs.*, 1976, **85**, 184 442).

the rate constant of exchange of $[Hedta]^{3-}$ is nearly equal to the value for $[edta]^{4-}$.[730] ^1H N.m.r. relaxation data have been used to determine the composition and formation constants for Cu^{II} with ethylenediamine or sulphosalicylate.[731] The ^1H n.m.r. spectra of the Cu^{II}–edta and Cu–cydta indicate that there is an equilibrium between five- and six-co-ordination of $[Cu(edta)]^{2-}$.[732] The decrease in chemical shift of the vinylic protons of cyclohexene with increasing Ag^+ concentration indicates that the extent of olefin π-complexation to silver salts decreases in the order $AgO_3SCF_3 > AgSbF_6 > AgBF_4$. The ^1H n.m.r. spectra show that alkynes form π-complexes nearly as well as olefins, and stability constants were measured.[733] ^{13}C N.m.r. spectroscopy has been used to show that $[Au_4(SR)_7]^{3-}$ (HSR = *N*-acetyl-L-cysteine) undergoes fast exchange with $[RS]^-$ but not with RSH.[734]

When $[Zn(phen)]^{2+}$ or $[Zn(bipy)]^{2+}$ is added to $[Me_3SiCH_2CH_2CH_2SO_3]^-$ there is a shift of the methyl protons. No such shift is found if Zn^{2+}, phen, or bipy is used alone. Complex formation due to hydrophobic interactions with a formation constant of 10 l mol^{-1} was suggested.[735] ^1H N.m.r. spectroscopy has been used to examine Cd^{II} complexing by thiosemicarbazide.[736] $\delta(^1H)$ and $^2J(^1H, ^{199}Hg)$ resulting from co-ordination have been measured as a function of pH and $[MeHg]^+$ concentration. With compounds containing thiol groups (cysteine, penicillamine, and glutathione), the co-ordination number is 1 with a 1 : 1 $[MeHg]^+$: ligand ratio, and the resulting RSHgMe formed is not affected by pH. At higher $[MeHg]^+$ concentration and low pH, $[RS(HgMe)_2]^+$ is formed. At higher pH, the second $[MeHg]^+$ is shifted to the amino-group and is partially dissociated due to MeHgOH formation.[737] The stability constant for $[MeHg]^+$ complexation to $[SO_4]^{2-}$, $[SeO_4]^{2-}$, $[CO_3]^{2-}$, $[SO_3]^{2-}$, $[SeO_3]^{2-}$, $[SeO_3H]^-$, $[SCN]^-$, and $[SeCN]^-$ has been determined from the pH dependence of the ^1H chemical shift and $^2J(^1H, ^{199}Hg)$ for the methyl group of $[MeHg]^+$ in solutions containing $[MeHg]^+$ and X^-. $\delta(^1H)$ and $^2J(^1H, ^{199}Hg)$ were deduced for each complex. The ligand donor atom in each complex was identified using formation constants. $[SeO_3]^{2-}$ co-ordinates *via* the oxygen atom not the selenium atom, but $[SO_3]^{2-}$ co-ordinates *via* the sulphur atom.[738] ^{13}C N.m.r. spectra of (91) and its Hg^{II} complex have been recorded as a function of pH. It was found that mercury co-ordinates to sulphur.[739] Hindered rotation of the carbon–nitrogen bond has been examined in metal complexes containing uracil. The parameters of rotation were calculated by using the Eyring and Arrhenius equation and total lineshape analysis. $HgCl_2(uracil)_2$ has two conformations and their rate constants,

730 R. K. Forcé, R. S. Marianelli, and J. D. Carr, *Inorg. Chem.*, 1976, **15**, 2653.
731 M. S. Shapnik, A. N. Gil'manov, V. E. Ermakova, and I. Kh. Muzeev, *Zhur. neorg. Khim.*, 1975, **20**, 3117.
732 J. G. Kloosterboer, 'Proceedings of the 16th International Conference on Coordination Chemistry', 1974, 2.29b (*Chem. Abs.*, 1976, **85**, 37 819).
733 G. S. Lewandos, D. K. Gregston, and F. R. Nelson, *J. Organometallic Chem.*, 1976, **118**, 363.
734 A. A. Isab and P. J. Sadler, *J.C.S. Chem. Comm.*, 1976, 1051.
735 P. R. Mitchell and H. Sigel, *Angew. Chem. Internat. Edn.*, 1976, **15**, 548.
736 P. Trinderup and E. Larsen, 'Proceedings of the 16th International Conference on Coordination Chemistry', 1974, R32 (*Chem. Abs.*, 1976, **85**, 69 026).
737 D. L. Rabenstein, M. T. Fairhurst, C. A. Evans, M. C. Tourangeau, and S. Libich, 'Proceedings of the 16th International Conference on Coordination Chemistry', 1974, 5.2 (*Chem. Abs.*, 1976, **85**, 191 640).
738 D. L. Rabenstein, M. C. Tourangeau, and C. A. Evans, *Canad. J. Chem.*, 1976, **54**, 2517.
739 N. Motohashi, I. Mori, and Y. Suguira, *Chem. and Pharm. Bull. Japan*, 1976, **24**, 1737.

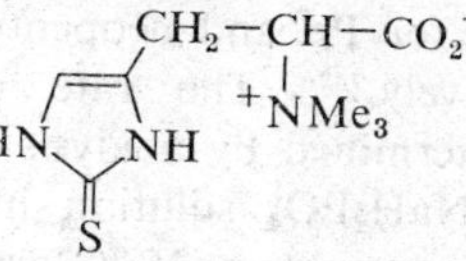

(91)

coalescence temperatures, $\Delta H^{\ddagger}$, and $\Delta S^{\ddagger}$ were calculated. Mercury co-ordination hindered carbon–nitrogen bond rotation.[740]

Equilibria among Uncharged Species. The hydrogen-bond energy, E_H, of weak proton donors in 1 : 1 complexes has been related to the change ($\Delta\delta$) of chemical shift of the hydrogen-bonding proton on formation of the complex by the equation $-E_H/\text{kcal mol}^{-1} = 1.15\Delta\delta/\text{p.p.m.}$[741] ^{1}H N.m.r. spectroscopy has been used to investigate the complexation of (92) with transition metal salts.[742]

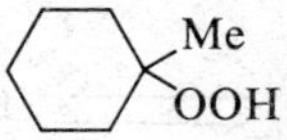

(92)

Exchange of organo-phosphorus ligands in 3*d*, 4*f*, and 5*f* element complexes has been investigated using ^{31}P n.m.r. relaxation data.[743]

Titanium, vanadium, niobium, and tungsten. Exchange of alkyl groups between $TiBz_4$, Ti([$^2H_8$]-Bz)$_4$, or $TiMe_4$ with $AlMe_3$ or $AlBz_3$ takes place *via* $[TiR_3]^+$-$[AlR_4]^-$. Decomposition then follows slowly with redistribution.[744] A variable-temperature n.m.r. investigation of a chloroform solution of MeCXYZ (X, Y, or Z = CH_2OMe, CH_2SMe, or CH_2NMe_2) with $TiCl_4$ shows that all four ligands act only as bidentate ligands towards titanium and that exchange can take place between free and co-ordinated ligand atoms. Nitrogen was found to bond more strongly than oxygen or sulphur to titanium. Oxygen co-ordination is preferred to sulphur co-ordination when nitrogen is already bound whereas sulphur co-ordination is preferred to oxygen co-ordination when oxygen already occupies a site on titanium.[745] $TiCl_4$ as well as the lanthanides has proved to be useful in inducing shifts in the ^{1}H n.m.r. spectra of some β-lactams.[746] Variable-temperature ^{13}C n.m.r. studies have been carried out on $Ti(OR)_4$ (R = Et, Pr^n, Bu^n, Pr^i, Bu^i, neopentyl, or Bu^t) and $Nb(OEt)_5$. The coexistence of trimers and trimer aggregates was suggested for the straight-chain alkoxides of titanium. For the branched-chain derivatives, monomer–dimer–trimer equilibria were proposed for Bu^i and monomer–dimer equilibria for Pr^i and neopentyl. Use of previously published ^{1}H n.m.r. data permitted the estimation of enthalpy and

740 T. Sakaguchi and T. Fujita, *Yakugaku Zasshi*, 1976, **96**, 1137 (*Chem. Abs.*, 1976, **85**, 184 462).
741 O. P. Yablonskii, *Zhur. fiz. Khim.*, 1975, **49**, 2538.
742 A. M. Syroezhko, *V Sb., Issled. V Obl. Khim. i Tekhnol. Produktov Pererabotki Goryuchikh Iskopaemykh.*, 1975, 49 (*Chem. Abs.*, 1976, **85**, 113 254).
743 A. A. Vashman and I. S. Pronin, *V Sb., XI Mendeleevsk. S'Ezd Po Obshchei i Priklad Khim. Ref. Dokl. i Soobshch.*, 1975, **1**, 219 (*Chem. Abs.*, 1976, **84**, 157 721).
744 L. S. Bresler, V. P. Ivanova, and A. S. Khachaturov, *Zhur. strukt. Khim.*, 1976, **17**, 257.
745 A. J. McAlees, R. McCrindle, and A. R. Woon-Fat, *Inorg. Chem.*, 1976, **15**, 1065.
746 A. K. Bose, M. S. Manhas, P. R. Srinivasan, H. P. S. Chawla, B. Dayal, and D. A. Foley, *Org. Magn. Resonance*, 1976, **8**, 151.

entropy of dimer formation for Pr^{i} and neopentyl as *ca.* -14 kcal mol^{-1} and -50 cal K^{-1} mol^{-1}, respectively.[747] The rate of loss of $[H_2PO_4]^-$ from $[OV(O_4PH_2)_2]^+$ has been determined by analysis of the broadening of the ^{31}P n.m.r. resonance of 1.0M-NaH_2PO_4 solution in the presence of $[VO]^{3+}$ at pH 2.0. The aquation rate constant at 25 °C was found to be $(2.5 \pm 0.3) \times 10^4$ s^{-1} and the corresponding anation rate constant estimated from the formation constant of the bis complex was 2×10^6 l mol^{-1} s^{-1}. The activation energy for the aquation process is 52 kJ mol^{-1}. The linewidths of metal-free phosphate solution at pH 2 were found to be linear in T/η and are thus apparently due to spin–rotation interaction.[748] Alcohol exchange with $(RO)_3AsO$ and $(RO)_3VO$ is rapid, *via* a five-co-ordinate transition state. $\Delta H^{\ddagger}$ and $\Delta S^{\ddagger}$ were derived.[749] Equilibrium constants have been determined between 1 : 1 adducts of nitrogen bases and OV(acac)$_2$ and Cu(acac)$_2$ using ^{1}H n.m.r. spectroscopy.[750] ^{1}H N.m.r. spectroscopy has been used to investigate the solvolysis of $VOCl_3$, and the equilibrium constant for

$$VOCl_3 + EtOH \rightleftharpoons EtOVOCl_2 + HCl$$

has been determined.[751] Stepwise complex formation between $NbOCl_3$ and $(Me_2N)_3PO$ has been investigated by ^{1}H n.m.r. spectroscopy.[752]

^{1}H N.m.r. spectroscopy has been used to determine the rate of *anti–gauche* isomerization of $[W_2(CH_2SiMe_3)_2(NEt_2)_4]$.[753]

Iron and osmium. PPh$_2$NMeCHMePh exchange in $[(-)_{365}$-$(\eta^5$-$C_5H_5)$Fe(CO)-(PPh$_2$NMeCHMePh)I] has been investigated.[754] Equilibrium constants for complex formation between substituted imidazoles and pyridines and various Fe^{III} porphyrins have been determined and the complexes have been shown to be the low-spin bis-complexes by use of susceptibility measurements.[755] The ^{13}C n.m.r. spectrum of a *cis–trans* mixture of (93) has been investigated as pyridine was added to break up the dimers.[756]

Cobalt, rhodium, and iridium. ^{13}C N.m.r. spectroscopy has been used to investigate pyridine exchange on Co^{II} and Ni^{II}. Activation energies, 16 kcal mol^{-1}

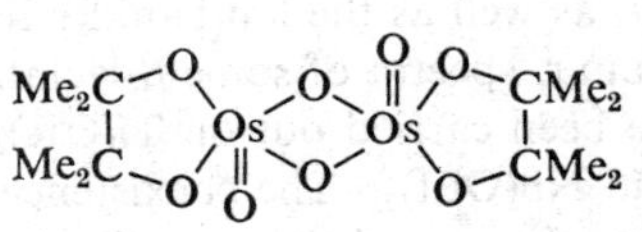

(93)

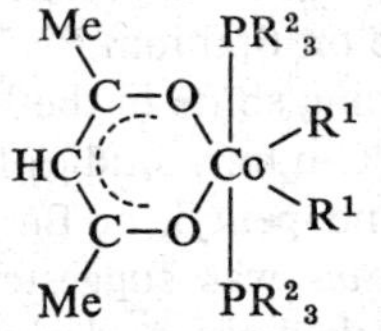

(94)

747 C. E. Holloway, *J.C.S. Dalton*, 1976, 1050.
748 M. W. Kendig and P. H. Rieger, *Inorg. Chim. Acta*, 1976, **17**, 175.
749 P. J. White, M. J. Kaus, J. O. Edwards, and P. H. Rieger, *J.C.S. Chem. Comm.*, 1976, 429.
750 E. Kwiatkowski and J. Trojanowski, *J. Inorg. Nuclear Chem.*, 1976, **38**, 131.
751 S. Ya. Kuchmii, V. K. Pogorelyi, Z. A. Tkachenko, A. V. Korzhak, and A. I. Kryukov, *Teor. i Eksp. Khim.*, 1976, **12**, 56.
752 L. G. Hubert-Pfalzgraf, R. C. Muller, M. Postel, and J. G. Riess, *Inorg. Chem.*, 1976, **15**, 40.
753 M. H. Chisholm and M. W. Extine, *J. Amer. Chem. Soc.*, 1976, **98**, 6395.
754 H. Brunner and F. Rackl, *J. Organometallic Chem.*, 1976, **118**, C19.
755 J. D. Salterlee, G. N. La Mar, and J. S. Frye, *J. Amer. Chem. Soc.*, 1976, **98**, 7275.
756 L. G. Marzilli, B. E. Hanson, T. J. Kistenmacher, L. A. Epps, and R. C. Stewart, *Inorg. Chem.*, 1976, **15**, 1661.

for Ni^{II} and 22 kcal mol^{-1} for Co^{II}, and the metal–pyridine distance, 3 Å, were determined.[757] The ^1H, ^{13}C, and ^{31}P n.m.r. spectra of (94) have been determined in several solvents. The co-ordinated PR_3 ligands are partially dissociated in solution. For (94; R^1 = Et, R^2_3 = Me_2Ph) $\Delta H^0 = 4.2$ kcal mol^{-1}, $\Delta G^0 = 4.0 \pm 0.2$ kcal mol^{-1}, and $\Delta S^0 = 0 \pm 2$ e.u. It is thought that the PR_3 ligand is replaced by solvent.[758, 759] The Evans method has been used to study the position of the high-spin–low-spin equilibrium

$$[CoCl_2\{PEt_n(OEt)_{3-n}\}_2] + PEt_n(OEt)_{3-n} \rightleftharpoons [CoCl_2\{PEt_n(OEt)_{3-n}\}_3]$$

ΔH^0 and ΔS^0 were determined. It was found that the determination of equilibrium constants from spectrophotometric data was less accurate than the n.m.r. method.[760] ^1H and ^{31}P n.m.r. spectroscopies have been used to follow the reaction

$$[RhCl(PAr_3)_3] + RCOCl \longrightarrow [RhCl_2(COR)(PAr_3)_2]$$

and to determine ΔH^0 and ΔS^0 for the equilibrium

$$[RhCl_2(COR)(PAr_3)_2] \rightleftharpoons [Rh(CO)Cl_2(PAr_3)_2R]$$ [761]

^1H and ^{13}C n.m.r. spectroscopies have been used to show that the various isomers of *trans*-$[(Ph_3P)_2(CO)M(RN{=}S{=}NR)]^+$ (M = Rh or Ir) and *cis*- and *trans*-$[PtL_2(RN{=}S{=}NR)X]^+$ interconvert *via* intra- and inter-molecular exchange processes.[762]

Nickel, palladium, and platinum. ^1H and ^{31}P n.m.r. spectroscopies have been used to investigate the dissociative exchange of PEt_3 for $[HNi(PEt_3)_3CN]$ with $\Delta G_{300}^{\ddagger} = 12.0$ kcal mol^{-1}. When bulkier tertiary phosphines were used, only $[HNi(PR_3)_2CN]$ is formed.[763] The ^1H n.m.r. spectra of $[NiR_2(bipy)]$ and $[NiEt_2(bipy)(\eta^2\text{-olefin})]$ have been studied over the range from -70 to 80 °C. The temperature-dependent shifts of the ethyl and 2,2′-bipyridyl protons of $[NiEt_2(bipy)]$ in non-aromatic solvents were analysed on the basis of the formation of a 1 : 1 solvent–solute complex, and K, ΔG^0, ΔH^0, and ΔS^0 were given. The n.m.r. spectra of $[NiR_2(bipy)]$ in aromatic solvents show induced solvent shifts and the chemical shifts of the CH_2 and CH_3 group of $[NiEt_2(bipy)]$ coincide at 76 °C in benzene to give a singlet. The ^{13}C n.m.r. spectra were also recorded. The formation constant, ΔG^0, ΔH^0, and ΔS^0 were determined for $[NiEt_2(bipy)(\eta^2\text{-olefin})]$. The chemical shift difference changes in the separation of the CH_2 and CH_3 protons on olefin co-ordination were attributed to an increase in electronegativity of the nickel due to back donation from nickel to the olefin.[764] The dynamic behaviour of the ligands on $[NiR(acac)(PPh_3)_n]$ has been examined in solution. PPh_3 exchange occurs rapidly by an S_N2 mechanism. The ^{31}P n.m.r. spectrum of $[NiEt(acac)(PPh_3)]$ shows that in pyridine, the pyridine replaces

757 P. Mignot and J. P. Grivet, *Chem. Phys.*, 1976, **15**, 261.
758 T. Ikariya, Y. Nakamura, and A. Yamamoto, *J. Organometallic Chem.*, 1976, **118**, 101.
759 T. Ikariya and A. Yamamoto, *J. Organometallic Chem.*, 1976, **120**, 257.
760 I. B. Joedicke, H. V. Studer, and J. T. Yoke, *Inorg. Chem.*, 1976, **15**, 1352.
761 D. Egglestone and M. C. Baird, *J, Organometallic Chem.*, 1976, **113**, C25.
762 J. Kuyper, L. G. Hubert-Pfalzgraf, P. C. Keijzer, and K. Vrieze, *J. Organometallic Chem.*, 1976, **108**, 271.
763 J. D. Druliner, A. D. English, J. P. Jesson, P. Meakin, and C. A. Tolman, *J. Amer. Chem. Soc.*, 1976, **98**, 2156.
764 T. Yamamoto, Y. Nakamura, and A. Yamamoto, *Bull. Chem. Soc. Japan*, 1976, **49**, 191.

PPh_3. The acetylacetonate undergoes rapid exchange with E_a = 10 kcal mol^{-1}. On standing, the CH_2 and CD_3 groups of [Ni(CH_2CD_3)(acac)(PPh_3)] scramble.[765] ^{31}P N.m.r. spectroscopy has been used to study exchange reactions of [Ni(*o*-tolyl)-Br(PPh_3)] with PR_3. The observed ligand exchange equilibria are dominated by steric factors.[766] Pyridine exchange on the pyridine adducts of [(salicylaldehydato)$_2$Ni] is relatively slow and its activation energy increases with decreasing electron-donor properties of the in-plane ligands. Changes in spin delocalization are small.[767] The ^{31}P n.m.r. spectrum of [$NiX_2(PMe_3)_3$] (X = Cl, Br, or I) is AB_2 which is consistent with (95). For [$Ni(CN)_2(PMe_3)_2$], PMe_3

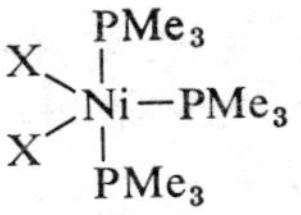

(95)

exchanges dissociatively with $\Delta H^{\ddagger}$ = +13.7 ± 0.5 kcal mol^{-1} and $\Delta S^{\ddagger}$ = +10.4 e.u.[768] ^{31}P N.m.r. spectroscopy has been used to show PPh_3 dissociation from [NiCl(PPh_3)(S_2CNEt_2)].[769] For *trans*-[$PdCl_2$(RN=S=NR)L] intra- and inter-molecular exchange have been studied by ^{1}H n.m.r. spectroscopy.[770] Variable-temperature ^{1}H n.m.r. spectroscopy has shown that on addition of EEt_2 to [$MX_2(EEt_2)_2$] (M = Pd or Pt; E = S, Se, or Te), fast exchange occurs except for [$PtX_2(SEt_2)_2$]. The rate of exchange decreases in the orders $TeEt_2 \gg SeEt_2 > SEt_2$ and Pd > Pt. Coalescences due to chalcogen inversion are quite distinct from the exchange coalescences in the systems SEt_2–[$PdX_2(SEt_2)_2$] and $SeEt_2$–[$PtX_2(SeEt_2)_2$].[771] The ratio of N- and S-bonded isomers of *trans*-[PtH(NCS)(PBu^n_3)$_2$] has been determined as a function of solvent.[772]

Copper and zinc. Low-temperature ^{1}H n.m.r. spectroscopy of MeLi–MeCu mixtures at various ratios in Me_2O shows the presence of $LiCuMe_2$, $LiCu_2Me_3$, and Li_2CuMe_3.[773] ^{1}H N.m.r. spectroscopy indicates that the stability of the complex Zn_2Et with L decreases in the order L = THF > propylene oxide > propylene sulphide > Et_2O.[774] Complex formation in [Zn(*meso*-trifluoroacetoxyoctaethylporphyrin)] has been studied using ^{1}H and ^{13}C n.m.r. spectroscopies. The large concentration dependence of the ^{1}H chemical shifts was analysed in terms of a monomer–dimer equilibrium to give $K \sim 3.01$ mol^{-1}, and monomer–dimer shifts of up to 4.2 p.p.m. (for the γ-*meso*-hydrogen atom). The dimers are immediately dissociated upon the addition of methanol. The analogous acetate showed no dependence on concentration or methanol. The

765 T. Yamamoto, T. Saruyama, Y. Nakamura, and A. Yamamoto, *Bull. Chem. Soc. Japan*, 1976, **49**, 589.
766 Y. Nakamura, K.-I. Maruya, and T. Mizoroki, *J. Organometallic Chem.*, 1976, **104**, C5.
767 M. Iwaizumi, K. Migita, and T. Isobe, *Bull. Chem. Soc. Japan*, 1976, **49**, 3325.
768 P. Meier, A. E. Merbach, M. Dartiguenave, and Y. Dartiguenave, *J. Amer. Chem. Soc.*, 1976, **98**, 6402.
769 J. A. McCleverty and N. J. Morrison, *J.C.S. Dalton*, 1976, 541.
770 J. Kuyper, P. I. Van Vliet, and K. Vrieze, *J. Organometallic Chem.*, 1976, **108**, 257.
771 R. J. Cross, T. H. Green, R. Keat, and J. F. Paterson, *J.C.S. Dalton*, 1976, 1486.
772 M. W. Adlard and G. Socrates, *J. Inorg. Nuclear Chem.*, 1976, **38**, 531.
773 E. C. Ashby and J. J. Watkins, *J.C.S. Chem. Comm.*, 1976, 784.
774 V. M. Denisov and Yu. P. Kuznetsov, *Izvest. Akad. Nauk S.S.S.R., Ser. khim.*, 1975, **12**, 2710.

large complexation shifts of the trifluoroacetate were explained on the basis of a dimer structure in which the O_2CCF_3 group plays no part in association which was presumed to be due to a novel metal–porphyrin interaction.[775] ^{1}H N.m.r. spectroscopy has been used to examine mercaptide anion exchange in R^1HgSR^2–R^1HgSR^3 and exhibits second-order kinetic behaviour. Evidence was presented that anion exchange proceeds through a four-centre bridged intermediate with total exclusion of the ionic pathway.[776]

Boron and aluminium. The dependence of spectral parameters of triple resonance $^1H(^{11}B, ^{14}N)$ magnetic resonance on spin–lattice relaxation times, solution, concentration, acidity, and impurities has been studied to determine rates and mechanism of hydrogen exchange between BNH_2 and NNH_2 groups in $BH_3N_2H_4$ in N_2H_4. The stripping of H^+ from a BNH_2 group by a N_2H_4 molecule is rate-limiting and the activation energy is 11.4 ± 1.0 kcal mol^{-1}.[777] ^{1}H N.m.r. lineshape analysis has been used to determine exchange parameters for Me_3BL. The reactions are dissociative in CH_2Cl_2 at 27 °C and the basicity towards BMe_3 decreases in the order $NMe_3 \sim PMe_3 > PMe_2Bu^t \sim PMe_2(NMe_2) > PMe_2Ph \gg PBz_3 > PMe_2CF_3 \sim PMe_3O \sim PMePh_2$.[778] ^{1}H and ^{11}B n.m.r. spectroscopies have been used to investigate scrambling reactions between BX_3 and BBr_3 and have shown that mixed species on equilibrium are not uniformly favoured. The effects of Lewis acidity and steric hindrance were discussed.[779]

$Me_2P(S)NRBX_2$ exists in solution as monomers and dimers in equilibrium. On the basis of ^{11}B n.m.r. spectroscopy it was suggested that sulphur–boron co-ordination occurs rather than nitrogen–boron co-ordination.[780] BX_3 adducts of $(Me_2N)_2CO$ and $(Me_2N)_2C{=}CH_2$ show similar co-ordination shifts and restricted rotation of the central carbon–nitrogen bonds. The rotation becomes more hindered as the Lewis acid strength increases $BF_3 < BCl_3 < BBr_3$.[781] A study of BF_3 complexes of C_4—C_{10} cycloalkanones and nine alkyl ketones has been completed using low-temperature ^{13}C n.m.r. spectroscopy. At temperatures between -85 and -125 °C the exchange is slow enough on the n.m.r. time-scale to permit the observation of separate ^{13}C n.m.r. signals for the bound and free ligand. Chemical shift displacements induced by complex formation were *ca.* 24 p.p.m. to higher frequency for the CO group and about an order of magnitude less for the other carbon signals. Doublets were observed for signals of the bound ligands with several members of both series, a feature which was interpreted in terms of a slow exchange of the BF_3 group between the two sp^2 orbitals of the carbonyl oxygen atom. Some evidence was presented for the much greater stability of only one of these *cis–trans* isomers in four of the alkyl ketone cases.[782] ^{1}H and ^{19}F n.m.r. spectroscopies have been used to investigate tetramethylthiourea adducts of BF_3 which partially ionize to $[L_2BF_2]^+$, $[BF_4]^-$, and $[B_2F_7]^-$. Below -70 °C all the methyl groups are inequivalent.[783]

[775] R. J. Abraham, G. H. Barnett, G. E. Hawkes, and K. M. Smith, *Tetrahedron*, 1976, **32**, 2949.
[776] R. D. Bach and A. T. Weibel, *J. Amer. Chem. Soc.*, 1976, **98**, 6241.
[777] G. V. Lagodzinskaya, N. G. Yunda, G. N. Nechiporenko, and G. B. Manelis, *Zhur. fiz. Khim.*, 1975, **49**, 2244.
[778] K. J. Alford, E. O. Bishop, and J. D. Smith, *J.C.S. Dalton*, 1976, 920.
[779] J.-P. Costes, G. Cros, and J.-P. Laurent, *Canad. J. Chem.*, 1976, **54**, 2996.
[780] G. Muckle, H. Nöth, and W. Storch, *Chem. Ber.*, 1976, **109**, 2572.
[781] J. S. Hartman and R. R. Yetman, *Canad. J. Chem.*, 1976, **54**, 1130.
[782] A. Fratiello, R. Kubo, and S. Chow, *J.C.S. Perkin II*, 1976, 1205.
[783] J. S. Hartman, G. J. Schrobilgen, and P. Stilbs, *Canad. J. Chem.*, 1976, **54**, 1121.

^{1}H N.m.r. studies on mixtures of Me_6Al_2 and (cyclopropyl)$_6Al_2$ have demonstrated that the cyclopropyl groups preferentially occupy the bridging position. Me_5(cyclopropyl)Al_2 and Me_4(cyclopropyl)$_2Al_2$ have been identified by low-temperature n.m.r. studies. Bridge–terminal methyl group exchange in Me_5(cyclopropyl)Al_2 was postulated to be intramolecular in toluene solution proceeding with an Arrhenius activation energy of 15.8 ± 0.7 kcal mol^{-1}. A two-site averaging process of terminal methyl groups in Me_4(cyclopropyl)$_2Al_2$ was studied. It was proposed that this averaging process results from rotation of the bridging cyclopropyl groups with $E_a \sim 11$ kcal mol^{-1}.[784] The equilibrium (96) has been suggested on the basis of ^{1}H n.m.r. data.[785] ^{1}H N.m.r. spectroscopy

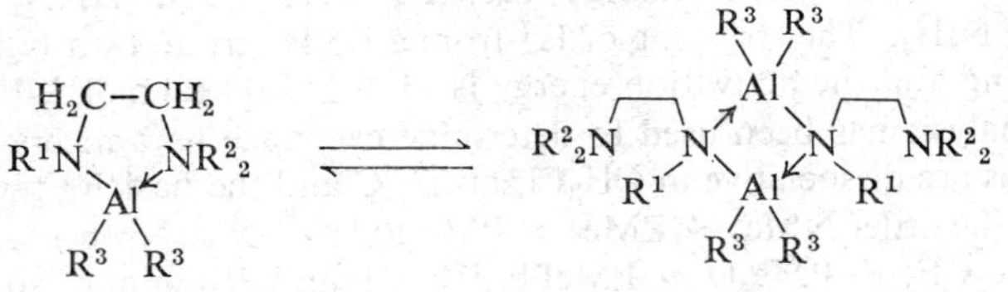

(96)

has been used to show that methacrylonitrile and methyl methacrylate in CH_2Cl_2 form binary complexes with $EtAlCl_2$.[786] The interaction of $[Al(OPr^i)_3]_4$ with ethylenediamine in benzene has been studied using ^{1}H, ^{13}C, and ^{27}Al n.m.r. spectroscopies. The initial products appear to be a linear diadduct and a polymer, but other unidentified aluminium compounds may also be formed. There was no evidence to support the suggestion that significant amounts of complexes are formed containing five-co-ordinate aluminium.[787] ^{1}H N.m.r. spectroscopy has been used to investigate equilibria of $[Al(OH)_nCl_{3-n}]$.[788]

Silicon and tin. The n.m.r. spectra of solutions containing both $Li^+[ESiH_3]^-$ and $E(SiH_3)_2$ (E = O, S, or Se) show that exchange occurs which is rapid on the n.m.r. time-scale unless E = O. N.m.r. data were given for ^{29}Si, ^{31}P, and ^{77}Se n.m.r. spectroscopies. Activation parameters, ΔH^0, and ΔS^0 were given for equilibrium (97).[789]

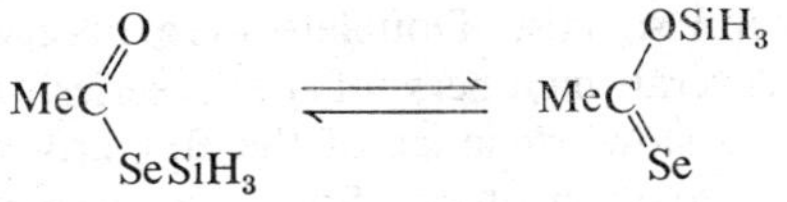

(97)

^{1}H and ^{13}C n.m.r. spectroscopy has been used to demonstrate that trace amounts of water in *N*-trimethylsilyl-imidazole and -2-methylimidazole are responsible for rapid intermolecular silyl exchange. Exchange can be slowed down by further purification or by the addition of an excess of triethylamine.

784 D. A. Sanders, P. A. Scherr, and J. P. Oliver, *Inorg. Chem.*, 1976, **15**, 861.
785 O. T. Beachley, jun. and K. C. Racette, *Inorg. Chem.*, 1976, **15**, 2110.
786 J. Furukawa, E. Kobayashi, and S. Nagata, *J. Polymer Sci., Polymer Chem.*, 1976, **14**, 237 (*Chem. Abs.*, 1976, **84**, 90 650).
787 J. W. Akitt and R. H. Duncan, *J.C.S. Dalton*, 1976, 1119.
788 A. P. Shut'ko, I. Ya. Mulik, and B. A. Geller, *Ukrain. khim. Zhur.* (*Russ. Edn.*), 1976, **42**, 993 (*Chem. Abs.*, 1976, **85**, 200 316).
789 S. Cradock, E. A. V. Ebsworth, D. W. H. Rankin, and W. J. Savage, *J.C.S. Dalton*, 1976, 1661.

A mechanism was proposed which accounts for hydrolysis and intermolecular silyl exchange.[790] The kinetics of hydrogen exchange between Me_3SiOH and Me_3SiONa have been determined by ^1H n.m.r. spectroscopy. The average silanol lifetime during hydrogen exchange with silanolate was *ca.* 100 times that of silanol in 1 : 1 association with silanolate.[791] Kinetic data and activation parameters for

$$(R^1{}_3Si)_2NOSiR^2{}_3 \rightleftharpoons (R^1{}_3Si)(R^2{}_3Si)NOSiR^1{}_3$$

indicate that the rearrangement probably involves a four-centre transition state.[792] Low-temperature n.m.r. spectra of the equilibrium (98) reveal both

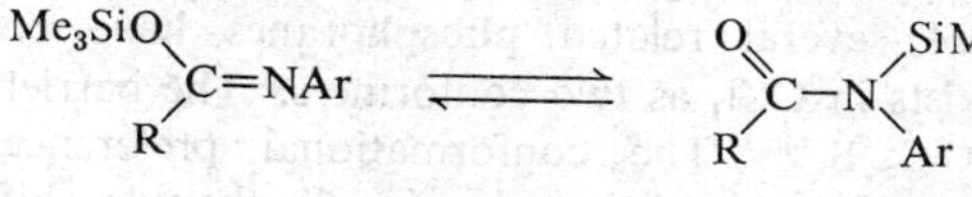

(98)

hindered rotation and exchange processes for the formanilides and benzanilides and exchange for acetanilides. Isomer ratios and free energies of activation were reported.[793] ^1H N.m.r. spectroscopy has been used to determine $\Delta G^0 = 1.7$ kcal mol^{-1} and $\Delta G^{\ddagger} = 21$—27 kcal mol^{-1} for the equilibrium (99).[794]

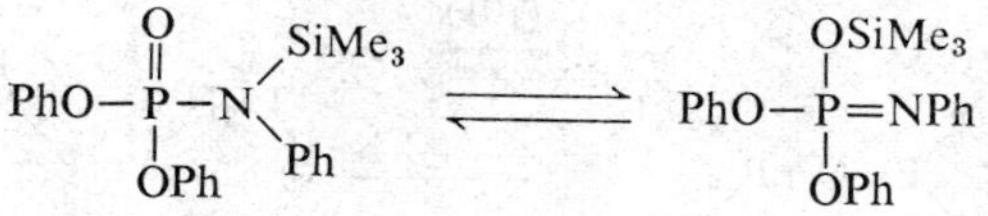

(99)

Equilibrium constants of a proton-silyl exchange reaction between RCO_2SiMe_3 and $BzCO_2H$ in CCl_4 have been determined by ^1H n.m.r. spectroscopy. There is a rapid equilibrium and the mechanism was discussed.[795] Activation parameters for the rearrangement of $C_5Me_5GeR^1{}_2R^2$ have been determined.[796] Variable-temperature ^{13}C n.m.r. spectra of $C_6H_{11}MMe_3$ (M = Sn or Pb) have been recorded at -69 °C to show the equilibrium between equatorial and axial isomers. Me_3Sn favours the equatorial position by 1.06 ± 0.14 kcal mol^{-1} while for $PbMe_3$ the value is 0.67 ± 0.06 kcal mol^{-1}.[797] ^1H and ^{119}Sn n.m.r. spectroscopies have been used to demonstrate the equilibrium (100).[798] The dependence of ^{13}C chemical shifts and spin–lattice relaxation times of ^{13}C nuclei of EtOAc, EtOH, and $C_5H_{11}OH$ in $SnCl_4$ or $SbCl_3$ solution on composition and temperature has been studied. The carbonyl group of EtOAc and the OH group

[790] R. E. Wasylishen, G. S. Birdi, and A. F. Janzen, *Inorg. Chem.*, 1976, **15**, 3054.
[791] K. A. Andrianov, A. I. Chernyshev, V. A. Temnikovskii, V. V. Yastrebov, and V. M. Kopylov, *Teor. i eksp. Khim.*, 1975, **11**, 678.
[792] P. Nowakowski and R. West, *J. Amer. Chem. Soc.*, 1976, **98**, 5616.
[793] C. H. Yoder and A. D. Belber, *J. Organometallic Chem.*, 1976, **114**, 251.
[794] P. K. G. Hodgson, R. Katz, and G. Zon, *J. Organometallic Chem.*, 1976, **117**, C63.
[795] J. Pola, A. El Attar, and V. Chvalovský, *Coll. Czech. Chem. Comm.*, 1976, **41**, 1772.
[796] V. S. Shpiro, Yu. A. Strelenko, Yu. A. Ustynyuk, N. N. Zemlyanskii, and K. A. Kocheshkov, *Doklady Akad. Nauk S.S.S.R.*, 1976, **228**, 1128.
[797] W. Kitching, D. Doddrell, and J. B. Grutzner, *J. Organometallic Chem.*, 1976, **107**, C5.
[798] J.-M. Brocas, B. de Jeso, and J.-C. Pommier, *J. Organometallic Chem.*, 1976, **120**, 217.

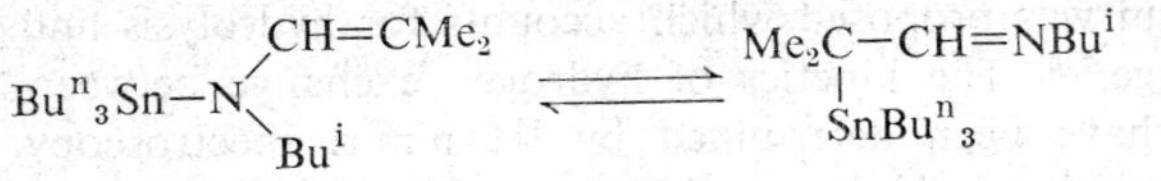

(100)

of alcohols serve as electron donors. The temperature and concentration dependence of T_1 were compared with the viscosity of the solution.[799] The relative amounts of *cis*- and *trans*-isomers of $[\{R^1R^2P(O)\}_2SnCl_4]$ have been shown to depend on the size of R^1 and R^2.[800]

Phosphorus, arsenic, and antimony. ^{31}P N.m.r. spectroscopy has demonstrated tellurium exchange between R_3PTe and R_3P.[801] The rates of ionization of $MeP(OPh)_4$ and of several related phosphoranes have been measured.[802] MePO(EPr)OPh exists in CCl_4 as two conformers. The barriers increase in the order $E = N < O < S$.[803] The conformational preference of R in the equilibrium (101) has been investigated using lanthanide shift reagents. The P=O is mainly axial when R = Me, Bz, Ph, Bu^t, CPh_3, or NMe_2, and is mainly

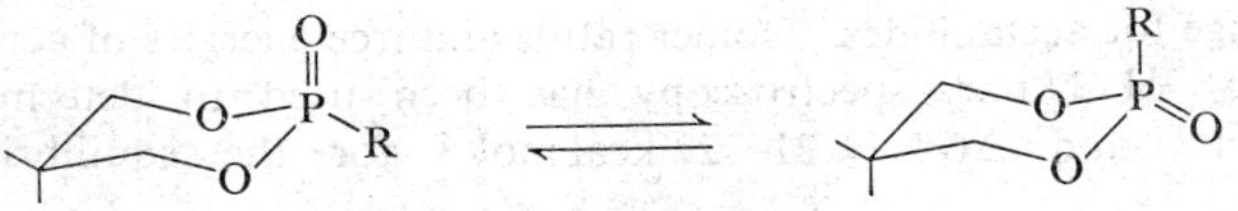

(101)

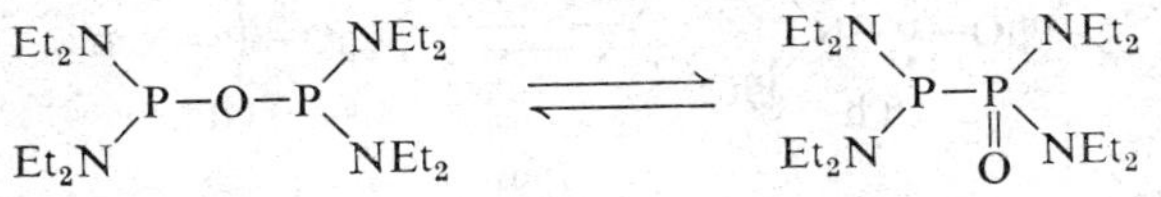

(102)

equatorial when R = OMe, OBu^t, OPh, SCH_3, Cl, or F.[804] ^1H and ^{31}P n.m.r. spectroscopies have been used to demonstrate the tautomerism (102).[805] The effect of base on the equilibrium (103) has been investigated by ^{31}P n.m.r.

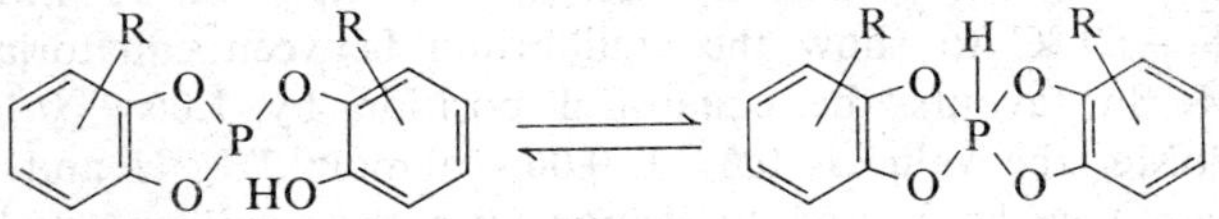

(103)

spectroscopy.[806] For ring inversion of $MeAs(CH_2)_5$ (cyclic), E_a = 7.8 kcal mol^{-1} and

799 B. P. Nikolaev and O. F. Bezrukov, *Teor. i eksp. Khim.*, 1976, **12**, 212.
800 A. V. Aganov, A. A. Musina, I. Ya. Kuramshin, Yu. Yu. Samitov, and A. A. Muratova, *Tezisy Dokl.-Vses. Chugaevskoe Soveshch. Khim. Kompleksn. Soedin.*, 12*th*, 1975, **2**, 159 (*Chem. Abs.*, 1976, **85**, 171 014).
801 W.-W. Du Mont and H.-J. Kroth, *J. Organometallic Chem.*, 1976, **113**, C35.
802 D. I. Phillips, I. Szele, and F. H. Westheimer, *J. Amer. Chem. Soc.*, 1976, **98**, 184.
803 L. A. Ashkinazi, P. M. Zavlin, and B. I. Ionin, *Zhur. obshchei Khim.*, 1976, **46**, 925.
804 A. J. Dale, *Acta Chem. Scand.*, 1976, **30B**, 255.
805 V. L. Foss, Yu. A. Veits, N. V. Lukashev, and I. F. Lutsenko, *J. Organometallic Chem.*, 1976, **121**, C27.
806 A. Munoz, G. Gençe, M. Koenig, and R. Wolf, *Compt. rend.*, 1975, **280**, *C*, 485.

$\Delta G_{145}^{\ddagger}$ = 6.8 kcal mol^{-1}. There are equal populations of axial and equatorial isomers at −140 °C. ^{1}H N.m.r. spectra of ^{15}N-labelled Me_2AsNMe_2–$HNMe_2$ and $MeAs(NMe_2)_2$–$HNMe_2$ show rapid exchange of the NMe_2 group.[808] ^{1}H N.m.r. spectra of some π- and p-electron donors were determined in cyclohexane containing $AsBr_3$. Both 1 : 1 and 1 : 2 compounds were formed and formation constants were given.[809] Stability constants were determined for AsX_3 with donors.[810] ^{19}F N.m.r. spectra of ligand exchange between SbF_5 and $SbCl_5$ in MeCN have revealed the existence of a series of geometric isomers of $[SbF_nCl_{5-n}(NCMe)]$.[811]

Course of Reactions.—*Lithium and Magnesium.* CIDNP (^{1}H and ^{19}F n.m.r. spectra) has been observed during the reaction of butyl-lithium with R^1R^2CHCl,[812] and the formation of Grignard reagents,[813] and from Grignard reagents undergoing iron-catalysed halogen–metal exchange, for which the mechanism was discussed.[814] Pulse Fourier transform has been used to study ^{1}H CIDNP arising from thermolysis of Bu^t_2Mg, Bu^t_2Zn, and both thermolysis and photolysis of Bu^t_2Hg in deuteriated hydrocarbons. Decomposition gives the metal-polarized isobutane and isobutene at rate k_d and unpolarized 2,2,3,3-tetramethylbutane at rate k_c. Analysis of the ratio of k_d/k_c as well as activation entropies for decomposition established a radical mechanism for the main process. In the decomposition of Bu^t_2Mg and Bu^t_2Zn, an A/E multiplet polarization of the isobutane and isobutene was observed. The opposite to E/A polarization of these products arose during the thermolysis of Bu^t_2Hg. Photolysis of Bu^t_2Hg, however, leads to an A/E polarization. The different phases of the multiplets indicate that decomposition of Bu^t_2Hg is a cage reaction. Bu^t_2Mg and Bu^t_2Zn decompose in a free-radical reaction of uncorrelated Bu^t pairs.[815] CIDNP has been observed from Bu^tMgCl with CCl_4 or $BrCCl_3$.[816] An n.m.r. spectroscopic and chromatographic study has been carried out on the thermolysis of calcium complexonate in aqueous solutions.[817]

Vanadium and Molybdenum. ^{31}P CIDNP data indicate that the reaction of hydroperoxides with phosphites catalysed by vanadylacetylacetonate is non-radical.[818] ^{1}H N.m.r. spectroscopy has been used to follow the isomerization of (104) due to rotation about the carbon–naphthyl bond.[819] ^{1}H N.m.r. spectroscopy has been used to monitor the concentration-independent but water-dependent

807 J. B. Lambert and H.-N. Sun, *J. Organometallic Chem.*, 1976, **117**, 17.
808 O. Adler and F. Kober, *Chem.-Ztg.*, 1976, **100**, 235 (*Chem. Abs.*, 1976, **85**, 143 216).
809 M. I. Vakhrin and V. S. Grechishkin, *Zhur. fiz. Khim.*, 1976, **50**, 314.
810 M. I. Vakhrin and E. S. Sbcherbakova, *Zhur. fiz. Khim.*, 1975, **49**, 2726.
811 E. G. Il'in, L. I. Klyuev, and Yu. A. Buslaev, *Doklady Akad. Nauk S.S.S.R.*, 1975, **223**, 95.
812 A. V. Podoplelov, T. V. Leshina, R. Z. Sagdeev, and M. A. Kamkha, *Zhur. org. Khim.*, 1976, **12**, 495.
813 B. J. Schaart, H. W. H. J. Bodewitz, C. Blomberg, and F. Bickelhaupt, *J. Amer. Chem. Soc.*, 1976, **98**, 3712.
814 R. G. Lawler and P. Livant, *J. Amer. Chem. Soc.*, 1976, **98**, 3710.
815 R. Benn, *Chem. Phys.*, 1976, **15**, 369.
816 V. I. Savin, A. G. Abul'khanov, and Yu. P. Kitaev, *Zhur. org. Khim.*, 1976, **12**, 484.
817 O. I. Martynova, E. P. Miklashevskaya, S. V. Rykov, E. D. Glinka, and B. M. Larin, *Trudy Moskov Energ. In-Ta*, 1975, **238**, 124 (*Chem. Abs.*, 1976, **84**, 49 701).
818 D. G. Pobedimskii, A. D. Pershin, Sh. A. Nasybullin, and A. L. Buchachenko, *Izvest. Akad. Nauk S.S.S.R., Ser. khim.*, 1976, **1**, 79.
819 H. Brunner and J. Wachter, *Angew. Chem. Internat. Edn.*, 1976, **15**, 316.

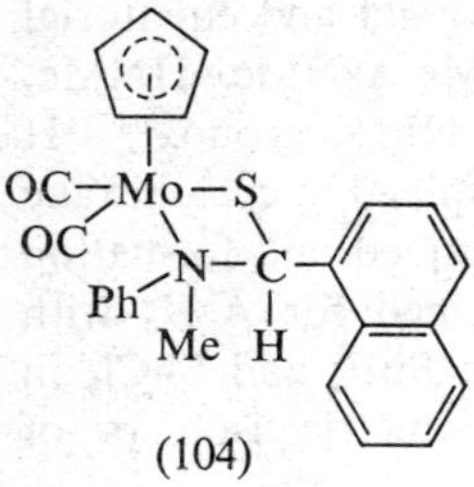

(104)

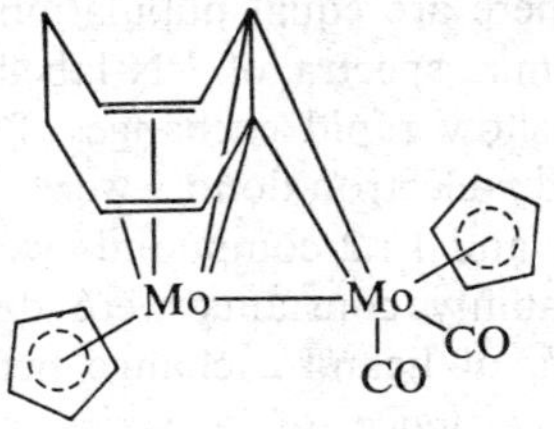

(105)

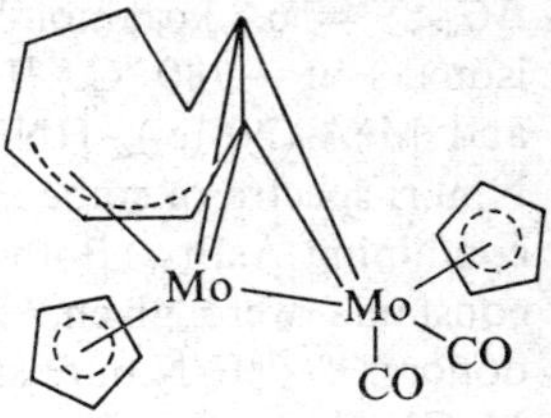

(106)

isomerization of (105) to (106). ^{13}C N.m.r. data were also reported.[820] ^{19}F N.m.r. spectroscopy has been used to follow the oxidation of $[(\eta^5\text{-}C_5H_5)Mo(CO)\text{-}(\eta^2\text{-}CF_3C{\equiv}CCF_3)SCF_3]$ to $[(\eta^5\text{-}C_5H_5)MoO(\eta^2\text{-}CF_3C{\equiv}CCF_3)SCF_3]$.[821]

Cobalt and Iridium. The decomposition of $[(\eta^5\text{-}C_5H_5)CoMe(\mu\text{-}CO)]_2$ has been monitored by ^{1}H n.m.r. spectroscopy.[822] ^{13}C N.m.r. spectroscopy has been used to show that thermal reaction between *trans*-$[Co(CN)_4(SO_3)_2]^{5-}$ and $[^{13}CN]^-$ leads to stereospecific incorporation to give *trans*-$[Co(^{13}CN)(CN)_4SO_3]^{4-}$. Thermal aquation at pH 7.5 at 60 °C gives complete scrambling.[823] ^{1}H N.m.r. spectroscopy has been used to measure the rate of isotopic deuterium exchange in some Co^{III} amine complexes in D_2O. The substituents $[CN]^-$, $[NO_2]^-$, and possibly $[SO_3]^{2-}$ decelerate the exchange of the amine protons *trans* to these ligands, leading in some cases to a smaller rate for these protons than for the *cis* ones. On the other hand, the *trans*-proton labilizing ligands include F^-, Cl^-, Br^-, H_2O, $[CO_3]^{2-}$, $[OCONH_2]^-$, $[N_3]^-$, CH_3NH_2, $[MeCO_2]^-$, [malonate]$^{2-}$, [fumarate]$^{2-}$, and [oxalate]$^{2-}$.[824] Similar work has shown that (i) the rates for *trans* (to O) amines are greater than for *cis*, (ii) the rates in carbonato complexes, especially for *trans* amines, are smaller than those in the oxalato and malonato analogues, (iii) the difference between rates for *cis* and *trans* amines is smaller for bis-diamine complexes than for tetra-ammine complexes.[825] The kinetics of proton exchange of β-*cis*-$[Co\{(R,R)(S,S)\text{-}2,3,2\text{-tet}\}Cl_2]^+$ (2,3,2-tet = 1,9-diamino-3,7-diazanonane) have been studied. One primary and one secondary amine proton exchange rapidly and another primary amine proton somewhat more slowly: presumably the three protons on the nitrogen atoms *trans* to chlorine atoms. No exchange of the other three protons occurs until the complex undergoes base hydrolysis. Activation parameters were determined.[826] ^{1}H N.m.r. spectroscopy has been used to show that $[Ir(\eta^4\text{-cyclo-octadiene})(PMePh_2)_2]^+\text{-}[PF_6]^-$ in CD_2Cl_2 at -80 °C reacts with H_2 to form (107), but on warming to room temperature hydrogen is lost.[827]

Nickel and Platinum. An *in situ* n.m.r. method of directly studying the rate and mechanism of photochromic reactions including both the forward, light-induced

[820] R. Goddard, S. A. R. Knox, F. G. A. Stone, M. J. Winter, and P. Woodward, *J.C.S. Chem. Comm.*, 1976, 559.
[821] P. S. Braterman, J. L. Davidson, and D. W. A. Sharp, *J.C.S. Dalton*, 1976, 241.
[822] N. E. Schore, C. Ilenda, and R. G. Bergman, *J. Amer. Chem. Soc.*, 1976, **98**, 7436.
[823] K. F. Miller and R. A. D. Wentworth, *Inorg. Chem.*, 1976, **15**, 1467.
[824] U. Sakaguchi, K. Maeda, and H. Yoneda, *Bull. Chem. Soc., Japan*, 1976, **49**, 397.
[825] T. Taura, U. Sakaguchi, and H. Yoneda, *Bull. Chem. Soc. Japan*, 1976, **49**, 1867.
[826] E. Ahmed and M. L. Tobe, *Inorg. Chem.*, 1976, **15**, 2635.
[827] R. H. Crabtree, H. Felkin, and G. E. Morris, *J.C.S. Chem. Comm.*, 1976, 716.

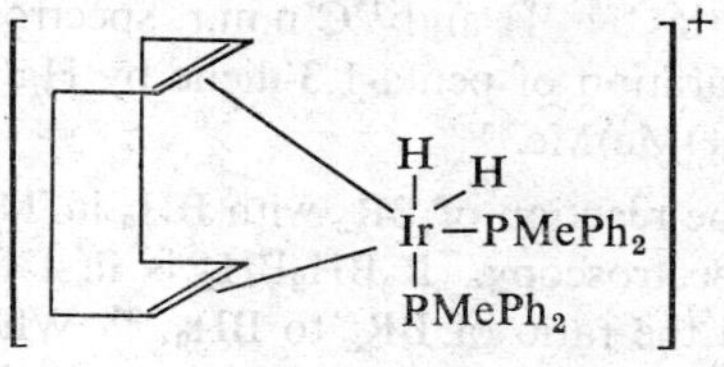

(107)

reactions and the return of the system to equilibrium in the dark has been developed and applied to $[(Bz_2NCS_2)_3Ni]^+Br^-$.[828] ^{13}C N.m.r. spectroscopy has been used to show the ready equilibration (108).[829] The reaction of *trans*-$[Pt(C{\equiv}CH)_2L_2]$ with anhydrous HCl to give *trans*-$[PtCl(C{\equiv}CH)L_2]$ *via* reactions

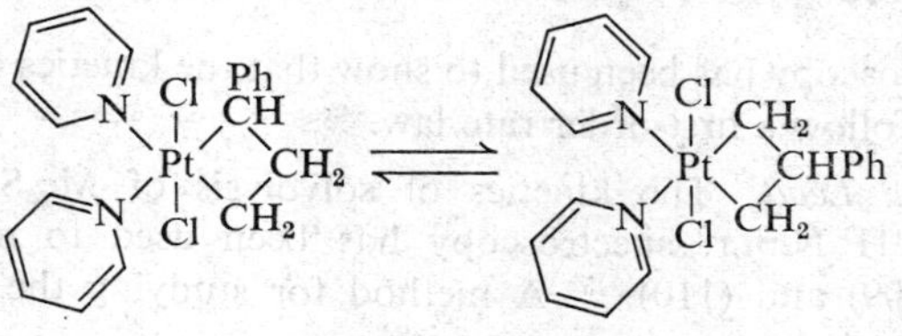

(108)

involving *trans*-$[Pt(C{\equiv}CH)(CCl{=}CH_2)L_2]$, *trans*-$[Pt(CCl{=}CH_2)_2L_2]$, and *trans*-$[PtCl(CCl{=}CH_2)L_2]$ has been monitored by ^1H n.m.r. spectroscopy.[830] The reaction of $[PtCl_2(\eta^2\text{-olefin})]_2$ with pyridine to give $[PtCl_2(py)(\eta^2\text{-olefin})]$ and then with $P(OPh)_3$ to give $[PtCl_2(py)\{P(OPh)_3\}]$ has been investigated by ^1H n.m.r. spectroscopy.[831] When *cis*-$[Pt(DMSO)_2Cl_2]$ reacts with various pyridines, the initially formed *trans*-$[Pt(py)(DMSO)Cl_2]$ isomerizes to a *cis–trans* mixture.[832]

Gold, Zinc, and Mercury. The decomposition of *cis*-$[Me_2AuL_2]^+$ has been monitored by ^1H n.m.r. spectroscopy. The rates depend on L: $PPh_3 > PMePh_2 > PMe_2Ph > PMe_3$ (attributed to bulk), and $SbPh_3 > AsPh_3 > PPh_3$. The decomposition is slow with chelating ligands and inhibited by an excess of ligand. The anion has only a small effect.[833] $MeAuPPh_3$ reacts with $CHCl_3$ under u.v. irradiation to give methyl and solvent-derived radicals in triplet pairs showing CIDNP. An S_H2 substitution by methyl radicals at the gold atoms was observed.[834] ^1H N.m.r. spectroscopy has been used to identify the site of zinc binding to L-aspartyl-L-phenylalanine methyl ester, aspartate, and L-phenylalanine-methyl ester, and to determine the relative rates of ester hydrolysis.[835] The photolysis of symmetric and asymmetric organomercury compounds has been studied by CIDNP. No indications of the existence of RHg· radicals was found. Depending on the solvent, reactions occur from excited states with

[828] D. P. Schwendiman and J. I. Zink, *J. Amer. Chem. Soc.*, 1976, **98**, 1248.
[829] R. J. Puddephatt, M. A. Quyser, and C. F. H. Tipper, *J.C.S. Chem. Comm.*, 1976, 626.
[830] R. A. Bell and M. H. Chisholm, *J.C.S. Chem. Comm.*, 1976, 200.
[831] W. Partenheimer, *J. Amer. Chem. Soc.*, 1976, **98**, 2779.
[832] P.-C. Kong, D. Iyamuremye, and F. D. Rochon, *Canad. J. Chem.*, 1976, **54**, 3224.
[833] P. Lawrence and R. S. Tobias, *J. Organometallic Chem.*, 1976, **122**, 429.
[834] P. W. N. M. Van Leeuwen, R. Kaptein, R. Huis, and C. F. Roobeck, *J. Organometallic Chem.*, 1976, **104**, C44.
[835] M. L. D. Touche and D. R. Williams, *J.C.S. Dalton*, 1976, 2001.

different spin multiplicities.[836] ^1H and ^{13}C n.m.r. spectroscopy has been used to show that 1,4-oxymercuration of penta-1,3-diene by $Hg(NO_3)_2$ in MeOH yields $ClHgCH_2CH{=}CHCH(OMe)Me$.[837]

Boron and Thallium. The reaction of BR_3 with BH_3 in THF has been studied by means of ^{11}B n.m.r. spectroscopy. $R_2BH_2BH_2$ is first formed and subsequent steps depend on R and the ratio of BR_3 to BH_3.[838] When R_3B reacts with air, CIDNP has been observed.[839] ^{19}F N.m.r. spectroscopy has been used to determine the kinetics of hydrolysis of $[BF_3OMe]^-$.[840] The removal of the first fluoride ion from the inner sphere of the boron is rate limiting and a S_N1cb mechanism was suggested.[841] Stopped-flow ^1H n.m.r. spectroscopy has been used to determine the rate for [842]

$$[Al(DMSO)_6]^{3+} + [^2H_6]DMSO \longrightarrow [Al([^2H_6]DMSO)_6]^{3+}$$

^1H N.m.r. spectroscopy has been used to show that the kinetics of decomposition of $MeTl(OAc)_2$ follow a first-order rate law.[843]

Silicon, Tin, and Lead. The kinetics of solvolysis of Me_3SiCX_3 have been determined.[844] ^1H N.m.r. spectroscopy has been used to monitor *cis–trans* isomerism in (109) and (110).[845] A method for studying the kinetics of low-

(109) (110)

temperature vulcanization of siloxane rubbers by $(EtO)_4Si$ is based on the use of ^1H n.m.r. spectroscopy to determine the concentration of EtOH formed during vulcanization.[846] ^1H N.m.r. spectroscopy has been used to follow the reaction of Me_6Sn_2 and Me_3PbCl and Me_3SnBu^t and Me_3PbCl. The rates of these reactions were determined,[847] and the thermal decomposition of Me_6Pb_2 was investigated.[848]

Phosphorus and Selenium. ^1H, ^{13}C, and ^{31}P n.m.r. spectroscopies have been used to investigate the reaction of Ph_3P with CCl_4 to give compounds such as $[Ph_3PCH_nCl_{3-n}]^+$ and $[Ph_3PCHPPh_3]^+$.[849] ^{31}P N.m.r. spectroscopy has been

836 F. J. J. de Kanter, *Org. Magn. Resonance*, 1976, **8**, 129.
837 A. J. Bloodworth, M. G. Hutchings, and A. J. Sotowicz, *J.C.S. Chem. Comm.*, 1976, 578.
838 B. Wrackmeyer, *J. Organometallic Chem.*, 1976, **117**, 313.
839 H. Friebolin and R. Rensch, *Org. Magn. Resonance*, 1976, **8**, 576.
840 V. N. Plakhotnik, V. V. Evsikov, and E. V. Yanchenko, *Koord. Khim.*, 1976, **2**, 855.
841 V. N. Plakhotnik and V. V. Evsikov, *Doklady Akad. Nauk S.S.S.R.*, 1976, **228**, 1112.
842 A. J. Brown, D. A. Couch, O. W. Howarth, and P. Moore, *J. Magn. Resonance*, 1976, **21**, 503.
843 U. Phol and F. Huber, *J. Organometallic Chem.*, 1976, **116**, 141.
844 W. Stańczyk and J. Chojnowski, *J. Organometallic Chem.*, 1976, **117**, 219.
845 F. K. Cartledge, B. G. McKinnie, and J. M. Wolcott, *J. Organometallic Chem.*, 1976, **118**, 7.
846 V. V. Severnyi, R. M. Minas'yan, V. V. Sokolov, and A. V. Kisin, *Kauch. Rezina*, 1976, **2**, 14 (*Chem. Abs.*, 1976, **84**, 181 390).
847 D. P. Arnold and P. R. Wells, *J. Organometallic Chem.*, 1976, **108**, 345.
848 D. P. Arnold and P. R. Wells, *J. Organometallic Chem.*, 1976, **111**, 269.
849 R. Appel, F. Knoll, W. Michel, W. Morbach, H.-D. Wihler, and H. Veltmann, *Chem. Ber.*, 1976, **109**, 58.

used to investigate the reactions of R^1R^2PCl with $MeCH(OEt)_2$,[850] $Cl_2PN(Me)$-$P(O)Cl$ with Bu^tNH_2 to give (111),[851] $P\{P(O)(OEt)_2\}_3$ with piperidine,[852] Ph_4P^+Cl

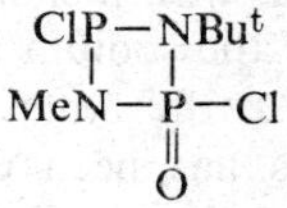

(111)

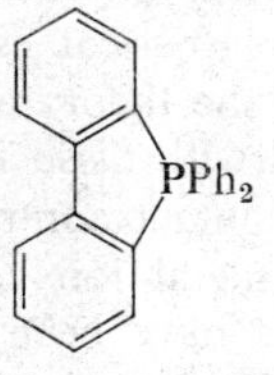

(112)

with $LiNR^1R^2$ to give (112) and Ph_4PNR_2,[853] and $[Ph_3PCXCl_2]^+$ with $Ph_2P{=}C{=}PPh_2$ to give $Ph_3P{=}Cl_2$ and $[Ph_3P{-}CX{-}PPh_3]^+$, which also had ^{13}C n.m.r. spectra recorded.[854] ^1H N.m.r. spectroscopy has been used to follow the equilibria [855]

$$PhSeCH_2C{\equiv}CH \rightleftharpoons PhSeCH{=}C{=}CH_2 \rightleftharpoons PhSeC{\equiv}CCH_3$$

4 Paramagnetic Complexes

In this section, compounds of *d*-block transition elements will be considered first and then those of the lanthanide and actinide elements. Papers concerning the use of paramagnetic complexes as 'shift reagents' are usually omitted. A number of paramagnetic complexes have been referred to in references 157, 259, 289, 290, 375, 413, 471, 667, 671—675, 708—715, 719—721, 723—726, 729—732, 750, 757, 760, 767, 951, 952, 961, 1008, 1012—1024, 1027, 1028, 1031, 1035—1038, 1040—1045, 1049, 1112, 1114, 1126, 1411, and 1451.

A number of reviews have appeared discussing various aspects of n.m.r. shift reagents.[856–860] The case of the effect of slow motion on nuclear relaxation processes of paramagnetic complexes has been analysed.[861]

The theory of the pseudocontact contribution to n.m.r. shifts of paramagnetic systems has been developed,[862] and methods to study spin–lattice relaxation in paramagnetic substances have been investigated.[863] The ^1H spin–lattice relaxation times have been measured from 100 to 180 K for both pure NH_3 and

[850] M. B. Gazizov, D. E. Sultanova, A. I. Razumov, T. V. Zykova, N. A. Anoshina, and R. A. Salakhutdinov, *Zhur. obshchei Khim.*, 1975, **45**, 1704.
[851] G. Bulloch and R. Keat, *J.C.S. Dalton*, 1976, 1113.
[852] D. Weber and E. Fluck, *Inorg. Nuclear Chem. Letters*, 1976, **12**, 515.
[853] N. A. Nesmeyanov, O. A. Rebrova, V. V. Mikul'shina, P. V. Petrovsky, V. I. Robas, and O. A. Reutov, *J. Organometallic Chem.*, 1976, **110**, 49.
[854] R. Appel, F. Knoll, and H. Veltmann, *Angew. Chem. Internat. Edn.*, 1976, **15**, 315.
[855] G. Pourcelot and J.-M. Cense, *Bull. Soc. chim. France*, 1976, 1578.
[856] C. C. Hinckley, *Proc. Rare Earth Res. Conf.*, 10*th*, 1 *Conf*-730402-*P*1, 1973, 362 (*Chem. Abs.*, 1976, **85**, 37 682).
[857] K. Tori, *Kagaku To Kogyo* (*Tokyo*), 1975, **28**, 642 (*Chem. Abs.*, 1976, **84**, 81 827).
[858] M. R. Willcott and R. E. Davis, *Proc. Rare Earth Res. Conf.*, 10*th*, 1 *Conf*-730402-*P*1, 1973, 405 (*Chem. Abs.*, 1976, **85**, 62 268).
[859] I. Mohyla, Z. Ksandr, and M. Hajek, *Chem. Listy*, 1975, **69**, 1148 (*Chem. Abs.*, 1976, **84**, 36 768).
[860] M. Tasumi, *Kagaku* (*Kyoto*), *Zokan*, 1975, **66**, 227 (*Chem. Abs.*, 1976, **84**, 97 188).
[861] A. J. Vega and D. Fiat, *Mol. Phys.*, 1976, **31**, 347.
[862] R. M. Golding, R. O. Pascual, and L. C. Stubbs, *Mol. Phys.*, 1976, **31**, 1933.
[863] B. Sozanieki, *Radiospektrosk. Ciala Stalego*, 1975, 233 (*Chem. Abs.*, 1976, **85**, 168 901).

a 2.5 atom % solution of sodium in NH_3. Both liquids had identical spin–lattice relaxation times except in the interval 140—160 K. In the dielectric region, *ca.* 150 K, a deep minimum of T_1 was observed for the sodium–liquid ammonia solutions. Numerical estimates of T_1 and of the activation energy suggest the validity of the theory as a metal with metallic sodium atoms in the conducting region and with dissociation of the sodium atoms at 150 K.[864] The theory of the pseudocontact contribution to the n.m.r. screening constants in t_2^1 and t_2^5 transition-metal ion complexes has been developed.[865] ^1H Spin–lattice relaxation times have been determined as a function of magnetic field strength for a series of paramagnetic transition-metal complexes chosen so that for some the electron spin relaxation times (τ_e) fall in the Redfield limit ($\tau_e \gg \tau_r$) while for others $\tau_e \ll \tau_r$ (τ_e = rotational correlation time). When $\tau_e \gg \tau_r$, τ_r dominates nuclear relaxation and the experimental results can be readily explained by Redfield theory. When $\tau_e \gg \tau_r$, current theory predicts that the nuclear T_1 values do not get longer as H_0 decreases. This was not found but the discrepancy was explained. $[Fe(S_2CNR_2)_2X]$, $[V(acac)_3]$, $[Mo(acac)_3]$, and $[Mn(acac)_3]$ were chosen as examples.[866] ^{13}C Spin–lattice relaxation times have been measured for $[M(acac)_n]$. The results were interpreted as due to the relaxation process being dominated by dipolar interactions with the unpaired electron spin residing on the ligand. A simple analysis of ligand-centred dipolar relaxation was used to quantify its importance and it was demonstrated that for spin densities of greater than 10^{-3}, unpaired electrons may be significant.[867] Further work, also using spin–spin relaxation data, examined the effects of fast methyl group internal motion and good agreement was obtained between theory and experiment. In all cases electron–nuclear dipolar coupling dominates T_1 whereas the hyperfine contribution can be important for T_2.[868]

An n.m.r. study has been carried out on the state of Ti^{III} and V^{II} ions in aqueous solutions of inorganic acids.[869] The Evans method has been applied to V^{IV} gluconate complexes.[870] ^{13}C N.m.r. data have been reported for $[(\eta^5\text{-Bu}^nC_5H_4)_2M]$ (M = V, Cr, Co, or Ni) in order to investigate hyperconjugation in the n-butyl group shifted by unpaired electrons. The shifts were compared with $J(^{13}C, ^{199}Hg)$.[871] A ^1H chemical shift range of 500 p.p.m. was found for $[(\eta^5\text{-1,3-R,Ph-}C_5H_3)_2M]$ (M = V, Cr, Co, or Ni; R = Me or Ph).[872] ^1H and ^{13}C n.m.r. spectra have been recorded for some mono- and di-alkylated metallocenes of V, Cr, Co, and Ni. Experimental conditions as well as new principles of peak assignment were discussed. This leads to a full assignment and a systematic arrangement of ^1H and ^{13}C chemical shifts which allows a

[864] V. M. Berezov, V. D. Korepanov, V. S. Romanov, and B. M. Khabibullin, *Pis'ma Zhur. eksp. i teor. Fiz.*, 1976, **24**, 138 (*Chem. Abs.*, 1976, **85**, 169 324).
[865] R. M. Golding, R. O. Pascual, and J. Vrbancich, *Mol. Phys.*, 1976, **31**, 731.
[866] D. T. Pegg, D. M. Doddrell, M. R. Bendall, and A. K. Gregson, *Austral. J. Chem.*, 1976, **29**, 1885.
[867] D. M. Doddrell, D. T. Pegg, M. R. Bendall, H. P. W. Gottlieb, A. K. Gregson, and M. Anker, *Chem. Phys. Letters*, 1976, **39**, 65.
[868] D. M. Doddrell, D. T. Pegg, M. R. Bendall, and A. K. Gregson, *Chem. Phys. Letters*, 1976, **40**, 142.
[869] A. N. Glebov and Yu. I. Sal'nikov, *Sb. Aspirantsk. Rabot. Kazan. Un-T. Estestv. N. Biol. Khimiya*, 1975, 191 (*Chem. Abs.*, 1976, **85**, 101 766).
[870] R. Good and D. T. Sawyer, *Inorg. Chem.*, 1976, **15**, 1427.
[871] K. Eberl, F. H. Köhler, and L. Mayring, *Angew. Chem. Internat. Edn.*, 1976, **15**, 554.
[872] F. H. Köhler and G. E. Matsubayashi, *Chem. Ber.*, 1976, **109**, 329.

convenient identification of paramagnetic complexes in solution and the determination of their structure.[873]

The use of [Cr(acac)$_3$] has proved to be capable of probing the solution structure of organic liquids. The theory of describing relaxation effects has been presented. It was assumed that the nuclear spin–lattice relaxation mechanism is due to the modulation of electron–nuclear dipole–dipole interaction. It was found that the jump diffusion model for translational motion explains the experimental data with jump distances corresponding to the diameter of the solvent molecules.[874] ^{1}H Spin–lattice relaxation times have been measured for several solute and solvent molecules in the presence of [Cr(acac)$_3$]. Two models were necessary to explain the results. In the first, T_1 is determined by the different correlation times of solute molecules. This corresponds to a case in which there is no detectable second co-ordination sphere and leads to the prediction that T_1 will be independent of concentration. Acetone, $CHCl_3$, and CH_2Cl_2 show this behaviour. The second model involves molecules in a second discrete co-ordination sphere and leads to the prediction that above a certain concentration, T_1 will vary linearly with solute concentration. MeOH shows this behaviour. Calculations showing good agreement with the observed T_1 were found for a model with a co-ordination number of eight, a Cr—Me separation of 7 Å and an equilibrium constant for displacing solvent ($CHCl_3$) of *ca.* 10. It was suggested that these results lead to a criterion for second co-ordination sphere formation compatible with the second co-ordination sphere requirements for an 'interchange' type of ligand exchange mechanism.[875] ^{13}C Spin–lattice relaxation times of organic compounds in the presence of β-diketonates of chromium, manganese, iron, nickel, and gadolinium have been reported and discussed. Hydrogen-bonding metal–ligand co-ordination and electrostatic effects modulated by contributing steric effects were indicated as causes for substrate–chelate complexation. When no specific complex occurs, the observed relaxation results from electron–nuclear dipole–dipole interaction modulated by translational motion and also, under some conditions, electronic relaxation. Specific complexation or orientation of the substrate leads to enhanced relaxation.[878] The ^{11}B n.m.r. signal of [Cr(BH$_4$)$_2$,-2THF] is at δ 96.5 and 375 Hz broad.[879]

From a comparison of ^{1}H and ^{13}C T_1 and T_2 in the presence of Mn^{2+} it was concluded that in many cases the dipolar term is not the dominant contributor to line broadening. Hence selective broadening experiments to determine the site of Mn^{2+} binding to small ligands based on an r^{-6} dependence between Mn^{2+} and the affected nucleus are not generally applicable. Binding sites and distances may be established from selective T_1 measurements, but in this case it must be established that the predominant dipolar interaction contributing to relaxation is that between the paramagnetic ion and the affected nucleus and that other closer interactions from the unpaired spin density on the ligand do not contribute importantly.[880] ^{13}C Relaxation rates have been measured for aqueous solutions

[873] F. H. Köhler, *J. Organometallic Chem.*, 1976, **110**, 235.
[874] J. G. Hexem, U. Edlund, and G. C. Levy, *J. Chem. Phys.*, 1976, **64**, 936.
[875] S. O. Chan and D. R. Eaton, *Canad. J. Chem.*, 1976, **54**, 1332.
[878] G. C. Levy, U. Edlund, and C. E. Holloway, *J. Magn. Resonance*, 1976, **24**, 375.
[879] H. Nöth and M. Seitz, *J.C.S. Chem. Comm.*, 1976, 1004.
[880] W. G. Espersen and R. B. Martin, *J. Phys. Chem.*, 1976, **80**, 161.

of imidazole at 31 °C at various pH values. The influence of Mn^{2+} and Cu^{2+} on these relaxation rates was examined. The results indicate that ^{13}C transverse relaxation rates of imidazole in the presence of these metals are completely dominated by the scalar mechanism and hence do not give the geometry.[881] The effect of Mn^{2+} on the relaxation of non-polar molecules, *e.g.* benzene, has been examined.[882] Correlation times and electron spin relaxation parameters for iron and manganese(II) probes have been discussed. A dynamic model which takes into account the various temperature-dependent processes was suggested. The correlation times for the processes which modulate IS interactions were analysed with reference to ^{13}C and ^{1}H relaxation rates. The limiting case, in which the longitudinal relaxation is determined by dipolar interaction whereas the transverse relaxation depends on scalar interaction, was examined. The treatment showed a distribution of τ_S values and the effect of this on T_1 and T_2. The model took into account the ligand (and solvent) exchange mechanism, the presence of solvent-separated species, different binding sites, and the influence of unpaired spin density. Experimental ^{1}H and ^{13}C T_1 and T_2 data were discussed in terms of the proposed model.[883] The ^{1}H n.m.r. spectra of *mer*-$[TcCl_3(PMe_2Ph)_3]$ and its rhenium analogues have been compared.[884]

The ^{1}H n.m.r. spectra of $[Bu^n_3PBz]^+[Fe(CN)_6]^{3-}$ have been analysed to determine the degree of association.[885] The ^{1}H n.m.r. spectra of co-ordinated axial imidazole ligands in low-spin ferric bisimidazole complexes with synthetic porphyrins have been resolved and assigned. The dominant dipolar shifts for certain porphyrin positions have allowed the quantitative separation of the observed imidazole shifts into their dipolar and contact contributions. The contact shifts were shown to arise from directly delocalized π spin densities. Comparison with Hückel MO calculations suggests that these shifts arise from π^* charge transfer from iron to imidazole.[886] The ^{1}H chemical shifts of the dicyanide complex of some haemins have been determined in various solvents. Simultaneous study of the high- and low-spin signals revealed a slow exchange, probably related to cyanide complexing.[887] The ^{1}H n.m.r. spectra of co-ordinated axial imidazoles in a series of low-spin ferric bisimidazole complexes with natural porphyrins have been assigned. The imidazole spectra for the bis complexes are essentially the same as those for synthetic porphyrins, with the hyperfine shifts exhibiting comparable contributions from dipolar and contact interactions. The contact contribution reflects spin transfer into a vacant imidazole π-orbital. The spectra of mono- and bis-imidazole complexes concur in predicting that only the 2-H and 5-CH_2 signals of an axial histidine are likely to resonate clearly outside the diamagnetic 0—10 p.p.m. from the Me_4Si region in haemoproteins.[888] ^{1}H Chemical

[881] R. E. Wasylishen and M. R. Graham, *Canad. J. Chem.*, 1976, **54**, 617.
[882] Yu. I. Gladkin, V. A. Rogov, and Yu. N. Molin, *Izvest. Akad. Nauk S.S.S.R., Ser. khim.*, 1975, **10**, 2330.
[883] F. Laschi, E. Tiezzi, and G. Valensin, *Gazzetta*, 1976, **106**, 359.
[884] U. Mazzi, G. de Paoli, P. di Bernardo, and L. Magon, *J. Inorg. Nuclear Chem.*, 1976, **38**, 721.
[885] S. Papp, P. Kvintovics, F. Nagy, jun., and A. Vertes, *Acta Chim. Acad. Sci. Hung.*, 1975, **87**, 343 (*Chem. Abs.*, 1976, **84**, 171 855).
[886] J. D. Satterlee and G. N. La Mar, *J. Amer. Chem. Soc.*, 1976, **98**, 2804.
[887] W. Schoessler, K. Pommerening, R. Hintsche, and P. Mohr, *Abh. Akad. Wiss. DDR Int. Symp. Strukt. Funkt. Erythrozyten, 7th* (1973), 1975, 165 (*Chem. Abs.*, 1976, **85**, 1473).
[888] G. N. La Mar, J. S. Frye, and J. D. Satterlee, *Biochim. Biophys. Acta*, 1976, **428**, 78.

shifts, T_1 and T_2, have been measured for low-spin complexes of Fe^{III}protoporphyrin(IX) dimethyl ester and Fe^{III}tetraphenylporphyrin in $CHCl_3$ between 203 and 253 K. Deviations between measured and calculated chemical shifts were observed for some protons (*e.g.* methyl group) in Fe^{III}protoporphyrin but not for the pyrrole protons in Fe^{III}tetraphenylporphyrin. These deviations were analysed theoretically and it was found that higher-order terms of the pseudocontact shift are too small to explain the deviations observed. Instead a considerable temperature dependence of the hyperfine coupling constant, *A*, for the protons of Fe^{III}-protoporphyrin had to be assumed in order to explain the deviations observed, whereas *A* was found to be independent of temperature for Fe^{III}tetraphenylporphyrin. This result was confirmed by independent relaxation time measurements and the values for hyperfine coupling constants extracted from the two methods were found to be in excellent agreement.[889] The ^1H n.m.r. spectrum of $[Fe(X\text{-}sal_2trien)]^+$ has been reported.[890] ^{13}C Chemical shifts and ^1H relaxation measurements have been measured for $[(Et_2NCS_2)_2FeX]$ and the results interpreted in terms of current relaxation theory. For X = Cl, the zero field splitting parameters are sufficiently small to yield electronic relaxation times longer than the overall molecular correlation time, and good agreement was obtained between experimental and theoretical relaxation times. The agreement was poor for X = Br or I. Although the electronic relaxation is a complex process, it was shown that one electronic relaxation time may dominate nuclear relaxation processes. ^1H and ^{13}C contact shifts are in good agreement with the proposal that it is the π-electronic framework of the ligand that is responsible for the transmission of unpaired spin density.[891, 892] The ^1H n.m.r. spectra of $[Fe_2S_2(SR)_2]^{2-}$ has been monitored as a function of temperature and the shifts were found to correlate with susceptibility.[893]

The temperature dependence of the ^1H and ^{13}C n.m.r. spectra of $[(\eta^5\text{-}Pr^iC_5H_4)_2Co]$ does not follow the Curie law, and it was suggested that two forms of the complex are present.[894] The ^1H n.m.r. line broadening and shift effects have been investigated for 5,7,7,12,14,14-hexamethyl-1,4,7,11-tetra-azacyclodeca-4,11-dienecobalt(II) in H_2O, MeOH, and Me_2NCHO. None of the systems show chemical exchange effects on the ^1H n.m.r. linewidth and shifts. The observations were quantitatively accounted for by the sum of inner-sphere dipolar, scalar, and outer-sphere dipolar relaxation mechanisms. For the copper analogue the temperature dependence of the shift requires a five–six co-ordination equilibrium.[895]

The porphyrin ^1H n.m.r. shifts of low-spin tetra-*p*-tolylporphinatocobalt(II) have been investigated upon the addition of various aromatic π-acceptors and π-donors. The interactions were described by the formation of 1 : 1 adducts and in the case of trinitrobenzene, thermodynamic parameters were obtained. Sizeable changes in dipolar shifts upon complex formation were observed with

889 E. V. Goldammer, H. Zorn, and A. Daniels, *J. Magn. Resonance*, 1976, **23**, 199.
890 M. F. Tweedle and L. J. Wilson, *J. Amer. Chem. Soc.*, 1976, **98**, 4824.
891 D. A. Doddrell, M. R. Bendall, and A. K. Gregson, *Austral. J. Chem.*, 1976, **29**, 55.
892 D. T. Pegg and D. M. Doddrell, *Austral. J. Chem.*, 1976, **29**, 1869.
893 W. O. Gillum, R. B. Frankel, S. Foner, and R. H. Holm, *Inorg. Chem.*, 1976, **15**, 1095.
894 F. H. Köhler, *J. Organometallic Chem.*, 1976, **121**, C61.
895 L. L. Rusnak and R. B. Jordan, *Inorg. Chem.*, 1976, **15**, 709.

the magnetic anisotropy increasing markedly in the presence of π-acceptors and decreasing significantly in the presence of a donor. The increase in magnetic anisotropy requires that the adducts are peripheral π-compounds. Similar changes in porphyrin linewidths were also recorded. The changes in magnetic anisotropy and ^1H relaxation times were analysed in detail in terms of the changes in electronic structure of the metal and in metal–porphyrin bonding. The changes in electronic structure of the cobalt were shown to arise primarily from changes in the metal–porphyrin σ-bond.[896] Subsequently attention was focused on the trinitrobenzene and the induced shift was separated into its ring current and dipolar contributions by determining the shift for the related complex with a diamagnetic nickel(II) porphyrin. By assuming parallel π-planes and scaling the known dipolar shift and linewidth for the *ortho*-hydrogen of the porphyrin by the known ratio of geometric factors to the computed geometric factors in various configurations, the solution structure of the 1 : 1 adduct was uniquely determined. This determination required the use of both the shift and relaxation geometric factors. The structure is consistent with a peripheral complex indicating purely π–π-contacts with the interplane spacing 3.2 ± 0.2 Å and with the trinitrobenzene centred approximately over a pyrrole.[897] Weak intermolecular complexes of cobalt(II) and nickel(II) mesoporphyrin dimethyl ester with 1,4-naphthoquinone, vitamin K_3, vitamin K_1, and ubiquinone 30 have been detected by ^1H n.m.r. spectroscopy. The relative orientation of components of the complexes was discussed.[898]

The complete ^1H n.m.r. spectrum of [Co(salen)] has been reported. Evidence was presented that the pattern of isotropic shifts obtained arises from a $(3d_{z^2})^1$ ground-state electronic configuration. Comparison was made with previous n.m.r. work from which a $(3d_{yz})^1$ ground state was postulated. The temperature dependence of the isotropic shifts for each proton was presented. These suggest a possible spin equilibrium for [Co(salen)] at room temperature in solution.[899] The ^{13}C n.m.r. spectra of [(pyridine *N*-oxide)$_6$M] (M = Co or Ni) have been recorded and the isotropic shifts compared with the ^1H n.m.r. isotropic shifts of the same complexes. Whereas the ^1H contact shift pattern is the same for the two metals, the ^{13}C contact shift pattern is not the same and cannot be unequivocally related to the ^1H n.m.r. contact shifts.[900] ^1H and ^{19}F n.m.r. spectra of [M(PhCOCH$_2$COCF$_3$)$_2$(pyridine)$_2$] (M = Co or Ni) have been reported.[901] The Evans method has been applied to [Co{P(OEt)$_3$}$_4$].[902] ^1H N.m.r. contact shifts of [M(S$_2$PR$_2$)$_2$L] (M = Co or Ni; L = pyridine *N*-oxide or Me$_2$NCHO) have been reported. The results indicate that a π-mechanism is operative, and comparison of the experimental data with the theoretically calculated spin densities suggests that the highest bonding orbital is involved in the spin-transfer mechanism. The shift pattern of [Co(S$_2$PR$_2$)$_2$L] is influenced by a dipolar contribution.[903] The

[896] G. P. Fulton and G. N. La Mar, *J. Amer. Chem. Soc.*, 1976, **98**, 2119.
[897] G. P. Fulton and G. N. La Mar, *J. Amer. Chem. Soc.*, 1976, **98**, 2124.
[898] L. Ford, H. A. O. Hill, B. E. Mann, P. J. Sadler, and R. J. P. Williams, *Biochim. Biophys. Acta*, 1976, **430**, 413.
[899] C. Srivanavit and D. G. Brown, *J. Amer. Chem. Soc.*, 1976, **98**, 4447.
[900] I. Bertini, C. Luchinat, and A. Scozzafava, *Inorg. Chim. Acta*, 1976, **19**, 201.
[901] A. P. Gulya, D. G. Batyr, and E. V. Iorga, *Zhur. neorg. Khim.*, 1976, **21**, 1038.
[902] G. Schiavon, S. Zecchin, G. Zotti, and G. Pilloni, *Inorg. Chim. Acta*, 1976, **20**, L1.
[903] R. N. Mukherjee, M. S. Venkateshan, M. D. Zingde, and M. M. Dhingra, *J. Inorg. Nuclear Chem.*, 1976, **38**, 689.

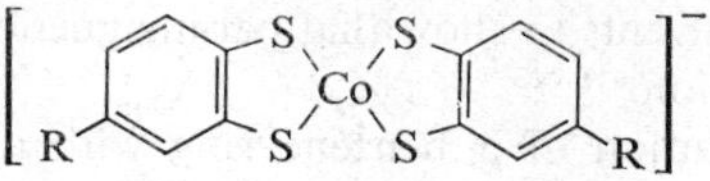

(113)

^{1}H n.m.r. spectra of (113; R = H or Me) have been measured. Comparison of the hyperfine interaction constants with MO calculations support a ground state $(a_g)^1(b_{3g})^1$ in which the a_g orbital is mainly metal-based d_{z^2} and the b_{3g} orbital is half metal-based d_{yz} and half delocalized into the π-system of the aromatic ligands.[904]

The Evans method has been applied to the mixture of Ni^{II} and $Et_3Al_2Cl_3$.[905] The distribution of unpaired electron-spin density in a series of ring-substituted nickelocenes, specifically designed to test spin delocalization mechanisms, has been measured by ^{1}H n.m.r. methods. Contact shifts of the cyclopentadienyl ring hydrogen atoms are almost constant (250 ± 19 p.p.m.) over the series of seven variously substituted nickelocenes, while in contrast the β-hydrogen atoms on the α-carbon atom directly attached to the cyclopentadienyl rings exhibit contact shifts which range from 0 to −312 p.p.m. These β contact shifts were shown to vary as ($\cos^2 \phi$), where ϕ is the dihedral angle between the cyclopentadienyl p_z orbital and the C_α—H_β bond. In two nickelocenes with rigid substituents, large contact shifts for a hydrogen atom remote from the cyclopentadienyl ring were observed when the W arrangement of the remote C—H bond and a cyclopentadienyl p_z orbital is satisfied. For one such hydrogen atom, a remote shift of −95.6 p.p.m. was observed while its geminal counterpart experienced no significant contact shift.[906]

Electron spin distribution and hyperfine coupling parameters for the α- and β-carbon atoms of the alkyl group in [alkylamineNi(acac)$_2$] complexes have been determined using ^{1}H, ^{13}C, and ^{14}N contact shift measurements. T_1 Measurements were used to determine T_{1e} of *ca.* 2×10^{-10} s and a reorientation correlation time of 4×10^{-11} s.[907, 908] The overall spin density distribution in [Ni(acac)$_2$] complexes with pyridines determined by ^{1}H and ^{13}C n.m.r. spectroscopies occurs mainly by a σ-electron mechanism. The pronounced contribution of the π-electron delocalization of unpaired electron density was also found. The correlation between proton hyperfine coupling constants in paramagnetic compounds and spin–spin coupling constants in isostructural molecules is incorrect for ^{13}C nuclei.[909] The ^{1}H n.m.r. spectra of paramagnetic complexes of Ni^{2+} with *N*-hydroxypropylsalicylaldimine have been measured and assigned.[910] The measurement of ^{1}H n.m.r. spectra of [(*N*-n-alkylsalicylaldiminato)$_2$Ni] as a function of solvent, concentration, and temperature have given information

904 P. J. Van der Put and J. Reedijk, 'Proceedings of the 16th International Conference on Coordination Chemistry', 1974, R78 (*Chem. Abs.*, 1976, **85**, 54 250).
905 R. T. Pennington and J. E. Howell, *J. Organometallic Chem.*, 1976, **113**, 75.
906 M. T. Scroggins, M. F. Rettig, and R. M. Wing, *Inorg. Chem.*, 1976, **15**, 1381.
907 C. Chachaty, A. Forchioni, and J. C. Ronfard-Haret, *Report CEA-CONF*-3074, 1975 (*Chem. Abs.*, 1976, **85**, 85 146).
908 C. Chachaty, A. Forchioni, and J. C. Ronfard-Haret, *Mol. Phys.*, 1976, **31**, 325.
909 A. A. Obynochnyi, R. Z. Sagdeev, Yu. N. Molin, and A. I. Rezvukhin, *Izvest. Sib. Otd. Akad. Nauk S.S.S.R., Ser. khim. Nauk*, 1975, **6**, 114 (*Chem. Abs.*, 1976, **84**, 113 889).
910 H. A. O. Hill and N. Zarb-Adami, *J. Inorg. Nuclear Chem.*, 1976, **38**, 2189.

on equilibria in inert solvents to show that paramagnetic tetrahedral species may also be formed in solution.[911]

The effect of substitution of a benzene ring with a pyrazole ring in paramagnetic [Ni(*S*-thiopyrazolonate)$_2$] on the configuration of the co-ordination junction as well as the distribution of unpaired electron density was investigated by contact shifts in ^{1}H n.m.r. spectra. Substitution with a pyrazole ring resulted in a strong stabilization of the tetrahedral configuration of the complex. It was concluded that along the benzene ring, $\frac{1}{15}$ to $\frac{1}{30}$ of an unpaired electron was delocalized.[912] The temperature dependence of paramagnetic shifts in complexed pyridine and β- and γ-picoline with Ni^{II} ethylxanthate or [{(EtO)$_2$PS$_2$}$_2$Ni] has been measured. The constants of isotropic electron–nucleus spin–spin hyperfine splitting and electron densities have been calculated. A small dependence of the degree of spin-density delocalization on the nature of the chelate and amine was observed.[913] The ^{1}H n.m.r. spectra of [Ni{(EtO)$_2$PS$_2$}$_2$(3-aminopyridine)] have been measured. The results indicate spin distribution through the σ-structure. There does not seem to be any correlation between the spin density on the α-protons and the basicity of the substituted pyridine ligand.[914]

When $H_2N(CH_2)_nNH_2$ (n = 2—9) interacts with Cu^{2+} all the ^{13}C resonances are affected.[915] Selective broadening of ^{1}H and ^{13}C ligand resonances in n.m.r. spectra by Cu^{II} has been evaluated as a method of determining Cu^{II} binding sites in aliphatic and aromatic nitrogen and carboxylate ligands. For many chelating ligands, such as peptides, the complexes present at the very high ligand to Cu^{II} molar ratios of a selective broadening experiment are different from those found at stoicheiometric concentration. Measurements of T_1 and T_2 indicate that neither ^{1}H nor ^{13}C resonances are broadened predominantly by a dipolar mechanism so that interpretations based on an r^{-6} dependence are invalid. For protons the broadening is determined mainly by scalar interaction and increases in importance as the donor group is varied from carboxylate oxygen to π-bonding amines to aliphatic amines.[916] In a related study, earlier interpretations of ^{1}H n.m.r. line-broadening effects in Cu^{II} amino-acid and peptide systems were challenged as follows: (i) the Cu^{II}-containing entities involved in the line broadening have not been properly identified; (ii) the exchange lifetime broadening has not been adequately distinguished from other relaxation effects; and (iii) the relative contributions of contact and dipolar interactions to the relaxation have not been established. The interaction of Cu^{II} with glycine was examined and various species were found. A model was developed to fit the data and slow exchange was found. It was concluded that ^{1}H n.m.r. broadening is not generally useful for obtaining structural information about such compounds in solution.[917] Isotropic n.m.r. chemical shifts have been measured for imidazole

[911] J. Csaszar, T. Szabo, and G. Dombi, *Acta Chim. Acad. Sci. Hung.*, 1976, **88**, 1 (*Chem. Abs.*, 1976, **84**, 157 737).
[912] L. E. Nivorozhkin, L. E. Konstantinovskii, V. I. Minkin, O. A. Osipov, A. D. Garnovskii, V. P. Kurbatov, and I. Ya. Kvitko, *Zhur. neorg. Khim.*, 1975, **20**, 3012.
[913] N. V. Mel'nikova and A. T. Pilipenko, *Koord. Khim.*, 1975, **1**, 1703.
[914] A. T. Pilipenko, N. V. Mel'nikova, and A. L. Rozenfel'd, *Ukrain. khim. Zhur.* (*Russ. Ed.*), 1976, **42**, 981 (*Chem. Abs.*, 1976, **85**, 200 315).
[915] W. Voelter, G. J. Strobel, and U. Weser, *Bioinorg. Chem.*, 1975, **5**, 183.
[916] W. G. Espersen and R. B. Martin, *J. Amer. Chem. Soc.*, 1976, **98**, 40.
[917] J. K. Beattie, D. J. Fensom, and H. C. Freeman, *J. Amer. Chem. Soc.*, 1976, **98**, 500.

and derivatives with various copper acetylacetonates. The pseudocontact contribution to the isotropic shifts is negligibly small and the Fermi contact interaction is dominant in these complexes. The normal temperature effect and the appreciable spin delocalization suggest that the imidazole and substituted imidazole adducts with copper(chelate)$_2$ are all six-co-ordinate complexes. The magnitude of the contact shifts [Cu(hfac)$_2$] > [Cu(tfac)$_2$] > [Cu(acac)$_2$] follows the stability constants order.[918] ^{19}F N.m.r. chemical shifts of CuF_2,4MeCN,4IF_5 shows broad shifted signals due to paramagnetic Cu^{II}.[919]

Compounds of the Lanthanides and Actinides.—Current theories of the origin of lanthanide-induced n.m.r. shifts have been examined and compared with one another and with experimental results in an attempt to understand the temperature dependence of the shifts. It was concluded that shifts arising from a pseudo-contact mechanism should in general show a T^{-2} behaviour, as predicted by Bleaney. In most situations with a fixed molecular geometry, theory does not predict a substantial T^{-1} component of the shift, but in some instances a fortuitous combination of crystal field parameters can lead to additional T^{-n} components ($n \geqslant 3$) which cause the overall shift to appear to be a linear function of T^{-1} with a large intercept at $T^{-1} = 0$. Extension of Bleaney's method to the positive T^{-3} term shows it to be negligible compared to that in T^{-2} for systems with $J = \frac{3}{2}$ and relatively small (*ca.* 10% for larger J). In the absence of detailed experimental studies of crystal field parameters in solution it is likely that for the majority of lanthanide adduct systems, the most realistic description of ^{1}H shifts is in terms of the well-established T^{-2} dependence.[920, 921] The errors in calculation of lanthanide-induced n.m.r. shifts have been evaluated as a function of substrate–paramagnetic ion distance.[922] The effects of rotational and site averaging on shifts and relaxation rates induced by lanthanides have been examined.[923] The limitations of lanthanide shift reagents for the determination of conformations of non-rigid molecules have been detailed.[924] The sources of error in computer analysis of induced axial pseudocontact shifts involves assessing the goodness of fit between calculated and observed shifts.[925] The use of shift reagents in polyfunctional systems has been examined.[926]

The ^{1}H n.m.r. spectra of [$(C_5H_5)_3MCNC_6H_{11}$] (M = Ce, Th, Nd, U, Sm, Pu, Eu, or Am) have been measured. It was found that the large dipolar shifts of the cyclohexyl ring decrease in the order Eu > Nd > Ce > Sm, whereas the spectra of the actinide complexes are poorly shifted. Only the uranium complex displays a spectrum, which is partially first order, similar to the samarium complex. The large susceptibility and shifting power of europium result from the close

[918] Sung-Mao Wang and Jung-Tso Chen, *J. Chin. Chem. Soc.* (*Taipei*), 1975, **22**, 179 (*Chem. Abs.*, 1976, **84**, 10 718).
[919] J. A. Berry, D. W. A. Sharp, and J. M. Winfield, *Inorg. Nuclear Chem. Letters*, 1976, **12**, 869.
[920] E. W. Stout, jun. and H. S. Gutowsky, *J. Magn. Resonance*, 1976, **24**, 389.
[921] E. W. Stout, jun. and H. S. Gutowsky, *U.S. NTIS, Ad Rep. AD-A*023232, 1976, 26 pp. (*Chem. Abs.*, 1976, **85**, 133 595).
[922] A. J. Rafalski and J. Barciszewski, *Uniw. Adama Mickiewicza Poznaniu, Wydz. Mat., Fiz. Chem.* (*Pr.*), *Ser. Chem.*, 1975, **18**, 161 (*Chem. Abs.*, 1976, **84**, 120 635).
[923] R. E. Lenkinski and J. Reuben, *J. Amer. Chem. Soc.*, 1976, **98**, 4065.
[924] G. R. Sullivan, *J. Amer. Chem. Soc.*, 1976, **98**, 7162.
[925] C. C. Hinckley and W. C. Brumley, *J. Amer. Chem. Soc.*, 1976, **98**, 1331.
[926] J. K. M. Sanders, *Proc. Rare Earth Res. Conf.*, 10*th*, 1 *Conf*-730402-*P*1, 1973, 380 *Chem. Abs.*, 1976, **85**, 62 344).

energetic proximity of the paramagnetic state $^7F_1$ (350 cm^{-1}) which is partially populated in the temperature region of n.m.r. measurements and from second-order Zeeman interaction with higher excited states. For the actinide complexes, contrary to the very little shifted C_6H_{11} signals, shifts for the cyclopentadienyl hydrogens are large and comparable to those of the lanthanide complexes. Thus a large contact contribution must be present also in actinide complexes.[927]

N.m.r. spectra and shift reagent capabilities have been studied for *meso*-tetra-arylporphyrin complexes of the entire series of tervalent lanthanide complexes in non-aqueous solution and of lanthanide complexes of water-soluble porphyrins which show potential for use as shift reagents in aqueous solution. The Yb^{III} tetraphenylporphyrin complex has *ca.* 2.2 times the dipolar shifting ability of the prosthetic group in metmyoglobin cyanide. A study of the temperature dependence of the resonance shifts provides another exception to the Bleaney theory, but they are consistent with other theories. The Yb^{III} tetraphenylporphyrin acetylacetonate complex acts as a downfield shift reagent towards alcohols and heterocyclic amines. Lu^{III} tetraphenylporphyrin acetylacetonate serves as a diamagnetic ring-current shift reagent. For the porphyrin, the aryl proton resonance shifts are consistent with a dominant dipolar interaction with the metal atom situated significantly out of the porphyrin plane by 1.8 Å (Eu) and 1.6 Å (Yb).[928, 929] The n.m.r. spectra of Pr^{III} complexes with the terdentate donors, diethylenetriamine, *N*-(2-aminoethyl)-1,3-diaminopropane, and 3,3′-diaminodipropylamine, and the quadridentate donor, triethylenetetramine, showed two methylene proton resonances for each non-equivalent methylene group in the free ligand. A model was proposed to explain methylene proton non-equivalence. The non-equivalence demonstrated by the two methylene protons in polydentate N-donor chelates was accounted for by using the axial model with allowance for rapid exchange of several geometrical and/or conformational isomers, providing that the magnetic (symmetry) axis of each isomer is defined by the orientation of the amine nitrogen atoms in the co-ordination sphere.[930] The isotropic shifts of $[Ln(tren)_n]^{3+}[ClO_4]_3^-$ (Ln = La, Pr, Nd, Gd, Er, or Y) have been examined as a function of temperature. For $[Nd(tren)_2]^{3+}$ the isotropic shift decreases rather than increases with decreasing temperature.[931] ^1H and ^{13}C lanthanide-induced shift n.m.r. data obtained for alanine indicated that the ^1H n.m.r. shifts contain a significant, though not dominant, contact contribution. The calculated contact and pseudocontact contributions to ^1H n.m.r. shifts were presented for several rare earth ions. Even the qualitative use of Gd^{3+} as a line-broadening agent must be used with caution in view of the fact that the contact shift of the α-carbon atom is nearly twice as large as that for the much closer carbonyl carbon in alanine.[932]

927 R. Von Ammon and B. Kanellakopulos, 'Proceedings of the 16th International Conference on Coordination Chemistry', 1974, 2.42A (*Chem. Abs.*, 1976, **85**, 39 060).
928 W. De W. Horrocks, jun, C.-P. Wong, R. F. Venteicher, and E. G. Hove, *Proc. Rare Earth Res. Conf.*, 12*th*, 1976, **1**, 241 (*Chem. Abs.*, 1976, **85**, 169 325).
929 W. De W. Horrocks, jun. and C.-P. Wong, *J. Amer. Chem. Soc.*, 1976, **98**, 7157.
930 J. H. Forsberg and G. Liu, *Proc. Rare Earth Res. Conf.*, 12*th*, 1976, **1**, 140 (*Chem. Abs.*, 1976, **85**, 133 625).
931 M. F. Johnson and J. H. Forsberg, *Inorg. Chem.*, 1976, **15**, 734.
932 A. D. Sherry, *Proc. Rare Earth Res. Conf.*, 12*th*, 1976, **1**, 250 (*Chem. Abs.*, 1976, **85**, 133 626).

The suitability of edta chelates of tervalent lanthanides as aqueous shift and relaxation reagents for carboxylic substrates has been established in a study of their effects on the ^{1}H n.m.r. spectrum of acetate. The Pr^{III} chelate forms mainly a 1 : 1 acetato complex with a downfield intrinsic shift of 1.83 p.p.m. at 39 °C and a dissociation constant of the order of 0.2 mol l^{-1}, the latter depending on the ionic medium. The useful pH range is between 6 and 10, confined on the acidic side by the formation of acetic acid and on the basic side by the formation of hydroxy complexes of the chelates. Similar behaviour was observed for $[Yb(edta)]^{-}$ which induces upfield shifts (intrinsic shift 3.5 p.p.m.) and $[Gd(edta)]^{-}$ which causes line broadening.[933, 934] The lanthanide-induced shifts in the ^{1}H and ^{13}C n.m.r. spectra of the axially symmetric tris-complexes of pyridine-2,6-dicarboxylate and tris-(4-methyl-2,6-pyridinedicarboxylate) were compared with the purely dipolar models using geometric factors calculated from crystallographic data. The ^{1}H n.m.r. chemical shifts are consistent with the solid-state structure. The ^{13}C chemical shifts contain deviations that cannot be explained if Fermi contact contributions are neglected. The latter were evaluated by difference and correlate well with Golding's $\langle S_z \rangle$ factors. Contact shift contributions are important even for carbon atoms five bonds removed from the lanthanide. Variable-temperature experiments also substantiate the relative contributions of the contact and dipolar shifts computed for the various ^{1}H and ^{13}C nuclei *via* the difference procedure based on the solid-state structure.[935] The differences between the calculated and observed behaviour of lanthanide shift reagents for isoquinoline and *endo*-norborneol have been recorded and interpreted as contact shifts. It was found that $Yb(dpm)_3$ is the best reagent for pseudocontact shifts.[936] The effects of reagent concentration, resonance frequency, and sample temperature on the line broadenings induced by the lanthanide shift reagent $Yb(fod)_3$ in the ^{1}H n.m.r. spectrum of $Bu^{t}COMe$ have been investigated and analysed in detail. For comparison, the effects of other members of the $Ln(fod)_3$ series were also examined. It was found that a term containing the modulation by chemical exchange of the chemical shift difference between co-ordinated and free substrate molecules makes a significant contribution to the observed line broadening, leading to their non-linear dependence on the fraction of complexed substrate. Also the ratio between line broadenings of two different proton-containing groups is a non-linear function of this fraction. Analysis of the results yielded the relaxation rates of protons in the co-ordinated state and their residence time, as well as the electron spin relaxation time of the central lanthanide ion.[937]

The structure of the 1 : 1 ion pairs $[Et_4N]^{+}[Ln(S_2CNR_2)_4]^{-}$ (Ln = Pr, Tb, Dy, Ho, Er, Tm, or Yb) in acetonitrile was determined using n.m.r. shift and relaxation ratios. Evidence of considerable ion clustering in solvents of lower electric permittivity was found.[938] Subsequently, the temperature dependence of

[933] J. Reuben and G. A. Elgavish, *Proc. Rare Earth Res. Conf.*, 12*th*, 1976, **1**, 25 (*Chem. Abs.*, 1976, **85**, 133 624).
[934] G. A. Elgavish and J. Reuben, *J. Amer. Chem. Soc.*, 1976, **98**, 4755.
[935] J. F. Desreux and C. N. Reilley, *J. Amer. Chem. Soc.*, 1976, **98**, 2105.
[936] O. A. Gansow, P. A. Loeffler, R. E. Davis, R. E. Lenkinski, and M. R. Willcott, *J. Amer. Chem. Soc.*, 1976, **98**, 4250.
[937] R. E. Lenkinski and J. Reuben, *J. Magn. Resonance*, 1976, **21**, 47.
[938] H. A. O. Hill, R. G. Roberts, D. Williams, and N. Zarb-Adami, *J.C.S. Faraday II*, 1976, **72**, 1267.

the shifts was investigated and interpreted in terms of both a T^{-1} and T^{-2} contribution. It was shown that a treatment due to Bleaney satisfactorily accounts for the T^{-2} contribution. It was also suggested that there is a significant contact contribution which is reflected in the T^{-1} term.[939]

The ^{13}C n.m.r. shift of $[(C_5H_5)_3UCl]$ in THF for $160 \leqslant T \leqslant 320$ K consists of a Curie–Weiss part with a Weiss temperature of 140 ± 10 K and a temperature-independent part of +59 p.p.m. with respect to TMS. The contact shift is -309 ± 120 p.p.m. at 290 K and is larger than the pseudocontact shift by about a factor of two.[940, 941] The ^1H n.m.r. chemical shift of [114; X = CH_2,

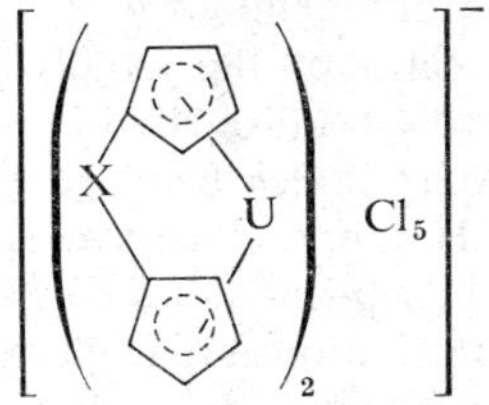

(114)

$SiMe_2$, or $(CH_2)_3$] has been recorded.[942] The ^1H n.m.r. chemical shift of $[U(NEt_2)_4]$ is remarkable in that in pentane and benzene, the CH_2 group is shifted downfield by *ca.* 12 p.p.m. but upfield by *ca.* 12 p.p.m. in ether and THF.[943] The ^{13}C n.m.r. spectrum of $[UCl_2\{HB(pyrazolyl)_3\}_2]$ shows two sets of signals in the ratio 2 : 1 indicating that the $[HB(pyrazolyl)_3]^-$ group is bidentate. There are no large chemical shifts consistent with the major contribution to the proton chemical shift being due to the pseudocontact mechanism which depends largely on distance. The spectrum of $[U(HBPz_3)_4]$ was assigned on the basis of a ten-co-ordinate uranium atom with two bidentate and two terdentate $[HB(pyrazolyl)_3]^-$ groups in the molecule.[944] In CD_3CN, the ^{13}C n.m.r. spectrum of $[U(NCS)_8]^{4-}$-$[NEt_4]_4{}^+$ shows two sets of $[N^{13}CS]^-$ signals and contact shifts with an electron density on the $[SCN]^-$ of -3.0×10^{-4}.[945] U^{IV} has been found to induce shifts in the ^1H n.m.r. spectra of some amino-acids.[946] The ^1H n.m.r. chemical shifts of $[(CF_3COCHCO_2Et)_4U]$ are sensitive to solvent and were interpreted in terms of contact shifts.[947]

5 Solid-state N.M.R. Spectroscopy

This section consists of three parts: 'Motions in Solids', 'Structure of Solids', and 'Molecules Sorbed onto Solids'.

939 H. A. O. Hill, D. Williams, and N. Zarb-Adami, *J.C.S. Faraday II*, 1976, **72**, 1494.
940 E. Fukushima and S. D. Larsen, *Chem. Phys. Letters*, 1976, **44**, 285.
941 H.-D. Amberger, *J. Organometallic Chem.*, 1976, **116**, 219.
942 C. A. Secaur, V. W. Day, R. D. Ernst, W. J. Kennelly, and T. J. Marks, *J. Amer. Chem. Soc.*, 1976, **98**, 3713.
943 J. G. Reynolds, A. Zalkin, D. H. Templeton, and N. M. Edelstein, *Inorg. Chem.*, 1976, **15**, 2498.
944 K. W. Bagnall, J. Edwards, and F. Heatley, *Transplutonium* 1975, *Proc. Int. Transplutonium Elem. Symp., 4th*, 1976, 119 (*Chem. Abs.*, 1976, **85**, 85147).
945 G. Folcher, H. Marquet-Ellis, P. Rigny, E. Soulié, and G. Goodman, *J. Inorg. Nuclear Chem.*, 1976, **38**, 747.
946 C. Neveu, G. Folcher, and A. M. Laurent, *J. Inorg. Nuclear Chem.*, 1976, **38**, 1223.
947 J. Paris and G. Folcher, *Canad. J. Chem.*, 1976, **54**, 1433.

Books and reviews published this year include 'Nuclear Magnetic Resonance in Solids', by J. Ebert and G. Seifert,[948] 'Metallic Shifts in N.M.R.', by G. C. Carter, L. H. Bennett, and D. J. Kahan,[949] 'Magnetic Resonance Studies of Hydrogen Bonding in Solids', by R. Blinc,[950] 'Spin Relaxation in Paramagnetic Crystals', by J. Stankowski,[951] 'Spin–Lattice Relaxation Theory in Paramagnetic Ionic Crystals', by L. K. Aminov,[952] 'N.M.R. Study of Intermolecular Interactions in Ion-exchange Resins', by V. V. Mank and O. D. Kurilenko,[953] 'The N.M.R. Study of Proton Exchange between Adsorbed Species and Oxides and Silicate Surfaces', by J. J. Fripiat,[954] 'An N.M.R. Study of Surface Complexes under the Conditions of Fast Exchange between Physisorbed and Chemisorbed Molecules', by V. Yu. Borovkov, G. M. Zhidomirov, and V. B. Kazanskii,[955] 'Adsorption Phenomena in Zeolites as Studied by Nuclear Magnetic Resonance', by H. Pfeifer,[956] 'Study of the Ion Exchanger-solvent System by a Nuclear Magnetic Resonance Method', by G. S. Bystrov, G. A. Grigor'eva, and N. I. Nikolaev,[957] 'Surface Analysis by means of Magnetic Resonance Techniques', by J. H. Lunsford,[958] 'N.M.R. Investigations on the Structure and Sorption Problems of Faujasite-type Zeolites', by H. Lechert,[959] and 'Study of Adsorptional Self-diffusion on Zeolites using the N.M.R. Field Gradient Pulse Technique', by J. Kaerger, H. Pfeifer, and M. Buelow.[960] In a review of 'Techniques of Temperature Measurements by N.M.R. in Solids', by J. Vanier, it was pointed out that n.m.r. is a good tool for temperature measurements in ferro- and antiferri-magnetic materials below 40 K.[961]

Motions in Solids.—The spin–lattice relaxation time of ^7Li in Li_2O has been investigated up to 780 °C at various Larmor frequencies. In pure Li_2O, the relaxation by dipole–dipole interaction prevails. The corresponding maximum relaxation rate observed between 600 and 750 °C depends on the Larmor frequency. The mean jump frequency of Li^+ ions above 600 °C is given by

$$\nu(\text{Li}) = 2.5 \times 10^{13} \exp\left(-\frac{1.1 \pm 0.05\ \text{eV}}{kT}\right) \text{s}^{-1}$$

[948] J. Ebert and G. Seifert, 'Nuclear Magnetic Resonance in Solids', Hilger, London, 1975.
[949] G. C. Carter, L. H. Bennett, and D. J. Kahan, 'Progress in Materials Science, Vol. 20: Metallic Shifts in NMR', Pergamon, Oxford, 1976 (*Chem. Abs.*, 1976, **85**, 184 746).
[950] R. Blinc, *Hydrogen Bond*, 1976, **2**, 831 (*Chem. Abs.*, 1976, **85**, 168 946).
[951] J. Stankowski, *Radiospektrosk. Ciala Stalego*, 1975, 211 (*Chem. Abs.*, 1976, **85**, 113 856).
[952] L. K. Aminov, *Paramagn. Rezon.*, 1974, **8**, 3 (*Chem. Abs.*, 1976, **85**, 168 908).
[953] V. V. Mank and O. D. Kurilenko, 'NMR Study of Intermolecular Interactions in Ion-exchange Resins', Naukova Dumka, Kiev, 1976, 79 pp. (*Chem. Abs.*, 1976, **85**, 78 901).
[954] J. J. Fripiat, 'Magnetic Resonance in Colloid and Interface Science' (Symposium No. 34), The American Chemical Society, Washington, 1976, p. 261 (*Chem. Abs.*, 1976, **85**, 198 663).
[955] V. Yu. Borovkov, G. M. Zhidomirov, and V. B. Kazanskii, in 'Magnetic Resonance in Colloid and Interface Science' (Symposium No. 34), The American Chemical Society, Washington, 1976, p. 233 (*Chem. Abs.*, 1976, **85**, 198 662).
[956] H. Pfeifer, in 'Magnetic Resonance in Colloid and Interface Science' (Symposium No. 34), The American Chemical Society, Washington, 1976, p. 36 (*Chem. Abs.*, 1976, **85**, 198 660).
[957] G. S. Bystrov, G. A. Grigor'eva, and N. I. Nikolaev, *Uspekhi Khim.*, 1976, **45**, 1621.
[958] J. H. Lunsford, *CRC Crit. Rev. Solid State Sci.*, 1976, **6**, 337 (*Chem. Abs.*, 1976, **85**, 167 039).
[959] H. Lechert, *Catal. Rev.-Sci. Eng.*, 1976, **14**, 1 (*Chem. Abs.*, 1976, **85**, 99 547).
[960] J. Kaerger, H. Pfeifer, and M. Buelow, *Z. Chem.*, 1976, **16**, 85.
[961] J. Vanier, *Temp.: Its Meas. Control Sci. Ind.*, No. 4, Pt. 2, 1972, 1197 (*Chem. Abs.*, 1976, **84**, 76 190).

By doping MgO and LiF, two different atomic defects are produced which both give rise to another maximum in the relaxation rate. This type of relaxation was attributed to quadrupolar interactions. The temperatures of the corresponding relaxation maxima are lower than for the dipole–dipole relaxation. In both cases the defects lead to an increase of the Li^+ jump frequency. The defects were identified as cation vacancies and cation interstitials. The activation energies of migration are 0.55 ± 0.05 and 0.3 ± 0.05 eV, respectively. It was concluded that the intrinsic disorder in Li_2O is of the Frenkel type with a formation energy of 1.3 ± 0.2 eV.[962] Spin–spin relaxation measurements of ^7Li have permitted the precise study of diffusion. A discontinuity in the diffusion processes was observed and a possible explanation proposed.[963] Wideline n.m.r. spectra of $LaNi_5H_x$ ($2.1 \leqslant x \leqslant 6.8$) and $LaNi_5D_x$ ($3.4 \leqslant x \leqslant 6.7$) have been measured at 3—100 MHz for ^1H and 16 MHz for ^2H nuclei at 4.2—380 K. The temperature dependence of the shape, linewidth, and second moment of the ^1H n.m.r. spectrum and nuclear quadrupole interaction parameters of the ^2H n.m.r. spectrum reflects the effects of rapid diffusive motion of hydrogen species at temperatures above 140 K. At low temperatures, the proton resonance shape is distinctly non-Gaussian with a second moment of 12.78 ± 0.26 Oe^2, independent of composition and the deuterium spectrum may be characterized by two or three quadrupole interactions which are also independent of composition. Both observations indicate the existence of a single hydride phase possessing a narrow range of stoicheiometry. At temperatures above 270 K, the deuterium spectrum shows only a single weak, although well-resolved, quadrupole interaction. Evidence of a hydrogen surface layer from 4.2 to 100 K was found.[964] The ^{19}F n.m.r. spectrum of $NOUF_7$ has shown different motions in the complex fluoride.[965]

Spin–lattice relaxation times on $Ti_{0.50}Nb_{0.50}H_{0.900}$ give an activation energy for the migration of the protons as 0.27 eV. The dependence of the T_1 minimum values on concentration and comparison with theoretical ones give the best agreement with Hogg's model for the incorporation of protons in the lattice.[966] Wideline n.m.r. spectra of ^1H and ^{15}N nuclei have been observed in the intercalation complexes $TaS_2(NH_3)$ and $TiS_2(NH_3)$ at temperatures from 5 to 300 K. Changes in the dynamics of intercalated NH_3 molecules were followed by determining the temperature dependence of linewidths and second moments of ^1H resonances. Above 160 K, the molecular motions average proton dipolar interactions to zero, which requires that molecular diffusion and reorientation occur at rates $>10^5$ s^{-1}. The residual ^1H shifts reflect the magnetic properties of the complexes. Around 125 K a strong line broadening indicates that translation and reorientation have lowered to this rate. Although reorientation and translation are correlated, the data suggest that they are not strongly coupled to one another. The molecules continue to spin about the three-fold (C_3) symmetry axes even at 5 K, and saturation studies of the resonance line suggest

[962] Y. S. Oei and H. Richtering, *Ber. Bunsengesellschaft phys. Chem.*, 1976, **80**, 1007.
[963] J. P. Motte, A. El Maslout, and P. Granger, *J. Solid State Chem.*, 1975, **15**, 253 (*Chem. Abs.*, 1976, **84**, 35 518).
[964] R. G. Barnes, W. C. Harper, S. O. Nelson, D. K. Thorne, and D. R. Torgeson, *J. Less-Common Met.*, 1976, **49**, 483 (*Chem. Abs.*, 1976, **85**, 200 317).
[965] P. Charpin, J. Michel, and P. Rigny, *Inorg. Nuclear Chem.* (Herbert H. Hyman Memorial Volume), 1976, 131.
[966] O. J. Zogal, B. Nowak, G. Berg, and D. Pohl, *Cryst. Lattice Defects*, 1975, **6**, 117 (*Chem. Abs.*, 1976, **85**, 54 239).

that the rate for this spinning is still *ca.* 10^8 s^{-1} at this temperature. In contrast to solid NH_3, the NH_3 molecules in these intercalation complexes are much more mobile in spite of comparable molecular densities. Apparently the two-dimensional character of the NH_3 arrays precludes the establishment of directed hydrogen bonds that inhibit motion in solid NH_3.[967] A related ^2H n.m.r. study was carried out on $TaS_2(ND_3)$ and $TiS_2(ND_3)$. The magnetic and temperature dependence of the observed electric quadrupole coupling constant $(e^2qQ/h)_{eff}$ have provided information about the configuration of intercalated ND_3 and their dynamics. The similarity of $(e^2qQ/h)_{eff}$ to previously reported values for solid ND_3 suggests that intercalation entails only small changes in molecular geometry or electronic configuration. Quadrupolar interaction does not vanish at 300 K, showing that reorientation about the C_3 axis is not isotropic, and the form of the motion was deduced from the high-temperature value of the coupling constant. A central component of the spectrum was observed at low temperatures and attributed to ND_3 molecules residing near vacancies in the layers.[968]

Motion of the cyclo-octatetraene moiety in the series of fluxional organometallic molecules $[(C_8H_8)Fe(CO)_3]$, $[(C_8H_8)Fe_2(CO)_5]$, and $[(C_8H_8)_2Ru_3(CO)_4]$ in the solid state has been confirmed by measurements of T_1 and $T_{1\rho}$ and accurate values were obtained for the activation energies. The results were discussed in relation to the structure and bonding in the compounds and to the results from high-resolution n.m.r. measurements in solution and from *X*-ray crystallographic investigations in the solid state.[969] Proton relaxation has been measured for two near-faujasite hydrogen zeolites with different iron contents. The motion of the hydrogen atoms was investigated and it was concluded that the activation energy jump is 19 kcal mol^{-1}.[970]

Diffusion of hydrogen in the α' phase of palladium and palladium–silver alloys has been studied by ^1H n.m.r. spectroscopy. The spin–lattice relaxation times and the diffusion coefficients from the pulsed field gradient method are consistent with random jumps between tetrahedral sites. The activation energy increases from 0.23 ± 0.01 to 0.28 ± 0.02 eV on the addition of 17 atom % Ag.[971] ^{109}Ag N.m.r. lines in a sample of polycrystalline AgF broaden substantially when the ^{19}F n.m.r. signal is irradiated near the magic angle in the rotating frame. This is due to a reduction in the ^{19}F–^{19}F dipolar coupling which normally causes fluctuations in ^{19}F–^{109}Ag interactions inducing an exchange narrowing analogous to classical motional narrowing. The ^{109}Ag linewidths obtained over the entire motional range at different ^{19}F frequencies were compared with those calculated exactly from the ratio of second and fourth moments.[972]

Motion of the organic moiety of fluxional organometallic compounds in the series $[(\eta^1\text{-}C_5H_5)HgX]$ (X = η^1-C_5H_5, Cl, Br, or I) in the solid state was confirmed by the temperature dependence of T_1 and $T_{1\rho}$ and values for the activation

[967] B. G. Silbernagel, M. B. Dines, F. R. Gamble, L. A. Gebhard, and M. S. Whittingham, *J. Chem. Phys.*, 1976, **65**, 1906.
[968] B. G. Silbernagel and F. R. Gamble, *J. Chem. Phys.*, 1976, **65**, 1914.
[969] A. J. Campbell, C. E. Cottrell, C. A. Fyfe, and K. R. Jeffrey, *Inorg. Chem.*, 1976, **15**, 1321.
[970] M. M. Mestdagh, W. E. E. Stone, and J. J. Fripiat, *J.C.S. Faraday I*, 1976, **72**, 154.
[971] P. P. Davis, E. F. W. Seymour, D. Zamir, W. D. Williams, and R. M. Cotts, *J. Less-Common Met.*, 1976, **49**, 159 (*Chem. Abs.*, 1976, **85**, 184 454).
[972] M. Mehring, G. Sinning, and A. Pines, *Z. Phys. (B)*, 1976, **24**, 73.

energy for this process were obtained. The results were discussed in relation to the molecular structure of the compounds and to the results from high-resolution n.m.r. spectroscopy.[973] ^{1}H Relaxation measurements in $[Cd(NH_3)_4(O_2CMe)_2]$ and $[Zn(NH_3)_2(O_2CMe)_2]$ at 100—300 °C yielded 12 and 3 for the number of protons effective in relaxation, *i.e.* all four NH_3 groups and only one NH_3 group, respectively, participated in the relaxation. The experimental data were explained by strong spin-diffusion bonds between the NH_3 groups in the complex compounds in which the NH_3 groups were co-ordinated to the central ion, whereas in the case of NH_3 groups not belonging to the same complex each group participated in the relaxation independently.[974]

The ^{11}B n.m.r. spectrum of $[B_3H_8]^-$ shows phase transitions at 270 and 316 K.[975] ^{1}H and ^{11}B n.m.r. spectroscopies have been used to characterize molecular reorientation in crystalline 9-Br-*meta*-carbaborane. A structural phase transition was found at 344 ± 2 K. In the high-temperature (α) phase reorientation is rapid and apparently isotropic with $E_A = 22.4 \pm 1.3$ kJ mol^{-1}. In the low-temperature (β) phase reorientation is anisotropic with an apparent activation energy of 47.0 ± 6.3 kJ mol^{-1} for temperatures above the proton T_1 minimum (300—344 K) and 14.7 ± 1.3 kJ mol^{-1} for temperatures below the proton T_1 minimum (230—290 K). The temperature-dependent linewidths and shapes in the β-phase were tentatively interpreted to indicate a more complicated reorientation than was observed in *ortho*-carbaborane.[976] ^{1}H N.m.r. spin–lattice relaxation times have been measured for Me_3NBH_3 from 120 to 380 K, a few degrees above the melting point. Minima in T_1 at 157 and 259 K were attributed to a three-fold reorientation of each of the three methyl groups and the BH_3 group and to the three-fold reorientation of the whole molecule about the boron–nitrogen axis, respectively. Activation energies for these processes were found to be 3.3 and 6.7 kcal mol^{-1}. Abrupt changes in T_1 at 350 and 360 K correspond exactly with heat capacity transitions observed by other workers. The time constant for the free induction decay decreases by two orders of magnitude at 360 K. Having a value of 3 ms above 360 K, it shows that there must be rapid diffusion as well as molecular tumbling in the highest temperature solid phase.[977] Self-diffusion processes in EuB_6 have been studied using the pulsed gradient spin-echo n.m.r. technique. At 1.7 K the zero field value of the self-diffusion parameter was (66 ± 4) × 10^{-6} μs^{-3}, while with a 5 kG polarizing field the value was (3 ± 0.6) × 10^{-6} μs^{-3}.[978]

A ^{23}Na n.m.r. investigation of a single crystal of β-Al_2O_3 has been made. The low-temperature rotation patterns show that Na^+ occupies several independent sites in the lattice. At temperatures above 110 K sodium atom jumps occur. At higher temperatures only one line is observed and the activation energy for the jump is 0.1 eV.[979] The ^{27}Al n.m.r. spectrum of ruby has been measured at

[973] A. J. Campbell, C. E. Cottrell, C. A. Fyfe, and K. R. Jeffrey, *Inorg. Chem.*, 1976, **15**, 1326.
[974] J. R. Goslar, J. Stankowski, and S. K. Hoffmann, *Uniw. Adama Mickiewicza Poznaniu, Wydz. Mat., Fiz. Chem. (Pr.), Ser. Fiz.*, 1975, **19**, 337 (*Chem. Abs.*, 1976, **84**, 187 162).
[975] R. T. Paine, E. Fukushima, and S. B. W. Roeder, *J. Magn. Resonance*, 1975, **32**, 566.
[976] A. J. Leffler, M. N. Alexander, and P. L. Sagalyn, *J. Chem. Phys.*, 1976, **65**, 2650.
[977] T. T. Ang and B. A. Dunell, *Canad. J. Chem.*, 1976, **54**, 1087.
[978] M. M. Bajaj, *Phys. Letters (A)*, 1976, **56**, 390 (*Chem. Abs.*, 1976, **85**, 25 637).
[979] W. Bailey, S. Glowinkowski, H. Story, and W. L. Roth, *J. Chem. Phys.*, 1976, **64**, 4126.

temperatures below that of liquid helium. The temperature dependence of the observed linewidths was discussed in terms of the spin-diffusion barrier model.[980]

Proton Zeeman and dipolar spin–lattice relaxation times and second moments have been measured in polycrystalline SiH_4 at 24 MHz between 90 and 3 K. Down to 25 K the classical model for molecular motion applies; below 25 K quantum effects have to be taken into account.[981] Longitudinal n.m.r. relaxation of polysiloxanes has been used to examine crystalline processes in cross-linked polymers and for comparison of segmental motion at various distances of the cross-link knots in polymer networks. The frequency of molecular motion corresponded to the frequency of the n.m.r. relaxation at the temperature minimum of the relaxation time, T_1. The temperature minimum of crystalline polysiloxanes prepared by crosslinking poly(dimethylsiloxanes) with $[MeViSiO]_3$ was observed at temperatures higher than for amorphous polymers. The temperature minimum of T_1 for amorphous polysiloxanes was shifted to lower temperatures, with an increase in distance between the cross-link knots of the network which was related to an increased segmental mobility.[982] ^{19}F N.m.r. spectra, second moments, and T_1 have been measured for solid SiF_4 and the results indicate a hindered molecular motion and piezoelectricity. A value of 24 kJ mol^{-1} was obtained for the activation energy of reorientational motion of the SiF_4 tetrahedra around the three-fold symmetry axis. The values of T_1 measured from 150 to 183 K by the n.m.r. pulse method at 30 MHz decrease with increasing temperature from 40 to 0.3 s.[983] The coupled ^1H and ^{19}F nuclear spin–lattice relaxation mechanism has been studied in polycrystalline $MSiF_6,6H_2O$ (M = Mg or Zn) at 30 MHz over the range 130—420 K. The results indicate that below 280 K in the magnesium complex and below 390 K in the zinc complex reorientation of the $[SiF_6]^{2-}$ ions is the dominant relaxation mechanism. Reorientation of individual water molecules has not to be taken into account as an effective relaxation mechanism. The phase transition in $MgSiF_6,6H_2O$ at 298 K is reflected by a change in $(T_1)^{-1}$. In $MgSiF_6,6H_2O$ the relaxation behaviour becomes purely exponential above the phase transition temperature, whereas in $ZnSiF_6,6H_2O$ there is a non-exponential relaxation behaviour over the whole temperature range investigated. The activation energies and frequency factors obtained from $(T_1)^{-1}$ measurements are consistent with data obtained by use of the Waugh–Fedin relation.[984]

Fluoride ion motion in β-PbF_2 doped with NaF has been studied using ^{19}F T_2, T_1, and T_{1r} measurements. Activation energies of 0.205 ± 0.01, 0.29 ± 0.02, and 0.27 ± 0.01 eV per ion were found. The pressure dependence of T_1 and T_2 yielded activation volumes of $\leqslant 0.2$ and 1.76 ± 0.05 cm^{-3} $gmol^{-1}$ respectively.[985] ^{19}F N.m.r. relaxation times and linewidths have been measured in polycrystalline $CsPbF_3$ in the range from −150 to +300 °C. The observed motional narrowing of the n.m.r. linewidth at low temperatures and the temperature dependence of

980 M. Takeuchi, T. Tatsukawa, M. Inoue, and H. Yagi, *Fukui Daigaku Kogakubu Kenkyu Hokoku*, 1975, **23**, 271 (*Chem. Abs.*, 1976, **84**, 67 503).
981 G. Janssens, P. Van Hecke, and L. Van Gerven, *Chem. Phys. Letters*, 1976, **42**, 445.
982 K. A. Andrianov, A. A. Zhdanov, V. M. Litvinov, and B. D. Lavrukhin, *Doklady Akad. Nauk S.S.S.R.*, 1976, **226**, 85.
983 J. Makela and L. Niemela, *Phys. Letters* (*A*), 1976, **56**, 307 (*Chem. Abs.*, 1976, **85**, 12 109).
984 H. D. Jannek and H. Rager, *Z. Naturforsch.*, 1975, **30a**, 1615.
985 T. Y. Hwang, I. J. Lowe, K. F. Lau, and R. W. Vaughan, *J. Chem. Phys.*, 1976, **65**, 912.

relaxation times were analysed in terms of a high vacancy diffusion of F^- ions. A phase transition was found at -93 °C from n.m.r. lineshapes.[986]

The spin–lattice relaxation times for ^1H and ^2H nuclei have been measured in polycrystalline $[NH_4]^+{}_2[TiF_6]^{2-}$ and $[ND_4]^+{}_2[TiF_6]^{2-}$ between liquid helium and room temperature at 7.0 MHz. The activation energy of reorientation of the $[NH_4]^+$ and $[ND_4]^+$ ions, respectively, and the reorientation of $[NH_4]^+$ ions by classical and by tunnelling rotation about one C_3 axis have been discussed.[987] The ^1H n.m.r. T_1, $T_{1\rho}$, and T_{1D} of $[NH_4]^+[ReO_4]^-$ have been reported as a function of temperature over the range 60—300 K. The temperature dependence of T_1 was described in terms of dipolar interaction between the protons within an $[NH_4]^+$ ion made time-dependent by rapid classical reorientation at a rate which may be described by an Arrhenius relation. The activation energy for the molecular reorientation is 2.2 kcal mol^{-1}. Over limited regions of temperature this same model can also be used to explain $T_{1\rho}$ and T_{1D} results. However, at low temperatures, the $T_{1\rho}$ data suggest the importance of quantum mechanical tunnelling. At high temperatures the dipolar coupling between the ^1H and rhenium spins, rendered temperature-dependent by the rapid rhenium spin–lattice relaxation, dominates both $T_{1\rho}$ and T_{1D}. Due to a strong angular dependence of relaxation rates associated with the hydrogen–rhenium interaction, non-exponential decay occurs for a powdered sample as used in these experiments. From the $T_{1\rho}$ and T_{1D} results it was concluded that the rhenium T_1 is dominated by quadrupolar Raman spin–phonon interaction. There is no evidence in any of the measurements for the previously postulated order–disorder phase transition in the vicinity of 200 K and involving the $[NH_4]^+$ ions.[988] Spin–lattice relaxation times of ^1H and ^{35}Cl nuclei in α-$[NH_4]^+[HgCl_3]^-$ have been reported. The ^1H n.m.r. T_1 data were explained in terms of C_3 reorientation and S_4 flips of the $[NH_4]^+$ ions. The latter mode is also largely responsible for the ^{35}Cl quadrupolar relaxation rate. A marked change in the ^1H n.m.r. T_1 at 55 K indicated a phase transition ascribable to ordering of the $[NH_4]^+$ ions.[989] The magnetic field dependence of the proton Zeeman relaxation in polycrystalline $[NH_4]^+{}_2[PbCl_6]^{2-}$ has been studied at 20 and 43 K. The relaxation rate *versus* frequency curve exhibited tunnelling maxima at 22 and 44 MHz (at 20 K) and 21 MHz (at 43 K). The interpretation of the time dependence of the magnetization as the sum of two exponentials showed that the time constants differed from each other much more at 20 than at 43 K.[990] An extension of this work suggested that the ^1H n.m.r. free induction decay consists of two components with temperature-independent second moments of 0.7 and 11.5 G^2. The temperature dependence of the corresponding weights leads to a decrease in the total second moment from 6 G^2 at 4.2 K to 2 G^2 at 50 K. The results were explained in terms of a model which assumes that the $[NH_4]^+$ tunnelling splittings are much larger than $\hbar\Delta\omega$, and that the rates k_2 and k_3 of 180° and 120° reorientation of the $[NH_4]^+$ ion, respectively, obey the equation $k_2 \gg \Delta\omega$ and $k_3 \ll \Delta\omega$, where $\Delta\omega$ is the line-

986 V. M. Bouznik, Yu. N. Moskvich, and V. N. Voronov, *Chem. Phys. Letters*, 1976, **37**, 464.
987 L. K. E. Niemelä and P. H. Oksman, *Chem. Phys. Letters*, 1976, **41**, 174.
988 R. L. Armstrong, J. A. J. Lourens, and K. R. Jeffrey, *J. Magn. Resonance*, 1976, **23**, 115.
989 K. Sagisawa, H. Kiriyama, and R. Kiriyama, *Chem. Letters*, 1975, **12**, 1285 (*Chem. Abs.*, 1976, **84**, 51 902).
990 M. Punkkinen, J. E. Tuohi, and E. E. Ylinen, *Chem. Phys.*, 1976, **13**, 265.

width.[991] Nuclear magnetic second moment and spin–lattice relaxation time studies have been performed over the range 77—388 K on ferroelectric NH_4HSO_4 and antiferroelectric $(NH_4)_3H(SO_4)_2$. The results suggest a rapid general reorientation of the $[NH_4]^+$ ions over the whole range in both materials brought about by the superposition of a three-fold reorientation on two less hindered two-fold reorientations. The NH_4HSO_4 shows an additional mechanism absent from the $(NH_4)_3H(SO_4)_2$ results which was attributed to a loss of distortion of the $[NH_4]^+$ tetrahedron at high reorientation frequencies.[992] ^1H N.m.r. spectroscopy has also been used to examine molecular motion in ammonium tartrate.[993] The frequency dependence of the Zeeman spin–lattice relaxation of ^1H in $[NH_4]^+[ClO_4]^-$ has been observed at various temperatures below 20 K. Plots of relaxation times against resonance-frequency curves give minima which yield detailed information about the temperature dependence of low-symmetry tunnelling splitting of the $[NH_4]^+$ group, but it depends on the quality of the crystal.[994]

The ^1H n.m.r. spin–lattice relaxation times at 20 and 60 MHz and the second moment of the ^1H n.m.r. resonance have been determined for $[MeNH_3]^+{}_2$-$[PtCl_6]^{2-}$, its partially deuteriated derivative, and $[MeNH_3]^+{}_2[SnCl_6]^{2-}$ at various temperatures between 4.2 and 400 K. The fully protonated complexes show narrow ^1H n.m.r. signals having a second moment of *ca.* 7 G^2 even at 77 K, indicating that the CH_3 and NH_3 groups in $[CH_3NH_3]^+$ ions undergo rapid reorientation about the triad axis. The partially deuteriated analogues $[CH_3ND_3]^+{}_2[PtCl_6]^{2-}$ and $[CD_3NH_3]^+{}_2[PtCl_6]^{2-}$ give rise to a single T_1 minimum at nearly the same temperature attributable to the reorientation of cations about their respective triad axes. On the other hand, the fully protonated complexes yield two T_1 minima, one of which, appearing at a lower temperature, was deep while the other, appearing below the transition point, was shallow. The deep minimum was assigned to reorientation of the cation as a whole and the shallow minimum to reorientation of the CH_3 and NH_3 groups. The potential barrier to internal reorientation of the $[CH_3NH_3]^+$ ion is 8 kJ mol^{-1}.[995] Wideline ^1H n.m.r. spectra of monocrystalline Me_2NH_2Cl have been recorded near and above room temperature. It was concluded from a comparison between experimental and theoretical second moments that methyl group reorientation is rapid and that the cation as a whole reorientates around two different axes. One axis is close to the symmetry axis of the free cation and the other is parallel to the carbon–carbon direction. Some of the relaxation constants, time factors, and activation barriers of the different motions in the three phases exhibited by the complex were reported.[996] This work was extended to the iodide, bromide, and hydrogenoxalate over the range 100—400 K. Relaxation constants, time factors, and activation barriers for different phases of various compounds were determined and discussed. It was concluded that methyl groups as well as cations as a whole reorientate in all the complexes.[997]

991 M. Punkkinen, J. E. Tuohi, and E. E. Ylinen, *J. Magn. Resonance*, 1976, **22**, 527.
992 A. Watton, E. C. Reynhardt, and H. E. Petch, *J. Chem. Phys.*, 1976, **65**, 4370.
993 Z. M. El Saffar, *J. Chem. Phys.*, 1976, **64**, 1854.
994 M. Punkkinen and J. P. Pyy, *Phys. Fenn.*, 1975, **10**, 215 (*Chem. Abs.*, 1976, **84**, 187 196).
995 R. Ikeda, Y. Kume, D. Nakamura, Y. Furukawa, and H. Kiriyama, *J. Magn. Resonance*, 1976, **24**, 9.
996 R. Sjoblom, *Chem. Scripta*, 1976, **9**, 127 (*Chem. Abs.*, 1976, **85**, 142 182).
997 R. Sjoblom, *Chem. Scripta*, 1976, **9**, 133 (*Chem. Abs.*, 1976, **85**, 142 183).

The librational amplitudes in $[NMe_4]^+$ halides have been investigated.[998] The spin–lattice relaxation times of ^1H nuclei in $[Me_4E]^+I^-$ (E = N, P, As, or Sb) have been investigated in the range 100—460 K at 34 MHz. The experiments show that the reorientational motion of the methyl groups at low temperatures and the reorientation of the whole cation at high temperatures are responsible for the relaxation maxima. The relaxation behaviour of the ^1H n.m.r. spin system is purely exponential in the temperature range investigated. Application of the Waugh–Fedin model to the ^1H n.m.r. results and the far i.r. and Raman spectra show that at high temperatures a 180° reorientation step of the cation is probable.[999] Spin–lattice relaxation times for $[Me_4P]^+X^-$ (X = Cl, Br, or I) have been measured between 100 and 500 K, and as in the previous paper two minima in T_1 were found for each compound which were assigned to methyl group reorientation and whole-cation tumbling. The second moments also indicate that the cations are tumbling isotropically at n.m.r. frequencies in the upper half of this temperature range. It was suggested that librational oscillation of the whole cation occurs at a frequency of at least 10^5 s^{-1} near 150 K. The energy barriers for both methyl group reorientation and isotropic tumbling decrease from Cl to Br but increase from Br to I.[1000]

T_{1D} Measurements have been used to investigate the motion of defects in high-purity hexagonal ice. The n.m.r. correlation times τ agree rather well with both the Schottky H_2O diffusion times and the deuteron correlation times in D_2O, suggesting that Schottky rather than interstitial diffusion dominates spin–lattice relaxation in both H_2O and D_2O ice.[1001] Spin–lattice measurements have been made on pure, NH_3-doped, and HF-doped ice crystals. Three relaxation processes were distinguished: a low-temperature process and an extrinsic and an intrinsic process at high temperatures with activation energies of 0.08, 0.35, and 0.62 eV, respectively. Vested vacancies were proposed as the defects effective in the low-temperature process. The extrinsic high-temperature processes are due to vacancies which are formed by shifting a HF molecule to an interstitial site. Schottky defects are responsible for the intrinsic high-temperature relaxation mechanism.[1002] ^1H N.m.r. spectroscopy has been used to investigate the motion of the cage water molecules of the clathrate hydrate of xenon.[1003]

The ^1H n.m.r. T_1 and $T_{1\rho}$ in single crystals of polycrystalline $SnCl_2,2H_2O$ have been measured. $T_{1\rho}$ exhibits a minimum at *ca.* 270 K. Both T_1 and $T_{1\rho}$ are dominated by three-fold reorientation of water molecules with an activation energy of 49 kJ mol^{-1}. The ^2H n.m.r. spectra suggest the combined motion of a 180° flip and a three-fold rotation. Near T_{tr} = 218 K the $\ln T_1$ against T^{-1} curve shows a dip as a critical phenomenon.[1004]

Structure of Solids.—Single crystals of $KHCO_3$ and $KHSO_4$ have been investigated by solid-state high-resolution ^1H n.m.r. techniques. These compounds

998 M. Polak, *J. Chem. Phys.*, 1976, **65**, 2785.
999 H. Rager and A. Weiss, *Ber. Bunsengesellschaft phys. Chem.*, 1976, **80**, 138.
1000 T. T. Ang and B. A. Dunell, *Canad. J. Chem.*, 1976, **54**, 1985.
1001 R. Blinc, H. Graenicher, G. Lahajnar, and I. Zupancic, *Z. Phys.* (*B*), 1975, **22**, 211 (*Chem. Abs.*, 1976, **84**, 67 404).
1002 J. H. Bilgram, J. Roos, and H. Graenicher, *Z. Phys.* (*B*), 1976, **23**, 1 (*Chem. Abs.*, 1976, **84**, 82 193).
1003 P. A. Bottomley, J. M. Pope, and B. A. Cornell, *Mol. Phys.*, 1976, **31**, 1277.
1004 H. Kiriyama, O. Nakamura, and R. Kiriyama, *Chem. Letters*, 1976, **7**, 689 (*Chem. Abs.*, 1976, **85**, 101 782).

were considered to be model systems for compounds that contain abundant quadrupolar nuclei (^{39}K) with a small magnetogyric ratio in addition to the protons. Despite considerable line broadening resulting from ^{1}H–^{39}K dipole–dipole coupling, all four possible proton shielding tensors in $KHSO_4$ and an averaged shielding tensor for the two different proton sites in $KHCO_3$ could be measured accurately. The results were analysed to determine the degree of ionization of the protons in the three types of hydrogen bonds occurring in $KHCO_3$ and $KHSO_4$.[1005] Measurement of the ^{14}N quadrupole coupling constant, asymmetry parameter, inverse linewidth parameter T_2^*, T_2, T_{1S}, and $T_{1\rho}$ have been reported for $NaNO_2$ over the range 77—467 K. The relevant spin–lattice relaxation times and the transition probabilities W_+, W_Δ, and W_- were deduced for a spin one system and transition probability were calculated from T_1 data. The results indicate the existence of short-range order above the ferroelectric phase transition.[1006] The effect of plastic deformation in the NaCl-type crystals on the broadening of the n.m.r. line has been studied by the Monte Carlo method for a number of different dislocations. The calculations were compared with experimental data.[1007] The ^{1}H n.m.r. relaxation time in $Cu_xZn_{1-x}Cs_2(SO_4)_2,6H_2O$ has been measured at liquid helium temperatures. The results were explained by use of a thermodynamic model which distinguishes the various electronic Zeeman systems and a dipolar interaction system strongly coupled to a proton spin system. Theory gives a quantitively satisfactory explanation of the complicated behaviour of the ^{1}H spin relaxation time as a function of magnetic field and temperature.[1008]

The ^{31}P chemical shielding tensors have been measured in a crystal of α-$Ca_2P_2O_7$. The principal values with respect to 85% H_3PO_4 were found to be −42, 32, and 65 p.p.m. {phosphorus(1)} and −42, 44, and 67 p.p.m. {phosphorus(2)}. The probable orientations of the tensors were determined.[1009] In the ^{1}H n.m.r. spectra of synthetic gypsum and the hemihydrates α- and β-$CaSO_4,0.5H_2O$ a sharp line due to water, probably adsorbed on the crystal surface, was found.[1010] The ordering of nuclear spin at low temperatures has been examined with special reference to the antiferromagnetic state of CaF_2.[1011]

The ^{31}P n.m.r. spectrum of $Tb_xY_{1-x}P$ has been measured from 4 to 300 K. Broad lines were found and the uniform spin polarization model was found to be inapplicable.[1012] The ^{153}Eu spin–spin relaxation times for two spherical EuO single crystals have been investigated,[1013] and their dependence on stoicheiometry was found to be very marked.[1014] ^{1}H N.m.r. spectra of MLnedta,nH_2O (M = Na,

1005 H. Feucht, U. Haeberlen, M. Pollak-Stachura, and H. W. Spiess, *Z. Naturforsch.*, 1976, **31a**, 1173.
1006 G. Petersen and P. J. Bray, *J. Chem. Phys.*, 1976, **64**, 522.
1007 B. A. Grinberg and Yu. M. Plishkin, *Deposited Pub. Viniti*, 1972, 4278 (*Chem. Abs.*, 1976, **85**, 54 219).
1008 G. M. Van Den Heuvel, R. J. J. Brak, L. S. J. M. Henkens, W. T. Wenckebach, and N. J. Poulis, *Physica B + C (Amsterdam)*, 1976, **81**, 53 (*Chem. Abs.*, 1976, **84**, 157 666).
1009 S. J. Kohler, J. D. Ellett, jun., and M. P. Klein, *J. Chem. Phys.*, 1976, **64**, 4451; **65**, 2922.
1010 C. Ostrowski, *Cem.-Wapno-Gips*, 1975, **29**, 177 (*Chem. Abs.*, 1976, **94**, 94 582).
1011 S. F. J. Cox, *Trudy Mezhdunar. Konf. Magn.*, 1973, **3**, 14 (*Chem. Abs.*, 1976, **84**, 53 019).
1012 R. K. Quinn and H. T. Weaver, *J. Solid State Chem.*, 1976, **16**, 197.
1013 H. G. Bohn, R. R. Arons, H. Luetgemeier, and K. J. Fischer, *J. Magn. Magn. Mater.*, 1976, **2**, 67 (*Chem. Abs.*, 1976, **85**, 151 484).
1014 R. R. Arons, *Report AED-CONF*-75-449-001, 1975 (*Chem. Abs.*, 1976, **85**, 85 154).

K, or Cs; Ln = La, Pr, Nd, Eu, Gd, Dy, Er, or Yb) have been determined. The co-ordinated water gives a broad signal while outer-sphere water gives a broad and a narrow line. The number of water molecules in the inner sphere is three for the light lanthanides and decreases to zero or one for ytterbium.[1015] The contact contribution to the isotropic shift has been estimated from the experimental shifts of gadolinium oxysalts, which were assumed to be purely contact. $\delta(^{14}N)$, $\delta(^{31}P)$, and $\delta(^{51}V)$ were measured for the nitrate, phosphate, and vanadate and used to obtain the dipolar shifts.[1016]

A reformulation of the covalent model was developed to explain the ^{19}F hyperfine interactions observed for Yb^{3+} and Tm^{2+} in cubic and tetragonal sites of alkaline earth fluorides. The same model did not explain the data for Ce^{3+}, Gd^{3+}, or Eu^{2+}, indicating that the spin polarization model is the dominant mechanism in the first part of the rare earth series but that the covalent model becomes dominant in the latter half of the series. This analysis showed that the distortion from cubic symmetry in the tetragonal site is not as great as an earlier n.m.r. analysis suggested, but it is still large. It was further suggested that the interstitial ion is closer to the rare earth ion than most observers have assumed.[1017] In a further application of this work, the good agreement between theory and experiment was taken as strong evidence for the covalent model being correct for Yb^{3+}. The calculations revealed that the main contribution to the isotropic shift does not come from the Fermi contact interaction, as has been generally assumed, but rather from the second-order Zeeman term.[1018]

Chemical shifts and spin–lattice relaxation times have been measured for ^{13}C and ^{15}N nuclei in $ThC_{1-x}N_x$ and $UC_{1-x}N_x$. The band character of the $5f$ electrons is confirmed at least in low nitrogen content uranium carbonitrides. The increase of the magnetic susceptibility with nitrogen content in $UC_{1-x}N_x$ is due to an increase of spin susceptibility enhancement for $x < 0.5$.[1019] The use of ^1H n.m.r. spectroscopy to determine the moisture content of TiO_2 has been investigated.[1020] Samples of non-stoicheiometric Na_xTiS_2 have been examined by n.m.r. spectroscopy and the Knight shift and quadrupolar coupling constants were determined.[1021] Similar measurements have been carried out on Li_xTiS_5 for $0 \leqslant x \leqslant 1$. The results suggest the donation of an appreciable fraction of the lithium $2s$ electron density to the TiS_2 layers, which decreases as x increases.[1022]

Niobium deuterides of composition $NbD_{0.55}$ to $NbD_{0.88}$ have been investigated by means of ^2H and ^{93}Nb n.m.r. spectroscopies over the range 4.2—420 K. The β to α phase transition in the range 340—415 K is sharply defined by the extinction of the ^2H quadrupolar interaction as well as by the appearance of the

1015 K. I. Popov, L. I. Martynenko, V. F. Chuvaev, N. I. Pechurova, and V. I. Spitsyn, *Izvest. Akad. Nauk S.S.S.R., Ser. khim.*, 1976, **4**, 725.

1016 M. Bose, M. Bhattacharya, and S. Ganguli, *Proc. Nuclear Phys. Solid State Phys. Symp.*, 1974, **17C**, 420 (*Chem. Abs.*, 1976, **84**, 51 862).

1017 B. R. McGarvey, *J. Chem. Phys.*, 1976, **65**, 955.

1018 B. R. McGarvey, *J. Chem. Phys.*, 1976, **65**, 962.

1019 J. L. Boutard and C. H. De Novion, *Plutonium* 1975 *Other Actinides, Proc. Internat. Conf., 5th*, 1976, 505 (*Chem. Abs.*, 1976, **85**, 54 291).

1020 V. S. Il'Enko, I. M. Karpovskii, and O. M. Shapovalova, *Nauch. Trudy Vses. N.-I. I. Proekt. In-T Titana*, 1975, **11**, 120 (*Chem. Abs.*, 1976, **84**, 152 297).

1021 B. G. Silbernagel and M. S. Whittingham, *Mater. Res. Bull.*, 1976, **11**, 29 (*Chem. Abs.*, 1976, **84**, 82 209).

1022 B. G. Silbernagel and M. S. Whittingham, *J. Chem. Phys.*, 1976, **64**, 3670.

central component of the ^{93}Nb spectrum when the cubic γ phase is formed. At 4.2 K, e^2qQ = 33.0 kHz for ^2H. The temperature-dependent behaviour of the ^{93}Nb resonance spectrum indicates that the boundary between the α and β regions of the Nb–H(D) phase diagram must pass between the compositions $NbD_{0.78}$ and $NbD_{0.82}$ at 300 K. N.m.r. spectroscopy did not furnish positive support for the occurrence of the ξ- and ε-phases nor for the occurrence of a cubic γ-phase but they are not incompatible with the occurrence. They merely reflect the intrinsic insensitiviiy of ^2H n.m.r. signals to long-range ordering.[1023] Tantalum deuterides in the composition range $TaD_{0.43}$ to $TaD_{0.76}$ have been investigated by ^2H n.m.r. spectroscopy over the range 4.2—400 K. The β to α phase transition in the range 320—370 K for these compounds is sharply defined by the extinction of the deuterium quadrupolar interaction when the cubic α-phase is reached. No other significant change in the n.m.r. spectrum was found down to 42 K at which temperature the quadrupolar coupling constants are equal within experimental error having the value $e^2qQ/h = 33.9 \pm 0.7$ kHz independent of compound. The asymmetry parameter values, however, tend towards zero with increasing deuterium content. The thermal width of the β to α transition is increased by a high oxygen content of the tantalum metal.[1024]

The similarity of ^{51}V n.m.r. signals in V_2O_5 and V_2O_5,nH_2O has been taken as evidence that the surroundings of the vanadium atoms in these compounds are the same.[1025] The temperature dependence of the n.m.r. spectra of polycrystalline $NaVO_3,2H_2O$ and KVO_3,H_2O has been studied. Above −50 °C, diffusion of the water molecules along regular vacancies of the Schottky type has been observed. In KVO_3,H_2O diffusion even occurs at −150 °C. Protons in H_2O molecules and OH groups in $NaVO_3,2H_2O$ are in the ratio 0.98 : 0.02. From ^{51}V and ^{23}Na n.m.r. spectra the following quadrupolar coupling constants were determined: 4.05 ± 0.05 and *ca.* 2.1 for ^{51}V and *ca.* 0.84 for ^{23}Na.[1026] The identification and qualitative analysis of mixtures of solid phases can be performed when signals of all the phases appear in the n.m.r. spectrum. The size and variation of isotropic shifts are not a serious problem, as shown by the ^{51}V n.m.r. spectrum of $K_{0.33}V_2O_5$ and for the Fe_2O_3–V_2O_5 system. Where signals of only one phase are observed the analysis is based on the dependence of the integral intensities of the spectrum and second moment of the individual components on concentration. Boundary conditions of the solid solution $x = 0.04$ and $x = 0.22$ were found for the $Li_xV_2O_5$ system. Similarly the inflection on a plot of peak separation for the V_2O_5–$FeVO_4$ system against concentration gave good evidence for the existence of solid solutions of $Fe_xV_{2-x}O_4$ in the regions $0 < x < 0.10$ and $0.10 < x < 0.25$. These phases were not distinguished previously by use of *X*-ray techniques so that in this case the n.m.r. method is more effective.[1027] A ^{51}V n.m.r. spectroscopic study of the reaction of

1023 R. G. Barnes, K. P. Roenker, and H. R. Brooker, *Ber. Bunsengesellschaft phys. Chem.*, 1976, **80**, 875.
1024 K. P. Roenker, R. G. Barnes, and H. R. Brooker, *Ber. Bunsengesellschaft phys. Chem.*, 1976, **80**, 470.
1025 R. N. Pletnev, A. A. Ivakin, V. V. Gorshkov, and A. K. Chirkov, *Doklady Akad. Nauk S.S.S.R.*, 1975, **224**, 106.
1026 R. N. Pletnev, V. V. Gorshkov, and A. A. Ivakin, *Zhur. neorg. Khim.*, 1976, **21**, 331.
1027 R. N. Pletnev and A. A. Fotiev, *Doklady Akad. Nauk S.S.S.R.*, 1976, **226**, 361.

V_2O_5 with $SrCO_3$ in a hydrogen atmosphere has indicated that the reaction proceeds fastest at 650 °C with the formation of SrV_2O_4. Unchanged V_2O_5 was reduced by hydrogen to V_2O_3 and a significant amount of $Sr_3(VO_4)_2$ was also formed. V_2O_3 reacted with $SrCO_3$ and SrO to give SrV_2O_4.[1028] In the ^1H n.m.r. spectra of $Ca_2V_2O_7,2H_2O$ at −55 °C and $Sr_2V_2O_7,1.3H_2O$ at −159 °C the width and size of the second moment are typical for a substance having water of crystallization. Narrow lines in the former spectrum are due to highly mobile water and increase with temperature up to 150 °C. Water is progressively lost between 115 and 200 °C. The water is much more mobile in $Sr_2V_2O_7,1.3H_2O$.[1029] Niobium hydroxide, $Nb_2O_7,6H_2O$, and $Nb_2O_9,6H_2O$ have been investigated by ^1H n.m.r. spectroscopy. All the compounds contain 0.4 to 0.5 OH^- groups per niobium, a value which did not change after two months in air.[1030] The ^{51}V n.m.r. spectrum of $Ta_{1-x}V_xS_2$ when $x < 0.8$ shows only one resonance, but when $x = 0.9$ a second ^{51}V resonance appears.[1031]

The ^2H n.m.r. spectrum of $[ND_4]^+[Cr_2O_7]^{2-}$ shows four phase transitions.[1032] ^1H N.m.r. spectroscopy has been used to study hydrogen bonding in $Na_2Cr_2O_7$,-$2H_2O$ crystals,[1033] and to investigate $K_{10}[W_{12}(OH)_2(H_2O)_2O_{40}],6H_2O$. Good agreement was found between theoretical and experimental spectra for the structure being $K_{10}[W_{12}(OH)_2O_{40}],8H_2O$.[1034] ^{53}Cr and ^{199}Hg n.m.r. spectra of $HgCr_2S_4$ have confirmed the zero field spin structure previously proposed.[1035] The ^{53}Cr n.m.r. spectrum of polycrystalline Cr_3Te_4 has been measured at 4.2 and 67—140 K and yielded two resonance lines due to Cr^{3+} and Cr^{2+} at 45.3 and 57.2 MHz at 4.2 K and at 41.8 and 54.0 MHz at 77 K, corresponding to local fields of 188 and 237 kOe at 4.2 K and 173 and 224 kOe at 77 K. A curve of frequency against temperature had a drop at 88 K which was explained by a phase transition involving formation of an antiferromagnetic component.[1036]

^{19}F N.m.r. spectroscopy has been used to investigate the spin flop phase of MnF_2 near the spin flop–paramagnetic phase boundary.[1037] The ^{19}F n.m.r. spectrum of K_2MnF_4 shows two different ^{19}F environments. One of the fluorine environments in the KF planes is sensitive to only one magnetic ion, and the other fluorine is surrounded by two symmetrically placed Mn^{2+} ions in the MnF_2 planes.[1038]

[1028] R. N. Pletnev, A. A. Fotiev, A. K. Slepukhin, and V. L. Volkov, *Zhur. fiz. Khim.*, 1975, **49**, 2717.
[1029] V. V. Gorshkov, R. N. Pletnev, and I. G. Chufarova, *Zhur. neorg. Khim.*, 1976, **21**, 566.
[1030] V. A. Titova, I. G. Slatinskaya, and V. G. Pitsyuga, *Zhur. neorg. Khim.*, 1975, **20**, 2699.
[1031] N. Karnezos, M. W. Shafer, G. V. S. Rao, G. V. Subba, and L. B. Welsh, *AIP Conf. Proc.* 29 *Magn. Mater., Annu. Conf., 21st*, 1976, 402 (*Chem. Abs.*, 1976, **85**, 55 715).
[1032] S. D. Damle, L. C. Gupta, and R. Vijayaraghavan, *Chem. Phys. Letters*, 1976, **37**, 614.
[1033] G. R. Bulka and V. M. Vinokurov, *Magnit. i Optich. Spektroskopiya Mineralov i Gorn. Porod*, 1974, 63 (*Chem. Abs.*, 1976, **84**, 51 880).
[1034] J. Chojnacka, E. Hodorowicz, and S. Sagnowski, *J. Inorg. Nuclear Chem.*, 1976, **38**, 1811.
[1035] T. J. Burch, S. B. Berger, and J. I. Budnick, *Trudy Mezhdunar. Konf. Magn.*, 1974, **3**, 402 (*Chem. Abs.*, 1976, **84**, 24 152).
[1036] E. Jedryka, S. Nadolski, A. Pajaczkowska, M. Wojcik, and W. Zbieranowski, *Uniw. Adama Mickiewicza Poznaniu, Wydz. Mat., Fiz. Chem. (Pr.), Ser. Fiz.*, 1975, **19**, 391 (*Chem. Abs.*, 1976, **84**, 187 163).
[1037] A. R. King and H. Rohrer, *AIP Conf. Proc.* 29 *Magn. Magn. Mater., Annu. Conf., 21st*, 1975, 456 (*Chem. Abs.*, 1976, **85**, 114 234).
[1038] C. Bucci and G. Guidi, *Trudy Mezhdunar. Konf. Magn.*, 1974, **1**, (Pt. 2), 333 (*Chem. Abs.*, 1976, **85**, 27 139).

The ^1H n.m.r. spectra of single crystals of ferrocene have been studied by multiple pulse techniques at room temperature. The data were analysed to yield $\Delta\sigma = \sigma_{\parallel} - \sigma_{\perp} = -6.5 \pm 0.1$ p.p.m. and $\sigma_{iso} = -4.2 \pm 0.5$ p.p.m.[1039] ^1H N.m.r. spectroscopy has been used to determine the distribution pattern of iron ions in the cation positions of the octahedral layers in the protolithionite, zinnwaldite, and polytypic modifications. The distribution pattern of the octahedral cations was controlled by the condition of the mica.[1040] An n.m.r. investigation of ordering in Fe_3O_4 shows eight Fe^{3+} ions on A sites and eight Fe^{2+} and eight Fe^{3+} ions on B sites.[1041] The ^{57}Fe n.m.r. spectrum of $DyFeO_3$ has been measured. The strongly decreasing resonance frequency dependence on temperature has been correlated with nuclei in the centres of domain boundaries and with nuclei on the peripheral of the domain.[1042] ^{19}F N.m.r. spectroscopy has been used to demonstrate that the fluorine is on Al^{III} or Fe^{III} in $Fe^{II}Fe^{III}F_5,7H_2O$, $Fe^{II}Al^{III}F_5,7H_2O$, $Zn^{II}Fe^{III}F_5,7H_2O$, and $Zn^{II}Al^{III}F_5,7H_2O$.[1043, 1044]

The spin configuration and magnetic ordering of $CoCl_3(Me_3NH),2H_2O$ have been studied by analysing the n.m.r. results on ^1H and $^{35,37}Cl$ nuclei. The results show that the compound orders magnetically in the magnetic space group Pnm^1a^1.[1045]

Polycrystalline samples of glycine complexes of platinum and palladium have been investigated by ^1H n.m.r. spectroscopy over the range from −196 to 200 °C. In order to shorten T_1, the samples were irradiated with γ-rays. For *cis*- and *trans*-$[Pt(O_2CCH_2NH_2)_2]$ and *cis*-$[Pd(O_2CCH_2NH_2)_2]$ the ^1H n.m.r. spectrum is determined by magnetically isolated NH_2CH_2 groups. *cis*- and *trans*-$[Pt(H_2NCH_2CO_2H)_2Cl_2]$, *trans*-$[PdCl_2(H_2NCH_2CO_2H)_2]$, and *cis*-$[PtCl(H_2NCH_2CO_2)(H_2NCH_2CO_2H)]$ show, together with a broad line due to H_2NCH_2, a relatively sharp line due to the protons in a $-CO_2H\cdots Cl-$ grouping. The exothermic process changing the shape and second moment of the ^1H n.m.r. signal has been observed for the chlorinated *cis* complexes.[1046] The broad-line ^{19}F n.m.r. spectrum of $M_n[PtF_6]_m$ has been investigated between room temperature and −100 to −150 °C. Anomalous chemical shifts were found. A tetragonal distortion of $[PtF_6]^{2-}$ was found which is related to six inequivalent platinum–fluorine bonds.[1047]

1039 H. W. Spiess, H. Zimmermann, and U. Haeberlen, *Chem. Phys.*, 1976, **12**, 123.
1040 A. M. Kalinichenko, V. I. Pavlishin, and I. V. Matyash, *Konst. Svoistva Miner.*, 1975, **9**, 48 (*Chem. Abs.*, 1976, **85**, 23 611).
1041 S. Iida, K. Mizushima, M. Mizoguchi, J. Mada, S. Umemura, K. Nakuo, and J. Yoshida, *AIP Conf. Proc.* 29 *Magn. Magn. Mater., Annu. Conf.*, 21*st*, 1976, 388 (*Chem. Abs.*, 1976, **85**, 55713).
1042 A. V. Zalesskii, A. M. Savvinov, I. S. Zheludev, A. N. Lobachev, A. N. Ivashchenko, and O. K. Mel'nikov, *Trudy Mezhdunar. Konf. Magn.*, 1974, **3**, 416 (*Chem. Abs.*, 1976, **84**, 51 844).
1043 T. Sakai and T. Tominaga, *Radiochem. Radioanalyt. Letters*, 1975, **23**, 329 (*Chem. Abs.*, 1976, **84**, 82 228).
1044 T. Sakai and T. Tominaga, *Bull. Chem. Soc. Japan*, 1976, **48**, 3168.
1045 R. D. Spence and J. L. Knirk, *Trudy Mezhdunar. Konf. Magn.*, 1973, **3**, 297 (*Chem. Abs.*, 1976, **84**, 53 057).
1046 V. S. Bondarenko, S. P. Gabuda, G. D. Mal'chikov, and N. A. Shestakova, *Zhur. strukt. Khim.*, 1976, **17**, 479.
1047 S. V. Zemskov, S. P. Gabuda, Yu. I. Nikonorov, V. A. Shipachev, and G. I. Zharkova, *Izvest. Sib. Otd. Akad. Nauk S.S.S.R., Ser. khim. Nauk*, 1976, **3**, 73 (*Chem. Abs.*, 1976, **85**, 101 744).

Spin–lattice relaxation time of the nuclear dipolar energy in a CF_3COOAg single crystal has been studied as a function of temperature and crystal orientation. The results were compared with the predictions of density matrix theory.[1048] ^1H and ^2H n.m.r. spectra have been studied for single crystals of $LiCuCl_3,2H_2O$ and $LiCuCl_3,2D_2O$, respectively, at room temperature. ^1H N.m.r. spectra were used to derive the orientation of the hydrogen–hydrogen vectors in the unit cell and the hydrogen–hydrogen distance in the water molecule and compared with a neutron diffraction study. ^2H N.m.r. studies showed that the H_2O molecules are flipping fast at room temperature. N.m.r. and *X*-ray data were combined to determine hydrogen atom positions.[1049] The suggestion that bromide vacancies in AgBr might be the cause of the observed increase at high temperatures of the ^{81}Br n.m.r. linewidths over the magnetic width *via* a quadrupole relaxation mechanism (lifetime broadening) was examined and confirmed.[1050] N.m.r. spectroscopy has proved to be a convenient and simple method to provide direct evidence for one-dimensional polymorphism in AgI. By studying quadrupole splitting on the ^{127}I resonance, the existence of polytypes was found.[1051]

The effect of temperature on the width and the second moment of the n.m.r. spectrum of some carbaborane-containing polymers has been examined; various phase transitions were observed which moved to higher temperatures with increasing crystallinity.[1052] ^{11}B N.m.r. spectroscopy has been used to investigate the structure and the nature of the bond in hexaborides[1053] and amorphous materials.[1054] Quadrupolar-broadened n.m.r. spectra have been measured for ^{17}O, ^{10}B, and ^{11}B in B_2O_3 glasses using samples specially enriched in ^{17}O and ^{10}B. Lineshape and ^{17}O n.m.r. spectra indicate the presence of two oxygen sites. Site 1 has a quadrupolar coupling constant; $Q_{cc} = 4.69$ MHz, and an asymmetry parameter $\eta = 0.58$ corresponding to oxygen atoms in boroxyl rings. Site 2 is characterized by $Q_{cc} = 5.6$ MHz and a large distribution of η about a value of 0.2 corresponding to oxygen-connecting boroxol rings. From one of the features of the ^{10}B n.m.r. spectrum, the distribution of ^{10}B coupling constants was obtained centred about the value $Q_{cc} = 5.51$ MHz (corresponding to $Q_{cc} = 2.64$ MHz for ^{11}B). A distribution about $\eta = 0.12$ was postulated for the boron asymmetry parameter which fits the remainder of the spectrum. These distributions agree with the ^{11}B spectrum.[1055]

The spin–lattice relaxation times of ^{11}B and ^{23}Na have been measured in B_2O_3, $Na_2B_4O_7$, and $(Na_2O)_{0.3}(SiO_2)_{0.7}$. In all cases the relaxation had been shown previously to be dominated by a strong quadrupolar mechanism. T_1 decreased

1048 M. Punkkinen and L. P. Ingman, *Chem. Phys. Letters*, 1976, **38**, 138.
1049 N. C. Vizia, P. N. Murty, C. R. K. Murty, and V. Nagarajan, *J. Magn. Resonance*, 1976, **22**, 439.
1050 L. G. Conti, *Gazzetta*, 1976, **106**, 393.
1051 D. Brinkmann and W. Freudenreich, *Z. Krist.*, 1976, **143**, 67 (*Chem. Abs.*, 1976, **84**, 128 448).
1052 V. S. Matochkin, A. I. Kalachev, E. I. Lyamenkova, P. M. Valetskii, S. V. Vinogradova, Yu. V. Zelenev, and V. V. Korshak, *Mater, Vees. Soveshch. Relaksatsionnym Yavleniyam Polim.*, *2nd*, 1974, **2**, 118 (*Chem. Abs.*, 1976, **85**, 63 521).
1053 M. L. Afanas'ev, E. M. Dudnik, V. K. Ermolaev, Yu. B. Paderno, and V. N. Shcherbakov, *V Sb., Khim. Svyaz' V Kristallakh i Ikh. Fiz. Svoistva.*, 1976, **1**, 154 (*Chem. Abs.*, 1976, **85**, 133 644).
1054 G. E. Peterson, C. R. Kurkjian, and A. Carnevale, *J. Non-crystalline Solids*, 1976, **21**, 283 (*Chem. Abs.*, 1976, **85**, 114 317).
1055 L. W. Panek, G. E. Jellison, jun., and P. J. Bray, *AIP Conf. Proc.* 31 *Struct. Excitations Amorphous Solids*, 1976, 79 (*Chem. Abs.*, 1976, **85**, 151 481).

between 1.2 and 273 K and is field independent up to 55 kG. In view of the available experimental results, the various models which have been proposed to account for n.m.r. relaxation in glasses are now unsatisfactory.[1056] The spin–lattice relaxation times of ^{11}B have been measured in crystalline and amorphous B_2O_3 over the ranges 100—500 K and 10—30 MHz. The results indicate a minimum in T_1 as a function of temperature near 300 K and an increase in T_1 thereafter. In the limited temperature range 200—100 K, T_1^{-1} is proportional to T^2. A theory was developed which explains many of the experimental properties of the nuclear T_2 relaxation, and it was postulated that low-frequency phonons exist whose lifetime is limited by structural relaxation of the two-level defects in the glass which have an average activation energy equivalent to 3150 K.[1057] ^{11}B and ^{27}Al n.m.r. spectra of Na_2O–B_2O_3–Al_2O_3 glasses have been investigated. The fraction of four-co-ordinate BO_4 units was determined by computer analysis. The ^{27}Al n.m.r. spectrum indicates that there is no abrupt change in the average aluminium environment. Certain linear relationships were found which yield detailed information on the competing processes of BO_3, BO_4, and AlO_4 formation and the formation of triclusters consisting of three tetrahedra having one oxygen atom in common.[1058] ^{11}B N.m.r. spectroscopy has been used to show three states of tetrahedral boron atoms in sodium boron silicate glasses.[1059] ^{11}B and ^{207}Pb n.m.r. spectra have been used to study the structure of glasses in the system PbO–B_2O_3–SiO_2. The results indicate that the glasses do not contain an appreciable number of $[BO_3]^-$ units and that the network-forming lead atoms play an important role in compositional behaviour. It was concluded that the Pb^{2+} atoms are shared evenly between the BO and SiO networks.[1060] ^{11}B N.m.r. spectra of $[AsR_nCl_{4-n}]^+[BCl_4]^-$ have also been reported.[1061]

A combination of conventional and multiple-pulse n.m.r. measurements has been used to examine the hydroxy protons in a series of high surface area silica aluminas. The protons remaining after calcination at 500 °C exist as immobile and widely separated hydroxy-groups at room temperature. The concentrations of SiOH and AlOH groups were reported as a function of adsorbant compound. In samples of high silica content no AlOH groups were detectable, but in those containing 50% or less SiO_2 most of the surface protons were present as AlOH.[1062] The temperature and pressure dependence of two of the n.m.r. transition frequencies of ^{27}Al in $LiAl(SiO_3)_2$ have been measured over the ranges 4.2—300 K and 0—7 kbar. Reasonable agreement was found between experiment and theory.[1063] The second moment of the ^{19}F n.m.r. line is reproduced by calculation if fluoride is assumed to replace oxygen atoms in the lattice of muscovite, in contrast to the usual assumption that it substitutes in an OH group

1056 J. Szeftel and H. Alloul, *AIP Conf. Proc.* 31 *Struct. Excitations Amorphous Solids*, 1976, 74 (*Chem. Abs.*, 1976, **85**, 133 656).
1057 M. Rubinstein and H. A. Resing, *Rep. NRL Prog.*, 1975, 24 (*Chem. Abs.*, 1976, **84**, 142 836).
1058 R. Gresch, W. Mueller-Warmuth, and H. Dutz, *J. Non-crystalline Solids*, 1976, **21**, 31 (*Chem. Abs.*, 1976, **85**, 98 276).
1059 S. P. Zhdanov and G. Schmidel, *Fiz. Khim. Stekla*, 1975, **1**, 452 (*Chem. Abs.*, 1976, **84**, 183 937).
1060 K. S. Kim, P. J. Bray, and S. Merrin, *J. Chem. Phys.*, 1976, **64**, 4459.
1061 K. B. Dillon, R. J. Lynch, and T. C. Waddington, *J.C.S. Dalton*, 1976, 1478.
1062 L. B. Schreiber and R. W. Vaughan, *J. Catalysis*, 1975, **40**, 226 (*Chem. Abs.*, 1976, **84**, 22 548).
1063 S. F. Harley, C. D. Williams, and R. F. Tipsword, *J. Chem. Phys.*, 1976, **64**, 4815.

position.[1064] It has been shown, using powdered $H_2AlP_3O_{10},2H_2O$ and related compounds, that magnetic configurations either with two protons or with three protons at the vertices of an equilateral triangle are usually inadequate to calculate ^1H n.m.r. curve-shape functions of powdered hydrated hydrogeno salts. It is better to use instead the isosceles triangle magnetic configuration model proposed by Andrew and Finch.[1065] The thermal decomposition of $KAl_2(OH)(PO_4)_2,2H_2O$, $K_2Al_3(OH)_2(PO_4)_3,1.5H_2O$, $RbAl_2(OH)(PO_4)_2,2H_2O$, $RbAl_2(OH)_3(HPO_4)_2$, $NH_4Al(OH)(PO_4)_2,2H_2O$, and $Li_4Al(OH)(PO_4)_2$ have been investigated by n.m.r. spectroscopy.[1066] Samples of micas have been examined by ^1H and ^{19}F n.m.r. spectroscopies and doublets found due to F–F_A and F–H_A pairs. The number of F–F_A pairs was always more than was calculated from chance distribution.[1067]

The temperature-dependent n.m.r. spectrum of opal has been studied. Tentative assignments of lines to chemically bonded OH groups and water in cavities were made.[1068] ^7Li N.m.r. spectroscopy has been used to study the effect of chemical elements on the arrangement of lithium ions in silicate glasses.[1069, 1070] In the ^{29}Si n.m.r. spectra of Na_2O–SiO_2 and K_2O–SiO_2 it has been shown that the local magnetic fields are *ca.* 10 times smaller than those deduced from ^{23}Na n.m.r. spectroscopy, suggesting that the glasses have a microstructure which consists of different regions with either much higher or lower than average alkali content.[1071] ^1H N.m.r. spectroscopy has been used to investigate sodium carbonate–SiO_2 mixtures.[1072] The ^{19}F n.m.r. spectra of PbF_2–SiO_2–PbO melts show cluster formation.[1073] ^{207}Pb N.m.r. signals have been observed in β-PbF_2 using heteronuclear polarization transfer.[1074]

The temperature dependence and thermal hysteresis of ^1H and ^2H spin–lattice relaxation times at high fields, and a study of the recovery of ^1H and ^2H magnetization in ammonium halides, have been reported. The anomalies occurring near the order–disorder phase transition were related to structural, dielectric, and Raman data. The hysteresis effect and the establishment of long-range order were shown to be related.[1075] The spin–lattice relaxation time of polycrystalline NH_4Br has been measured as a function of temperature and pressure to complete a phase diagram.[1076] ^{11}B and ^{31}P n.m.r. spectra of a number of solid Me, Et, and

[1064] A. M. Belongov, A. G. Grammakov, N. M. Mikhailov, N. B. Strakhov, and Yu. L. Chernov, *Izvest. Leningr. Elektrotekh. Inst.*, 1974, **141**, 12 (*Chem. Abs.*, 1976, **85**, 195 331).
[1065] C. Dorémieux-Morin, *J. Magn. Resonance*, 1976, **21**, 419.
[1066] I. I. Boldog, A. M. Golub, and A. M. Kalinichenko, *Zhur. neorg. Khim.*, 1976, **21**, 684.
[1067] J. Sanz and W. E. E. Stone, *Compt. rend.*, 1976, **282**, *B*, 361.
[1068] A. M. Belonogov, A. G. Grammakov, N. M. Mikhailov, and N. B. Strakhov, *Izvest. Leningr. Elektrotekh. Inst.*, 1974, **141**, 3 (*Chem. Abs.*, 1976, **85**, 110 829).
[1069] M. L. Zlatogorskii, T. A. Sidorov, and V. S. Grechishkin, *V Sb., Radiospektroskopiya*, 1974, **8**, 25 (*Chem. Abs.*, 1976, **84**, 128 459).
[1070] M. L. Zlatogorskii, T. A. Sidorov, and V. S. Greschishkin, *V Sb., Radiospektroskopiya*, 1974, **8**, 17 (*Chem. Abs.*, 1976, **84**, 128 460).
[1071] B. D. Mosel, W. Mueller-Warmuth, and H. Dutz, *Phys. Chem. Glasses*, 1974, **15**, 154 (*Chem. Abs.*, 1976, **84**, 34 610).
[1072] I. M. Kiselev, A. G. Brekhunets, and V. V. Mank, *V Sb., Fiz.-Khim. Issled. Neorgan. Soedin.*, 1975, **1**, 84 (*Chem. Abs.*, 1976, **85**, 27 121).
[1073] I. A. Novokhatskii, V. K. Prokopenko, and V. I. Arkharov, *Svar. Flyusy Shlaki, Dokl. Mezhdunar. Semin. Vopr. Issled., Proizvod. Primen. Flyusov Shlakov Svarki Spets. Elektrometall.*, 1974, 27 (*Chem. Abs.*, 1976, **84**, 49 095).
[1074] M. E. Stoll, W.-K. Rhim, and R. W. Vaughan, *J. Chem. Phys.*, 1976, **64**, 4808.
[1075] A. R. Sharp and M. M. Pintar, *Chem. Phys.*, 1976, **15**, 431.
[1076] W. Mandema and N. J. Trappeniers, *Physica B + C (Amsterdam)*, 1976, **81**, 285 (*Chem. Abs.*, 1976, **84**, 187 199).

PhCl phosphoranes and their ionic 1 : 1 adducts with Lewis acids such as PCl_5, $SbCl_5$, BCl_3, $AlCl_3$, or ICl have been recorded. The results show that the compounds PMe_nCl_{5-n}, PEt_2Cl_3, PEt_3Cl_2, and PPh_3Cl_2 have ionic structures of the type $[PR_nCl_{4-n}]^+Cl^-$ in the solid state whereas PPh_2Cl_3 and $PPhCl_4$ are molecular. The adducts all have structures of the type $[PR_nCl_{4-n}]^+$(Lewis acid)Cl^-.[1077] N.m.r. spectra of a series of single crystals and powders of natural and synthetic apatites and phosphonites have been examined before and after calcination at 1000 °C.[1078] The n.m.r. spectrum of the transition between the lowest levels of an $I = \frac{5}{2}$ nucleus (^{121}Sb in Sb_2O_3 powder) subject to a strong quadrupole interaction of zero asymmetry and a weaker Zeeman interaction has been examined. Powder patterns were simulated in first-order Zeeman perturbation with emphasis on dipole broadening on the edge discontinuity near $\nu = 3\nu_0$. Numerical evaluation of the exact theory was used to determine the transverse chemical shift component $\sigma_\perp$ and the strength ν_Q of the quadrupole interaction from the field dependence of the resonance frequency.[1079] ^1H N.m.r. spectroscopy was also used to study the hydration of sulphide minerals.[1080]

^1H and ^{19}F wideline dipolar spectra of a single crystal of KHF_2 have been recorded at various orientations to the magnetic field. Due to the small hydrogen–fluorine distance in the bifluoride ion, the proton intramolecular splitting is unusually large (maximum splitting is 293 kHz) yielding exceptionally well-resolved five-line spectra. The splitting allows an unusually precise determination of the hydrogen–fluorine distance of 1.153 ± 0.005 Å, which is in good agreement with neutron diffraction data and confirms that the proton occupies a symmetric position between the hydrogen-bonded fluorine atoms. The calculated proton intermolecular broadening fully accounts for the angular dependence of the width of the various lines in the dipolar spectrum. The wideline ^{19}F spectrum shows a clear asymmetry due to the anisotropy of the magnetic shielding. From the experimental spectra, the following principal values of the ^{19}F shielding tensor relative to C_6F_6 were deduced: $\sigma_{xx} = -59.7 \pm 5$, $\sigma_{yy} = -43.4 \pm 5$, and $\sigma_{zz} = +32.5 \pm 5$ p.p.m.[1081] In a related ^1H n.m.r. study, the ^1H magnetic shielding tensor of the hydrogen bond in $[FHF]^-$ in KHF_2 was determined at 93 MHz by applying a fluorine-decoupled proton multiple pulse technique on the central line of the dipolar spectrum. The principal values with respect to Me_4Si of the nearly axially symmetric tensor are $\sigma_{xx} = -39.7 \pm 1.0$, $\sigma_{yy} = -32.4 \pm 1.0$, and $\sigma_{zz} = +8.8 \pm 1.0$ p.p.m. The most deshielded direction is the bond axis. The anisotropy of 44.8 p.p.m. is the largest proton anisotropy so far reported. The results were analysed using Gierke and Flygare's atom–dipole approximation of the magnetic shielding tensor. Theory and experiment were brought into close agreement by introducing empirically a reduced mean square electron cloud radius which accounts for the overlap of the shielded nucleus

1077 K. B. Dillon, R. J. Lynch, R. N. Reeve, and T. C. Waddington, *J.C.S. Dalton*, 1976, 1243.
1078 G. Knubovets and L. D. Kislovskii, *Kristallokhim. Miner. Geol. Probl.*, 1975, 212 (*Chem. Abs.*, 1976, **84**, 182 543).
1079 E. von Meerwall, R. B. Creel, and C. F. Griffin, *J. Chem. Phys.*, 1976, **65**, 4370.
1080 M. S. Shelestov, R. A. Skormina, E. M. Li, A. A. Petrov, and B. A. Savostin, *Trudy N.-I. Proekt. In-T Po Obogashch. Red Tsvet. Met. Kazmekhanobr Sb.*, 1973, **11**, 136 (*Chem. Abs.*, 1976, **84**, 93 192).
1081 P. Van Hecke, H. W. Spiess, U. Haeberlen, and S. Haussühl, *J. Magn. Resonance*, 1976, **22** 93.

by the electrons of the neighbouring nuclei. The analysis of the hydrogen shielding tensor suggests that the hydrogen–fluorine bond in the $[FHF]^-$ is strongly ionic.[1082] The hydrogen–fluorine distance in $Cs^+[FHF]^-$ has been determined by temperature-dependent n.m.r. studies. The phase transition from α- to β-modifications in $CsHF_2$ is accompanied by the appearance of isotropic reorientation of the $[FHF]^-$ ions in interstitial sites.[1083] The n.m.r. spectrum of $TlHF_2$ has been measured and the fluorine co-ordinates determined. The thallium–fluorine distance is 2.56 Å, which is shorter than the sum of the ionic radii, owing to the partial covalent character of the thallium–fluorine bond. A phase transition was detected at 83 °C, the structure of the high-temperature phase being assigned to the CsCl type. The dynamics of $[HF_2]^-$ was examined and at high-temperature diffusion was observed.[1084] The high-resolution ^1H n.m.r. spectrum of solid $[(n\text{-hexyl})_4N]^+[ClO_4]^-$ was examined and four well-defined resonances were found showing that the chain moves freely.[1085]

Molecules Sorbed onto Solids.—This section is subdivided into two subsections: water sorbed onto solids, and other molecules sorbed onto solids.

Water Sorbed onto Solids. Important differences have been observed between the ^1H n.m.r. spectrum of free water and spectra taken with thin adsorption films of solution on the (001) face of $KClO_3$, the (100) face of $NaNO_2$, and the (010) face of KNO_3.[1086] The n.m.r. spectra of water adsorbed on apatites and human teeth, $M_{10}(PO_4)_6X_2$ (M = Ca, Sr, or Ba; X = OH or F), have been observed. The linewidth of the adsorbed water varied as a function of the amount of water adsorbed.[1087] The ^1H n.m.r. spectra of water adsorbed onto TiO_2 indicate the presence of two types of OH groups with interprotonic distances of 2.7 and 4.0 Å, respectively, as well as two types of adsorbed water.[1088] ^1H N.m.r. spectroscopy has been used to study water adsorbed on ThO_2 catalyst.[1089] The ^1H n.m.r. spectra of dry nickelates exhibit a sharp singlet of a weakly bound water, which is removed at 120 °C, and asymmetric lines which were assigned to $[OH]^-$ and $[H_3O]^+$ groups. With $Co(OH)_2$, the symmetric ^1H n.m.r. spectrum is changed upon oxidation to an asymmetric cobalt hydroperoxide line.[1090]

The ^1H n.m.r. line-broadening mechanisms of adsorbed species in zeolites have been examined. The dominant mechanism originates from inhomogeneous magnetic fields inside zeolite cavities due to bulk susceptibility effects, although this is partially narrowed by diffusion over interparticulate dimensions.

1082 P. Van Hecke, H. W. Spiess, and U. Haeberlen, *J. Magn. Resonance*, 1976, **22**, 103.
1083 B. D. Zil'berman, T. D. Fedotova, and S. P. Gabuda, *Zhur. strukt. Khim.*, 1976, **17**, 273.
1084 B. D. Zil'berman, N. K. Moroz, M. Khaitova, D. D. Ikrami, and S. P. Gabuda, *Zhur. strukt. Khim.*, 1975, **16**, 765.
1085 E. M. Barrall, T. T. Horikawa, and J. T. S. Andrews, *Thermochim. Acta*, 1976, **14**, 99 (*Chem Abs.*, 1976, **84**, 82 216).
1086 V. V. Sipyagin, A. A. Chernov, E. I. Fedin, and P. O. Okulevich, *Kristallografiya*, 1976, **21**, 370 (*Chem. Abs.*, 1976, **85**, 114 228).
1087 K. Umezawa, M. Hiramatsu, N. Takai, T. Yamabe, Y. Umezawa, T. Nishizawa, and M. Chiba, *Bunseki Kagaku*, 1975, **24**, 79 (*Chem. Abs.*, 1976, **84**, 67 401).
1088 V. S. Il'enko and A. V. Uvarov, *Kolloid. Zhur.*, 1975, **37**, 1161 (*Chem. Abs.*, 1976, **84**, 50 213).
1089 W. S. Brey, B. W. Martin, B. H. Davis, and R. B. Gammage, in 'Magnetic Resonance in Colloid and Interface Science' (Symposium No. 34), The American Chemical Society, Washington, 1976, p. 216 (*Chem. Abs.*, 1976, **85**, 198 782).
1090 I. S. Shamina, I. K. Kuchkaeva, P. N. Bityutskii, and L. A. Chetovskaya, *Spektrosk. EE Primen. Geofiz. Khim.*, 1975, 338 (*Chem. Abs.*, 1976, **84**, 67 124).

Chemical exchange reactions among ammonium ions, water molecules, and lattice hydroxy-groups are then differentiated and discussed semiquantitatively. Lattice OH groups were shown to be strongly acidic and apparently lead to structure degradation above *ca.* 16 per unit cell. Proton exchange between $[NH_4]^+$ and H_2O has been shown to occur *via* deammoniation of $[NH_4]^+$ at measurable rates. In contrast to aqueous solutions, rapid quadrupolar relaxation of ^{14}N in $[NH_4]^+$ occurs inside the zeolite cavities. The results provide a powerful method of obtaining structural, kinetic, and dynamic information about simple chemical processes inside zeolite cavities in comparison with aqueous solutions.[1091]

The ^1H chemical shifts of H_2O measured in swollen ion exchanger KU-2 charged with H^+, Cs^+, Rb^+, K^+, Na^+, Li^+, $[NH_4]^+$, Ba^{2+}, Sr^{2+}, Ca^{2+}, Mg^{2+}, Be^{2+}, Zn^{2+}, Cd^{2+}, Hg^{2+}, Al^{3+}, Y^{3+}, La^{3+}, Sc^{3+}, Ga^{3+}, and In^{3+} and in the ion exchanger AV-17 containing F^-, Cl^-, Br^-, and I^- have been correlated with the radii and charges of the ions. Anomalous behaviour of exchangers containing Rb^+, Cs^+, and ions with outer *d*-orbitals was discussed.[1092] Linewidths and chemical shifts of Li^+, F^-, $[NH_4]^+$, and H^+ in aqueous dispersions of a number of ion-exchange resins and aluminosilicates in solutions of high molecular weight compounds have been measured. The shift of the water signals against temperature is smaller for electrolytes than water, and hydration numbers were determined. The activation energy for molecular motion was calculated from the temperature dependence of the linewidth as being 8—16 kcal mol^{-1}, in good agreement with previous work.[1093]

^{23}Na Spin–lattice relaxation times have been determined at 10 MHz on Dowex 50W-XZ ion exchange resin as a function of the dielectric constant of the solvent. The relaxation time varied from 24.8 ms in distilled water (D = 78.5) to values too short to be determined in water–ethanol mixtures below an electric permittivity of 35. Similar studies were conducted on 0.25M-$NaHSO_3$ in a series of aqueous–non-aqueous mixtures. The results indicate that T_1 may be expressed as a function of the macroscopic electric permittivity for any binary solvent system. For systems with an electric permittivity greater than 65, the relationship is given by $T_1 = a\exp(bD)$, where a and b are constant for a particular solvent system.[1094] N.m.r. spectroscopy has been used to study the state of hydrogen (H^+, $[OH]^-$, and $[OH_3]^+$) in hydrated NaScY and NaZrY zeolites, and to show that the interproton distances in water molecules in zeolites increase with the formation of a water–Sc or water–Zr co-ordinate bond. The hydrated zeolites contain less than 5 mole % water in the form of $[H_3O]^+$. The zeolites contain isolated $[OH]^-$ and H^+ which give narrow n.m.r. bands and which are not embedded into the zeolite skeleton structure and are considered as inert mechanical impurities. The intensity of these narrow bands increases with the increasing number of repeated hydration–dehydration cycles. The dehydrated zeolites do not contain H^+ centres.[1095] Nuclear magnetic relaxation

1091 E. B. Whipple, P. J. Green, M. Ruta, and R. L. Bujalski, *J. Phys. Chem.*, 1976, **80**, 1350.
1092 V. V. Mank, V. D. Grebenyuk, N. P. Gnusin, and F. D. Ovcharenko, *Fiz. Tverd. Tela*, 1974, 305 (*Chem. Abs.*, 1976, **85**, 184 422).
1093 O. D. Kurilenko and V. V. Mank, *Croat. Chem. Acta*, 1976, **48**, 505.
1094 M. Goldsmith, D. Hor, and R. Damadian, *J. Chem. Phys.*, 1976, **65**, 1708.
1095 M. M. Selim, S. P. Gabuda, A. A. Panich, Z. V. Grayaznova, and N. K. Moroz, *Doklady Akad. Nauk S.S.S.R.*, 1975, **224**, 1119.

of water protons in zeolites of the faujasite type has been measured. Two regions, A and B, of relaxation were found, the intensities and relaxation times of which were determined at different silicon : aluminium ratios, contents of Fe^{3+}, coverages, temperatures, and D_2O dilutions. Region A is given by mobile water in the large cavities and region B seems to consist of water molecules inside the cubo-octahedra. Intensity and spin–spin relaxation times of region B are independent of silicon : aluminium ratio and content of Fe^{3+} and are produced by hydrogen–hydrogen interaction. Exchange between regions A and B was not observed and can only take place on a time-scale of seconds. The equilibrium of sorption is shifted by temperature in favour of the region A. At low coverages, region B is occupied predominantly.[1096] These observations have two important consequences. (i) The number of water molecules per cubo-octahedron can be determined by n.m.r. spectroscopy. (ii) The results derived from n.m.r. relaxation measurements are valid only for water molecules adsorbed in the supercages of the zeolites since the short phase does not contribute to the signal under certain conditions. The transverse relaxation time is much less than that of the water molecules adsorbed in the supercages. It corresponds to a correlation time τ_{ck} which is, at room temperature, *ca.* 500 times greater than the correlation time of water molecules in the supercages at higher pore-filling factors.[1097] ^1H and ^2H n.m.r. spectra have been reported for H_2O/D_2O adsorbed on NaX zeolite. A quadrupolar relaxation mechanism dominates ^2H relaxation. The coupling constant is 225 kHz for ^2H in physically adsorbed water. This was proposed as a method for estimating the quadrupolar coupling constant of a liquid. The Al—O—D angle is $121.5 \pm 0.02°$ and the aluminium–oxygen distance is 1.68 ± 0.20 Å in the complex.[1098] The temperature dependence of the ^1H n.m.r. relaxation time of water adsorbed on various faujasite-type zeolites has been determined at a high pore-filling factor from -50 to 200 °C. Possible structures of the hydrated zeolites were based on these studies and related studies on water and aqueous salt solutions.[1099] The interaction of water with aluminosilicates containing Li^+ and other cations has been studied. The aluminosilicates were montmorillonite, vermiculite, phlogopite, zeolites A and X, and mordenite. Silica gel was also studied. The adsorbed water molecules can be described by a two-phase mode: molecules with anisotropic rotation and molecules with a random reorientation in the interpocket space. The water–aluminosilicate interaction depends on pore size and the chemical nature of the surface and may be either chemi- or physi-sorption.[1100] Sodium Llano vericulite has been studied by ^1H n.m.r. spectroscopy and a doublet and a singlet were found. The doublet splitting is orientation dependent between 0 and 50 °C and orientation independent below -60 °C. An octahedral distribution of water around Na^+ fits the data; the six rotation axes around which the water molecules are spinning rapidly

1096 W. D. Basler, H. Lechert, and H. Kacirek, *Ber. Bunsengesellschaft phys. Chem.*, 1976, **80**, 451.
1097 H. Pfeifer, *Surface Sci.*, 1975, **52**, 434 (*Chem. Abs.*, 1976, **84**, 24 173).
1098 H. A. Resing, *J. Phys. Chem.*, 1976, **80**, 186.
1099 A. Gutsze, *Uniw. Adama Mickiewicza Poznaniu, Wydz. Mat., Fiz. Chem.* (*Pr.*), *Ser. Fiz.*, 1975, **19**, 321 (*Chem. Abs.*, 1976, **84**, 185 345).
1100 F. D. Ovcharenko, V. V. Mank, and A. G. Brekhunets, *Spektrosk. At. Mol., Tezisy Dokl. Sess. Kom. Opt. Spektrosk. Akad. Nauk Ukrain. S.S.R.*, 1974, 29 (*Chem. Abs.*, 1976, **85**, 167 127).

being tilted by 65 °C with respect to the C* axis. The cation hydration shell is characterized by rotational diffusional motion with an activation energy of 8.5 kcal mol^{-1}.[1101] High-resolution ^1H n.m.r. spectra of water on the surface of the H^+, Mg^{2+}, and Na^+ forms of kaolinite with different water contents have been recorded. Five water retentions, differing in mobility of the water molecules, were detected in the dispersed cation-exchanged kaolinite. The energy state of the surface-bound water differed with the type of exchangeable cation. At high concentrations of the mineral and at low moisture contents, Mg^{2+} and H^+ have approximately the same effect on water molecules. The Mg^{2+} form of kaolinite bonds water more strongly than the Na^+ form over the entire range of mineral concentrations.[1102] N.m.r. spectroscopy has also been used to study the differential porosity of aerated structural materials.[1103]

A spin-echo n.m.r. study of adsorbed water on carbon has been investigated from −175 to 95 °C. The longitudinal relaxation at temperatures below −50 °C is due to dipolar proton–proton interactions and at higher temperatures to proton–electron interactions. The latter did not contribute to transversal relaxation.[1104] On the basis of a study of ^1H and ^{19}F n.m.r. spectra in hydrated and fluorinated samples of SiO_2, conclusions have been reached regarding the character of the surface of the hydrated coatings. The role of hydration in the creation of proton acid centres on the surface was investigated.[1105] A study of the surface electrical conductivity and surface mobility of water films adsorbed on quartz as a function of the number of monolayers has shown that the increasing mobility of protons when clusters of adsorbed water coalesce is mainly responsible for the increase in conductivity with increasing surface coverage. N.m.r. spectra of adsorbed water films on hydrophilic and hydrophobic quartz surfaces, on colloidal silica particles, and on an aqueous dispersion of colloidal silica indicate that the temperature dependence of the surface conductivity is very sensitive to the appearance of a mobile phase of water on the thawing of a disperse system.[1106] The temperature dependence (180—310 K) of the linewidth and chemical shift of high-resolution n.m.r. spectra of methanol, acetonitrile, and water molecules adsorbed on non-pressed Aerosil has been investigated. The dependence of the OH linewidth of MeOH and H_2O on sample temperature was conditioned by H^+ exchange between the surface silanol groups and OH groups of adsorbed molecules. The character of the MeOH and MeCN molecular motions on the Aerosil surface changed at *ca.* 123 to *ca.* 203 K and at *ca.* 240 K, respectively. The analysis of the chemical shifts indicated that the association of adsorbed methanol molecules increased with decreasing temperature and increasing surface filling.[1107]

[1101] J. Hougardy, W. E. E. Stone, and J. J. Fripiat, *J. Chem. Phys.*, 1976, **64**, 3840.
[1102] A. D. Voronin, V. F. Babanin, V. I. Malinovskii, and Yu. E. Shapiro, *Pochvovedenie*, 1976, **4**, 156 (*Chem. Abs.*, 1976, **85**, 49 304).
[1103] I. Ya. Kiselev, *Trudy Nii Stroit. Fiz. Gosstroi S.S.S.R.*, 1975, **10**, 3 (*Chem. Abs.*, 1976, **85**, 129 524).
[1104] R. Hummel, *Z. Phys. Chem.* (*Leipzig*), 1976, **257**, 81.
[1105] G. F. Golovanova, *Zhur. fiz. Khim.*, 1976, **50**, 2036.
[1106] A. B. Kurzaev, S. N. Kozlov, and V. F. Kiselev, *Doklady Akad. Nauk S.S.S.R.*, 1976, **228**, 877.
[1107] G. Winde, A. V. Volkov, A. V. Kiselev, and V. I. Lygin, *Zhur. fiz. Khim.*, 1975, **49**, 2913.

Other Molecules Sorbed onto Solids. The spin–lattice relaxation times of ^3He on porous material are anomalously low with respect to the minimum allowable value for internuclear dipolar interaction.[1108] A method of correcting chemical shifts of chemisorbed molecules of bulk susceptibility using physisorbed gases has been presented.[1109]

Wideline ^1H n.m.r. spectroscopy has been used to support and extend the i.r.-based distinction between chemisorbed and physisorbed species. Physisorbed pyridine or thiophen on molybdena–alumina show liquid-like n.m.r. behaviour. Chemisorbed pyridine is much more rigidly held and only undergoes C_2 motion.[1110] The n.m.r. spectrum of propylene adsorbed on a sample of MoO_3 on MgO was recorded. Displaced lines were detected for the protons of the propylene in a weak field which is produced by surface complexes of propylene.[1111] The n.m.r. spectra of benzene adsorbed on Co^{2+} could not be explained by pseudocontact interactions. A rapid exchange of physisorbed molecules with molecules entering the co-ordination sphere of Co^{2+} was observed. The effective number of paramagnetic centres was determined.[1112] ^1H and ^{63}Cu n.m.r. spectroscopies have been used to investigate the absorption of H_2 on Cu. The activation energy for surface migration of hydrogen atoms is 2.0 ± 0.3 kcal mol^{-1}.[1113]

^1H N.m.r. spectroscopy has been used to follow the rate of *ortho–para* conversion in liquid hydrogen in contact with nickel–alumina catalysts.[1114] N.m.r. pulse techniques have been used to determine the temperature dependence of T_1 and T_2 of n-butene and n-butane molecules adsorbed on CaNaA zeolites with different contents of Ca^{2+} ions. The observed diminution of correlation times with rising degree of Ca^{2+} exchange could only be explained if translational motions dominate proton spin relaxation of the adsorbates.[1115] n-Hexane on NaX zeolites have been examined by ^1H n.m.r. spectroscopy for self-diffusion.[1116] Using n.m.r. pulsed field gradient techniques, the temperature and pore-filling factor dependence of self-diffusion coefficients of cyclohexane and cyclohexene adsorbed on NaX-zeolites has been measured. Whereas the activation energies for the diffusion of adsorbed cyclohexane molecules (2.1—3.0 kcal mol^{-1}) are comparable with the corresponding values in the free liquid, the activation energies for the specifically adsorbed cyclohexane molecules are considerably higher (4.9 kcal mol^{-1}). In contrast to cyclohexene, the self-diffusion of cyclohexane molecules is characterized by a sharp decrease at higher pore-filling factors ($\theta > 0.6$). This fact is in accordance with results of n.m.r. relaxation time measurements which indicate that with increasing pore-filling factor, the adsorbed molecules hinder mutually and consequently possess a shorter mean free jump length.[1117] The ^1H n.m.r. spectra of butenes adsorbed on NaX zeolites have been reported. The rotational frequency of the adsorbed molecules

1108 N. S. Sullivan, *J. Low Temp. Phys.*, 1976, **22**, 313 (*Chem. Abs.*, 1976, **84**, 113 884).
1109 J. L. Bonardet and J. P. Fraissard, *J. Magn. Resonance*, 1976, **22**, 543.
1110 L. Petrakis and F. E. Kiviat, *J. Phys. Chem.*, 1976, **80**, 606.
1111 V. M. Mastikhin, A. A. Davydov, M. A. Fedotov, and A. A. Budneva, *Zhur. fiz. Khim.*, 1976, **50**, 458.
1112 W. D. Hoffmann, *Kinetika i Kataliz*, 1976, **17**, 718.
1113 T. Ito and T. Kadowaki, *Japan J. Appl. Phys.*, 1975, **14**, 1673 (*Chem. Abs.*, 1976, **84**, 10 722).
1114 J. Tourneur, A.-P. Legrand, and D. Marchand, *Compt. rend.*, 1975, **281**, *C*, 841.
1115 D. Michel and V. Roessiger, *Surface Sci.*, 1976, **54**, 463 (*Chem. Abs.*, 1976, **84**, 113 967).
1116 J. Kärger, M. Bülow, and N. van Phat, *Z. phys. Chem.* (*Leipzig*), 1976, **257**, 1217.
1117 J. Kärger, P. Lorenz, H. Pfeifer, and M. Bülow, *Z. phys. Chem.* (*Leipzig*), 1976, **257**, 209.

was calculated from the minimum in the temperature dependence of the longitudinal proton spin–lattice relaxation time and was compared with an intracrystal self-diffusion coefficient as found by n.m.r. pulsed field gradient techniques. It was found that the *cis*-isomer interacts more strongly than *trans*-but-2-ene with Na^+.[1118] n-Butenes adsorbed on NaAgX and NaTlX have been studied by ^1H spin-echo and pulsed field gradient techniques. Ag^+ reduces diffusion, and ^{13}C n.m.r. spectroscopy has been used to demonstrate butene binding. Tl^+ increases mobility.[1119] ^{13}C N.m.r. spectra of alkenes adsorbed on NaX, NaY, or TlX faujasites and on Al_2O_3–Na_2O indicate the presence of alkene–Na^+ bonds. Na^+ did not interact with the alkenes in solution. ^{13}C N.m.r. spectra indicate that Ag^+ interacts with alkenes in solution and in the adsorbed states. The temperature dependence of the ^{13}C n.m.r. spectrum of but-1-ene over NaCaY indicates that universal delocalization of π-electrons occurred over the whole volume, ruling out carbonium and carbanion intermediates.[1120, 1121] The mobility of benzene molecules in cation-exchanged synthetic zeolites has been studied *via* n.m.r. linewidths. The mobility remained high for Ca^{2+} and La^{3+} forms and the Na^+ forms with Si : Al ratio <2.43, but decreased sharply at *ca.* 5 molecules of benzene per cavity for Na^+ zeolites with Si : Al ratio >2.43. The higher mobility in La^{3+}-exchanged zeolites containing cavities with nearly no cations can account for the increased catalytic activity regardless of the catalyst active centres.[1122] Wideline ^1H n.m.r. spectroscopy has been used at 77 K in the study of benzene adsorbed on NaX and NaY zeolites. The second moment of benzene does not depend on the pore-filling factor in the range θ = 0.2—0.85. This was explained by anisotropic rotation of the molecules around the six-fold axis of symmetry and by interaction between Na^+ ions and the aromatic π-electrons.[1123] A linear increase of the benzene relaxation time with increasing concentration has been observed for aluminosilicate catalysts containing Fe^{3+} or Mn^{2+}, or washed free of metal ions.[1124] ^{13}C N.m.r. spectra of several amines, especially pyridine and aniline, adsorbed on SiO_2, Al_2O_3, and mixed SiO_2–Al_2O_3 have been measured. On SiO_2 only weak interactions with surface hydroxy-groups were observed. An anomalous case is *NN*-diethylaniline, which shows a large C-2 chemical shift. On Al_2O_3 stronger interactions were observed. There was no evidence however of large chemical shifts such as occur on protonation of corresponding amines in aqueous acids. The strong acid sites on Al_2O_3 are sterically hindered. On SiO_2–Al_2O_3, chemical shifts due to protonation were observed. ^{13}C Chemical shifts can be used to estimate the concentration of surface proton acids.[1125]

1118 J. Kärger and D. Michel, *Z. phys. Chem.* (*Leipzig*), 1976, **257**, 983.
1119 J. Kärger, D. Michel, A. Petzold, J. Caro, H. Pfeifer, and R. Schöllner, *Z. phys. Chem.* (*Leipzig*), 1976, **257**, 1009.
1120 D. Michel, W. Meiler, and H. Pfeifer, *Catal., Proc. Internat. Symp.*, 1975, 199 (*Chem. Abs.*, 1976, **84**, 120 838).
1121 D. Michel, W. Meiler, and H. Pfeifer, *J. Mol. Catalysis*, 1976, **1**, 85 (*Chem. Abs.*, 1976, **84**, 163 687).
1122 H. Lechert, W. Haupt, and K. P. Wittern, *J. Catalysis*, 1976, **43**, 356 (*Chem. Abs.*, 1976, **85**, 99 750).
1123 W. D. Hoffmann, *Z. phys. Chem.* (*Leipzig*), 1976, **257**, 315.
1124 M. D. Danchev, V. V. Erchenkov, and G. M. Lanchenkov, *Neft Gaz*, 1974, 152 (*Chem. Abs.*, 1976, **84**, 16 427).
1125 I. D. Gay and S. Liang, *J. Catalysis*, 1976, **44**, 306 (*Chem. Abs.*, 1976, **85**, 182 747).

A pulse n.m.r. study of benzene adsorbed on graphitized carbon black has been carried out and T_1 and T_2 were measured. The benzene molecules are adsorbed at fixed sites with planes perpendicular to the graphite basic planes. The molecules undergo reorientational jumps through 60° to the next position. The adsorbed benzene layers have two-dimensional ordered symmetry.[1126] T_1 and T_2 of benzene on SiO_2 were found to be 100—200 ms and are about an order of magnitude longer than with γ-Al_2O_3. With SiO_2 saturated with $Al(NO_3)_3$, the relaxation time decreased to 15—30 ms. ^1H and ^{13}C high-resolution n.m.r. spectra of benzene adsorbed on Co^{2+}-modified aerosil have been measured. The temperature dependence of the ^1H paramagnetic shift is in accord with Curie's law, while the behaviour of the ^{13}C line-shift contradicts the theory. This unusual effect was explained by a change of mean distance between the ions and benzene molecules.[1128]

6 Group III Compounds

Information concerning complexes of these elements can be found at the following sources: $[Ir(BH_4)H_2(PRBu^t{}_2)_2]$,[318] $[HB(3,5\text{-}Me_2C_3HN_2)_3Mo(NO)(CO)_2Br(OPr^i)]$,[183] $[PtMe\{(pyrazolyl)_3BH\}(\eta^2\text{-olefin})]$,[397] $RBCR^1{=}CR^2BRNRNR$,[173] $[(\eta^5\text{-}C_5H_5)(\eta^5\text{-}C_5H_4BRX)Fe]$,[296] $[M(C_5H_5BR)_2]$ (M = Ru or Os),[298] $[(\eta^5\text{-}C_5H_5)Fe(EtCEtCBMeSBMe)Fe(\eta^5\text{-}C_5H_5)]$,[299] boron–lead compounds,[1245] *mer*-$[(BF_4)Cr(CO)_3(PMe_3)CMe]$,[134] $[(\eta^5\text{-}C_5H_5)_2MMe_2AlMe_2]$ (M = Sc or Y),[101] $[MO_2Al_2(OPr^i)_4]$ (M = Mo or Zn),[207] [In(tetraphenylporphyrin)Cl],[306] and [Tl(octaethylporphyrin)] derivatives.[102]

Boron Hydrides and Carbaboranes.—^{10}B and ^{11}B spin–lattice and spin–spin relaxation times of some small boron hydride derivatives have been reported. Measurements of T_1 and T_2 relaxation rates in magnetically dilute ^{10}B species confirm that the quadrupolar mechanism dominates T_2 as well as T_1 relaxation in small boron hydrides. It was found that T_1 equals T_2, and unresolved boron–boron coupling constants were estimated from lineshape analyses employing measured T_1 relaxation rates. The magnitude of $J(^{11}B, ^{11}B)$ was related to bonding between the coupled nuclei, and a correlation of observed boron–boron coupling constants with current localized MO views of boron hydrides was demonstrated.[1129] A linear relationship has been observed between calculated charge densities $q_B{}^{total}$ or $q_B{}^p$ at the boron and $\delta(^{11}B)$ for a considerable number of boron compounds. The term $p^2{}_{B(2s)H}$ from the density matrix correlates linearly with $^1J(^1H, ^{11}B)$. The results obtained by CNDO/S calculations were confirmed by *ab initio* calculations. By correlating $^1J(^{11}B, ^{11}B)$ with $p^2{}_{B(2s)B(2s)}$ reasonable values for $^1J(^{11}B, ^{11}B)$ in two-centre two-electron bonds were estimated.[1130] An extensive study of ^{11}B and ^{13}C shielding values of icosahedral heteroatom boranes has been described. Shielding effects on polyhedral atoms are due to

1126 B. Boddenberg and J. A. Moreno, in 'Magnetic Resonance in Colloid and Interface Science' (Symposium No. 34), The American Chemical Society, Washington, 1976, p. 189 (*Chem. Abs.*, 1976, **85**, 198 760).
1127 M. D. Danchev, *Neft Gaz*, 1974, 151 (*Chem. Abs.*, 1976, **84**, 16 426).
1128 W.-D. Hoffmann, *Z. phys. Chem.* (*Leipzig*), 1976, **257**, 817.
1129 E. J. Stampf, A. R. Garber, J. D. Odom, and P. D. Ellis, *J. Amer. Chem. Soc.*, 1976, **98**, 6550.
1130 J. Kroner and B. Wrackmeyer, *J.C.S. Faraday II*, 1976, **72**, 2283.

variations in (i) directly bonded exopolyhedral groups, (ii) *ortho* cage groups, (iii) *meta* cage groups, and (iv) *para* cage groups. A linear relationship between ^{11}B and ^{13}C shielding values of isoelectronic and isostructural icosahedral molecules was demonstrated.[1131] $^{1}J[^{11}B(1), ^{11}B(3)]$ for B_4H_{10}, $^{1}J[^{11}B(1), ^{11}B(4)]$ for CB_5H_9, and a number of other compounds have been determined. From both $J(^{1}H, ^{11}B)$ and $J(^{11}B, ^{11}B)$ values for B_2H_6, B_4H_{10}, and B_5H_9, a set of s_B-orbital populations was derived and found to compare favourably to those obtained from PRDDO wavefunctions. Also $^{1}J(^{1}H, ^{11}B)$ was shown to be a function of bond distance, and from this a more reasonable bond length of 1.30 Å was suggested for the unique bridge hydrogen bonded to the apex boron atom of B_5H_{11}. Values of $^{1}J[^{1}H(\mu\text{-}2,3), ^{11}B(2)] = 14\ (\pm 10\%)$ Hz and $^{1}J[^{1}H(\mu\text{-}2,3), ^{11}B(3)] = 41\ (\pm 10\%)$ Hz for B_5H_{11} and $^{1}J[^{1}H(\mu), ^{11}B(5)] = 24\ (\pm 10\%)$ Hz for 2,3-$C_2B_4H_8$ were predicted.[1132]

In the ^{1}H and ^{27}Al n.m.r. spectra of $Al(BH_4)Et_2$ and related compounds, the signals are often in excess of 2000 Hz wide.[1133] The ^{1}H, ^{11}B, and $^{1}H\{^{27}Al\}$ n.m.r. spectra of $Al(BH_4)_3NH_2R$ have been investigated. The signals were shifted to high field compared to $Al(BH_3)_3$. The strongest changes of shielding were observed for ^{27}Al due to symmetry changes from planar in $Al(BH_3)_3$ to tetrahedral in complexes. The donor strengths of the ligands are $NMe_3 > NHMe > NH_2Me > NH_3$. No proof of the existence of a bridge structure in the complexes was obtained. The observed spectra show an equivalence of all the hydrogen atoms on the boron. At −80 °C, a narrow singlet was observed showing equivalence of all the hydrogen atoms in $[BH_4]^-$ and a thermal decoupling of boron and aluminium.[1134] The ^{1}H n.m.r. spectrum of $Al(BH_4)_3,OEt_2$ has been measured over the range from −80 to +30 °C in toluene and from −110 to +30 °C in ether. The temperature lowering brings about ^{11}B and ^{27}Al decoupling. Even at −100 °C, the BH_4 group has equivalent hydrogen atoms. The energy of activation for molecular rotation and the quadrupolar coupling constant were determined.[1135] ^{11}B N.m.r. data have been reported for $B_5H_{10}BeBH_4$.[1136]

The broadening of the NH signals in $H_2NNH_2BH_3$ at high temperatures was attributed to quadrupolar relaxation of the ^{14}N nuclei and proton exchange.[1137] ^{1}H, ^{11}B, ^{13}C, and ^{31}P n.m.r. spectra have been reported for $[(MeO)_3PBH_3]_2^{2-}$; this was believed to contain a phosphorus–phosphorus bond, but the spectra showed none of the complexity predicted if $^{1}J(^{31}P, ^{31}P) > 10$ Hz.[1138] An n.m.r. study of a few XYZP → BH_3 complexes was undertaken in order to compare some parameters, *e.g.* $^{1}J(^{1}H, ^{31}P)$, $^{1}J(^{11}B, ^{31}P)$, $\delta(^{11}B)$, with those derived from the enthalpies of the reaction of B_2H_6 with PXYZ. No overall correlation was found between $|^{1}J(^{11}B, ^{31}P)|$ and the stability of the complexes.[1139] ^{11}B N.m.r. data

[1131] L. J. Todd, A. R. Siedle, G. M. Bodner, S. B. Kahl, and J. P. Hickey, *J. Magn. Resonance*, 1976, **23**, 301.
[1132] T. Onak, J. B. Leach, S. Anderson, M. J. Frisch, and D. Marynick, *J. Magn. Resonance.* 1976, **23**, 237.
[1133] P. R. Oddy and M. G. H. Wallbridge, *J.C.S. Dalton*, 1976, 2076.
[1134] G. N. Boiko, Yu. I. Malov, K. N. Semenenko, and S. P. Shilkin, *Koord. Khim.*, 1976, **2**, 686.
[1135] G. N. Boiko, *Izvest. Akad. Nauk S.S.S.R., Ser. khim.*, 1976, **6**, 1215.
[1136] D. F. Gaines and J. L. Walsh, *J.C.S. Chem. Comm.*, 1976, 482.
[1137] G. V. Lagodzinskaya, N. G. Yunda, L. N. Erofeev, A. A. Zhirnov, G. N. Nechiporenko, and G. B. Manelis, *Zhur. fiz. khim.*, 1975, **49**, 2238.
[1138] L. A. Peacock and R. A. Geanangel, *Inorg. Chem.*, 1976, **15**, 244.
[1139] C. Jugie, C. Jouany, L. Elegant, J. F. Gal, and M. Azzaro, *Bull. Soc. chim. France*, 1976, 1.

have also been reported for R_3NBH_3, B_3H_7,DMF,[1140] and $PhPF_2 \rightarrow BH_3$ (including ^{19}F and ^{11}B n.m.r. data).[1141]

For Me_3NBH_2NC, the ^{11}B n.m.r. spectrum shows the presence of two isomers at δ 28.5 and δ 33.3.[1142] For LBH_2X (L = py, Me_3N, or Me_3P; X = F, Cl, Br, I, NCS, or NCSe) there is a correlation between $^1J(^1H, ^{11}B)$, the electronegativity of X, E_X, and the estimated boron–X bond distance, D_X. For L = Me_3N,

$$^1J(^1H, ^{11}B) = 3.336E_X + 36.426D_X - 47.233$$

and for L = py

$$^1J(^1H, ^{11}B) = 34.583E_X + 18.116D_X - 10.328$$

while for L = PMe_3 the changes are negligible.[1143] ^{11}B N.m.r. data have also been reported for $H_2CBH_2NMe_2BH_2NMe_2$ (cyclic, C–N bonded),[1144] $Me_3NBH_2CONHEt$ (including ^{13}C n.m.r. data),[1145] $[HXBNMe_2]_2$ (X = H, Cl, or Br),[1146] BH_2Cl,[1147] and $BHCl_2$.[1148]

The ^{11}B n.m.r. spectra of $Me_{3-n}NH_nB_3H_7$ have been reported. On going from $H_3NB_3H_7$ to $Me_3NB_3H_7$, the chemical shift of B-1 (nitrogen attached) decreases while the chemical shift of B-2,3 increases. Thus for $Me_3NB_3H_7$ all the boron nuclei have the same shift.[1149] $Me_4C_4B_8H_8$ has two isomers, one of which is fluxional. ^{11}B N.m.r. data have also been given for $[(Me_2C_2B_4H_4)_2FeH_2]$, $[(\eta^5\text{-}C_5H_5)Co(Me_2C_2B_3H_3)Fe(CO)_3]$, $[(Me_2C_2B_4H_4)_2Fe]^-$, and $[(Me_4C_4B_8)\text{-}FeCo(\eta^5\text{-}C_5H_5)]$.[1150] The electronic structure of the triple-decked metallocarbaborane complexes 1,7,2,3- and 1,7,2,4-$[(\eta^5\text{-}C_5H_5)_2Co_2C_2B_3H_5]$ has been examined *via* ^1H and ^{11}B n.m.r. spectra for a series of derivatives containing substituents on the central $[C_2B_3H_5]^{4-}$ or end $[C_5H_5]^-$ rings. It was suggested that the 1,7,2,4 complex contains a highly electron-delocalized metallocene-like central ring while the 1,7,2,3 system is dominated by strong local π-interactions between the metals and an ethylenic carbon–carbon bond in the central ring. In both isomers the substitution of Me, Ph, or $SiMe_3$ at a carbaboranyl ring carbon atom or of Br or I at B-5 produces moderate to large changes in chemical shift (relative to the parent molecule) at all boron positions and at all hydrogen positions in each ring. End-ring substitution by Me, Et, or $SiMe_3$ in each isomer generates strong chemical shift effects in the central ring, and the $SiMe_3$ species also shows a small effect in the unsubstituted cyclopentadienyl ring. A direct *trans*-polyhedral electronic interaction between the metal atoms was proposed to account for the antipodal effects.[1151] ^{11}B N.m.r. data have also been

[1140] P. J. Dolan, J. H. Kindsvater, and D. G. Peters, *Inorg. Chem.*, 1976, **15**, 2170
[1141] C. Jouany, G. Jugie, G. V. Roschenthaler, and O. Stelzer, *J. Inorg. Nuclear Chem.*, 1976, **38**, 1383.
[1142] J. L. Vidal and G. E. Ryschkewitsch, *J.C.S. Chem. Comm.*, 1976, 192.
[1143] M. L. Denniston, M. A. Chiusano, and D. R. Martin, *J. Inorg. Nuclear Chem.*, 1976, **38**, 979.
[1144] B. R. Gragg and G. E. Ryschkewitsch, *Inorg. Chem.*, 1976, **15**, 1201.
[1145] B. F. Spielvogel, L. Wojnowich, M. K. Das, A. T. McPhail, and K. D. Hargrave, *J. Amer. Chem. Soc.*, 1976, **98**, 5702.
[1146] O. T. Beachley, jun. and B. Washburn, *Inorg. Chem.*, 1976, **15**, 725.
[1147] H. C. Brown and N. Ravindran, *J. Amer. Chem. Soc.*, 1976, **98**, 1785.
[1148] H. C. Brown and N. Ravindran, *J. Amer. Chem. Soc.*, 1976, **98**, 1798.
[1149] A. R. Dodds and G. Kodama, *Inorg. Chem.*, 1976, **15**, 741.
[1150] W. M. Maxwell, V. R. Miller, and R. N. Grimes, *Inorg. Chem.*, 1976, **15**, 1343.
[1151] R. Weiss and R. N. Grimes, *J. Organometallic Chem.*, 1976, **113**, 29.

reported for [(η^5-C_5H_5)$_3$$Co_3B_3H_5$], [($\eta^5$-$C_5H_5$)$_4$$Co_4B_4H_4$],[1152] 2-Cl-1,5-$C_2B_3H_4$,[1153] [$B_5H_3(CO)_2Fe(CO)_3$],[1154] [$B_5H_9Fe(CO)_3$] (including ^{13}C n.m.r. data),[1155] [(Me_2-$C_2B_4H_4$)$_2$CoH], [(η^5-C_5H_5)Co($Me_2C_2B_3H_3$)CoH($Me_2C_2B_3H_3$)Co(η^5-C_5H_5)],[1156] [(η^5-C_5H_5)$_3$$Ni_3$($CB_5H_5Me$)],[1157] 2-Cl-1,6-$C_2B_4H_5$,[1158] and Me_3P-2,4-$C_2B_4H_6$.[1159]

The ^{11}B n.m.r. spectra of *B*-methyl derivatives of 2,4-$C_2B_5H_7$ show an antipodal chemical shift effect for the 1,7-positions.[1160] The low-temperature ^1H and $^{11}B\{^1H\}$ n.m.r. spectra of B_8H_{14} and iso-B_9H_{15} have been obtained. It was suggested that B_8H_{14} contains four unique terminal and two bridging hydrogen atoms around the open face of the B_8 cage while iso-B_9H_{15} has six asymmetric bridging hydrogen atoms around the open face of the B_9 cage of C_{3v} symmetry.[1162] Halogenothiadecaborane has been examined by ^{11}B n.m.r. spectroscopy and spin–lattice relaxation times of up to 4 s.[1163] ^{11}B N.m.r. data have also been reported for 3,3′-(2,4-$C_2B_5H_6$)$_2$,[1164] 4-CB_8H_{14}, [6-CB_9H_{12}]$^-$,[1165] [6,6-(Ph_3P)$_2$-6,2,3-$RuC_2B_7H_9$], [6,6-(R_3P)$_2$-6-H-6,2,3-$MC_2B_7H_9$] (M = Rh or Ir),[1166] [(η^5-C_5H_5)Co(CB_9H_{10})]$^-$,[1167] [2-(Ph_3P)$_2$Ir-Cl-1-SB_9H_8],[1168] [($B_{10}H_{12}$)Ir(CO)\{$C_2S_2(CN)_2$\}-(PPh_3)$_2$]$^-$,[1169] [\{(Ph_3P)$_2$Cu\}$_2$$B_{10}H_{10}$],[1170] 6-$Me_3SiOB_{10}H_{13}$, $Ph_2SnB_{10}H_{12}$,[1171] [(η^5-C_5H_5)Co(η-7-$B_{10}H_{10}$PR)], [7-$B_{10}H_{11}$PR]$^-$,[1172] *arachno*-9-L-6-CB_9H_{13},[1173] SB_9H_{11},[1174] 2,2′- or 2,7′-(1-SB_9H_8)(1′-$SB_{11}H_{10}$),[1175] 2-R-7-$SB_{10}H_{11}$,[1177] [(1,2-$C_2B_9H_{11}$)Re(CO)$_3$]$^-$, [(1,2-$C_2B_9H_{11}$)Pt(PPh_3)$_2$],[1178] 2,1,7-[(Ph_3P)$_2$$RuC_2B_9H_{11}$] (including ^{31}P n.m.r. data),[1179] 8,8′-[RE($C_2B_9H_{10}$)$_2$Co] (E = O, Se, or Te),[1180] and *closo*-3,8-[(Ph_3P)$_2$-3-X-3,1,2-$NiC_2B_9H_{10}$] (X = H or Cl).[1181]

1152 V. R. Miller and R. N. Grimes, *J. Amer. Chem. Soc.*, 1976, **98**, 1600.
1153 R. C. Dobbie, E. W. Distephano, M. Black, J. B. Leach, and T. Onak, *J. Organometallic Chem.*, 1976, **114**, 233.
1154 J. A. Ulman and T. P. Fehlner, *J.C.S. Chem. Comm.*, 1976, 632.
1155 T. P. Fehlner, J. Ragaini, M. Mangion, and S. G. Shore, *J. Amer. Chem. Soc.*, 1976, **98**, 7085.
1156 W. M. Maxwell, V. R. Miller, and R. N. Grimes, *J. Amer. Chem. Soc.*, 1976, **98**, 4818.
1157 C. G. Salentine, C. E. Strouse, and M. F. Hawthorne, *Inorg. Chem.*, 1976, **15**, 1832.
1158 J. R. Spielman, R. G. Warren, D. A. Bergquist, J. K. Allen, D. Marynick, and T. Onak, *Synth. React. Inorg. Metal-Org. Chem.*, 1975, **5**, 347.
1159 L. Lew, G. Haran, R. Dobbie, M. Black, and T. Onak, *J. Organometallic Chem.*, 1976, **111**, 123.
1160 J. F. Ditter, E. B. Klusmann, R. E. Williams, and T. Onak, *Inorg. Chem.*, 1976, **15**, 1063.
1162 D. C. Moody and R. Schaeffer, *Inorg. Chem.*, 1976, **15**, 233.
1163 W. L. Smith, B. J. Meneghelli, N. McClure, and R. W. Rudolph, *J. Amer. Chem. Soc.*, 1976, **98**, 624.
1164 J. S. Plotkin and L. G. Sneddon, *J.C.S. Chem. Comm.*, 1976, 95.
1165 B. Štíbr, K. Baše, S. Heřmánek, and J. Plešek, *J.C.S. Chem. Comm.*, 1976, 150.
1166 C. W. Jung and M. F. Hawthorne, *J.C.S. Chem. Comm.*, 1976, 499.
1167 J. Dolanský, K. Baše, and B. Štíbr, *Chem. and Ind.*, 1976, 853.
1168 D. A. Thompson and R. W. Rudolph, *J.C.S. Chem. Comm.*, 1976, 770.
1169 A. R. Siedle and L. J. Todd, *Inorg. Chem.*, 1976, **15**, 2838.
1170 G. G. Outterson, jun., V. T. Brice, and S. G. Shore, *Inorg. Chem.*, 1976, **15**, 1456.
1171 R. E. Loffredo, L. F. Drullinger, J. A. Slater, C. A. Turner, and A. D. Norman, *Inorg. Chem.*, 1976, **15**, 478.
1172 J. L. Little, *Inorg. Chem.*, 1976, **15**, 114.
1173 K. Baše, S. Heřmánek, and B. Štíbr, *Chem. and Ind.*, 1976, 1068.
1174 W. R. Pretzer and R. W. Rudolph, *J. Amer. Chem. Soc.*, 1976, **98**, 1441.
1175 W. R. Pretzer and R. W. Rudolph, *Inorg. Chem.*, 1976, **15**, 1779.
1177 D. A. Thompson, W. R. Pretzer, and R. W. Rudolph, *Inorg. Chem.*, 1976, **15**, 2948.
1178 H. M. Colquhoun, T. J. Greenhough, and M. G. H. Wallbridge, *J.C.S. Chem. Comm.*, 1976, 1019.
1179 E. H. S. Wong and M. F. Hawthorne, *J.C.S. Chem. Comm.*, 1976, 257.
1180 J. Plešek, S. Heřmánek, K. Baše, L. J. Todd, and W. F. Wright, *Coll. Czech. Chem. Comm.*, 1976, **41**, 3509.
1181 S. B. Miller and M. F. Hawthorne, *J.C.S. Chem. Comm.*, 1976, 786.

The ^1H and ^{11}B n.m.r. spectra of *o*-, *m*-, and *p*-carbaborane and its derivatives have been correlated with structure. The substituent effects are not describable in terms of classical reaction constants but are made up of a number of components.[1182] In other carbaboranes, it has been noted that the effect of the substituent is not on the surrounding atoms but on the atom directly attached to it and at the opposite vertex of the icosahedron.[1183] In the ^1H, ^{11}B, and ^{13}C n.m.r. spectra of dicarba-*closo*-dodecaboranes(12), the signals can be assigned by use of an antipodal shift effect and a vicinal shift effect.[1184] The effect of *C*-methyl substitution on ^{11}B and ^{13}C n.m.r. spectra and on the heats of formation of *o*-, *m*-, and *p*-carbaborane(12) has been related to the interaction of electrons in the methyl group with electrons in the carbaborane MOs and to the redistribution of charge density in the carbaboranes.[1185] The ^{13}C n.m.r. spectra of 1,2-diphenyl-*o*-carbaborane, 1,7-diphenyl-*m*-carbaborane, 1,12-diphenyl-*p*-carbaborane, their mono- and di-anions, and 1,2-dimethyl-*o*-carbaborane and its dianion have been examined. The spectra of all the monoanions and some of the dianions were identified.[1186, 1187] N.m.r. data have also been published for $[M(C_2Me_2B_{10}H_{10})_2]^{2-}$ (M = Ti, Zr, or Hf; ^{11}B n.m.r. data),[1188] $[(Me_4C_4B_8H_8)FeCo(\eta^5\text{-}C_5H_5)]$ (^{11}B n.m.r. data),[1189] and $(Me\text{-}o\text{-}CB_{10}H_{10}C)_3P$.[1190]

The ^{11}B n.m.r. spectra of normal and isotopically labelled (^2D and ^{10}B) $B_{13}H_{19}$ and $KB_{13}H_{18}$ have been discussed and related to structure.[1191]

Other Compounds of Boron.—He(I) Photoelectron spectra of $Me_{3-n}BX_n$ have been discussed and compared with BX_3 and BMe_3 on the basis of *ab initio* calculations. Correlations of orbital ionization energies, π-charge densities, and ^{11}B chemical shifts as well as calculated and measured dipole moments gave insight into the electronic structure of three-co-ordinate boron compounds.[1192] The ^1H, ^{11}B, and ^{14}N n.m.r. spectra of $R_{3-n}BX_n$ (X = NMe_2, Cl, or Br) have been reported. Comparison of R = alkyl with R = aryl points to increased π-back-bonding from the ring to boron in the series Ph < 3-thienyl < 2-thienyl < 1-Me-2-pyrrolyl < 2-furyl, which is comparable with the decrease of 'aromaticity' of the cyclic system. The influence of a conjugated double-bond system leads to a high-field shift of the ^{11}B resonance relative to that of vinyl boranes.[1193] The ^1H and ^{13}C n.m.r. spectra of (115) have shown that the α-hydrogen and carbon

1182 T. A. Babushkina, T. P. Klimova, G. A. Anorova, V. V. Khrapov, and V. I. Stanko, *Zhur. obshchei Khim.*, 1976, **46**, 1064.
1183 V. I. Stanko, T. A. Babushkina, T. P. Klimova, Yu. V. Gol'tyapin, A. I. Klimova, A. M. Vasil'ev, A. M. Alymov, and V. V. Khrapov, *Zhur. obshchei Khim.*, 1976, **46**, 1071.
1184 S. Heřmánek, V. Gregor, B. Štíbr, J. Plešek, Z. Janoušek, and V. A. Antonovich, *Coll. Czech. Chem. Comm.*, 1976, **41**, 1492.
1185 V. I. Stanko, Yu. V. Gol'tyapin, V. K. Pavlovich, and G. L. Gal'chenko, *Doklady Akad. Nauk S.S.S.R.*, 1975, **224**, 1360.
1186 L. I. Zakharkin, V. N. Kalinin, V. A. Antonovich, and E. G. Rys, *Izvest. Akad. Nauk S.S.S.R., Ser. khim.*, 1976, **5**, 1036.
1187 L. I. Zakharkin, V. N. Kalinin, and N. I. Kobel'kova, *Synth. React. Inorg. Metal-Org. Chem.*, 1976, **6**, 65.
1188 C. G. Salentine and M. F. Hawthorne, *Inorg. Chem.*, 1976, **15**, 2872.
1189 W. M. Maxwell, E. Sinn, and R. N. Grimes, *J. Amer. Chem. Soc.*, 1976, **98**, 3490.
1190 V. I. Bregadze, N. N. Godovikov, A. N. Degtyarev, and M. I. Kabachnik, *J. Organometallic Chem.*, 1976, **112**, C25.
1191 J. C. Huffman, D. C. Moody, and R. Schaeffer, *Inorg. Chem.*, 1976, **15**, 227.
1192 H.-O. Berger, J. Kroner, and H. Nöth, *Chem. Ber.*, 1976, **109**, 2266.
1193 B. Wrackmeyer and H. Nöth, *Chem. Ber.*, 1976, **109**, 1075.

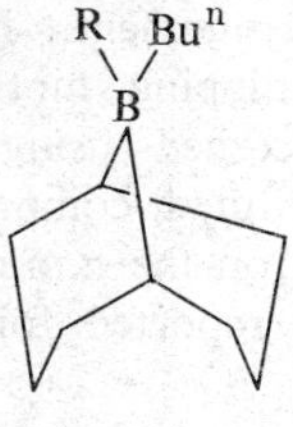

(115)

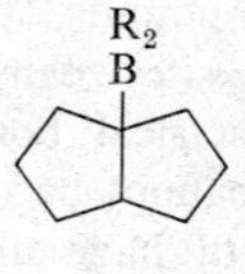

(116)

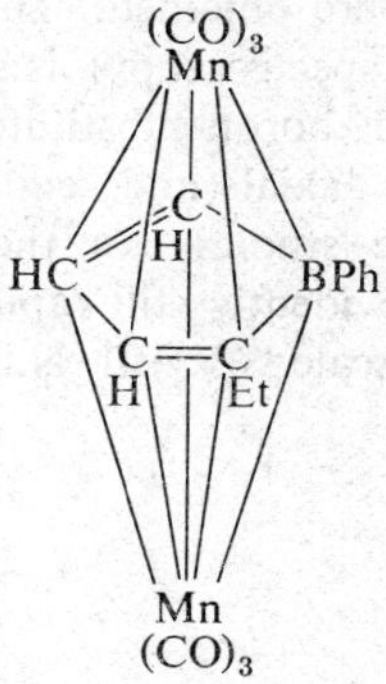

(117)

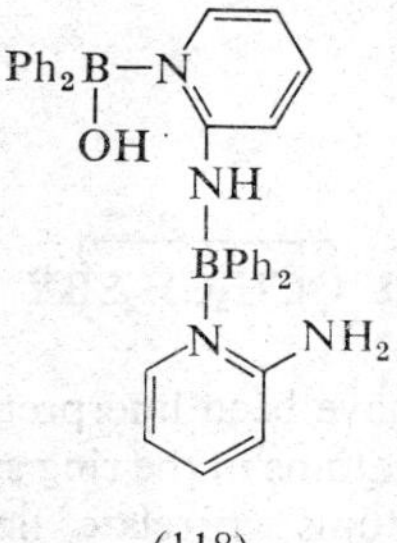

(118)

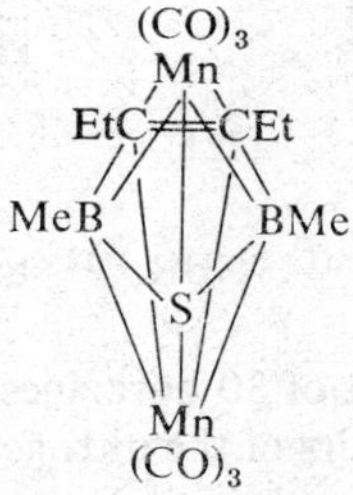

(119)

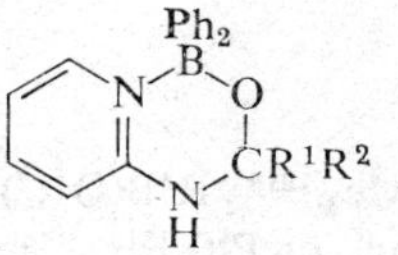

(120)

atoms are shielded in comparison to R_3B. The results indicate a hydridic character for the α-hydrogen atoms in the boronate complexes.[1194] ^{11}B N.m.r. data have also been reported for $(Me_3Sn)MeC{=}C(BMe_2)(SnMe_3)$,[1195] $Me_nB(C_2F_3)_{3-n}$ (including ^{19}F n.m.r. data),[1196] (116),[1197] (117),[1198] (118),[1199] $(CF_3)_2CNBMe_2$ (only ^{13}C n.m.r. data),[1200] (119),[1201] $[(Ph_3P)_2PtBMe_2Br_3]$,[1202] $BBu^n_2(OCH_2Ph)$,[1203] $(\overline{CH_2CHMeCH_2CH_2B})_nN(SnMe_2)_{3-n}$ (including ^{14}N n.m.r. data),[1204] $(\overline{CH_2CHMeCH_2CH_2B})_nNR_2$ (including ^{14}N n.m.r. data),[1205] R_2BOR,[1206] and (120).[1207]

A multinuclear investigation of magnetic resonance parameters of 14 *P*-hexamethylcyclotriboraphosphanes and *P*-octamethylcyclotetraboraphosphane has been carried out. ^1H, ^{11}B, ^{13}C, ^{19}F, and ^{31}P chemical shifts and coupling

1194 Y. Yamamoto, H. Toi, A. Sonoda, and S. Murahashi, *Chem. Letters*, 1975, **11**, 1199 (*Chem. Abs.*, 1976, **84**, 16 629).
1195 B. Wrackmeyer and H. Nöth, *J. Organometallic Chem.*, 1976, **108**, C21.
1196 E. J. Stampf and J. D. Odom, *J. Organometallic Chem.*, 1976, **108**, 1.
1197 G. W. Kramer and H. C. Brown, *J. Amer. Chem. Soc.*, 1976, **98**, 1964.
1198 G. E. Herberich, J. Hengesbach, U. Kölle, G. Huttner, and A. Frank, *Angew. Chem. Internat. Edn.*, 1976, **15**, 433.
1199 B. R. Gragg and K. Niedenzu, *J. Organometallic Chem.*, 1976, **117**, 1.
1200 B. R. Gragg and K. Niedenzu, *Synth. React. Inorg. Metal.-Org. Chem.*, 1976, **6**, 275.
1201 W. Siebert and K. Kinberger, *Angew. Chem. Internat. Edn.*, 1976, **15**, 434.
1202 M. Fishwick, H. Nöth, W. Petz, and M. G. H. Wallbridge, *Inorg. Chem.*, 1976, **15**, 491.
1203 K. Smith and K. Swaminathan, *J.C.S. Dalton*, 1976, 2297.
1204 W. Storch and H. Nöth, *Angew. Chem. Internat. Edn.*, 1976, **15**, 235.
1205 H. Nöth and W. Storch, *Chem. Ber.*, 1976, **109**, 884.
1206 W. Fenzl, H. Kosfeld, and R. Köster, *Annalen*, 1976, 1370.
1207 B. R. Gragg, R. E. Handshoe, and K. Niedenzu, *J. Organometallic Chem.*, 1976, **116**, 135.

constants were obtained. The results show that n.m.r. spectroscopy (especially ^{11}B n.m.r. spectroscopy) is invaluable for structure assignment. In the unsymmetrical boron-substituted compounds, the ring-flipping motion which equilibrates axial and equatorial sites can be discerned using variable-temperature studies. In the symmetrical boron-substituted compounds, the motion is evidently still rapid and cannot be discerned on the n.m.r. chemical shift time-scale.[1208] ^{11}B N.m.r. data have also been reported for (121),[1209]

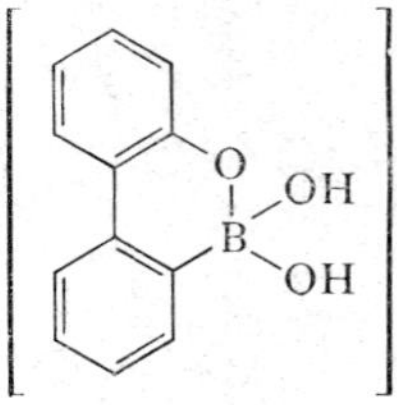

(121)

$X_2BC_5H_4SiMe_3$.[1210] $R^1B(OR^2)_2$,[1211] X_2BR,[1212] and $\overline{OCH_2CH_2SBX}$ (X = SEt, NEt_2, $NHBu^n$, or Ph).[1213]

The ^{14}N and ^{11}B n.m.r. spectra of 30 borazines have been interpreted in terms of delocalization of p_z electron pairs of the nitrogen atoms in the ring system. The calculated π-charge densities at the nitrogen atoms correlate linearly with $\delta(^{14}N)$.[1214] The ^{13}C n.m.r. spectra have been measured for Me_3N adducts of $BX_{3-n}Y_n$. $\delta(^{13}C)$ moves downfield on substitution with the heavier halides, the shifts reflecting the order of Lewis acidity. A linear correlation was found between $\delta(^{13}C)$ and $\delta(^1H)$, $\delta(^{11}B)$, and $^1J(^{11}B, ^{19}F)$, showing that the ^{13}C chemical shifts reflect changes in the boron–nitrogen bond. No coupling $^2J(^{11}B, ^{13}C)$ was found. Both ^1H and ^{13}C shifts show pairwise additivity with respect to halogen substitution at boron.[1215] Studies of ^{11}B n.m.r. spectra of adducts formed between BX_3 and Ph_nEX_{3-n} (E = P or As) have shown that both ^{11}B chemical shift increments and $^1J(^{11}B, ^{31}P)$ are sensitive to the strength of the donor–acceptor bond. The coupling constants were discussed in terms of Fermi-contact interactions.[1216] The ^{31}P n.m.r. spectra of $(4\text{-}XC_6H_4)_nPCl_{3-n}$, $(3\text{-}XC_6H_4)_n$-PCl_{3-n} (X = F, Cl, or H), and complexes with BCl_3 and BBr_3 have been examined. The chemical shift increments on complexation were interpreted in terms of an increased charge on the phosphorus atom and of concomitant charge delocalization, $\pi(p\text{–}p)$ Ar$\leftarrow$P and P$\rightarrow$Cl, and $\pi(p\text{–}d)$ Ar$\rightarrow$P and P$\leftarrow$Cl, in general, and of mesomeric contributions in the particular case of $(4\text{-}FC_6H_4)_n$-PCl_{3-n} compounds and their complexes.[1217] N.m.r. data have also been reported

1208 M. A. Sens, J. D. Odom, and M. H. Goodrow, *Inorg. Chem.*, 1976, **15**, 2825.
1209 B. M. Mikhailov and M. E. Kuimova, *J. Organometallic Chem.*, 1976, **116**, 123.
1210 P. Jutzi and A. Seufert, *Angew. Chem. Internat. Edn.*, 1976, **15**, 295.
1211 R. Köster and W. V. Dahlhoff, *Annalen*, 1976, 1925.
1212 P. Paetzold and B. Boeke, *Chem. Ber.*, 1976, **109**, 1011.
1213 R. H. Cragg, J. Dudman, and J. P. N. Husband, *J. Organometallic Chem.*, 1976, **116**, 281.
1214 B. Wrackmeyer and H. Nöth, *Chem. Ber.*, 1976, **109**, 3480.
1215 J. M. Miller and T. R. B. Jones, *Inorg. Chem.*, 1976, **15**, 284.
1216 E. Muylle, G. P. Van der Kelen, and E. G. Claeys, *Spectrochim. Acta*, 1976, **32A**, 1149.
1217 E. Muylle and G. P. Van der Kelen, *Spectrochim. Acta*, 1976, **32A**, 599.

for (122; X = BMe_2, NMe_2, or OMe) (^{11}B n.m.r. data),[1218] $B_6(NMe_2)_{12}Al_6Me_{12}$ (^{11}B and ^{13}C n.m.r. data),[1219] $BX_n(NCS)_{3-n}$ (X = Cl, Br, or I; ^{11}B n.m.r. data),[1220] $Me_3PBX_n(NCS)_{3-n}$ (X = Cl, Br, or I; ^{11}B n.m.r. data).[1221] $Me_3NCl_2BN(SCF_3)_2$ (^{11}B and ^{19}F n.m.r. data),[1222] $BF_2NH(PF_2)$ (^{19}F and ^{31}P n.m.r. data),[1223] (123)

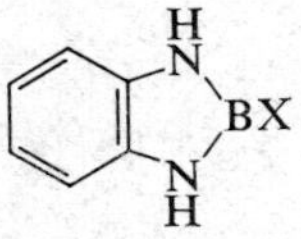
(122)

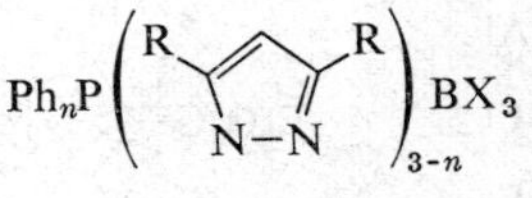
(123)

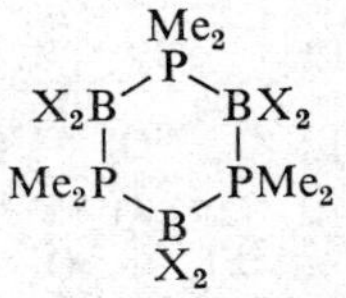
(124)

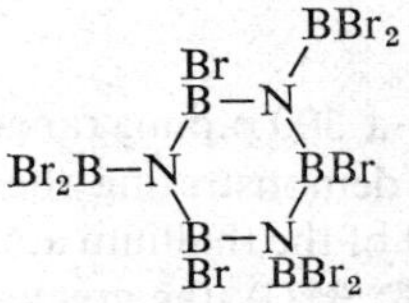
(125)

(^{19}F and ^{31}P n.m.r. data; X = F or Cl),[1224] (124) (^{11}B n.m.r. data),[1225] $[BF_n(OH)_{4-n}]^-$ (^{19}F n.m.r. data),[1226] and (125), (^{11}B n.m.r. data).[1227]

Complexes of other Group III Elements.—The ^{27}Al n.m.r. spectrum of $LiAlH_4$ shows $^1J(^1H, ^{27}Al) = 175.2 \pm 1.2$ Hz.[1228] The ^{27}Al n.m.r. spectra of $LiAlH_4$, AlH_3, and $LiAlH(ER)_3$ (E = O or S) have been recorded and $\delta(^{27}Al)$, $^1J(^1H, ^{27}Al)$, and linewidths were discussed.[1229]

From a comparison of the low-temperature ^1H n.m.r. spectra of $(Et_2Al)_2O$ and Et_3Al, a bridging structure (126) was suggested.[1230] In strong acids, $^2J(^1H, M)$ (M = Sn, Tl, or Pb) of some Me_2M complexes have been found to be smaller and the ^1H resonances of the methyl groups attached to the metal atom were shifted to lower magnetic field. From the correlation of $^2J(^1H, M)$ and $\delta(^1H)$, the decrease of $^2J(^1H, M)$ of these compounds with increasing acid strengths were considered to be mainly due to the increase of the mean singlet–triplet excitation energy in the Fermi-contact term. The lowering of the metal–carbon bond strengths in strong acids was revealed in ν(M, C) and may be responsible

1218 R. Goetze and H. Nöth, *Chem. Ber.*, 1976, **109**, 3247.
1219 B. A. Amero and E. P. Schram, *Inorg. Chem.*, 1976, **15**, 2842.
1220 J. Dazord, H. Mongeot, H. Atchekzai, and J. P. Tuchagues, *Canad. J. Chem.*, 1976, **54**, 2135.
1221 H. Mongeot, J. Dazord, H. F. Atchekzai, and J. P. Tuchagues, *Synth. React. Inorg. Metal.-Org. Chem.*, 1976, **6**, 191.
1222 A. Haas, M. Häberlein, and C. Krüger, *Chem. Ber.*, 1976, **109**, 1769.
1223 D. E. J. Arnold, E. A. V. Ebsworth, and D. W. H. Rankin, *J.C.S. Dalton*, 1976, 823.
1224 S. Fischer, J. Hoyano, and L. K. Peterson, *Canad. J. Chem.*, 1976, **54**, 2710.
1225 M. H. Goodrow and R. I. Wagner, *Inorg. Chem.*, 1976, **15**, 2830.
1226 L. Maya, *Inorg. Chem.*, 1976, **15**, 2179.
1227 K. Dehnicke and V. Fernandéz, *Chem. Ber.*, 1976, **109**, 488.
1228 R. E. Santini and J. B. Grutzner, *J. Magn. Resonance*, 1976, **22**, 155.
1229 J. Huet, J. Durand, and Y. Infarnet, *Org. Magn. Resonance*, 1976, **8**, 382.
1230 M. Bolesławski, S. Pasynkiewicz, A. Kunicki, and J. Serwatowski, *J. Organometallic Chem.*, 1976, **116**, 285.

for the decrease in $^1J(^1H, M)$.[1231] ^1H, ^{13}C, and ^{205}Tl n.m.r. spectra have been determined for Me_2TlX (X = NO_3, I, or OPh) as a function of solvent and concentration. The n.m.r. parameters most sensitive to changes in the environment of the thallium nucleus are the ^{205}Tl chemical shift and $^1J(^{13}C, ^{205}Tl)$. The

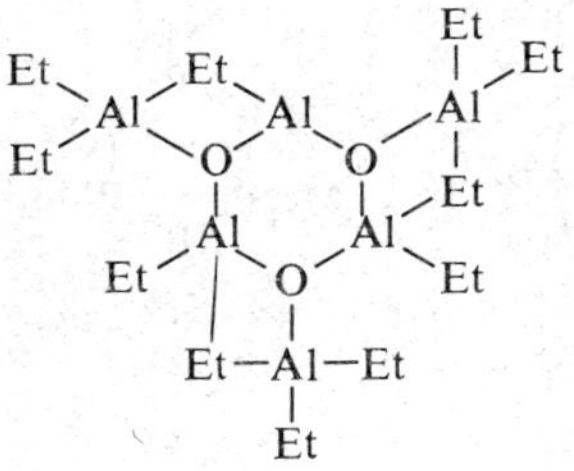

(126)

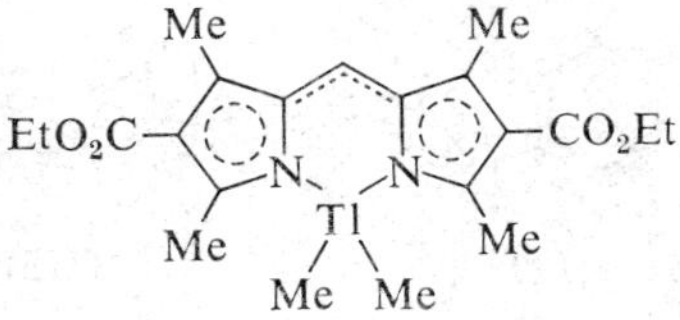

(127)

^{205}Tl chemical shift varies within a 300 p.p.m. range as the anion, solvent, and concentration are changed, thus demonstrating the potential of this parameter as an indicator of the environment of the thallium atom. The nature of the solvent is the strongest influence on $^1J(^{13}C, ^{205}Tl)$, the greatest change being observed for Me_2TlNO_3 between water and pyridine. The solvent dependence of $^1J(^{13}C, ^{205}Tl)$ and $^2J(^1H, ^{205}Tl)$ was discussed in terms of the approximate expression for the Fermi-contact mechanism for spin–spin coupling constant.[1232] The ^1H and ^{13}C n.m.r. spectra of (127) have been reported and long-range thallium coupling observed.[1233]

For $PhTl(O_2CCF_3)_2$, $^1J(^{13}C, ^{203}Tl)$ = 10 718 Hz and $^1J(^{13}C, ^{205}Tl)$ = 10 623 Hz.[1234] The ^{205}Tl resonance frequency has been measured in a series of substituted aryl $Tl(O_2CCF_3)_2$ complexes. The resonance frequency of ^{205}Tl is remarkably sensitive to the substituent, with a correlation existing between the resonance frequency and Hammett σ parameter. A correlation between the resonance frequency and $^2J(^1H, ^{205}Tl)$ was also found, as were very large solvent effects.[1235] ^{27}Al N.m.r. spectra of aluminium ketoenolate complexes and alkoxides in the condensed state have been reported. With aluminium ethylacetoacetate, possible isomers with non-equivalent aluminium–oxygen bonds were discussed.[1236] ^{31}P N.m.r. data have also been reported for $(Me_3Si)_2NPN(SiMe_3)AlCl_3$, {$\delta(^{31}P)$ = 450.4},[1237] $Cl_3AlP(NMeNMe)_3PAlCl_3$ (including ^{14}N and ^{27}Al n.m.r. data),[1238] and $(Me_2N)_3PMe_2NPClAlCl_3$.[1239]

7 Compounds of Silicon, Germanium, Tin, and Lead

Information concerning complexes of these elements can be found at the following sources: *trans*-$[PtH(YSiH_3)(PEt_3)_2]$,[402] *trans*-$[PtI(H_2SiNHSiH_3)$-

1231 Y. Kawasaki and M. Aritomi, *J. Organometallic Chem.*, 1976, **104**, 36.
1232 P. J. Burke, R. W. Matthews, and D. G. Gillies, *J. Organometallic Chem.*, 1976, **118**, 129.
1233 A. T. T. Hsieh, C. A. Rogers, and B. O. West, *Austral. J. Chem.*, 1976, **29**, 49.
1234 J. Y. Lallemand and M. Duteil, *Org. Magn. Resonance*, 1976, **8**, 328.
1235 J. F. Hinton and R. W. Briggs, *J. Magn. Resonance*, 1976, **22**, 447.
1236 L. A. Mai and V. A. Shcherbakov, *Latv. PRS Zinat. Akad. Vestis, Kim. Ser.*, 1975, **5**, 548 (*Chem. Abs.*, 1976, **84**, 82 144).
1237 E. Niecke and R. Kröher, *Angew. Chem. Internat. Edn.*, 1976, **15**, 692.
1238 H. Nöth and R. Ullmann, *Chem. Ber.*, 1976, **109**, 1942.
1239 C. W. Schultz and R. W. Parry, *Inorg. Chem.*, 1976, **15**, 3046.

$(PEt_3)_2]$,[434] $C_6H_4(NSiMe_3)_2BNX$,[1218] $[(Me_3SiCH_2)_4M(CSiMe_3)_2]$ (M = Nb, Ta, or W),[111] 6-$Me_3SiOB_{10}H_{13}$, $Ph_2SnB_{10}H_{12}$,[1171] $(Me_3Si)_2NP(SiMe_3)(AlCl_3)$,[1237] $X_2BC_5H_4SiMe_3$,[1210] $R_3PC(SiMe_3)X$,[1366] 1,7,2,4-$[(\eta^5\text{-}C_5H_5)_2Co_2C_2B_3H_4SiMe_3]$,[1151] $[(\eta^3\text{-}CH_2CHSiMe_2)Fe(SiMe_3)(CO)_3]$,[260] $(Me_3Si)_2NP(NSiMe_3)_2PN(SiMe_3)_2$,[1246] $(Me_3Si)_2NP(S)(NSiMe_3)_2PN(SiMe_3)_2$,[1436] $Me_3SiN{=}P\{N(SiMe_3)_2\}_2N{=}C{=}NSi$-$Me_3$,[1435] $Me_2NP(NSiMe_3)_2NMe_2$,[1437] $(Me_2CHO)_2PONBzMR_3$ (M = Si, Ge, or Sn; R = Me or Ph),[1431] $Me_3P{=}N{-}SiMe_2N{=}PMe_3$,[1373] $CH_2CMeHCH_2CH_2BN$-$(SnMe_3)_2$,[1204] $(Me_3Sn)MeC{=}C(BMe_2)(SnMe_3)$,[1195] $[Me_3Sn]^+$, $[Me_3Pb]^+$,[483] $[(\eta^5\text{-}C_5H_5)W(CO)_2(PR_3)MMe_3]$ (M = Ge or Pb),[149] $[(Et_3Si)_2\{(C_6H_{11})_3P\}_2Pt_2$-$(\mu\text{-}H)_2]$,[389] $[C_6H_4(SiMe_2)_2Pt(PPh_3)_2]$,[420] $[R_nX_{m-n}Sn\{Co(CO)_4\}_{4-m}]$,[329] $[(OC)_5CrC$-$(NMe_2)(SiPh_3)]$,[128] $[(OC)_3Re(dppe)COSiPh_3]$,[218] $Bu^t{}_2(Me_2ClM)P$ (M = Ge or Sn),[189] $[Bu^t{}_2(Me_2MCl)PNi(CO)_3]$ (M = Ge or Sn),[436] *cis*-$[Pt(PPh_3)_2(SnMe_n$-$Ph_{2-n}Cl)Ph]$,[414] Me_2MX_2 (M = Sn or Pb),[1231] SiF_4,[1495] and [Pb(octaethylporphyrin)].[102]

A review entitled 'Nuclear Magnetic Resonance Spectroscopy of Organosilicon Compounds' has appeared.[1240] The use of ^{13}C, ^{14}N, ^{19}F, and ^{29}Si n.m.r. spectroscopies for the control of peptide synthesis and structure determination of polypeptides has been described.[1241] The ^{29}Si n.m.r. spectra of six silanols and four silylamines have been examined in several solvents of varying electron-pair donor ability. A linear correlation was found between the silanol ^{29}Si chemical shift and solvent donor ability. The silylamine ^{29}Si chemical shifts were considerably less sensitive to solvent. The effect was attributed to hydrogen bonding between the hydroxyl proton of the silanol and the electron pair of the solvent.[1242] ^{31}P N.m.r. spectroscopy has shown that the inversion barrier at the phosphorus is low in 1,3-diphosphoranes and 1,3-diphosphorinanes with a Group IV atom (C, Si, Ge, Sn) in the 2-position.[1243] The ^{13}C n.m.r. spectra of organometallic compounds containing tin or lead bonded to furan or thiophen rings have been reported. The ^{13}C chemical shifts and $J(^{13}C, M)$ were interpreted in terms of $(d\text{-}p)\pi$-bonding.[1244] $^1J(^{207}Pb, X)$ (X = H, B, C, N, P, Se, Sn, or Pb) and other n.m.r. parameters have been reported for model lead compounds. Comparison of the reduced coupling constants $^1K(^{207}Pb, X)$ and $^1K(Sn, X)$ in analogous compounds shows no clear pattern, and it was concluded that these couplings depend significantly upon β_{PbX} and β_{SnX}, the *s* overlap integrals between lead or tin and the element X, in addition to the expected factors $\alpha_S{}^2$ and $\psi^2(0)$.[1245]

^{29}Si N.m.r. spectroscopy has been used with ^1H decoupling to study a number of cyclic siloxanes with asymmetric units. Observed chemical shifts were discussed in relation to the configuration of the various possible isomers. In the case of $[MePhSiO]_4$ the four possible isomers were examined separately. The neighbouring effects on chemical shift are markedly non-additive, in contrast

1240 D. E. Williams, *Chem. Anal.* (*N.Y.*) 41 *Anal. Silicones*, 1974, 287 (*Chem. Abs.*, 1976, **85**, 136 934).
1241 E. Bayer, G. Jung, E. Breitmaier, P. Hunziker, W. Koenig, and W. Voelter, *Pept.*, *Proc. Eur. Pept. Symp.*, 11*th*, 1973, 265 (*Chem. Abs.*, 1976, **84**, 25 588).
1242 E. A. Williams, J. D. Cargioli, and R. W. Larochelle, *J. Organometallic Chem.*, 1976, **108**, 153.
1243 A. Hauser, A. Zschunke, K. Issleib, and W. Boettcher, *Phosphorus*, 1975, **5**, 261.
1244 G. Barbieri, *Atti Soc. Nat. Mat. Modena*, 1974, **105**, 127 (*Chem. Abs.*, 1976, **85**, 27 071).
1245 J. D. Kennedy, W. McFarlane, and B. Wrackmeyer, *Inorg. Chem.*, 1976, **15**, 1299.

to the situation for $[MeHSiO]_4$ isomers.[1246] ^{13}C and ^{29}Si n.m.r. spectra of the fractions chromatographically isolated from the product of Cl_2MeSiH and $Cl_2MeSiCH{=}CH_2$ cohydrolysis agree with the structures proposed previously on the basis of mass and i.r. spectra. One of the non-geminal divinylhexamethylcyclotetrasiloxane fractions has been shown by ^{29}Si n.m.r. spectroscopy to consist of two compounds. It was concluded that all four possible isomers are formed in cohydrolysis, not just three as thought previously.[1247] Spin–spin and spin–lattice relaxation measurements have been made on some triorganylsilanes.[1248]

The ^{13}C n.m.r. spectra have been reported for some Me_3Si-, Me_3Ge-, Me_3Sn-, and Me_3Pb-substituted 1,1-cyclohexanes, 2,2-norbornanes, and (128), and a

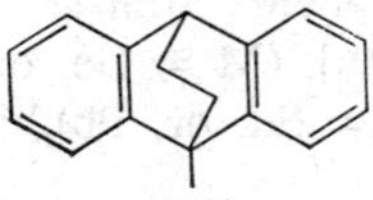

(128)

γ-effect, dependent on the geometry, was found.[1249] ^2J(H, C, M) for Me_3MR (M = C, Si, Sn, or Pb) has been analysed in terms of the *s*-character of the metal–carbon bond and distortion of C—M—R angle.[1250] ^{19}F N.m.r. data have been reported for $C_6F_5C{\equiv}CMR_3$ (M = Si, Ge, or Sn; R = Me or Et).[1251] The ^{13}C n.m.r. spectra of $ArMMe_3$ (M = C, Si, Ge, Sb, or Pb; Ar = Ph, naphthyl, biphenyl, or 9-anthryl) have been recorded and assigned, and $^1J(^{13}C, M)$ determined. For these directly bonded metalloid systems, the substituent chemical shifts at formally conjugated positions have been considered in terms of mesomeric electron withdrawal, associated with orbitals of π-symmetry and X, and opposing electron donation resulting from the aryl carbon–X bond polarization. For silicon, the former effect predominates, but for germanium, tin, and lead, opposing electron-donating mechanisms assume importance. For 1-naphthyl systems, the effects of molecular distortion on ^{13}C chemical shifts are critical at certain positions. The substituent effects of MMe_3 in aromatic systems are very feeble in comparison with the conventional substituents in organic chemistry.[1252] ^{19}F N.m.r. spectra of $C_6F_5MMe_3$ (M = Si, Ge, or Sn) have been examined. The n.m.r. parameters were correlated with the σ-constants of the MMe_3 groups. Correlations which are valid for non-organometallic C_6F_5Y compounds do not apply, thus revealing the particular nature of the C_6F_5—M bond.[1253]

1246 R. K. Harris, B. J. Kimber, M. D. Wood, and A. Holt, *J. Organometallic Chem.*, 1976, **116**, 291.
1247 J. Souček, G. Engelhardt, K. Stránský, and J. Schraml, *Coll. Czech. Chem. Comm.*, 1976, **41**, 234.
1248 I. P. Biryukhov, A. F. Babkin, Yu. I. Khudobin, N. P. Kharitonov, N. A. Andreeva, and O. S. Aidanova, *V Sb., Issled. V Obl. Fiz. i Khimii Kauchukov i Rezin.*, 1975, 57 (*Chem. Abs.*, 1976, **84**, 179 159).
1249 W. Kitching, M. Marriott, W. Adcock, and D. Doddrell, *J. Org. Chem.*, 1976, **41**, 1673.
1250 G. Barbieri and R. Benassi, *Atti Soc. Nat. Mat. Modena*, 1974, **105**, 39 (*Chem. Abs.*, 1976, **85**, 62 311).
1251 V. S. Zavgorodnii, E. T. Bogoradovskii, V. L. Maksimov, V. B. Lebedev, B. I. Rogozev, and A. A. Petrov, *Zhur. obshchei Khim.*, 1975, **45**, 2466 (*Chem. Abs.*, 1976, **84**, 42 722).
1252 M. Bullpitt, W. Kitching, W. Adcock, and D. Doddrell, *J. Organometallic Chem.*, 1976, **116**, 161.
1253 J. C. Maire and Y. Limouzin, *J. Organometallic Chem.*, 1976, **122**, 21.

The ^{13}C and ^{29}Si n.m.r. spectra of $Me_{4-n}(OCH_2CH_2NR_2)_nSi$ (R = H or Me) and related compounds indicate a decrease in the degree of $(p-d)\pi$-interaction of the silicon–oxygen bond of 2-aminoethoxysilanes with respect to the corresponding alkoxysilanes.[1254] The ^{13}C n.m.r. spectra of $X_nSi(OR)_{4-n}$ (X = Me, Ph, or H) have been recorded and assigned. The chemical shifts of the α-carbon of the alkoxy-groups were shown to depend on both the nature of the alkoxy-group and the number and type of substituents on the silicon. Regression analysis of the data gave empirical substitution chemical shift parameters for silyl substituents. The β-carbon resonances were shown to be dependent on the presence of a silyl group but not on the specific silyl substituents.[1255] The ^{29}Si n.m.r. chemical shifts of $Me_{4-n}SiX_n$ (X = NMe_2, OMe, or SMe) move upfield for X = NMe_2 or OMe with increasing values of *n*. This was accounted for by increasing silicon–oxygen and silicon–nitrogen back-donation. van der Waals interaction was proposed to explain certain n.m.r. parameters. A linear correlation was found between $^1J(^1H, ^{13}C)$ and the product of electronegativity and van der Waals radii in Me_3SiX. This provided an explanation for the lower coupling constant of Me_3SiOMe and Me_3SiNMe_2 compared with those in $SiMe_4$.[1256] 1*s* Carbon binding energies have been correlated with ^{13}C chemical shifts for Me_nMCl_{4-n} (M = Si or Ge).[1257]

^1H, ^{13}C, ^{19}F, and ^{29}Si n.m.r. spectra have been measured for $R_3Si(CH_2)_mF$ derivatives. According to the data, the α-derivatives CFH_2SiMe_3 and CH_2FBu^t appear to be different from all the other fluoroalkylsilanes and fluoroalkanes. An interpretation of the observed trends in structural terms was offered.[1258] ^1H and ^{13}C n.m.r. data have been presented for six different compounds containing the fragment $PhCCH_2SiMe_3$. It was found unexpectedly that the $SiMe_3$ ^1H chemical shift is upfield of Me_4Si. This behaviour was attributed to ring-current effects due to a preferred conformation of the $SiMe_3$, which is flat to the phenyl group.[1259] For (129), the ^1H nuclear Overhauser enhancement has been used to determine the stereochemistry.[1260] ^{13}C and ^{29}Si n.m.r. spectra of Me_3Si benzenes substituted at different positions by methyl and Me_3Si groups have been investigated with special reference to steric interactions between *ortho*-substituents. The steric effects as measured by ^{13}C and ^{29}Si chemical shifts are generally smaller in Me_3Si- than in Bu^t-substituted compounds. Both nuclei follow the same general trends in benzene derivatives.[1261] The ^1H chemical shifts of some organosilicon derivatives of thiophen are linearly related to charge-transfer bands with $(NC)_2C{=}C(CN)_2$.[1262] ^{13}C N.m.r. data have been

[1254] T. Pehk, E. Lippmaa, E. Lukevics, and L. I. Simchenko, *Zhur. obshchei Khim.*, 1976, **46**, 602.
[1255] P. E. Rakita, L. S. Worsham, and J. P. Srebro, *Org. Magn. Resonance*, 1976, **8**, 310.
[1256] E. V. Van den Berghe and G. P. Van der Kelen, *J. Organometallic Chem.*, 1976, **122**, 329.
[1257] J. E. Drake, C. Riddle, and L. Coatsworth, *Canad. J. Chem.*, 1975, **53**, 3602.
[1258] J. Schraml, J. Včelák, G. Engelhardt, and V. Chvalovsky, *Coll. Czech. Chem. Comm.*, 1976, **41**, 3758.
[1259] B. L. Shapiro and T. W. Proulx, *Org. Magn. Resonance*, 1976, **8**, 40.
[1260] H. Bouas-Laurent, M. Daney, R. Lapouyade, C. Lang, N. Platzer, and J.-J. Basselier, *J. Organometallic Chem.*, 1976, **118**, 233.
[1261] J. Schraml, V. Chvalovsky, M. Mägi, E. Lippmaa, R. Calas, J. Dunoguès, and P. Bourgeois, *J. Organometallic Chem.*, 1976, **120**, 41.
[1262] V. A. Kuznetsov, A. N. Egorochkin, E. A. Chernyshev, V. I. Savushkina, and V. Z. Anisimova, *Tezisy Dokl. Nauchn. Sess. Khim. Tekhnol. Org. Soedin. Sery Sernistykh Neftei*, 13*th*, 1974, 276 (*Chem. Abs.*, 1976, **85**, 142 464).

determined for $Me_3Si(CH_2)_nOX$ (including ^{19}F n.m.r. data),[1263] (130) (including ^{19}F n.m.r. data),[1264] $R_3Si(CH_2)_nF$ (only ^{19}F n.m.r. data),[1265] (131) [including $\delta(^{29}Si^1) = -106.2$],[1266] $H_3CC(O)SiMe_3$,[1267] $Me_3SiC{\equiv}CC{\equiv}CMe$,[1268] $Me_3SiC{\equiv}CCH{\equiv}CHR$,[1269] (132),[1270] and (133) (including ^{29}Si n.m.r. data).[1271]

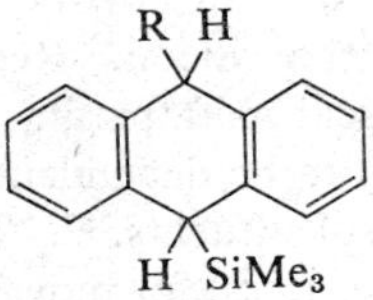
(129)

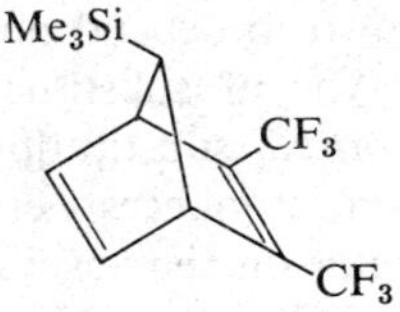
(130)

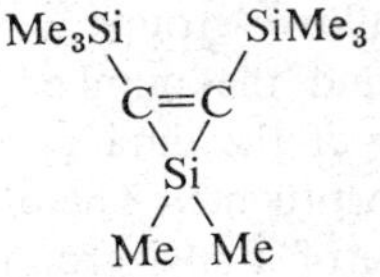
(131)

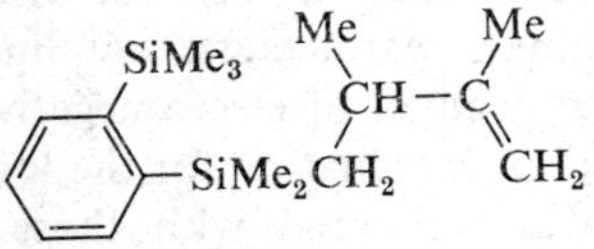
(132)

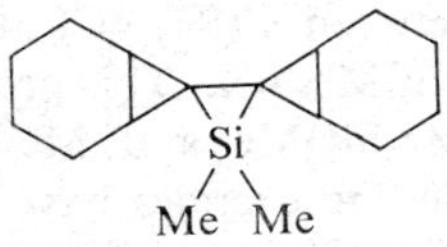
(133)

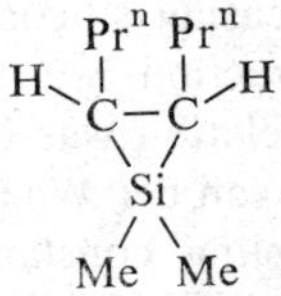
(134)

$\delta(^{29}Si) = 59.8$ for (134) is consistent with the silacyclopropane structure.[1272] The ^{13}C chemical shifts of pentamethylene heterocycles, $\overline{\lfloor (CH_2)_5X \rfloor}$ *e.g.* X = MeAs, Me_2Ge, Me_2Si, Se, $AsMe_2^+$, $SeMe^+$, and $TeMe^+$ have been reported. The chemical shifts of the α- and γ-carbon atoms are governed almost entirely by the electronegativity of the hetero-group at the 1-position with a small perturbation due to substituents on the heteroatom. The chemical shift of the β-carbon atom also depends on the heteroatom electronegativity but also on the orientation of the substituent in the 1-position. Increased heteroatom electronegativity causes a downfield shift for the α-carbon atom but an upfield shift for the β- and γ-carbon atoms in the ratio 20 : −1 : −2. Thus the polarization of the charge apparently reverses sign between the first (α) and second (β) carbon atom in agreement with the Pople–Gordon theory.[1273] Subsequently it was found that the relaxation rate of the α-carbon atom is proportional to the molecular weight. The anisotropic tumbling about a preferred axis of rotation from the heteroatom to C-4 was indicated for heteroatoms from Groups IV and V.[1274] ^{13}C

[1263] J. Pola and V. Chvalovský, *Coll. Czech. Chem. Comm.*, 1976, **41**, 581.
[1264] A. R. L. Bursico, M. Murray, and F. G. A. Stone, *J. Organometallic Chem.*, 1976, **111**, 31.
[1265] J. Včelák, V. Chvalovský, M. G. Voronkov, V. B. Pukhnarevich, and V. A. Pestunovich, *Coll. Czech. Chem. Comm.*, 1976, **41**, 386.
[1266] D. Seyferth, D. C. Annarelli, and S. C. Vick, *J. Amer. Chem Soc.*, 1976, **98**, 6382.
[1267] E. M. Dexheimer and L. Spialter, *J. Organometallic Chem.*, 1976, **107**, 229.
[1268] M. T. W. Hearn, *Austral. J. Chem.*, 1976, **29**, 2315.
[1269] M. T. W. Hearn, *J. Magn. Resonance*, 1976, **22**, 521.
[1270] M. Ishikawa, T. Fuchikami, and M. Kumada, *J. Organometallic Chem.*, 1976, **118**, 139.
[1271] D. Seyferth, R. L. Lambert, jun., and D. C. Annarelli, *J. Organometallic Chem.*, 1976, **122**, 311.
[1272] D. Seyferth and D. C. Annarelli, *J. Organometallic Chem.*, 1976, **117**, C51.
[1273] J. B. Lambert, D. A. Netzel, H.-N. Sun, and K. K. Lilianstrom, *J. Amer. Chem. Soc.*, 1976, **98**, 3778.
[1274] J. B. Lambert and D. A. Netzel, *J. Amer. Chem. Soc.*, 1976, **98**, 3783.

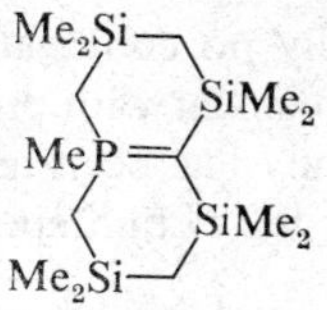

(135)

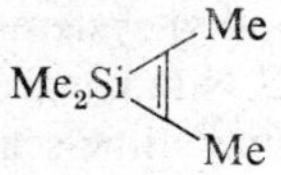

(136)

N.m.r. data have also been reported for (135) (including ^{29}Si and ^{31}P n.m.r. data),[1275] (136),[1276] and Me_nGeF_{4-n} (only ^{19}F n.m.r. data).[1277]

^{13}C N.m.r. parameters for 33 organotin compounds with a variety of structural features have been investigated in order to obtain information about the relationship between their structure and ^{13}C n.m.r. parameters. It was found that the substitution of a proton by a trialkyl tin group generally produces an upfield shift for the directly bonded carbon atom. The γ-nuclei usually resonate at lower fields except where there is appreciable steric strain, while the β-carbon atoms undergo a relatively constant shift of *ca.* 3.5—4.5 p.p.m. to lower fields. The magnitude of $^{1}J(^{13}C, ^{119}Sn)$ is influenced by hybridization of the tin atom and the directly attached carbon atoms. In rigid organotin compounds, the *vic* coupling constants show a Karplus type variation. In aliphatic organotin compounds, the values of the *vic* $J(^{13}C, ^{119}Sn)$ indicate a flexible molecular framework with a clear-cut preference for certain conformations.[1278] ^{13}C and ^{119}Sn n.m.r. spectra of 33 organotin compounds of the type $RSnMe_nCl_{3-n}$ and related types have been discussed. The $SnMe_3$ group causes a small upfield shift of the carbon atom attached to it while the other groups cause downfield shifts. The shifts show a monotonic change on replacing methyl groups in Me_3Sn by chlorine atoms. The effects on carbon atoms further removed from the tin atom were discussed. Variation in R causes little change in $^{n}J(^{13}C, ^{117,119}Sn)$ or $\delta(^{119}Sn)$.[1279] Trends in the ^{1}H chemical shifts and $^{2}J(^{1}H, C, ^{119}Sn)$ of $Me_{3-n}Br_nSn(CH_2)_mBr$ have been discussed in terms of inductive effects. Intramolecular co-ordination between the ω-bromine atom and the tin atom could not be demonstrated.[1280] The ^{119}Sn n.m.r. chemical shifts of a series of isomeric norbornenyltrimethyltin compounds have been measured in order to assess the effect of isolated bonds on the magnetic shielding of the tin nucleus. It was concluded that changes in C—Sn—C interbond angles may have to be invoked to account for observed changes in ^{119}Sn shielding.[1281] ^{13}C N.m.r. spectra of $ArCH_2SnMe_3$ (Ar = Ph, naphthyl, or biphenyl) have been recorded and essentially completely assigned by specific deuteriation, the magnitude of $J(^{13}C, ^{117,119}Sn)$, and other criteria. Comparison was made with other appropriate neopentyl compounds. At conjugated positions there is substantial shielding and $\Delta\delta$ (the substitution chemical shift) correlates well with Δq (the SCF-π-MO change in π-charge density on going from ArMe to $[ArCH_2]^-$) as would be expected for a conjugative interaction.

1275 H. Schmidbaur and M. Heimann, *Angew. Chem. Internat. Edn.*, 1976, **15**, 367.
1276 R. T. Conlin and P. P. Gaspar, *J. Amer. Chem. Soc.*, 1976, **98**, 3715.
1277 S. C. Pace, *Synth. React. Inorg. Metal-org. Chem.*, 1975, **5**, 373.
1278 H. G. Kuivila, J. L. Considine, R. H. Sarma, and R. J. Mynott, *J. Organometallic Chem.*, 1976, **111**, 179.
1279 T. N. Mitchell, *Org. Magn. Resonance*, 1976, **8**, 34.
1280 E. J. Bulten, H. F. M. Gruter, and H. F. Martens, *J. Organometallic Chem.*, 1976, **177**, 329.
1281 J. D. Kennedy, *J. Organometallic Chem.*, 1976, **104**, 311.

These data indicate that there is essentially no conjugative transmission to the 5-position in the 1-naphthyl system in agreement with previous reports based on ^{19}F n.m.r. data.[1282] ^{13}C N.m.r. spectroscopy has been used to demonstrate that (137) is static on the n.m.r. time-scale.[1283] ^{13}C N.m.r. spectra have been reported

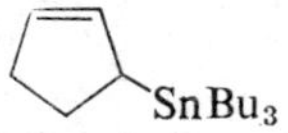

(137)

and discussed for 32 simple $R^1{}_nSnR^2{}_{4-n}$ compounds (n = 0—4) in which the tin atom is four-co-ordinate. It was found that the value of $^1J(^{13}C, ^{119}Sn)$ in the compounds R_4Sn depends on the effective nuclear charge at tin. In mixed organotin compounds, Bent's postulate was apparently verified and in a number of cases coupling constants were correlated with Taft's σ^*-constants for the alkyl groups. The values of $^1J(^{13}C, ^{119}Sn)$ in benzyltin compounds are anomalously small. Long-range $J(^{13}C, ^{117,119}Sn)$ coupling constants were discussed. The chemical shifts of the carbon atoms in methyl groups bound to tin provide a second useful probe for estimating the electron-withdrawing or electron-releasing power of alkyl and other groups bound to the tin atom.[1284] ^{119}Sn Chemical shifts have been used to show that $Me_{4-n}Sn(OBu^t)_n$ polymerize in solution.[1285] $^1J(^{13}C, ^{207}Pb)$ of $Et_nPb(OAc)_{4-n}$ is markedly dependent on n, being 200 Hz for n = 4, 235 Hz for n = 3, and 490 Hz for n = 2.[1286] ^{13}C Chemical shifts and $J(^{13}C, ^{207}Pb)$ have been reported for $Ar_nMe_{4-n}Pb$. The value of $J(^{13}C, ^{207}Pb)$ was shown to depend on the number and nature of the aryl groups.[1287]

INDO has been used to calculate $J(^{13}C, ^{29}Si)$ in Me_3SiX compounds and generally good agreement was found between theory and experiment.[1288] $[MeHg]^+$, $[Me_3Sn]^+$, and $[Me_3Pb]^+$ have been examined by ^1H n.m.r. spectroscopy in lyotropic nematic media. The metal–carbon distance was determined by assuming a tetrahedral configuration at carbon and a carbon–hydrogen distance of 1.096 Å. This procedure revealed a systematic trend of shorter metal–carbon distances being determined by the n.m.r. method as compared to the best recent electron diffraction results. No allowance appears to have been made for vibrational averaging.[1289] The non-equivalence of the diastereotopic isopropyl methyl groups in $Pr^i{}_2PhSiLi$ up to 185 °C indicates that the barrier to inversion at silicon in the silyl anion is in excess of 24 kcal mol^{-1}.[1290] ^{29}Si Chemical shifts and $^1J(^{29}Si, ^{29}Si)$ have been measured for a wide variety of silicon compounds. Values ranging from 52.5 Hz for $(Me_3Si)_4Si$ to 186 Hz for $(Cl_3Si)_2SiCl_2$ have been measured. A good correlation was found between $^1J(^{29}Si, ^{29}Si)$ and

1282 M. Bullpitt, W. Kitching, W. Adcock, and D. Doddrell, *J. Organometallic Chem.*, 1976, **116**, 187.
1283 U. Schröer and W. P. Neumann, *J. Organometallic Chem.*, 1976, **105**, 183.
1284 T. N. Mitchell and G. Walter, *J. Organometallic Chem.*, 1976, **121**, 177.
1285 J. D. Kennedy, *J. Mol. Structure*, 1976, **31**, 207.
1286 D. C. van Beelen, D. de Vos, G. J. M. Bots, L. J. van Doom, and J. Wolters, *Inorg. Nuclear Chem. Letters*, 1976, **12**, 581.
1287 D. de Vos, *J. Organometallic Chem.*, 1976, **104**, 193.
1288 K. D. Summerhays and D. A. Deprez, *J. Organometallic Chem.*, 1976, **118**, 19.
1289 L. W. Reeves, M. Suzuki, and J. A. Vanin, *J. Organometallic Chem.*, 1976, **15**, 1035.
1290 J. B. Lambert, M. Urdaneta-Perez, and H.-N. Sun, *J.C.S. Chem. Comm.*, 1976, 806.

the sum of the electronegativities of the substituents was found.[1291] Heteronuclear magnetic double-resonance experiments have been used to determine the signs and magnitudes of $^1J(^{119}Sn, ^{119}Sn)$ and other coupling constants in $Li[Sn(SnMe_3)_3]$, 3THF, $[Fe(\eta^5\text{-}C_5H_5)\{P(OPh)_3\}_2Sn(SnMe_3)_3]$, and $SnMe_2(SnMe_3)_3$. Values of *ca.* −5000 Hz were found for the first of the compounds and the results indicate that in general tin–tin coupling constants are determined mainly by changes in β_{SnSn}, the *s*-overlap integral for the tin–tin bond, which are brought about by changes in substituent at the tin atom.[1292] The suggestion that the assignment of $^1J(^{13}C, ^{207}Pb)$ and $^2J(^{13}C, ^{207}Pb)$ in Me_6Pb_2 should be reversed has been shown to be invalid.[1293]

^{13}C and ^{29}Si chemical shifts, coupling constants, and relaxation times have been measured for Me_3SiMe and $(Me_3Si)_3N$.[1294] Benzene-induced shifts have been measured for some organometallic 8-quinolinolato (L) complexes, *e.g.* $LMMe_3$ and LMe_3SbO_2CR (M = Si, Ge, or Sn). A large upfield shift (0.5—0.7 p.p.m.) of the 2-proton of the 8-quinolinolato ligand was found in benzene when the nitrogen of the ligand was co-ordinated but is small (0.0—0.3 p.p.m.) in the absence of nitrogen co-ordination.[1295] ^{31}P Chemical shifts and $^1J(^{31}P, ^{117,119}Sn)$ have been reported for $Bu^t{}_2(Me_{3-n}Cl_nSn)P$. $^1J(^{31}P, ^{117}Sn)$ is markedly dependent on the compound, being 927 Hz for $Me_2Sn(PBu^t{}_2)_2$ and 1720 Hz for $Cl_2Sn(PBu^t{}_2)_2$.[1296] N.m.r. data have also been reported for $Bu^t(Bu^tNH)SiFNBu^tSiMe_3$ (^{19}F n.m.r. data),[1297] (138), (^{13}C and ^{31}P n.m.r. data),[1298] $(Me_3Si)_2NCR^1{=}CHCR^2(OSiMe_3)CR^2(OSiMe_3)CH{=}CR^1N(SiMe_3)_2$

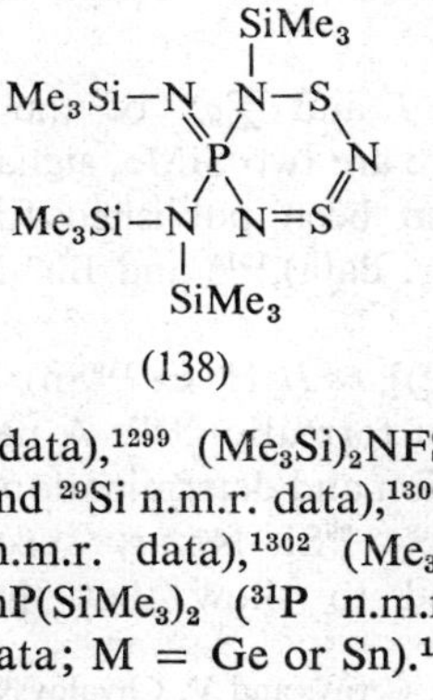

(138)

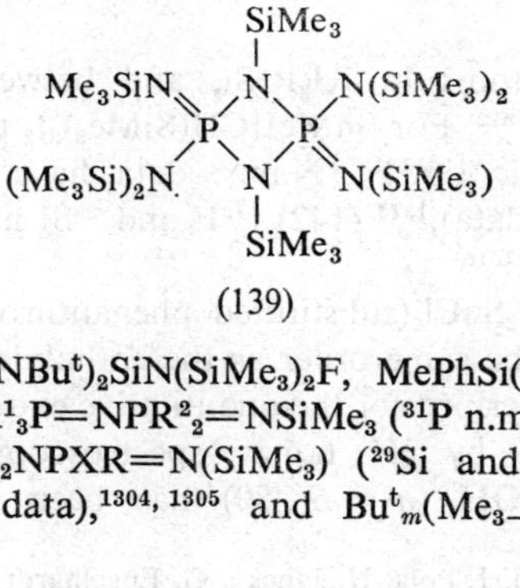

(139)

(^{13}C n.m.r. data),[1299] $(Me_3Si)_2NFSi(NBu^t)_2SiN(SiMe_3)_2F$, $MePhSi(NBu^t)(NR)$-$SiMe_2$ (^{19}F and ^{29}Si n.m.r. data),[1300] $R^1{}_3P{=}NPR^2{}_2{=}NSiMe_3$ (^{31}P n.m.r. data),[1301] (139) (^{31}P n.m.r. data),[1302] $(Me_3Si)_2NPXR{=}N(SiMe_3)$ (^{29}Si and ^{31}P n.m.r. data),[1303] $PhP(SiMe_3)_2$ (^{31}P n.m.r. data),[1304, 1305] and $Bu^t{}_m(Me_{3-n}MCl_n)_{3-m}P$ (^{31}P n.m.r. data; M = Ge or Sn).[1306]

1291 K. G. Sharp, P. A. Sutor, E. A. Williams, J. D. Cargioli, T. C. Farrar, and K. Ishibitsu, *J. Amer. Chem. Soc.*, 1976, **98**, 1977.
1292 J. D. Kennedy and W. McFarlane, *J.C.S. Dalton*, 1976, 1219.
1293 W. McFarlane, *J. Organometallic Chem.*, 1976, **116**, 315.
1294 R. K. Harris and B. Lemarié, *J. Magn. Resonance*, 1976, **23**, 371.
1295 Y. Kawasaki, *Bull. Chem. Soc. Japan*, 1976, **49**, 2319.
1296 H. Schumann, W.-W. du Mont, and H.-J. Kroth, *Chem. Ber.*, 1976, **109**, 237.
1297 U. Klingebiel and A. Meller, *Angew. Chem. Internat. Edn.*, 1976, **15**, 619.
1298 R. Appel and M. Halstenburg, *Angew. Chem. Internat. Edn.*, 1976, **15**, 696.
1299 T. N. Mitchell and B. Kleine, *Tetrahedron Letters*, 1976, 2173.
1300 U. Klingebiel and A. Meller, *Angew. Chem. Internat. Edn.*, 1976, **15**, 312.
1301 W. Wolfsberger and W. Hager, *J. Organometallic Chem.*, 1976, **118**, C65.
1302 R. Appel and M. Halstenberg, *J. Organometallic Chem.*, 1976, **121**, C47.
1303 E. Niecke and W. Bitter, *Chem. Ber.*, 1976, **109**, 415.
1304 R. Appel and K. Geisler, *J. Organometallic Chem.*, 1976, **112**, 61.
1305 O. Dahl, *Acta Chem. Scand.*, 1976, **30B**, 799.
1306 H. Schumann, W.-W. du Mont, and B. Wöbke, *Chem. Ber.*, 1976, **109**, 1017.

^{13}C and ^{29}Si n.m.r. spectra of a series of compounds containing the Me_3SiOC fragment have been presented. It was shown that the ^{29}Si chemical shift varies more in this type of compound than in compounds with a $Me_3SiC(sp^3)$ fragment. The substituent effect of the chlorine atom in Me_3SiOCH_2Cl is diamagnetic. The large sensitivity of the Me_3SiO ^{29}Si chemical shift to substituent can be analytically utilized when a compound is isolated or prepared as a silylated derivative. ^{29}Si N.m.r. spectra would be especially helpful in the analysis of polyfunctional compounds, and have been applied to a sugar and some steroids.[1307] Similarly, the ^{29}Si chemical shifts in a series of *para*- and *meta*-substituted phenyloxytrimethylsilanes are twice as sensitive to benzene ring substitution as are the shifts in corresponding $ArSiMe_3$ compounds, in which the silicon atom is one bond closer to the substituent.[1308] ^{29}Si N.m.r. spectroscopy has been used to show that a compound is (140) rather than (141).[1309] A nuclear magnetic spin-echo study of triorganylsiloxybenzenes has been carried out. A linear relationship was found between the nuclear spin-echo intensities, *I*, and the number of

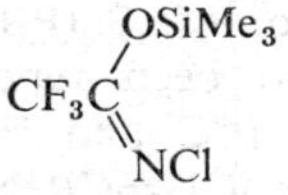
(140)

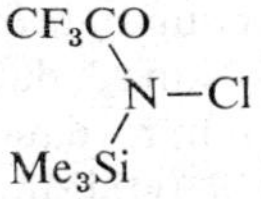
(141)

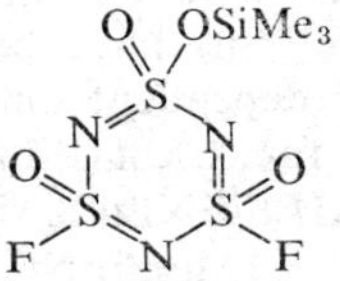
(142)

hydrogen atoms in $R^1_2R^2Si$, and between *I* and $\sum\sigma^*$ of the R^1 and R^2 groups.[1310–1312] For $SnMeI\{CH(SiMe_3)_2\}_2$ there are two $SiMe_3$ signals due to the diastereotopic tin.[1313] N.m.r. data have also been published for Me_3SiOR, (^{13}C n.m.r. data),[1314] (142) (^{19}F and ^{29}Si n.m.r. data),[1315] and $Bu^n_2Bu^tSnCl$ (^{13}C n.m.r. data).[1316]

For $[Me_2SnCl_2$(substituted phenanthroline)], $^2J(^1H, C, ^{119}Sn)$ values fall roughly in the same order as the enthalpies of formation.[1317] A new procedure has been described for the quantitative extraction and determination of dimethylpolysiloxane by ^1H n.m.r. spectroscopy.[1318] ^{29}Si N.m.r. spectroscopy of $H(OSiMe_2)_nOH$ (n = 6—80) has been used to show that the degree of

[1307] J. Schraml, J. Pola, H. Jancke, G. Engelhardt, M. Černý, and V. Chvalovský, *Coll. Czech. Chem. Comm.*, 1976, **41**, 360.
[1308] J. Schraml, P. Koehler, K. Licht, and G. Engelhardt, *J. Organometallic Chem.*, 1976, **121**, C1.
[1309] W. Lidy and W. Sundermeyer, *Chem. Ber.*, 1976, **109**, 2542.
[1310] I. P. Biryukov, A. F. Babkin, Yu. P. Shumilov, Yu. I. Khudobin, and N. P. Kharitonov, *V Sb., Issled. V Obl. Fiz. i Khimii Kauchukov i Rezin.*, 1975, 60 (*Chem. Abs.*, 1976, **84**, 179 158).
[1311] I. P. Biryukov, Yu. I. Khudobin, N. P. Kharitonov, and B. Z. Shterenberg, *Zhur. fiz. Khim.*, 1976, **50**, 522.
[1312] I. P. Biryukov, A. F. Babkin, Yu. I. Khudobin, and N. P. Kharitonov, *Zhur. fiz. Khim.*, 1976, **50**, 755.
[1313] J. D. Cotton, P. J. Davidson, and M. F. Lappert, *J.C.S. Dalton*, 1976, 2275.
[1314] M. J. Buchanan, R. H. Cragg, and A. Steltner, *J. Organometallic Chem.*, 1976, **120**, 189.
[1315] D.-L. Wagner, H. Wagner, and O. Glemser, *Chem. Ber.*, 1976, **98**, 1424.
[1316] A. G. Davies, B. Muggleton, B. P. Roberts, M.-W. Tse, and J. N. Winter, *J. Organometallic Chem.*, 1976, **118**, 289.
[1317] W. D. Honnick, M. C. Hughes, C. D. Schaeffer, jun., and J. J. Zuckerman, *Inorg. Chem.*, 1976, **15**, 1391.
[1318] M. L. Anhoury, P. Crooy, R. De Neys, and A. Laridant, *J. Pharm. Sci.*, 1976, **65**, 590.

polymerization is given by $n = \eta - 26$, where η is the bulk viscosity at 25 °C.[1319] N.m.r. data have also been reported for (143; E = C, Ge, Sn, As, N, S, or P) (^{31}P n.m.r. data),[1320] $Bu^t_2FSiNHBu^t$ (^{19}F and ^{29}Si n.m.r. data),[1321] $Me(CF_3)_2$-$CNSiMe_2N\{C(CF_3)_2Mr\}SiMe_2$ (^{19}F n.m.r. data),[1322] $R^1R^2SiFNHBu^t$ (^{19}F n.m.r. data),[1323] (144) (^{19}F n.m.r. data),[1324] and $F(Me_2SiO)_nMe_2SiF$ (^{19}F n.m.r. data).[1325]

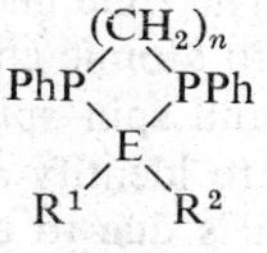

(143)

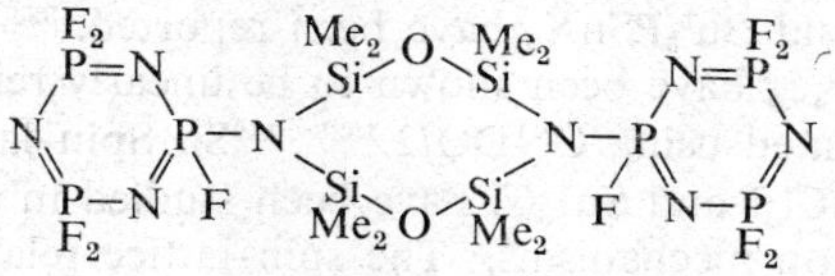

(144)

The ABC spectrum of (α-phenyl)phenylethyltrichlorosilane has been analysed.[1326] The ^{13}C n.m.r. spectra of a series of 38 substituted $PhSiX_3$ compounds have been recorded and assigned. The phenyl resonances were correlated in terms of five to nine empirical substituent chemical shift parameters. The *m* and *p* resonances, which are highly correlated, exhibit a linear dependence on the nature of the substituents. The other two phenyl resonances show a more complex behaviour requiring second-order terms for a successful correlation.[1327] The ^1H n.m.r. spectrum of the 2,2′-bipyridyl adduct of Bu^nSnCl_3 indicates the non-equivalence of the two pyridyl rings. Consequently the n-butyl group is in the SnN_2 plane.[1328] In the ^{19}F n.m.r. spectra of $SiF_3SiF_2CF_3$ and related compounds linear relationships have been found, *e.g.* $^3J(^{19}F_3CCX^{19}F_2)$ against $^3J(^{19}F_3CSiX^{19}F_2)$.[1329] N.m.r. data have also been reported for (145) (^{13}C n.m.r.

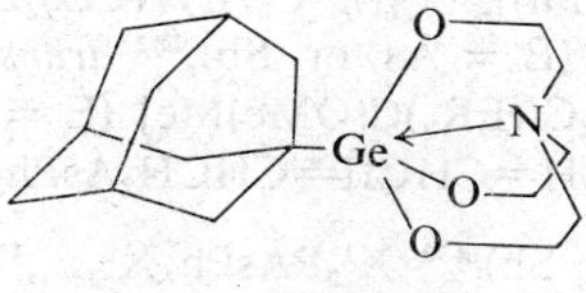

(145)

data),[1330] $(CF_3)_2NOCHClCF_2SiF_3$ (^{19}F n.m.r. data),[1331] and $Cl_3SiCHFCFHCF_2CF_2$ (^{19}F n.m.r. data).[1332]

1319 R. W. Larochelle, J. D. Cargioli, and E. A. Williams, *Macromolecules*, 1976, **9**, 85 (*Chem. Abs.*, 1976, **84**, 106 254).
1320 K. Issleib and W. Böttcher, *Synth. React. Inorg. Metal.-org. Chem.*, 1976, **6**, 179.
1321 U. Klingebiel and A. Meller, *Angew Chem. Internat. Edn.*, 1976, **15**, 313.
1322 K. E. Peterman and J. M. Shreeve, *Inorg. Chem.*, 1976, **15**, 743.
1323 U. Klingebiel and A. Meller, *Chem. Ber.*, 1976, **109**, 2430.
1324 H. W. Roesky and B. Kuhtz, *Chem. Ber.*, 1976, **109**, 3958.
1325 S. C. Pace and J. G. Riess, *J. Organometallic Chem.*, 1976, **121**, 307.
1326 Yu. F. Sigolaev, A. M. Evdokimov, V. S. Tskhovrebashvili, and V. O. Reikhsfel'd, *Issled. V Obl. Fiz. i Khimii Kauchukov i Rezin.*, 1975, 50 (*Chem. Abs.*, 1976, **84**, 179 162).
1327 P. E. Rakita, J. P. Srebro, and L. S. Worsham, *J. Organometallic Chem.*, 1976, **104**, 27.
1328 G.-E. Matsubayashi and T. Tanaka, *J. Organometallic Chem.*, 1976, **120**, 347.
1329 K. G. Sharp, S. Li, and R. B. Johannesen, *Inorg. Chem.*, 1976, **15**, 2295.
1330 V. F. Mironov, T. K. Gar, and A. V. Kisin, *Khim. Elementoorg. Soedin.*, 1976, 33 (*Chem. Abs.*, 1976, **85**, 192 835).
1331 R. N. Haszeldine, D. J. Rogers, and A. E. Tipping, *J.C.S. Dalton*, 1976, 1056.
1332 C. J. Attridge, M. G. Barlow, W. I. Bevan, D. Cooper, G. W. Cross, R. N. Haszeldine, J. Middleton, M. J. Newlands, and A. E. Tipping, *J.C.S. Dalton*, 1976, 694.

In aqueous sodium silicate solutions containing 0.37—2.0M-SiO_2 with Na to Si ratios ranging from 40 : 1 to 3 : 1, fast ^{29}Si spin–lattice relaxation has been observed (0.5—4.5 s) which decreased with increasing SiO_2 concentration. Thus ^{29}Si spin–lattice relaxation times for monosilicate, linear, oligomer, and cyclic silicates with 3 : 1 Na : Si were 3.5, 1.4, and 4.6 s, respectively, for 0.9M-SiO_2, and 1.2, 0.5, and 1.7 s, respectively, for 2M-SiO_2.[1333] The ^{31}P n.m.r. spectra of R_3PGeX_2 and $Bu^t{}_3PSnX_2$ have been reported.[1334, 1335] The ^{77}Ge chemical shifts of $GeCl_nBr_{4-n}$ have been shown to be linearly related to atomic charges which were calculated using CNDO/2.[1336] ^{119}Sn Spin–lattice and spin–spin relaxation times in $SnCl_3I$ and SnI_3Cl have been studied in order to identify and separate the relaxation mechanisms. The spin–lattice relaxation is due to a mixture of scalar and spin–rotation mechanisms while the spin–spin relaxation is due to the scalar relaxation mechanism. The T_1 of both $SnICl_3$ and $SnCl_3I$ show evidence for a maximum in $(T_1)^{-1}{}_{scalar}$ when $T_2(^{127}I)$ is equal to the difference in the ^{119}Sn and ^{127}I Larmor frequency. $^1J(^{118}Sn, ^{127}I)$ = 1638 Hz for $SnCl_3I$ while for SnI_3Cl, $^1J(^{119}Sn, ^{127}I)$ = 1097 Hz and $^1J(^{119}Sn, ^{35}Cl)$ = 421 Hz. The rotational correlation times and halogen relaxation times were measured as a function of temperature. No T_1 component due to chemical shift anisotropy was found at or below 13.3 kG.[1337] ^{31}P N.m.r. data have been reported for $[Bu^t{}_3PX]^+[MX_3]^-$ (X = Cl or Br; M = Ge or Sn).[1338, 1339]

8 Compounds of Group V Elements

Information concerning complexes of these elements can be found at the following sources: $(C_6F_4)_3E_2$ (E = As or Sb),[498] $(CH_2)_5AsMe$,[1273, 1274] $[Ph_4As]^+$,[1169] *trans*-$[PtCl_2L(Ph_3AsCHCOPh)]$,[404] $[(\eta^5\text{-}C_5H_5)V(CO)_3EPh_3]$ (E = As or Sb),[113] $[XCr(CO)_3(EMe_3){\equiv}CMe]$ (E = As or Sb),[132] *trans*-$[Cr(CO)_4(EMe_3){\equiv}CMe]^+$ (E = As or Sb),[133] $[(OC)_4Cr(ER_3)C(OMe)Me]$ (E = As or Sb),[136] $[Re_2(CO)_8(AsCF_3)_2]$,[226] $[(\eta^4\text{-}MeCH{=}CHCH{=}CHCH_2AsPh_3)Fe(CO)_3]^+$,[276] $[PtX_n(EMe_3)_{4-n}]^{2-n}$ (E = As or Sb),[459] $X^1{}_3PAsPh_nX^2{}_{3-n}$,[1216] $PhP(CH_2)_nPPhAsR$,[1320] $[(PhAs)_2Mo_6O_{24}]^{4-}$,[204] Me_3SbO_2CR,[1295] $HN(PPh_2N)PRS$,[452] $[(OC)_5M^1SbF\{M^2(CO)_3(\eta^5\text{-}C_5H_5)\}_2]$ (M^1 = Cr or W; M^2 = Mo or W),[150] $Me_3SiNSN{=}S{=}NP({=}NSiMe_3)N(SiMe_3)_3$,[1298] $N(PF_2N)_2PFN(SiMe_2OSiMe_2)_2NPF(NPF_2)_2N$,[1324] $[AsF_6]^-$,[1490] and $[Sb_2F_{11}]^-$.[1485]

The ^{15}N spin–lattice relaxation time has been measured in pure liquid nitrogen over the range 63.1—77.3 K and in the β-solid phase down to 38 K, using a sample enriched to 99.5% in $^{15}N_2$. These measurements were compared with previous work on $^{13}N_2$. The relaxation of $^{14}N_2$ is due to electric quadrupolar interaction and was used to determine the orientational correlation time, τ_Q, of individual molecules. Spin-rotation interaction dominates $^{15}N_2$ spin–lattice relaxation and the angular momentum correlation time, τ_{SR}, was deduced. The

1333 G. Engelhardt, *Z. Chem.*, 1975, **15**, 495 (*Chem. Abs.*, 1976, **84**, 82 184).
1334 W.-W. du Mont and G. Rudolph, *Chem. Ber.*, 1976, **109**, 3419.
1335 W.-W. du Mont, B. Neudert, G. Rudolph, and H. Schumann, *Angew. Chem. Internat. Edn.*, 1976, **15**, 308.
1336 P. Geerlings and C. Van Alsenoy, *J. Organometallic Chem.*, 1976, **117**, 13.
1337 R. R. Sharp and J. W. Tolan, *J. Chem. Phys.*, 1976, **65**, 522.
1338 W.-W. du Mont, B. Neudert, and H. Schumann, *Angew. Chem. Internat. Edn.*, 1976, **15**, 308.
1339 W.-W. du Mont, H.-J. Kroth, and H. Schumann, *Chem. Ber.*, 1976, **109**, 3017.

temperature dependence of τ_Q exhibits a discontinuity at the triple point which implies that molecular reorientation is much slower in the liquid than in the solid. The Hubbard relationship

$$\tau_Q \tau_{SR} = I_0/6kT$$

where I_0 is the molecular moment of inertia, which was developed for the rotational diffusion of spherical molecules in a liquid is approached in liquid nitrogen for $T < 85$ K and may hold in the solid.[1340] A nematic liquid crystal has been used as a solvent in order to observe hydrogen–hydrogen coupling in $[NH_4]^+$.[1341]

The nuclear Overhauser enhancement between hydrogen and phosphorus and the ^{31}P spin–lattice relaxation time for phosphorus-containing compounds have been determined. The dominant mechanism of phosphorus relaxation is dipole–dipole interaction between phosphorus and hydrogen. From the nuclear Overhauser enhancement and spin-relaxation data, phosphorus–hydrogen distances were determined. A nuclear Overhauser enhancement of 1 : 2 was found to be common.[1342] $^1J(^1H, ^{31}P)$ and $^1J(^{13}C, ^{31}P)$ have been calculated using the finite perturbation method.[1343] $^1J(^{31}P, ^{31}P)$ has been calculated using the results obtained from an MO-SCF *ab initio* calculation on P_2H_4 in four different conformations. $^1J(^{31}P, ^{31}P)$ is highly dependent on the rotation angle, being −238.0 Hz in the eclipsed conformation and 10.9 Hz in the staggered one. The shape of the theoretical $^1J(^{31}P, ^{31}P)$ plot seems to be in good agreement with experiment.[1344] For the two isomers of $(CF_3PH)_2$, the ^1H and ^{31}P n.m.r. spectra are very temperature dependent, *e.g.* $^1J(^{31}P, ^{31}P)$ for one isomer changes from −135.2 Hz at 35 °C to −114.8 Hz at −100 °C, the corresponding values for the other isomer being −183.7 Hz and −203.4 Hz. This behaviour was attributed to differing rotamer populations.[1345] The assignment of n.m.r. signals due to the racemic and *meso* forms of PhPHPHPh has been achieved by using an optically active solvent at 40 and 111 MHz for ^{31}P n.m.r. spectroscopy.[1346] ^{19}F N.m.r. data have been reported for *trans*-$[H_2PF_4]^-$.[1347] ^1H, ^{15}N, ^{19}F, and ^{31}P n.m.r. spectra have been reported for $PF_2H(NH_2)_2$, and show that NH_2 rotation is rapid at room-temperature but slows on cooling. The signs of the coupling constants were determined.[1348] ^{75}As Fourier transform n.m.r. spectra of $[AsH_4]^+$ shows $^1J(^1H, ^{75}As)$ of 555 ± 6 Hz at 0 °C.[1349]

The temperature dependence of ^{31}P chemical shifts of a variety of phosphorus compounds has been examined and shifts of up to 0.0647 p.p.m. per degree were found. Compounds of the type $RP(OMe)_2$ were found to be the most susceptible.[1350] ^{31}P Magnetic shielding anisotropies $\sigma_\parallel - \sigma_\perp$ of ±6, +210, and

1340 L. M. Ishol, T. A. Scott, and M. Goldblatt, *J. Magn. Resonance*, 1976, **23**, 313.
1341 K. Radley and A. Saupe, *Mol. Phys.*, 1976, **32**, 1167.
1342 H. Beierbeck, R. Martins, J. K. Saunders, and C. Benezra, *Canad. J. Chem.*, 1976, **54**, 1918.
1343 T. A. Albright, *Org. Magn. Resonance*, 1976, **8**, 489.
1344 J. P. Albrand, H. Faucher, D. Gagnaire, and J. B. Robert, *Chem. Phys. Letters*, 1976, **38**, 521.
1345 J.-P. Albrand, J.-B. Robert, and H. Goldwhite, *Tetrahedron Letters*, 1976, **12**, 949.
1346 J.-P. Albrand and J.-B. Robert, *J.C.S. Chem. Comm.*, 1976, 876.
1347 K. O. Christe, C. J. Schack, and E. C. Curtis, *Inorg. Chem.*, 1976, **15**, 843.
1348 D. E. J. Arnold and D. W. H. Rankin, *J.C.S. Dalton*, 1976, 1130.
1349 R. Gut, *Inorg. Nuclear Chem. Letters*, 1976, **12**, 149.
1350 M. D. Gordon and L. D. Quin, *J. Magn. Resonance*, 1976, **22**, 149.

+127 p.p.m. have been reported for Me_3P, Me_3PO, and Me_3PS, respectively. Liquid crystals and INDOR were used.[1351] ^{13}C and ^{31}P n.m.r. spectra have been obtained for a number of $(R^1R^2P)_2$ and $(R^1R^2PS)_2$ compounds. The satellite resonances in the ^{31}P n.m.r. spectrum due to ^{13}C-containing isotopomers were examined, and the data were treated together with ^1H n.m.r. studies where appropriate to obtain $J(^{13}C, ^{31}P)$ and $^1J(^{31}P, ^{31}P)$ as well as ^{13}C and ^{31}P chemical shifts. These n.m.r. parameters were discussed in terms of steric interactions and stereoisomerisms. The values of $^1J(^{31}P, ^{31}P)$ for both series of compounds increase markedly as the size of the substituent increases and this effect correlates with an increase in $\delta(^{31}P)$.[1352] ^{31}P Spin–lattice relaxation times and $^{31}P\{^1H\}$ nuclear Overhauser enhancements have been reported for three diphosphines and eight diphosphine disulphides. In three instances, the work was carried out at several temperatures. The ($^{31}P^1H$) dipolar contribution to relaxation varies with substituent bulk and temperature, particularly for diphosphines. It is probable that spin–rotation interactions provide the other major relaxation mechanism.[1353] ^1H and ^{13}C n.m.r. spectroscopies have been used to show that Ar_5Sb is fluxional even at −130 °C. The ^{13}C n.m.r. chemical shifts of Ar_3Sb and Ar_5Sb have been discussed in terms of group trends. A linear plot was found between the ^{13}C chemical shift of Ph_3E and the electronegativity of E (E = N, P, As, or Sb).[1354] ^1H, ^{13}C, and ^{31}P n.m.r. spectra of phosphabenzene, arsabenzene, and stibabenzene have been measured and the ^1H spectra analysed by the use of spectral simulation. The α-hydrogen is highly deshielded due to the diamagnetic anisotropy of the adjacent heteroatom. The ^{31}P nuclei of phosphabenzene are highly deshielded and all $^1J(^1H, ^{13}C)$ are close to 159 Hz.[1355] The ^1H and ^{13}C n.m.r. spectra of (146; E = P or As) have been recorded. For (146; E = P),

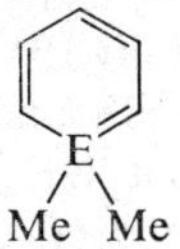

(146)

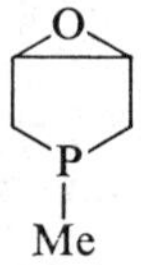

(147)

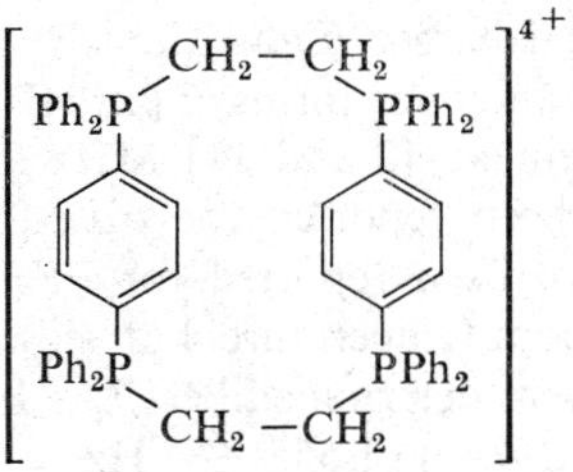

(148)

$\delta(^{13}C\text{-}2,6)$ = 67.5 with $^1J(^{13}C, ^{31}P)$ = 94 Hz.[1356] N.m.r. data have also been reported for (147) (^{13}C and ^{31}P n.m.r. data),[1357] neopentyl poly(tertiary phosphines) (^{13}C and ^{31}P n.m.r. data),[1358] cyclohexyl tertiary phosphines (^{13}C n.m.r. data),[1359]

1351 J. D. Kennedy and W. McFarlane, *J.C.S. Chem. Comm.*, 1976, 666.
1352 S. Aime, R. K. Harris, E. M. McVicker, and M. Fild, *J.C.S. Dalton*, 1976, 2144.
1353 R. K. Harris and E. M. McVicker, *J.C.S. Faraday II*, 1976, **72**, 2291.
1354 G. L. Kuykendall and J. L. Mills, *J. Organometallic Chem.*, 1976, **118**, 123.
1355 A. J. Ashe, R. R. Sharp, and J. W. Tolan, *J. Amer. Chem. Soc.*, 1976, **98**, 5451.
1356 A. J. Ashe and T. W. Smith, *J. Amer. Chem. Soc.*, 1976, **98**, 7861.
1357 C. Symmes, jun. and L. D. Quin, *Tetrahedron Letters*, 1976, 1853.
1358 R. B. King, J. C. Cloyd, jun., and R. H. Reimann, *J. Org. Chem.*, 1976, **41**, 972.
1359 M. D. Gordon and L. D. Quin, *J. Org. Chem.*, 1976, **41**, 1690.

$(CF_3)_2PCF{=}C(CF_3)_2$ (^{19}F and ^{31}P n.m.r. data),[1360] $Me_2AsC{=}CCF_2CF_2C{=}CAs$-$Me_2CF_2CF_2$ (^{19}F n.m.r. data),[1361] (148) (^{31}P n.m.r. data; X = H, Cl, or Me),[1362] $Bu^n{}_3PC(S)NR$ (^{31}P n.m.r. data),[1363] and phosphorus-containing heterocycles.[1364]

^{13}C and ^{31}P n.m.r. data for phosphorus ylides have been compared with data from CNDO/2 and MINDO/3 MO calculations. Evidence was found for the ylide carbon atom to be trigonal with carbon–phosphorus π-bonding. Charge delocalization from the phosphorus atom into the π-system was found.[1365] Atomic charges obtained using CNDO calculations have a linear correlation with ^{13}C chemical shifts for R_3PCXY.[1366] ^{13}C Chemical shift assignments have been given for a series of arsenic ylides with a *p*-substituted phenacyl group linked to the carbanion. The chemical shifts of the ylide carbon atoms show a linear dependence on the dual substituent parameters of Swain and Lupton, the resonance term being predominant. The chemical shifts of the carbonyl carbon atoms do not exhibit an analogous correlation. The chemical shift of carbon atoms in phenyl groups linked to arsenic indicate a limited interaction between the aromatic ring and the arsenic in $Ph_3As{=}CHCOC_6H_4X$.[1367, 1368] The ylide carbon of Me_2AsCHN_2 is at δ 24.1.[1369] The upper and lower limits of $^3J(^{31}P, ^{31}P)$ have been found for *trans*-$P^VCH{=}CHP^V$-containing compounds.[1370] ^{31}P N.m.r. data have been reported for $Ph_3P{=}CHC(OEt)CHCO_2Et$.[1371]

$Me_3PN{=}PMe_2{=}CH_2$ shows only one ^1H, ^{13}C, and ^{31}P n.m.r. signal, and consequently the dynamic equilibrium (149) was postulated.[1372] ^1H N.m.r.

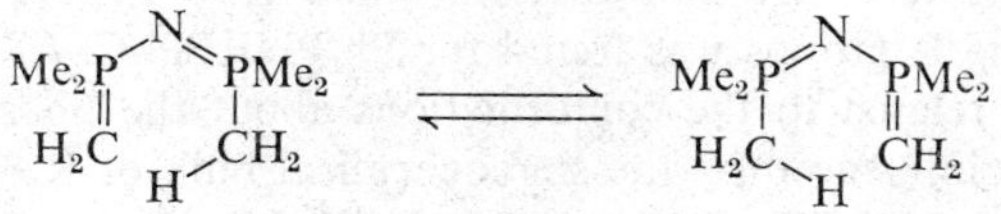

(149)

spectroscopy has been used to show that 1 : 1 complexes of $Me_3P{=}NSiMe_2$-$N{=}PMe_3$ show non-rigid behaviour.[1373] The electron distribution and conformation of $Ph_3P{=}NR$ and some phosphinazines have been discussed with reference to ^{13}C and ^{31}P n.m.r. spectroscopy and CNDO/2 MO calculations.[1374]

1360 K. G. Sharp and I. Schwager, *Inorg. Chem.*, 1976, **15**, 1697.
1361 W. R. Cullen, A. W. Wu, A. R. Davis, F. W. B. Einstein, and J. D. Hazlett, *Canad. J. Chem.*, 1976, **54**, 2871.
1362 S. D. Venkataramu, M. El-Deek, and K. D. Berlin, *Tetrahedron Letters*, 1976, 3365.
1363 K.-Y. Akiba, T. Yoneyama, H. Hamada, and N. Inamoto, *Bull. Chem. Soc. Japan*, 1976, **49**, 1970.
1364 G. A. Gray, S. E. Cremer, and K. L. Marsi, *J. Amer. Chem. Soc.*, 1976, **98**, 2109.
1365 T. A. Albright, M. D. Gordon, W. J. Freeman, and E. E. Schweizer, *J. Amer. Chem. Soc.*, 1976, **98**, 6249.
1366 K. A. Ostoja Starzewski and H. Bock, *J. Amer. Chem. Soc.*, 1976, **98**, 8486.
1367 P. Frøyen and D. G. Morris, *Acta Chem. Scand.* (*B*), 1976, **30**, 435.
1368 P. Frøyen and D. G. Morris, *Acta Chem. Scand.* (*B*), 1976, **30**, 790.
1369 P. Krommes and J. Lorberth, *J. Organometallic Chem.*, 1976, **110**, 195.
1370 Yu. Yu. Samitov, E. A. Berdnikov, F. R. Tantasheva, B. Ya. Margulis, and E. G. Kataeva, *Zhur. obshchei Khim.*, 1975, **45**, 2130.
1371 H.-J. Bestmann and R. W. Saalfrank, *Chem. Ber.*, 1976, **109**, 403.
1372 H. Schmidbaur and H.-J. Füller, *Angew. Chem. Internat. Edn.*, 1976, **15**, 501.
1373 W. Wolfsberger and H. Schmidbaur, *J. Organometallic Chem.*, 1976, **122**, 5.
1374 T. A. Albright, W. J. Freeman, and E. E. Schweizer, *J. Org. Chem.*, 1976, **41**, 2716.

^{13}C and ^{31}P n.m.r. spectra of Ph_3PSe and $Ph_3P(OAc)_2$ have been measured. For Ph_3PSe, $^1J(^{31}P, ^{77}Se) = 736.2$ Hz.[1375] The ^1H n.m.r. spectra of $Et_nPh_{3-n}AsXY$ have been interpreted in terms of the electron density on arsenic.[1376] The ^{31}P n.m.r. spectra of PPh_nCl_{5-n} (n = 0—3) gives two signals due to $[PPh_nCl_{4-n}]^+$ and PPh_nCl_{5-n}. The chlorine contributions are additive and change linearly as a function of the number of chlorine atoms in the compound. The data were discussed in terms of the nuclear magnetic shielding by chlorine atoms, the nature of the phosphorus–chlorine bond and the co-ordination number of the phosphorus.[1377] ^{31}P N.m.r. data have been reported for $CH_2(PPh_2N)_2SO_2$,[1378] $Ph_nP(N\text{-pyrrole})_{3-n}$,[1379] $Ph_3PNRCH_2CH(CH_2OPh)O$,[1380] $(R^1O)R^2_4E$ (including ^{13}C n.m.r. data; E = P or As),[1381] (150) (including ^{19}F n.m.r. data),[1382] and $Ar_3Sb(CF_3CO_2CH_2COR)$ (only ^{19}F n.m.r. data).[1383]

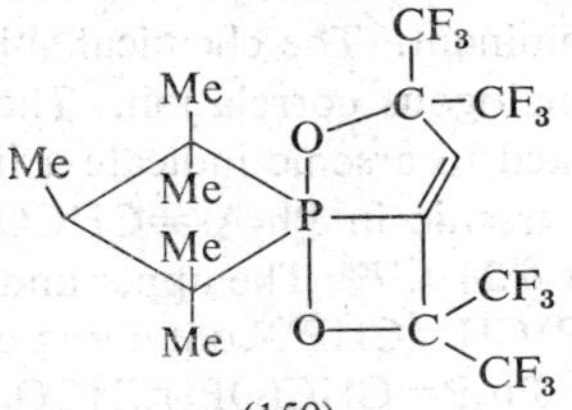

(150)

$^1H\{^{31}P\}$ and ^{31}P n.m.r. measurements of $(Ph_2PNR)_2PPh$ show that $^2J(^{31}PN^{31}P)$ is +280 Hz when R = Me and 25.1 Hz when R = Et. A similar marked dependence on the R groups was found for $Ph_2PNRPPhCl$ (+334 to −35 Hz) and this may be related to the conformations about the phosphorus–nitrogen bonds, which are influenced by the stereochemical bulk of R. $^1H\{^{31}P\}$ INDOR experiments on $Ph_2PNRPPh_2$ indicate that $^2J(^{31}P, N, ^{31}P)$ is much greater when R = Me than when R = Et or Pr^i.[1384] Regression analysis of the dependence of ^{19}F chemical shifts in R^1R^2POF on the inductive σ^* Taft constants of the substituents was performed. The equation

$$\delta(^{19}F) = b_0 + b_1 \sum\sigma^*$$

was suggested for a quantitative description of the dependence. The contribution $\delta(^{19}F^p)$ for the phosphorus–fluorine bond polarity to $\delta(^{19}F)$ values was found. $\delta(^{19}F^m)$, due to $(p\text{–}d)\pi$-bonding, was derived.[1385] Subsequently, this work was extended to ^{31}P n.m.r. spectra. $\delta(^{31}P)$ is linearly dependent on the sum of the

1375 C. Glidewell, *J. Organometallic Chem.*, 1976, **116**, 199.
1376 B. E. Abalonin, Yu. F. Gatilov, T. V. Zykova, G. I. Vasilenko, Z. M. Izmailova, and N. A. Zhikhareva, *Doklady Akad. Nauk S.S.S.R.*, 1976, **226**, 1323.
1377 G. G. Arzoumanidis, *Prakt. Panelleniou Chem. Synedrious*, 4th, 1972, 46 (*Chem. Abs.*, 1976, **85**, 191 982).
1378 R. Appel, I. Ruppert, and M. Montenark, *Chem. Ber.*, 1976, **109**, 71.
1379 S. Fischer, J. Hoyano, I. Johnson, and L. K. Peterson, *Canad. J. Chem.*, 1976, **54**, 2706.
1380 R. Appel and M. Halstenberg, *Chem. Ber.*, 1976, **109**, 814.
1381 H. Schmidbaur and P. Holl, *Chem. Ber.*, 1976, **109**, 3151.
1382 H. A. E. Aly, J. H. Barlow, D. R. Russell, D. J. H. Smith, M. Swindles, and S. Trippett, *J.C.S. Chem. Comm.*, 1976, 449.
1383 F. Ebina, T. Uehiro, T. Iwamoto, A. Ouchi, and Y. Yoshino, *J.C.S. Chem. Comm.*, 1976, 245.
1384 R. J. Cross, T. H. Green, and R. Keat, *J.C.S. Dalton*, 1976, 1424.
1385 B. D. Ryzhikov and E. T. Gainullina, *Zhur. strukt. Khim.*, 1975, **16**, 754.

Taft substituent constant. Positive mesomeric effects of the substituent on the tetrahedral phosphorus atom decrease with increasing sum of Taft substituent constants and with respect to the size of decrease in value of the Taft substituent constant due to $(p\text{–}d)\pi$-interaction.[1386] ^{31}P N.m.r. spectroscopy has been used to optimize the preparation of Me_2PCl from red phosphorus and methyl chloride.[1387] The ^{13}C chemical shift of the *para*-carbon atom of Ph_2SbX (X = Cl, Br, OAc, SC_6H_4F-4, or Ph) has been determined, as has the inductive σ-effect which is given by[1388]

$$\delta(^{13}C) = (3.78 \pm 0.24)\sigma + 128.2 \pm 0.1$$

^{31}P N.m.r. data have also been reported for (151),[1389] (152) (including ^{19}F n.m.r. data),[1390] (153),[1391] $R^1R^2P(O)(OR^3)$,[1392] and (154) (only ^{13}C n.m.r. data).[1393]

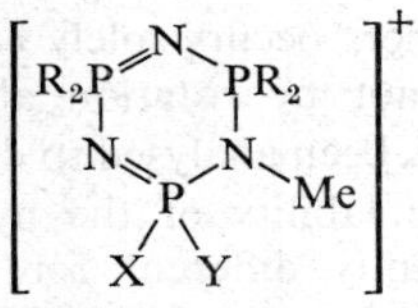

(151)

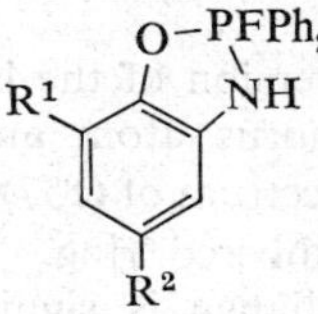

(152)

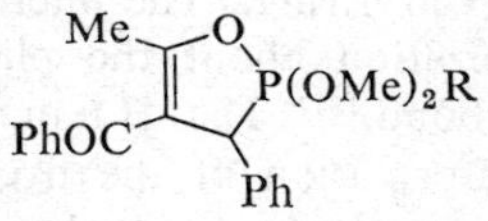

(153)

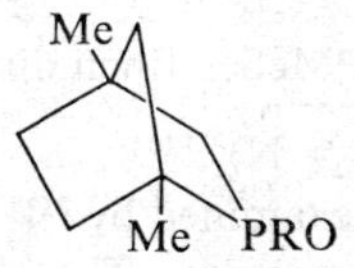

(154)

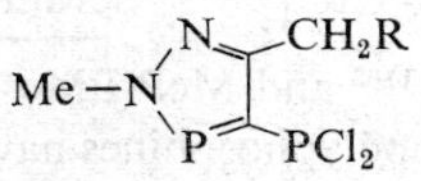

(155)

The ^{31}P n.m.r. spectrum of (155) shows an AB pattern.[1394] The ^{19}F{^{1}H} and ^{31}P{^{1}H} n.m.r. spectra of $(RF_2PNMe)_2$ have been recorded and analysed both at a temperature high enough for *gauche* ⇌ *trans* exchange to be rapid on the n.m.r. time-scale and at a low temperature where exchange is slow. Some ^{1}H n.m.r. data, including an analysis of an $[A(X)_2M_3]_2$ spin system (R = Me), have been given. Many signs of coupling constants were obtained either from spectral analysis or from double-resonance experiments. There is a very large isomer effect on $^2J(^{31}P, ^{31}P)$.[1395] For $R^1AsNR^2CH_2CH_2XR^3$ and PhOAsOCHMeCHMeO, the ^{1}H n.m.r. spectra have been fully analysed to give chemical shifts and coupling constants. Ring torsional angles were derived by

[1386] E. T. Gainullina and M. K. Baranaev, *Zhur. fiz Khim.*, 1976, **50**, 1951.
[1387] F. W. Parrett and M. S. Sun, *Synth. React. Inorg. Metal-Org. Chem.*, 1976, **6**, 115.
[1388] S. I. Pombrik, E. D. Korniets, D. N. Kravtsov, L. A. Fedorov, and E. I. Fedin, *Izvest. Akad. Nauk S.S.S.R., Ser. khim.*, 1976, **7**, 1667.
[1389] A. Schmidpeter, *Chem. Ber.*, 1976, **109**, 2340.
[1390] H. B. Stegmann, H. V. Dumm, and K. B. Ulmschneider, *Tetrahedron Letters*, 1976, 2007.
[1391] B. A. Arbuzov, A. A. Musina, A. V. Aganov, R. M. Aminova, N. A. Polezhaeva, and Yu. Yu. Samitov, *Doklady Akad. Nauk S.S.S.R.*, 1976, **228**, 865.
[1392] R. Appel and U. Warning, *Chem. Ber.*, 1976, **109**, 805.
[1393] Y. Kashman and A. Rudi, *Tetrahedron Letters*, 1976, 2819.
[1394] J. Luber and A. Schmidpeter, *J.C.S. Dalton*, 1976, 887.
[1395] R. K. Harris, M. I. M. Wazeer, O. Schlak, and R. Schmutzler, *J.C.S. Dalton*, 1976, 17.

means of the *R* value method. The data were adequately explained on the basis of rapid interconversion of twist envelope conformations.[1396] Low and variable temperature ^1H and ^{31}P n.m.r. studies of (156) show the presence of unequally populated diastereoisomeric forms corresponding to two possible orientations

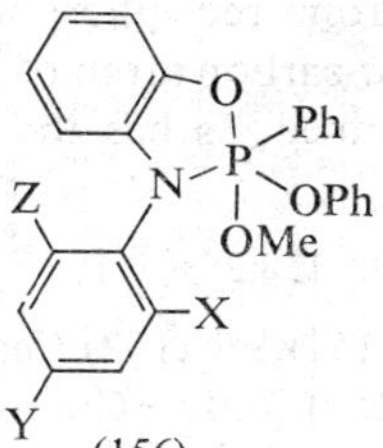

(156)

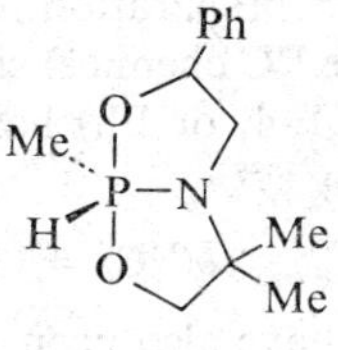

(157)

of the *N*-aryl ring. The interconversion of the isomers occurs solely *via* ligand reorganization about the phosphorus atom and not by rotation about the N–aryl bond.[1397] The ^1H n.m.r. spectrum of (157) has been analysed to determine the conformation of the five-membered ring. The lability of the hydrogen–phosphorus bond to proton exchange is significantly different for the two enantiomers.[1398] ^{31}P N.m.r. data have also been reported for $R^1PF_3NR^2R^3$ (including ^{19}F n.m.r. data),[1399] $FSO_2NCOP(NMe_2)_3$ (including ^{19}F n.m.r. data),[1400] $[P{=}NR^1NHCR^2{=}CR^3]^+$,[1401] $SeMeP(NR)_2PMeSe$ [including $^1J(^{31}P, ^{77}Se)$],[1402] $Bu^tP(NMe)_2PBu^t$,[1403] and $MeP(PMe{=}NPPh_2{=}N)_2$.[1404]

The $(RP)_n$ cyclopolyphosphines have been examined by ^{31}P n.m.r. spectroscopy and give AB_2, AA′BB′C, or ABCDE spin systems. The shift is indicative of ring size.[1405] The analysis of the ^{31}P n.m.r. spectrum of $(PPh)_6$ oriented in a nematic phase has been reported. The ring conformation obtained is in good agreement with *X*-ray data. The $^1J(^{31}P, ^{31}P) = -106$ Hz supports the rotational dependence around the phosphorus–phosphorus bond previously suggested from experimental and theoretical studies.[1406] The weak resonances at +22 and +49 p.p.m. in the ^{31}P n.m.r. spectrum of melts or solutions of $(PPh)_5$ are caused by $(PPh)_4$ and $(PPh)_6$, respectively. $(PEt)_4$ and mixed $P_4Et_{4-n}Ph_n$ species exhibit similar chemical shifts as $(PPh)_4$ and similar behaviour was found for $P_6Et_{6-n}Ph_n$ and $(PPh)_6$. Generally cyclophosphanes establish an equilibrium between $(PR)_n$ species of different ring size by a multi-centre mechanism.[1407] The ^{31}P n.m.r. spectrum of $PhAs\{P(O)Me(OMe)\}_2$ shows two singlets and an AB pair due to the three isomeric forms.[1408]

[1396] D. W. Aksnes, F. A. Amer, and K. Bergesen, *Acta Chem. Scand.* (*A*), 1976; A**30**, 109.
[1397] J. I. G. Cadogan, R. S. Strathdee, and N. J. Tweddle, *J.C.S. Chem. Comm.*, 1976, 891.
[1398] J. Devillers, D. Houalla, T. Mouheich, and R. Wolf, *Org. Magn. Resonance*, 1976, **8**, 558.
[1399] A. V. Fokin, G. I. Drozd, and M. A. Landau, *Zhur. strukt. Khim.*, 1976, **17**, 385.
[1400] H. W. Roesky and G. Sidiropoulos, *Angew. Chem. Internat. Edn.*, 1976, **15**, 693.
[1401] J. Luber and A. Schmidpeter, *Angew. Chem. Internat. Edn.*, 1976, **15**, 111.
[1402] O. J. Scherer and G. Schnabl, *Chem. Ber.*, 1976, **109**, 2996.
[1403] O. J. Scherer and G. Schnabl, *Angew. Chem. Internat. Edn.*, 1976, **15**, 772.
[1404] A. Schmidpeter, J. Högel, and F. R. Ahmed, *Chem. Ber.*, 1976, **109**, 1911.
[1405] L. R. Smith and J. L. Mills, *J. Amer. Chem. Soc.*, 1976, **98**, 3852.
[1406] J. P. Albrand, A. Cogne, and J. B. Robert, *Chem. Phys. Letters*, 1976, **42**, 498.
[1407] M. Baudler, B. Carlsohn, W. Boehm, and G. Reuschenbach, *Z. Naturforsch.*, 1976, **31b**, 558.
[1408] K. M. Abraham and J. R. Van Wazer, *J. Organometallic Chem.*, 1976, **113**, 265.

The ^{13}C chemical shifts and $J(^{13}C, ^{31}P)$ have been reported for $RP(O)(OMe)_2$ of known geometry and the possibility of a Karplus type relationship for $^3J(^{13}C, C, C, ^{31}P)$ was discussed.[1409] The signs of $J(^1H, ^{31}P)$ and $J(^{13}C, ^{31}P)$ have been determined for $MePECl_2$ and related compounds. For $EtPOCl_2$ and $Cl(CH_2)_2POCl_2$, $^2J(^1H, ^{31}P)$ and $^3J(^1H, ^{31}P)$ have opposite signs.[1410] A simplified topological approach to the computer simulation of lanthanide-induced shifts has been applied to 2-substituted 5-Bu^t-2-oxo-1,3,2-dioxaphosphorinanes.[1411] A total analysis of the ^1H, ^{13}C, and ^{31}P n.m.r. spectra of $RP(SCH_2)_2CH_2$ has been carried out. $^3J(^1H, ^{31}P)$, $^4J(^1H, ^{31}P)$, and $^4J(^{13}C, ^{31}P)$ are highly sensitive to the bond orientation around phosphorus and can be used as a tool for stereochemical assignment.[1412] ^1H, ^{13}C, and ^{31}P n.m.r. spectra have been used to show that when $(EtO)_2P(O)CHRCO_2Me$ is treated with Bu^{n}Li, then (158) is formed,

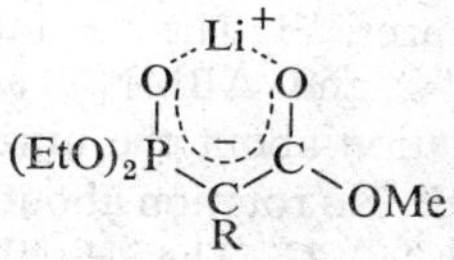

(158)

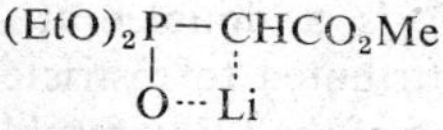

(159)

whereas treatment with Bu^{t}OLi yields (159).[1413] The ^1H n.m.r. spectra of 15 six-membered arsenites have been analysed. The n.m.r. data were explained in terms of two chair conformations differing essentially in configuration at the arsenic. The predominant isomers of all compounds except those possessing *trans*-4,6-dimethyl groups appear to exist almost entirely in one chair conformation with axially and equatorially oriented substituents at arsenic and carbon, respectively. In *trans*-4,6-dimethyl-containing compounds, there is strong evidence for a conformational equilibrium between the two chair forms.[1414]

The low-temperature ^{13}C n.m.r. spectrum of $PhAs(OCMe_2CMe_2O)_2$ shows one CO and two methyl resonances, and it was therefore concluded that the structure is square-pyramidal rather than trigonal-bipyramidal.[1415] The ^{13}C and ^{19}F n.m.r. spectra of (1-adamantyl)PF_4 shows rapid fluorine exchange even at −100 °C.[1416] N.m.r. data have also been reported for O=C=NSOF (^{19}F n.m.r. data),[1417] $MeP(O)(OR^1)(SR^2)$ (^{31}P n.m.r. data),[1418] $H_2C{=}CMeCH_2P(O)(OEt)_2$ (^{31}P n.m.r. data),[1419] $R(X)PYCH_2CH_2CHMeY$ (^{31}P n.m.r. data),[1420] and

[1409] G. W. Buchanan and C. Benezra, *Canad. J. Chem.*, 1976, **54**, 231.
[1410] V. I. Zakharov, Yu. V. Belov, Yu. L. Kleiman, N. V. Morkovin, and B. I. Ionin, *Zhur. obshchei Khim.*, 1976, **46**, 1415.
[1411] P. Finocchiario, A. Recca, W. G. Bentrude, H.-W. Tan, and K. C. Lee, *J. Amer. Chem. Soc.*, 1976, **98**, 3537.
[1412] J. Martin, J. B. Robert, and C. Taieb, *J. Phys. Chem.*, 1976, **80**, 2417.
[1413] T. Bottin-Strzalko, J. Seyden-Penne, and M.-P. Simonnin, *J.C.S. Chem. Comm.*, 1976, 905.
[1414] D. W. Aksnes, J. Andersen, and K. Bergesen, *Acta Chem. Scand.* (*A*), 1976, **30**, 327.
[1415] H. Goldwhite, J. Grey, and R. Teller, *J. Organometallic Chem.*, 1976, **113**, C1.
[1416] J.-V. Weiss and R. Schmutzler, *J.C.S. Chem. Comm.*, 1976, 643.
[1417] H.-J. Krannich and W. Sundermeyer, *Angew. Chem. Internat. Edn.*, 1976, **15**, 110.
[1418] A. A. Abduvakhabov, A. A. Sadykov, Kh. A. Aslanov, N. N. Godovikov, and A. S. Sadykov, *Doklady Akad. Nauk Uzbek. S.S.R.*, 1974, **31**, 30.
[1419] L. Maier, *Phosphorus*, 1975, **5**, 223 (*Chem. Abs.*, 1976, **84**, 59 652).
[1420] J. A. Mosbo, *Tetrahedron Letters*, 1976, 4789.

$R^1P\{P(O)(OR^2)_2\}_2$ (^{31}P n.m.r. data).[1421] The signs and magnitudes of $^1J(^{15}N, ^{31}P)$, $^2J(^1H, N, ^{31}P)$, $^3J(^1H, C, P, ^{15}N)$ and certain other coupling constants and ^{31}P and ^{15}N chemical shifts have been determined in 24 organophosphorus compounds with phosphorus–nitrogen bonds enriched in ^{15}N. The coupling constant $^1K(^{15}N, ^{31}P)$ has substantial negative values in phosphorus(III) compounds and becomes more positive with values near zero when the phosphorus atom is four-co-ordinate. The behaviour of $^2J(^1H, N, ^{31}P)$ is consistent with the hybridization of the nitrogen atom being close to sp^2 in these species, and this was attributed to $(p\text{–}d)\pi$-overlap with phosphorus d-orbitals and/or $(p\text{–}p)\pi$-overlap with N-phenyl groups.[1422] ^{13}C, ^{15}N, and ^{17}O n.m.r. spectra have been measured for $(Me_2N)_3P$ [$^1J(^{15}N, ^{31}P) = +59.1$ Hz], $(Me_2N)_3PS$ [$^1J(^{15}N, ^{31}P) = \pm 6.0$ Hz], and $(Me_3N)_3PO$ [$^1J(^{15}N, ^{31}P) = -26.9$ Hz, $^1J(^{17}O, ^{31}P) = 145$ Hz].[1423] The ^{31}P chemical shifts of PXY(CN) are in good agreement with predictions based on theories of Letcher and Van Wazer.[1424] The ^{19}F n.m.r. spectrum of $F_4PNMeNMePF_4$ is a doublet at +80 °C and ABMNX at −80 °C. This behaviour was attributed to restricted rotation about the nitrogen–phosphorus bond and stopping of the PF_4 scrambling.[1425] No rotation about the phosphorus–nitrogen bond is observed in (160) even at 240 °C. The ^1H, ^{31}P, and ^{29}Si n.m.r.

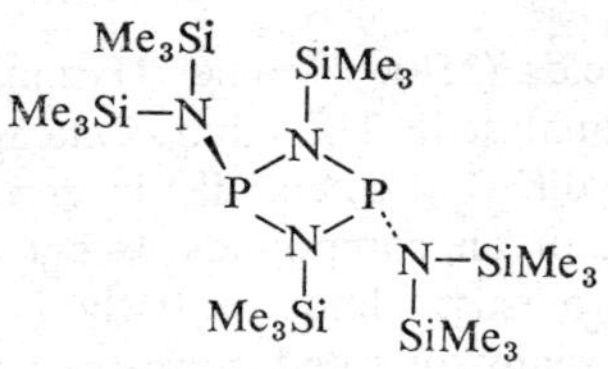

(160)

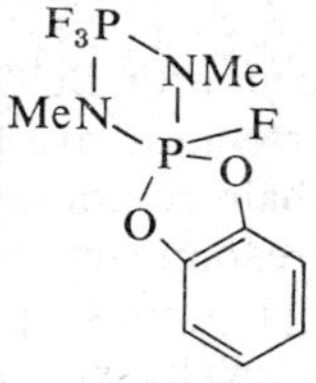

(161)

spectra were recorded.[1426] Analysis of the ^{19}F and ^{31}P n.m.r. spectra of (161) has shown that the fluorine atom is in the equatorial position rather than the axial position.[1427] The ^{31}P n.m.r. spectra of an extensive series of primary and secondary amine derivatives of $N_3P_3Cl_6$ and $N_4P_4Cl_8$ have been recorded and analysed. $^1H\{^{31}P\}$ INDOR studies show that $^2J(^{31}P, N, ^{31}P)$ for cis-$N_3P_3Cl_3(NMe_2)_3$, $N_3P_3(NMe_2)_6$, and $N_4P_4(NMe_2)_8$ are positive. Trends in ^{31}P chemical shifts and P—N—P coupling constants were discussed and compared with similar data for chloro(phenyl)- and amino(phenyl)-cyclophosphazenes. The appearance of 'virtual coupling' effects in the ^1H n.m.r. spectra of some of these derivatives and the origins of broadening effects in ^{31}P n.m.r. spectra were discussed.[1428] The magnitude of $^4J(^{31}P, ^{31}P)$ for phosphazenylcyclophosphazenes seems to be related to the conformation adopted by the phosphazenyl group relative to the phosphazene ring.[1429] On the basis of ^1H n.m.r. data it has been

1421 K. M. Abraham and J. R. Van Wazer, *Inorg. Chem.*, 1976, **15**, 2323.
1422 W. McFarlane and B. Wrackmeyer, *J.C.S. Dalton*, 1976, 2351.
1423 G. A. Gray and T. A. Albright, *J. Amer. Chem. Soc.*, 1976, **98**, 3857.
1424 K. B. Dillon, M. G. C. Dillon, and T. C. Waddington, *J. Inorg. Nuclear Chem.*, 1976, **38**, 1149.
1425 R. Goodrich-Haines and J. W. Gilje, *Inorg. Chem.*, 1976, **15**, 470.
1426 E. Niecke, W. Flick, and S. Pohl, *Angew. Chem. Internat. Edn.*, 1976, **15**, 309.
1427 R. K. Harris, M. I. M. Wazeer, O. Schlak, and R. Schmutzler, *J.C.S. Dalton*, 1976, 306.
1428 R. Keat, R. A. Shaw, and M. Woods, *J.C.S. Dalton*, 1976, 1582.
1429 M. Biddlestone, R. Keat, H. Rose, D. S. Rycroft, and R. A. Shaw, *Z. Naturforsch.*, 1976, **31b**, 1001.

shown that for $Ph_2NAsOCH_2CH_2O$ there is a twist envelope conformation with slow inversion of the arsenic.[1430] The ^{13}C and ^{31}P n.m.r. spectra of $(Me_2CHO)_2$-$PONBzR^1$ [R^1 = H or MR^2_3 (M = Si, Ge, or Sn; R^2 = Me or Ph)] have been recorded. The ^{13}C inequivalence of the isopropyl methyl carbon atoms was attributed to slow rotation of the carbon–oxygen bond rather than molecular asymmetry.[1431] ^{31}P N.m.r. data have also been reported for $PF_2(NMe_2)_3$ (including ^{19}F n.m.r. data),[1432] *cis-* and *trans*-$NP(NHBu)_2(NSOF)_2$ (including ^{19}F n.m.r. data),[1433] $PCl_{3-m}F_mNMePCl_{3-n}F_nNMe$ (including ^{19}F n.m.r. data),[1434] (162),[1435] $(Me_3Si)_2N(S)P\{N(SiMe_3)\}_2PN(SiMe_3)_2$ (including ^{19}F n.m.r. data),[1436] (163),[1437] $N\{C(CX_3)N\}_2PCl_2$ (including ^{19}F n.m.r. data),[1438] $NPCl_2NPCl_2NPOCH_2$-

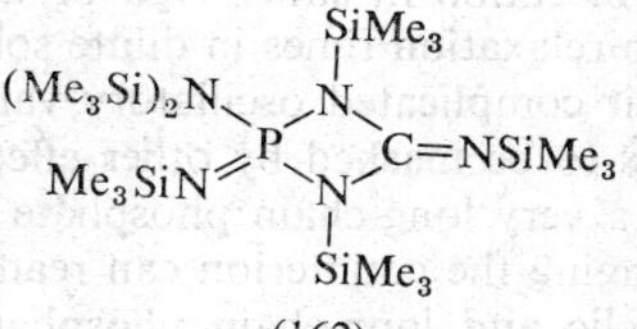

(162)

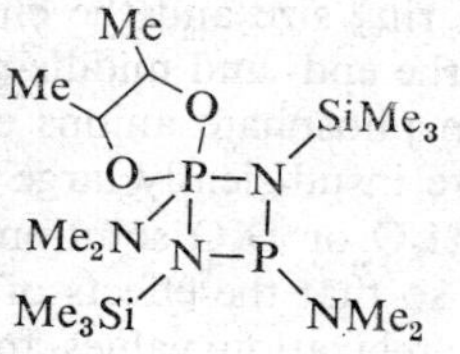

(163)

$CF_2CH_2CF_2CH_2O$,[1439] $N_nP_nCl_{2n-m}F_m$ (including ^{19}F n.m.r. data),[1440] $N_4P_4Cl_6$-$(NHEt)_2$,[1441] (164) (only ^{19}F n.m.r. data),[1442] $OCR^1{}_2CR^1{}_2OP(NHPh)OCH_2$-$CH_2NR^2$,[1443] $OCH_2CH_2CH_2P(O)N(CH_2CH_2Cl)_2$ (only ^{13}C n.m.r. data),[1444] 1,4,2-oxazaphosphorinanes,[1445] and $(R^1O)_2P(Se)NMeP(OR^2)_2$.[1446]

^{31}P Spin–lattice relaxation times of various esters of orthophosphoric acid have been reported and discussed.[1447] The ^{31}P n.m.r. spectrum of $[H_nPO_4]^{n-3}$ shows

[1430] O. Adler and F. Kober, *Z. Naturforsch.*, 1976, **31b**, 472.
[1431] C. Glidewell, *J. Organometallic Chem.*, 1976, **108**, 335.
[1432] H. Oberhammer and R. Schmutzler, *J.C.S. Dalton*, 1976, 1454.
[1433] J. B. van den Berg, E. Klei, B. de Ruiter, J. C. van de Grampel, and C. Kruk, *Rec. Trav. chim.*, 1976, **95**, 206.
[1434] R. K. Harris and M. I. M. Wazeer, *J.C.S. Dalton*, 1976, 302.
[1435] R. Appel and M. Halstenberg, *J. Organometallic Chem.*, 1976, **116**, C13.
[1436] E. Niecke and W. Flick, *J. Organometallic Chem.*, 1976, **104**, C23.
[1437] W. Zeiss, *Angew. Chem. Internat. Edn.*, 1976, **15**, 555.
[1438] G. Schöning and O. Glemser, *Chem. Ber.*, 1976, **109**, 2960.
[1439] V. N. Sharov, V. N. Prons, V. V. Korol'ko, A. L. Klebanskii, and G. P. Kondratenkov, *Zhur. obshchei Khim.*, 1975, **45**, 1953.
[1440] N. L. Paddock and D. J. Patmore, *J.C.S. Dalton*, 1976, 1029.
[1441] S. S. Krishnamurthy, A. S. Sau, A. R. Vasudeva Murthy, R. Keat, R. A. Shaw, and M. Woods, *J.C.S. Dalton*, 1976, 1405.
[1442] H. W. Roesky and E. Janssen, *Angew. Chem. Internat. Edn.*, 1976, **15**, 39.
[1443] M. Sanchez, J.-F. Brazier, C. Houalla, A. Munoz, and R. Wolf, *J.C.S. Chem. Comm.*, 1976, 730.
[1444] W. Egan and G. Zon, *Tetrahedron Letters*, 1976, 813.
[1445] Yu. Yu. Samitov, M. A. Pudovik, L. K. Kibardina, and A. E. Pudovik, *Zhur. obshchei Khim.*, 1975, **45**, 2134 (*Chem. Abs.*, 1976, **84**, 43 158).
[1446] N. P. Grechkin, I. A. Nuretdinov, L. K. Nikonorova, and E. I. Loginova, *Zhur. obshchei Khim.*, 1976, **46**, 1753.
[1447] T. Glonek and J. R. Van Wazer, *J. Phys. Chem.*, 1976, 80, 636.

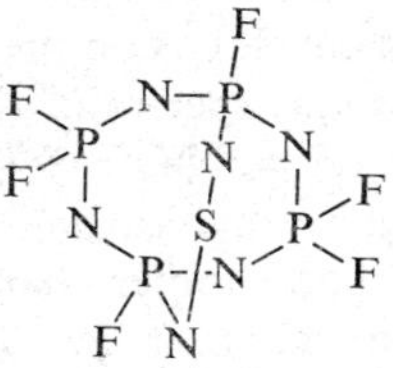

(164)

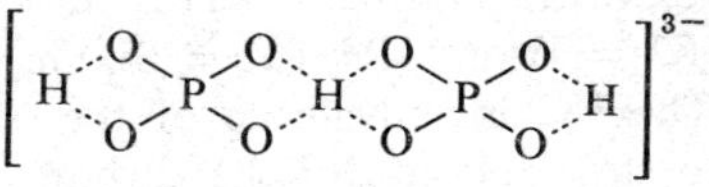

(165)

a nuclear Overhauser enhancement of 2.2 when $n = 1.5$ and it is thought that there is a species (165) present in solution.[1448] The ^{31}P spin–lattice relaxation times of dilute solutions of cyclic metaphosphate anions exhibiting three to eight phosphorus atoms per molecule have been found to be independent of pH but to vary with ring size and the choice of cation in either H_2O or D_2O. On the other hand, the end- and middle-group relaxation times in dilute solution of the shorter chain phosphate anions exhibit complicated oscillatory variations with pH which are insufficiently large so as to be masked by other effects. The ^{31}P T_1 in dilute H_2O or D_2O solutions of a very long-chain phosphate is again pH independent so that the effects of changing the counterion can readily be determined. The relaxation values for cyclic and long-chain phosphates in dilute aqueous solutions are quantitatively interpreted in terms of solvent ^{1}H–^{31}P and intramolecular ^{31}P–^{31}P dipole–dipole relaxation as well as spin-rotation relaxation. Preliminary data were presented for the adenosine di- and tri-phosphates and acid-dissociation constants were calculated for the pyro- and tri-polyphosphoric acids from the variations of the ^{31}P chemical shifts and $^{1}J(^{31}P, ^{31}P)$ with pH.[1449] Use has been made of the ^{31}P chemical shifts of polyphosphates with the terminal group around 9.5—12 p.p.m. and the middle groups around 20—27 p.p.m. in order to determine chain lengths.[1450] Europium and praesodymium shift reagents have been applied to the ^{1}H, ^{13}C, and ^{31}P chemical shifts of compounds such as $(EtO)_3P{=}O$. The ^{1}H shifts are predominantly pseudo-contact while the ^{13}C shifts are both pseudo and contact and the ^{31}P shifts are predominantly contact.[1451] ^{31}P N.m.r. data have also been reported for $(MeO)_5P$, $[(MeO)_6P]^-$ (including ^{13}C n.m.r. data),[1452] $[(PhO)_6P]^-$, $(PhO)_5P$,[1453] $[(\overline{OC_6H_4O)_3P}]^-$,[1454] and $(RO)_3\overline{POOO}$.[1455]

The ^{19}F n.m.r. spectrum of $OSFN{=}S(O)F_2$ is ABX, showing that the fluorine atoms of the $NSOF_2$ are inequivalent.[1456] ^{19}F N.m.r. data have also been reported for $\overline{SN{=}SN{-}S}{=}NSO_2F$,[1457] $FSO_2N(SO_2)_2NSO_2F$,[1458] and (166) (including ^{31}P n.m.r. data).[1459]

1448 T. Glonek, *J. Amer. Chem. Soc.*, 1976, **98**, 7090.
1449 T. Glonek, P. J. Wang, and J. R. Van Wazer, *J. Amer. Chem. Soc.*, 1976, **98**, 7968.
1450 T. Umegaki, Y. Nakayama, and T. Kanazawa, *Bull. Chem. Soc. Japan*, 1976, **49**, 2105.
1451 T. A. Gerken and W. M. Ritchey, *J. Magn. Resonance*, 1976, **24**, 155.
1452 D. B. Denney, D. Z. Denney, and C. F. Ling, *J. Amer. Chem. Soc.*, 1976, **98**, 6755.
1453 C. L. Lerman and F. H. Westheimer, *J. Amer. Chem. Soc.*, 1976, **98**, 179.
1454 J. Gloede and H. Gross, *Tetrahedron Letters*, 1976, 917.
1455 H. J. Bestmann, L. Kisielowski, and W. Distler, *Angew. Chem. Internat. Edn.*, 1976, **15**, 298.
1456 H.-J. Krannich and W. Sundermeyer, *Angew. Chem. Internat. Edn.*, 1976, **15**, 311.
1457 H. W. Roesky, G. Holtschneider, H. Wiezer, and B. Krebs, *Chem. Ber.*, 1976, **98**, 1358.
1458 K.-D. Schmidt, R. Mews, and O. Glemser, *Angew. Chem. Internat. Edn.*, 1976, **15**, 614.
1459 T.-P. Lin and O. Glemser, *Chem. Ber.*, 1976, **109**, 3537.

^{19}F and ^{31}P magnetic shielding anisotropies of PSF_3 have been determined at a single temperature and in the smectic liquid crystal phase. The F—P—F bond angle was determined. The absolute sign of the indirect phosphorus–fluorine coupling constant was found to be negative and the anisotropy of this quantity was assumed to be negligible. In addition, the ^{31}P n.m.r. spectrum of PF_3 dissolved in the smectic liquid crystalline solvent was also investigated. The value of $\Delta\sigma$ in the smectic phase for PF_3 was found to be in disagreement with the previous value measured in the nematic liquid crystalline phase.[1460] P_4S_4 exists in two forms, α and β. The ^{31}P n.m.r. spectrum of the α-form is a singlet, which is consistent with (167), but the β-form has an AMX_2 ^{31}P n.m.r. spectrum as

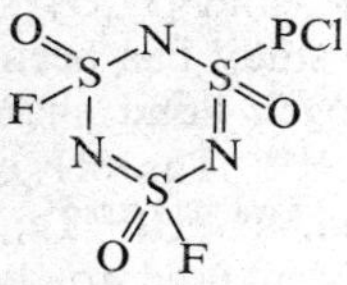

(166)

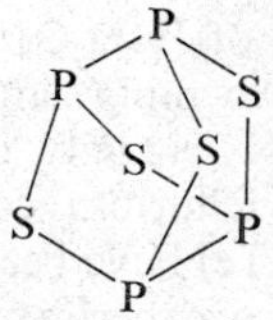

(167)

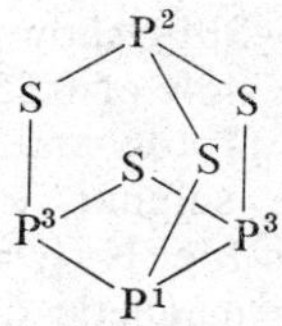

(168)

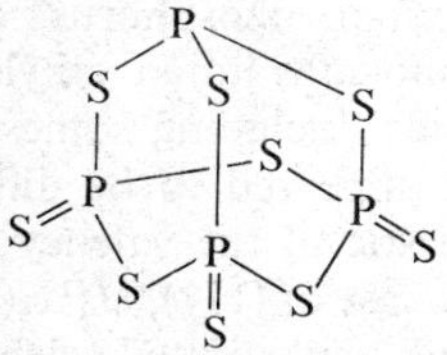

(169)

(170)

expected for (168), with $^2J[^{31}P(1), ^{31}P(2)] = 50.4$ Hz, $^1J[^{31}P(1), ^{31}P(3)] = 168.4$ Hz and $^2J[^{31}P(2), ^{31}P(3)] = 18.2$ Hz.[1461] ^{31}P N.m.r. spectra have been reported for polycrystalline P_4S_9 and for CS_2 solutions of P_4S_9 and P_4S_{10}. A solution of P_4S_9 gives a well-resolved AB_3 spectrum with $J(^{31}P, ^{31}P) = 96$ Hz, which is consistent with (169). The spectrum of solid P_4S_9 was determined using high-speed magic angle rotation techniques but was incompletely resolved. P_4S_{10} in CS_2 gave a singlet.[1462] The ^{31}P chemical shift of $(MeSe)_3P$ is 107 p.p.m., with $^1J(^{31}P, ^{77}Se)$.[1463] For $[NF_4]^+$, $^1J(^{14}N, ^{19}F) = 230$ Hz.[1464] The ^{31}P chemical shift of $[PCl_{6-n}F_n]^-$ has been interpreted theoretically and reasonable agreement found with experiment.[1465] The ^1H and ^{19}F n.m.r. spectrum of MeF–SbF_5 in SO_2F_2 at -145 °C shows signals due to (170).[1466]

1460 A. J. Montana, N. Zumbulyadis, and B. P. Dailey, *J. Chem. Phys.*, 1976, **65**, 4756.
1461 A. M. Griffin, P. C. Minshall, and G. M. Sheldrick, *J.C.S. Chem. Comm.*, 1976, 809.
1462 E. R. Andrew, W. Vennart, G. Bonnard, R. M. Croiset, M. DeMarcq, and E. Mathieu, *Chem. Phys. Letters*, 1976, **43**, 317.
1463 L. Maier, *Helv. Chim. Acta*, 1976, **59**, 252.
1464 K. O. Christe, C. J. Schack, and R. D. Wilson, *Inorg. Chem.*, 1976, **15**, 1275.
1465 R. Radeglia, *Z. phys. Chem.* (*Leipzig*), 1976, **257**, 199.
1466 J.-Y. Calves and R. J. Gillespie, *J.C.S. Chem. Comm.*, 1976, 506.

9 Compounds of Groups VI, VII, Krypton, and Xenon

Information concerning complexes of these elements can be found at the following sources: $[(OC)_4BrMSeHPPh_2]$ (M = Mn or Re),[230] $[MeHg(HSeCH_2CH_2NH_3)]^{2+}$,[488] $\overline{(CH_2)_5E}$ (E = Se or Te),[1274] $[\overline{(CH_2)_5E}Me]^+$ (E = Se or Te),[1273] 8,8′-$[RE(C_2B_9H_{10})_2Co]$ (E = Se or Te),[1180] $[M(CO)_n(SeCF_3)]_2$ (M = Mn, n = 3; M = Fe, n = 2),[227] $[PtX_n(EMe_2)_{4-n}]^{2-n}$ (E = Se or Te),[459] *trans*-$[PtH(PEt_3)_2(SeSiH_3)]$,[393] $[(R_3P)_2Ag(Se_2CNR_2)]$,[476] lead–selenium compounds,[1245] $(MeSe)_3P$,[1463] $[(\eta^5\text{-}C_5H_5)(OC)_3MP(Se)Cl_2]$ (M = Cr or W),[158] Ph_3PSe,[1375] $SeMeP(NR)_2PMeSe$,[1402] $(R^1O)_2P(Se)NMeP(OR^2)_2$,[1446] and $[F_nU(OTeF_5)_{6-n}]$.[103]

The position of the ^1H n.m.r. resonance of water has been measured as a function of temperature and pressure.[1467] The ^1H n.m.r. spectrum of D_2O in HSO_3F–SbF_5 shows the presence of $[HD_2O]^+$, $[H_2DO]^+$, and $[H_3O]^+$ with an isotope shift of 0.055 p.p.m.[1468] An n.m.r. and CNDO study has been carried out of proton and electron donor capacity in a hydrogen bond shared with oxygen, sulphur, selenium, and tellurium compounds.[1469] The ^1H chemical shifts of MeEH (E = O, S, Se, or Te) have been discussed.[1470] $^1H\{^{125}Te\}$ Heteronuclear magnetic double-resonance experiments have been used to determine tellurium chemical shifts for 15 organotellurium compounds.[1471]

A relationship has been found between $^3J(^1H, ^1H)(trans)$, $^3J(^1H, ^1H)(cis)$ and the lengths of carbon–halogen, carbon–carbon, carbon–phosphorus, carbon–sulphur, and carbon–selenium bonds in some mono-substituted ethylenes.[1472] ^{77}Se N.m.r. spectra of 26 organoselenium compounds, including some with an endocyclic selenium atom, have been recorded. It appeared to be difficult to define a range of chemical shifts which is characteristic of the valency state of selenium or of its endo- or exo-cyclic position. In the case of (171), $^4J\{^1H(4), ^{77}Se\}$ was observed.[1473] $J(^1H, ^1H)$ and $J(^1H, ^{77}Se)$ of 2- and 3-substituted selenophens, whose signs have been determined by double-resonance experiments, have been correlated with the reactivity constants, F and R, of Swain and Lupton, by means of the linear equations

$$J = i + fF + rR$$

The relative inductive and mesomeric contributions of the coupling constants were discussed. Substituent effects on $J(^1H, ^1H)$ and $J(^1H, ^{77}Se)$ were found to be additive in 24 disubstituted selenophens. In agreement with experimental results, $^3J(^1H, ^{77}Se)$ is negative in 2,4-dinitroselenophen. Evidence was given from long-range coupling constants that the 2-formyl derivative is almost exclusively in the Se,O *cis* conformation and the 3-formyl derivative in the Se,O *trans* conformation.[1474] In (172), $\delta(^{13}CO)$ = 204.9 and $\delta(^{77}Se)$ = 584 relative to Me_2Se.[1475]

1467 J. W. Linowski, N.-I. Liu, and J. Jonas, *J. Chem. Phys.*, 1976, **65**, 3383.
1468 V. Gold, J. L. Grant, and K. P. Morris, *J.C.S. Chem. Comm.*, 1976, 397.
1469 V. K. Pogorelyi, *V Sb., XI Mendeleevsk. S'Ezd Po Obshch. i Prikl. Khimii. Ref. Dokl. I Soobshch.*, 1975, **3**, 77 (*Chem. Abs.*, 1976, **84**, 157 726).
1470 K. Hamada and H. Morishita, *Jap. J. Appl. Phys.*, 1976, **15**, 748 (*Chem. Abs.*, 1976, **84**, 171 863).
1471 W. McFarlane, F. J. Berry, and B. C. Smith, *J. Organometallic Chem.*, 1976, **113**, 139.
1472 S. V. Yalymova, Yu. Yu. Samitov, and A. Sh. Agishev, *Zhur. strukt. Khim.*, 1975, **16**, 991.
1473 J. Christiaens, J.-L. Piette, L. Laitem, M. Baiwir, J. Denoel, and G. Llabres, *Org. Magn. Resonance*, 1976, **8**, 354.
1474 M. P. Simonnin, M. J. Pouet, J. M. Cense, and C. Paulmier, *Magn. Resonance*, 1976, **8**, 508.
1475 M. Evers, R. Weber, Ph. Thibaut, L. Christiaens, M. Renson, A. Croisy and P. Jacquignon, *J.C.S. Perkin I*, 1976, 2452.

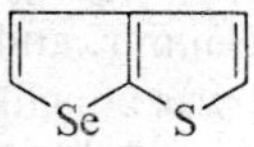

(171)

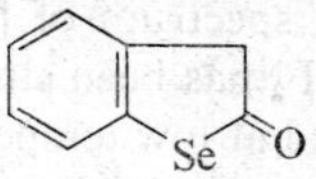

(172)

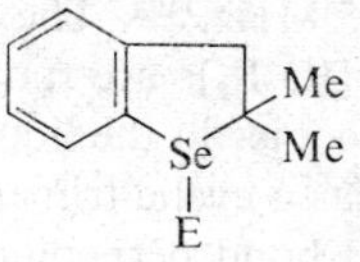

(173)

Geminal $J(^{13}C, ^{77}Se)$ and vicinal $J(^{1}H, ^{77}Se)$ coupling constants are stereospecific in [173; X = lone pair, O, Me^+, and $C(CO_2Me){=}C(CO_2Me)O$]. The coupling constant is largest to the nuclei nearest to the selenium lone pair(s).[1476] In (174), $^{1}J(^{13}C, ^{125}Te)$ = 134.7 Hz.[1477] N.m.r. data have also been reported for $(CF_3)_2S{=}NH$ (^{19}F n.m.r. data),[1478] vinylarylselenides (^{13}C and ^{77}Se n.m.r.

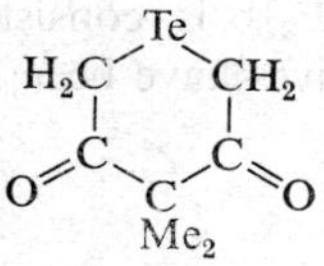

(174)

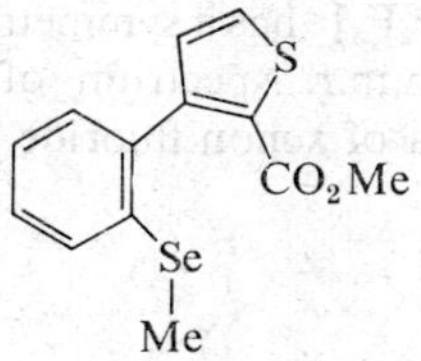

(175)

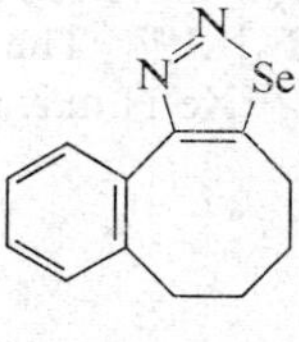

(176)

data),[1479] (175) (^{13}C n.m.r. data),[1480] 2-substituted tellurophens (^{13}C n.m.r. data),[1481] CF_3SF_4X (^{19}F n.m.r. data),[1482] $(CF_3S)_2CFSSSeCN$ (^{19}F n.m.r. data),[1483] (176) (^{13}C n.m.r. data),[1484] and $[Se(C_2F_5)I_2][Sb_2F_{11}]$ (^{19}F n.m.r. data).[1485]

The ^{19}F n.m.r. spectrum of $HOSF_5$ is AB_4.[1486] The A_2BC ^{19}F n.m.r. spectra of *cis*-$[(RO)(MeO)TeF_4]$ and *cis*-$[(RO)(Me_2N)TeF_4]$ have been recorded and analysed. $^{2}J(^{19}F, ^{19}F)$ range from 137 to 174 Hz, and trends in ^{19}F chemical shifts permit complete assignment of the resonances. Stereospecific coupling between ^{19}F and the ^{1}H of the NMe_2 group was observed.[1487] ^{19}F N.m.r. data have also been reported for $SF_5OOCF_2NHCF_3$,[1488] $(TeF_4O)_2$, and IF_3O_2.[1489]

Natural abundance ^{17}O n.m.r. spectra have been studied in ClO_3F, $[ClO_4]^-$, $ClOF_3$, ClO_2F, and $[ClO_2]^+[AsF_6]^-$ as pure ligands or in solution. In ClO_3F, a lineshape analysis and longitudinal relaxation time measurement gave $^{1}J(^{17}O, ^{35}Cl)$ = 108 ± 5 Hz and the quadrupolar interaction of ^{17}O, e^2qQ/h =

1476 H. J. Reich and J. E. Trend, *J.C.S. Chem. Comm.*, 1976, 310.
1477 J. C. Dewan and J. Silver, *Inorg. Nuclear Chem. Letters*, 1976, **12**, 647.
1478 S. D. Morse and J. M. Shreeve, *J.C.S. Chem. Comm.*, 1976, 560.
1479 D. F. Kusharev, G. A. Kalabin, T. G. Mannafov, V. A. Mullin, M. F. Larin, and V. A. Pestunovich, *Zhur. org. Khim.*, 1976, **12**, 1482.
1480 F. B. Iteke, L. Christiaens, and M. Renson, *Tetrahedron*, 1976, **32**, 689.
1481 F. Fringuelli, S. Gronowitz, A.-B. Hörnfeldt, I. Johnson, and A. Taticchi, *Acta Chem. Scand. (B)*, 1976, **30**, 605.
1482 S.-L. Yu and J. M. Shreeve, *Inorg. Chem.*, 1976, **15**, 14.
1483 A. Haas and M. Yazdanbakhsch, *Chem. Ber.*, 1976, **109**, 1976.
1484 H. Meier, M. Layer, W. Combrink, and S. Schniepp, *Chem. Ber.*, 1976, **109**, 1650.
1485 J. Passmore and P. Taylor, *J.C.S. Dalton*, 1976, 804.
1486 K. Seppelt, *Angew. Chem. Internat. Edn.*, 1976, **15**, 44.
1487 J. Dalton, W. McFarlane, C. R. Wickens, G. W. Fraser, and J. B. Miller, *Org. Magn. Resonance*, 1976, **8**, 522.
1488 E. R. Falardeau and D. D. DesMarteau, *J. Amer. Chem. Soc.*, 1976, **98**, 3529.
1489 I. Beattie, R. Crocombe, A. German, P. Jones, C. Marsden, G. V. Schalkwyk, and A. Bukovszky, *J.C.S. Dalton*, 1976, 1380.

14 ± 1.5 MHz.[1490] The ^{19}F n.m.r. spectrum of $[BrF_4O]^-$ is a singlet,[1491] while IF_5 and $[IO_2F_2]^-$ are static.[1492] IO_2F_3 has been studied by ^{19}F n.m.r. and Raman spectroscopies in the liquid state and at low temperatures. It was concluded that it exists as a cyclic trimer with *cis*-oxygen bridges in the boat conformation with the I=O bond perpendicular to the plane containing the bridge.[1493] ^{19}F N.m.r. spectroscopy has been used to show that IX, IF_5, 3py (X = F, Cl, or Br) contains I···F···I bridges.[1494]

^{129}Xe Chemical shifts have been measured in a mixture of xenon and methane, CF_4, and SiF_4 as a function of temperature and density to investigate the amount of interaction and the well depth for the pairs.[1495] Similarly $\delta(^{129}Xe)$ has been measured in NO gas as a function of temperature and density of NO. A greater interaction was found for NO than O_2.[1496] ^{19}F N.m.r. spectroscopy has been used to show that in BrF_5, $[Kr_2F_3]^+$ has a symmetrically fluorine-bridged structure like $[Xe_2F_3]^+$.[1497] The ^{19}F n.m.r. spectrum of $[XeOF_3]^+$ is consistent with (177).[1498] ^{129}Xe N.m.r. spectra of xenon fluoride derivatives have been reported.

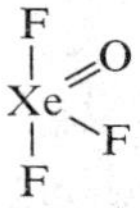

(177)

A linear relationship exists between ^{19}F and ^{129}Xe chemical shifts, which gives a method of determining the degree of covalency of the xenon–fluorine bond.[1499]

10 Appendix

This appendix contains a list of papers where the use of nuclei other than ^1H has been described. The nuclei are ordered by increasing mass.

^2H 56, 700, 964, 968, 987, 1001, 1004, 1023, 1024, 1032, 1049, 1075, and 1098.
^6Li 643.
^7Li 56, 65, 644, 689, 690, 699—701, 962, 963, 1022, 1069, 1070, and 1093.
^9Be 56 and 656.
^{10}B 56, 1055, and 1129.
^{11}B 56, 299, 318, 602, 604, 777, 779, 780, 838, 879, 975, 976, 978, 1053—1061, 1077, 1129—1132, 1134, 1136, 1138—1142, 1144—1185, 1188, 1191—1193, 1196—1199, 1201—1216, 1218—1222, 1225, 1227, and 1245.
^{13}C 10, 17, 56—64, 66, 81, 88, 89, 92, 93, 97, 99—101, 106, 108, 109, 111, 124, 125, 127—149, 152, 154, 155, 157, 162, 163, 165—169, 172, 175—177,

1490 J. Virlet and G. Tantot, *Chem. Phys. Letters*, 1976, **44**, 296.
1491 R. J. Gillespie and P. Spekkens, *J.C.S. Dalton*, 1976, 2391.
1492 J. B. Milne and D. M. Moffett, *Inorg. Chem.*, 1976, **15**, 2165.
1493 R. J. Gillespie and J. P. Krasznai, *Inorg. Chem.*, 1976, **15**, 1251.
1494 E. Lehmann, D. Naumann, and M. Schmeisser, *J. Fluorine Chem.*, 1976, **7**, 135 (*Chem. Abs.*, 1976, **84**, 113 953).
1495 C. J. Jameson. A. K. Jameson, and S. M. Cohen, *J. Chem. Phys.*, 1976, **65**, 3401.
1496 C. J. Jameson, A. K. Jameson, and S. M. Cohen, *J. Chem. Phys.*, 1976, **65**, 3397.
1497 R. J. Gillespie and G. J. Schrobilgen, *Inorg. Chem.*, 1976, **15**, 22.
1498 R. J. Gillespie, B. Landa, and G. J. Schrobilgen, *Inorg. Chem.*, 1976, **15**, 1256.
1499 J. H. Holloway, G. J. Schrobilgen, P. Granger, and C. Brevard, *Compt. rend.*, 1976, **282**, *C*, 519.

183, 188, 192, 193, 195, 196, 198, 207, 211—215, 218—223, 225, 228, 235—238, 242, 245, 246, 249, 250, 252, 254—258, 260, 261, 263—266, 269—273, 275—288, 291—293, 296, 300, 301, 303—306, 314, 317, 327, 333, 342, 344—346, 348, 349, 352, 354, 355, 359, 361, 363, 366, 369—371, 376—380, 382, 387, 397—399, 405—410, 412, 421, 424—428, 432, 433, 435, 438, 442, 444—449, 451, 456, 460, 468, 472—474, 479, 485—487, 489, 493— 495, 499, 500, 502, 506, 507, 513, 516, 517, 519, 521—523, 525, 526, 528, 529, 531, 539—544, 546, 549—556, 558—560, 562—565, 567—570, 572, 573, 575, 585, 586, 588, 595, 597, 599, 605, 606, 608, 610, 612, 619, 626—631, 693, 696, 697, 716, 718, 734, 739, 747, 756—758, 762, 764, 782, 787, 790, 797, 799, 820, 823, 829, 837, 849, 854, 867, 868, 871, 873, 874, 878, 880, 881, 883, 891, 894, 900, 907—909, 915, 916, 932, 935, 940, 944, 945, 1019, 1119—1121, 1125, 1126, 1128, 1131, 1138, 1145, 1155, 1184—1187, 1194, 1200, 1208, 1215, 1219, 1232—1234, 1241, 1244, 1245, 1247, 1249, 1252, 1254, 1255, 1257—1259, 1261, 1263, 1264, 1266—1271, 1273, 1275, 1276, 1278, 1279, 1282—1284, 1286—1288, 1292—1294, 1298, 1299, 1307, 1314, 1316, 1327, 1330, 1343, 1352, 1354—1359, 1364—1369, 1372, 1374, 1375, 1381, 1388, 1393, 1409, 1410, 1412, 1413, 1415, 1416, 1423, 1431, 1444, 1451, 1452, 1475—1477, 1479—1481, and 1484.

^{14}N 56, 667, 777, 907, 908, 1006, 1016, 1193, 1204, 1205, 1214, 1238, 1241, and 1245.

^{15}N 56, 110, 422, 444, 967, 1019, 1340, 1422, and 1423.

^{17}O 56, 73, 83, 203—205, 655, 667, 675, 1055, 1423, and 1490.

^{19}F 4, 72, 84, 86, 87, 103—105, 121, 122, 130, 150, 154, 156, 159, 170, 171, 174, 186, 191, 197, 199, 206, 208, 216, 217, 226, 227, 243, 247, 248, 251, 253, 262, 267, 302, 315, 337—340, 351, 353, 358, 361, 362, 391, 394, 415, 463, 464, 469, 478, 496, 497, 507, 524, 578, 581—583, 587, 591, 623, 626, 632, 633, 636, 647, 702, 703, 717, 783, 811, 812, 821, 840, 841, 901, 919, 965, 972, 983—986, 1011, 1017, 1018, 1037, 1038, 1043, 1044, 1047, 1048, 1064, 1067, 1073, 1081, 1083, 1084, 1093, 1105, 1141, 1196, 1208, 1222—1224, 1226, 1241, 1251, 1253, 1258, 1263—1265, 1277, 1297, 1300, 1315, 1321—1325, 1329, 1331, 1332, 1347, 1360, 1361, 1382, 1383, 1385, 1390, 1395, 1399, 1400, 1416, 1417, 1425, 1427, 1432—1434, 1436, 1438, 1440, 1442, 1456—1460, 1464, 1466, 1478, 1482, 1483, 1485—1489, 1491—1494, and 1497—1499.

^{23}Na 56, 65, 94, 95, 646, 647, 651, 652, 691, 695, 699—701, 979, 1007, 1021, 1026, 1056, 1071, and 1094.

^{25}Mg 56 and 657.

^{27}Al 56, 787, 980, 1058, 1063, 1133, 1134, 1228, 1229, 1236, and 1238.

^{29}Si 28, 393, 422, 610, 789, 1071, 1241, 1242, 1246, 1247, 1254, 1256, 1258, 1261, 1271, 1272, 1275, 1288, 1291, 1294, 1300, 1303, 1307—1309, 1315, 1319, 1321, 1333, and 1426.

^{31}P 72, 112, 115—120, 132, 134, 149, 151, 153, 154, 158, 160, 174, 178—182, 185—187, 189—191, 194, 199, 210, 224, 229—231, 233, 234, 238—241, 274, 277, 307—313, 318—326, 330—336, 341, 343, 347, 350, 353, 356, 357, 360, 361, 383—386, 388, 389, 392—394, 396, 401, 402, 410, 411, 414, 416—418, 420—422, 431, 433, 434, 436, 447, 452—454, 456—458, 460—

462, 464—470, 472, 475—477, 480—482, 492, 495, 507, 558, 576, 577, 584, 589—592, 622—624, 626, 634, 680, 704, 727, 728, 748, 758, 759, 761, 763, 765, 766, 768, 769, 789, 801, 805, 818, 849—854, 1009, 1012, 1016, 1077, 1138, 1139, 1141, 1179, 1190, 1208, 1217, 1223, 1224, 1237—1239, 1243, 1245, 1275, 1298, 1301—1306, 1320, 1334, 1335, 1338, 1339, 1342—1346, 1350—1353, 1355, 1357, 1358, 1360, 1362, 1363, 1365, 1370—1372, 1374, 1375, 1377—1382, 1384, 1386, 1387, 1389—1392, 1394, 1395, 1397, 1399—1408, 1410, 1412, 1413, 1418—1422, 1424, 1426—1429, 1431—1441, 1443, 1445—1455, 1459—1463, and 1465.

^{33}S 56 and 203.
^{35}Cl 56, 83, 636, 661, 667, 688, 690, 701, 989, 1007, 1045, and 1337.
^{37}Cl 56, 636, 688, 701, 1007, and 1045.
^{39}K 56, 65, 200, and 701.
^{51}V 56, 83, 112—114, 118, 120, 1016, 1025—1028, and 1031.
^{53}Cr 56, 83, 200, 1035, and 1036.
^{55}Mn 56, 83, and 209.
^{57}Fe 56, 1041, and 1042.
^{59}Co 56, 329, and 381.
^{63}Cu 56, 480, and 113.
^{69}Ga 56 and 685.
^{71}Ga 56 and 685.
^{73}Ge 56 and 1336.
^{75}As 56 and 1346.
^{77}Se 56, 230, 393, 639, 789, 1245, 1473, 1475, and 1479.
^{79}Br 56 and 701.
^{81}Br 56, 701, and 1050.
^{85}Rb 56, 65, and 701.
^{93}Nb 56 and 1023.
^{95}Mo 56, 83, 201, and 202.
^{97}Mo 56, 201, and 202.
^{109}Ag 56, 639, 676, and 972.
^{113}Cd 56, 503, 505, 639, and 679.
^{119}Sn 56, 619, 639, 1245, 1279, 1281, 1285, 1292, and 1337.
^{121}Sb 56 and 1079.
^{125}Te 56, 393, 639, and 1471.
^{127}I 56, 701, 1051, and 1337.
^{129}Xe 56, 1495, 1496, and 1499.
^{133}Cs 56, 65, 96, 650, 694, 700, and 701.
^{139}La 56, 660, 661, and 707.
^{153}Eu 56, 1013, and 1014.
^{183}W 56, 153, and 639.
^{195}Pt 56, 393—395, 422, 437, 453, 459, 469, and 639.
^{199}Hg 56, 484, 490, 491, 639, and 1035.
^{205}Tl 56, 686, 1232, and 1235.
^{207}Pb 56, 1060, 1074, 1245, and 1293.

2

Nuclear Quadrupole Resonance Spectroscopy

BY R. J. LYNCH

1 Introduction

This Report is concerned with pure nuclear quadrupole resonance (n.q.r.) in solids. Quadrupole coupling constants determined from solid-state n.m.r., Mössbauer, and microwave spectroscopy are not included in this Chapter, but n.q.r. frequencies which have been detected by double-resonance techniques are included.

The arrangement of the material closely follows that used for previous Reports. The Main-group elements are discussed first, in their Groups of the Periodic Table. These are followed by transition-metal resonances, in order of increasing atomic number.

Resonances are reported for 23 nuclides of 17 elements, in order of increasing atomic number:

$I = 1$: ^{2}H, ^{14}N

$I = \frac{3}{2}$: ^{23}Na, ^{35}Cl, ^{37}Cl, ^{63}Cu, ^{65}Cu, ^{69}Ga, ^{71}Ga, ^{75}As, ^{79}Br, ^{81}Br

$I = \frac{5}{2}$: ^{17}O, ^{27}Al, ^{121}Sb, ^{127}I, ^{185}Re, ^{187}Re

$I = \frac{7}{2}$: ^{59}Co, ^{123}Sb

$I = \frac{9}{2}$: ^{93}Nb, ^{115}In, ^{209}Bi

Included in this Report are papers from the December 1975 issue of the *Bulletin of the Academy of Sciences of the U.S.S.R., Physical Series*, which was exclusively concerned with n.q.r. A translated edition appeared in 1976.

A second volume of 'Advances in Nuclear Quadrupole Resonance' has been published, which includes reviews on molecular orientation, optical detection of quadrupole interactions in excited states, double resonance, polyiodides, *a priori* field-gradient calculations, and phase transitions in perovskites and antifluorites.[1] An introduction to n.q.r. and its applications,[2] a review of n.m.r.–n.q.r. studies of hydrogen bonding,[3] and an article on the role of n.q.r. in theoretical chemistry have also appeared.[4] A discussion of the applications of high-frequency spectroscopy to the chemical industry deals with n.q.r., e.s.r., and dielectric spectroscopy (published in German).[5]

[1] 'Advances in Nuclear Quadrupole Resonance', ed. J. A. S. Smith, Heyden, London, 1975, Vol. 2.
[2] R. J. Lynch, in 'Spectroscopy', ed. B. P. Straughan and S. Walker, Chapman and Hall, London, 1976, Vol. 1, p. 175.
[3] R. Blinc, in 'The Hydrogen Bond', ed. P. Schuster, G. Zundel, and C. Sandorfy, North-Holland, Amsterdam, 1976, Vol. 2, p. 831.
[4] P. T. Narasimhan, *J. Sci. Ind. Res.*, 1976, **35**, 11.
[5] H. G. Fitzsky, *Git. Fachz. Lab.*, 1976, **20**, 341, 344.

2 Instrumentation

Papers describing instrumentation of varying levels of sophistication have appeared this year. An easily constructed n.q.r. spectrometer of the self-quenched super-regenerative oscillator type has been reported.[6] The detection uses a ratemeter to monitor the pulse repetition rate, which varies when a resonance occurs. A ^{14}N n.q.r. Fourier Transform instrument has been described [7] which uses a commercially available pulsed n.m.r. system plus minicomputer. Sensitivity to ^{14}N resonances in succinonitrile appears to be superior to that of continuous-wave r.f. systems. Other pulsed n.q.r. instrumentation used for the study of ^{14}N has also been described,[8, 9] and also for higher frequencies, 2—20 MHz [10] and up to 1000 MHz.[11] N.q.r. has also been observed using a superconducting magnetometer.[12] Both n.q.r. transitions from ^{27}Al in sapphire (Al_2O_3) were observed in this way at 4.2 K. Adiabatic fast-passage conditions were used, and a weak magnetic field was present to remove the degeneracy of the nuclear states. The technique can be applied to nuclei which have long relaxation times. An observation on the use of pulse methods for n.q.r. detection pointed out that the r.f. pulse width for greatest amplitude of the free induction decay is not the same as the width required for sample saturation, when the n.q.r. sample is poly-crystalline.[13] A thermometer which uses the temperature dependence of ^{63}Cu quadrupole resonance in copper oxide has been reported.[14] Temperatures were measured up to 870 K with a precision of 0.1—0.01 K.

3 Main-group Elements

Group I (Deuterium, Sodium-23).—Double-resonance techniques have greatly assisted deuterium n.q.r. Resonance has been detected in $Sb(C_6D_5)_3$, using the $^{121}Sb(\pm\frac{5}{2} \leftrightarrow \pm\frac{3}{2})$ resonance as the indicator line.[15] The ^2H frequency of 135.5 kHz did not appear to change over the range 77—125 K (η = 0.044).

The n.q.r. spectra of ^2H and ^{17}O have been obtained at 77 K in isotopically enriched ice II, a proton-ordered high-pressure phase of ice.[16] Eight ^2H lines were observed between 157 and 177 kHz, consistent with four inequivalent deuterons. By observation of double quadrupole transitions (simultaneous transitions of two deuterons which share an appreciable dipole coupling), the lines were assigned to two inequivalent water molecules, which is consistent with *X*-ray and neutron diffraction data. The values of e^2Qq deduced are 225.8, 219.8, 226.1, 220.3 kHz, and η values are 0.127, 0.134, 0.100, and 0.102, respectively. The corresponding values for normal ice Ih are 213.4 kHz and 0.112.[17] There appears to be an

6 E. P. A. Sullivan, *Rev. Sci. Instr.*, 1976, **47**, 134.
7 A. Colligiani and R. Ambrosetti, *Gazzetta*, 1976, **106**, 439.
8 G. L. Petersen, *Diss. Abs. Internat.* (*B*), 1976, **37**, 317.
9 E. O. Azizov and Yu. M. Luganskii, *Izvest. Akad. Nauk S.S.S.R., Ser. fiz.*, 1975, **39**, 2627.
10 B. N. Pavlov, *Izvest. Akad. Nauk S.S.S.R., Ser. fiz.*, 1975, **39**, 2620.
11 B. G. Ignatov, A. L. Aleksandrov, L. Z. Pososhenko, and G. K. Semin, *Izvest. Akad. Nauk S.S.S.R., Ser. fiz.*, 1975, **39**, 2630.
12 T. Jach, *Diss. Abs. Internat.* (*B*), 1976, **36**, 4567; T. Jach, *Appl. Phys. Letters*, 1976, **28**, 49.
13 A. H. Brunetti, *J. Phys.* (*E*), 1976, **9**, 164.
14 A. A. Boguslavskii, Yu. N. Bobkov, and R. Sh. Lotfullin, *Zavodskaya Lab.*, 1976, **42**, 461.
15 J. Barocas, *Diss. Abs. Internat.* (*B*), 1976, **37**, 243.
16 D. T. Edmonds, S. D. Goren, A. L. Mackay, and A. A. L. White, *J. Magn. Resonance*, 1976, **23**, 505.
17 D. T. Edmonds and A. L. Mackay, *J. Magn. Resonance*, 1975, **20**, 515.

approximately linear correlation between e^2Qq and the O···O bond distance (Figure 1). A correlation has also been shown for N—D···O bonds in amino-acids.[18] There is a lack of accurate information of the O—D bond length in ice II. The coupling constants suggest a slightly shorter O—D bond in ice II than in ice Ih, but not by as much as 0.03 Å, as the neutron-diffraction data

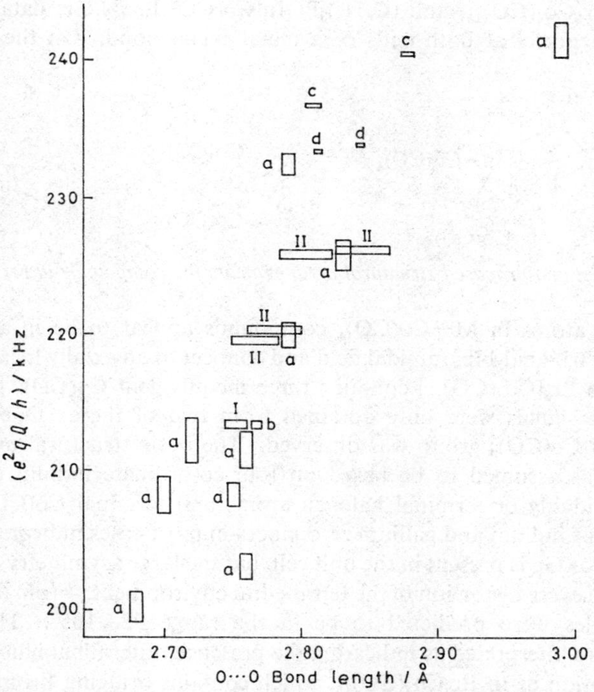

Figure 1 *A plot of the* ^{2}H *quadrupole coupling constant* e^2qQ/h *against the* O···O *hydrogen-bond length for deuteriated ice II (II), ice Ih (I), and the waters of crystallization of copper sulphate pentahydrate (a), ammonium oxalate hydrate (b),* L*-serine hydrate (c), and* L*-asparagine hydrate (d)*
(Reproduced by permission from *J. Magn. Resonance*, 1976, **23**, 505)

indicate. From the n.q.r. data, the two water molecules in ice II are thought to have similar pairs of O—D bonds, with a much greater difference between O—D bond lengths within one water molecule. The ^{17}O ($I = \frac{5}{2}$) n.q.r. spectrum of 10% enriched, fully protonated ice II showed three resonances, at 1695, 1888, and 3575 kHz (77 K), which gave e^2Qq = 6983 kHz, $\eta = 0.865$.[16] Corresponding values for ice Ih are 6525 kHz and $\eta = 0.925$.[19] A close similarity of the two water molecules must exist, since only one set of lines was observed.

The ^{23}Na n.q.r. frequency in $NaBrO_3$ has also been found, using a double-resonance technique.[19] At 77 K the frequency is 440.5 ± 1 kHz and the temperature coefficient is -2.0×10^{-4} K^{-1} at 273 K.

[18] M. J. Hunt and A. L. Mackay, *J. Magn. Resonance*, 1976, **22**, 295.
[19] D. T. Edmonds and A. Zussman, *Phys. Letters (A)*, 1972, **41**, 167.

Group III (Aluminium-27, Gallium-69, Gallium-71, and Indium-115).—Both ^{27}Al n.q.r. transition frequencies have been observed from sapphire (Al_2O_3) at 359 kHz ($\frac{1}{2} \leftrightarrow \frac{3}{2}$) and 717 kHz ($\frac{3}{2} \leftrightarrow \frac{5}{2}$) at 4.2 K.[12]

Brill and Miller [20] have measured ^{69}Ga, ^{71}Ga, ^{115}In, ^{59}Co, ^{35}Cl, ^{79}Br, and ^{81}Br n.q.r. frequencies in a series of four-co-ordinate gallium and indium compounds of the form $XM[Co(CO)_4]_2$, $XM[Co(CO)_4]_2$,THF (M = Ga or In; X = Cl, Br, or acac), $In_3Br_3Co_4(CO)_{15}$, and $(C_6H_5)_3POIn[Co(CO)_4]_3$. N.q.r. data were thus obtained from nuclei at both ends of a metal–metal bond. On the basis of η

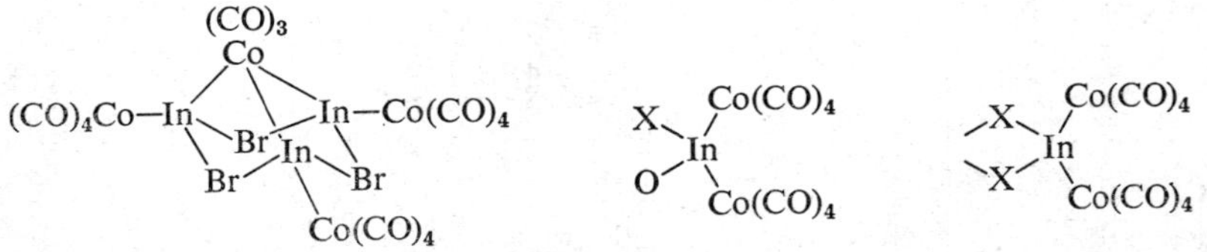

Figure 2 *Known and proposed structural arrangements for compounds in ref.* 20

values, cobalt atoms in M—$Co(CO)_4$ compounds appear to lie in an approximately regular trigonal-bipyramidal field and connect to an axially located indium or gallium. $In_3Br_3[Co_4(CO)_{15}]$ contains three inequivalent $Co(CO)_4$ groups (see Figure 2), but signals were only obtained from two of these. One resonance from the axial $Co(CO)_3$ group was observed. The basic structural unit of these compounds was assumed to be based on four-co-ordinate indium or gallium, with either bridging or terminal halogen atoms and terminal $Co(CO)_4$ groups. The single set of indium and gallium resonances in most cases indicated that only one type of In or Ga is present in the unit cell, but the large asymmetry parameters for ^{115}In imply severe distortion of the tetrahedral environment. From the η values, Br–In–Br angles were predicted to be in the range 92—105°. The halogen resonances were interpreted as indicating the presence of terminal halogen atoms, with the exception of $In_3Br_3Co_4(CO)_{15}$, which contains bridging bromine atoms; this is confirmed by a crystal structure determination.[21] By comparison of values of the ^{59}Co quadrupole coupling constants with analogous Co_2SnX_2 compounds, the In—Co bond was thought to be more ionic than the Sn—Co bond. Substitution of indium for gallium or germanium for tin produced little difference in the orbital population distribution of cobalt. π-Bonding in the In—Co and Ga—Co bonds was thought not to be an important factor in these four-co-ordinate complexes. The trend of e^2Qq for ^{59}Co of $Co_2MX_2 > Co_2MOX > Co_2MO_2$ (M = Ga or In; X = Cl or Br) was interpreted in terms of a decreasing X(O)–In–(O)X angle, and was supported by the observed shifts in the ^{115}In coupling constants.

An additive electric-field-gradient model has been proposed [22] for the interpretation of ^{115}In quadrupole coupling constants in four-co-ordinate and six-co-ordinate indium compounds. The method appears to be consistent with trends observed in Mössbauer quadrupole splittings in similar Sn^{IV} compounds.

[20] T. B. Brill and D. C. Miller, *Inorg. Chem.*, 1976, **15**, 2553.
[21] P. D. Cradwick and D. Hall, *J. Organometallic Chem.*, 1970, **22**, 203.
[22] T. B. Brill, *Inorg. Chem.*, 1976, **15**, 2558.

The ^{115}In n.q.r. spectra of Me_2InX (where X = F, Cl, InI_4, ReO_4, Br, I, or AcO) have been studied[23] and a correlation has been shown between the C–In–C valence angle and the asymmetry parameter. The indium atoms of solid Me_2InReO_4 were found to occupy equivalent positions, and the C–In–C angle was calculated to be 174°.

Gallium and aluminium halide donor–acceptor complexes have been the subject of three studies. ^{69}Ga and ^{71}Ga n.q.r. frequencies were observed at 14.450 and 9.105 MHz (274 K) from benzophenone,$GaCl_3$ and at 23.202 and 14.606 MHz from 4-chlorophenol,$GaCl_3$.[24] Charge transfer was found to increase in the order $4\text{-}ClC_6H_4OH < GaCl_4^- < C_6H_5COC_6H_5 < Cl^-$. In complexes of substituted benzophenone and benzoyl chloride with $AlCl_3$ and $AlBr_3$, ^{35}Cl and ^{81}Br data indicated that the benzoyl chlorides were weaker donors than benzophenones, despite the fact that the shifts in the i.r. carbonyl stretching frequency on formation of the complexes are less for benzophenones.[24] For the benzoyl chloride–$AlCl_3$ adducts, a positive correlation was found between the shift in the ^{35}Cl n.q.r. frequency of $^{35}Cl{-}C$ and the shift of the C—Cl i.r. stretching frequency. The n.q.r. data also confirmed that these complexes are covalent rather than ionic ($[ArCO]^+[AlCl_4]^-$). The absence of a large separation of bromine frequencies in $(4\text{-}ClC_6H_4OAlBr_2)_2$ leads to the structure (1) being proposed.

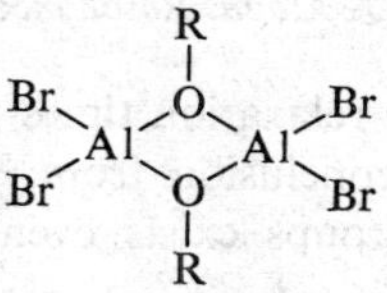

(1)

Variations in the ^{35}Cl n.q.r. frequencies[25] of $GaCl_3$ complexes with donors R_2O, R_2S, R_3P, and R_3As (R = Me or Et) indicated that the order of charge transfer is $R_3P > R_3As > R_2S > R_2O$. Substitution of methyl for ethyl caused less than 100 kHz shift in the chlorine frequency. Methyl derivatives showed two chlorine resonance of intensity ratio 2 : 1, whereas ethyl derivatives gave three resonances. The former complexes therefore probably have a mirror plane of symmetry.

^{79}Br and ^{81}Br n.q.r. data showed that *o*- or *p*-xylene–Al_2Br_6 π-complexes have the Al_2Br_6 in a dimeric stricture.[26] Mesitylene and durene form complexes of L,$AlBr_3$, the n.q.r. spectra of which show no lines corresponding to bridged bromine atoms. Based on n.q.r. frequency shifts, the stability of the complexes appeared to increase with increasing methyl substitution of the donor molecule.

Group V (Nitrogen-14, Arsenic-75, Antimony-121, Antimony-123, and Bismuth-209).—A study of $^{14}N_2$ molecules enclosed in a β-quinone clathrate (Figure 3) showed three well-resolved major components at 4.2 K, between 3.577 and

[23] H. Schmidbauer and D. Koth, *Naturwiss.*, 1976, **63**, 482.
[24] T. Deeg and A. Weiss, *Ber. Bunsengesellschaft phys. Chem.*, 1976, **80**, 2.
[25] J. C. Carter, R. Haran, and G. Jugie, *Compt. rend.*, 1976, **282**, *C*, 623.
[26] A. F. Volkov, I. P. Roman, E. N. Gur'yanova, and K. A. Kocheskov, *Izvest. Akad. Nauk S.S.S.R., Ser. khim.*, 1976, **25**, 1365.

3.596 MHz, each with partially resolved fine structure.[27] The spectrum was explained by assuming that the filling of a cavity produces a slight lattice distortion, which extends to nearest-neighbour cavities arranged along a crystal axis. The molecular libration frequency was deduced to be 50.7 cm^{-1} at 0 K. The static value of e^2Qq (5.37 MHz) is close to that found for solid nitrogen (5.39 MHz), so that incorporation in the clathrate lattice appears to cause little perturbation of the nitrogen molecule. The n.q.r. spectrum of solid α nitrogen[28] has been reinterpreted as the superposition of spectra of ortho- and para-N_2.[29]

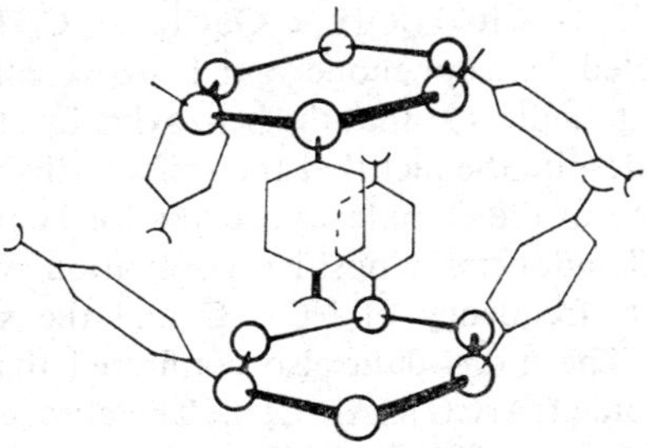

Figure 3 *The crystal structure of β-quinone clathrate* $3C_6H_4(OH)_2,xM$, *where x is the fraction of cavities occupied by constituent* M
(Reproduced by permission from *J. Magn. Resonance*, 1976, **24**, 103)

A study of ^{14}N quadrupolar relaxation times in $NaNO_2$ over the range 77—467 K supported a previous conclusion from ^{23}Na n.m.r. studies that short-range ordering of the NO_2 groups exists even above the ferroelectric phase transition.[30]

All three ^{14}N quadrupole resonance frequencies (ν_+, ν_-, ν_0) have been obtained at 77 K in a complex of 3,5-dichloropyridine with iodine monochloride.[31] Values of e^2Qq and η are 2.257 MHz and 0.208, respectively. ^{35}Cl n.q.r. frequencies were also observed, at 23.54 and 35.74 MHz (at 296 K), indicating that the degree of charge transfer is almost the same as in the corresponding pyridine and 4-methylpyridine complexes, which is estimated to be about 0.3*e*.

^{14}N n.q.r. data have been reported[32] for glyoximes $(HONCX)_2$ (X = Cl, Ph, or Me) and for bis(glyoximato)metal(II) complexes $M(HO_2N_2C_2X_2)_2$ (X = Me, Ph, or Cl when M = Pd; X = Me when M = Ni). The nitrogen e^2Qq values were reduced substantially on complex formation, *e.g.* the dimethylglyoxime value (5002 kHz) was reduced to 3580—3709 kHz on complexing with Pd and to 2822—3089 kHz with Ni. Two ^{14}N transition frequencies were found in each case, as required by two non-equivalent nitrogen atoms, which is as shown in *X*-ray structures. The nitrogens appear to donate more charge to Ni than to Pd, which is unexpected since Pd^{II} is generally regarded as more electron-withdrawing than Ni^{II}. However, the strong hydrogen bonding in these compounds makes the

27 A. A. V. Gibson, R. Goc, and T. A. Scott, *J. Magn. Resonance*, 1976, **24**, 103.
28 J. R. Brookeman, M. M. McEnnan, and T. A. Scott, *Phys. Rev.* (*B*), 1971, **4**, 3661.
29 S. Hirokawa, *J. Phys. Soc. Japan*, 1976, **41**, 1417.
30 G. Petersen and P. J. Bray, *J. Chem. Phys.*, 1976, **64**, 522.
31 G. A. Bowmaker, *J.C.S. Faraday II*, 1976, **72**, 1964.
32 Y. N. Hsieh, P. S. Ireland, and T. L. Brown, *J. Magn. Resonance*, 1976, **21**, 445.

situation rather complicated. Taking the data for all of the complexes as a whole, the orbital populations of the nitrogen lone pair are in the range 1.60—1.75*e*, giving effective metal charges of +0.40*e* to +1.00*e*. These values imply that the amount of charge transfer is approximately similar to that found in the pyridine ICl complex.[31]

In a study of $K_2Pt(SCN)_6$ and $K_2Hg(SCN)_6$ complexes [33] the ^{14}N asymmetry parameters which were found (η = 0.39 and 0.20) suggested an asymmetric π-electron distribution about a CN bond axis. (The η value for KSCN is 0.0285.) Pt—SCN and Hg—SCN are therefore thought to be non-linear, with the degree of ionic character of the metal–ligand bonds being estimated at 28% and 58%, respectively. *X*-Ray data for these complexes show no positions of light atoms, but a bent Hg—S—C unit with a sulphur angle of *ca.* 100° is anticipated from crystal data of analogous compounds.

Calculation of the charge transfer in RCN···Cl—Cl and $RCN\cdots HCCl_3$ complexes amounted to 0.02—0.04*e*, as indicated by ^{14}N and ^{35}Cl n.q.r.[34] In the chlorine complexes the charge is transferred to the end chlorine atom.

A ^{14}N–H double-resonance study [35] of $(MeNH_3)_2CdCl_4$ determined the ^{14}N quadrupole coupling constants and η values in three phases: 726 kHz and η = 0.21 for the orthorhombic room-temperature phase; 790 kHz and η = 0.1 in the low-temperature ($173 < T/K < 279$) tetragonal phase; 880 kHz, η = 0.20 in the monoclinic (<173 K) phase. The corresponding values for $MeNH_3Cl$ are 913 kHz and η = 0.109. The low e^2Qq values in all phases indicated an almost undistorted trigonal symmetry for the $MeNH_3$, some distortion of which might be expected to occur as a result of hydrogen bonding. The structure of $(MeNH_3)_2CdCl_4$ consists of layers of $CdCl_6$ octahedra with shared corner chlorine sites. The orthorhombic room-temperature phase has one N—H—Cl bond directed to an unshared chlorine and two N—H—Cl bonds to the corner-sharing Cl atoms. In the monoclinic, low-temperature phase, two hydrogen bonds are directed to unshared chlorine atoms and one to the corner-sharing site. The observed increase in e^2Qq on going from the orthorhombic phase to the tetragonal low-temperature phase, which is coupled with a simultaneous decrease in η, followed from the partial freezing-in of the dynamic disorder in the C—N bond directions. The data for the monoclinic phase are consistent with a frozen-in C—N bond in a deformed lattice, with N—H—Cl bonds of different length.

The study of aziridines (RN◁) has been extended [36] with the reporting of ^{14}N frequencies of C_4H_9N◁, P(N◁)$_3$, P(N◁^{Me_2})$_3$, DN◁, and Me_2Si(N◁)$_2$. The variations of the nitrogen orbital populations with substituents agree in general with the scheme of three-centre bonding for aziridine.

N.q.r. frequencies of ^{14}N in compounds containing N—Si bonds have been measured [37, 38] in Me_2NSiMe_3, Et_2NSiMe_3, $(Me_2SiNH)_3$, and Me_3SiN_3. The values of e^2Qq/kHz and (η) at 77 K are 4854.85 (0.447), 4804.57 (0.464), 2669.23

[33] T. Asaji, R. Ikeda, and D. Nakamura, *Z. Naturforsch.*, 1976, **31b**, 1483.
[34] B. V. Dresvyankin, V. S. Grechishkin, and V. I. Lunegov, *Izvest. Akad. Nauk S.S.S.R., Ser. fiz.*, 1975, **39**, 2584.
[35] J. Seliger, R. Blinc, H. Arend, and R. Kind, *Z. Phys. (B)*, 1976, **25**, 189.
[36] D. Ya. Osokin, I. A. Safin, and I. A. Nuretdinov, *Mol. Phys.*, 1976, **32**, 979.
[37] E. Schempp and M. Chao, *J. Phys. Chem.*, 1976, **80**, 193.
[38] M. Chao, *Diss. Abs. Internat. (B)*, 1976, **36**, 3454.

(0.645), 4026.18 or 4041.42 (0.095 or 0.084), respectively. The last compound gave different frequencies depending on the crystallization history of the sample. $(Me_2SiNH)_3$ has a cyclic structure, and the spectrum showed a three-fold splitting of the nitrogen lines, consistent with a distortion of the ring that makes the three nitrogens inequivalent.

A Townes–Dailey treatment indicated that N—Si bonding appears to be characterized by strong polarization of the N—Si σ-bond and rather weak p_π–d_π bonding, weaker at least than the p_π–p_π bonding in amino-pyridines. The low coupling constant found for $(Me_2SiNH)_3$ (2.67 MHz), which is only about half that in Me_2NSiMe_3, indicated the readily polarizable nature of the three bonds to nitrogen in the former compound. The ring has a slight degree of aromatic character, with a silicon *d*-orbital occupancy of 0.25*e*.

Two studies of ^{75}As n.q.r. of arsenic metal have reported e^2Qq values of 47.360 MHz (from extrapolation to 0 K)[39] and 47.123 MHz at 4.2 K.[40] The ionic contribution to the field gradient accounted for only 15% of the total. The conduction electrons therefore appear to be responsible for a major proportion of the field gradient, even though arsenic is not particularly metallic.[40] In bismuth metal, ^{209}Bi resonances were observed at 6.10 MHz ($\frac{5}{2} \leftrightarrow \frac{7}{2}$) and 8.11 MHz ($\frac{7}{2} \leftrightarrow \frac{9}{2}$) (300 K), which gave a value of 51.2 MHz for e^2Qq.[39] This agrees with previous values obtained by n.m.r. and low-temperature specific heat measurements.

The temperature and pressure dependences of ^{75}As, $^{121, 123}$Sb, and ^{35}Cl n.q.r. frequencies in $AsCl_3$ and $SbCl_3$ have received attention.[41] Intermolecular bonding increases in the order $AsCl_3 < SbCl_3 < BiCl_3$, and this trend is reflected in the relative importance of volume effects in determining the n.q.r. frequencies of these solids. Pressure dependences were explained in terms of negative static and positive dynamic contributions to the pressure coefficient. The static contribution increases in magnitude as intermolecular bonding increases.

The ^{121}Sb, ^{123}Sb n.q.r. spectra[42] of ClF_3,SbF_5 and BrF_3,SbF_5 indicated a larger value for η in the latter compound (0.23) than for the former (0.056), which was consistent with structural data showing fluorine bridges in a *cis*-configuration in BrF_3,SbF_5 and a *trans*-configuration in ClF_3,SbF_5. An order of increasing donor ability of fluorine halides was proposed: $ClF_5 < BrF_5 < ClF_3$. A tetrameric chain structure with a *cis* arrangement of fluorine bridges was proposed for SbF_5, presumably similar to NbF_5,SbF_5. Two types of Sb atom were found in $ClF_3,2SbF_5$ and $ClF_5,2SbF_5$, and their large η values indicated that bridging fluorines were *cis*-located. In the corresponding 1 : 4 compounds the two end Sb sites and the two inner Sb sites of the $Sb_4F_{21}^-$ anion are equivalent. The ^{35}Cl n.q.r. frequency of ClF_5 was reported as 20.572 MHz (77 K).[42]

The ^{121}Sb coupling constants of $M_2SbC_2O_4F_3,2H_2O$ (M = K, Rb, or NH_4) were found to be very similar to those of M_2SbF_5, indicating similar square-pyramidal structures for $[SbC_2O_4F_3]^{2-}$ and $[SbF_5]^{2-}$.[43] The double salt formu-

[39] T. J. Bastow and H. J. Whitfield, *Solid State Comm.*, 1976, **18**, 955.
[40] S. N. Sharma, *Phys. Letters* (*A*), 1976, **57A**, 379.
[41] G. C. Gillies and R. J. C. Brown, *Canad. J. Chem.*, 1976, **54**, 2266.
[42] I. A. Kuz'min, V. I. Shpanko, V. P. Kazakov, V. F. Sukhoverkov, B. E. Dzevitskii, and E. V. Bryukhova, *Izvest. Akad. Nauk S.S.S.R., Ser. fiz.*, 1975, **39**, 2555.
[43] L. A. Zemnukhova and R. L. Davidovich, *Izvest. Akad. Nauk S.S.S.R., Ser. fiz.*, 1975, **39**, 2545.

lation $M_2C_2O_4,2SbF_3,nH_2O$ for the complexes $M_2Sb_2C_2O_4F_6,nH_2O$ (M = Na, K, Rb, Cs, or NH_4; n = 0 or 4), which had been suggested from i.r. measurements, was rejected because the e^2Qq values differed greatly from that of SbF_3.

^{209}Bi, ^{35}Cl, and ^{37}Cl spectra of $RCN,BiCl_3$ (R = Me, CH_2Cl, Et, Ph, or BrC_6H_4) and $RCN,2BiCl_3$ (R = Me, Ph, or NCC_6H_4) have been observed.[44] Chlorine frequencies decreased on formation of the 1 : 1 complex, as is usually found in Group V trihalide complexes, but the ^{209}Bi quadrupole coupling constant increased. In similar As and Sb complexes the ^{75}As e^2Qq is reduced, and ^{121}Sb e^2Qq is not significantly changed upon complex formation. Large changes in bond lengths and angles take place in the arsenic and antimony complexes, and these are greater for Sb than As. Presumably the trend continues to the bismuth complexes, resulting in the increased e^2Qq values. In the 1 : 2 complexes two chemically independent bismuth atoms were found, and structures such as $MeCN \rightarrow BiCl_3 \rightarrow BiCl_3$ were proposed to explain this.

^{209}Bi resonance was observed in $Bi_4(GeO_4)_3$ and $Bi_4(SiO_4)_3$.[45] All four transitions were observed ($I = \frac{9}{2}$), and values of e^2Qq were calculated to be 490.8 MHz for the germanate and 470.4 MHz for the silicate. In both cases $\eta = 0$, implying axially symmetric field gradients. The purely ionic model was inadequate in explaining the magnitudes of the coupling constants.

Group VI (Oxygen-17).—^{17}O N.q.r. frequencies have been obtained,[46] by a double-resonance technique, from $Ba(ClO_3)_2,H_2{}^{17}O$. Three resonance lines were observed, each showing some fine structure. The ^{17}O quadrupole coupling constant was calculated to be −7.61 MHz, and $\eta = 0.94$. An estimate of the O—H bond length is 0.99 ± 0.02 Å and ∠HOH = 110 ± 2°.

For other ^{17}O n.q.r. data, see under Group I.

Group VII (Chlorine-35, Chlorine-37, Bromine-79, Bromine-81, and Iodine-127).— The four ^{35}Cl quadrupole coupling constants in $NaAlCl_4$ have been determined by single-crystal Zeeman n.q.r. spectra,[47] and lie in the range 21.6—23.0 MHz, while η varies from 0.18 to 0.32 for the four independent chlorine sites that are present. The small spread in e^2Qq is due to the small deviations of the bond lengths from the mean value of 2.13 Å (the crystal structure was redetermined in this paper [47]). The considerable spread of η values results from the influence of external charges on the efg tensor, from the interactions between chlorines within one $AlCl_4^-$ tetrahedron, and from the anisotropic thermal vibrations of the chlorine atoms. An angular distortion of the tetrahedron of 4° was observed in the *X*-ray structure. In $NaAlCl_4$ all the nuclei have a quadrupole moment, making this a 'total spin system'. The quadrupole coupling constants of ^{23}Na and ^{27}Al were determined by n.m.r., and gave the values 1.1117 and 1.4895 MHz, respectively, the latter value arising from the distortion of $AlCl_4^-$ from T_d symmetry.

An analysis of the temperature dependence of the ^{35}Cl n.q.r. frequency in sodium chloroacetate [48] gave two molecular torsional frequencies of 110 and

44 L. A. Lobanova, E. N. Guryanova, and A. F. Volkov, *Izvest. Akad. Nauk S.S.S.R., Ser. fiz.*, 1975, **39**, 2574.
45 K. V. Gopalakrishnan and L. C. Gupta, *Pramana*, 1976, **6**, 343.
46 M. Shporer and A. Achlama, *J. Chem. Phys.*, 1976, **65**, 3657.
47 W. Scheinert and A. Weiss, *Z. Naturforsch.*, 1976, **31a**, 1354.
48 R. Chandramani, N. Devaraj, and V. S. S. Sastry, *Austral. J. Chem.*, 1976, **29**, 2363.

40 cm^{-1}, both of which have negative temperature coefficients. Unfortunately, these values were not compared with vibrational spectroscopic data.

The hydrogen bonding in a wide range of acid salts $M^+[RCO_2 \cdots H \cdots O_2CR]^-$ has been studied by observation of the ^{35}Cl n.q.r. spectra of chlorine atoms situated in the organic group R (M = K, Rb, Cs, or NH_4; R = CH_2Cl, $CHCl_2$, CCl_3, or *o*-, *m*-, or *p*-ClC_6H_4).[49] Symmetrical O—H—O bonds were indicated for $MH(CCl_3COO)_2$ salts and for $CsH(CH_2ClCOO)_2$, while for the $H(CHCl_2COO)_2^-$ ion an asymmetric hydrogen bond was predicted in all the compounds examined. In most cases the deuteriated analogues were also studied; the asymmetrically hydrogen-bonded species gave significant ^{35}Cl frequency shifts (up to 0.9%), whereas the symmetric species showed little or no shift. $MH(p\text{-}ClC_6H_4COO)_2$ and $KH(m\text{-}ClC_6H_4COO)_2$ were thought to contain O—H—O bonds of the symmetric type, in agreement with *X*-ray data, whereas the corresponding *o*-chloro-compound has the proton asymmetrically located. Agreement with neutron diffraction data was obtained in the case of potassium hydrogen chloromaleate, and a similar symmetric bond was predicted from the n.q.r. data in potassium hydrogen dichloromaleate. This method appears to be generally applicable for determining hydrogen-bond symmetry.

Available n.q.r. data for the Group IV tetrahalides have been reviewed, and these were used to classify them into the five structural types.[50] CCl_4 at 77 K probably has the same structure as monoclinic CBr_4, with 32 molecules in the unit cell. The n.q.r. spectra of $SiCl_4$, $GeCl_4$, $GeBr_4$, $SnCl_4$, $TiCl_4$, and $TiBr_4$ show isomorphism with $SnBr_4$, with four independent bromine atoms. In this structure there is a weak co-ordination of one of the bromine atoms with an adjacent molecule. $GeBr_4$, GeI_4, $SiBr_4$, SiI_4, $TiBr_4$, TiI_4, and another form of $SnBr_4$ all have analogous spectra to SnI_4, which has two independent iodine atoms. CI_4,$PbCl_4$, a third form of $SnBr_4$, and a second form of TiI_4 all show a single halogen resonance line. The final group ($HfCl_4$, $ZrBr_4$, $HfBr_4$, ZrI_4, and HfI_4) are all isostructural with $ZrCl_4$, which contains two independent chlorine atoms, one of which is bridging. Their n.q.r. spectra therefore showed two widely separated lines.

Halogen resonances of diphenylhalonium ions have been measured in $(C_6H_5)_2X^+ BF_4^-$ and $[(p\text{-}ClC_6H_4)X(C_6H_5)]^+ BF_4^-$ (X = Cl, Br, or I).[51] Two resonances were observed for $(C_6H_5)_2Cl^+$, at 36.08 and 36.28 MHz, respectively, while values for the corresponding ^{79}Br- and ^{127}I-containing ions are 291.88 and 344.2 MHz, respectively, at 77 K [^{127}I frequency is for the ($\frac{1}{2} \leftrightarrow \frac{3}{2}$) transition]. Frequencies were essentially similar in $(C_6H_5)_2X^+ X^-$ complexes.[52] These exist as dimers, with the four halogen atoms in an approximately square configuration. By means of a Townes–Dailey analysis, the halogen charges were evaluated: Ph_2I^+, $+1.0e$; Ph_2Br^+, $+0.5e$; and Ph_2Cl^+, $+0.3e$. Significant

[49] R. J. Lynch, T. C. Waddington, T. A., O'Shea, and J. A. S. Smith, *J.C.S. Faraday II*, 1976, **72**, 1980.

[50] S. I. Kuznetsov, I. M. Alymov, E. V. Bryuchova, and G. K. Semin, *Izvest. Akad. Nauk S.S.S.R., Ser. fiz.*, 1975, **39**, 2523.

[51] A. N. Nesmeyanov, V. V. Saat-sazov, T. L. Khotsyanova, S. I. Kuznetsov, I. N. Lisichkina, and T. P. Tolstaya, *Doklady Akad. Nauk S.S.S.R.*, 1976, **227**, 670.

[52] T. G. Khotsyanova, T. A. Babushkina, and V. V. Saat-sazov, *Izvest. Akad. Nauk S.S.S.R., Ser. fiz.*, 1975, **39**, 2530; *Koord. Khim.*, 1976, **2**, 1567.

delocalization of charge in the latter two cases was indicated by a shift of +2 MHz in the ^{35}Cl frequency of the *p*-chloro-group of chlorophenyl compounds.

A study of the ^{35}Cl spectra of chlorovinyl- and (chloromethyl-)silanes gave no evidence for p_π–d_π interaction.[53] N.q.r. frequencies gave a linear correlation with Taft parameters.

In a study of phosphorus amide complexes with $SnCl_4$,[54] a single ^{35}Cl resonance was observed for $[Ph_2P(O)NMe_2]_2SnCl_4$ at 17.66 MHz (77 K), consistent with a *trans*-structure (D_{4h}) for the complex with four equivalent chlorine atoms in the crystal. $[Ph_2P(S)NMe_2]_2SnCl_4$ was also thought to be the *trans*-complex (ν_{Cl} = 17.79 and 18.35 MHz), and similarly $[Et_2P(S)NMe_2]_2SnCl_4$ (ν_{Cl} = 18.03 MHz). These conclusions were reinforced by i.r. and tin Mössbauer data, and were included in a wider study[55] of 35 phosphoryl, thiophosphoryl, and phosphorus amide complexes of $SnCl_4$. For complexes with ligands of similar size it was found that, as the donor ability of the ligand increased, *trans* structures rather than *cis* were preferred. Bulky ligands, *e.g.* the phosphorus amides, were generally found to form *trans*-complexes, as expected. The ionic character of the Sn—Cl bond was observed to increase with the nature of the donor, in the order $P < P{=}S \simeq P{=}O$. The ^{14}N n.q.r. of the complexes $[R_2P(O)NMe_2]_2SnCl_4$ has also been examined;[55] on complex formation a decrease in η was observed, and a slight decrease in the population of the nitrogen lone-pair orbital. A larger decrease was noted in the N—P σ-bond population, typically $-0.05e$, while only slight changes in nitrogen–phosphorus p_π–d_π conjugation were noted.

The ^{35}Cl n.q.r. spectra of a series of substituted chlorophosphoranes PR_nCl_{5-n} (R = Me, Et, or Ph) have been studied, together with the corresponding 1 : 1 adducts with PCl_5, $SbCl_5$, BCl_3, $AlCl_3$, or ICl.[56] $PMeCl_4$, PMe_2Cl_3, PMe_3Cl_2, PEt_2Cl_3, PEt_3Cl_2, and PPh_3Cl_2 were shown to have ionic structures of the type $[PR_nCl_{4-n}]Cl$, whereas PPh_2Cl_3 and $PPhCl_4$ are molecular in the solid state, with phenyl groups located equatorially in the trigonal-bipyramidal structures. The adducts all have ionic structures $[PR_nCl_{4-n}][XCl]$, where X represents the Lewis acid. Confirmatory evidence was provided by solid-state ^{31}P and ^{11}B n.m.r. chemical shifts. A ^{35}Cl n.q.r. study of analogous arsenic compounds[57] indicated that $AsPhCl_4$ and $AsPh_2Cl_3$ are also molecular in the solid state, and are probably isostructural with their phosphorus analogues. A contrast was found between $AsPh_3Cl_2$ and PPh_3Cl_2; the former compound was found to have a trigonal-bipyramidal structure, with axial chlorine atoms, whereas the latter has an ionic structure $[PPh_3Cl]Cl$. In contrast with their phosphorus analogues, $AsMe_2Cl_3$ and $AsMe_3Cl_2$ are molecular. Lewis acid adducts gave ^{35}Cl frequencies consistent with the ionic structures $[AsR_nCl_{4-n}]$-

[53] V. P. Feshin, M. G. Voronkov, V. F. Mironov, L. S. Romanenko, and T. K. Gar, *Teor. i eksp. Khim.*, 1976, **12**, 260.
[54] A. N. Pudovik, I. Ya, Kuramshin, N. R. Safiullina, A. A. Muratova, N. P. Morozova, and E. G. Yarkova, *Zhur. obshchei Khim.*, 1976, **46**, 765.
[55] A. I. Andreeva, I. Ya. Kuramshin, A. A. Muratova, D. Ya. Osokin, A. N. Pudovik, and I. A. Safin, *Koord. Khim.*, 1976, **2**, 683; *Izvest. Akad. Nauk S.S.S.R., Ser. fiz.*, 1975, **39**, 2590.
[56] K. B. Dillon, R. J. Lynch, R. N. Reeve, and T. C. Waddington, *J.C.S. Dalton*, 1976, 1243.
[57] K. B. Dillon, R. J. Lynch, and T. C. Waddington, *J.C.S. Dalton*, 1976, 1478.

[XCl], where X = BCl_3, $SbCl_5$, or ICl. ^{35}Cl Frequencies were also reported for Ph_3CCl, $SiPh_3Cl$, $GePh_2Cl_2$, and $GePh_3Cl$, and a comparison was made between Group V and Group IV MR_nCl_{4-n} systems. Less π-bonding was indicated in As—Cl bonds than in corresponding P—Cl bonds, and similarly $\pi_{Ge} < \pi_{Si}$. An order of decreasing π-bonding was proposed: $[PCl_4]^+ > PCl_3O > [PPhCl_3]^+ > [PPh_2Cl_2]^+ > [PPh_3Cl]^+$. A variable-temperature study of the ^{35}Cl n.q.r. frequency and of T_1 of $Cl_3PNPO(CCl_3)_2$ showed that the rate of reorientation of CCl_3 was significantly lower than that of the PCl_3 group at the same temperature.[58] A similar study on $Cl_3PNSO_2(p\text{-}ClC_6H_4)$ estimated the barrier to reorientation of the PCl_3 group to be 5.4 ± 0.2 kcal mol^{-1}.[59] In PCl_4CCl_3 the barrier to reorientational motion was found to be 15.2 kcal mol^{-1} from both ^{35}Cl–C and ^{35}Cl–P measurements.[60] This motion was not thought to be associated with the CCl_3 group, but with the molecule as a whole. No similar reorientation relaxation process was found for $PCl_3(CCl_3)_2$.[60]

The ^{35}Cl spectra of $SbCl_5$,L charge-transfer compounds, where L = MeOH, EtOH, Pr^{n}OH, Me_2NOCH, $(CH_2)_2OHCl$, $O(CH_2)_4$, H_3PO_4, $CH_2ClCOCN$, ClC_5-H_4N, or $(Me_2N)_3PO$, indicated an increase in charge on the chlorine bonded to antimony of 0.3—0.5*e*.[61] In complexes for which ^{121}Sb data were obtained, it was found that, upon charge transfer, the electron density at Sb decreased rather than increased, and that charge was transferred to the chlorine atoms.

A single-crystal Zeeman study[62] of sodium chlorite, $NaClO_2,3H_2O$, gave the value for e^2Qq of 98.382 MHz and $\eta = 0.54 \pm 0.02$. The *z* component of the field gradient was found to lie nearly perpendicular to the ClO_2^- plane, while the *y*-axis lay along the bisector of the OClO angle. These directions differ markedly from those in ClO_2. The crystal structure, which was also determined,[17] gave Cl—O bond lengths of 1.546(1) and 1.557(1) Å in the trihydrate and 1.564(3) Å in the anhydrate, with OClO angles of 108.23(6)° and 108.2(1)°, respectively.

Solid solutions of $NaClO_3$–$KClO_3$ showed ^{35}Cl linewidths up to 35 kHz in samples containing up to 10% $NaClO_3$ or 10% $KClO_3$,[63] while line multiplicities have been observed in solid solutions of chlorobenzene–bromobenzene and of methyl bromide–methyl iodide.[64] In the former case three ^{79}Br resonances were observed, at 77 K, the intensities of which were dependent upon the relative concentrations. Only a single line spectrum is normally observed from pure bromobenzene at 77 K. Electrostatic intermolecular interactions were said to be responsible for the observed splittings.

Single halogen n.q.r. resonances were observed from Ph_2SeX_2 (X = Cl or Br) and Ph_2TeX_2 (X = Cl, Br, or I) compounds, indicating the presence of a molecular two-fold axis.[65] These compounds all appear to have trigonal-

58 V. A. Mokeeva, I. A. Kyuntsel, and G. B. Soifer, *Zhur. strukt. Khim.*, 1976, **17**, 366.
59 V. A. Mokeeva, I. A. Kyuntsel, and G. B. Soifer, *Fiz. Tverd. Tela*, 1976, **18**, 3120.
60 V. A. Mokeeva, I. V. Izmest'ev, and I. A. Kyuntsel, *V. Sb., Radiospektroskopiya*, 1975, **9**, 52, 59.
61 O. Kh. Poleshchuk, and Yu. K. Maksyutin, *Izvest. Akad. Nauk S.S.S.R., Ser. fiz.*, 1975, **39**, 2579.
62 C. Tarimci, *Diss. Abs. Internat.* (*B*), 1976, **36**, 6199.
63 J. Hales, M. Rakos, and J. Lasanda, *Czech. J. Phys.*, 1976, **B26**, 936.
64 G. K. Semin and I. M. Alymov, *Phys. Status Solidi* (*A*), 1976, **33**, K63, K111; I. A. Alymov, G. K. Semin, T. L. Khotsyanova, and T. A. Babushkina, *Izvest. Akad. Nauk S.S.S.R., Ser. fiz.*, 1975, **39**, 2457.
65 V. V. Saat-sazov, T. L. Khotsyanova, V. P. Kazakov, and E. V. Bryukhova, *Izvest. Akad. Nauk S.S.S.R., Ser. fiz.*, 1975, **39**, 2535.

bipyramidal structures, with equivalent halogen atoms in the axial positions and a lone pair in the equatorial position.

Two ^{35}Cl resonances have been observed from $(MeNH_3)_2MnCl_4$ at 4.56 and 7.71 MHz (room temperature),[66] corresponding to two crystallographically independent chlorine atoms. The perovskitic layer structure, which consists of sheets of $MnCl_6$ octahedra sharing corners, undergoes an order–disorder phase transition at 393.7 K. The n.q.r. frequencies are somewhat lower than in other perovskite-type structures; 7.7 MHz in $CsPbCl_3$ [67] and 11.2 MHz in $RbCdCl_3$,[68] and two lines were found in $CsCdCl_3$ (D_{6h}^4) at 11.765 and 6.50 MHz (24 °C).[69] The latter compound also has a cubic phase (O_h^1), formed under high pressure, from which only a single resonance at 11.958 MHz was observed (24 °C). Zeeman experiments on $(MeNH_3)_2MnCl_4$ showed that the field gradient tensor for the lower line was axially symmetric, with the *z*-axis perpendicular to the layers, so that this resonance was assigned to chlorines situated above and below the Mn plane, *i.e.* non-bridging chlorines. The other bridging chlorines were found to have an η value of 0.7. Here the assignment appears to be the converse of that usually found for chlorines at high n.q.r. frequencies, *i.e.* ν(bridging Cl) < ν(terminal Cl). In this case the situation is complicated by the very ionic nature of the bonding and also by N—H···Cl hydrogen bonding, which is thought to involve both types of chlorine atom. A jumpwise rotation of the $MeNH_3^+$ ion was deduced to be responsible for the observed phase transition at 393.7 K. The C—N bond is said to move on a conical surface between four potential wells which exist because of hydrogen bonding. Above the phase transition, all wells are equally populated, but below 393.7 K one of them becomes favoured.[66] A similar model explained the ^{14}N results of $(MeNH_3)_2CdCl_4$ (see under Group V).

In a study of *trans*-$PtCl_2(C_2H_4)L$ complexes (L = substituted pyridines, MeNC, or Me_2NH), the i.r. ν(C=C) and ν(Pt—C_2) frequencies and ^{13}C n.m.r. shifts showed some dependence upon L (*i.e. trans* effect), but the ^{35}Cl n.q.r. frequencies which occurred in the range 20.38—21.45 MHz were not found to exhibit any *cis*-effect.[70]

Bowmaker and Whiting[71] observed ^{35}Cl n.q.r. frequencies at 17.58 and 17.70 MHz from $[Et_4N]AuCl_2$, 17.55 MHz from $[Bu_4N]AuCl_2$, and 17.80 MHz from $[(Bu_2NCS_2)_2Au]AuCl_2$ (all frequencies at 77 K). The singlet resonances are consistent with the proposed linear symmetrical structure of $AuCl_2^-$, while the doublet from Et_4NAuCl_2 was attributed to crystallographic splitting. From a Townes–Dailey analysis the Au—Cl bond was found to be 68% ionic; a value of 82% was previously obtained for $CuCl_2^-$.[72] There is some similarity in the frequencies of $AuCl_2^-$, $PtCl_4^{2-}$, and $PdCl_4^{2-}$ (17—19 MHz) and also of $AuCl_4^-$, $PtCl_6^{2-}$, and $PdCl_6^{2-}$ (26—28 MHz). The net metal atom charges all lie in the region 0—0.6*e*, and this has been taken as support for the view that the degree

[66] R. Kind and J. Roos, *Phys. Rev. (B)*, 1976, **13**, 45.
[67] N. J. Tovborg-Jensen, *J. Chem. Phys.*, 1969, **50**, 559.
[68] R. Kind and J. Roos in Proceedings of the 18th Ampère Congress, ed. P. S. Allen, E. R. Andrew, and C. A. Bates, Univ. of Nottingham, 1974, p. 297.
[69] K. S. Aleksandrov, A. K. Moskalev, V. L. Serebrennikov, V. M. Buznik, and B. V. Beznosikov, *Izvest. Akad. Nauk S.S.S.R., Ser. fiz.*, 1975, **39**, 2474.
[70] M. A. M. Meester, D. J. Stufkens, and K. Vrieze, *Inorg. Chim. Acta*, 1976, **16**, 191.
[71] G. A. Bowmaker and R. Whiting, *Austral. J. Chem.*, 1976, **29**, 1407.
[72] G. A. Bowmaker, L. D. Brockliss, and R. Whiting, *Austral. J. Chem.*, 1973, **26**, 29.

of covalence of a metal–halogen bond is determined by the requirement for the metal charge to remain quite low, regardless of the formal charge, *i.e.* oxidation state, of the metal.

Two studies of mercuric halide complexes have appeared. Wulfsberg [73] examined the halogen n.q.r. frequencies of 1 : 1 polyether adducts $RO(CH_2$-$CH_2O)_{n-1}R$ (R = Me or Me_3C, n = 2—5 or 7) and the crown ether $(CH_2CH_2O)_6$ with $HgCl_2$, $HgBr_2$, and HgI_2. A distinct correlation was shown to hold between the ^{35}Cl n.q.r. frequency and the Hg—Cl i.r. stretching frequency. ^{35}Cl and ^{81}Br frequencies were also correlated for pairs of related compounds, and the proportionality constant came close to the ratio of their atomic quadrupole coupling constants. It was found that each Cl—Hg intermolecular interaction, as judged from molecular models or crystal structures, lowered the ^{35}Cl n.q.r. frequency by 0.7—1.5 MHz; while each polyether donor lowered the frequency by 0.6 MHz. Five-co-ordinate mercury was predicted for HgX_2(diglytbu) (R = Me_3C, n = 3; X = Br or I) and a co-ordination number of seven was expected in HgI_2(triglyme) (R = Me, n = 4), which contains a bridging iodine atom. A phase transition in $HgBr_2$(triglyme) near 308 K between dimer and monomer forms was indicated.

Fichtner and Weiss [74] have examined oxygen, nitrogen, and sulphur donors with $HgCl_2$ (donors: dioxan, 1,3,5-trioxan, 1,3,5-trithian, tetrahydrothiophen, cyclohexane-1,4-dione, collidine, benzaldehyde, and *o*-, *m*-, and *p*-chloro-benzaldehyde). Apart from the sulphur compounds, an approximately octahedral co-ordination at mercury was indicated. $HgCl_2$ formed a dimer with *o*-Cl-benzaldehyde and a trimer with *p*-Cl-benzaldehyde. Only small shifts from the $HgCl_2$ n.q.r. frequency were found for dioxan, benzaldehyde and *o*-Cl-benzaldehyde adducts, indicating only weak donor interaction. The structure suggested for the *o*-Cl-benzaldehyde adduct is (2; D = donor). A positive temperature coefficient was found for the lower frequency ^{35}Cl resonance, and this was assigned to a bridging chlorine atom. Bridging chlorines were also detected in the case of $3HgCl_2$, 2(*p*-Cl-benzaldehyde) by anomalous temperature coefficients. The suggested structure is (3). If the central mercury atom in (3)

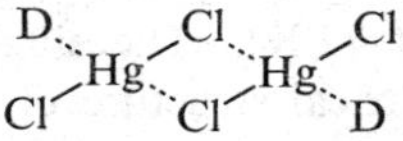

(2)

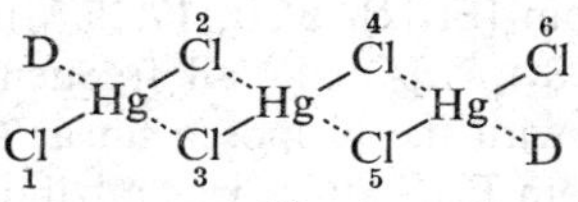

(3)

is at a centre of inversion, then Cl-1 and Cl-6, Cl-2 and Cl-5, and Cl-3 and Cl-4 should be crystallographically equivalent. Only one ^{35}Cl–C signal was found; therefore the two donor molecules should also be equivalent. Chlorines 3 and 4 have the lowest n.q.r. frequency, owing to the *trans*-effect of donor D.

The structure proposed for $HgCl_2$, 2(*m*-Cl-benzaldehyde) is (4), while $HgCl_2$,-2(dioxan) was thought to have a structure analogous to the *X*-ray structure of the $HgBr_2$ adduct (5). For K_2HgCl_4,H_2O and Rb_2HgCl_4,H_2O only a single ^{35}Cl resonance was found in each case, indication that the double-salt formulation $2KCl,HgCl_2,H_2O$ is justified.[74]

[73] G. Wulfsberg, *Inorg. Chem.*, 1976, **15**, 1791.
[74] W. Fichtner and A. Weiss, *Z. Naturforsch.*, 1976, **31b**, 1626.

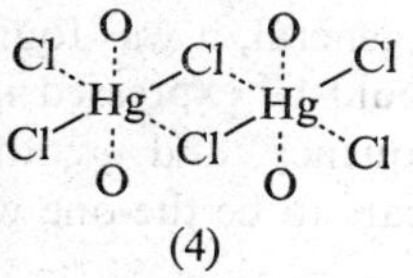

(4)

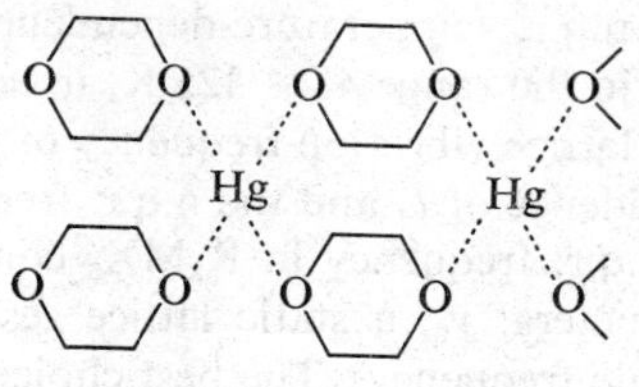

(5)

In a ^{79}Br n.q.r. investigation[75] of dibromo(dialkylamino)boranes Br_2BNR_2 (R = methyl, ethyl, n-propyl, n-butyl, or phenyl), the n.q.r. frequency decreased as the length of the alkyl group on the nitrogen atom increased. The B—N bond was found to exhibit considerable double-bond character.

^{79}Br n.q.r. frequencies have been observed in a number of tribromides (Table 1).[76] NH_4Br_3, $RbBr_3$, and $CsBr_3$ appear to contain asymmetric tribromide ions, while in the tetra-n-butylammonium salt, Br_3^- appears to be symmetric. The tetra-n-propylammonium salt is weakly asymmetric. There appears to be some

Table 1 *Bromine n.q.r. frequencies in tribromides*

Compound	^{79}Br *n.q.r. frequencies*/MHz *at* 77 K
NH_4Br_3	42.66, 264.83, 345.11
$RbBr_3$	97.09, 215.67, 345.60
$CsBr_3$	104.09, 211.62, 344.21
$Pr^n{}_4NBr_3$	156.60, 167.80, 340.68
$Bu^n{}_4NBr_3$	158.03, 346.66

similarity with I_3^- compounds[77] and with HCl_2^- compounds.[78] In both cases an increase in the size of the cation, from Me_4N^+ to Et_4N^+, results in a change from an asymmetric to a symmetric anion. The charge on the central bromine atom remained unaffected by the change in symmetry, and was constant at $+0.06e$ for all the tribromides. The lower line in NH_4Br_3 showed a positive temperature dependence, thought to be due to the involvement in weak hydrogen bonding and the effect of neighbouring cations.[76]

Positive ^{81}Br and ^{127}I n.q.r. temperature coefficients were found for $CdBr_2$ and CdI_2 (but not for PbI_2).[79] These were accounted for by the combination of oppositely signed ionic and covalent contributions to the electric field gradient, but their calculated relative magnitudes did not account for the observed n.q.r. frequencies. Electric field experiments on single crystals of the same compounds gave values for electric field constants, and it was estimated that the contribution from electron polarizability, due to the polarization of bonds by the electric field, was an order of magnitude smaller than the contribution of ionic polarizability.[80]

75 E. V. Bryuchova, V. P. Kazakov, and E. A. Balabanova, *Izvest. Akad. Nauk S.S.S.R., Ser. khim.*, 1976, **7**, 1614.
76 Y. Kume and D. Nakamura, *J. Magn. Resonance*, 1976, **22**, 235.
77 H. Harada, D. Nakamura, and M. Kubo, *J. Magn. Resonance*, 1974, **13**, 56.
78 C. J. Ludman, R. J. Lynch, J. A. Salthouse, J. A. S. Smith, and T. C. Waddington, *Chem. Comm.*, 1970, 405.
79 D. L. Lyfar, V. E. Goncharuk, and S. M. Ryabchenko, *Phys. Status Solidi (B)*, 1976, **76**, 183.
80 A. V. Bondar, A. Ya. Gordon, and D. L. Lyfar, *Fiz. Tverd. Tela*, 1976, **18**, 1235.

A ^{79}Br n.q.r. temperature-dependence study of Rb_2PtBr_6 showed no phase transition in the range 4.2—325 K, in contrast to the potassium and ammonium salts.[81] A lattice vibration frequency of 41 cm^{-1} was predicted from the temperature dependence of T_1 and the n.q.r. frequency. In general, it was found that the halogen n.q.r. frequency in R_2MX_6 compounds could be expressed in terms of two parameters: ν_0, a static lattice resonance frequency, and ω_e, the effective lattice mode frequency. The best choice of ν_0 appears to be the one which leads to the simplest temperature variation of ω_e.[82]

A single-crystal Zeeman study of β-$LiIO_3$ led to a determination of the field gradient axes in the crystal, and a value for η of 0.027.[83] Application of a two-frequency Bayer–Kushida model to halogen n.q.r. temperature-dependence data for α- and β-$LiIO_3$, $NaBrO_3$, $NaClO_3$, and $KClO_3$ predicted frequencies which matched bands in their vibration spectra.[84] The ^{127}I e^2Qq values were also determined[85] for a number of simple and complex iodates: M^IIO_3, $M^{II}(IO_3)_2$, $Bi(IO_3)_2$, $M^I{}_2Sn(IO_3)_6$, $M^I{}_2Ge(IO_3)_6$, and $Mg(CrIO_6)_2,6H_2O$, (M^I = Li, Na, K, Rb, Cs, NH_4^+, Ag, or Tl; M^{II} = Mg, Ca, Sr, Ba, Zn, Cd, Hg, Cu, or Pb). The value of e^2Qq was found to decrease as the cation changed from Li to Cs and from Mg to Ba, due to a decrease in the polarizing capability of the cation, and a linear correlation was shown with ionic radius. The average values of e^2Qq and η also increased linearly with the valency of the element X participating in the X—O(IO_2) bond, implying a steady increase in its covalency.

In a study of complex iodomercurates[86] NMe_4HgI_3, $KHgI_3,H_2O$, Cs_2HgI_4, Cs_3HgI_5, and $Cs_2Hg_3I_8$, the ^{127}I frequency correlated with the length of the Hg—I bond, and bridging iodide ions were clearly identified by their large η values. Bridging iodine atoms had e^2Qq values in the range 553—623 MHz and η = 0.48—0.74, while terminal iodines had e^2Qq values of 813—1019 MHz, η = 0.01—0.13. All these compounds contain HgI_4^{2-} tetrahedra; these are isolated in Cs_2HgI_4, linked at their vertices into chains of HgI_3^- in $KHgI_3,H_2O$ and NMe_4HgI_3, and linked in planar networks of $Hg_3I_8^{2-}$ in $Cs_2Hg_3I_8$. Cs_3HgI_5 corresponds to the double salt Cs_2HgI_4,CsI. The iodide ion lies nearly at the centre of an octahedron of Cs ions. The lowest frequency of the ^{127}I spectrum was assigned to this iodide ion at 99.6 MHz ($\frac{1}{2} \leftrightarrow \frac{3}{2}$) and the other three resonances to ^{127}I–Hg at 103.5, 108.3, and 121.5 MHz. If this assignment is correct, it implies that the charges on all the iodine atoms in the molecule are very similar, and that the Hg—I bonds must be significantly ionic.

Other halogen n.q.r. data appear under Groups III and V.

4 Transition Metals

Cobalt-59.—Two ^{59}Co resonances were observed in $(PhCH_2)_3SnCo(CO)_4$ at 14.24 MHz ($\pm\frac{3}{2} \leftrightarrow \pm\frac{5}{2}$) and 21.51 MHz ($\pm\frac{5}{2} \leftrightarrow \pm\frac{7}{2}$) at 300 K, giving e^2Qq =

[81] R. G. C. McElroy and R. L. Armstrong, *J. Magn. Resonance*, 1976, **24**, 165.
[82] R. G. C. McElroy and R. L. Armstrong, *J. Magn. Resonance*, 1976, **23**, 231.
[83] D. F. Baisa, A. I. Barabash, and G. A. Puchkovskaya, *J. Mol. Structure*, 1976, **31**, 247.
[84] D. F. Baisa, A. I. Barabash, G. A. Puchkovskaya, and Yu. A. Frolkov, *Izvest. Akad. Nauk S.S.S.R., Ser. fiz.*, 1975, **39**, 2487.
[85] V. A. Shiskin, V. S. Grechishkin, T. G. Balicheva, G. A. Petrova, and L. A. Cheburina, *Izvest. Akad. Nauk S.S.S.R., Ser. fiz.*, 1975, **39**, 2537.
[86] B. I. Pakhomov, P. M. Fedorov, I. M. Alymov, I. N. Ivanova-Korfini, and G. K. Semin, *Izvest. Akad. Nauk S.S.S.R., Ser. fiz.*, 1975, **39**, 2519.

100.4 MHz and $\eta = 0.15$.[87] For comparison, the values for the corresponding trimethyl and triphenyl complexes are $e^2Qq = 96.8$, 104.1 MHz and $\eta = 0.03$, 0.05, respectively. A correlation was suggested between e^2Qq and the barrier to axial–radial interchange of CO groups, as determined by ^{13}C n.m.r. in solution.

For other ^{59}Co resonances see under Group III.

Copper-63 and Copper-65.—^{63}Cu and ^{65}Cu n.q.r. frequencies have been observed in [Cu(tetramethylthiourea)$_3$]BF_4 at 27.390 and 25.341 MHz, and in the ClO_4^- salt at 27.390 and 25.340 MHz (296 K).[88] The Cu^I complex has a distorted trigonal planar structure, unlike the tetrahedral Cu(thiourea)$_3$$BF_4$, which is a sulphur-bridged dimer. Some involvement of the copper $3d_{z^2}$ electron in back donation to the π-orbital of the ligands was suggested to explain the magnitude of the field gradient.

^{63}Cu and ^{65}Cu n.q.r. frequencies and spin–lattice relaxation times have been measured in Cu_3SbS_3, Cu_2S, CuS,CuS_2, K[Cu(CN)$_2$], and Cu_2O.[89] For compounds containing similar bonds (Cu—O or Cu—S) the frequency spectrum (16—26 MHz) is narrower than that found for antimony compounds. In most cases the resonances faded above room temperature, due in some cases to the tendency of the Cu atoms to diffuse. Measured T_1 values at 77 K were in the range 270—500 ms, with the exception of Cu_2O, for which $T_1 \approx 2$ s.

Niobium-93.—The linear temperature dependence of the ^{93}Nb($\pm\frac{1}{2} \leftrightarrow \pm\frac{3}{2}$) frequency in $NbSe_2$ from 2.585 MHz at 77 K to 2.47 MHz at 376 K contrasts with the approximate $T^{\frac{3}{2}}$ dependence found for non-cubic metals. The layered, quasi-two-dimensional structure is thought to be responsible for this behaviour.[90]

Rhenium-185 and Rhenium-187.—Studies of perrhenates have continued (see last year's Report) with an investigation of the pressure dependences of ^{187}Re frequencies in NH_4ReO_4, $KReO_4$, $NaReO_4$, and $AgReO_4$.[91] All had positive pressure coefficients, but only the ammonium salt showed an increasing value as the temperature was lowered to 221 K. Below this temperature the resonance disappeared. *X*-Ray diffraction of NH_4ReO_4 over the range 143—371 K showed that the molar volume remained practically constant. Since this implies a positive value for $(\partial\nu/\partial T)_v$, torsional lattice modes are unlikely to influence the n.q.r. frequency, as these normally produce a negative $(\partial\nu/\partial T)_v$. Variations in the quadrupole frequency were therefore explained by changes in the lattice parameters or in the atomic position parameters as the temperature or pressure is changed. The n.q.r. frequency showed a linear correlation with the unit-cell parameter *a* for the perrhenates studied.[91] $TlReO_4$ has also been investigated.[92] In the range 280—360 K the ^{187}Re temperature coefficient is an order of magnitude greater than that found below 200 K. In the upper range, an η value of 0.3 was measured, and the distorted ReO_4 tetrahedra are expected to have up to C_{2v} symmetry. For the low-temperature range, $\eta = 0$; therefore hardly any

[87] D. L. Lichtenberger and D. R. Kidd, *J. Amer. Chem. Soc.*, 1976, **98**, 629.
[88] Y. Hijama, N. Watanabe, and E. Nika, *Bull. Chem. Soc. Japan*, 1976, **49**, 2986.
[89] R. S. Abdullin, I. A. Pen'kov, and I. A. Safin, *Izvest. Akad. Nauk S.S.S.R., Ser. fiz.*, 1975, **39**, 2543.
[90] D. R. Torgeson and F. Borsa, *Phys. Rev. Letters*, 1976, **37**, 956.
[91] R. J. C. Brown, J. G. Smeltzer, and R. D. Heyding, *J. Magn. Resonance*, 1976, **24**, 269.
[92] P. K. Burkert and F. M. Hutter, *Z. Naturforsch.*, 1976, **31b**, 145.

distortion from T_d symmetry is expected. Sixteen resonances due to both ^{185}Re and ^{187}Re have been observed from $LiReO_4,2H_2O$ at room temperature, consistent with four independent rhenium sites.[16]

Two covalent perrhenates, $Me_3SiOReO_3$ and $Me_3GeOReO_3$, have also been examined.[93] Below 195 K, two ^{187}Re and two ^{185}Re resonances were observed from the silicon compound, but above 294 K only one resonance from each isotope was found, with much greater negative temperature dependence. The presence of only one type of rhenium atom at room temperature is consistent with the reported *C*2/*c* space group. At low temperature two inequivalent rhenium atoms were indicated. In $Me_3GeOReO_3$ the two ^{187}Re resonance lines ($\frac{1}{2} \leftrightarrow \frac{3}{2}$) and ($\frac{3}{2} \leftrightarrow \frac{5}{2}$) were found to have non-linear negative and positive temperature dependences respectively. Consequently, the value of η varied from 0.68 at 77 K to 0.358 at room temperature. There was evidence for several phase transitions between room temperature and 123 K, which probably accounts for the curious temperature-dependence data.

[93] P. K. Burkert, D. Koth, and H. Schmidbaur, *Z. Naturforsch.*, 1976, **31b**, 149.

3
Microwave Spectroscopy

BY A. P. COX

1 Introduction

This chapter signposts recent publications of high-resolution studies of inorganic and organometallic molecules in the gaseous phase. Microwave spectroscopy continues to be the most valuable spectroscopic tool for structure determination in the gaseous phase, not only for molecular geometry and internal motions, but for such properties as dipole moments, quadrupole coupling, and magnetic constants. Many microwave papers about the study of rotational relaxation have been published; some significant papers are mentioned below. Regarding instrumentation, the most impressive feature is the combination of microwave spectroscopy with other techniques (particularly laser techniques), considerably extending the range of molecular information. A pulsed microwave Fourier transform spectrometer has been described [1] that is capable of detecting ions and short-lived species. A specific analytical application [2] of microwave spectroscopy is the study of exchange reactions of deuterium-labelled propene on catalytic surfaces of rutile and magnesia.

Several new microwave species have been reported. Some, such as OC_3S and the phospha-alkenes, have been characterized for the first time through their microwave spectra. Others, such as N_2H^+ and HNC, have given new spectra in the laboratory, and are of particular interest because of their previous detection in interstellar space. Other new species include the van der Waals molecules; two review articles [3,4] were recently published on their spectroscopy and structural properties.

An important compilation of structural parameters has appeared,[5] being a critically reviewed list of gas-phase structures determined by electron diffraction and spectroscopic techniques. A review article by Lide gives a lucid account [6] of structural determination from spectroscopic data. A compilation of current microwave and electron-diffraction abstracts is provided by the Sektion für Strukturdokumentation (Dr. B. Starck) of the Universität Ulm. Some research groups have their own literature surveys from computer scanning, such as that circulated by Professor Sasada *et al.* on behalf of several Japanese laboratories.

1 J. Ekkers and W. H. Flygare, *Rev. Sci. Instr.*, 1976, **47**, 448.
2 C. S. John, C. Kemball, R. Dickinson, and J. K. Tyler, *J.C.S. Faraday I*, 1976, **72**, 1782.
3 G. E. Ewing, *Canad. J. Phys.*, 1976, **54**, 487.
4 B. J. Howard, in International Review of Science Physical Chemistry Series 2, Vol. 2: 'Molecular Structure and Properties', ed. A. D. Buckingham, Butterworth, Sevenoaks, 1975, Ch. 3.
5 'Structure Data of Free Polyatomic Molecules', ed. K.-H. and A. M. Hellwege, New Series, Group II, Vol. 7, Landolt–Börnstein, Springer-Verlag, Berlin, 1976.
6 D. R. Lide, in ref. 4, Ch. 1.

2 Diatomic Molecules

A double-oven technique has allowed the bismuth monohalides, which are high-temperature radicals, to be studied[7–9] under optimum conditions. Spectra observed between 55 and 100 GHz for BiCl, BiBr, and BiI yield very precise molecular constants for the 0^+ electronic ground state (see Table 1). Transitions of BiF have also been observed. Good agreement was found with band spectroscopic studies for the chloride and bromide. A striking similarity in Dunham constants was found between the monohalides of ^{209}Bi and ^{205}Tl, and the quadrupole coupling constants at both ^{209}Bi and ^{35}Cl were satisfactorily interpreted on the basis of similar chemical bonding.

The microwave spectrum of BaS has been measured[10] in an absorption cell at 2000 K, using a saturation modulation spectrometer. The molecular constants (see Table 1) agree well with the less precise optical values. Comparison with isoelectronic CsCl shows BaS to have a much larger bond force constant, but smaller anharmonic constants. The vibrational dependence of the dipole moment was calculated as $\mu_1 = 0.022(5)$ D, from earlier molecular beam electric resonance (m.b.e.r.) work, similar to BaO.

A critical review[11] of microwave spectral data of SiS has been published for application to radio astronomy (see Table 1).

A study of several isotopic species of GeS and GeSe has demonstrated[12] the breakdown of the Born–Oppenheimer approximation for the rotational constant Y_{01}. The adiabatic correction has been extracted for both molecules, results being different from those obtained in earlier work on PbS. The same group has studied[13] quadrupole fine structure of the $J = 3 \rightarrow 4$ transition of ^{85}Rb^{127}I, to give coupling constants at both nuclei (see Table 1), thus completing the study of the set of rubidium halides (apart from RbBr) and the alkali-metal iodides.

Hyperfine structure of ^{7}Li^{127}I in the $v = 0$, $J = 1, 2$ states has been closely examined[14] by m.b.e.r. techniques. Precise electric quadrupole coupling constants, at both nuclei, plus spin–rotation and spin–spin constants were determined. No evidence was obtained for magnetic octupole or electric hexadecapole interactions for the iodine nucleus. The r.f. m.b.e.r. spectrum of LiF has been re-examined[15] and found to be in substantial agreement with earlier work.

A novel method for determining the sign of dipole moments has been reported.[16] The structure and dipole moment of the weakly bound 'antihydrogen-bonded' complex HFClF has been determined by m.b.e.r., using supersonic expansion techniques (see Section 4). These results have been used, on the basis of little charge transfer, to show that the sign of the dipole moment for ClF is $^+$ClF$^-$. This result contradicts the earlier result obtained from molecular *g* values, but is in accord with chemical intuition and *ab initio* calculations.

7 P. Kuijpers, T. Törring, and A. Dymanus, *Chem. Phys.*, 1976, **18**, 401.
8 P. Kuijpers and A. Dymanus, *Chem. Phys. Letters*, 1976, **39**, 217.
9 P. Kuijpers, T. Törring, and A. Dymanus, *Chem. Phys.*, 1976, **12**, 309.
10 E. Tiemann, Ch. Ryzlewicz, and T. Törring, *Z. Naturforsch.*, 1976, **31a**, 128.
11 E. Tiemann, *J. Phys. Chem. Ref. Data*, 1976, **5**, 1147.
12 W. U. Stieda, E. Tiemann, T. Törring, and J. Hoeft, *Z. Naturforsch.*, 1976, **31a**, 374.
13 E. Tiemann and J. Hoeft, *Z. Naturforsch.*, 1976, **31a**, 236.
14 A. R. Jacobson and N. F. Ramsey, *J. Chem. Phys.*, 1976, **65**, 1211.
15 A. J. Hebert and C. D. Hollowell, *J. Chem. Phys.*, 1976, **65**, 4327.
16 K. C. Janda, W. Klemperer, and S. E. Novick, *J. Chem. Phys.*, 1976, **64**, 2698.

Table 1 *Some molecular constants for diatomic molecules*

Molecule [a]	B_e/MHz	r_e/Å	μ_e/D	*Nucleus*	eqQ/MHz	*Ref.*
$^{209}Bi^{35}Cl(0^+)$	2761.8538(13)	2.471 55(7)	—	^{209}Bi	−1027(12)	7
				^{35}Cl	−34.5(28)	7
$^{209}Bi^{79}Br(0^+)$	1295.5609(12)	2.609 53(7)	—	—	—	8
$^{209}Bi^{127}I(0^+)$	816.119 43(13)	2.800 53(8)	—	—	—	9
$^{138}Ba^{32}S$	3097.2847(12)	2.507 350 8(5)	10.852(14)	—	—	10
$^{28}Si^{32}S$	9099.5369(15) [b]	—	1.73(6) [c]	^{33}S	10.90(20)	11
$^{74}Ge^{32}S$	$\mu_r B_e$ = 124 836.875(52) [b] a.m.u. MHz	—	—	—	—	12
$^{74}Ge^{80}Se$	$\mu_r B_e$ = 110 913.18(6) [b] a.m.u. MHz	—	—	—	—	12
$^{85}Rb^{127}I$	$B_0 + 32Y_{02}$ = 982.6574(15)	—	—	^{85}Rb	−40.40(30)	13
				^{127}I	−59.89(30)	13
$^{7}Li^{127}I$	—	—	—	^{7}Li	0.171 000(30) [d]	14
				^{127}I	−198.478 627(10) [d]	14
$^{7}Li^{19}F$	—	—	6.2839(10)	^{7}Li	0.4156(4) [c]	15
$^{14}N^{16}O(X^2\Pi_{\frac{1}{2}})$	50 844.97(16) [c]	—	—	—	—	18
$^{14}N^{16}O(X^2\Pi_{\frac{3}{2}})$	50 850.92(18) [c]	—	—	—	—	18
$^{14}N^{16}O(A^2\Sigma^+)$	—	—	1.10(3) [e]	^{14}N	−6.77(40) [e]	19
$^{16}OD(A^2\Sigma^+)$	—	—	—	^{2}D	0.18(10)	19
$^{14}NH(X^3\Sigma^-)$	489 960(6) [c]	1.037 22(2)	—	—	—	20
$^{14}ND(X^3\Sigma^-)$	263 263(2) [c]	—	—	—	—	20
$^{32}S^{16}O(X^3\Sigma^-)$	21 609.552(21)	1.481 083(1)	—	—	—	21
$^{32}S^{16}O(a^1\Delta)$	21 295.405(24) [c]	1.491 967(1) [c]	—	—	—	21

[a] All electronic states are $X^1\Sigma$ unless otherwise noted. [b] Including Born–Oppenheimer corrections. [c] $v = 0$ state. [d] $v = 0, J = 1$. [e] $v = 3$.

A gas-phase electron paramagnetic resonance investigation of six isotopic species of NO has been carried out in the *X*-band.[17] A perturbation analysis for the $^2\Pi_{\frac{3}{2}}$ state gave better agreement with experiment than previously obtained. Almost simultaneously, a paper[18] by Meerts gave an impressive analysis of the rotational and hyperfine Λ-doubling spectrum of ^{14}NO, by diagonalization of the secular matrix containing all couplings between different *J*-states. There were 22 constants required to fit 91 pieces of microwave and m.b.e.r. data. A note[19] from Woods and Dixon also concerns NO (in the $A^2\Sigma^+$ state). They have corrected the quadrupole coupling constant and magnetic hyperfine parameters previously obtained by Zare and co-workers, using optical–r.f. double resonance. Similar corrections have been made for the $A^2\Sigma^+$ state of OD, thus improving agreement with *ab initio* calculations for both molecules.

At submillimetre wavelengths the laser magnetic resonance spectrum of imine (NH) and its isotopes has been measured.[20] More precise values of the rotational constants and the equilibrium bond distance were derived than those obtained from optical spectroscopy. Hyperfine parameters, measured for the first time, were compared with *ab initio* calculations.

The wide spectral coverage afforded by submillimetre microwave techniques has enabled[21] considerable refinement of, and addition to, molecular constants of SO, both in its $^1\Delta$ excited electronic state and its $^3\Sigma$ ground state (see Table 1).

The magnetic resonance spectrum of the van der Waals molecule KAr has been studied[22] in a supersonic molecular beam. The paramagnetic molecule behaves largely as the valence electron of an alkali metal interacting with two impenetrable cores. The spin–rotational constant was determined as $\gamma(v = 0, N = 0) = 240(10)$ kHz, and its variation with internuclear distance was examined. An interatomic potential calculated for KAr was discussed.

Several papers have been published on microwave lines of OH, particularly related to relaxation effects, because of its importance as an interstellar source. Population inversion between Λ-doublet states was observed,[23] using the reaction of H atoms with NO_2 in a beam maser and by m.b.e.r. In another paper, line-width measurements of pressure-broadened OH lines are reported,[24] and the difficulties involved described. CO is another important interstellar source. One paper relates[25] emission from $J = 1 \rightarrow 0$ with visual extinction in cloud L134, and thereby supports the use of ^{13}CO column densities as a measure of the abundance of interstellar molecular hydrogen.

3 Triatomic Molecules

The study of vibration–rotation interaction in KOH has been extended[26] by Pearson *et al.* A detailed analysis of *l*-type resonance in the millimetre region

[17] F. H. Jarke, N. A. Ashford, and I. J. Solomon, *J. Chem. Phys.*, 1976, **64**, 3097.
[18] W. L. Meerts, *Chem. Phys.*, 1976, **14**, 421.
[19] R. C. Woods and T. A. Dixon, *J. Chem. Phys.*, 1976, **64**, 5319.
[20] F. D. Wayne and H. E. Radford, *Mol. Phys.*, 1976, **32**, 1407.
[21] W. W. Clark and F. C. De Lucia, *J. Mol. Spectroscopy*, 1976, **60**, 332.
[22] R. R. Freeman, E. M. Mattison, D. E. Pritchard, and D. Kleppner, *J. Chem. Phys.*, 1976, **64**, 1194.
[23] J. J. ter Meulen, W. L. Meerts, G. W. M. van Mierlo, and A. Dymanus, *Phys. Rev. Letters*, 1976, **36**, 1031.
[24] R. Bustreel, J. L. Destombes, and C. Marlière, *Chem. Phys. Letters*, 1976, **42**, 154.
[25] K. D. Tucker, R. L. Dickman, P. J. Encrenaz, and M. L. Kutner, *Astrophys. J.*, 1976, **210**, 679.
[26] E. F. Pearson, B. P. Winnewisser, and M. B. Trueblood, *Z. Naturforsch.*, 1976, **31a**, 1259.

was interpreted in terms of an almost harmonic bending potential; thus KOH is described as linear, not quasi-linear. Equilibrium bond distances were derived (see Table 2) and compared with those of the heavier alkali-metal hydroxides.

Table 2 *Structural parameters for triatomic molecules*

Molecule [a]	*Type*	*r*(A—B)/Å	*r*(B—C)/Å	∠ABC/°	μ/D	*Ref.*
KOH	r_e	2.196	0.960	180	—	26
NaOH	r_0	1.95	—	180	—	27
HNC	r_e	0.993 57(117)	1.164 53(28)	180	3.05	34, 31
	r_s	0.986 07(9)	1.171 68(22)			
HCN	r_e	1.065 49(24)	1.153 21(5)	180	—	34
SCl_2	r_0	2.014(5)	—	102.8(2)	0.36(1)	35
$SO_2(C^1B_2)$	r_0	1.560	—	104.3	—	36
$HO_2(X^2A'')$	r_0/r_s	0.977(10)	1.335(5)	104.1(15)	—	41

[a] All electronic states are $X^1\Sigma$ unless otherwise stated.

Despite severe experimental difficulties, further measurements have been made [27] on NaOH at 1000 K in the 100 GHz region. Analysis showed vibration–rotation interaction to be more severe than for the heavier alkali-metal hydroxides. A comparison was made of the bonding properties of the alkali-metal hydroxides and fluorides. The same authors, using equally severe conditions, have detected [28] the first microwave spectrum of an alkali-metal cyanide, KCN, at 600 K. A spectrum of surprising complexity was obtained between 85 and 107 GHz, giving rise to a *B*-constant which the authors interpret as being too large for a linear structure. Instead they tentatively invoke Clementi's polytopic bond, with CN^- rotating more or less freely about K^+.

The $J = 0 \rightarrow 1$ quadrupole multiplet of the N_2H^+ ion, recently identified in the interstellar medium, has now been observed in the laboratory by Saykally *et al.*[29] in glow discharges of mixtures of nitrogen and hydrogen. The determination of a precise structure for the ion now becomes feasible, as well as its velocity distribution in astronomical sources.

Using the same glow-discharge spectrometer, Woods and his co-workers [30] have also detected the $J = 0 \rightarrow 1$ transition of HNC. Of three different gas mixtures used, nitrogen and acetylene gave the best results, and detection of the U90.663 line in the laboratory under these conditions left no doubt as to the original interstellar assignment. Two other groups have been actively engaged in observations of HNC. Brown and co-workers [31] have detected HNC, $HN^{13}C$, and DNC, using thermal isomerization of the corresponding cyanide at 1000 K. A parallel-plate spectrometer was used, enabling the dipole moment to be determined (see Table 2) from measurements of the Stark effect. In addition, they have observed [32] $J = 1 \rightarrow 0$ emission lines of $HN^{13}C$, DNC, and $H^{15}NC$ from

27 P. Kuijpers, T. Törring, and A. Dymanus, *Chem. Phys.*, 1976, **15**, 457.
28 P. Kuijpers, T. Törring, and A. Dymanus, *Chem. Phys. Letters*, 1976, **42**, 423.
29 R. J. Saykally, T. A. Dixon, T. G. Anderson, P. G. Szanto, and R. C. Woods, *Astrophys. J.*, 1976, **205**, L101.
30 R. J. Saykally, P. G. Szanto, T. G. Anderson, and R. C. Woods, *Astrophys. J.*, 1976, **204**, L143.
31 G. L. Blackman, R. D. Brown, P. D. Godfrey, and H. I. Gunn, *Nature*, 1976, **261**, 395.
32 R. D. Brown, P. D. Godfrey, J. W. V. Storey, and F. O. Clark, *Nature*, 1976, **262**, 672; R. D. Brown (personal communication).

various interstellar sources. Column densities for DNC and $H^{15}NC$ and a narrow linewidth for $HN^{13}C$ indicate that there is enhanced isotopic enrichment, due to kinetic isotope effects on the rates of formation. For ^{13}C and ^{15}N, localized enrichment is implied. Winnewisser and co-workers have studied [33, 34] all eight stable isotopic species of HNC and HCN in the millimetre-wave region. In this work HNC was generated by the reaction of methyl bromide or iodide with active nitrogen, formed in a r.f. discharge. Accurate structures have been derived for both molecules (see Table 2) that are in close agreement with *ab initio* calculations, but with the possibility of quasi-linear behaviour of HNC remaining to be investigated.

The microwave spectrum of SCl_2 has been observed [35] in the 9—36 GHz region. An accurate structure has been obtained, together with the dipole moment (see Table 2) and the quadrupole coupling constants for $S^{35}Cl_2$ [$\chi_{aa} = -38.98(10)$, $\chi_{bb} = -8.87(10)$, and $|\chi_{ab}| = 64(10)$ MHz]. The χ_{ab} value shows that the S—Cl bond lies very close to the axis of the field gradient tensor.

A detailed rotational analysis [36] of the 235 nm system of SO_2 has given accurate rotational constants and an averaged structure for the $^1B_2$ excited electronic state. From the vibrational structure, definite evidence for an unsymmetrical equilibrium geometry (Δr = 0.12 Å) was obtained, corresponding to a double-minimum potential barrier of $\sim$100 cm^{-1} in the antisymmetric stretching co-ordinate. Comparison with the ground-state structure indicates a sharp increase in average bond length (0.125 Å) and a large decrease in bond angle (−15.0°), in remarkable agreement with configuration interaction calculations.

Two other papers concerning SO_2 have appeared. Absorption coefficients of SO_2 lines up to $J = 50$ and to 200 GHz have been calculated [37] for application to high-sensitivity monitoring of environmental SO_2 or the study of reaction kinetics. Measured coefficients at 70 GHz in this work, and earlier measurements at lower frequencies, show 5—10% agreement with calculation. Aliev and Watson have derived simplified perturbation expressions [38] for the dependence of the sextic centrifugal distortion constants on the cubic anharmonic potential constants. Numerical results computed for SO_2 from a published anharmonic force field gave good agreement with observed distortion constants. Less good agreement was obtained from the force fields of NH_3 and CH_4. Other force-constant calculations,[39, 40] this time harmonic, pertaining to microwave data have been carried out for NSCl and NSF, and for SeO_2.

[33] R. A. Creswell, E. F. Pearson, M. Winnewisser, and G. Winnewisser, *Z. Naturforsch.*, 1976, **31a**, 221.
[34] E. F. Pearson, R. A. Creswell, M. Winnewisser, and G. Winnewisser, *Z. Naturforsch.*, 1976, **31a**, 1394.
[35] J. T. Murray, W. A. Little, jun., Q. Williams, and T. L. Weatherly, *J. Chem. Phys.*, 1976, **65**, 985.
[36] J. C. D. Brand, P. H. Chin, A. R. Hoy, and H. D. Bist, *J. Mol. Spectroscopy*, 1976, **60**, 43.
[37] W. F. Kolbe, B. Leskovar, and H. Buscher, *J. Mol. Spectroscopy*, 1976, **59**, 86.
[38] M. R. Aliev and J. K. G. Watson, *J. Mol. Spectroscopy*, 1976, **61**, 29.
[39] A. Müller, N. Mohan, S. J. Cyvin, N. Weinstock, and O. Glemser, *J. Mol. Spectroscopy*, 1976, **59**, 161.
[40] T. R. Ananthakrishnan, C. P. Girijavallabhan, and G. Aruldhas, *J. Mol. Spectroscopy*, 1976, **60**, 239.

The microwave spectrum of the $N = 0 \rightarrow 1$ transition of DO_2 has been observed[41] near 60 GHz. The data have been combined with earlier results on HO_2 to compute the structure, given in Table 2, which is in good agreement with the latest of several theoretical estimates. Prior to this experimental data on DO_2, some collective consideration of the structures of HCO, HNO, and HO_2 had been given by Ogilvie.[42]

The $6_{16} \rightarrow 7_{07}$ microwave transition of NO_2 in the $^2B_2$ excited electronic state has been observed[43] by microwave optical double-resonance spectroscopy (m.o.d.r.) with a tunable dye laser near 593 nm. Spin–rotation and hyperfine interactions are discussed, together with earlier results. The Fermi-contact constant was obtained and found to be $\sim$50 MHz from the $K_a = 0$ sub-levels, much smaller than that for the $^2A_1$ ground state, consistent with the unpaired spin density being essentially in a b_2 molecular orbital where $|\psi(0)|^2$ would be zero. The microwave spectrum of NO_2 that is in the ground-electronic state but in the $v_2 = 1$ state has been studied[44] in a free-radical cavity spectrometer. Twenty hyperfine components of the $8_{08} \rightarrow 7_{17}$ and $9_{19} \rightarrow 10_{010}$ rotational transitions have been fitted to hyperfine coupling constants, not significantly different from $v = 0$, but with a fine-structure constant A_s which changes by 10%.

M.o.d.r. has also been used to detect[45] *magnetic* dipole transitions in the ground vibronic state of NH_2. Analysis of transitions between two J components of the same rovibronic state gave improved ^{14}N hyperfine constants for the X^2B_1 state over the saturation spectral study published earlier[46] by the same authors, which also gave coupling constants for an excited electronic state. Comparison of the magnetic hyperfine constants with other π-radicals NF_2 and ClO_2 indicated the unpaired electron to be predominantly p (out-of-plane). Similarity in the electronic structures of NF_2 and NH_2 does not seem to extend to the quadrupole coupling constants.

Van der Waals bonding in the wide-amplitude bender ArHCl has been further investigated,[47] using rotational spectroscopy on a supersonic beam. Centrifugal distortion constants D_J were reported for four isotopic species, plus quadrupole coupling constants at the chlorine nucleus ($\chi_{aa} = -36.241$ MHz for $ArD^{35}Cl$). The distortion constants were found to be substantially different for ArHCl and ArDCl. The difference was interpreted as due to different effective radial potential functions. The equilibrium structure is now believed to be linear.

Infrared microwave double resonance and double-photon transitions have been studied[48] for FCN in a conventional microwave spectrometer. Using three different types of information, it was shown that the $R(2)$ transition of ν_1 for FCN occurs at 1079.5871 $\pm$ 0.0005 cm^{-1}, *i.e.* lower by 104 MHz than the $R(20)$, 9.4 μm CO_2 laser line.

[41] Y. Beers and C. J. Howard, *J. Chem. Phys.*, 1976, **64**, 1541.
[42] J. F. Ogilvie, *J. Mol. Structure*, 1976, **31**, 407.
[43] T. Tanaka and D. O. Harris, *J. Mol. Spectroscopy*, 1976, **59**, 413.
[44] L. W. Hrubesh and R. F. Curl, *J. Mol. Spectroscopy*, 1976, **61**, 144.
[45] G. W. Hills, J. M. Cook, R. F. Curl, jun., and F. K. Tittel, *J. Chem. Phys.*, 1976, **65**, 823.
[46] G. W. Hills, D. L. Philen, R. F. Curl, jun., and F. K. Tittel, *Chem. Phys.*, 1976, **12**, 107.
[47] S. E. Novick, K. C. Janda, S. L. Holmgren, M. Waldman, and W. Klemperer, *J. Chem. Phys.*, 1976, **65**, 1114.
[48] H. Jones, *Z. Naturforsch.*, 1976, **31a**, 1614.

A submillimetre-wave spectrometer, using BWO sources with acoustic detection, has been used by Krupnov and co-workers [49] to extend greatly the microwave data for N_2O. Over 40 rotational and rovibrational constants were determined for 22 spectroscopically different species.

There are six papers reporting rotational relaxation studies on OCS. Tanaka and Hirota obtained [50] a significant difference between T_1 and T_2 from the nutation signals of the $J = 1 \rightarrow 2$ transition of OCS, finding $(2\pi T_2 p)^{-1} - (2\pi T_1 p)^{-1} = 0.94(38)$ MHz Torr^{-1}. Schwendeman and co-workers,[51] by very careful examination of the $J = 1 \rightarrow 2$ line-width, obtained $(2\pi T_2 p)^{-1} = 6.03(5)$ MHz Torr^{-1}, and suggested the probability that T_1/T_2 is slightly greater than 1. A T_1 value has also been obtained for this transition from an adiabatic rapid-passage microwave double-resonance technique.[52] Hoke *et al.*, from transient measurements,[53] obtained $T_1 = T_2$ within 6% for both the $J = 0 \rightarrow 1$ and $J = 1 \rightarrow 2$ transitions. Two further microwave double resonance studies [54, 55] of OCS, involving separate *M*-states, both confirmed the importance of higher than first-order dipole collisional interactions. Two papers bearing on the vibrational potential function of OCS have been published.[56, 57] Millimetre-wave spectra of various species, isotopic and excited-state, have yielded centrifugal distortion data for comparison with an earlier theoretical treatment. Laser Stark measurements by Maki and Freund of excited states of ν_2 for OCS gave precise dipole-moment changes and band centres that were accurate to several MHz.

4 Tetra-atomic Molecules

A critical review of microwave data for five isotopic species of HNCO has appeared.[58] Some of the charge properties are quoted in Table 3. The structure

Table 3 *Dipole components and quadrupole coupling constants for tetra-atomic molecules*

Molecule	*Dipole*/D	*Nucleus*	$eQ(q)_{aa}$/MHz	$eQ(q)_{bb}$/MHz	*Ref.*
HNCO	$\mu_a = 1.575(5)$	^{14}N	2.0527(10)	−0.473(7)	58
	$\mu_b = 1.35(10)$	^{2}H	0.0536(2)	—	
$^{81}Br_2CO$	—	^{81}Br	248.6(18)	−62.3(15)	64
ClFCO	—	^{35}Cl	−73.04(2)	44.71(2)	65
$NH_2D(\nu_2)$	$\mu_c = 1.09(4)$	^{14}N	2.52(24)	1.46(36)	68
$HF^{35}ClF$	$\mu_a = 2.313(1)$	^{35}Cl	−144.746(25)	—	67
$DF^{35}ClF$	$\mu_a = 2.295(1)$	^{35}Cl	−144.548(25)	—	67

has been determined to be r(N—H) = 0.986, r(N—C) = 1.209, r(C—O) = 1.166 Å, ∠HNC = 128.0°, and ∠NCO = 180°. Anomalies in earlier microwave

49 B. A. Andreev, A. V. Burenin, E. N. Karyakin, A. F. Krupnov, and S. M. Shapin, *J. Mol. Spectroscopy*, 1976, **62**, 125.
50 K. Tanaka and E. Hirota, *J. Mol. Spectroscopy*, 1976, **59**, 286.
51 R. A. Creswell, S. R. Brown, and R. H. Schwendeman, *J. Chem. Phys.*, 1976, **64**, 1820.
52 T. Kasuga, T. Amano, and T. Shimizu, *Chem. Phys. Letters*, 1976, **42**, 278.
53 W. E. Hoke, D. R. Bauer, J. Ekkers, and W. H. Flygare, *J. Chem. Phys.*, 1976, **64**, 5276.
54 L. J. Retallack and R. M. Lees, *J. Chem. Phys.*, 1976, **65**, 3793.
55 H. Mäder, W. Schrepp, and H. Dreizler, *Z. Naturforsch.*, 1976, **31a**, 1419.
56 J. G. Smith, *J.C.S. Faraday II*, 1976, **72**, 2298.
57 A. G. Maki and S. M. Freund, *J. Mol. Spectroscopy*, 1976, **62**, 90.
58 G. Winnewisser, W. H. Hocking, and M. C. L. Gerry, *J. Phys. Chem. Ref. Data*, 1976, **5**, 79.

data of HCNO have been interpreted[59] in terms of gross anharmonicity in the two-dimensional C—H bending vibration. Data for HCNO and DCNO have been treated in terms of a single potential surface with a minimum at a slightly bent HCN configuration (~170°) with a small maximum to linearity of ~2 cm^{-1}. This model successfully accounts for the anomalous pattern of vibration–rotation constants, and leads to a more normal value of 1.060 Å for the C—H bond length.

The first microwave study of $AsBr_3$ has been reported.[60] Data for the symmetric-top species $As^{79}Br_3$ and $As^{81}Br_3$ gave an r_z structure of r(As—Br) = 2.324(3) Å and ∠BrAsBr = 99.8(2)°, in excellent agreement with electron diffraction.

Work on $SbCl_3$ has been extended[61] to the millimetre-wave range for six isotopic molecules. An r_s structure was determined as r(Sb—Cl) = 2.3217(16) Å and ∠ClSbCl = 97.19(13)°. Centrifugal distortion data, mean amplitudes from electron diffraction, plus vibrational frequencies have been combined to determine a unique harmonic force field.

Ambiguity in the in-plane force field of NO_2F has been resolved[62] by *ab initio* Hartree–Fock calculations. Good agreement was obtained with a new experimental force field refined by the inclusion of isotopic shifts of the vibrational fundamentals.

The microwave spectrum of $^{37}ClFC^{34}S$ has been measured, allowing structural refinement from eight rotational constants.[63] An r_z structure was determined to be r(C—S) = 1.5957(2), r(C—Cl) = 1.7188(2), r(C—F) = 1.3265(2) Å, ∠SCCl = 127.00(1)°, and ∠SCF = 123.99(2)°; this confirms the planarity and, apart from the C—F bond, agrees reasonably well with the $r_\alpha{}^0$ electron-diffraction structure. An accurate structure has also been reported[64] for Br_2CO from spectra analysed for ^{79}Br and ^{81}Br isotopes in the 25—165 GHz range. The following r_0 structure was obtained: r(C—O) = 1.172(3), r(C—Br) = 1.917(1) Å, and ∠BrCBr = 112.3(1)°. Quadrupole coupling constants at bromine (except for χ_{ab}) have been obtained from hyperfine splittings (see Table 3), while centrifugal-distortion data have been combined with the vibrational frequencies to give a harmonic force field. Transformation of the coupling constants into the bond-axis system indicated appreciable π-bonding in the C—Br bond, leading to a valence-bond picture which is not fully supported by either the structure or the valence force constants.

The rotational Zeeman effect of ClFCO has been measured.[65] The chlorine nuclear g_I factor, the molecular g factor, and the magnetic susceptibility anisotropies have been determined. The bulk susceptibility has been calculated to be −57.3(67) erg G^{-2} mol^{-1}. Rotational constants and quadrupole coupling constants at chlorine have been refined (see Table 3). The molecular quadrupole

59 J. A. Duckett, A. G. Robiette, and I. M. Mills, *J. Mol. Spectroscopy*, 1976, **62**, 19.
60 A. G. Robiette, *J. Mol. Structure*, 1976, **35**, 81.
61 G. Cazzoli and W. Caminati, *J. Mol. Spectroscopy*, 1976, **62**, 1.
62 K. Molt, W. Sawodny, P. Pulay, and G. Fogarasi, *Mol. Phys.*, 1976, **32**, 169.
63 R. Hamm, H. J. Kohrmann, H. Günther, and W. Zeil, *Z. Naturforsch.*, 1976, **31a**, 594.
64 J. H. Carpenter, J. G. Smith, I. Thompson, and D. H. Whiffen, *J.C.S. Faraday II*, 1977, **73**, 384.
65 F. Scappini and A. Guarnieri, *Z. Naturforsch.*, 1976, **31a**, 369.

moments have been determined with rather large error; the out-of-plane moment Q_{cc} has a high positive value [19.4(73) $\times 10^{-26}$ e.s.u. cm^2], similar in trend to F_2CO and FHCO. A further paper, on F_2CO, refers to pressure-broadening studies.[66]

A second paper[67] on the m.b.e.r. study of the HFClF complex gives more details of the structure and bonding (see Table 3). The atomic arrangement is HFClF, with the three heavy atoms collinear, and the proton off axis by 55°. The van der Waals bond length is $r(\mathrm{F}\cdots\mathrm{Cl}) = 2.76$ Å. The structure and bonding are strikingly similar to those of the hydrogen-bonded HF dimer, and both complexes are described in simple M.O. terms as Lewis acid–base structures.

The first direct observations of ^{14}N quadrupole splittings in the optical region have been observed,[68] using Lamb dip spectroscopy on the $v_2 = 0 \rightarrow 1$ vibrational band of NH_2D. Stark-effect tuning brought molecular resonances ($4_{04} \rightarrow 5_{14}$, $4_{04} \rightarrow 5_{05}$) into coincidence with the CO_2 laser lines at 10.53 and 10.59 μm, respectively. Splittings of less than 1 MHz were resolved to give nuclear quadrupole coupling constants for the excited vibrational state, which show that the threefold symmetry of electronic charge is broken in the excited state, due to differences in the H,D amplitudes. The dipole moment (see Table 3) is also much smaller than that of the ground state, similar to the situation in NH_3.

The ν_2 band of NH_3 (and $^{15}NH_3$) has also been studied under high resolution by i.r.–microwave two-photon spectroscopy.[69] The tunable microwave frequency was added to or subtracted from i.r. radiation obtained from several lines of CO_2 or N_2O lasers. The waveguide cell is either outside the laser cavity, for straightforward spectroscopy, or inside, to allow two-photon Lamb dips to be observed. New approaches to non-rigid vibration–rotation theory will be necessary to analyse the accurate data given for ammonia. Such an approach has been reported by Špirko *et al.*[70] They use a non-rigid bender Hamiltonian to treat centrifugal distortion and Coriolis interactions in the ground and excited vibration–inversion states of ammonia. A harmonic force field and inversion potential function for (NH_3, $^{15}NH_3$) and for (ND_3, NT_3) were determined.

Kano *et al.*[71] have used the coincidence of the P(13) laser line of N_2O with the $\nu_2 a^qQ(8,7)$ line of NH_3 to study energy transfer by i.r.–microwave double resonance. Microwave inversion lines of both $^{15}NH_3$ and NH_3, in mixtures of the two gases, showed double-resonance signals, even though $^{15}NH_3$ is not pumped by the laser. Thus earlier results, which seemed to contradict collisional selection rules forbidding ortho–para conversion, are now shown to arise through energy transfer to collision partners, and not by collision-induced transitions. Amano has used a similar i.r.–microwave double-resonance system, but in reverse, to measure T_1 for NH_3.[72] Transient effects on the $\nu_2 a^qQ(8,7)$ transition have been observed due to rapid passages applied to the $J = 8$, $K = 7$ microwave transition. The value of T_1 is found to be much larger than T_2, in strange

[66] G. P. Srivastava and D. Kumar, *J. Phys.* (*B*), 1976, **9**, 651.
[67] S. E. Novick, K. C. Janda, and W. Klemperer, *J. Chem. Phys.*, 1976, **65**, 5115.
[68] E. W. Van Stryland and R. L. Shoemaker, *J. Chem. Phys.*, 1976, **64**, 4968.
[69] S. M. Freund and T. Oka, *Phys. Rev.* (*A*), 1976, **13**, 2178.
[70] V. Špirko, J. M. R. Stone, and D. Papoušek, *J. Mol. Spectroscopy*, 1976, **60**, 159.
[71] S. Kano, T. Amano, and T. Shimizu, *J. Chem. Phys.*, 1976, **64**, 4711.
[72] T. Amano, *Chem. Phys. Letters*, 1976, **37**, 585.

contrast to the transient experiments of Flygare and co-workers[53] and a very recent paper by Amano and Schwendeman,[73] where T_2/T_1 for a range of transitions in $^{15}NH_3$ lies between 1.1 and 1.6.

Two other papers have been published on rotational energy transfer in NH_3. McMahon[74] has included the effect of orientating collisions of dipolar molecules to account for the inversion bandshape of NH_3 and ND_3. Srivastava and Nahar[75] have studied the collision-induced shift of an inversion transition in NH_3 by microwave–microwave double resonance.

5 Inorganic and Organometallic Molecules containing Five or More Atoms

Three reports on carbaboranes have appeared this year, including a fuller version[76] of the microwave determination of CB_5H_7 reported two years ago. Accurate skeletal parameters have been determined[77] for 1,6-$C_2B_7H_9$, the largest carbaborane studied in detail by microwave spectroscopy. The structure is a tricapped trigonal prism, with B—B bond lengths correlating well with the simple inductive rule. The dipole moment was determined to be 2.14(17) D. Beaudet and co-workers have also determined[78] accurate B—B bond lengths for 1,7-$C_2B_6H_8$, a dodecahedron structure distorted towards a square antiprism. The determination supports earlier *X*-ray studies, but there is evidence for compression of molecular parameters in the crystal. The dipole moment was determined to be 1.40(1) D.

The microwave spectrum of OC_3S, possibly a new chemical species, has been detected.[79] The dipole moment was found to be 0.634 D. A preliminary study of the three degenerate bending modes shows it to be a well-behaved linear molecule, unlike carbon suboxide, C_3O_2. The latter is quasi-linear, and in a recent paper[80] vibration–rotation data have been successfully interpreted in terms of a strongly anharmonic potential for the lowest degenerate vibration, ν_7, with the same effective Hamiltonian used for HCNO and SiH_3NCO. The bending potential for ν_7 in C_3O_2 has a minimum at a bending angle of *ca.* 24° at the central carbon, with a maximum of *ca.* 23 cm^{-1} at the linear configuration. In the microwave spectrum of SiH_3NCO, vibration–rotation constants assigned[81] out as far as $v_{10} = 3$ (the lowest bending mode) show remarkable variation with quantum number. These variations have been explained in terms of an anharmonic potential surface with a minimum at a bent SiNC configuration (158°) and a maximum of 20 cm^{-1}, at the linear configuration, close to the ground-vibrational level. This analysis was aided considerably by an earlier electron-diffraction study. Substitution of S for O in both these species appears to restore normal linear behaviour. The planar asymmetric rotor NCNCO also displays evidence

[73] T. Amano and R. H. Schwendeman, *J. Chem. Phys.*, 1976, **65**, 5133.
[74] D. R. A. McMahon, *Phys. Rev. (A)*, 1976, **14**, 768.
[75] G. P. Srivastava and N. R. M. Nahar, *Phys. Letters*, 1976, **58A**, 387.
[76] G. L. McKown, B. P. Don, R. A. Beaudet, P. J. Vergamini, and L. H. Jones, *J. Amer. Chem. Soc.*, 1976, **98**, 6909.
[77] K. K. Lau and R. A. Beaudet, *Inorg. Chem.*, 1976, **15**, 1059.
[78] H. N. Rogers, K. K. Lau, and R. A. Beaudet, *Inorg. Chem.*, 1976, **15**, 1775.
[79] M. Winnewisser and J. J. Christiansen, *Chem. Phys. Letters*, 1976, **37**, 270.
[80] J. A. Duckett, I. M. Mills, and A. G. Robiette, *J. Mol. Spectroscopy*, 1976, **63**, 249.
[81] J. A. Duckett, A. G. Robiette, and I. M. Mills, *J. Mol. Spectroscopy*, 1976, **62**, 34.

of quasi-linearity in its microwave spectrum.[82] The molecule is bent, with $\angle$CNC $\simeq$ 140°, but there is a moderately low barrier to linearity ($\sim$500 cm^{-1}), as suggested by the behaviour of the rotational constants for the lowest vibrational mode. The dipole moment was determined to be 2.533(11) D; large, on account of the cyanide group.

A linear hydrogen-bonded complex, HCN···HF, has been detected by microwave spectroscopy.[83] A preliminary value of r(N···F) = 2.796 Å was obtained that was quite close to *ab initio* predictions. The dipole moment was measured as 5.59(2) D, an enhancement of 0.78 D over the sum of monomer values. The monomer dipole moments and the enhancement are over-estimated by the *ab initio* calculations, but, as shown for recent calculations on the HF dimer,[84] the general features of the molecular data are well represented.

Microwave studies of two amino-compounds have been reported. Rotational transitions of cyanamide, NH_2CN, in the ground and the first excited inversion state have been extended to 120 GHz,[85] principally to confirm identification of cyanamide signals from Sagittarius B2. Several of the transitions were perturbed, due to Coriolis resonance between these two states, which, from the analysis, were determined to be 49.2(1) cm^{-1} apart. Spectra of the three species of [1,2-$^2H_2$] hydrazine have been measured[86] up to 105 GHz. Effects due to inversion were found to be small. The structure was determined to be r(N—H) = 1.008(8), r(N—N) = 1.447(5) Å, $\angle$HNH = 113.3(30), and $\angle$NNH = 109.2(8)°, with the dihedral angle = 88.9(15)°. An average dipole moment was determined to be 1.90(8) D.

The microwave spectrum of chlorine nitrate, $ClONO_2$, has been reported[87] in the ground and first excited vibrational states. The molecule is shown to be planar, with dipole components μ_a = 0.72(7) and μ_b = 0.28(2) D. Quadrupole coupling constants for ^{35}Cl were determined to be $\chi_{aa} = -83.9(3)$ and χ_{bb} = 18.8(10) MHz. The chemical reactivity of $ClONO_2$ was discussed in terms of its suggested role in stratospheric ozone chemistry.

The elusive microwave spectrum of CF_3NO has now been measured[88] for three isotopic species. The dipole components were found to be even smaller than expected, μ_a = 0.145(1) and μ_b = 0.10(1) D. The barrier to internal rotation of the CF_3 top was obtained as V_3 = 770(50) cal mol^{-1}. The structure was found to have a long C—N bond, in line with other structural studies. For a series of NO compounds, Mason has demonstrated[89] a correlation between ^{14}N quadrupole coupling constants and the product of the n.m.r. shielding constant and an excitation energy that is approximated by the $n \rightarrow \pi^*$ u.v.–visible band.

Vibrationally excited states of CF_3Br have been studied by a combination of conventional microwave and i.r.–microwave double resonance.[90] Coincidences

82 W. H. Hocking and M. C. L. Gerry, *J. Mol. Spectroscopy*, 1976, **59**, 338.
83 A. C. Legon, D. J. Millen, and S. C. Rogers, *Chem. Phys. Letters*, 1976, **41**, 137.
84 L. A. Curtiss and J. A. Pople, *J. Mol. Spectroscopy*, 1976, **61**, 1.
85 D. R. Johnson, R. D. Suenram, and W. J. Lafferty, *Astrophys. J.*, 1976, **208**, 245.
86 S. Tsunekawa, *J. Phys. Soc. Japan*, 1976, **41**, 2077.
87 R. D. Suenram, D. R. Johnson, L. C. Glasgow, and P. Z. Meakin, *Geophys. Res. Letters*, 1976, **3**, 611, 758; R. D. Suenram and D. R. Johnson, *J. Mol. Spectroscopy*, 1977, **65**, 239.
88 P. H. Turner and A. P. Cox, *Chem. Phys. Letters*, 1976, **39**, 585.
89 J. Mason, *J.C.S. Faraday II*, 1976, **72**, 2064.
90 H. Jones, F. Kohler, and H. D. Rudolph, *J. Mol. Spectroscopy*, 1976, **63**, 205.

of two lines of the CO_2 laser with transitions of the $\nu_6 \rightarrow (\nu_6 + \nu_1)$ band allowed pumping of microwave vibrational satellites. The i.r. band centre was determined to be 1083.530(1) cm^{-1}, and vibration–rotation constants were obtained for ν_6, $(\nu_6 + \nu_1)$, ν_1, and ν_3.

Four papers have appeared on rotational properties in tetrahedral molecules. The rotational magnetic-moment spectra of CH_4, SiH_4, GeH_4, and the corresponding fluorides have been studied [91] by molecular beam magnetic resonance. The rotational g factor and average spin–rotation constant, C_a, for H or F were determined for each molecule, signs being determined (apart from those for the C_a values of the fluorides). These constants are of chemical interest because they provide tests for *ab initio* calculations from electronic wavefunctions and because of the importance of their interactions in the role of nuclear spin relaxation. The centrifugal distortion dipole moment of SiH_4 in the ground vibronic state has been measured [92] from the Stark effect of a previously assigned [93] Q-branch transition, the $E(2) \rightarrow E(3)$ of $J = 18$. A value of $\mu_Q(18) = 3.34(4) \times 10^{-5}$ D was obtained, in agreement with earlier work on spectral intensities of R-branch lines. A further paper [94] gives a new formulation of the vibration–rotation Hamiltonian for the F fundamentals of XY_4 molecules.

Three papers have appeared on the silyl halides. The microwave spectra of SiH_3F and SiD_3F for the ground, v_2, v_3, v_5, and $v_6 = 1$ excited states have been measured.[95] X–Y Coriolis resonance between excited states of ν_2 and ν_5, separated by 29 and 8 cm^{-1} for the two isotopes, has been analysed in detail. Molecular constants agree well with those from an earlier i.r. analysis. The value of the dipole moment of SiH_3F has been revised to $\mu = 1.298(4)$ D. Microwave data [96] for SiH_2DF and $SiHD_2F$ in the ground vibrational state have been combined with earlier data to give an r_z structure, r(Si—H) = 1.480(2), r(Si—F) = 1.596(2) Å, and ∠HSiF = 108.4(1)°. A general harmonic force field was determined. Three isotopic species of SiH_2Cl_2 have been measured,[97] yielding the 'substitution' parameters r(Si—Cl) = 2.033(3), r(Si—H) = 1.480(15) Å, ∠ClSiCl = 109.7(3), and ∠HSiH = 111.3(7)°. The dipole moment has been measured as 1.13(2) D, and the quadrupole coupling constants for $SiH_2{}^{35}Cl_2$ have been found to be $\chi_{aa} = -21.13(35)$ and $\chi_{cc} = 21.00(23)$ MHz. The coupling constants show little asymmetry about the Si—Cl bond; remarkable in view of the suspected $(p\text{-}d)\pi$ bonding. A systematic shortening of the Si—Cl bond is found with progressive Cl substitution.

Two organosilicon compounds have been studied. The V_6 internal barrier of [α-$^2H_3$]phenylsilane has been determined [98] as 15.93 cal mol^{-1}, 11% less than an earlier result for the SiH_3 top. The microwave spectra of the *cis*- and skew isomers of methylvinylsilane have been measured.[99] The V_3 barriers have been obtained from A, E splittings to be 1800(20) and 1585(15) cal mol^{-1}, respectively.

[91] I. Ozier, S. S. Lee, and N. F. Ramsey, *J. Chem. Phys.*, 1976, **65**, 3985.
[92] R. H. Kagann, I. Ozier, and M. C. L. Gerry, *J. Chem. Phys.*, 1976, **64**, 3487.
[93] I. Ozier, R. M. Lees, and M. C. L. Gerry, *Canad. J. Phys.*, 1976, **54**, 1094.
[94] A. G. Robiette, D. L. Gray, and F. W. Birss, *Mol. Phys.*, 1976, **32**, 1591.
[95] C. Georghiou, J. G. Baker, and S. R. Jones, *J. Mol. Spectroscopy*, 1976, **63**, 89.
[96] A. G. Robiette, C. Georghiou, and J. G. Baker, *J. Mol. Spectroscopy*, 1976, **63**, 391.
[97] R. W. Davis and M. C. L. Gerry, *J. Mol. Spectroscopy*, 1976, **60**, 117.
[98] W. Caminati and G. Cazzoli, *Chem. Phys. Letters*, 1976, **38**, 218.
[99] M. Hayashi and M. Imachi, *Chem. Letters*, 1976, 121.

The dipole moments were determined to be 0.727(8) D for the *cis*-form and 0.68 D for the skew isomer. These properties, and likely structural differences between the isomers, appear similar to the isomers of but-1-ene.

The microwave spectrum of CH_3GeH_2F has been studied by two groups.[100, 101] Both determined the barrier and dipole moment, which may be quoted as V_3 = 940(20) cal mol^{-1} and μ = 2.53(15) D. In addition, Roberts *et al.* determined accurate structural parameters from twelve isotopic species, r(C—Ge) = 1.925(2), r(Ge—F) = 1.751(5), r(C—H) = 1.094(5) Å, ∠HCH = 108.9(3), and ∠FGeC = 106.3(3)°. The CH_3 group is tilted (1.9°) towards the Ge—F bond. The shortening of r(C—Ge) and the inclination of the dipole moment away from the Ge—F bond (23°) was taken as some indication of $(p\text{–}d)\pi$ bonding in the C—Ge bond. The internal barrier is 300 cal mol^{-1} lower than that in CH_3GeH_3, similar to the silicon case, but different from the carbon series. A fuller version[102] of the microwave study of $ClCH_2GeH_3$ has been published. The barrier to internal rotation of the GeH_3 group is 1740(30) cal mol^{-1}, now reported from *A*, *E* splittings, 500 cal mol^{-1} higher than that in CH_3GeH_3. The structural and quadrupole data are the same as those reported last year.

Several microwave studies of phosphorus compounds have appeared. Ground-state rotational spectra of OPF_3 and $^{18}OPF_3$ have been recorded[103] up to the millimetre-wave region, and lines for all excited vibrational states below 1000 cm^{-1} of the main species located. A detailed analysis is given for v_5 = 1 and v_6 = 1. For the latter, a value of the *A* rotational constant, rarely determined, has been obtained from an accidental near degeneracy. The three rotational constants gave the following structure: r_0(P—F) = 1.523, r_0(P—O) = 1.436 Å, and ∠FPF = 101.3°, in excellent agreement with electron diffraction. An improved harmonic force field was calculated. A similar study has been made of SPF_3.[104] Coriolis constants for the v_5 = 1, v_6 = 1 states, centrifugal distortion data, and vibrational isotopic shift data (from i.r. matrix studies) have been used to determine the most general harmonic force field. The structure has been determined, with an assumed FF distance, as r(P—F) = 1.538(10), r(P—S) = 1.849(3) Å, and ∠FPF = 98.5(20)°, in fair agreement with a more recent electron-diffraction study.

The microwave, i.r., and Raman spectra of CH_3POF_2 have been recorded.[105] Excited-state splittings in the microwave spectrum gave a barrier of 3.58(4) kcal mol^{-1}, consistent with torsional frequencies observed in the vibrational studies. The dipole moment was determined as 3.62(20) D. Approximate structural parameters and a greater barrier, relative to CH_3PF_2 (2.30 kcal mol^{-1}), were discussed in terms of co-ordination of the lone pair of electrons.

The microwave spectra of ethylphosphine and some deuteriated forms have been recorded.[106] The *gauche*- and *trans*-conformers were identified, with the

[100] L. C. Krisher and J. A. Morrison, *J. Chem. Phys.*, 1976, **64**, 3556.
[101] R. F. Roberts, R. Varma, and J. F. Nelson, *J. Chem. Phys.*, 1976, **64**, 5035.
[102] J. Nakagawa and M. Hayashi, *Bull. Chem. Soc. Japan*, 1976, **49**, 3441.
[103] J. G. Smith, *Mol. Phys.*, 1976, **32**, 621.
[104] J. G. Smith and I. Thompson, *Mol. Phys.*, 1976, **32**, 1247.
[105] J. R. Durig, K. S. Kalasinsky, and V. F. Kalasinsky, *J. Mol. Structure*, 1976, **34**, 9.
[106] J. R. Durig and A. W. Cox, jun., *J. Chem. Phys.*, 1976, **64**, 1930.

trans being more stable by 200(100) cm^{-1}. The dipole moments were measured as 1.22(7) and 1.226(5) D, respectively. From limited isotopic data, the following structural parameters could be estimated: $\angle CCP = 110.1(2)°$ and $r(C—P) = 1.880$ Å; $\angle CCP = 115.2(1)°$ and $r(C—P) = 1.876$ Å for the two forms, respectively. The microwave and Raman spectra of SiH_3PD_2 have been studied.[107] From *A*, *E* splittings, the barrier has been determined to be 1512(26) cal mol^{-1}, and the dipole moment obtained was 0.60(1) D. Moments of inertia were combined with earlier microwave data to derive structural parameters that imply little contribution of 3*d* orbitals of silicon to the Si—P bond. The barrier is 0.45 kcal mol^{-1} smaller than that in methylphosphine.

The microwave spectra of several unstable phospha-alkenes (CF_2=PH, CH_2=PCl, and CH_2=PH) have been detected;[108] the compounds are formed by pyrolysis of CF_3PH_2, CH_3PCl_2, and $(CH_3)_2PH$, respectively. These are rare examples involving $(p–p)\pi$ multiple bonding to phosphorus. Preliminary structure calculations give $r(C=P) = 1.67$ Å for all three molecules, intermediate between $r(C—P) = 1.848$ Å in $(CH_3)_2PH$ and 1.5421 Å for $r(C\equiv P)$ in HCP. In a second paper by Hopkinson *et al.*, the phosphorus analogue of CH_3CN, *viz.* CH_3CP, has been identified by microwave spectroscopy[109] in the pyrolysis products of $CH_3CH_2PCl_2$. CH_3CP is reasonably stable, having a half-life of up to several hours at low gas pressures.

Millimetre-wave measurements on ClF_5 by Jurek and co-workers[110] have led to the revision of their earlier values of D_{JK} and *eqQ*. Force constants have been calculated using the centrifugal distortion information, including R_6. A more recent paper by Bodenseh *et al.*[111] confirms the revised constants, giving $eqQ(^{35}ClF_5) = 43.93(8)$ MHz. The dipole moment was determined to be 0.536(10) D. The determinable structural parameter $\angle F_{ax}ClF_{eq}$ was obtained as 86.5(10)° from both studies, similar to BrF_5. Microwave studies[112] of BrF_5, the first molecule for which C_{4v}, *K*-type doubling was observed, have been extended to excited vibrational states, where the first observation of *l*-doubling of $Kl = -1$ states has been made. This effect, predicted by de Heer in 1951, has been measured for $v_9(E) = 1$, the analysis of which is complicated by strong Coriolis resonance with $v_5(B_1) = 1$, the other state studied. Preliminary vibration–rotation constants have been derived. Vibrational satellites in the microwave spectrum of the C_{2v} molecule SF_4 have been studied[113] for the two lowest bending modes ν_4 and ν_9. Relative intensities and Coriolis considerations were used to distinguish between two proposed assignments from vibrational spectra.

Electron-diffraction and microwave data have been combined to determine the structures of two hexaco-ordinated compounds of Main-group elements, IOF_5 and SF_5Cl. For IOF_5, the derived structural parameters[114] were $r_g(I—O) = 1.715(4)$, $r_g(I—F_{eq}) = 1.817(2)$, $r_g(I—F_{ax}) = 1.863(4)$ Å, and $\angle OIF_{eq} = 98.0(3)°$.

[107] J. R. Durig, Y. S. Li, M. M. Chen, and J. D. Odom, *J. Mol. Spectroscopy*, 1976, **59**, 74.
[108] M. J. Hopkinson, H. W. Kroto, J. F. Nixon, and N. P. C. Simmons, *J.C.S. Chem. Comm.*, 1976, 513.
[109] M. J. Hopkinson, H. W. Kroto, J. F. Nixon, and N. P. C. Simmons, *Chem. Phys. Letters*, 1976, **42**, 460.
[110] P. Goulet, R. Jurek, and J. Chanussot, *J. Phys.*, 1976, **37**, 495.
[111] H. K. Bodenseh, W. Hüttner, and P. Nowicki, *Z. Naturforsch.*, 1976, **31a**, 1638.
[112] P. N. Brier, S. R. Jones, and J. G. Baker, *J. Mol. Spectroscopy*, 1976, **60**, 18.
[113] H. Inoue, A. Naruse, and E. Hirota, *Bull. Chem. Soc. Japan*, 1976, **49**, 1260.
[114] L. S. Bartell, F. B. Clippard, and E. J. Jacob, *Inorg. Chem.*, 1976, **15**, 3009.

The axial I—F bond is longer than the equatorial bond, contrary to the primary deformation expected from Gillespie–Nyholm ideas. Arguments have therefore been proposed for a secondary relaxation effect, reversing the difference in axial and equatorial bond lengths. The S—F axial bond length was also found to be the larger in SF_5Cl, also indicating a secondary effect, but following 'steric' deformation.[115] The r_z structure was found to be r(S—Cl) = 2.045(3), $r(S—F_{eq})$ = 1.564(3), $r(S—F_{ax})$ = 1.586(9) Å, and $\angle ClSF_{eq}$ = 90.7(2)°. The inclusion of microwave data was necessary to establish the difference between axial–equatorial bond lengths for both these molecules.

Microwave studies of two organoselenium compounds have been reported. The ^{13}C species of tetrahydroselenophen have been investigated,[116] allowing a substitution ring structure to be derived; r(C-2—Se) = 1.963, r(C-2—C-3) = 1.549, r(C-3—C-4) = 1.527 Å, ∠C-5SeC-2 = 90.7, ∠SeC-2C-3 = 105.0, and ∠C-2C-3C-4 = 106.9°. The molecule has C_2 symmetry, with an angle of twist = 29.7°. The two-top molecule $(CD_3)_2Se$ has been examined in the ground[117] and first excited[118] torsional states. From the data for the excited state, the value of V_3 obtained was 1493(9), with the top–top coupling parameter V_{12}' = 28.4(3) cal mol^{-1}; the value of the angle between the top-axis and the molecular symmetry axis that was derived was 50.0(1)°.

The microwave spectrum of MnO_3F has been remeasured.[119] Very early vibrational-state data have been revised, and consistent data are now available for the ν_3, ν_5, and ν_6 modes.

The author wishes to thank Dr. Y. Kawashima for his assistance with the literature search.

[115] C. J. Marsden and L. S. Bartell, *Inorg. Chem.*, 1976, **15**, 3004.
[116] A. H. Mamleev, N. M. Pozdeev, and N. N. Magdesieva, *J. Mol. Structure*, 1976, **33**, 211.
[117] G. K. Pandey and H. Dreizler, *Z. Naturforsch.*, 1976, **31a**, 357.
[118] G. K. Pandey, H. Lutz, and H. Dreizler, *Z. Naturforsch.*, 1976, **31a**, 1413.
[119] J. Høg and T. Pedersen, *J. Mol. Spectroscopy*, 1976, **61**, 243.

4

Vibrational Spectra of Small Symmetric Species and of Single Crystals

BY D. M. ADAMS, P. N. GATES, AND P. GANS

1 General Introduction

By D. M. Adams

Spectroscopically speaking, 1976 was a year of steady advance rather than notable progress. With still tighter controls on publication space, it is becoming necessary to make unpalatable decisions as to the areas that are truly relevant to the four 'vibrational' Chapters of this Report. We have attempted to maintain the archival value of the Report by listing all appropriate publications, but it has become increasingly necessary to omit mention of many papers that list routine i.r. or Raman data for characterization only.

The core of the Report remains that material directly relevant to the needs of inorganic chemists, the emphasis being upon collection of well-based assignments. In this context we note and welcome the continued increase in the number of assignments established by single-crystal methods (Part 3 of this Chapter). Widespread use of metal and other isotopic substitution continues, often giving uniquely valuable data. It must, however, be strenuously affirmed that the metal-isotope technique is complementary to methods that admit of the determination of symmetry species, and never a substitute. Information from resonance Raman work, and from the new but potentially sizeable field of spectroscopy at high pressures, has been omitted or treated as of marginal relevance unless new species have been characterized or assignments of more general value established.

Some particularly welcome books have appeared this year. Maslowsky's 'Vibrational Spectra of Organometallic Compounds' will fulfil a real need if it really does cover the field of its title in depth.[1] Mineral spectroscopy now has a valuable source book [2] (unaccountably omitted last year), although it is uneven in depth of treatment. Several works of a theoretical nature are welcome, chiefly by virtue of the stature of their authors;[3–8] as are two compendia,[9, 10] and an

[1] E. Maslowsky, jun., 'Vibrational Spectra of Organometallic Compounds', Wiley, New York, 1976.

[2] 'The Infrared Spectra of Minerals', ed. V. C. Farmer, The Mineralogical Society, London, 1974.

[3] S. Califano, 'Vibrational States', Wiley, Chichester, 1976.

[4] C. J. H. Schutte, 'The Theory of Molecular Spectroscopy, Vol. 1: The Quantum Mechanics and Group Theory of Vibrating and Rotating Molecules', Elsevier, Amsterdam, 1976.

[5] H. Poulet and J. P. Mathieu (transl. A. Simievic), 'Vibration Spectra and Symmetry of Crystals', Gordon and Breach, New York, 1976.

[6] M. M. Suschinskii, 'Ramanspektren von Molekülen und Kristallen', Heyden, Rheine, 1976.

outstanding volume in Durig's series.[11] Proceedings of the XIIth European Congress on Molecular Spectroscopy are available.[12]

Fourier transform spectroscopy,[13–15] laser-Raman spectroscopy of matrix-isolated species,[16, 17] and the Raman effect in crystals,[18, 19] vibrational properties of amorphous solids,[20] actinide chelate complexes,[21] the hydrogen bond in ice,[22] and anion-substituted apatites [23] have been reviewed. Two particularly valuable accounts have been given of the vibrational properties of metal ammine complexes,[24] and of the principles of NCA, with particular reference to the use of isotopic data.[25] A pantechnicon of Raman applications, many of inorganic interest, has been discussed;[26] and some impressive examples have been given of the power of the rotating-sample method in inorganic Raman spectroscopy.[27] An alternative to Edgell's 'hidden symmetry' story has been proposed.[28]

2 Spectra of Small Symmetric Species

By P. N. Gates

Diatomic Species.—As in previous years, matrix isolation and physical studies comprise the bulk of this Section. Absorptions at 4248, 3727, and 2973 cm^{-1} have been assigned to Q branches in the i.r. spectra of H_2, HD, and D_2, respectively, in Ar matrices.[29] Solid crystalline N_2 containing traces of CO_2 shows the ν(N—N) absorption at 2328 cm^{-1} and also its combinations with the CO_2 Fermi doublet (ν_1, $2\nu_2$).[30] The experimentally observed shift, splitting, and intensity ratio of the vibrational Raman band in solid α-N_2 have been reproduced in an *a priori* calculation based on quadrupole moments and polarizabilities of N_2 in the gas phase.[31] Detailed analyses of photographically recorded pure rotational

7 D. C. Harris and M. D. Bertolucci, 'Symmetry and Spectroscopy: An Introduction to Vibrational and Electronic Spectroscopy', Oxford University Press, 1976.
8 E. G. Brame, jun. and J. G. Grasselli, 'Infrared and Raman Spectroscopy. Part A', Dekker, Basel, 1976.
9 'Absorption Spectra in the Infrared Region. Vol. 2', ed. L. Lang *et al.*, Butterworth, Sevenoaks, 1976.
10 C. N. R. Rao, 'Bibliography on Infrared Spectroscopy Through 1960, Parts 1, 2, and 3', N.B.S. Special Publication 428, N.B.S., Washington, D.C., 1976.
11 'Vibrational Spectra and Structure. Vol. 5', ed. J. R. Durig, Elsevier, Amsterdam, 1976.
12 'Molecular Spectroscopy of Dense Phases', ed. M. Grosmann, S. G. Elkomoss, and J. Ringeissen, Elsevier, Amsterdam, 1976.
13 E. Knözinger, *Angew. Chem. Internat. Edn.*, 1976, **15**, 25.
14 J. B. Bates, *Science*, 1976, **191**, 31.
15 R. Geick, *Topics Current Chem.*, 1975, **58**, 73.
16 L. Andrews, *Appl. Spectroscopy Rev.*, 1976, **11**, 125.
17 M. Mozkovits and G. A. Ozin, *Vib. Spectra Struct.*, 1975, **4**, 187.
18 L. Brindus, *Studii Cercetari Fiz.*, 1976, **28**, 769.
19 M. Y. Valakh, *Kvantovaya Elektron.* (*Kiev*), 1975, **9**, 108.
20 R. J. Bell, *Methods Comput. Phys.*, 1976, **15**, 215.
21 U. Casellato, M. Vidali, and P. A. Vigato, *Inorg. Chim. Acta*, 1976, **18**, 77.
22 E. Whalley, *Hydrogen Bond*, 1976, **3**, 1425.
23 R. G. Knubovets and L. D. Kislovskii, *Tr. Inst. Geol. Geofiz., Akad. Nauk S.S.S.R., Sibirsk. Otdel.*, 1975, **50**, 63.
24 K. H. Schmidt and A. Müller, *Coordination Chem. Rev.*, 1976, **19**, 41.
25 N. Mohan, S. J. Cyvin, and A. Müller, *Coordination Chem. Rev.*, 1976, **21**, 221.
26 J. E. D. Davies, *J. Mol. Structure*, 1976, **31**, 217.
27 H. Homborg and W. Preetz, *Spectrochim. Acta*, 1976, **32A**, 709.
28 M. Bigorgne, *Spectrochim. Acta*, 1976, **32A**, 1365.
29 G. R. Smith, J. A. Warren, and W. A. Guillory, *J. Chem. Phys.*, 1976, **65**, 1591.
30 B. Nelander, *Chem. Phys. Letters*, 1976, **42**, 187.
31 G. Zumofen and K. Dressler, *J. Chem. Phys.*, 1976, **64**, 5198.

Raman spectra have enabled the calculation (some new) of precise molecular constants for the following species: $^{16}O_2$, $^{18}O_2$, and $^{16}O^{18}O$,[32] $^{35}Cl_2$ and $^{37}Cl_2$,[33] and $^{19}F_2$.[34] Molecular constants for $^{12}C^{16}O$, $^{13}C^{16}O$, $^{12}C^{18}O$, and $^{13}C^{18}O$ have been derived from i.r. absorption[35] and emission[36] studies. Emission intensities for overtone bands of HF and DF from a chemical laser source have been used to determine dipole-moment functions and i.r. transition probabilities for these species.[37] An analysis of the temperature-dependence of the band envelopes in the Raman spectrum of liquid HF suggests that a model involving all polymer HF species may be more realistic than previous models emphasizing the hexamer.[38] A detailed theoretical investigation of R.R.E. scattering from homonuclear diatomics was tested by scattering experiments made on I_2.[39] The influence of non-rigidity of interacting molecules on Raman and i.r. bandshapes has been analysed theoretically and applied to vibrational linewidths of HCl and CO trapped in rare-gas crystals.[40] An analysis of the spectra of CO in rare-gas matrices suggests that CO can be used as an efficient probe to study the dynamics of rare-gas lattices.[41]

Far-i.r. peak frequencies, half-bandwidths, intensities, and spectral moments have been correlated quantitatively with various other molecular properties for a series of binary rare-gas mixtures at total pressures up to 100 atmospheres.[42] Values of ω_e/cm^{-1} are:

He–Ar	He–Kr	He–Xe	Ne–Ar	Ne–Kr	Ne–Xe	Ar–Kr	Ar–Xe
50.9	47.5	42.9	46.7	42.5	38.1	46.5	42.4

Andrews has completed the $M^+X_2^-$ (X = F, Cl, Br, or I) series with the Raman spectra for $M^+Br_2^-$ in Ar matrices. The wavenumbers/cm^{-1} are:

M:	Li	Na	K	Rb	Cs
ν(Br—Br):	149	115	160	159	157

The anomalously low value for $Na^+Br_2^-$ was accounted for on the basis of interaction with the interionic Na^+—Br_2^- mode at 194 cm^{-1}.[43] UN has been identified by i.r. spectra in an Ar matrix. Three bands (1000.9, 995.5, and 991.9 cm^{-1}) are assigned as $U^{14}N$ on different sites, with a further triplet at 969.2, 963.9, and 960.5 cm^{-1} when ^{15}N is used, in excellent agreement with the calculated isotope shifts.[44] Matrix-isolated ^{14}NI and ^{15}NI have i.r. absorptions at 590 and 573 cm^{-1}, respectively.[45]

32 H. G. M. Edwards, E. A. M. Good, and D. A. Long, *J.C.S. Faraday II*, 1976, **72**, 865.
33 H. G. M. Edwards, E. A. M. Good, and D. A. Long, *J.C.S. Faraday II*, 1976, **72**, 927.
34 H. G. M. Edwards, E. A. M. Good, and D. A. Long, *J.C.S. Faraday II*, 1976, **72**, 984.
35 Da-Wun Chen, K. N. Rao, and R. S. McDowell, *J. Mol. Spectroscopy*, 1976, **61**, 71.
36 T. R. Todd, C. M. Clayton, W. B. Telfair, T. K. McCubbin, and J. Plíva, *J. Mol. Spectroscopy*, 1976, **62**, 201.
37 R. N. Sileo and T. A. Cool, *J. Chem. Phys.*, 1976, **65**, 117.
38 I. Sheft and A. J. Perkins, *J. Inorg. Nuclear Chem.*, 1976, **38**, 665.
39 D. L. Rousseau and P. F. Williams, *J. Chem. Phys.*, 1976, **64**, 3519.
40 D. Robert and L. Galatry, *J. Chem. Phys.*, 1976, **64**, 2721.
41 H. Dubost, *Chem. Phys.*, 1976, **12**, 139.
42 E. Bar-Ziv and S. Weiss, *J. Chem. Phys.*, 1976, **64**, 2412.
43 C. A. Wight, B. S. Ault, and L. Andrews, *Inorg. Chem.*, 1976, **15**, 2147.
44 D. W. Green and G. T. Reedy, *J. Chem. Phys.*, 1976, **65**, 2921.
45 R. Minkwitz and F. W. Froben, *Chem. Phys. Letters*, 1976, **39**, 473.

Factor-group analysis for β-Na_2S_2 predicts four Raman-active fundamentals, assigned as follows:[46]

451.4	A_{1g}	$\nu(^{32}S^{32}S)$
442.1	A_{1g}	$\nu(^{32}S^{34}S)$
134.5	E_{1g}	libration of two S_2^{2-} groups in opposite directions
64.5	E_{2g}	out-of-phase oscillation of S_2^{2-}
27.0	E_{2g}	out-of-phase oscillation of Na^+

Several overtones were observed in the resonance Raman emission (R.R.E.) spectra of the S_2^- ion in alkali-metal halide crystals at 6 K. Comparison of Raman and emission data in MI (M = Na, K, or Rb) are consistent with there being a single S_2^- site with Π_g splitting, and, in some alkali-metal halides, a time dependence of the polarized emission intensity was analysed in terms of the reorientation kinetics of the optically excited S_2^- species.[47] γ-Irradiation of $NaClO_3$ produced a metastable O_2 species [$\nu(^{16}O_2)$ 1544 cm^{-1}, compared to 1558 cm^{-1} in stable O_2], identified by R.R.E. of isotopic variants, and for which the progression 1544 (ν), 3058 (2ν), 4536 (3ν), and 5990 cm^{-1} (4ν) was observed.[48, 49] Identification of the species Se_n (n = 2, 3, 5, 6, 7, or 8) and the novel InI [ν(InI) = 175 cm^{-1}] in the vapour phase during the preparation of In_2Se_3 by chemical vapour transport has been achieved by Raman spectroscopy.[50]

Triatomic Species.—Apart from various studies of H_2O, the majority of papers in this Section are concerned with matrix-isolated species and high-resolution work. In a study of the action of microwave discharges in matrix-isolation experiments, it was concluded that vacuum-u.v. photolysis is the mechanism by which energy is transmitted from the discharge to the cold window for the production of new species in the Ar–HCl–Cl_2 system, rather than direct transport of ions from the tube to the window.[51] I.r. spectra of $M^+HBr_2^-$ in Ar matrices are consistent with a linear centrosymmetric HBr_2^- species. Typically, wavenumbers/cm^{-1} are:

	ν_3	$(\nu_1 + \nu_3)$	ν_1
Na^+	670	893	168

where ν_1 is calculated from ν_3 and $(\nu_1 + \nu_3)$. The ion-pairs M^+HClBr^- (formed either by reactions of the type MCl + HBr or MBr + HCl) show absorption (typically) for Na^+HClBr^- at 1135 (ν_3) and 1355 cm^{-1} (ν_2).[52] Observation of both ν_1 and ν_3 (at 253 and 340 cm^{-1}, respectively, in $Rb^+Cl_3^-$) in the i.r. spectra of $M^+Cl_3^-$ ion-pairs suggests an asymmetric (probably $C_{\infty v}$) structure for the Cl_3^- ion,[53] whereas mutual exclusion appears to operate in the CsF_3 species with

[46] G. J. Janz, J. R. Downey, E. Roduner, G. J. Wasilczyk, J. W. Coutts, and A. Eluard, *Inorg. Chem.*, 1976, **15**, 1759.
[47] C. A. Sawicki and D. B. Fitchen, *J. Chem. Phys.*, 1976, **65**, 4497.
[48] J. B. Bates and H. D. Stidham, *J. Chem. Phys.*, 1976, **65**, 3901.
[49] J. B. Bates and H. D. Stidham, *Chem. Phys. Letters*, 1976, **37**, 20.
[50] J. K. Grinberg, V. A. Boniakova, V. K. Chevel'kov, R. Hillel, and J. Bouix, *J. Inorg. Nuclear Chem.*, 1976, **38**, 383.
[51] C. A. Wight, B. S. Ault, and L. Andrews, *J. Chem. Phys.*, 1976, **65**, 1244.
[52] B. S. Ault and L. Andrews, *J. Chem. Phys.*, 1976, **64**, 1986.
[53] B. S. Ault and L. Andrews, *J. Chem. Phys.*, 1976, **64**, 4853.

a linear $D_{\infty h}$ geometry.[54] Wavenumbers/cm^{-1} are:

	ν_1	ν_3		ν_1	ν_3
$RbCl_3$	253	340;	CsF_3	461	550

ν_3 has been assigned for products for reactions of the type MX + Y_2 (X, Y = Cl, Br):[53]

	$K^+ClBrCl^-$	$K^+BrClBr^-$	$K^+BrClCl^-$	$K^+ClBrBr^-$
ν_3/cm^{-1}	345	311	282	273

A linear symmetric I_3^- species is consistent with a single i.r. band (at 135 cm^{-1}) for $[Co(en)_2X_2](I_3)$ (X = NO_2 or Cl).[55]

In a Raman and i.r. Ar matrix study of MgX_2 (X = F, Cl, Br, or I) the ν_1 mode in MgF_2 was assigned to the 550 cm^{-1} band in the Raman, disagreeing with an earlier i.r. assignment at 478 cm^{-1}, which had been taken as evidence for non-linearity. The present assignments, consistent with linear species, are:[56]

	ν_1	ν_2	ν_3/cm^{-1}
MgF_2	550.0	249.0	841.8
$MgCl_2$	326.5	93.0	600.8
$MgBr_2$	197.9	81.5	497.1
MgI_2	147.6	55.8	444.9

Raman spectra of $MgCl_2$ at 1010 K indicate the presence of discrete tetrahedral $[MgCl_4]^{2-}$ species in equilibrium with polynuclear species of unknown structure. The assignment is ν_1 (A_1) 244; ν_2 (E) 100; ν_3 (T_2) 335; and ν_4 (T_2) 140.[57] Raman and i.r. spectra of the recently reported[58a] $[NEt_4][InX_2]$ (X = Cl, Br, or I) clearly indicate bent monomeric InX_2^- species. Thus $InCl_2^-$ shows Raman bands at 328 (ν_1), 177 (ν_2), and 291 (ν_3); 330 and 291 cm^{-1} in i.r. Raman scatter occurs at 236 and 197 cm^{-1} for $InBr_2^-$ and at 186 and 140 cm^{-1} for InI_2^-.[58b] A detailed i.r. Ar matrix-isolation study of $GeCl_2$ has led to assignments of ν_1 and ν_3 for the fifteen possible isotopic variants involving $^{70, 72, 73, 74, 76}Ge$ and $^{35, 37}Cl$. Typical of these are:

	ν_1	ν_3/cm^{-1}
$^{74}Ge^{35}Cl_2$	398.6	373.5
$^{76}Ge^{35}Cl_2$	397.0	377.5
$^{74}Ge^{35}Cl^{37}Cl$	395.2	369.7

A calculation led to an estimated bond angle of 99 ± 4°.[59] Force constants and mean-square amplitudes of vibration have been calculated for $GeCl_2$, using gas-phase Raman data for ν_2 and Ar matrix i.r. data for ν_1 and ν_3 from the literature.[60] Raman and i.r. spectra of CdX_2 and CdXY (X, Y = Cl, Br, or I) in Kr matrices have been assigned for a variety of isotopic variants of Cl and Br.

54 B. S. Ault and L. Andrews, *J. Amer. Chem. Soc.*, 1976, **98**, 1591.
55 L. V. Konovalov and O. S. Konovalenko, *Russ. J. Inorg. Chem.*, 1976, **21**, 624.
56 M. L. Lesiecki and J. W. Nibler, *J. Chem. Phys.*, 1976, **64**, 871.
57 C. H. Huang and M. H. Brooker, *Chem. Phys. Letters*, 1976, **43**, 180.
58 (*a*) J. J. Habeeb and D. G. Tuck, *J.C.S. Chem. Comm.*, 1975, 600; (*b*) J. J. Habeeb and D. G. Tuck, *J.C.S. Dalton*, 1976, 866.
59 A. K. Mal'tsev, V. A. Svyatkin, and O. M. Nefedov, *Doklady Chem.*, 1976, **227**, 344.
60 J. Brunvoll and I. Hargittai, *Acta Chem. Scand.* (*A*), 1976, **30**, 230.

A few typical assignments to ν_1 (cm^{-1}) are: $Cd^{37}Cl_2$ (320.4); $Cd^{79}Br^{81}Br$ (209.1); $Cd^{37}Cl^{79}Br$ (392.8), and $Cd^{35}ClI$ (385.7). Anomalies in the calculated stretching force constants and isotope shifts could be removed on the assumption of slightly non-linear geometries,[61] and a similar conclusion could be reached from calculations based on the Kr matrix spectra for many of the isotopic variants of the systems HgX_2 and HgXY (X, Y = Cl, Br, or I).[62, 63] Assignments made to dimeric species were consistent with the generally accepted centrosymmetric (probably D_{2h}) geometry.[62] Vibrational spectra of HgX_2 (X, Y = Cl, Br, or I) in a variety of donor solvents revealed weak activity of the (formally inactive) ν_3 mode in the Raman spectra for several (*e.g.* acetic acid, ethanol, acetone) implying a departure from linearity.[64] Bending frequencies of some first-row transition-metal dihalides calculated from electron-diffraction data are compared with literature values from matrix-isolated studies.[65] Bands at 1031, 1011, and 987 cm^{-1} have been assigned to the ν_3 mode of VO_2 trapped in an Ar matrix, with the bond angle estimated as 93 ± 5°.[66] The characteristic isotopic triplet pattern establishes a 'side-on' configuration for the $Au(O_2)$ species isolated from Au atoms and O_2/Ar mixtures at 10 K. Wavenumbers/cm^{-1} for the triplet are 1091.7 ($^{16}O_2$), 1029.8 ($^{18}O_2$), and 1060.6 ($^{16}O^{18}O$).[67] Assignments have been made for linear UO_2^+ species produced as the ion pairs $UO_2^+NO_2^-$ or $UO_2^+NO^-$ from i.r. spectra in Ar matrices. Wavenumbers/cm^{-1} are:[68]

	$U^{16}O_2^+$	$U^{16}O^{18}O^+$	$U^{18}O_2$
ν_1	—	802.4	—
ν_3	891.9	878.1	847.3

I.r. spectra of the isolated BO_2^- ion in twelve alkali-metal halide pressed discs[69] and as an impurity in single crystals[70] have been measured. Anharmonic force fields (approximately quartic) were calculated for each lattice, and correlations of force-constant shifts and fundamental frequencies made with lattice parameters.[69] The Raman spectrum of solid BaS_3 is consistent with a C_{2v} S_3^- ion at room temperature, but at higher temperatures the spectral differences are interpreted in terms of bond-angle changes and rotation of S_3^- in the solid.[71] A feature at 535 cm^{-1} has been assigned as the ν_1 (sym. stretch) of the S_3^- radical ion produced from the breakdown of polysulphide chains of Na_2S_4 in KSCN melts.[72]

Measurement of centrifugal distortion constants from microwave spectra for isotopic variants of the weakly bound Ar—HCl species enabled the calculation of ν(Ar—HCl) and force constants to be made by considering HCl as a rigid

[61] A. Strull, A. Givan, and A. Loewenschuss, *J. Mol. Spectroscopy*, 1976, **62**, 283.
[62] A. Givan and A. Loewenschuss, *J. Chem. Phys.*, 1976, **64**, 1967.
[63] A. Givan and A. Loewenschuss, *J. Chem. Phys.*, 1976, **65**, 1851.
[64] J. H. Smith and T. B. Brill, *Inorg. Chim. Acta*, 1976, **18**, 225.
[65] I. Hargittai and J. Tremmel, *Coordination Chem. Rev.*, 1976, **18**, 257.
[66] L. V. Serebrennikov and A. A. Mal'tsev, *Vestnik. Moskov. Univ., Khim*, 1975, **16**, 486.
[67] D. McIntosh and G. A. Ozin, *Inorg. Chem.*, 1976, **15**, 2869.
[68] D. W. Green, S. D. Gabelnick, and G. T. Reedy, *J. Chem. Phys.*, 1976, **64**, 1697.
[69] D. F. Smith, *Spectrochim. Acta*, 1976, **32A**, 319.
[70] I. I. Kondilenko, V. A. Pasechnyi, and Y. P. Tsyashchenko, *Zhur. priklad. Spektroskopii*, 1975, **23**, 1049.
[71] G. J. Janz, E. Roduner, J. W. Coutts, and J. R. Downey, *Inorg. Chem.*, 1976, **15**, 1751.
[72] H. H. Eysel and D. Nöthe, *Z. Naturforsch.*, 1976, **31b**, 411.

rod.[73] Calculated wavenumbers/cm^{-1} are:

	$Ar{-}H^{35}Cl$	$Ar{-}H^{37}Cl$	$Ar{-}D^{35}Cl$	$Ar{-}D^{37}Cl$
ω_e	32.4	31.9	34.4	33.7

Temperature dependence of the intensities of two i.r. absorptions (2887 and 2879 cm^{-1}) in the spectra of gaseous Ar–HCl mixtures at various pressures leads to a Ar–HCl bond-energy estimate of *ca.* 5 kJ mol^{-1}, which contrasts with the generally accepted value of *ca.* 1.7 kJ mol^{-1}.[74] A very thorough i.r. study of NSCl has enabled the first complete assignment for the gas-phase spectrum to be made. The fundamentals are at 1324.5 (ν_1), 414.8 (ν_2), and 271.6 cm^{-1} (ν_3). Force constants consistent with independent data such as isotope shifts and centrifugal distortion constants were calculated and the force constants for NSF redetermined.[75] A broad band centred at 1850 cm^{-1} in the emission from a carbon tube furnace has been interpreted as the ν_3 mode of the C_3 species.[76]

High-resolution studies and the calculation of molecular constants have been made for a variety of isotopic variants of HCN,[77] SO_2,[78] N_2O,[79–83] and NO_2.[84–86] Laser magnetic resonance of $^{14}NO_2$ and $^{15}NO_2$ near 1600 cm^{-1} yielded molecular constants in good agreement with conventional high-resolution i.r. studies.[87] Vibrational and electronic spectra of the O_3^- ion have been reviewed,[88] and anharmonic force constants have been calculated using spectroscopic data for six isotopic (^{16}O and ^{18}O) variants of O_3.[89] Other studies include preresonance Raman spectroscopy of SO_2, N_2O, and H_2S,[90] i.r. and Raman band profiles of OCS in various solvents,[91] and calculation of a general quartic valence force field for OCS from 117 pieces of experimental data.[92]

Interest continues unabated in a variety of studies of water in its various phases and states of aggregation. This area is reported here, even though some of the species are larger than triatomic. A detailed i.r. study appears to settle the argument in favour of an open, rather than cyclic, configuration for the water

[73] S. E. Novick, K. C. Janda, S. L. Holmgren, M. Waldman, and W. Klemperer, *J. Chem. Phys.*, 1976, **65**, 1114.
[74] A. W. Miziolek and G. C. Pimentel, *J. Chem. Phys.*, 1976, **65**, 4462.
[75] A. Muller, N. Mohan, S. J. Cyvin, N. Weinstock, and O. Glemser, *J. Mol. Spectroscopy*, 1976, **59**, 161.
[76] R. R. Treffers and D. P. Gilva, *Astrophys. J.*, 1975, **202**, 839.
[77] A. G. Maki and R. L. Sams, *J. Mol. Spectroscopy*, 1976, **60**, 57.
[78] R. van Riet, D. van Lerberghe, G. Steenbeckeliers, and A. Fayt, *Bull. Classe Sci., Acad. Royale Belg.*, 1975, **61**, 610.
[79] A. Valentin, M. F. Le Moal, P. Cardinet, and J.-P. Boissy, *J. Mol. Spectroscopy*, 1976, **59**, 96.
[80] C. Amiot and G. Guelachvili, *J. Mol. Spectroscopy*, 1976, **59**, 171.
[81] C. Amiot, *J. Mol. Spectroscopy*, 1976, **59**, 191.
[82] C. Amiot, *J. Mol. Spectroscopy*, 1976, **59**, 380.
[83] K. Altman, W. Klöckner, and G. Strey, *Z. Naturforsch.*, 1976, **31a**, 1311.
[84] A. Cabana, M. Laurin, W. J. Lafferty, and R. L. Sams, *Canad. J. Phys.*, 1975, **53**, 1902.
[85] J. L. Hardwick and J. C. D. Brand, *Canad. J. Phys.*, 1976, **54**, 80.
[86] A. Cabana, M. Laurin, and C. Pépin, *J. Mol. Spectroscopy*, 1976, **59**, 13.
[87] S. M. Freund, J. T. Hougen, and W. J. Lafferty, *Canad. J. Phys.*, 1975, **53**, 1929.
[88] A. I. Karelin, *Neorg. Perekisnye Soedin., Dokl. Vses Soveshch.*, 1975, p. 172.
[89] P. Hennig and G. Strey, *Z. Naturforsch.*, 1976, **31a**, 244.
[90] P. W. Low, *Sci. Tech. Aerospace Rep.*, 1975, **13**, Abstract No. N75–28157.
[91] D. Legay, J. P. Perchard, A. M. Goulay-Bize, and J. Soussen-Jacob, *J. Chim. phys.*, 1975, **72**, 863.
[92] K. A. Davis and J. Overend, *Spectrochim. Acta*, 1976, **32A**, 233.

dimer species in Ar matrices.[93–96] Limitations of space preclude an adequate description of this definitive work, but a variety of evidence is adduced in support of this conclusion. On the basis of variation of i.r. band intensities over a large range of matrix : water ratios, and spectral changes on annealing (which distinguishes between dimers and non-rotating monomers), the assignments of Table 1 were deduced.[93] After dosing the systems H_2O (D_2O, HDO)/HCl and

Table 1 *Vibrational wavenumbers*/cm^{-1} *for monomeric and dimeric* H_2O *in* Ar *at* 7 K

	H_2O	$(H_2O)_2$	D_2O	$(D_2O)_2$	HDO	$(HDO)_2$
ν_1	3638.0	{3574.5, 3634.1}	2657.7	{2614.8, 2654.8}	2710.0	{2639.0, 2705.7}
ν_2	1589.1	{1593.6, 1611.2}	1174.6	{1177.9, 1189.1}	1398.8	{1398.3, 1403.0}
ν_3	3734.3	3709.5	2746.4	2746.4	3687.3	3681.1

Me_2O/HCl with Xe (which enhances the intensity of absorptions due to non-rotating monomers), assignments to these systems were also made.[94] I.r. spectra for $^{18}OH_2$, $^{18}OD_2$, and ^{18}ODH and the various dimeric combinations were assigned, the relative simplicity of the spectra suggesting that dynamical coupling between the units of the dimer is low, thus supporting the open, rather than the cyclic, configuration.[95] Results from the use of double-derivative i.r. spectroscopy to study solutions of H_2O in CCl_4 support the view that the dimeric species is open rather than cyclic.[97]

Band origins for T_2O, DTO, and HTO (many of which have not been experimentally observed) have been calculated using variational vibrational methods.[98] Significant intensity decreases and bandshape changes occurring in the ν(OH) stretching region of the Raman spectrum of water as concentrations of dissolved lysozyme are increased have been explained on the basis of a model involving three different water species in association with protein, each having a different scattering efficiency.[99]

A re-examination of the i.r. spectra of aqueous HX (X = F, Cl, Br, or I) solutions using improved experimental conditions was interpreted as confirming an earlier assignment of broad bands at 2900, 1730, and 1200 cm^{-1} to fundamentals of the H_3O^+ ion, and suggesting that there was no evidence for $H_5O_2^+$ in these systems.[100] Assignments for the H_3O^+, H_2DO^+, HD_2O^+, and D_3O^+ fundamentals have been made from a Raman study of species of the type $H_3O^+Cl^-$ in liquid SO_2.[101] Typical wavenumbers/cm^{-1} are:

	H_3O^+	D_3O^+
ν_1	3385	2490
ν_3	{3470, 3400}	{2660, 2580}
ν_4	{1700, 1635}	1255

[93] G. P. Ayres and A. D. E. Pullin, *Spectrochim. Acta*, 1976, **32A**, 1629.
[94] G. P. Ayres and A. D. E. Pullin, *Spectrochim. Acta*, 1976, **32A**, 1641.
[95] G. P. Ayres and A. D. E. Pullin, *Spectrochim. Acta*, 1976, **32A**, 1689.
[96] G. P. Ayres and A. D. E. Pullin, *Spectrochim. Acta*, 1976, **32A**, 1695.

Hydrated salts continue to attract attention, with studies on crystal hydrates of MCl_2 and $M(NO_3)_2$ (M = Mg, Ca, Sr, or Ba),[102] $BaCl_2,2H_2O$,[103] $Ba(ClO_3)_2$,-H_2O,[104] MSO_4,H_2O (M = Zn or Cu),[105] α- and β-$CaSO_4,\frac{1}{2}H_2O$,[106] and $CuSO_4$,-$5H_2O$ and $CuSO_4,5D_2O$.[107] The separation between intermolecularly uncoupled ν_3 and ν_1 modes of H_2O (or D_2O) in non-aqueous solvents and hydrates varies linearly with ν(OH) [or ν(OD)], suggesting that this could be useful for predicting geometry changes of water in hydrates upon deuteriation.[108] Reasonable agreement between calculated and observed (low-temperature i.r.) wavenumbers was obtained for varyingly deuteriated samples of Li_2SO_4,H_2O.[109] The ν(OH) at 3606 cm^{-1} and ν(OD) at 2657 cm^{-1} modes of partially deuteriated $NaPF_6,H_2O$ occur at what appears to be the highest values so far recorded for HDO in crystalline hydrates.[110] Data from a large number of crystal hydrates have been used to plot the uncoupled ν(OH) value against r_0 (obtained by subtracting the H-bond O—H···Y length from the sums of the ionic radii of O and Y in the free state). It was inferred that M—O co-ordination in hydrates reduces the ionic size of the water oxygen, and that the effect on ν(OH) is opposite to that of H-bonding.[111]

Compounds containing the $H_5O_2^+$ species have received attention this year. In contrast to ref. 100, a band at 1200 cm^{-1} in the i.r. spectra of aqueous solutions of strong acids was assigned to an antisymmetric vibration of the central proton along the O—O axis of the symmetrical $[H_2O\cdots H\cdots OH_2]^+$ species.[112, 113] Increasing the concentration of a neutral salt in aqueous acid solutions results in a decrease in intensity of i.r. absorption due to the polarizable H-bond in the $H_5O_2^+$ group. It was suggested that the H-bonds are polarized by local fields due to neutral salt ions, with the H^+ becoming trapped at one of the H_2O ends of the $H_5O_2^+$ species.[114] An *ab initio* calculation was used to aid the assignment of i.r. bands of $H_5O_2^+$ $SbCl_6^-$ and $D_5O_2^+$ $SbCl_6^-$. Thus for $H_5O_2^+$:[115]

$\nu(OH_2)$	ν(OHO)	$\delta(OH_2)$	δ(OHO)/cm^{-1}
3425	1015	1695	2200

[97] J. H. Keighley and A. J. Whitaker, *Mol. Phys.*, 1976, **31**, 643.
[98] G. D. Carney, L. A. Curtiss, and S. R. Langhoff, *Appl. Spectroscopy*, 1976, **30**, 453.
[99] F. Cavatorta, M. P. Fontana, and A. Vecli, *J. Chem. Phys.*, 1976, **65**, 3635.
[100] P. A. Giguère and S. Turrell, *Canad. J. Chem.*, 1976, **54**, 3477.
[101] P. A. Giguère and C. Madec, *Chem. Phys. Letters*, 1976, **37**, 569.
[102] G. S. Karetnikov, O. V. Bazileva, and T. V. Gerzha, *Zhur. fiz. Khim.*, 1975, **49**, 815.
[103] Y. S. Jain, V. K. Kapoor, and H. D. Bist, *Appl. Spectroscopy*, 1976, **30**, 440.
[104] G. Brink, *Appl. Spectroscopy*, 1976, **30**, 630.
[105] V. A. Kolesova and A. N. Lazarev, *Zhur. neorg. Khim.*, 1975, **20**, 2846.
[106] V. N. Nosov, N. G. Frolova, and V. F. Karmyshov, *Zhur. priklad. Spektroskopii*, 1976, **24**, 713.
[107] B. K. Srivastava, D. P. Khandelwal, S. Khandelwal, and H. D. Bist, *Chem. Phys. Letters*, 1976, **43**, 557.
[108] J. Schiffer, M. Intenzo, P. Hayward, and C. Calabrese, *J. Chem. Phys.*, 1976, **64**, 3014.
[109] H. P. Hayward and J. Schiffer, *J. Chem. Phys.*, 1976, **64**, 3961.
[110] A. M. Heyns, *Chem. Phys. Letters*, 1976, **40**, 471.
[111] B. K. Srivastava, D. P. Khandelwal, and H. D. Bist, *Chem. Phys. Letters*, 1976, **39**, 366.
[112] N. B. Librovich, V. D. Maiorov, and V. A. Sevel'ev, *Doklady Akad. Nauk S.S.S.R.*, 1975, **225**, 1358.
[113] N. B. Librovich, V. D. Maiorov, and V. A. Sevel'ev, *Doklady Phys. Chem.*, 1976, **225**, 1383.
[114] D. Schioberg and G. Zundel, *Canad. J. Chem.*, 1976, **54**, 2193.
[115] R. Janoschek, R. Ortwein, and A. Schmidt, *Chem. Phys. Letters*, 1976, **37**, 396.

Assignments proposed for a series of $SbCl_6^-$ salts of $H_5O_2^+$, $H_7O_3^+$, and $H_9O_4^+$ (and D analogues) differ somewhat from those in the literature for the $H_7O_3^+$ species.[116] The $H_5O_2^+$ species has also been recognized in $HGaCl_4,2H_2O$, $HGaBr_4,2H_2O$,[117] and *trans*-[Co(en)$_2$Cl$_2$]Cl,HCl,2H_2O.[118]

Tetra-atomic Species.—Interest continues in the assignment and structure of matrix-isolated $AlCl_3$. In a previous investigation, an i.r. band at 382.2 cm^{-1} was assigned to the ν_1 mode of a pyramidally distorted species.[119] A Raman band at 393.5 cm^{-1} has now been assigned as ν_1, with the suggestion that such a discrepancy from the i.r. value (11 cm^{-1}) could hardly arise from the same molecule, and that the i.r. band at 382.2 cm^{-1} does not therefore imply distortion from planarity.[120] Further, independent i.r. investigations both indicate that the band at 382 cm^{-1} only appears if N_2 is present in the matrix, and this has been ascribed to an $AlCl_3N_2$ species.[121, 122] ν_3 modes for various isotopic variants of $AlCl_3$ and $AlBr_3$ in Ar, Ne, and N_2 matrices have also been reported.[122, 123] Raman spectra of gaseous monomeric GaX_3 (X = Cl, Br, or I), at high temperatures lead, with results from previous work, to a complete set of fundamental assignments (Table 2),[124–126] from which thermodynamic functions were computed.[124]

Table 2 *Vibrational wavenumbers*/cm^{-1} *for* GaX_3 (X = Cl, Br, *or* I)[124]

Mode	$GaCl_3$	$GaBr_3$	GaI_3
ν_1 (A_1')	381	232 [b]	162
ν_2 (A_2')	145 [a]	98 [a]	63 [a]
ν_3 (E')	457	343 [c]	276
ν_4 (E')	128	83 [c]	60

[a] Ref. 125. [b] Estimated from ref. 126. [c] Ref. 126.

Fundamental bands for the mixed CdX_2Y^- (X, Y = Cl, Br, or I) have been assigned on the basis of C_{2v} symmetry from Raman spectra of solid $Pr^n{}_4N^+$-CdX_2Y^-.[127] It has been suggested that Raman bands at 75 and 112 cm^{-1} in $Cd_4(AsF_6)_2$ (formed from Cd and AsF_5 in liquid SO_2) are the symmetric stretching modes of a symmetrical linear Cd_4^{2+} cation.[128]

Vibrational frequencies for Na_2S_4 from independent studies are in agreement, although differing in the detailed assignments for the S_4^{2-} species of C_2 symmetry.[46, 129] An assignment based on a Raman and i.r. study of polycrystalline and molten K_2S_4 is 485 (ν_1), 434 (ν_2), 220 (ν_3), 478 (ν_4), and 266 cm^{-1} (ν_5).[130]

116 R. Ortwein and A. Schmidt, *Z. anorg. Chem.*, 1976, **425**, 10.
117 G. Picotin, J. Rozière, J. Potier, and A. Potier, *Adv. Mol. Relaxation Processes*, 1975, **7**, 177.
118 J. Rozière and J. M. Williams, *Inorg. Chem.*, 1976, **15**, 1174.
119 M. L. Lesiecki and J. S. Shirk, *J. Chem. Phys.*, 1972, **56**, 4171.
120 I. R. Beattie, H. E. Blayden, and J. S. Ogden, *J. Chem. Phys.*, 1976, **64**, 909.
121 J. S. Shirk and A. E. Shirk, *J. Chem. Phys.*, 1976, **64**, 910.
122 H. Schnöckel, *Z. anorg. Chem.*, 1976, **424**, 203.
123 H. Schnöckel, *Z. Naturforsch.*, 1976, **31b**, 1291.
124 M. C. Drake and G. M. Rosenblatt, *J. Chem. Phys.*, 1976, **65**, 4067.
125 G. K. Selivanov and A. A. Mal'tsev, *Zhur. strukt. Khim.*, 1973, **14**, 943.
126 I. R. Beattie and J. R. Horder, *J. Chem. Soc. (A)*, 1969, 2655.
127 J. G. Contreras and D. G. Tuck, *Canad. J. Chem.*, 1976, **54**, 3641.
128 B. D. Cutforth, R. J. Gillespie, and P. K. Ummat, *Rev. Chim. minérale*, 1976, **13**, 119.
129 H. H. Eysel, G. Wieghardt, H. Kleinschmager, and G. Weddigen, *Z. Naturforsch.*, 1976, **31b**, 415.
130 G. J. Janz, J. W. Coutts, J. R. Downey, and E. Roduner, *Inorg. Chem.*, 1976, **15**, 1755.

An i.r. and Raman study of XCF_2 (X = Se or S) includes the new wavenumbers/cm^{-1} (assigned on the basis of C_{2v} symmetry):[131]

A_1	ν_1,	ν(CSe)	1280	B_1	ν_4,	$\nu_{asym}(CF_2)$	1208
	ν_2,	$\nu_{sym}(CF_2)$	710		ν_5,	$\delta_{asym}(CF_2)$	352
	ν_3,	$\delta_{sym}(CF_2)$	432	B_2	ν_6,	γ	560

I.r. spectra of matrix-isolated MgF_2 co-deposited with alkali-metal and alkaline-earth fluorides have been assigned assuming planar cyclic C_{2v} $MMgF_3$ and $M'MgF_4$ structures.[132] Assignment of the i.r. spectrum of IN_3 has been made on the basis of C_s symmetry, with a linear NNN and bent INN bonds. Wavenumbers/cm^{-1} in benzene solution are:[133] 2058, $\nu_{asym}(N_3)$; 1176, $\nu_{sym}(N_3)$; 578, $\gamma(N_3)$; 338, ν(NI). Assignments have been given for the T-shaped (C_{2v}) XeF_3^+ and pyramidal (C_s) XeO_2F from the Raman spectra of species such as $XeF_3^+SbF_6^-$, $XeF_3^+Sb_2F_{11}^-$, and $XeO_2F^+Sb_2F_{11}^-$.[134] Wavenumbers/cm^{-1} for XeF_3^+ are: 644, ν_1 (A_1); 583, ν_2 (A_1); 209, ν_3 (A_1); 618, ν_4 (B_1); and 305, ν_5 (B_1), and for XeO_2F^+: 867, ν_1 (A'); 580, ν_2 (A'); 334, ν_4 (A'); and 938, ν_5 (A'').

The ν_1 modes of $KBr^{18}O_3$, $KBr^{16}O^{18}O_2$, $KBr^{16}O_2{}^{18}O$, and $KBr^{16}O_3$ have been observed at 744, 753, 768, and 792 cm^{-1}, respectively,[135] and a detailed study of the isotope effects in the Raman spectra of $MClO_3$ and $MBrO_3$ (M = alkali metal) has been made.[136] R.R.E. and i.r. spectra of electron-irradiated $NaClO_3$ have been interpreted on the basis of the ClO_3 radical, for which a partial assignment is:[137]

807 (Raman) } ν_1 (A_1) 850 (i.r.) ν_3 (E)
812 (Raman, i.r.) }

The ν_3 and ν_4 regions in the i.r. spectra of NH_3 and the NH_3–Cl_2 complex in solid N_2 below 15 K show fine structure, interpreted in terms of a hindered rotation about the C_3 axis.[138] The observation of ν(NH) and the first and second overtones in the gas-phase i.r. spectrum of HNSO, when combined with earlier microwave data, enabled the calculation of molecular constants for the first three vibrational states. Wavenumbers/cm^{-1} for the fundamental and overtones are:[139] 3308.50 (ν_1); 6463.56 ($2\nu_1$); 9465.04 ($3\nu_1$). An *ab initio* calculation of force constants for HNCO strongly supports the assignment of the observed band at 777.1 cm^{-1} to an out-of-plane vibration.[140] High-resolution work includes studies on $^{14}NH_3$ and $^{15}NH_3$ [141] and $^{12}C_2HD$.[142]

Penta-atomic Species.—A conference report of preresonance and resonance Raman spectra of 14 tetrahedral molecular hydrides, and of 17 octahedral and

[131] A. Haas, B. Koch, N. Welcman, and H. Willner, *Spectrochim. Acta*, 1976, **32A**, 497.
[132] A. S. Kana'an, R. H. Hauge, and J. L. Margrave, *J.C.S. Faraday II*, 1976, **72**, 1991.
[133] K. Dehnicke, *Angew. Chem. Internat. Edn.*, 1976, **15**, 553.
[134] R. J. Gillespie, B. Landa, and G. J. Schrobilgen, *Inorg. Chem.*, 1976, **15**, 1256.
[135] E. W. Holtzscheiter and E. R. Johnson, *Internat. J. Appl. Radiation Isotopes*, 1975, **26**, 780.
[136] J. B. Bates and H. D. Stidham, *J. Phys. and Chem. Solids*, 1976, **37**, 183.
[137] J. B. Bates and H. D. Stidham, *Chem. Phys. Letters*, 1976, **37**, 25.
[138] L. Fredin, B. Nelander, and G. Ribbegard, *Chem. Phys.*, 1976, **12**, 153.
[139] M. Carlotti, G. Di Lonardo, G. Galloni, and A. Trombetti, *J. Mol. Spectroscopy*, 1976, **60**, 388.
[140] P. Botschwina, E. Nachbauer, and B. M. Rode, *Chem. Phys. Letters*, 1976, **41**, 486.
[141] C. Chassagne, M. L. Grenier-Besson, and D. Bargues, *J. Phys. (Paris)*, 1976, **37**, 71.
[142] A. Baldacci, S. Ghersetti, S. C. Hurlock, and K. N. Rao, *J. Mol. Spectroscopy*, 1976, **59**, 116.

7 square-planar species, in solution has appeared.[143] Bond-polarizability derivatives and bond anisotropies have been calculated from a detailed Raman study of vapour-phase intensities of MH_4 (M = C, Si, Ge, or Sn) and MF_4 (M = C, Si, or Ge) species.[144] Intensities of the ν_1 modes in the Raman spectra of PdX_4^{2-} (X = Cl, Br, or I) species excited in the vicinity of electronically forbidden ligand-field transitions showed *minima* at these frequencies, in contrast to the maxima usually observed near electronically allowed transitions. This effect was ascribed to interference between the weak scattering from the forbidden electronic state and strong preresonance scattering from allowed electronic states of higher energy.[145] O.V.F.F. calculations based on available data for square-planar MX_4^{2-} species have been used to obtain 'best values' for the inactive ν_5 (B_{2u}) modes as part of an electronic spectral study of a single crystal. Calculated wavenumbers/cm^{-1} are: 174 (K_2PtCl_4), 157 (K_2PdCl_4), and 110 (K_2PdBr_4).[146]

Assignments have been made for $C^{35}Cl_4$ and $C^{37}Cl_4$ from i.r. spectra in rare-gas matrices.[147] The fundamental and five hot bands were identified, and rotational constants were determined from a high-resolution (using a tunable thin film PbTe diode laser) study of the ν_4 band (1565—1610 cm^{-1}) of gaseous C_3O_2.[148] Investigations on SiH_4 include a high-resolution study of the ν_4 band[149] and a bandshape analysis and spectral-moment calculation on the collision-induced far-i.r. spectrum.[150] New Raman intensity and polarization data have been reported for $HSiCl_3$ and $DSiCl_3$,[151] and N.C.A.'s have been performed on CH_3X and SiH_3X (X = F, Cl, or Br).[152] Natural $GeCl_4$ is a mixture of 25 isotopic species with, theoretically, 45 i.r. bands in the ν_3 region. The ν_3 bands of GeX_4 (X = Cl, Br, or I) in Ar matrices have been interpreted with the aid of computer simulation, and assignments made to many of the isotopic species.[153] Not surprisingly, calculation of Coriolis coupling constants from *PR* separations is rendered very difficult by the multitude of isotopic species in such systems, but this has been overcome by using the isotopically pure species $^{116}Sn^{35}Cl_4$, $^{74}Ge^{35}Cl_4$, and $^{104}RuO_4$; for example, the ν_3 (T_2) mode of $GeCl_4$ is broad and unresolved, but the *PR* separation is 7 cm^{-1} in $^{74}Ge^{35}Cl_4$.[154]

The novel $OAsCl_3$ has been prepared in Ar or Kr matrices by condensation of the products of a microwave discharge on $AsCl_3$, O_2, and Ar or by u.v. photolysis of O_3 in Ar (Kr) matrices in the presence of $AsCl_3$. Raman and i.r. spectra of isotopic variants are consistent with the C_{3v} formulation. Wave-

[143] Y. M. Bosworth, R. J. H. Clark, M. L. Franks, and P. D. Mitchell, *Proc. Int. Conf. Coord. Chem. 16th*, 1974, 2.18b.
[144] R. S. Armstrong and R. J. H. Clark, *J.C.S. Faraday II*, 1976, **72**, 11.
[145] P. Stein, V. Miskowski, W. H. Woodruff, J. P. Griffin, K. G. Werner, B. P. Gaber, and T. G. Spiro, *J. Chem. Phys.*, 1976, **64**, 2159.
[146] T. H. Harrison, H. H. Paterson, and J. D. Godfrey, *Inorg. Chem.*, 1976, **15**, 1291.
[147] H. Schnoeckel and H. J. Becher, *Z. phys. Chem. (Frankfurt)*, 1975, **97**, 171.
[148] W. H. Weber, P. D. Maker, and C. W. Peters, *J. Chem. Phys.*, 1976, **64**, 2149.
[149] J. W. C. Johns and W. A. Kreiner, *J. Mol. Spectroscopy*, 1976, **60**, 400.
[150] A. Rosenberg and I. Ozier, *J. Chem. Phys.*, 1976, **65**, 418.
[151] V. S. Dernova, I. F. Kovalev, M. G. Voronkov, O. G. Yarosh, and R. G. Mirskov, *Doklady Phys. Chem.*, 1975, **222**, 448.
[152] R. Aroca, E. A. Robinson, and T. A. Ford, *J. Mol. Structure*, 1976, **31**, 177.
[153] D. Tevault, J. D. Brown, and K. Nakamoto, *Appl. Spectroscopy*, 1976, **30**, 461.
[154] A. Müller and F. Koniger, *J. Mol. Structure*, 1976, **30**, 195.

numbers/cm^{-1} for $^{16}OAsCl_3$ are:[155] $\nu_1\,(A_1)$ 997.0; $\nu_2\,(A_1)$ 373.1; $\nu_4\,(E)$ 434.3; $\nu_5\,(E)$ 247.1. Raman and i.r. wavenumbers/cm^{-1} for Na_2S_5 and K_2S_5 have been listed without detailed assignments.[46, 129] The absorption at 353 cm^{-1} in the spectrum of SF_4 has been shown, by matrix isolation in N_2, to consist of two bands, thus providing the hitherto unobserved $\nu_9\,(B_2)$ mode. Detailed vibrational spectra and force-constant calculations on SF_4 and ClF_4^+ (in $ClF_4^+SbF_6^-$) have resulted in revised assignments. Wavenumbers/cm^{-1} are:[156]

		SF_4	ClF_4^+			SF_4	ClF_4^+
A_1	ν_1	892	800	B_1	ν_6	730	795
	ν_2	558	571		ν_7	532	515
	ν_3	356	385	B_2	ν_8	867	829
	ν_4	228	250		ν_9	350	385
A_2	ν_5	474	475				

Some changes in assignments for SeF_4 and TeF_4 have been proposed on the basis of M.U.B.F.F. calculations, taking account of the unpaired electrons in Se and Te.[157] Assignments on the basis of C_{3v} symmetry have been made for two SeO_3F^- salts. Thus:[158]

	$\nu_1\,(A_1)$	$\nu_2\,(A_1)$	$\nu_3\,(A_1)$	$\nu_4\,(E)$	$\nu_5\,(E)$	$\nu_6\,(E)$/cm^{-1}
$K^+SeO_3F^-$	896	580, 603	392	968, 974	409	301
$NO^+SeO_3F^-$	903	607, 624	388	971, 979	406	292, 309

Vibrational assignments for the $BrOF_3$ [159] (Table 3) and the $BrF_2O_2^-$ [160] (Table 4) species have been given for the first time. Raman data and assignments for $XeOF_3^+$ in SbF_6^- and $Sb_2F_{11}^-$ salts are listed in Table 3.[134]

Table 3 *Raman wavenumbers*/cm^{-1} *for the C_s species* $BrOF_3$ [159] *and* $XeOF_3^+$ [134]

Mode		$BrOF_3$ [a]	$XeOF_3^+$ [b]	*Mode*		$BrOF_3$	$XeOF_3^+$
A'	ν_1	1002	942	A'	ν_6	202	149 [d]
	ν_2	621	635	A''	ν_7	{605, 613} [c]	617
	ν_3	489	334		ν_8	396	356
	ν_4	351	{601, 594}		ν_9	334	319
	ν_5	242	199				

[a] Liquid; [b] $Sb_2F_{11}^-$ salt; [c] I.r. (gas); [d] SbF_6^- salt.

The novel CrO_3I^- anion has been partially assigned from the i.r. and Raman spectra of $[Ph_4P]^+[CrO_3I]^-$:[161]

A_1	$\nu_1, \nu_{sym}(CrO)$	902	A_1	$\nu_3, \delta_{sym}(CrO_3)$	360
	$\nu_2, \nu(CrI)$	354	E	$\nu_4, \nu_{asym}(CrO)$	952

155 F. W. S. Benfield, A. J. Downs, G. P. Gaskill, and S. E. Staniforth, *J.C.S. Chem. Comm.*, 1976, 856.
156 K. O. Christe, E. C. Curtis, C. J. Schack, S. J. Cyvin, J. Brunvoll, and W. Sawodny, *Spectrochim. Acta*, 1976, **32A**, 1141.
157 M. G. K. Pillai and V. Kumar, *Indian J. Pure Appl. Phys.*, 1976, **14**, 226.
158 M. Černík and K. Dostál, *Z. anorg. Chem.*, 1976, **425**, 37.
159 R. Bougon and Tue Bui Huy, *Compt. rend.*, 1976, **283**, *C*, 461.
160 R. J. Gillespie and P. Spekkens, *J.C.S. Dalton*, 1976, 2391.
161 E. Königer-Ahlborn and A. Müller, *Z. Naturforsch.*, 1976, **31b**, 1429.

The first liquid-phase data on CrO_2F_2 are consistent with a primarily monomeric C_{2v} geometry (Table 4).[162]

Table 4 *Raman wavenumbers*/cm^{-1} *for the* C_{2v} *species* $BrO_2F_2^-$ [160] *and* CrO_2F_2 [162]

Mode		$BrO_2F_2^-$ [a]	CrO_2F_2 [b]	*Mode*		$BrO_2F_2^-$	CrO_2F_2
A_1	ν_1	884	995	B_1	ν_6	442	1045
	ν_2	424	708		ν_7	380	295
	ν_3	400	403	B_2	ν_8	910	770
	ν_4	197	220		ν_9	307	325
A_2	ν_5	338	275				

[a] Solid as $KBrO_2F_2$ salt; [b] Liquid.

Sharp i.r. bands resulting from substitution of oxyanions in host lattices of other oxyanions allow precise measurements of isotope shifts. Data obtained for ClO_4^- ($^{35,37}Cl$), SO_4^{2-} ($^{16,18}O$ and $^{32,34}S$), MoO_4^{2-} ($^{92,100}Mo$), NO_3^- ($^{14,15}N$), and CO_3^{2-} ($^{12,13}C$) were used to calculate interaction constants in the force fields of these species.[163] A Raman and i.r. emission study of MoO_4^{2-} and WO_4^{2-} in KCl melts is consistent with T_d symmetry for these species, and the earlier conclusion that the previously unobserved ν_4 (T_2) mode is too weak to be observed appears to be confirmed.[164]

Hexa-atomic Species.—In a gas-phase Raman study of PF_5, AsF_5, and VF_5, the details of the ν_7 (E) vibration were explained on the basis of a two-dimensional interchange potential having its absolute minima at the D_{3h} conformation. It was suggested that the fluorine interchange (pseudorotation) occurs *via* a C_{4v} intermediate by initial displacement of the equatorial F atoms followed by mixing in the axial F distortions as the interchange proceeds.[165] Preparation of the elusive $AsCl_5$ (by u.v.-irradiation of a solution of $AsCl_3$ in Cl_2 at -105 °C) has been described. Raman wavenumbers/cm^{-1} (without detailed assignments) were given: 437, 369 (pol), 295 (pol), 220, 213, 83.[166] Vibrational assignments made on the basis of C_{4v} symmetry for the novel $BrOF_4^-$ ion are essentially in agreement in two independent studies.[160, 167] Wavenumbers/cm^{-1} are:[167]

A_1	ν_1	932	E	ν_7	506, 483
	ν_2	525		ν_8	434, 421, 407, 398
	ν_3	312		ν_9	194, 164
B_1	ν_4	459			
	ν_5	236			
B_2	ν_6	248			

Codeposition of F_2 with UF_4 in Ar [168] and u.v. photolysis of UF_6 isolated in Ar, Xe, and CO matrices [169] both produced bands in the i.r. at 561 and 584 cm^{-1},

[162] S. D. Brown, G. L. Gard, and T. M. Loehr, *J. Chem. Phys.*, 1976, **64**, 1219.
[163] H. J. Becher, F. Friedrich, and H. Willner, *Z. anorg. Chem.*, 1976, **426**, 15.
[164] T. W. Cape, V. A. Maroni, P. T. Cunningham, and J. B. Bates, *Spectrochim. Acta*, 1976, 32A, 1219.
[165] L. S. Bernstein, S. Abramowitz, and I. W. Levin, *J. Chem. Phys.*, 1976, **64**, 3228.
[166] K. Seppelt, *Angew. Chem. Internat. Edn.*, 1976, **15**, 377.
[167] R. Bougon, Tue Bui Huy, P. Charpin, and G. Tantot, *Compt. rend.*, 1976, **283**, *C*, 71.
[168] K. R. Kunze, R. H. Hauge, D. Hamill, and J. L. Margrave, *J. Chem. Phys.*, 1976, **65**, 2026.
[169] R. T. Paine, R. S. McDowell, L. B. Asprey, and L. H. Jones, *J. Chem. Phys.*, 1976, **64**, 3081.

which were assigned to monomeric UF_5. In both studies the data were insufficient to decide between C_{4v} or D_{3h} geometries, but calculation of the fundamental frequencies, using force constants transferred from UF_6, suggests that the structure may be square-based pyramidal (C_{4v}), with the U atom above the fluorine atom equatorial plane.[170] Vibrational spectra of liquid CrF_5 are consistent with a C_{2v} *cis*-fluorine-bridged polymer, with no evidence for the monomer.[171] Wavenumbers/cm^{-1} for the S_6^{2-} ion in K_2S_6 are: 504, 495, 455, 373, 358 (stretches); 337, 254 (bends); 157, 132 (torsions).[129]

Hepta-atomic Species.—New vibrational data for octahedral MX_6 species are summarized in Table 5. A vibrational study of the compounds Cs_2KMF_6 (M = Sc, Y, La, Gd, or Yb) includes the assignment of lattice modes and MVFF calculations.[172] The preparation and vibrational data for some $X^+AuF_6^-$ species (X^+ = $Xe_2F_{11}^+$, XeF_5^+, K^+, Cs^+, O_2^+, NO^+, or IF_6^+)[173] and $[ReF_6]^+X^-$ (X^- = SbF_6^- or $Sb_2F_{11}^-$)[174] have been reported (Table 5[169, 172–175, 180]). A

Table 5 *Vibrational wavenumbers*/cm^{-1} *for octahedral species*

Species	ν_1 (A_{1g})	ν_2 (E_g)	ν_3 (T_{1u})	ν_4 (T_{1u})	ν_5 (T_{2g})	*Ref.*
Cs_2KScF_6	498	390	176	90	230	172
Cs_2KYF_6	476	382	160	74	194	172
Cs_2KLaF_6	443	334	130	63	171	172
Cs_2KGdF_6	473	380	140	72	185	172
Cs_2KYbF_6	491	370	156	70	196	172
$CsAuF_6$	595	530	—	—	225	173
Cs_2MoF_6	685	598	653	250	274	180
UF_6 [a]	665.8	529.6	619.3	183.5	199.7	169
ReF_6^+ [b]	797	734	783	353	359	174
ThI_6^{2-} [c]	131	—	143	—	68, 50	175
PaI_6^{2-} [c]	—	—	143	—	—	175
UI_6^{2-} [c]	—	—	143	—	—	175

[a] In Ar matrix; [b] As SbF_6^- salt; [c] As Et_4N^+ salt (see Chapter 6).

preliminary R.R.E. study of the mixed-valence compound Cs_2SbCl_6 (see Vol. 9) has now been extended to several other compounds in the same series and to some related lead compounds involving Pb^{II} and Pb^{IV}.[176] Anomalous polarization effects in the R.R.E. of $IrBr_6^{2-}$ have been explained group-theoretically by invoking vibronic states.[177] The ν_4 (T_{1u}) mode showed significant pressure sensitivity in a far-i.r. study of salts of SeX_6^{2-} (X = Cl or Br) and TeX_6^{2-} (X = Cl, Br, or I), which was accounted for on the basis of a pressure-dependent delocalization of inert-pair electrons throughout the lattice.[178]

Some vibrational assignments and MVFF calculations involving the $MoOF_5^-$, $MoOF_5^{2-}$, and WOF_5^- species of C_{4v} symmetry have been given[179] (Table 6).

170 B. J. Krohn, W. B. Person, and J. Overend, *J. Chem. Phys.*, 1976, **65**, 969.
171 S. D. Brown, T. M. Loehr, and G. L. Gard, *J. Chem. Phys.*, 1976, **64**, 260.
172 R. Becker, A. Lentz, and W. Sawodny, *Z. anorg. Chem.*, 1976, **420**, 210.
173 N. Bartlett and K. Leary, *Rev. Chim. minérale*, 1976, **13**, 82.
174 E. Jacob and M. Fähnle, *Angew. Chem. Internat. Edn.*, 1976, **15**, 159.
175 D. Brown, P. Lidster, B. Whittaker, and N. Edelstein, *Inorg. Chem.*, 1976, **15**, 511.
176 R. J. H. Clark and W. R. Trumble, *J.C.S. Dalton*, 1976, 1145.
177 H. Hamaguchi and T. Shimanouchi, *Chem. Phys. Letters*, 1976, **38**, 370.
178 D. M. Adams, J. D. Findlay, M. C. Coles, and S. J. Payne, *J.C.S. Dalton*, 1976, 371.
179 A. Beuter and W. Sawodny, *Z. anorg. Chem.*, 1976, **427**, 37.

A matrix-isolation study (i.r. and Raman) of SF_5Cl, SF_5Br, and S_2F_{10} includes the first vibrational assignment for SF_5Br (Table 6), and makes some changes to previous assignments for SF_5Cl.[180] An assignment for $TeClF_5$ (Table 6)and a

Table 6 *Vibrational data (wavenumbers/cm^{-1}) for some C_{4v} MX_5Y species*

Assignments		$TeClF_5$	SF_5Br	$[MoOF_5]^-$ [a]	$[WOF_5]^-$ [a]	$[MoOF_5]^{2-}$ [b]
A_1	ν_1	711	{843, 836}	973	987	990
	ν_2	662	685.9	662	686	590
	ν_3	317	591.7	492	507	445
	ν_4	{410.5, 404.0}	273.5	300	286	282
B_1	ν_5	651	614.9	580	594	—
	ν_6	—	222.0	—	—	—
B_2	ν_7	302	497.3	—	—	—
E	ν_8	726	886.7	580	608	590
	ν_9	324.5	572.9	324	329	—
	ν_{10}	259	418.8	252	242	255
	ν_{11}	167	—	—	—	—
Ref.		181	180	179	179	179

[a] Cs salt; [b] K salt.

MVFF calculation on this and $SClF_5$, $SeClF_5$, SF_6, SeF_6, and TeF_6 have been given.[181]

SF_6 has been subjected to a variety of physical studies. Measurement of i.r. combination bands has enabled the calculation of anharmonicity constants and an improved force field.[182] The appearance of the symmetry-forbidden Raman bands, ν_3 (T_{1u}) and ν_6 (T_{2u}) in the gas phase has been treated theoretically in terms of all significant single-molecule and pair contributions to the scattering cross-section involving the various hyperpolarizability tensors.[183] The i.r. double-resonance technique (excitation by a high-power laser pulse followed by probing the appropriate region with a C.W. tunable laser) was used to observe *V—V* energy transfer from the ν_3 mode to other vibration–rotation levels.[184] First-order Coriolis coupling in liquids of spherical tops, such as SF_6 and CF_4, has been treated theoretically. Comparison with the experimental Raman spectra indicates that the coupling accounts for the difference in linewidths between the depolarized lines.[185]

3 Single-crystal and other Solid-state Spectroscopy

By D. M. Adams

The emphasis in this Section remains on the acquisition of new data and well-substantiated assignments. Areas excluded are: mixed or doped crystals, alloys, second-order Raman scattering, mechanisms of phase transitions, and detailed investigations of lattice dynamics.

180 R. R. Smardzewski, R. E. Noftle, and W. B. Fox, *J. Mol. Spectroscopy*, 1976, **62**, 449.
181 W. V. F. Brooks, M. Eshaque, C. Lau, and J. Passmore, *Canad. J. Chem.*, 1976, **54**, 817.
182 R. S. McDowell, J. P. Aldridge, and R. F. Holland, *J. Phys. Chem.*, 1976, **80**, 1203.
183 R. Samson and A. Ben-Reuven, *J. Chem. Phys.*, 1976, **65**, 3586.
184 D. S. Frankel, *J. Chem. Phys.*, 1976, **65**, 1696.
185 M. Gilbert and M. Drifford, *J. Chem. Phys.*, 1976, **65**, 923.

Some of the publications listed at the start of this Chapter are relevant to this Section. In addition, we note the appearance of the Proceedings of the Campinas Conference on 'Light Scattering in Solids';[186] a general group-theoretical method of discussing changes in vibrational spectra due to Landau-type structural phase transitions;[187] a tabulation of the irreducible Raman scattering tensor operator for the 32 crystallographic point groups and its application to R.R.E. and electronic Raman spectra;[188] and a review of the normal-co-ordinate analysis of crystals.[189]

'Simple' Lattice Types.—The majority of work on the simpler lattice types is of a detailed physical nature, and there are few new assignments to report. Nine of the ten Raman-active modes of α-boron (D^5_{3d}) have been located:[190] 587, 693, 710, 776, 796, 872, 931, 1125, and 1185 cm^{-1}, and an assignment has been attempted, based upon force constants transferred from $[B_{12}H_{12}]^{2-}$. A series of lattice modes ranging from 170 to 1450 cm^{-1} was observed for β-rhombohedral boron, using i.r. reflectance,[191] and polarized Raman spectra of lanthanum and samarium hexaborides have been obtained.[192] Raman single-crystal experiments (77 K) on AlN (C^4_{6v}) yielded:[193] A_1 659.4, 897; E_1 672.4, 912 cm^{-1}. A most thorough i.r. reflectance and Raman single-crystal study has been made of further pyrite-type compounds MS_2, with the results:

M	A_g/cm^{-1}	T_g/cm^{-1}		
Ni	478	488	281	—
Cu	512	—	264	207
Zn	504	—	273	243

Combining these results with earlier ones on other 3*d* metal sulphides showed that the phonon modes correlate with a progressive increase in cell size from FeS_2 to ZnS_2 as a result of the addition of antibonding electrons.[194] N.C.A.s have been carried out for the pyrite-type solids MS_2 (M = Mn, Fe, Ru, or Os), $RuSe_2$, $RuTe_2$, and PtP_2.[195]

The complex unit cell of ZnP_2 (D^4_4, $Z = 8$) has been characterized by single-crystal Raman spectroscopy, with the results (wavenumbers/cm^{-1}):

A_1	55, 275, 323, 359, 425, 477
B_1	56, 105, 119, 128
B_2	109, 201, 234, 335, 362, 462, 472
E	85, 100, 122, 227, 249, 334, 462, 477

which compare with the theoretical requirement:[196]

$$\Gamma = 9A_1 + (8A_2) + 9B_1 + 9B_2 + 17E$$

186 'Proceedings of the Third International Conference on Light Scattering in Solids', ed. M. Balkanski, R. C. C. Leite, and S. P. S. Porto, Flammarion, Paris, 1976.
187 J. Petzelt and V. Dvorak, *J. Phys. Chem.*, 1976, **9**, 1587.
188 B. F. Goechter, *J. Mol. Spectroscopy*, 1976, **63**, 1.
189 J. Th. M. De Hosson, *Comput. Phys. Comm.*, 1975, **10**, 104.
190 W. Richter, W. Weber, and K. Ploog, *J. Less-Common Metals*, 1976, **47**, 85.
191 H. Binnenbruck and H. Werheit, *J. Less-Common Metals*, 1976, **47**, 91.
192 R. J. Betsch and W. B. White, *Proc. Rare Earth Res. Conference*, 1976, **2**, 534.
193 R. Tsu and R. F. Rutz, in ref. 186, p. 393.
194 E. Anastassakis and C. H. Perry, *J. Chem. Phys.*, 1976, **64**, 3604.
195 H. D. Lutz, P. Willich, and H. Hauseler, *Z. Naturforsch.*, 1976, **31a**, 847.
196 I. S. Gorban, V. A. Gorynya, V. I. Lugovoi, and A. P. Mokovetskaya, *Fiz. Tverd. Tela*, 1975, **17**, 1638.

Mixed Oxides, Fluorides, and other Ternary Phases.—No new vibrational assignments have appeared for perovskite fluorides, although interest was maintained in details of their phase transitions: $RbMnF_3$,[197–199] $RbCaF_3$.[200, 201] Fragmentary i.r. data have been given for $3KF,NdF_3$.[202]

I.r. and Raman spectra of ABO_2 were analysed on the basis of either D_{3d}^5 ($NaYO_2$, KYO_2) or D_{4h}^{19} (A = Li, Na, or K; B = Gd).[203] A most detailed re-investigation of the single-crystal Raman spectra of $LiTaO_3$ at room temperature led to the conclusion that results are best understood on the basis of C_3 and not C_{3v} symmetry, as previously thought. All nine (A_1 + E) modes were found:[204]

A_1	81, 186, 202, 250, 355, 458, 594, 661, 751
E	69, 141, 249, 316, 345, 382, 462, 591, 660 cm^{-1}

21 of the 26 optical phonons of $LiNbO_3$ have been located in an analogous study.[205] Raman investigations of the orthorhombic to rhombohedral phase transition,[206] and the polariton frequencies,[207] have been reported for $KNbO_3$. Qualitative i.r. and Raman data have been listed for $LaLnO_3$ (Ln = Lu, Yb, Tm, Er, or Y),[208] and a full Raman assignment has been established for ferroelectric (tetragonal, C_{4v}) $PbTiO_3$, which has hitherto been difficult to obtain in single-crystal form. All the expected modes were found:[209]

A_1		B_1		E	
TO	LO	TO	LO	TO	LO
147	189	289	289	88	128
359	465			220	439
646	796			505	723
				289	289

Two informative studies have been made of the spinel lattice. Polarized i.r. reflectance and transmission spectra of the 'superionics' Na- and Ag-β-alumina showed the vibrations of the spinel block to be almost decoupled from metal-ion motions for modes normal to the *c*-axis.[210] Raman and i.r. work with spinels having cation ordering on the octahedral sites revealed very complex spectra.[211]

The use of $^{46,50}Ti$ substitution in M_2TiO_5 (M = La to Dy) showed that i.r. and Raman modes >775 cm^{-1} correspond to local vibrations involving the

197 D. J. Lockwood and B. H. Torrie, *NATO Adv. Study Inst. Ser., Ser. E*, 1974, **1**, 147.
198 D. J. Lockwood and G. J. Coombs, *J. Phys. (C).*, 1975, **8**, 4062.
199 V. V. Eremenko, V. I. Fomin, and Yu. A. Popkov, in ref. 186, p. 905.
200 J. B. Bates, R. W. Major, and F. A. Modine, *Solid State Comm.*, 1975, **17**, 1347.
201 A. J. Rushworth and J. F. Ryan, *Solid State Comm.*, 1976, **18**, 1239.
202 B. S. Zakharova, V. A. Sipachev, and L. P. Reshetnikova, *Russ. J. Inorg. Chem.*, 1976, **21**, 201.
203 N. V. Porotnikov, Kh. T. Sharipov, K. I. Petrov, and I. A. Rozdin, *Russ. J. Inorg. Chem.*, 1976, **21**, 163.
204 A. F. Penna, A. Chaves, and P. da R. Andrade, *Phys. Rev. (B)*, 1976, **13**, 4907.
205 M. Nippus, *Z. Naturforsch.*, 1976, **31a**, 231.
206 M. P. Fontana, C. Razzetti, and A. M. Quittet, in ref. 186, p. 850.
207 R. Claus and F. X. Winter, *Ber. Bunsengesellschaft phys. Chem.*, 1975, **79**, 1034.
208 N. V. Porotnikov, S. S. Plotkin, and K. I. Petrov, *Russ. J. Inorg. Chem.*, 1976, **21**, 777.
209 R. A. Frey and E. Silbermann, *Helv. Phys. Acta*, 1976, **49**, 1.
210 A. S. Barker, jun. and J. A. Ditzenberger, *Phys. Rev. (B)*, 1976, **14**, 386.
211 V. G. Keramidas, B. A. Deangelis, and W. B. White, *J. Solid State Chem.*, 1975, **15**, 233.

shortest Ti—O bond.[212] Further qualitative data have been reported for Ln_2MO_6 (M = Mo or W) [213] and $ABTeO_6$ (A = Sr or Ba; B = Ge or Sn).[214] $Ca_2Nb_2O_7$ (C_2^2, $Z = 4$) has been characterized by single-crystal Raman and i.r. spectroscopy, about half of the theoretical $66(A + B)$ modes being located,[215] and qualitative i.r. and Raman data have been given for $La_2Sn_2O_7$, $Y_2Sn_2O_7$, $Y_2Zr_2O_7$, La_2MgSnO_6, Mg_2SnO_4, and Sr_2SnO_4.[216]

Two soft modes have been identified by a Raman study of the orthorhombic phase of $Cr_3B_7O_{13}Cl$.[217] The complete Raman spectra of both ferroelectric ($P3$) and paraelectric ($P\bar{6}$) phases of $Pb_5Ge_3O_{11}$ have been established (spectra given, but no frequencies) and several *A*-modes have been seen to disappear at the phase transition.[218] For the related $Bi_4Ge_3O_{12}$ (T_d^6, $Z = 2$) theory predicts $4A_1 + 5A_2 + 9E + 14T_1 + 14T_2$ modes, with LO/TO splitting expected (and observed) for the T_2 species. 25 of the 27 optically active modes were located and assigned, using single-crystal i.r. reflectance and Raman techniques.[219]

In a particularly important paper, the first Raman spectra have been reported for polymorphs of Nb_2O_5 and for some related oxides PNb_9O_{25}, VNb_9O_{25}, and $Ti_2Nb_{10}O_{29}$. These structures are all characterized by having crystallographic shear planes. The structures and spectra proved too complex to allow complete interpretation, but bands characteristic of edge-shared and corner-shared NbO_6 octahedral were identified.[220]

Sheet and Chain Structures.—Interest in one- and two-dimensional lattices has blossomed and become highly fashionable: Zallen has reviewed parts of the field.[221]

Between 5 and 300 K, the temperature variation of a line of A_1 symmetry, *ca.* 40 cm^{-1}, is consistent with its origin in an amplitude mode of the charge-density wave in $K_2Pt(CN)_4Br_{0\cdot3},3H_2O$.[222] Qualitative Raman and i.r. spectra for the same compound, and the chlorine analogue, show no coincidences, although previous *X*-ray measurements indicated the non-centrosymmetric group *P4mm*.[223] I.r. data have been collected for many related compounds and given partial assignments:[224] $M_2Pt(CN)_4Br_2,xH_2O$ (M = Na, K, or Ba/2) and $MPt(CN)_4,xH_2O$ (M = Na, K, or Mg/2).

In the SnS type (D_{2h}^{16}, $Z = 4$) structure of GeS there are two layers per cell. Each layer has C_{2v}^7 symmetry and, since inter-layer coupling is weak, there are

212 M. Th. Paques-Ledent, *Spectrochim. Acta*, 1976, **32A**, 1339.
213 N. N. Morozov, E. N. Murav'ev, L. Z. Gokhman, G. V. Lysanova, M. M. Ivanova, E. M. Reznik, E. G. Spiridonov, and G. I. Karpushkina, *Izvest. Akad. Nauk S.S.S.R., Neorg. Materialy*, 1975, **11**, 2005.
214 M. Robert and P. Tarte, *Compt. rend.*, 1976, **283**, *C*, 195.
215 S. D. Ross, *Spectrochim. Acta*, 1976, **32A**, 1331.
216 A. J. H. Macke and G. Blasse, *J. Inorg. Nuclear Chem.*, 1976, **38**, 1407.
217 D. J. Lockwood, *Solid State Comm.*, 1976, **18**, 115.
218 W. Taylor, D. J. Lockwood, and J. W. Arthur, *Ferroelectrics*, 1976, **12**, 113.
219 M. Couzi, J. R. Vignalou, and G. Boulon, *Solid State Comm.*, 1976, **20**, 461.
220 A. A. McConnel, J. S. Anderson, and C. N. R. Rao, *Spectrochim. Acta*, 1976, **32A**, 1067.
221 R. Zallen, *Proc. Int. Sch. Phys. 'Enrico Fermi'*, 1975, **55**, 159.
222 E. F. Steigmeier, R. Loudon, G. Hanbeke, H. Anderson, and G. Scheiber, *Solid State Comm.*, 1975, **17**, 1447.
223 J. R. Ferraro, L. J. Basile, and J. M. Williams, *J. Chem. Phys.*, 1976, **64**, 732.
224 J. R. Ferraro, L. J. Basile, and J. M. Williams, *J. Chem. Phys.*, 1976, **65**, 3025.

near-degeneracies in i.r. and Raman spectra.[225–227] From single-crystal i.r. and Raman experiments, the following assignment was deduced:[225]

Layer, C_{2v}	Cell, D_{2h}	
$3A_1$	A_g	111, 238, 269
	B_{1u}	117.5, 257.5, —
$2A_2$	B_{1g}	96, 251.5
	A_{2u}	inactive
$3B_1$	B_{2g}	114, 132, 242
	B_{3u}	105, 237, 280
B_2	B_{3g}	212
	B_{2u}	201

with lattice modes at 48 (A_g), 55 (B_{3g}), and 76 cm^{-1} (B_{2g}).

Several materials with the CdI_2 or MoS_2 structures have been investigated. For the CdI_2-type, selection rules require $A_{1g} + E_g + A_{2u} + E_u$.[228] From i.r. reflectance (wavenumbers/cm^{-1}):[229, 230]

	A_{2u}		E_u	
	TO	LO	TO	LO
SnS_2	340	356	205	314
$SnSe_2$	241	248	144	204

and from Raman scattering[231] on $SnSe_2$: A_{1g} 339, E_g 320 cm^{-1}. Powder data yielded (wavenumbers/cm^{-1}):[232]

	MoS_2	$MoSe_2$	$MoTe_2$
A_{2u}	466	348	287
E_{1u}	383	285	235

Charge-density waves in 2H-TaS_2[233] and 2H-$NbSe_2$[234] have been studied by Raman scattering. The 2H- and 4H-polytypes of PbI_2 could be distinguished by a small splitting of the E_g mode (79 cm^{-1}, 2H) to 75, 78.5 cm^{-1} (4H). No difference was found between their i.r. spectra.[235]

Improved single-crystal spectra have been obtained for the Ga–Ga-bonded layer-structured material β-GaS (D_{3h}), largely confirming earlier work. From the Raman data (wavenumbers/cm^{-1}) at 4 K:[236]

22	E'	shear	294	E''	ν(Ga—S)
74	E''	δ(GaGaS)	298	E'	ν(Ga—S)
190	A_1'	ν(Ga—Ga)	364	A_1'	ν(Ga—S)

[225] J. D. Wiley, W. J. Buckel, and R. L. Schmidt, *Phys. Rev.* (*B*), 1976, **13**, 2489.
[226] H. R. Chandrasekhar and V. Zwick, *Solid State Comm.*, 1976, **18**, 1509.
[227] D. I. Siapkas, D. S. Kyriakos, and N. A. Economou, *Solid State Comm.*, 1976, **19**, 765.
[228] Yu. M. Seidov and S. M. Seid-Rzaeva, *Fiz. Nizk. Temp.* (*Kiev*), 1976, **2**, 528.
[229] G. Lucovsky, J. C. Mikkelsen, and W. Y. Liang, *Phys. Rev.* (*B*), 1976, **14**, 1663.
[230] H. Koehler and C. R. Becker, *Phys. Status Solidi* (*B*), 1976, **76**, K15.
[231] O. P. Agnihotri, A. K. Garg, and H. K. Sehgal, *Solid State Comm.*, 1975, **17**, 1537.
[232] A. K. Garg, *Phys. Status Solidi* (*B*), 1976, **76**, 171.
[233] D. R. Karecki and B. P. Clayman, *Solid State Comm.*, 1976, **19**, 479.
[234] J. C. Tsang, J. E. Smith, and M. W. Shafer, *Phys. Rev. Letters*, 1976, **37**, 1407.
[235] A. Grisel and Ph. Schmid, *Phys. Status Solidi* (*B*), 1976, **73**, 587.
[236] G. Lucazeau, *Solid State Comm.*, 1976, **18**, 917.

The γ (C_{3v}^5), δ (C_{6v}^4), and ε (D_{3h}^1) polytypes of GaSe may be distinguished in the lattice-mode region by Raman scattering.[237] They show the following bands, which all arise from 'rigid layer' modes: δ 13.2, 37.2, 59.1; ε 19.3, 56.9, 59.1; γ 59.2 cm^{-1}.

The first single-crystal i.r. reflectance data for red HgI_2, together with another Raman study, yielded the following assignment, in which interactions between the two layers per cell are indicated:[238]

Layer, D_{2d}	*Crystal, D_{4h}*		*Layer*	*Crystal*	
A_1	A_{1g}	113	E	E_g	—
	B_{2u}	inactive		E_u	105
B_2	B_{1g}	144	E	E_g	29
	A_{2u}	132		E_u	25
B_2	B_{1g}	—	E	E_g	17
	A_{2u}	acoustic		E_u	acoustic

Similar Davydov splitting (i.r./Raman) has been reported from single-crystal studies of BiI_3,[239, 240] which also has two layers per cell. A reinvestigation of the i.r. reflectance spectra of single-crystal V_2O_5 gave data in good agreement with those of Gilson *et al.*,[241] but revealed major errors in their Kramers–Krönig analyses, and hence in the TO frequencies quoted. No new assignments resulted.[242]

Interest in layer structures containing arsenic and antimony remains. A single-crystal Raman study of claudetite, As_2O_3 (C_{2h}^5), revealed a large number of Davydov doublets due to inter-layer coupling ($C_s \xrightarrow{\times 2} C_{2h}$), and 'rigid layer' modes at 30, 38, and 49 cm^{-1}.[243] A complementary i.r. single-crystal study has been reported for As_2S_3;[244] the out-of-plane absorption is substantially weaker that that in the planes of the layers. The first single-crystal Raman spectra of BiSBr and BiSCl (both SbSI structure) have been reported;[245] and a soft mode has been found to accompany the displacive phase transition at 481 K in Sb_5O_7I.[246, 247] A very complete assignment of the complex spectra of proustite, Ag_2AsS_3, has been made by single-crystal methods,[248] and some qualitative data have been given for $AgAsS_2$.[249]

K_2CuF_4 and K_2NiF_4 have long been attributed the same D_{4h}^{17} layer structure. However, recent magnetic and, now, single-crystal Raman data show unequivocally that the cell must be doubled to D_{4h}^5 for the copper compound,

[237] A. Polian, K. Kunc, and A. Kuhn, *Solid State Comm.*, 1976, **19**, 1079.
[238] Y. Ogawa, I. Harada, and H. Matsuura, *Spectrochim. Acta*, 1976, **32A**, 49.
[239] V. V. Artamonov, L. I. Berezhinskii, M. Y. Valakh, B. K. Krulikovskii, and M. P. Lisitsa, *Fiz. Tverd. Tela*, 1975, **17**, 3621.
[240] T. N. Mel'nichenko, *Izvest. V.U.Z., Fiz.*, 1976, **19**, 155.
[241] T. R. Gilson, O. F. Bizri, and N. Cheetham, *J. Chem. Soc. (A)*, 1973, 291.
[242] P. Clauws and J. Vennik, *Phys. Status Solidi (B)*, 1976, **76**, 707.
[243] E. J. Flynn, S. A. Solin, and G. N. Papatheodorou, *Phys. Rev. (B)*, 1976, **13**, 1752.
[244] D. Treacy and P. C. Taylor, *Phys. Rev. (B)*, 1975, **11**, 2941.
[245] E. Furman, O. Brafman, and J. Makovsky, *Phys. Rev. (B)*, 1976, **13**, 1703.
[246] W. Prettl and K. H. Rieder, *Phys. Rev. (B)*, 1976, **14**, 2171.
[247] W. Prettl and K. H. Rieder, in ref. 186, p. 939.
[248] H. D. Riccius, P. R. Carey, and O. Siimann, *Phys. Status Solidi (B)*, 1975, **72**, K99.
[249] D. M. Bercha, I. I. Golovach, and V. Yu. Slivka, *Vestnik Sb., Khim. Svyaz' V Kristallakh I Ikh Fiz. Svoistva*, 1976, **1**, 125.

to accommodate 'orbital ordering'. Thus:[250]

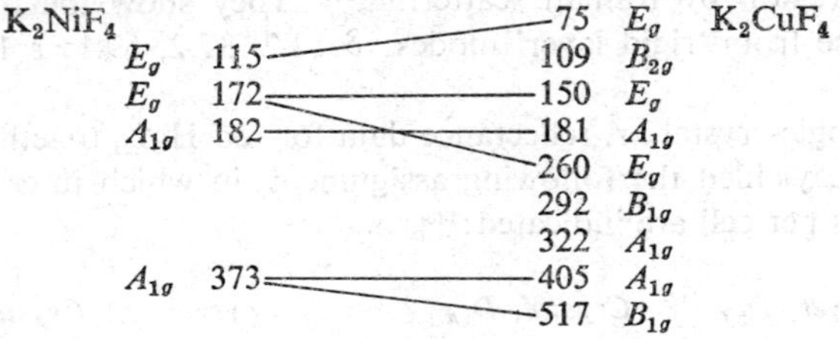

A complex series of layer structures $[RNH_3]_2[MCl_4]$ (M = Mn, Fe, Cu, or Cd) exists in which the individual layers are vibrationally isolated by layers of cations. Each of these structures undergoes one or more (up to four) phase transitions with temperature change. Their spectra are of great complexity, but evidence for rational assignments is slowly accumulating. We report far-i.r. reflectance data for $[MeNH_3]_2[MCl_4]$ (M = Mn or Fe)[251] and for $[C_nH_{2n-1}NH_3]_2[MCl_4]$ (M = Cu or Cd).[252] The latter study showed that at least part of the complexity is due to second-order absorptions.

Silicates.—Some very useful progress can be reported in this field, due to the increasing application of single-crystal methods. Qualitative studies abound, and will not be catalogued here. (See also Chapter 5.)

A very thorough single-crystal study of synthetic tephroite, Mn_2SiO_4 (olivine-type), complements earlier work on the closely similar forsterite, Mg_2SiO_4. The Raman-active components of the SiO_4^{4-} internal modes do not appear to interact with Mn motions, but the i.r. components of ν_2 and ν_4 are strongly mixed with lattice modes.[253] Another polarized-i.r. study of topaz seems to duplicate one by the same author reported last year.[254]

A full Raman assignment has been made to symmetry species for the pyrosilicate hemimorphite, $Zn_4(OH)_2Si_2O_7,H_2O$ (C_{2v}^{20}, $Z = 1$), and i.r. (powder) data have been obtained.[255] Detailed assignment to internal modes was not attempted, but could be approached by analogy with the excellent assignment now available for thortveitite.

Triorthosilicates $[Si_3O_{10}]^{8-}$ are rare in Nature, as they readily transform into either pyrosilicates or pyroxenes, $[Si_3O_9{}^{6-}]_n$. Spectra of six triorthosilicates have been given (for the 800—1100 cm^{-1} region only) and compared with the patterns from these structurally related types.[256] A detailed i.r. reflectance and Raman single-crystal investigation of benitoite, $BaTiSi_3O_9$, which contains a cyclic anion, allowed an almost complete assignment of the 41 optically active modes, as well as their LO components, many of which were directly observed.[257] Selection rules have been listed for the fibrous zeolite natrolite.[258]

[250] M. Kaneko, G. Kuuabara, and A. Misu, *Solid State Comm.*, 1976, **18**, 1085.
[251] N. Lehner, K. Strobel, R. Geick, and G. Heger, *J. Phys. (C).*, 1975, **8**, 4096.
[252] J. H. M. Stoelinga and P. Wyder, *J. Chem. Phys.*, 1976, **64**, 4612.
[253] H. D. Stidham, J. B. Bates, and C. B. Finch, *J. Phys. Chem.*, 1976, **80**, 1226.
[254] L. T. Kovaleva, *Tr. Inst. Geol. Geofiz., Akad. Nauk S.S.S.R., Sib. Otd.*, 1975, **271**, 26, 32.
[255] H. Poulet and J. P. Mathieu, *Bull. Soc. Fr. Minéral. Crist.*, 1975, **98**, 3.
[256] A. S. Povarennykh, B. N. Litvin, and M. Bednar, *Geol. Zhur. (Russ. Edn.)*, 1976, **36**, 76.
[257] D. M. Adams and I. R. Gardner, *J.C.S. Dalton*, 1976, 315.
[258] D. K. Arkhipenko, B. A. Orekhov, and I. A. Belitskii, *Trudy Inst. Geol. Geofiz., Akad. Nauk S.S.S.R., Sib. Otd.*, 1975, **271**, 19.

Single-crystal i.r. absorption or reflectance data have been reported for phlogopite,[259] muscovite (200—1400 cm^{-1}),[260, 261] and $Li_2Si_2O_5$.[262] The latter material is the simplest mica-like silicate, having only one layer per unit cell: i.r. reflectance spectra were obtained from all three orthorhombic directions, and form the basis of a theoretical discussion. Studies such as these hold out the promise of real advances in understanding the lattice dynamics of layer silicates. A comment on the OH direction in axinite is noted.[263]

Oxoanionic Crystals.—$Gd_2(MoO_4)_3$ has been studied by Raman scattering, revealing (at 80 K) 126 of the 201 theoretically allowed bands, and much LO–TO splitting.[264] Two A_1 modes (45, 51 cm^{-1} at 80 K) in the ferroelectric phase were found to be 'soft'.[265, 266]

Complex Cationic Salts.—Phase-transition studies of ammonium compounds continue to absorb the attention of many workers, and many new data for NH_4X (X = halogen,[267–270] NO_3,[271] SO_4, ClO_4, or BeF_4) have been reported, although few of these have resulted in new assignments of general concern. A Raman study of NH_4I up to 16.4 kbar revealed drastic changes at phase transitions, and allowed more accurate determination of the phase diagram.[270] Phases IV, V, and VII of NH_4NO_3 and ND_4NO_3 have been characterized by Raman methods.[271] Of the effort expended on $(NH_4)_2SO_4$,[272–276] the most useful for our purpose is the single-crystal Raman study of the paraelectric phase (ambient).[272] A very complete assignment was deduced and, by the use of ^{15}N substitution, a distinction was made between translational (152—223 cm^{-1}) and librational (358, 375 cm^{-1}) modes of NH_4^+. The discrepancy between these results and a recent neutron-diffraction estimate of the position of the librational mode (305 ± 25 cm^{-1}) was noted, but not explained. The ferroelectric phase of $(NH_4)_2SO_4$ has also been investigated by Raman scattering,[274, 275] as has the A.T.R. behaviour of the paraelectric material.[276] Raman spectra of NH_4ClO_4 in the 12—300 K region are consistent with retention of the D_{2h}^{16} structure throughout. A B_{1g} mode at 180 cm^{-1} was attributed to an NH_4^+ libration,[277]

259 D. K. Arkhipenko and N. A. Pal'chik, *Trudy Inst. Geol. Geofiz., Akad. Nauk S.S.S.R., Sib. Otd.*, 1975, **271**, 13.
260 V. I. Vettegren, K. N. Kaksenko, I. I. Novak, and A. E. Chmel, *Fiz. Tverd. Tela*, 1975, **17**, 3088.
261 E. Steger, *Spectrochim. Acta*, 1976, **32A**, 1125.
262 V. F. Pavinich, A. P. Mirgorodskii, and V. A. Kolesova, *Optika i Spektroskopiya*, 1976, **40**, 817.
263 Y. Takeuchi, *Z. Kristallogr., Kristallgeom., Kristallphys., Kristallchem.*, 1975, **141**, 471.
264 B. N. Ganguly, F. G. Ullman, R. D. Kirby, and J. R. Hardy, *Phys. Rev. (B)*, 1976, **13**, 1344.
265 T. Shigenari, Y. Takagi, and Y. Wakabayashi, *Solid State Comm.*, 1976, **18**, 1271.
266 Y. Takagi and Y. Wakabayashi, *J. Phys. Soc. Japan*, 1976, **41**, 719.
267 G. G. Mishin and V. S. Gorelik, *Fiz. Tverd. Tela*, 1975, **17**, 2889.
268 H. D. Hochheimer and T. Geisel, *J. Chem. Phys.*, 1976, **64**, 1586.
269 Y. Ebisuzaki, *Chem. Phys. Letters*, 1976, **37**, 127.
270 H. D. Hochheimer, E. Spanner, and D. Strauch, *J. Chem. Phys.*, 1976, **64**, 1583.
271 Z. Iqbal, *Chem. Phys. Letters*, 1976, **40**, 41.
272 R. L. Carter, *Spectrochim. Acta*, 1976, **32A**, 575.
273 Z. Iqbal and C. W. Christoe, *Ferroelectrics*, 1976, **12**, 177.
274 H. G. Unruh and O. Ayere, *Ferroelectrics*, 1976, **12**, 181.
275 Z. Iqbal and C. W. Christoe, *Solid State Comm.*, 1976, **18**, 269.
276 Y. S. Jain and H. D. Bist, *Proc. Nucl. Phys. Solid State Phys. Symp.*, 1974, **17C**, 284.
277 G. J. Rosasco and H. J. Prask, *Solid State Comm.*, 1975, **16**, 135.

but, in view of the above comments, is probably a translatory mode. The orthorhombic to cubic phase transition has also been investigated by the Raman technique.[278] Single-crystal Raman studies of $(NH_4)_2BeF_4$ have been reported for both ambient temperature[279–281] and 4.2 K.[281] NH_4^+ translatory modes were assigned in the range 145—200 cm^{-1}, but librational modes were not located.[281] The Raman bands at 183 and 198 cm^{-1} in $[K_{1-x}(NH_4)_x]_2$-$[CuCl_4(H_2O)_2]$, attributed to NH_4^+ librations, are, no doubt, translations.[282] The site symmetry of the ammonium ion in crystals can be probed using NH_3D^+, analogous to the use of HOD in hydrates: two studies of this kind have appeared dealing with NH_4ReO_4 and $[NH_4]_2[CuCl_4(H_2O)_2]$.[283, 284]

Further single-crystal Raman data for orthorhombic $N_2H_6SO_4$[285] complement an earlier study:[286] LO and TO modes were found. Studies of the phase behaviour of hexa-ammine complexes continue, with i.r.-profile measurements[287] of $[Co(NH_3)_6]I_3$ and a theoretical analysis of $[Ni(NH_3)_6]I_2$, which showed that the phase transitions may be attributed to coupling of a soft librational and a TA mode, and they result, at low temperatures, in freezing out of the hydrogen motions.[288]

Fragmentary data have been given for $Al(NO_3)_3,9H_2O$,[289] and fuller, but incompletely assigned, Raman data for $[M(H_2O)_9][BrO_3]_3$ (M = Nd or Gd)[290] and $[La(H_2O)_6][NO_3]_3$.[291]

Complex Anionic Salts.—KCN and NaCN continue to excite interest, due to the order–disorder phase transitions they undergo. In KCN, for example, there is complete order below 83 K, with antiparallel orientation of CN^-; then a phase in which the anions remain aligned but become disordered with respect to C and N; and, finally, above 168 K, a disordered phase in which the CN^- groups move in a multi-well potential that is of octahedral symmetry. A variety of spectroscopic manifestations of these changes have been reported.[292–296]

Many new data have been gathered, using single-crystal Raman methods, for MNCO (M = K, Rb, or Cs),[297] KNCS and RbNCS,[298] and KNCSe.[299] In these

[278] T. B. Brill and F. Goetz, *J. Chem. Phys.*, 1976, **65**, 1217.
[279] Y. S. Jain, *Indian J. Phys.*, 1976, **50**, 206.
[280] Y. S. Jain, *Proc. Nucl. Phys. Solid State Phys. Symp.*, 1974, **17C**, 295.
[281] Y. A. Popkov, V. F. Shabanov, V. V. Eremenko, and K. S. Aleksandrov, *Fiz. Nizk. Temp. (Kiev)*, 1975, **1**, 936.
[282] M. L. Bansal, V. C. Sahni, and A. P. Roy, *Proc. Nucl. Phys. Solid State Phys. Symp.*, 1974, **17C**, 250.
[283] I. A. Oxton, O. Knop, and M. Falk, *J. Phys. Chem.*, 1976, **80**, 1212.
[284] I. A. Oxton, O. Knop, and M. Falk, *Canad. J. Chem.*, 1976, **54**, 892.
[285] C. Caville, *Canad. J. Spectroscopy*, 1975, **20**, 123.
[286] M. Guay, J. Weber, and R. Savoie, *Canad. J. Spectroscopy*, 1974, **19**, 127.
[287] M. Goettel, J. M. Janik, and J. A. Janik, *Phys. Status Solidi (A)*, 1976, **35**, 675.
[288] A. R. Bates, I. L. Crick, and R. O. Davies, *J. Phys. (C)*., 1976, **9**, 3013.
[289] I. I. Kondilenko, P. A. Korotkov, and N. G. Golubeva, *Zhur. priklad. Spektroskopii*, 1976, **24**, 532.
[290] H. Poulet, J. P. Mathieu, D. Vergnat, and B. Vergnat, *Phys. Status Solidi (A)*, 1975, **32**, 509.
[291] A. Armengaud, M. Abenoza, and D. V. Luu, *Compt. rend.*, 1976, **283**, *B*, 155.
[292] W. Dultz, *J. Chem. Phys.*, 1976, **65**, 2812.
[293] W. Dultz, in ref. 186, p. 852.
[294] J. Daubert, K. Knorr, W. Dultz, H. Jex, and R. Currat, *J. Phys. (C)*., 1976, **9**, L389.
[295] D. Fontaine, M. Krauzman, and R. M. Pick, in ref. 186, p. 692.
[296] L. W. Winchester, *Diss. Abs. Internat. (B)*, 1975, **36**, 2336.
[297] S. S. Ti, S. F. A. Kettle, and O. Ra, *J. Cryst. Mol. Structure*, 1976, **6**, 43.
[298] S. S. Ti, S. F. A. Kettle, and O. Ra, *Spectrochim. Acta*, 1976, **32A**, 1603.
[299] S. S. Ti and S. F. A. Kettle, *Spectrochim. Acta*, 1976, **32A**, 1765.

studies, particular attention was paid to the analysis and provenance of the second-order features. For KNCSe only ν_1 and ν_2 showed any observable Davydov splitting (at ambient temperature); the following assignment was deduced (wavenumbers/cm^{-1}):[299]

ν_1	ν_2	ν_3
2068, 2076	419, 421, 427	563

The behaviour of $NaNO_2$ near the ferroelectric transition has been studied by both Raman[300] and single-crystal i.r. methods,[301] and a new band at *ca.* 825 cm^{-1} in the high-temperature phase has been associated with the onset of hindered rotation of the anions. Some partial, polarized, i.r. spectra have been reported for H_3PO_2 and D_3PO_2. ν(OH) shows a very large factor-group splitting.[302]

KNO_3 has been reinvestigated by the single-crystal Raman technique (ambient temperature), obtaining, for the first time, the full multiplet structure predicted by factor-group analysis for ν_3 and ν_2. No such structure was found for ν_1, ν_4. Thus:[303]

D_{3h}, *ion*	C_s, *site*	D_{2h}, *crystal*
ν_1, A_1'	A'	A_g, B_{3g} 1051
ν_2, A_2''	A'	A_g 827, B_{3g} 830
ν_3, E'	A'	A_g 1361, B_{3g} 1451
	A''	B_{1g} 1345, B_{2g} 1388
ν_4, E'		A_g, B_{1g}, B_{2g}, B_{3g} 716 cm^{-1}

Various Raman and i.r. studies have been made of the phase transitions in MNO_3 (M = Li, Na, or K)[304] and the ordered phases of $RbNO_3$ and $CsNO_3$[305] and of NH_4NO_3.[306]

A particularly thorough single-crystal Raman investigation of $AgNO_3$ at both ambient temperature and 77 K yielded 58 of the 60 predicted modes of the D_{2h}^{15}, $Z = 8$ cell, including a very long list of lattice modes. The data for the ν_1 and ν_3 regions are quoted, in illustration. All atoms are on general positions, and hence each internal mode will generate a complete D_{2h} vector of factor-group components:[307]

ν_1 A_1'	$\longrightarrow$	A_g	1046.9	B_{1g}	1045.4
		B_{2g}	1045.9	B_{3g}	1045.4
ν_3 E'	$\longrightarrow$	A_g	1308.1 and 1325.0		
		B_{1g}	1339.7 and 1365.7		
		B_{2g}	1331.0 and 1350.5		
		B_{3g}	1339.7 and 1347.6 cm^{-1}		

Most of the predicted Raman-active modes ($4A_g + 4E_g + 12T_g$) have also been located for the cubic (T_h^6, $Z = 4$) phase of $Sr(NO_3)_2$; only qualitative data were

300 I. N. Goncharuk and E. V. Chisler, *Fiz. Tverd. Tela*, 1975, **17**, 141.
301 E. A. Ivanova and E. V. Chisler, *Fiz. Tverd. Tela*, 1975, **17**, 2873.
302 S. Detoni, D. Hadzi, and B. Orel, *J. Mol. Structure*, 1976, **33**, 279.
303 R. Frech and J. P. Devlin, *Chem. Phys. Letters*, 1976, **38**, 79.
304 A. A. Artamanov, N. G. Episheva, G. Y. Keisanov, E. N. Fedorova, I. P. Belyaeva, and A. V. Sechkarev, *Spektrosk. Ee Primen. Geofiz. Khim.*, 1975, 163.
305 S. V. Karpov and A. A. Shulmin, *Fiz. Tverd. Tela*, 1975, **17**, 2868.
306 Z. Iqbal, *Chem. Phys. Letters*, 1976, **40**, 41.
307 C. H. Huang and M. H. Brooker, *Spectrochim. Acta*, 1976, **32A**, 1715.

obtained for the i.r. region.[308] Polycrystalline $M(NO_3)_2$ (M = Zn, Cd, or Ca) have been studied by both i.r. and Raman spectroscopy. At room temperature, $Ca(NO_3)_2$ has spectra compatible with the above $Sr(NO_3)_2$ structure, but it undergoes phase transitions at 88.5 and 287.8 K, whereas for M = Zn or Cd no phase changes were detected.[309]

Both $NaH_3(SeO_3)_2$ [310] and $NaD_3(SeO_3)_2$ [311] have been studied in the lattice-mode region by single-crystal Raman methods, and the modes have been classified as either A_g or B_g in C_{2h} (monoclinic). The only substantial frequency differences were in modes in the 140—280 cm^{-1} interval. Thus for H(D) the values of wavenumbers/cm^{-1} are:

A_g	278 (264)	B_g	156 (143)	B_g	112 (112)
B_g	252 (250)	A_g	133 (133)	B_g	92 (92)
A_g	206 (212)	A_g	125 (123)	A_g	70 (70)
B_g	159 (154)	A_g	113 (112)	B_g	57 (57)

An abstract has been published indicating that $MH_3(SeO_3)_2$ (M = Li, Na, K, or Cs) have all been studied as single crystals by Raman methods, both above and below their ferroelectric transition points.[312]

An assignment (single-crystal Raman) has been made of all the optical phonons of $AgClO_3$. For the C_{4h}^5, $Z = 4$ cell ($\Gamma = 8A_g + 8B_g + 7E_g$), all predicted modes were found, apart from three of A_g type. For the internal modes (wavenumbers/cm^{-1}):[313]

C_{3v}, *ion*	C_s, *site*	C_{4h}, *crystal*
ν_1, A_1	A'	A_g, B_g 895
ν_2		A_g, B_g 618
ν_3, E	$A' + A''$	A_g—, B_g 905, E_g 920
ν_4		A_g—, B_g 481, E_g 494

Qualitative Raman and i.r. data have been reported for $Ba(ClO_3)_2$, and are consistent with symmetry *Fd2d*,[314] and many isotopic variants (*e.g.* $^{35}Cl^{16}O_2{}^{18}O^-$, $^{37}Cl^{16}O_2{}^{18}O^-$) of chlorates and bromates have been characterized at low temperatures in single crystals MEO_3 (M = Na or K; E = Cl or Br). N.C.A. was used as a guide to assignment.[315] Some data have been given for $KBr^{18}O_3$.[316] Single-crystal Raman and i.r. data on $M(BrO_3)_3,9H_2O$ (M = Nd or Gd) are consistent with the space group C_{2v}^{12} for the low-temperature phase.[317] Details continue to be collected for $LiIO_3$, chiefly in respect of its phase transitions,[318–320] and an

308 M. Scrocco and L. Mattioli, *Phys. Status Solidi (B)*, 1975, **72**, K13.
309 M. H. Brooker, *Spectrochim. Acta*, 1976, **32A**, 369.
310 V. F. Shabanov and A. P. Fedotov, *Fiz. Tverd. Tela*, 1975, **17**, 553.
311 V. F. Shabanov and A. P. Fedotov, *Fiz. Tverd. Tela*, 1976, **18**, 1588.
312 C. A. Cody and R. K. Khanna, in ref. 186, p. 932.
313 G. M. Gualbento, A. F. Penna, C. A. Arguello, and H. Vargas, *Solid State Comm.*, 1975, **17**, 1481.
314 H. Vargas, J. Pelzl, J. Laugier, D. Dautieeppe, and J. P. Mattieu, *J. Magn. Resonance*, 1975, **20**, 269.
315 J. B. Bates and H. D. Stidham, *J. Phys. and Chem. Solids*, 1976, **37**, 183.
316 E. W. Holtzschieter and E. R. Johnson, *Internat. J. Appl. Radiation Isotopes*, 1975, **26**, 780.
317 H. Poulet, J. P. Mathieu, D. Vergnat, B. Vergnat, A. Hadni, and X. Gerbaux, *Phys. Status Solidi (A)*, 1975, **32**, 509.
318 K. Haller, E. Jarvekulg, and L. Rebane, *Fiz. Tverd. Tela*, 1976, **18**, 1409.
319 D. F. Baisa, A. I. Barabash, and G. A. Puchkovskaya, *J. Mol. Structure*, 1976, **31**, 247.
320 J. P. Misset and M. Remoissenet, *Solid State Comm.*, 1976, **20**, 443.

abstract indicates that i.r. and Raman single-crystal work has been done on $NaIO_3$.[321]

Some quite exceptionally beautiful single-crystal i.r. reflectance spectra have been obtained (at ambient temperature) for both HIO_3 and DIO_3, covering the entire range 30—3500 cm^{-1}. Optical constants were extracted by Kramers–Krönig analysis.[322] The crystals are orthorhombic (D_2^4, $Z = 4$). Apart from modes due to OH(D) motion, the spectra of the two materials were closely similar; these modes are (wavenumbers/cm^{-1}):

	H	D
B_1	2940, 1150, 564	—, 859, 416
B_2	2850, 1140, 570	2180, 854, —
B_3	3000, 1150, 571	2200, 862, 420

Single-crystal Raman assignments have been given for Na_2SO_4 (D_{2h}^{24}, $Z = 2$),[323] $LiKSO_4$,[324, 325] and of a series of alums,[326, 327] all studies yielding long lists of numbers, and assignments being in conformity with similar work by earlier authors. Vibrational normal co-ordinates of K_2SO_4 have been discussed.[328]

Some polarized i.r. spectra of very high quality have been given for the series $LuEO_4$ and YEO_4 (E = P, As, or V), which have the D_{4h}^{19}, $Z = 2$ structure. Typically, values of wavenumbers/cm^{-1} for $LuAsO_4$ are:[329]

	ν_1	ν_3	ν_4	ν (trans.)	ν (rot.)
A_{2u}	912	872	452	227	—
E_u	—	816	337	185	310

Further details of the lattice dynamics of the ferroelectric KH_2PO_4 [330] and KD_2PO_4 [330, 331] have been studied, including polarized i.r. transmission work.[330] Although not strictly within the area of this Chapter, we note a theoretical study of the torsional potential functions of OH and water in apatites which confirmed some earlier assignments.[332]

A single-crystal Raman study of Tl_3AsS_4 has been completed, but the authors report only the wavenumbers/cm^{-1} of the $14A_g$ modes for the D_{2h}, $Z = 4$ cell, *viz.* 46, 50, 55, 79, 85, 94, 111, 119, 187, 207, 222, 378, 402, and 420.[333]

A full assignment of the many site and factor-group components in the Raman spectrum of Cs_2ZnCl_4 has been deduced from single-crystal measurements.[334]

321 S. B. S. Sastry, R. B. Tripathi, and C. Ramasastry, *Proc. Nucl. Phys. Solid State Phys. Symp.*, 1974, **17C**, 247.
322 T. Ogawa, M. Ononichi, T. Arai, and K. Kudo, *Sci. Light (Tokyo)*, 1975, **24**, 97.
323 S. Montero, *Spectrochim. Acta*, 1976, **32A**, 843.
324 J. Hiraishi, N. Teniguchi, and H. Takahashi, *J. Chem. Phys.*, 1976, **65**, 3821.
325 T. R. Rao, M. L. Bansal, V. C. Sahni, and A. P. Roy, *Phys. Status Solidi (B)*, 1976, **75**, K31.
326 H. H. Eysel and J. Eckert, *Z. anorg. Chem.*, 1976, **424**, 68.
327 J. Eckert, H. H. Eysel, and G. L. Kampffmeyer, *Z. anorg. Chem.*, 1976, **424**, 81.
328 G.-J. Wu and R. Frech, *J. Chem. Phys.*, 1976, **64**, 4897.
329 A. Armbruster, *J. Phys. and Chem. Solids*, 1976, **37**, 321.
330 J. Baran and H. Ratajczak, 'Molecular Spectroscopy of Dense Phases; Proceedings of the European Congress on Molecular Spectroscopy', Elsevier, Amsterdam, 1975, p. 225.
331 R. Blinc, R. Pirc, and B. Zeks, *Phys. Rev. (B)*, 1976, **13**, 2943.
332 P. F. Gonzales-Diaz and A. Hidalgo, *Spectrochim. Acta*, 1976, **32A**, 1119.
333 V. A. Stefanovich, V. Yu. Slivka, Yu. V. Voroshilov, A. P. Zhdankin, M. I. Golovei, and I. Yu. Roman, *Fiz. Tverd. Tela*, 1975, **17**, 3643.
334 P. T. T. Wong, *J. Chem. Phys.*, 1976, **64**, 2186.

Although $[enH_2][PdCl_4]$ (C^5_{2h}, $Z = 2$) proved too dark for adequate Raman work, a most thorough investigation of its i.r. dichroic behaviour has been completed, and supported with work on both *C*- and *N*-deuteriated variants. The anion modes (100 K) were located at (wavenumbers/cm^{-1}):[335]

	ν(Pd—X)	δ(Pd—X)	π(Pt—X)
X = Cl	336	194	179
X = Br	256	142	137

The origin of some new features developed in the i.r. spectra of $M_2[PdCl_4]$ and $M_2[PtCl_4]$ (M = K, NH_4, or ND_4), near the ν(Pd—Cl) or ν(Pt—Cl) region, has been considered, but incompletely accounted for.[336]

Octahedral hexafluoride species have been quite fully studied this year, but the most important paper is unquestionably that by Forrest and Lane, in which single-crystal Raman and i.r. reflectance data are reported for K_2TiF_6, Rb_2ZrF_6, and Cs_2HfF_6.[337] These materials have the symmetry D^3_{3d}, $Z = 1$. Assignments are gathered in Table 7.[337, 338] Full unit-cell normal-co-ordinate analyses were

Table 7 *Single-crystal i.r. and Raman wavenumbers*/cm^{-1} *and assignments for some hexafluorometallates*(IV)

Mode		K_2TiF_6	Rb_2ZrF_6	Cs_2HfF_6	Cs_2TiF_6
ν_1	A_{1g}	618	589	572	597
ν_3	E_u	615	537	498	—
	A_{2u}	600	522	490	—
ν_2	E_g	(440) [a]	(416)	(389)	453
ν_5	A_{2g}	308	258	257	284
	E_g	300	244	247	283
ν_4	E_u	315	240	217	—
	A_{2u}	281	192	184	—
ν_6	E_u	(193)	(141)	(122)	—
ν_{LT}	E_u	136	98	87	—
ν_{LT}	A_{2u}	122	92	87	—
ν_{LR}	E_g	138	90	83	84
ν_{LT}	E_g	82	74	69	67
ν_{LT}	A_{1g}	85	73	58	58
Ref.		337	337	337	338

[a] Values in parentheses are calculated.

carried out. For all three compounds the E_u components of the octahedral ν_3 and ν_4 modes were at higher frequencies than their A_{2u} components. Placed alongside the results of calculations of the effect of D_{3d} distortion (elongation or flattening) upon normal modes of PtF_6 and WF_6, it was shown that, contrary to *X*-ray evidence, the distortions of the MF_6^{2-} octahedra in the three compounds are of the same type.

The group T_h 'is the only one that can account for the observed frequencies' of the anion in $[O_2][AsF_6]$, which are (wavenumbers/cm^{-1}):[339]

ν_1	A_g	689	ν_3	T_u	700	ν_5	T_g	375
ν_2	E_g	573	ν_4	T_u	385	ν_6	T_u	252

335 R. W. Berg, *Spectrochim. Acta*, 1976, **32A**, 1747.
336 D. M. Adams and R. W. Berg, *J.C.S. Dalton*, 1976, 52.
337 I. W. Forrest and A. P. Lane, *Inorg. Chem.*, 1976, **15**, 265.
338 W. von der Ohe, *J. Chem. Phys.*, 1976, **65**, 3575.
339 C. Naulin and R. Bougon, *J. Chem. Phys.*, 1976, **64**, 4155.

$\nu(O_2^+)$ is at 1858 cm^{-1}, and several of the anion modes split into two or three components on passing through an order–disorder phase transition at *ca.* 270 K. Vibrational data have been adduced in interpretation of the luminescence spectra of Cs_2MnF_6 and A_2MF_6:MnF_6 (A = K or Rb, M = Si; A = Cs, M = Si or Ge)$^{2-}$.[340] Raman and i.r. (powders) data were used to deduce probable space groups for $KSbF_6$ (I), D_{4h}^{10}, and $KSbF_6$ (II), T^3 or T^5.[341]

Low-temperature Raman and i.r. spectra (powders) for NO[$SbCl_6$], $[NO]_2$-[$SnCl_6$], and $[NO]_2[TiCl_6]$ showed pronounced differences from those obtained under ambient conditions, consistent with order–disorder phase changes. Bands at *ca.* 235 cm^{-1} were attributed to NO^+ librational motions.[342]

A new single-crystal Raman study of $K_2Hg(CN)_4$ improves on the assignments and data reported by others in the past two years and, in addition, notes that there is a phase change below 110 K to an unknown structure of lower symmetry, as indicated by splittings of the following bands at 4 K,[343] all of which are of T_g symmetry initially: 2149 → 2150.5, 2149; 331 → 331, 326; 137 → 145, 142 cm^{-1}. Some complex single-crystal Raman data for the Prussian Blue analogues $M_3[Co(CN)_6]_2,xH_2O$ (M = Mn or Cd) have been obtained, supplemented by related work on dehydration and ammoniation of the lattices, and discussed in terms of M—C bonding and other relevant parameters.[344] There is a phase change in $Cs_2Li[Cr(CN)_6]$ at low temperature, heralded by the appearance of several new bands in the 20—40 cm^{-1} region, including a soft mode which hardens to 31 cm^{-1} at 20 K.[345]

Although no mode descriptions were offered, the following assignment (wavenumbers/cm^{-1}) of the single-crystal Raman spectra of $LiCO_2H,H_2O$ has been given:[346]

A_1	83	196	594	561	1350
A_2	110	153	211	1047	1356
B_1	93	149	207	1052	1364
B_2	141	774			

Raman and i.r. spectra of $Nd(Cl_3CO_2)_3,3H_2O$ imply a C_{2h} factor group.[347]

$Rb_2[NiCl_4(H_2O)_2]$ occurs in the triclinic group $P\bar{1}$, with $Z = 1$: single-crystal studies are therefore unprofitable, but a rather thorough i.r. and Raman study yielded the assignment:[348] ν(Ni—O) 353 (i.r.); 342 (Ra); ν(Ni—Cl) 238, 215 (i.r.); 227, 200 (Ra) cm^{-1}.

Combining extant single-crystal i.r. data with new single-crystal Raman results allowed an assignment to be made for $K[B_5O_6(OH)_4],2H_2O$, assisted by N.C.A. The stretching vibrations could be interpreted in terms of BO_4 and BO_3 group vibrations, but the bending modes were heavily mixed.[349]

[340] S. L. Chodos, A. M. Black, and C. D. Flint, *J. Chem. Phys.*, 1976, **65**, 4816.
[341] A. M. Heyns and C. W. F. T. Pistorius, *Spectrochim. Acta*, 1976, **32A**, 535.
[342] D. M. Adams and R. Appleby, *J. Inorg. Nuclear Chem.*, 1976, **38**, 1601.
[343] W. Krasser and S. Hausshuel, *Solid State Comm.*, 1976, **20**, 191.
[344] B. I. Swanson, *Inorg. Chem.*, 1976, **15**, 253.
[345] B. I. Swanson and R. R. Ryan, in ref. 186, p. 944.
[346] M. Yamamoto, H. Ito, and H. Inaba, *Phys. Letters* (*A*), 1975, **55A**, 303.
[347] R. C. Naik, K. H. Ayyar, and N. A. Navasimham, *Proc. Nucl. Phys. Solid State Phys. Symp.*, 1974, **17C**, 254.
[348] G. E. Shankle and J. B. Bates, *J. Chem. Phys.*, 1976, **64**, 2539.
[349] V. Devarajan, E. Gräffe, and E. Funck, *Spectrochim. Acta*, 1976, **32A**, 1225.

Molecular Crystals.—Raman data have been collected for solid nitrogen in the ranges 0—220 K and 0—10 kbar. The best-defined spectra, from the γ-phase at 8 K, yielded the values 2331 cm^{-1} [$\nu(N{\equiv}N)$] and 58.4, 103.6 cm^{-1} (external modes).[350]

Ice VIII exists only under a combination of low temperature and high pressure. By using a pre-stressing technique, relatively large blocks of this phase were obtained at 100 K, and they yielded excellent single-crystal Raman spectra with very sharp features throughout the range 50—4000 cm^{-1}. Typically, in the $\nu(OH)$ region, the wavenumbers/cm^{-1} are:[351]

C_{2v}, *molecule*	D_{4h}^{19}, *crystal*
ν_1, A_1	A_{1g} (3358.8) + B_{1g} (3477.4)
ν_3, B_2	E_g (3447.6)

The Raman spectrum of ice Ih has been reinvestigated in the interval 20—350 cm^{-1}: vibrations below *ca.* 200 cm^{-1} are considered to be mainly disorder-allowed.[352]

An extremely detailed review of the electronic spectra of ReF_6 includes much detail on the vibrational fine-structure assignment for the solid state.[353]

Despite many detailed investigations of orthorhombic sulphur, the low-frequency i.r.-active lattice modes have never been directly observed. This omission has been repaired.[354] The polycrystalline material showed lattice modes (at 4.2 K) at 33.5, 44, 50, 60.5, and 96.5 cm^{-1}. A Davydov multiplet at 71, 74, 78 cm^{-1} was assigned to the internal mode ν_9. Monoclinic β-sulphur, studied by Raman scattering, shows internal modes close to those of the orthorhombic form. The external mode spectrum at 373 K is broad and typical of a partially disordered phase, but is gradually resolved on cooling (to 40 K) to yield a large number of lines.[355]

trans-[$Co(CH_3)(py)(dmg)_2$] (dmg = dimethylglyoximato) has the symmetry C_{2h}^5, $Z = 4$. Polarized i.r. spectra obtained from a (001) face showed some dichroic behaviour: an attempt was made to interpret the complex data using the oriented-gas model. Probably the most significant point is that the absorption at *ca.* 900 cm^{-1}, previously attributed to the out-of-plane O—H mode of the very short hydrogen bond, does not have the correct symmetry for this.[356] Finally, we note a single-crystal study of $CuCl_2(dmso)_2$ and $CuCl_2[^2H_6](dmso)_2$ at 77 K by i.r. reflectance and Raman spectroscopy.[357] An almost complete assignment was deduced. $\nu(Cu{-}O)_s$ and $\nu(Cu{-}O)_a$ are at 481 and 496 cm^{-1}, respectively; $\nu(Cu{-}Cl)_s$ and $\nu(Cu{-}Cl)_a$ are at 229 and 286 cm^{-1}.

4 Vibrational Spectra of Solutions

By P. Gans

Aqueous Solutions.—Near-i.r. studies of the water combination band ($\bar{\nu}_{1,3} + \nu_2$) for solutions of 16 electrolytes have indicated changes in solution

[350] F. D. Medina and W. B. Daniels, *J. Chem. Phys.*, 1976, **64**, 150.
[351] P. T. T. Wong and E. Whalley, *J. Chem. Phys.*, 1976, **64**, 2359.
[352] P. T. T. Wong and E. Whalley, *J. Chem. Phys.*, 1976, **65**, 829.
[353] E. R. Bernstein and G. R. Meredith, *J. Chem. Phys.*, 1976, **64**, 375.
[354] A. Anderson and P. G. Boczar, *Chem. Phys. Letters*, 1976, **43**, 506.
[355] G. Gautier and M. Debeau, *Spectrochim. Acta*, 1976, **32A**, 1007.
[356] A. Bigotto, E. Zangrando, and L. Randaccio, *J.C.S. Dalton*, 1976, 96.
[357] D. M. Adams and W. R. Trumble, *Inorg. Chem.*, 1976, **15**, 1968.

structure.[358] James and co-workers have extended their studies of the effects of solutes on the librational band of water (500—700 cm^{-1}). Upon examining solution films, of thickness 7 μm, between KRS 5 plates, of alkyl-ureas, thioureas, formamide, acetamide, acetone, and tetra-alkylammonium nitrates, they concluded that the unique solution behaviour of urea depends on its ability to hydrogen-bond in a pseudo-tetrahedral fashion.[359] A 'probable hydration number' of five was suggested for $LiNO_3$, $NaNO_3$, and KNO_3 in aqueous solution, but it was pointed out that the method used did not distinguish between cation and anion hydration.[360] The perchlorates $MClO_4$ (M = Li, Na, or NH_4) also had a hydration number of five, but, in contrast with previous studies, the libration band intensity was found to depend on the nature of the dissolved cation.[361] Downing and Williams have made further studies of the optical constants of aqueous electrolyte solutions, including the librational band.[362] They have also remeasured the reflectance spectra of aqueous HBr solutions, and have observed features which correspond with those of H_3O^+ observed in crystals.[363]

Aqueous solutions of strong acids showed a Raman-inactive i.r. absorption at *ca.* 1200 cm^{-1}, which was assigned to the species $H_5O_2^+$, and this was said to be distinct from the H_3O^+ species.[364] Raman spectra of aqueous solutions of nitric acid were deemed to be consistent with the presence of 'solvent-separated' ion-pairs.[365]

Raman spectra of zinc oxide dissolved in conc. KOH, NaOH, or NaOD indicated the possibility of both OH^- and H_2O as ligands in the zincate series. It was suggested that the major species in equilibrium in the solutions were $[Zn(OH)_3(H_2O)]^-$ and $[Zn(OH)_2(H_2O)_2]$.[366] Zr^{IV}/SO_4^{2-} complexes were investigated by the method of continuous variations;[367] details of observed wavenumbers are in Chapter 6. Lastly, but most opportunely, Irish and Brooker have written a very comprehensive review of the spectra of electrolyte solutions.[368]

Non-aqueous Solutions.—Solvent–solvent interactions were cited as the principal cause of 'anomalous' intensity patterns in the spectra of liquid and solid $POCl_3$ and $PSCl_3$. In dilute solution the ν_1 band shows the expected isotopic splitting pattern, with intensities in the ratio 27 : 27 : 9 : 1, but in the liquid this appears to be drastically altered.[369] (Similar behaviour was reported last year for BCl_3 by A. Loewenschuss, *Spectrochim. Acta*, 1975, **31A**, 679.) Buback has extended his previous work by examining Raman spectra of ammonia in the supercritical

358 S. Taniewska-Osinska and R. Grochowski, *Bull. Acad. polon. Sci., Sér. Sci. chim.*, 1976, **24**, 67.
359 D. W. James, R. F. Armishaw, and R. L. Frost, *J. Phys. Chem.*, 1976, **80**, 1346.
360 R. F. Armishaw and D. W. James, *J. Phys. Chem.*, 1976, **80**, 501.
361 D. W. James and R. F. Armishaw, *Inorg. Nuclear Chem. Letters*, 1976, **12**, 425.
362 H. D. Downing and D. Williams, *J. Phys. Chem.*, 1976, **80**, 1950.
363 H. D. Downing and D. Williams, *J. Phys. Chem.*, 1976, **80**, 1640.
364 N. B. Librovich, V. D. Maiorov, and V. A. Sevel'ev, *Doklady Akad. Nauk S.S.S.R.*, 1975, **225**, 1358.
365 V. N. Egorov and I. E. Kuzinets, *Russ. J. Inorg. Chem.*, 1975, **20**, 1762.
366 S. K. Sharma and M. D. Reed, *J. Inorg. Nuclear Chem.*, 1976, **38**, 1971.
367 F. G. Baglin and D. Breger, *Inorg. Nuclear Chem. Letters*, 1976, **12**, 173.
368 D. E. Irish and M. H. Brooker, 'Advances in Infrared and Raman Spectroscopy', ed. R. J. H. Clark and R. E. Hester, Heyden, London, 1976.
369 K. Olie and D. J. Stufkens, *Spectrochim. Acta*, 1976, **32A**, 469.

range, but no new assignments for the ν(N—H) region were proposed.[370] A previously unreported feature in the Raman spectrum of liquid ammonia, a broad band at *ca.* 1750 cm^{-1}, which could originate from associated solvent molecules, may also affect the ν(N—H) region by Fermi-resonant interaction.[371]

The cation–solvent stretching vibration from solutions of 12 metal cations was used to determine the ions' co-ordination number in ammonia; the only unexpected results were for Zn^{2+} and Ag^+, which appeared to be four-co-ordinate.[371] Cation solvation bands in propylene carbonate showed some anion dependence. For example, $LiClO_4$ gave a value of 398 ± 4, but LiO_2CCF_3 gave a value of 367 ± 3 cm^{-1}. For KO_2CCF_3 and RbO_2CCF_3 the bands were placed at 159 ± 6 and 118 ± 8 cm^{-1}, respectively.[372]

Evidence for cation–anion interactions came from far-i.r. studies of alkali-metal 2,4-dionates M[RCOCHCOR]$^-$ (R = Ph or Bu^t). Wavenumbers for ν(M—O) decreased linearly as a function of the reciprocal of the Pauling ionic radius of the cation, thus: Li (434) > Na (400) > K (339) > Cs (337) > (Bu_4N) (351 cm^{-1}), suggesting that ion-pairing was predominantly ionic.[373] Extensive ion-pairing has been found in solutions of nitrates, nitrites, azides, cyanides, and thiocyanates in liquid ammonia, and some spectra from sodium and potassium cyanide solutions were displayed to show the considerable differences between them.[374] Contact ion-pairing was proposed for solutions of $Na^+[Mn(CO)_4L]^-$ [L = CO, PMe_2Ph, PPh_3, or $P(OPh)_3$] in THF, between the cation and the oxygen atom of the equatorial carbonyl groups.[375] For example, for $Na[Mn(CO)_5]$, ion-pair bands were assigned at 1901, 1875, and 1829 cm^{-1}, the latter for ν(C—O···M). Addition of hexamethylphosphoramide, or the crown ether 15-crown-5, destroyed the ion-pairing, and left the 'free' anion with bands at 1894 and 1860 cm^{-1}. In a similar study, the gradual addition of the ether benzo 18-crown-6 to solutions of $Na^+[(\eta^5\text{-}C_5H_5)Fe(CO)_2]$ in THF or dimethoxyethane revealed the presence of at least three anionic species, with bands at 1877 and 1807 cm^{-1}, 1862 and 1770 cm^{-1}, and 1862 and 1786 cm^{-1}, respectively. Conjectures about the site of ion-pairing included one suggestion that there is direct Na$^+$–Fe interaction.[376] Alkali-metal cations interact more strongly with *cis*-1-benzyl-2,3-dibenzoylaziridine (1) than with the *trans*-isomer, for obvious reasons.[377] Spectra of (Pr_3MeP)(SFO_3) in $CDCl_3$ solution showed

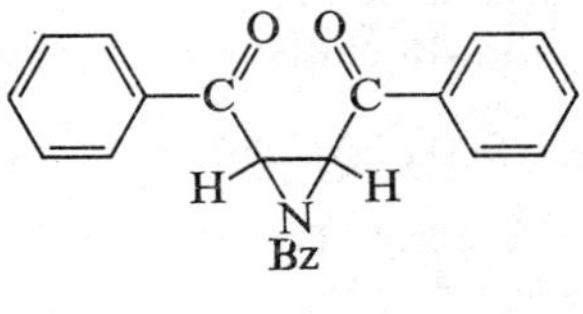

(1)

370 M. Buback and K. R. Schultz, *J. Phys. Chem.*, 1976, **80**, 2478.
371 P. Gans and J. B. Gill, *J.C.S. Dalton*, 1976, 779.
372 H. L. Yeager and H. Reid, *J. Phys. Chem.*, 1976, **80**, 850.
373 E. M. Arnett and V. M. Depalma, *J. Amer. Chem. Soc.*, 1976, **98**, 7447.
374 P. Gans, J. B. Gill, and M. Griffin, *J. Amer. Chem. Soc.*, 1976, **98**, 4661.
375 M. Y. Darensbourg, D. J. Darensbourg, D. Burns, and D. A. Drew, *J. Amer. Chem. Soc.*, 1976, **98**, 3127.
376 K. H. Parnell and D. Jackson, *J. Amer. Chem. Soc.*, 1976, **98**, 4443.
377 A. B. Norman, G. R. Smith, and H. P. Hopkins, *J. Phys. Chem.*, 1976, **80**, 25.

an unsplit strong band at 1285 cm^{-1}, assigned to ν_E(S—O), implying that any ion-pairing present did not cause perceptible effects. The band is split in *O*-bonded fluorosulphate derivatives.[378] In contrast, bands at 1050 cm^{-1} (free ion) and 1010 cm^{-1} (bonded ion) in the Raman spectra of $(Bu^n_4N)(HOSO_3)$ solutions in methyl cyanide were used to determine equilibrium constants.[379]

Two studies of complexes in non-aqueous media provided evidence for unusual bonding situations. An equilibrium between *C*- and *N*-bonded isomers of solvated silver cyanide was proposed for solutions of AgCN in liquid ammonia; possibly the first time that cyanide linkage isomers of a metal ion have been observed.[380] The corresponding ν(C—N) Raman shifts were at 2112 and 2126 cm^{-1}. Raman spectra of aluminium bromide in benzene and various methyl-benzenes at temperatures below 230 K did not contain the band at 210 cm^{-1}, characteristic of the doubly bridged dimer.[381] It was suggested that interaction with solvent caused the rupture of a single bridge bond, and that the resulting mono-bridged species had a ν(Al—Br—Al) mode at 200—217 cm^{-1}. The ν_{sym}(Hg—X) vibration in solutions of HgX_2 and MeHgX (X = Cl, Br, or I) in a wide range of solvents was found to decrease with increase of solvent donicity. For example, for $HgCl_2$, typical values were 340 cm^{-1} in benzene and 280 cm^{-1} in pyridine.[382] A Raman study of $(Bu^n_4N)Br$–$SbBr_3$ mixtures in CH_2Cl_2 led to the suggestion that the ion $[Sb_2Br_7]^-$ is formed at molar ratios less than 1 : 1.[383] Stability constants for the adducts of $SbCl_3$ with *N*-methyl- and *NN*-dimethyl-acetamide in chloroform and benzene have been determined.[384]

α–β isomerization in $[Mo_8O_{26}]^{4-}$ anions in solution in methyl cyanide was followed [385] by measuring their characteristic i.r. absorptions at 808 and 717 cm^{-1}. Raman spectra of liquid mixtures of MeCN and IF_5 indicated a weak inter-molecular interaction by changes in the $\nu(I—F_{ax})$ and ν(C≡N) values.[386] Evidence for the existence of the carbenium ion usually postulated as an inter-mediate in the Friedel–Crafts reaction was obtained by i.r., Raman, and n.m.r. studies of $Bu^tCl/AlCl_3$ mixtures in liquid hydrogen chloride.[387] Raman shifts of 317, 160, and 97 cm^{-1} identified the anion present as $[Al_2Cl_7]^-$.

Charge-transfer complexes of dihalogens and related species provide the basis of four reports this year. I.r. frequencies and integrated intensities of all three fundamentals of ICN have been measured in 28 solvents. As ν(I—C) decreased from 487 cm^{-1} (CCl_4) to 432 cm^{-1} (pyridine), the intensity increased by a factor of *ca.* 4. ν(C—N) and δ(ICN) vary between the ranges 2168—2157 and 319—333 cm^{-1}, respectively.[388] Dioxan formed complexes in which the ν(A—B) frequency decreased as follows:[389] I_2, 6; Br_2, 12; Cl_2, 12; IBr, 15; ICl, 37; BrCl, 36 cm^{-1}. Raman band intensities of complexes of Cl_2 or Br_2 with alkenes

378 C. Eaborn, N. Farrell, J. L. Murphy, and A. Pidcock, *J.C.S. Dalton*, 1976, 58.
379 Y. Lambean, G. Maes, and Th. Zeegers-Huyskens, *Bull. Soc. chim. belges*, 1976, **85**, 129.
380 P. Gans, J. B. Gill, and M. Griffin, *J.C.S. Chem. Comm.*, 1976, 169.
381 K. Schumann and H.-H. Perkampus, *Spectrochim. Acta*, 1976, **32A**, 1651.
382 K. Sone, M. Aritaki, K. Hiraoka, and Y. Fukuda, *Bull. Chem. Soc. Japan*, 1976, **49**, 2015.
383 G. C. Allen and R. F. McMeeking, *Inorg. Chim. Acta*, 1976, **20**, L55.
384 M. Van Cauteren and Th. Zeegers-Huyskens, *Inorg. Nuclear Chem. Letters*, 1976, **12**, 323.
385 W. G. Klemperer and W. Shum, *J. Amer. Chem. Soc.*, 1976, **98**, 8291.
386 J. A. Berry, D. W. A. Sharp, and J. M. Winfield, *Inorg. Nuclear Chem. Letters*, 1976, **12**, 869.
387 F. Kalchschmid and E. Mayer, *Angew. Chem. Internat. Edn.*, 1976, **15**, 773.
388 J. De Leeuw and Th. Zeegers-Huyskens, *Spectrochim. Acta*, 1976, **32A**, 617.
389 J. W. Anthonsen, *Spectrochim. Acta*, 1976, **32A**, 963.

have been discussed in terms of bonding and other features involved.[390] I.r. intensities of the internal modes of (substituted) pyridines were used to evaluate stability constants and ΔH_f values of pyridine–I_2 complexes.[391]

Raman and i.r. spectra of molten $Al(NO_3)_3$ and its hydrates have been studied, but apparently it was suggested that the nitrate ion was always unco-ordinated![392] Shifts in SO_4^{2-} and PO_4^{3-} modes as a function of temperature and cation radius were studied in Na, K, or Ca bromide or iodide melts.[393] New compounds were formed in the binary and quaternary molten mixtures of NH_4NO_3–$(NH_4)_4P_2O_7$–$(NH_4)H_2PO_4$–$(NH_4)_2HPO_4$; details were not given in the secondary source (*Chem. Abs.*, 1976, **85**, 83 843).[394]

390 Ya. M. Kimel'fel'd, A. B. Mostovoi, and L. M. Mostovaya, *Doklady Phys. Chem.*, 1975, **222**, 582.
391 F. M. Abdel Kerim and F. Abou E. Fotouh, *Appl. Spectroscopy*, 1976, **30**, 200.
392 I. I. Kondilenko, P. A. Korotkov, and N. G. Golubeva, *Zhur. priklad. Spektroskopii*, 1976, **24**, 532.
393 M. V. Smirnov, Yu. V. Yurinov, and A. P. Khaimenov, *Deposited Publ.*, 1973, VINITI 6146–73 (*Chem. Abs.*, 1976, **85**, 22 997).
394 A. P. Sizonenko, I. B. Gavrilenko, N. M. Smirnova, R. G. Lepolina, M. E. Pozin, and A. M. Shevyakov, *Zhur. priklad. Khim.*, 1976, **49**, 962.

5

Characteristic Vibrational Frequencies of Compounds containing Main-group Elements

BY P. GANS

The form of this chapter is as in Volume 9 of this series, although papers which give vibrational data only for characterization purposes are treated more briefly.

1 Group I Elements (including Hydrogen)

Hydrogen bonding continued to provide a seemingly inexhaustible field of study. Band-width phenomena in the i.r. transmission spectra of polycrystalline samples of KHF_2 and $CsHF_2$ have been accounted for in part by superposition of absorption and reflection processes and in part by anharmonic interactions between ν_3 and lattice modes.[1] With cobaltic acid, $HCoO_2$, much more pronounced effects were observed.[1]

In the hydrogendinitrate ion (1), the hydrogen atom is at the centre of a distorted tetrahedron of oxygen atoms. A group of bands near 1400 cm^{-1}

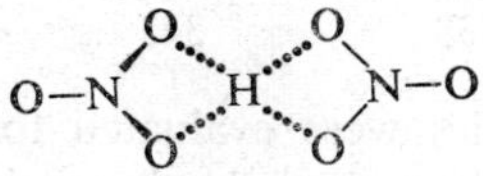

(1)

shifted to *ca.* 1010 cm^{-1} on deuteriation, and a strong broad feature at *ca.* 500 cm^{-1} was observed in the i.r. spectrum of this anion;[2] inelastic neutron scattering measurements on the caesium salt gave major peaks at 1240 ± 50 cm^{-1} and 450 ± 30 cm^{-1}. The latter was assigned to the O···H stretching vibration.[3] The systems CsF/RCO_2H (R = H, Me, Et, or Pr) were investigated by n.m.r. and i.r. spectroscopy. The latter revealed two broad features at *ca.* 2120 and 1550 cm^{-1} which were independent of the alkyl substituent, and arose from a strong F···H—OCOR hydrogen bond.[4] The uncomplexed formate ions in the compound $[NiL_6][HCOO]_2$ (L = imidazole) are stabilized through hydrogen bonding to the cation, and give five peaks in the 3150—2630 cm^{-1} region.[5] The i.r. spectra of some calcium aluminates of formula $[Ca_2Al(OH)_6]X^+$,-yH_2O were discussed mainly in terms of hydrogen bonding of the interlayer

[1] A. S. Gilbert and N. Sheppard, *Spectrochim. Acta*, 1976, **32A**, 923.
[2] J. M. Williams, N. Dowling, R. Gunde, D. Hadži, and B. Orel, *J. Amer. Chem. Soc.*, 1976, **98**, 1581.
[3] J. Roziere and C. V. Berney, *J. Amer. Chem. Soc.*, 1976, **98**, 1582.
[4] J. Emsley and O. P. A. Hoyte, *J.C.S. Dalton*, 1976, 2219.
[5] J. C. Jansen and J. Reedijk, *Rec. Trav. chim.*, 1976, **95**, 52.

water molecules,[6] and similar studies have been made of calcium apatites.[7] Adsorbed water molecules were the centre of interest in studies of the i.r. spectra of 'fluorosilicates' of K^+, Na^+, and Ca^{2+} and alkali halide surfaces.[8, 9]

Only one report concerning alkali metals has appeared this year. The vibrational spectra of a series of lithium derivatives $Li[E(SiH_3)_n]$ (E = O, S, or Se, for n = 1; E = P or As for n = 2) were described and bands at 465 cm^{-1} and 310 cm^{-1} were tentatively assigned to ν(Li—O) and ν(Li—S) respectively.[10]

2 Group II Elements

The vibrational spectra of the compound $Cs[Et_2BeN_3]$ were taken as indicative of a trimeric structure.[11] Proposed assignments include ν(Be—C) at 943 and 912 cm^{-1} and ν(Be—N) at 500 and 465 cm^{-1}. Polymeric structures and various group frequency assignments were suggested for other ethyl beryllium derivatives.[11] A number of assignments were proposed from the i.r. and Raman spectra of 1 : 1 adducts of $Bu^t{}_2Be$ with R_3N and R_3P (R = Me or Et),[12] and these, together with previous data on adducts of dibutyl beryllium, have been reviewed.[13] It was suggested that the basicity of the donor could be indicated by various skeletal frequencies in the adducts.[13]

Exhaustive isotopic substitution data (see also Chapter 6) on $[Mg(NH_3)_6]Cl_2$ confirmed the author's previous assignment of ν(Mg—N) modes,[14] as follows (values given in units of cm^{-1})

	$^{24}Mg^{14}N^1H$	$^{26}Mg^{14}N^1H$	$^{24}Mg^{14}N^2H$	$^{26}Mg^{14}N^2H$	$^{24}Mg^{15}N^1H$
A_{1g}	335	335	312	312	326
E_g	243	243	227	227	237
T_{1u}	363	358	343	340	357

Thermodynamic characteristics were evaluated for the monomer/dimer equilibrium of diethylmagnesium in diethyl ether using measured values for the intensities of the three Raman bands at 440 [δ(COC) of free Et_2O], 360 [$\nu(Mg_2C_2)$ of the dimer], and 410 cm^{-1} [$\nu_s(MgC_2)$ of the monomer].[15] Assignments have also been given for pure diethylmagnesium in the C_{4h} factor group.[16] The i.r. spectra of some chelated magnesium alkoxides such as $\overline{OCH_2CH_2OMg}$ and $\overline{OCH_2CH_2CH_2OMg}$ have been interpreted in terms of polymeric structures.[17] Magnesium trifluoroacetate shows the smallest splitting, $\Delta\nu$, between symmetric and asymmetric C—O stretching vibrations. The magnitude of $\Delta\nu$ decreases

[6] C. J. M. Houtepen and H. N. Stein, *Spectrochim. Acta*, 1976, **32A**, 1409.
[7] P. F. González-Diáz and A. Hidalgo, *Spectrochim. Acta*, 1976, **32A**, 631.
[8] V. V. Mank, S. P. Suleimanov, F. D. Ovcharenko, and N. G. Vasil'ev, *Doklady Phys. Chem.*, 1976, **224**, 1070.
[9] R. St. C. Smart and N. Sheppard, *J.C.S. Faraday II*, 1976, **72**, 707.
[10] S. Cradock, E. A. V. Ebsworth, D. W. H. Rankin, and W. J. Savage, *J.C.S. Dalton*, 1976, 1661.
[11] N. Atam and K. Dehnicke, *Z. anorg. Chem.*, 1976, **3**, 193.
[12] J. Mounier, B. Mula, and A. Potier, *J. Organometallic Chem.*, 1976, **105**, 289.
[13] J. Mounier, B. Mula, and A. Potier, *J. Organometallic Chem.*, 1976, **107**, 281.
[14] R. Plus, *Spectrochim. Acta*, 1976, **32A**, 263.
[15] J. Kress, *J. Organometallic Chem.*, 1976, **111**, 1.
[16] J. Kress, A. Novak, and J. Hervieu, *J. Organometallic Chem.*, 1976, **121**, 7.
[17] P. Maleki and M. J. Schwing-Weill, *J. Inorg. Nuclear Chem.*, 1976, **38**, 1783.

in the order Pb > Ba > Sr > Ca > Mg, which was said to conform with the expected trend.[18]

Bands at 415, 440, and 495 cm^{-1} were assigned to ν(Ca—N) modes in solid $(\eta^5\text{-}C_5H_5)_2Ca,NH_3$,THF and its indenyl analogue.[19] Anhydrous calcium nitrate appears to undergo phase transitions at 287.8 and 88.5 K. The i.r. and Raman spectra of the new phases were discussed, together with the vibrational spectra of the anhydrous nitrates of cadmium and zinc, thus extending the authors' previous work on $Sr(NO_3)_2$, $Ba(NO_3)_2$, and $Pb(NO_3)_2$.[20] The thermal decomposition of $Ba(ClO_3)_2,H_2O$ was followed by i.r. spectroscopy.[21]

3 Group III Elements

Compounds containing B—H Bonds.—A new dimension to vibrational spectroscopy has been created by the development of laser-induced chemical reactions.[22] Direct excitation of ν_{14} in $^{11}B_2H_6$ by a CO_2 laser tuned to 973.29 cm^{-1}, in the presence of H_2S, gave products quite different from those obtained by thermolysis. The new compound $HB(SH)_2$, which is stable in the gas phase but unstable as a liquid, was prepared by this route, and its vibrational frequencies, assigned with the aid of NCA, are in Table 1. In contrast, an investigation by g.l.c. and i.r.

Table 1 *Wavenumbers*/cm^{-1} *for the compound* $HB(SH)_2$ [a]

Mode	$H^{10}B(SH)_2$	$H^{11}B(SH)_2$
ν_1	~2620	~2620
ν_2	782	767
ν_4	1108	1089
ν_5	865	845
ν_6	~800	797

[a] Numbering of modes as for $HBCl_2$; see, for example, O. B. De Mandriola and J. F. Westerkamp, *Spectrochim. Acta*, 1964, **20**, 1633.

of a product alleged to be benzoyldibutylborane found it to be a mixture of benzyloxydibutylborane and benzylbenzoate.[23] The structure (2) was proposed for GaB_2H_9 based on i.r. and other data, but details were not supplied.[24] The compounds [25] $M(THF)_2(B_6H_9)_2$ (M = Mg or Zn) show $\nu(B—H)_t$ to be in the region 2490—2620 and $\nu(B—H)_b$ in the region 1800—2000 cm^{-1}, whereas the ferraborane analogue (3) of $B_6H_6^{2-}$ shows [26a] ν(B—H) to be at 2572 cm^{-1}. New molecular constants have been derived from a high resolution i.r. study of ν_7(E) (BH_3 rocking) of $^{10}BH_3CO$.[26b]

18 J. A. Faniran and K. S. Patel, *Spectrochim. Acta*, 1976, **32A**, 1351.
19 M. Kirilov, G. Petrov, and C. Angelov, *J. Organometallic Chem.*, 1976, **113**, 225.
20 M. H. Brooker, *Spectrochim. Acta*, 1976, **32A**, 369.
21 H. D. Lutz, H.-J. Klüppel, and H.-J. Kesterke, *Z. anorg. Chem.*, 1976, **423**, 83.
22 H. R. Bachmann, F. Bachmann, K. L. Kompa, H. Nöth, and R. Rinck, *Chem. Ber.*, 1976, **109**, 3331.
23 K. Smith and K. Swaminathan, *J.C.S. Dalton*, 1976, 2297.
24 A. J. Downs and P. D. P. Thomas, *J.C.S. Chem. Comm.*, 1976, 825.
25 D. L. Denton, W. R. Clayton, M. Mangion, S. G. Shore, and E. A. Meyers, *Inorg. Chem.*, 1976, **15**, 541.
26 (*a*) J. A. Ulman and T. P. Fehlner, *J.C.S. Chem. Comm.*, 1976, 632; (*b*) C. Pépin, L. Lambert, and A. Cabana, *J. Mol. Spectroscopy*, 1976, **59**, 63.

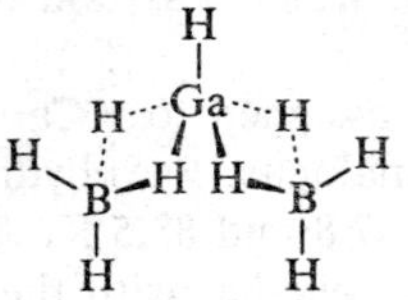

(2)

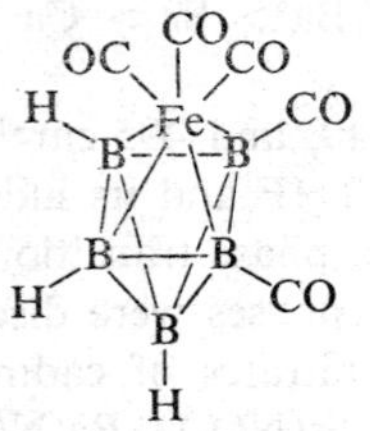

(3)

Hydridotris(pyrazol-1-yl) borato salts of the lanthanide elements, $M(HBpz_3)_3$ fall into two groups.[27] In the first (M = Er or Y) ν(B—H) is observed at 2465, 2437, and 2411 cm^{-1}, and in the second group (M = La, Ce, Pr, Sm, or Gd) ν(B—H) appears in the regions 2450—2440 and 2410—2400 cm^{-1}. The compound $(HBpz_3)TiCl_3$ shows ν(B—H) at 2519 cm^{-1}, falling to 2500 cm^{-1} in the Ti^{III} compound $(HBpz_3)TiCl_2$,THF.[28]

The borohydride $Al(BH_4)_2Me$ is covalent and monomeric in the gas phase, with double proton bridges, and i.r. absorptions at 2550, 2485 for $\nu(B—H)_t$ and 2035, 1950 cm^{-1} for $\nu(B—H)_b$. These frequencies [29] and those of $Al(BH_4)Me_2$ show that $\nu(B—H)_t$ decreases and $\nu(B—H)_b$ increases in the compounds $Al(BH_4)_xMe_{3-x}$ in the order $x = 3 > x = 2 > x = 1$. Formation of adducts of the type $Al(BH_4)_2MeL$ (L = NMe_3, NMe_2H, PMe_3, $AsMe_3$, OMe_2, OEt_2, or SMe_2) is accompanied by a drop of *ca.* 60 cm^{-1} in $\nu(B—H)_t$ and a rise of *ca.* 120 cm^{-1} in $\nu(B—H)_b$, depending on the extent to which the borohydride assumes ionic character.[29] A study of diethyl ether adducts of $Al(BH_4)_xR_{3-x}$ (R = alkyl) has shown that the ν(C—O—C) vibration can be used empirically to differentiate between alkyl aluminium compounds with a different number of alkyl groups.[30] A bidentate structure was also assigned to the borohydride in $[M(L)(BH_4)]$ (M = Co or Cu; L = $(Ph_2PCH_2)_3CMe$) on the basis of comparison with established assignments,[31] and in $(\eta^5-C_5H_5)_2Er(BH_4)THF$, although a tridentate borohydride group was proposed for the corresponding samarium compound.[32] $\nu(B—H)_t$ was found at 2400—2580, and $\nu(B—H)_b$ at 1960—2240 cm^{-1} in the Ti^{III} adducts $Ti(BH_4)_3L$ (L = Et_2O, $(THF)_2$, or $MeOCH_2CH_2OMe$), and structures were proposed accordingly.[33] The data from which Table 2 was selected was used to support the structure (4), of C_{2h} symmetry, for

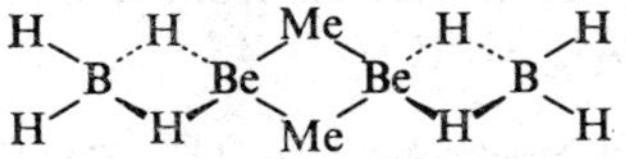

(4)

[27] K. W. Bagnall, A. C. Tempest, J. Takats, and A. P. Masino, *Inorg. Nuclear Chem. Letters*, 1976, **12**, 555.
[28] J. K. Kouba and S. S. Wreford, *Inorg. Chem.*, 1976, **15**, 2313.
[29] P. R. Oddy and M. G. H. Wallbridge, *J.C.S. Dalton*, 1976, 869.
[30] P. R. Oddy and M. G. H. Wallbridge, *J.C.S. Dalton*, 1976, 2076.
[31] P. Dapporto, S. Midollini, A. Orlandini, and L. Sacconi, *Inorg. Chem.*, 1976, **15**, 2768.
[32] T. J. Marks and G. W. Grynkewich, *Inorg. Chem.*, 1976, **15**, 1302.
[33] K. Franz, H. Fusstetter, and H. Nöth, *Z. anorg. Chem.*, 1976, **427**, 97.

Table 2 *Wavenumbers*/cm^{-1} *of methyl beryllium borohydride*

	$[CH_3Be(BH_4)]_2$		$[CH_3Be(BD_4)]_2$	
Vibration	*Raman*	*I.r.*	*Raman*	*I.r.*
$\nu_{as}(BH_2)_t$	2502	2493	1909	1897
$\nu_s(BH_2)_t$	2444	2437	1810	1793
$\nu_s(BH_2)_b$	2103	2097	1503	1498
$\nu_{as}(BH_2)_b$	2023	2017	*a*	*a*
$\nu(BeH_2)_b$	1615	1607	1117	1095
	1480	1468	1032	1032
$\tau(BH_2)_t$	1143	1143	819	—
$\delta(BH_2)_t$	1134	1138	843	873
$\rho(BH_2)_t$	1026	1029	789	788
$\nu_{as}(Be—C)$	937	924	945	942
$\nu(Be—B)$	—	740	—	730
$\nu_s(Be—C)$	526	594	524	588
$\nu(Be—Be)$	357	—	332	—

[a] Three bands observed in the deuterate.

methylberyllium borohydride.[34] Force constants have been calculated for B_2H_6 and X_2Y_6 (X = Al or Ga; Y = Cl, Br, or I) using a Green's function method.[35] Largely unassigned i.r. data have been given for various sila-,[36] carba-,[37] phospha-,[38] thia-,[39, 40] metalla-,[41–45] and other [46, 47] polyhedral boranes.

Compounds containing Al—H Bonds.—I.r. spectra have been presented for compounds formulated as $MAlH_{4-n}(OR)_n$ (M = Na or Li; R = Me, Et, Pr^i, or Bu^t). ν(Al—H—Al) was attributed to a broad band at *ca.* 1464 cm^{-1} on the basis of deuteriation.[48] The Al—H stretching vibrations were observed at 1830 and 1785 cm^{-1} in the *N*-alkyliminoalane (5) and at 1865 cm^{-1} in the cubane analogue $(EtNAlH)_4$.[49]

Compounds containing M—C Bonds (M = B, Al, In, or Tl).—The B—C vibrations in the compounds BR_3 (R = allyl, methallyl, or crotyl) were found at similar wavenumbers to those of triethylboron.[50] An i.r. absorption at

[34] L. J. Allamandola and J. W. Nibler, *J. Amer. Chem. Soc.*, 1976, **98**, 2096.
[35] K. Ramaswamy and G. Shanmugam, *J. Annamalai Univ. B*, 1972, **30**, 75.
[36] K. A. Andrianov and B. A. Astapov, *Doklady Chem.*, 1976, **237**, 302.
[37] L. A. Leites, L. E. Vinogradova, S. S. Bukalov, V. T. Aleksanyan, E. G. Rys, V. N. Kalinin, and L. I. Zakhankin, *Izvest. Akad Nauk S.S.R., Ser. khim.*, 1975, **9**, 1985.
[38] M. Mangion, R. K. Herz, M. L. Denniston, J. R. Long, W. R. Clayton, and S. G. Shore, *J. Amer. Chem. Soc.*, 1976, **98**, 449.
[39] W. R. Pretzer and R. W. Rudolph, *J. Amer. Chem. Soc.*, 1976, **98**, 1441.
[40] W. R. Pretzer and R. W. Rudolph, *Inorg. Chem.*, 1976, **15**, 1779.
[41] G. G. Oatterson, V. T. Brice, and S. G. Shore, *Inorg. Chem.*, 1976, **15**, 1456.
[42] W. M. Maxwell, V. R. Miller, and R. N. Grimes, *Inorg. Chem.*, 1976, **15**, 1343.
[43] J. L. Little, *Inorg. Chem.*, 1976, **15**, 114.
[44] A. R. Siedle and L. J. Todd, *Inorg. Chem.*, 1976, **15**, 2838.
[45] C. G. Salentine and M. F. Hawthorne, *Inorg. Chem.*, 1976, **15**, 2872.
[46] N. T. Kuznetsov, G. S. Klimchur, O. A. Kananaeva, and K. A. Solntsev, *Russ. J. Inorg. Chem.*, 1976, **21**, 505.
[47] R. E. Loffredo, L. F. Drullinger, J. A. Slater, C. A. Turner, and A. D. Norman, *Inorg. Chem.*, 1976, **15**, 478.
[48] J. Fusek, F. Hanousek, and F. Haruda, *Coll. Czech. Chem. Comm.*, 1976, **41**, 187.
[49] H. Nöth and P. Wolfgardt, *Z. Naturforsch.*, 1976, **31b**, 697.
[50] W. Brüser, S. Schröder, and K. Witke, *Z. anorg. Chem.*, 1976, **421**, 89.

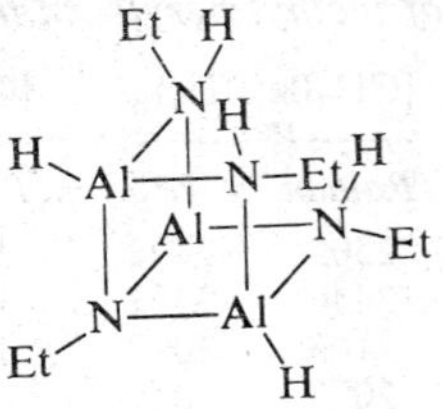

(5)

1330 cm^{-1}, characteristic of the B—Me group, was observed in two isomers of formula $[(\eta^5\text{-}C_5H_5)Ni]_3CB_5H_5Me$.[51]

Bands at 1009 and 660—691 cm^{-1} were assigned to $\nu_s(AlC_2)$ and $\nu_{as}(AlC_2)$ respectively in the compound $B_6(NMe_6)_{12},Al_6Me_{12}$ for which the structure (6)

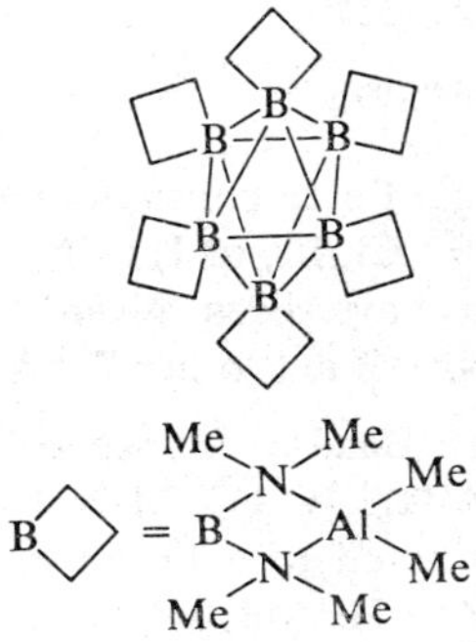

(6)

was suggested.[52] Bands at 940—950 cm^{-1} were tentatively assigned to cage-related vibrations, by reference to the i.r. spectra of boron subchlorides.[52] The dimer Al_2(cyclopropyl)$_6$ has i.r. absorptions at 786, 627, 608, 557, and 515 cm^{-1}, which were assigned [53] to ν(Al—C) modes by comparison with literature values for Al_2Me_6 and Al_2Ph_6.

ν_{as}(In—C) and ν_s(In—C) were assigned in the ranges 548—562 and 480—499 cm^{-1} in the compounds Me_2InX (X = F, Cl, Br, I, OAc, ReO_4, or InI_4).[54] The corresponding Tl—C modes were assigned ranges of 453—584 and 423—492 cm^{-1} in a wide variety of compounds of the type $(^nH_3C)_2TlNR_2$ (n = 1 or 2).[55] Insertion of various molecules into the Tl—N bonds of these compounds gave products formulated as Me_2Tl—YC(X)E (Y = O or S; X = O, S, PhN, or MeN; E = NMe or OEt) for which the ranges ν_{as}(Tl—C) 522—555 and ν_s(Tl—C) 473—499 cm^{-1} were assigned.[56] The spectra of the compound Me_2TlS_2COMe showed no evidence for ν(Tl—S), but ν(Tl—C) was found within

[51] C. G. Salentine, C. E. Strouse, and M. F. Hawthorne, *Inorg. Chem.*, 1976, **15**, 1832.
[52] B. A. Amero and E. P. Schram, *Inorg. Chem.*, 1976, **15**, 2842.
[53] D. A. Sanders, P. A. Scherr, and J. P. Oliver, *Inorg. Chem.*, 1976, **15**, 861.
[54] H. Schmidbauer and D. Koth, *Naturwiss.*, 1976, **63**, 482.
[55] B. Walther, A. Zschunke, B. Adler, A. Kobe, and S. Bauer, *Z. anorg. Chem.*, 1976, **427**, 137.
[56] B. Walther, R. Mahrwald, C. Jahn, and W. Klar, *Z. anorg. Chem.*, 1976, **423**, 144.

the usual ranges.[57] Two reports on dimethylthallium(I) compounds show that linear $[Me_2Tl]^+$ ions are present. In $(Me_2Tl)_2X$ ($X = S_2O_3^{2-}$ or $\alpha\omega$-dicarboxylates) ν_{as}(Tl—C) was observed at 555 cm^{-1} in the i.r. spectrum,[58] whereas in $(Me_2Tl)CN_2$ it was observed at 535 cm^{-1}, and a Raman shift of 485 cm^{-1} for ν_s(Tl—C) was also recorded.[59]

Compounds containing B—N or Al—N Bonds.—The spectroscopy of trigonal boron–nitrogen compounds has been reviewed.[60] Ammonia cyanoborane, $H_3NB(CN)H_2$, has an i.r. absorption at 714 cm^{-1} which was assigned to ν(B—N); other absorptions include ν_{as}(B—H) at 2430 and 2410, ν_s(B—H) at 2360, and a band at 2202 cm^{-1} assigned to the ν(C—N) vibration.[61] The methylamine analogue, $Me_3NB(CN)H_2$, however, exhibits two C—N stretching vibrations at 2135 and 2195 cm^{-1}, which were attributed to linkage isomers.[62] The Raman spectrum of borazine, $B_3N_3H_6$, in the gas phase has been reported.[63] Several ζ_i constants were derived from band contour measurements of e'' modes. Assignments are in Table 3. The i.r. spectrum of the compound $(DBNMe)_3$

Table 3 *Raman shifts*/cm^{-1} *for gaseous borazene*

Symmetry		*Mode*	*Wavenumber*/cm^{-1}
A_1'	ν_1	ν_s(N—H)	3472
	ν_2	ν_s(B—H)	2535
	ν_3	ν_s(B—N)	938
	ν_4	δ_s(BN)	849
E'	ν_{11}	ν_{as}(N—H)	3482
	ν_{12}	ν_{as}(B—H)	2512
	ν_{13}	ν_{as}(B—N)	1465
	ν_{14}	ν_{as}(B—N)	—
	ν_{15}	δ(BH)	1065
	ν_{16}	δ(NH)	985
	ν_{17}	δ(BN)	515
E''	ν_{18}	γ(BH)	960
	ν_{19}	γ(NH)	792
	ν_{20}	γ(BN)	275

was displayed, but ν(B—D) at 1875 cm^{-1} was the only assignment given.[64] Raman and i.r. spectra of the dibromoboryl borazine (7) were assigned in D_{3h} symmetry. The ν(B—N) ring breathing modes were found at 1450 and 1345 cm^{-1}, and bands at 1245 and 1045 cm^{-1} were assigned to symmetric and antisymmetric modes of the terminal B—N bonds.[65] Largely unassigned i.r. data have been given for the following compounds: $MeNH_2,B_3H_7$ and $Me_2NH,$-B_3H_7,[66] $[Me_3NBH_2NMe_2(CH_2)_nCO_2Et]^+$ and $[Me_3NBH_2O_2C(CH_2)_mNMe_3]^+$

[57] W. Schwarz, G. Mann, and J. Weidlein, *J. Organometallic Chem.*, 1976, **122**, 303.
[58] U. Knips and F. Huber, *J. Organometallic Chem.*, 1976, **107**, 9.
[59] P. Krommes and J. Lorberth, *J. Organometallic Chem.*, 1976, **120**, 131.
[60] Gmelin, New Supplement series, Vol. 23. 'Boron. Part 5. Boron-pyrazole derivatives and spectroscopy of trigonal boron–nitrogen compounds', Springer-Verlag, Weinheim, 1975.
[61] P. J. Bratt, M. P. Brown, and K. R. Seddon, *J.C.S. Dalton*, 1976, 353.
[62] J. L. Vidal and G. E. Ryschkewitsch, *J.C.S. Chem. Comm.*, 1976, 192.
[63] B. Roussel, A. Chaput, and G. Fleury, *J. Mol. Structure*, 1976, **31**, 371.
[64] A. J. DeStefano and R. F. Porter, *Inorg. Chem.*, 1976, **15**, 2569.
[65] K. Dehnicke and V. Fernández, *Chem. Ber.*, 1976, **109**, 488.
[66] A. R. Dodds and G. Kodama, *Inorg. Chem.*, 1976, **15**, 741.

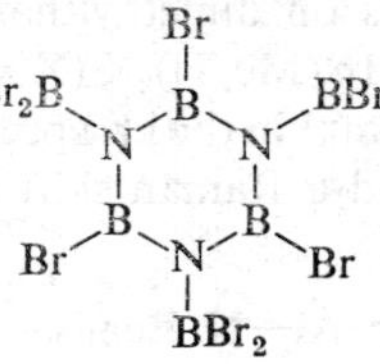

(7)

(n = 2 or 3; m = 1, 2, 3, or 4),[67] $(HXBNMe_2)_n$ (X = Cl or Br; n = 1 or 2) and $(H_2BNMe_2)_2$,[68] $H_2CN(Me_2)BH_2N(Me)_2B(H)L$ (L = Cl, Br, I, py, *etc.*),[69, 70] and $R^1{}_2AlNR^2(CH_2)_nNR^3{}_2$ (R^1 = Cl, H, Me, Et, or Ph; R^2 = Me or Et; R^3 = Me or Et; n = 2 or 3).[71]

Compounds containing B—P or Ga—As Bonds.—The adduct $(MeO)_3P,BH_3$ has an i.r. band at 794 cm^{-1} which was assigned to ν(B—P). Reduction of this compound gave a dimeric anionic species $[(MeO)_3(H_3B)P]_2{}^{2-}$ in which ν(B—P) was found at 810 cm^{-1} (i.r.) and 806 cm^{-1} (Raman); evidence for the dimeric structure of the anion includes the observation of a Raman shift of 499 cm^{-1}, which was assigned to ν(P—P), being some 60 cm^{-1} higher than that vibration in diphosphine.[72] I.r. data for the substituted phosphinoborines (8) remained

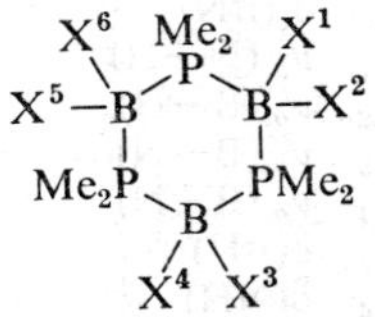

(8) X^1—X^6 = H or halogen

essentially unassigned, apart from one or two specific and a few general remarks concerning ν(B—H) vibrations.[73] Crystals of GaAs doped with oxygen atoms (50% ^{18}O enriched) have been studied in the 751—851 cm^{-1} region.[74]

Compounds containing M—O (M = B, Al, or Ga) Bonds.—I.r. (gas and solution) and Raman (solid and liquid) spectra of trimethylborate were found to be consistent with C_{3h} symmetry, and an NCA was carried out on the system.[75] A small assortment of ν(B—O—X) vibrations have been assigned as follows. For the compounds $M_2[RC(CH_2O)_3BOB(OCH_2)_3CR]$ (M = Na or K; R = Et or Bu) ν_{as}(B—O—B) was placed in the range 1255—1295 cm^{-1}, and other assign-

[67] N. E. Miller, *Inorg. Chem.*, 1976, **15**, 1735.
[68] O. T. Beachley and B. Washburn, *Inorg. Chem.*, 1976, **15**, 725.
[69] B. R. Gragg and G. E. Ryschkewitsch, *Inorg. Chem.*, 1976, **15**, 1201.
[70] B. R. Gragg and G. E. Ryschkewitsch, *Inorg. Chem.*, 1976, **15**, 1205.
[71] O. T. Beachley and K. C. Racette, *Inorg. Chem.*, 1976, **15**, 2112.
[72] L. A. Peacock and R. A. Geanangel, *Inorg. Chem.*, 1976, **15**, 244.
[73] M. H. Goodrow and R. I. Wagner, *Inorg. Chem.*, 1976, **15**, 2830.
[74] Z. L. Akkerman, L. A. Borisova, and A. F. Kravchenko, *Fiz. Tekh. Poluprovodn.*, 1976, **10**, 997.
[75] A. Rogstad, B. N. Cyvin, S. J. Cyvin, and J. Brunvoll, *J. Mol. Structure*, 1976, **35**, 121.

ments were also given.[76] For compound (9) ν(B—O) was assigned[77] to i.r. bands at 1155 and 808 cm^{-1}, and ν(B—F) was located at 955 cm^{-1}. For the compounds (10) ν(B—O—C) was found at 1073 cm^{-1} (R = H) and 1050 cm^{-1} (R = Me).[78] For the compounds (11) ν(B—O) was placed[79] in the much higher

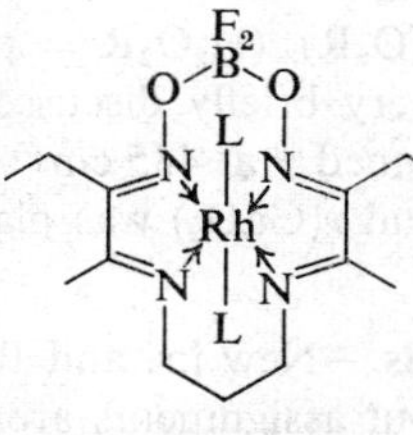

(9) L = SePh

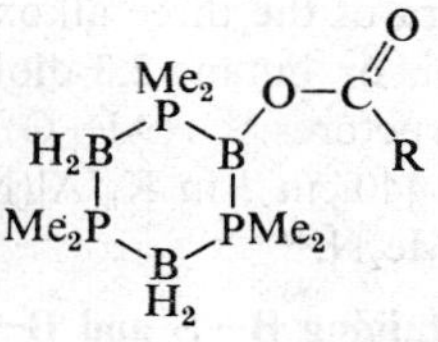

(10) R = H or Me

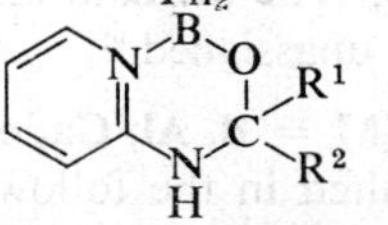

(11) R^1 = H, R^2 = Me or Ph
R^1 = R^2 = Me or $(CH_2)_5$

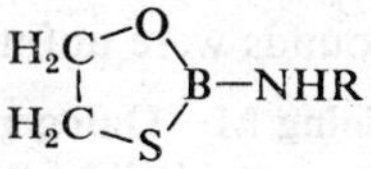

(12) R = Bu^n or Bu^t

region of 1360—1380 cm^{-1}. Finally, ν(B—O) was placed in the similar region 1330—1295 cm^{-1} in the compounds (12); the assignments of ν(B—N) at 1515—1470 cm^{-1} and ν(B—S) at 1035—970 cm^{-1} were also given for these compounds.[80]

Amongst papers on spectra of borates few have appeared which make a serious attempt on the assignment problem. Single crystal Raman data for $K[B_5O_6(OH)_4],2H_2O$ were combined with existing i.r. data to perform a NCA on the pentaborate ion, but few conclusions resulted apart from the mundane one that it is possible to distinguish between vibrations of the BO_4 and BO_3 fragments.[81] Other polymeric borate anions were studied in solution by Raman spectroscopy, and it was claimed that the species $[B_5O_6(OH)_4]^-$, $[B_3O_3(OH)_4]^-$, $[B_4O_5(OH)_4]^{2-}$, and $[B_3O_3F_6]^{3-}$ had been characterized.[82] Other systems studied include alkaline earth metaborates,[83] borosilicate glasses,[84] and GeO_2—PbO—B_2O_3 glasses.[85]

Raman spectra of the ionic conductor 'β-Al_2O_3' and several of its cation isomorphs show low frequency vibrations due to displacements of the mobile

[76] E. M. Shvarts, I. Kalve, A. Terauda, and A. Ievins, *Latvijas PSR Zinaltnu Akad. Vestis, Kim. Ser.*, 1976, 3.
[77] J. P. Collman, R. K. Rothrock, J. P. Sen, T. D. Tullius, and K. O. Hodgson, *Inorg. Chem.*, 1976, **15**, 2947.
[78] M. H. Goodrow and R. I. Wagner, *Inorg. Chem.*, 1976, **15**, 2836.
[79] B. R. Gragg, R. E. Handshoe, and K. Niedenzu, *J. Organometallic Chem.*, 1976, **116**, 135.
[80] R. H. Cragg, J. Dudman, and J. P. N. Husband, *J. Organometallic Chem.*, 1976, **116**, 281.
[81] V. Devarajan, E. Gräfe, and E. Funck, *Spectrochim. Acta*, 1976, **32A**, 1225.
[82] L. Maya, *Inorg. Chem.*, 1976, **15**, 2179.
[83] Z. I. Shishkina, S. G. Sushkova, and G. E. Gabova, *Tr. Ural'sk. Nauchn Issled. Khim. Inst.*, 1973, **32**, 109.
[84] W. L. Konijnendijk, *Bol. Soc. Esp. Ceram. Vidrio*, 1975, **14**, 517.
[85] I. A. Afonskaya and V. N. Morozov, *Zhur. priklad. Spektroskopii*, 1976, **24**, 733.

cations in a direction parallel to the conducting planes.[86] Raman studies of solutions of pure Al_2O_3 in molten cryolite and other melts showed that oxides could be detected at concentrations as low as 2 mole %. New bands were observed at 530, 460, 310, and 185 cm^{-1}, but the only structural conclusion was that terminal Al—O bonds were not observed.[87]

The i.r. spectra of the three alkoxides $Al_2(O_2R)_3$ (H_2O_2R = propane-1,3-diol, propane-1,2-diol, or butane-2,3-diol) were very briefly discussed in relation to the proposed structures.[88] ν(Al—O) was assigned[89] at 435 cm^{-1} in $K_3[Al(NO_3)_6]$ and at 455 and 440 cm^{-1} in $K_2[Al(NO_3)_5]$, and $\nu(GaO_2)$ was placed at 463 cm^{-1} in $Me_2GaO_2C_2Me_2N$.[90]

Compounds containing B—S and B—Se Bonds.—New i.r. and Raman data, and NCA have been obtained for $(MeS)_2BMe$, but assignments are similar to those of H. Vahrenkamp (*J. Organometallic Chem.*, 1971, **28**, 181) with the exception of the following new ones:[91] 382 cm^{-1}: (?) combination; 331 cm^{-1}: δ_{as}(BSC); 292 cm^{-1}: $\delta_{as}(CBS_2)$; 280 cm^{-1}: (?) τ(SMe). The i.r. data for $(CF_3Se)_3B(NMe_3)$ and related compounds were unfortunately unassigned.[92]

Compounds containing M—Halogen Bonds (M = B, Al, Ga, or In).—A thorough study[93] of the boron subhalide B_4Cl_4 resulted in the following assignments in the point group T_d: ν_{a_1} (B—B) 1302, ν_e (B—B) 692, ν_{a_1} (B—Cl) 358, ν_{t_2} 514 and 978, and bending modes at 117, 343, and 110 cm^{-1}. NCA showed extensive mixing of ν(B—B) and ν(B—Cl) for the modes of t_2 symmetry and f(B—B) at 230 ± 10 N m^{-1}, which is considerably lower than the value of 370 N m^{-1}, obtained previously for the more oxidized subhalide B_2Cl_4. The icosahedral subhalides $[B_{12}X_{12}]^{2-}$ (X = Cl, Br, or I) have been prepared and the assignments which were made are given in Table 4.[94] The mixed compounds $BX_2(NCS)$ and

Table 4 *Infrared wavenumbers*/cm^{-1} *for the anions* $[B_{12}X_{12}]^{2-}$

	ν(B—B)	ν(B—X)
X = Cl	1040	530, 460
X = Br	1000	430, 380
X = I	940	410, 370

$BX(NCS)_2$ (X = Cl, Br, or I) have been isolated in spite of rapid redistribution reactions in solution.[95] Reasonably complete Raman and i.r. data were obtained for the monothiocyanato derivatives, and a selection of these are given in Table 5.

A microwave, i.r., Raman, and NCA study of vinyldifluoroborane (13) gave results which indicated, somewhat inconclusively, little conjugation in the π-system;[96] selected fundamentals are given in Table 6. NCA on the series of

[86] C. H. Hao, L. L. Chase, and G. D. Mahan, *Phys. Rev., B*, 1976, **13**, 4306.
[87] B. Gilbert, G. Mamantov, and G. M. Begun, *Inorg. Nuclear Chem. Letters*, 1976, **12**, 415.
[88] P. Maleki and M. J. Schwing-Weill, *J. Inorg. Nuclear Chem.*, 1976, **38**, 1787.
[89] G. N. Shirokova, S. Ya. Zhuk, and V. Ya. Rosolovskii, *Russ. J. Inorg. Chem.*, 1976, **21**, 527.
[90] B. Eberwein, F. Sille, and J. Weidlein, *Z. Naturforsch.*, 1976, **31B**, 689.
[91] S. Lindøy, H. M. Seip, and R. Seip, *Acta Chem. Scand.*, 1976, **A30**, 54.
[92] A. Haas, B. Koch, and N. Welcman, *Z. anorg. Chem.*, 1976, **427**, 114.
[93] F. R. Brown, F. A. Miller, and C. Sourisseau, *Spectrochim. Acta*, 1976, **32A**, 125.
[94] N. T. Kuznetsov, L. N. Kulikova, and S. T. Zhukov, *Russ. J. Inorg. Chem.*, 1976, **21**, 51.
[95] J. Dazord, H. Mongeot, H. Atchekzai, and J. P. Tuchagues, *Canad. J. Chem.*, 1976, **54**, 2135.
[96] J. R. Durig, L. W. Hall, R. O. Carter, C. J. Wurrey, V. F. Kalasinsky, and J. R. Odom, *J. Phys. Chem.*, 1976, **80**, 1188.

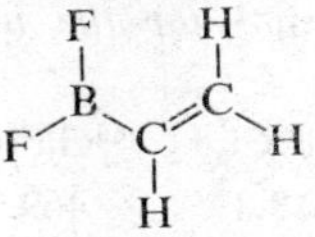

(13)

compounds Me_3NBX_3 (X = F, Cl, Br, or I) based on data from ^2H, ^{10}B, ^{11}B, and ^{15}N isotopic variants gave values for the force constant f(B—N) in the order X = Cl > X = F > X = Br > X = I.[97] ν(B—X) in the series of compounds $(\eta^5\text{-}C_5H_5)(\eta^5\text{-}C_5H_4BX_2)Fe$ was assigned as follows:[98] X = F, 1330 cm^{-1}; X = Cl, 950 and 905 cm^{-1}; X = Br, 852 and 836 cm^{-1}; X = I, 743 and 710 cm^{-1}. Other substituents gave the assignments ν(B—Me) at 1260, ν(B—SMe) at 900, and $\nu(B\text{—}NEt_2)$ at 1208 cm^{-1}. Sketchy assignments have also been given for various alkyl boron halides,[99] and compounds empirically formulated as $(chel)(BX_3)_2$ [chel = phen, bipy, or 1,8-bis(dimethylamino)naphthalene; X = Cl or F] were shown to have the ionic structure $[(chel)BX_2][BX_4]$.[100] Spectral data for complexes of BF_3 with 2′,4′-dihydroxychalcones have been presented.[101]

Table 5 *Selected Raman and i.r. wavenumbers*/cm^{-1} *for the compounds* $BX_2(NCS)$ (X = Cl, Br, or I)

Assignment	$BCl_2(NCS)$	$BBr_2(NCS)$	$BI_2(NCS)$
ν(C—N)	2045	2050	2040
ν(B—N)	1290,[a] 1260 [b]	1292,[a] 1265 [b]	1284,[a] 1259 [b]
ν_{as}(B—X)	—	850,[a] 812 [b]	742,[a] 712 [b]
ν(C—S)	800	766	753
ν_s(B—X)	434	325	—

[a] Attributed to $^{10}BX_2$ (NCS); [b] Attributed to $^{11}BX_2$ (NCS).

Table 6 *Selected vibrational wavenumbers*/cm^{-1} *for vinyldifluoroborane*

Assignment	F_2BCHCH_2	F_2BCDCD_2
$\nu_{as}(^{10}B\text{—}F)$	1425	1418
$\nu_{as}(^{11}B\text{—}F)$	1375	1369
$\nu_s(^{10}B\text{—}F)$	1338	1352
$\nu_s(^{11}B\text{—}F)$	1318	1315
ν(B—C)	765	720
$\delta(BF_2)$	538	495
$\rho_r(BF_2)$	427	397
$\rho_w(BF_2)$	198	185
$\tau(BF_2)$	103	95

Some new matrix isolation studies of group III trihalides [102] inevitably cover much old ground. In the present case the assignments in Table 7 have been

[97] P. H. Laswick and R. C. Taylor, *J. Mol. Structure*, 1976, **34**, 197.
[98] W. Ruf, T. Renk, and W. Siebert, *Z. Naturforsch.*, 1976, **31b**, 1028.
[99] W. Haubold and J. Weidlein, *Z. anorg. Chem.*, 1976, **420**, 251.
[100] D. D. Axtell, A. C. Cambell, P. C. Keller, and J. V. Rund, *J. Coordination Chem.*, 1976, **5**, 129.
[101] A. A. Efimov, O. I. Betin, A. I. Panasenko, and S. P. Starkov, *Zhur. priklad. Spektroskopii*, 1975, **23**, 1054.
[102] I. R. Beattie, H. E. Blayden, S. M. Hall, S. N. Jenny, and J. S. Ogden, *J.C.S. Dalton*, 1976, 666.

Table 7 *Selected wavenumbers/*cm^{-1} *for the in-plane vibrations of monomeric group III trihalides*

Symmetry	$Al^{35}Cl_3$	$^{69}Ga^{35}Cl_3$	$In^{35}Cl_3$
E'	618.8	469.3	400.5
A_1'	393.5	386.2	359.0
E'	150	132	119

offered for the monomeric compounds on the basis of D_{3h} selection rules and isotope splitting patterns, and are supported by an approximate NCA. Data for Al—halogen and other vibrations are in Table 8.[103–105] If the assignments in refs.

Table 8 *Miscellaneous Al—halogen and other vibrations*

Compound	*Assignments/*cm^{-1}		*Ref.*
$(\eta^5\text{-}C_5H_5)_2SnAlCl_3$	ν(Al—Cl)	485, 428	
	δ($AlCl_2$)	294, 274	103
$(\eta^5\text{-}C_5H_5)_2SnAlBr_3$	ν(Al—Br)	391, 333	103
Me_3ClAl_2O,*n*-PhCN	ν(Al—Cl)	443, 417	
	ν(Al—C)	695, 560	104
{[(Me_2N)AlCl]$_2$NMe}$_2$	ν(Al—Cl)	556	105
{[(Me_2N)AlBr]$_2$NMe}$_2$	ν(Al—Br)	380, 370	105
{[(Me_2N)AlI]$_2$NMe}$_2$	ν(Al—I)	355	105

102 and 103 are accepted there are clearly some very large shifts when the $(\eta^5\text{-}C_5H_5)_2Sn$ adducts of $AlCl_3$ and $AlBr_3$ are formed. Data for the compound $CuAl_2Cl_8$ are given in Chapter 6.

cis[Dichlorobis(1,10-phenanthroline)gallium]methyl trichlorogallate [106] displays ν(Ga—Cl) at 330 and 315 cm^{-1}, and ν(Ga—Me) at 570 cm^{-1}. Salts of the anion $[Ga_2I_7]^-$ display ν(Ga—I—Ga) as an intense Raman shift of 137 cm^{-1}, and a very weak i.r. absorption at *ca.* 145 cm^{-1}. On the basis of this and other evidence a structure of D_{3d} symmetry with a linear Ga—I—Ga bridge has been proposed for this anion.[107] This is in marked contrast to the chloro anion, $[Ga_2Cl_7]^-$, which is isoelectronic and isosteric with Cl_2O_7.

The first unambiguous compounds of indium(II) to be prepared are salts of the anions $[In_2X_6]^{2-}$. The assignments given in Table 9 were based on NCA and

Table 9 *Wavenumbers/*cm^{-1} *for the compounds* $(Bu_4N)_2[In_2X_6]$ (X = Cl, Br, or I) *assigned in* D_{3d} *symmetry*

Vibration	*Symmetry*	X = Cl	X = Br	X = I
ν(In—X)	A_g	289	202	152
ν(In—In)	A_g	174	139	108
ν(In—X)	E_g	325	244	210
ν(In—X)	A_{2u}	292	185	164
ν(In—X)	E_u	332, 284 [a]	235, 207 [a]	186

[a] Fermi doublet; unperturbed values estimated at 308 and 221 cm^{-1}.

103 P. G. Harrison and J. A. Richards, *J. Organometallic Chem.*, 1976, **108**, 35.
104 S. Pasynkiewicz, M. Bolesławski, and A. Sadowik, *J. Organometallic Chem.*, 1976, **113**, 303.
105 H. Nöth and P. Wolfgardt, *Z. Naturforsch.*, 1976, **31b**, 1447.
106 A. T. McPhail, R. W. Miller, C. G. Pitt, G. Gupta, and S. C. Scrivastava, *J.C.S. Dalton*, 1976, 1657.
107 D. Mascherpa-Corral and A. Potier, *J. Inorg. Nuclear Chem.*, 1976, **38**, 211.

analogy with $[Ga_2X_6]^{2-}$ spectra.[108] Unassigned Raman data were listed for the 'dihalides' of indium.[109] Raman data for the trihalides $InCl_3$, $InBr_3$ and InI_3 have been obtained for crystals and melts.[110] On melting, $InBr_3$ is transformed from an $InCl_3$-like layer structure to a dimeric form, and the spectra were assigned in D_{2h} symmetry. In_2I_6 is also present in this form in both crystalline and molten phases. Very fragmentary information of an $H[MI_4],2Et_2O$ (M = Ga or In) was presented[111] and mean amplitudes of vibration for $GaCl_3,NH_3$, $AlBr_3,NH_3$, and $GaBr_3,NH_3$ have been calculated.[112]

4 Group IV Elements

Compounds containing Si—H or Ge—H Bonds.—Only isolated reports of ν(Si—H) vibrations have appeared this year. In the compounds $Me_2HSiO(Me_2SiO)_nSiHMe_2$ (n = 0—3) the range 2143—2148 cm^{-1} was quoted.[113] The compound $MeCH{=}CHCH_2CH_2CH{=}CHCH_2SiHMe_2$ displays[114] ν(Si—H) at 2140 cm^{-1}, whereas in $PhCH_2SiHMeOCH_2C{\equiv}CH_2$ it was found[115] at 2090 cm^{-1}. The position of ν(Si—H) in various triorganosilanes was found to correlate with the electronic effects in the substituents but not with reactivity to hydrosilation of hex-1-yne.[116] Raman and i.r. spectra of trimethylvinylsilane were used as a basis for NCA[117] and Raman spectra of some ethynyl silanes have also been reported.[118]

The presence of ν(Si—H) at 2147 and $\delta(SiH_3)$ at 960 cm^{-1} in the i.r. spectrum of the compound $H_3SiV(CO)_6$ is not inconsistent with a seven-co-ordinate structure with SiH_3 capping one face of the octahedron.[119] See also ref. 10 for data pertaining to SiH_3 groups.

The square-root of the integrated intensity of the ν(Ge—H) absorption in compounds of the type R_3GeH was found to correlate linearly with the sum of the Taft σ^* constants of the substituents;[120] the presence of d_π-p_π bonding disturbs this correlation.

Compounds containing M—C Bonds (M = C, Si, or Ge).—ν(C≡C) vibrations have been observed in a variety of metallated alkynes. The wholly lithiated propyne derivative Li_4C_3, formulated as $Li[Li_2C{-}C{\equiv}CLi]$, shows ν(C≡C) at 1675 cm^{-1}. This frequency rises to 1770 and then 1850 cm^{-1} as the lithium atoms

108 B. H. Freeland, J. L. Hencher, D. G. Tuck, and J. G. Contreras, *Inorg. Chem.*, 1976, **15**, 2144.
109 B. H. Freeland and D. G. Tuck, *Inorg. Chem.*, 1976, **15**, 475.
110 W. Bues, Z. Akhras, and G. Okon, *Z. anorg. Chem.*, 1976, **425**, 193.
111 D. Raptis, D. Kaminaris, J. Kounis, and A. G. Galinos, *Monatsh.*, 1976, **107**, 345.
112 S. J. Cyvin and I. Hargittai, *Spectroscopy Letters*, 1976, **2**, 131.
113 I. Souchek, K. A. Andrianov, L. M. Khananashvili, and V. M. Myasina, *Doklady Chem.*, 1976, **227**, 172.
114 V. P. Yur'ev, I. M. Salimyareeva, O. Zh. Zhembarov, G. A. Tolstikov, and S. R. Rafikov, *Doklady Chem.*, 1976, **224**, 609.
115 K. A. Andrianov, I. A. Shikhiev, S. Z. Alieva, and S. F. Karaev, *Doklady Chem.*, 1976, **226**, 21.
116 M. G. Voronkov, V. B. Pukhnarevich, L. I. Kopylova, V. A. Nestunovich, E. O. Tsetlina, B. A. Trofimov, I. Pola, and V. Khvalovskii, *Doklady Chem.*, 1976, **227**, 165.
117 L. V. Khristenko and Yu. A. Pentin, *Vestn. Mosk. Univ. Khim.*, 1976, **17**, 16.
118 T. M. Shidakov, I. F. Kovalev, O. G. Yarosh, and M. G. Voronkov, *Izvest. sibirsk. Otdel. Akad. Nauk, Ser. khim. Nauk*, 1975, 122.
119 J. S. Allinson, B. J. Aylett, and H. M. Colquhoun, *J. Organometallic Chem.*, 1976, **112**, C7.
120 A. N. Egorochkin, S. Ya. Khorshev, N. S. Ostasheva, E. I. Sevastyanova, J. Satgé, P. Privière, and J. Barrau, *J. Organometallic Chem.*, 1976, **105**, 311.

attached to the trigonal carbon atom are replaced successively by methyl groups.[121] I.r. data have also been reported for various silyl allenynes, di-ynes, cumulenes, and propynes.[122, 123] An i.r. band at 2180 cm^{-1} was said to be characteristic of the $-C{\equiv}CSi{<}$ grouping.[124] $\nu(C{\equiv}C)$ was listed [125] for various compounds of the type $R^1{}_3MC{\equiv}CNR^2R^3$ (M = Si, Ge, or Sn; various R's) and the simpler alkynes $R_3MC{\equiv}CH$ (M = Si, Ge, or Sn).[126] Integrated intensities of the latter compounds have also been studied.[127] Two reports concerning organosilicon compounds are of marginal interest.[128, 129]

Equilibria between cyano and isocyano isomers among trialkyl cyanosilanes seem to be well established. New measurements in various solvents, including those on ^{13}C variants support this hypothesis.[130] For example, Me_3SiCN gives two i.r. bands in CCl_4 solution, at 2191.5 and 2096.5 cm^{-1} (*cf.* ref. 62). Vibrational data for the compounds $C_6D_5SiX_3$ (X = H, Cl, or B) have been reported.[131]

The preparation of a sila-ethene, prefigured in last year's Report by a theoretical prediction of spectrum, has at last been achieved. Photolysis of matrix isolated trimethylsilyldiazomethane at 8 K yielded 1,1,2-trimethylsilaethene.[132] Unfortunately $\nu(Si{=}C)$ could not be identified in the i.r. spectrum. A second, identical, preparation of this compound and its deuteriate showed that the absorption at 641 cm^{-1} is due to the out-of-plane deformation of the lone hydrogen atom.[133] A second sila-ethene derivative (14) has a band at 1315 cm^{-1} which was tentatively assigned to $\nu(Si{=}C)$.[134] The structure (14) was established by n.m.r. methods, and $\nu(C{-}O)$ at 2025, 1995, and 1975 cm^{-1} correspond 'nicely' to those of π-allyltricarbonyliron halides.

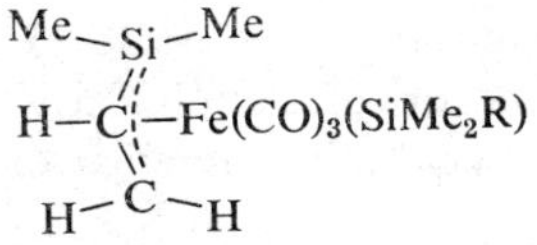

(14) R = Me or $CH{=}CH_2$

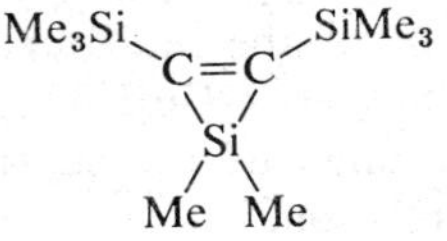

(15)

The first stable silacyclopropene (15) had i.r. absorptions at 1505 and 1500 cm^{-1} which are possibly due to $\nu(C{=}C)$. The values are lower than normally found

121 W. Priester, R. West, and T. Ling Chwang, *J. Amer. Chem. Soc.*, 1976, **98**, 8413.
122 W. Priester and R. West, *J. Amer. Chem. Soc.*, 1976, **98**, 8426.
123 W. Priester and R. West, *J. Amer. Chem. Soc.*, 1976, **98**, 8421.
124 K. A. Andrianov, I. A. Shikhiev, S. F. Karaev, and A. K. Khabibova, *Doklady Chem.*, 1976, **223**, 464.
125 G. Himbert, D. Frank, and M. Regitz, *Chem. Ber.*, 1976, **109**, 370.
126 M. G. Voronkov, N. I. Shergina, V. B. Pukhnarevich, L. V. Sherstyannikova, L. I. Kopylova, and B. A. Trofimov, *Doklady Phys. Chem.*, 1976, **224**, 888.
127 N. I. Shergina, L. V. Sherstyannikova, V. B. Pukhnarevich, L. I. Kopylova, B. A. Trofimov, and M. G. Voronkov, *Doklady Phys. Chem.*, 1976, **225**, 1359.
128 W. Lidy and W. Sundermeyer, *Chem. Ber.*, 1976, **109**, 2542.
129 K. A. Andrianov, É. M. Movsum-Zade, I. A. Shikhiev, M. G. Mamedov, and Ya. F. Nasirov, *Doklady Chem.*, 1976, **223**, 498.
130 J. A. Seckar and J. S. Thayer, *Inorg. Chem.*, 1976, **15**, 501.
131 F. Höfler, *Monatsh.*, 1976, **107**, 411.
132 O. L. Chapman, C.-C. Chang, M. E. Jung, J. A. Lowe, T. J. Barton, and M. L. Tumey, *J. Amer. Chem. Soc.*, 1976, **98**, 7844.
133 M. R. Chedekel, M. Skoglund, R. L. Kreeger, and H. Shechter, *J. Amer. Chem. Soc.*, 1976, **98**, 7846.
134 H. Sakurai, Y. Kamiyama, and Y. Nakadaira, *J. Amer. Chem. Soc.*, 1976, **98**, 7453.

in cyclopropenes, but comparable to those of straight-chain silylethenes such as $(Me_3Si)_2C{=}CHSiMe_3$ (1499 cm^{-1}).[135] Disilacyclobutanes have also been synthesized.[136]

A strong, polarized, line appeared at 300 cm^{-1} in the Raman spectrum of the unusual bivalent group IV metal alkyl $Ge\{CH(SiMe_3)_2\}_2$, and is presumably due to ν(Ge—C) vibrations.[137] In the solid state this compound and its Sn and Pb congeners dimerize.[138] Raman and i.r. data have been recorded for $(^nH_3C)_3$-Ge(CN) (n = 1 or 2) and all fundamentals except the internal torsions were assigned.[139]

Compounds containing M—C Bonds (M = Sn or Pb).—The Raman spectra of the compounds Me_2SnO, $Me_2Pb(NO_3)_2$, and $Me_2Tl(NO_3)$ showed only a single band (at 525, 471, and 490 cm^{-1} respectively) as expected for a linear cation $[MeMMe]^+$.[140] Assignments [141] of the vibrational spectra of the stannacyclohexanes $(CH_2)_5SnX_2$ (X = Me or Et), made with the aid of a comparison with silacyclohexanes, include ν_s(Sn—C) at 518 and ν_{as}(Sn—C) at 507 cm^{-1}. A veritable avalanche of papers concerning dimethyltin derivatives, set off no doubt by an explosion in Mössbauer studies, has appeared.[142–148] The ν(Sn—C) modes appear remarkably insensitive to changes of the other substituents. Discussion of compounds of the types $\{(Me_3Si)_2CH\}_2SnXR$, $\{(Me_3Si)_2N\}_2SnXR$, and $CH_2\{Sn(X)Y_2\}_2$ (X = Cl, Br, or I; R = Et, Me, Ph, *etc.*), for which the greater range 460—600 cm^{-1} for ν(Sn—C) has been proposed, must await fuller publication.[149]

Previous assignments for $SnPh_4$ were confirmed with the aid of pure ^{116}Sn and ^{124}Sn isotopes.[150] Some modified assignments were proposed for $SnPh_3I$ including ν(Sn—I) at 159 (^{116}Sn) and 154 cm^{-1} (^{124}Sn), in disagreement with a previous Raman study.[150] ν_{as}(Sn—Ph) modes were identified in the region 260—270 cm^{-1}. A similar value of 268 cm^{-1} was proposed for the five-co-ordinate complex $Ph_3Sn(OPPh_3)(NO_3)$,[151] and an even higher value of 290 cm^{-1} for the cation $[Ph_2Sn(NO_3)(DMSO)_3]^+$.[152] In a very lengthy paper on mono- and di-substituted aryl trimethyltin compounds, frequencies assigned to ν(Sn—Ph) were correlated with mass differences rather than Hammett σ constants of the substituent groups

135 D. Seyferth, D. C. Annarelli, and S. C. Vick, *J. Amer. Chem. Soc.*, 1976, **98**, 6382.
136 R. T. Conlin and P. P. Gaspar, *J. Amer. Chem. Soc.*, 1976, **98**, 868.
137 D. E. Goldberg, D. H. Harris, M. F. Lappert, and K. M. Thomas, *J.C.S. Chem. Comm.*, 1976, 261.
138 P. J. Davidson, D. H. Harris, and M. F. Lappert, *J.C.S. Dalton*, 1976, 2268.
139 F. Watari, *J. Mol. Structure*, 1976, **32**, 285.
140 Y. Kawasaki and M. Aritomi, *J. Organometallic Chem.*, 1976, **103**, 39.
141 G. Davidovics, M. Guiliano, J. Chouteau, R. Ouaki, and J. C. Maire, *Spectrochim. Acta*, 1976, **32A**, 301.
142 H. Schumann, W.-W. du Mont, and H.-J. Kroth, *Chem. Ber.*, 1976, **109**, 237.
143 R. G. Goel, H. S. Prasad, G. M. Bancroft, and T. K. Sham, *Canad. J. Chem.*, 1976, **54**, 711.
144 V. G. Kumar Das, *J. Inorg. Nuclear Chem.*, 1976, **38**, 1241.
145 N. V. Kozlova, I. F. Kovalev, Y. I. Khudobin, and M. G. Voronkov, *Doklady Akad. Nauk S.S.S.R.*, 1975, **223**, 1401.
146 R. B. Leblanc and W. H. Nelson, *J. Organometallic Chem.*, 1976, **113**, 257.
147 K. Tanaka, S. Araki, and T. Tanaka, *Inorg. Chim. Acta*, 1976, **16**, 107.
148 H. Schumann, W.-W. du Mont, and B. Wöbke, *Chem. Ber.*, 1976, **109**, 1017.
149 M. J. S. Gynane, M. F. Lappert, S. J. Miles, and P. P. Power, *J.C.S. Chem. Comm.*, 1976, 256.
150 N. S. Dance, W. R. McWhinnie, and R. C. Poller, *J.C.S. Dalton*, 1976, 2349.
151 M. Nardell, C. Pelizzi, and G. Pelizzi, *J. Organometallic Chem.*, 1976, **112**, 263.
152 L. Coghi, C. Pelizzi, and G. Pelizzi, *J. Organometallic Chem.*, 1976, **114**, 53.

attached to the phenyl rings. It was also implied that Sn—Ph π-bonding was probably unimportant in these systems.[153]

Compounds containing Si—Si or Sn—Sn Bonds.—I.r. and Raman spectra of the compounds $(^nH_3C)_3Si\ Si(C^nH_3)_3$ (n = 1 or 2) have been interpreted on the basis of $G^{\dagger}_{36}$ symmetry with free rotation about the Si—Si bond.[154] ν(Si—Si) was assigned at 336 cm^{-1} in 3-silylpentasilane,[155] and at 400 cm^{-1} in some 1,2-bis-(dimethylalkylamino)disilanes.[156] Vibrational spectra of cyclopentasilane, $Si_5{}^nH_{10}$ (n = 1 or 2) were used to show that a D_{5h} structure can be ruled out on the basis of i.r./Raman coincidences. Of interest are the ring breathing modes observed in the Raman, 382 (pol) and 464, and i.r., 406 cm^{-1}, spectra.[157] Data have also appeared for other cyclosilanes.[158, 159] Formation of the cyclic structure $(Ph_2P)_2\overline{Pt(SiPh_2)_3Si}Ph_2$ from Si_4Ph_8 is accompanied by an increase in ν(Si—Si) from *ca.* 335 to 368 cm^{-1}; the increase was said to reflect the anticipated increase in Si—Si bond order.[160, 161] Vibrational data for compound (16) clearly rule out

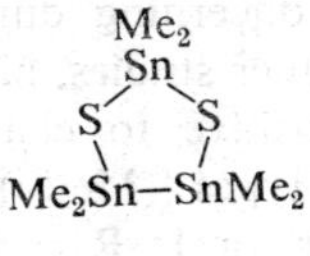

(16)

a planar ring conformation;[162] tentative assignments for this compound include ν(Sn—Sn), 190; $\nu_s(Sn_2S)$, 325; and $\nu_{as}(Sn_2S)$, 359 cm^{-1}.

Compounds containing M—N Bonds (M = Si, Ge, or Sn).—A model of D_{2d} symmetry sufficed to explain the observed Raman and i.r. data for $N_2(Si^nH_3)_4$ (n = 1 or 2) in all states of aggregation; the N—NSi_2 fragments are both planar and mutually perpendicular, indicating the presence of N → Si p_π–d_π bonding.[163] The i.r. spectra of methyleneamino derivatives of formula $M(N{=}CPH_2)_4$ (M = Si, Ge, or Sn) are consistent with bent C=N—M structures.[164] Tentative assignments for ν(M—N) were: M = Si, *ca.* 860; M = Ge, 700—741; M = Sn, 645 cm^{-1}. The formulation[165] of the compound $(Me_3Si)_2N{-}N{\equiv}C$ was supported by spectral assignments of ν(N≡C) at 2100, ν(N—N) at 1032, δ(NNC) at 437, $\nu_s(Si_2N)$ at 460 and $\nu_{as}(Si_2N)$ at 928 cm^{-1}. $\nu_{as}(Si_2N)$ was assigned in the similar range 905—955 cm^{-1} for the related compounds $(Me_3Si)_2NX$ (X = $SCMe_3$, SPh, or O_2SPh).[166] ν_{as}(SiN) was observed at *ca.* 850 cm^{-1} in

[153] M. E. Bishop, C. D. Schaeffer, and J. J. Zuckerman, *Spectrochim. Acta*, 1976, **32A**, 1519.
[154] F. Höfler, *Monatsh.*, 1976, **107**, 893.
[155] F. Feher and I. Fischer, *Z. anorg. Chem.*, 1976, **421**, 9.
[156] U. Wannagat, H. Autzen, and M. Schlingmann, *Z. anorg. Chem.*, 1976, **419**, 41.
[157] F. Höfler, G. Bauer, and E. Hengge, *Spectrochim. Acta*, 1976, **32A**, 1435.
[158] E. Hengge and F. Lunzer, *Monatsh.*, 1976, **107**, 371.
[159] U. Wannagat and M. Schlingmann, *Z. anorg. Chem.*, 1976, **419**, 48.
[160] M. F. Lemanski and E. P. Schram, *Inorg. Chem.*, 1976, **15**, 1489.
[161] M. F. Lemanski and E. P. Schram, *Inorg. Chem.*, 1976, **15**, 2515.
[162] B. Matiasch, *J. Organometallic Chem.*, 1976, **122**, 345.
[163] J. R. Durig, K. S. Kalasinsky, and V. F. Kalasinsky, *J. Mol. Structure*, 1976, **35**, 201.
[164] J. Keable, D. G. Othen, and K. Wade, *J.C.S. Dalton*, 1976, 1.
[165] N. Wiberg and G. Hübler, *Z. Naturforsch.*, 1976, **31b**, 1317.
[166] A. Blaschette, D. Rinne, and H. C. Marsmann, *Z. anorg. Chem.*, 1976, **420**, 55.

various silylamines,[167] but in the incredibly low region 425—476 cm^{-1} for some others.[168] A similarly low value of 598 cm^{-1} was suggested[169] for ν(Si—N) in the cyclic compound $HSi(OCH_2CH_2)_3N$. Raman and i.r. measurements[170, 171] on a wide range of cyclic compounds of the types (17) and (18), and many related acyclic compounds,[171a] resulted in the following ranges: ν(Si—N), 828—950, ν(Si—C), 757—775 and 642—695, ν(Si—Si), 378—463 cm^{-1}. Vibrational, m.s., and n.m.r. data were used to characterize the compounds Si_5N_4,

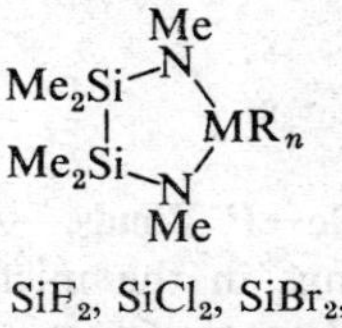

MR_n = SiF_2, $SiCl_2$, $SiBr_2$, $SiMe_2$, PMe, PMe_2^+, $GeCl_2$, SO, AsCl, SbCl

(17)

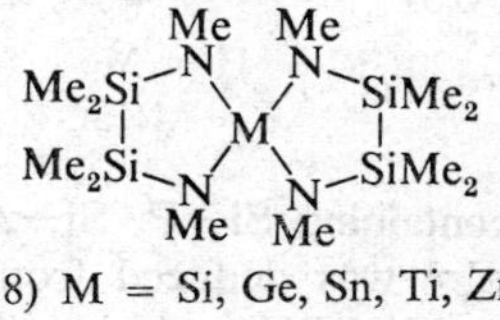

(18) M = Si, Ge, Sn, Ti, Zr

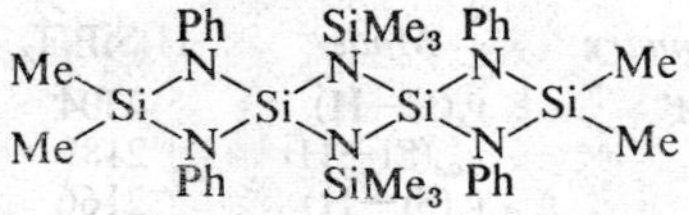

(19)

Si_5N_5, Si_5N_4O, and Si_7N_8, but the wavenumbers were simply listed without apparent assignment.[172] Unassigned i.r. data have also been published for the new dispirotetrasilazane (19),[173] some cyclic siloxazanes,[174] Schiff-base complexes of silicon,[175] and various compounds of which $\{Me_2SiClCF_3)_2\}_2$ is a typical example.[176]

The i.r. spectra of α and β Ge_3N_4 have been compared.[177] ν(Ge—N) was found at 670—690 cm^{-1} in the adducts $Ge(NCS)_4,4L$ (L = $R^1C_6H_4CH{=}NC_6H_4R^2$, R^1, R^2 = various alkyl, H, *etc.*).[178] The bismethylgermanium carbodi-imides $(Me_nH_{3-n}GeN{:})_2C$ have strong, polarized Raman shifts at *ca.* 425 cm^{-1}, which were assigned to ν(Ge—N); ν_s(Ge—C) was assigned in the region 576—611 cm^{-1}, but assignment of the ν(N=C=N) vibration was made complicated by a Fermi resonance.[179] Quite detailed vibrational assignments

[167] U. Klingebiel and A. Meller, *Chem. Ber.*, 1976, **109**, 2430.
[168] K. A. Andrianov, M. I. Shkol'nok, Zh. S. Syrtsova, K. I. Petrov, V. M. Kopylov, M. G. Zaitseva, and E. A. Koroleva, *Doklady Chem.*, 1976, **223**, 406.
[169] C. Eaborn, K. J. Odell, A. Pidcock, and G. R. Scollary, *J.C.S. Chem. Comm.*, 1976, 317.
[170] H. Bürger, M. Schlingmann, and G. Pawelke, *Z. anorg. Chem.*, 1976, **419**, 121.
[171] M. Schlingmann and U. Wannagat, *Z. anorg. Chem.*, 1976, **419**, 115.
[171a] D. Schmid, M. Schlingmann, and U. Wannagat, *Z. anorg. Chem.*, 1976, **419**, 32.
[172] M. Schlingmann and U. Wannagat, *Z. anorg. Chem.*, 1976, **419**, 108.
[173] H. Rosenberg, Tsu-Tzu Tsai, W. W. Adams, M. T. Gehatia, A. V. Frantini, and D. R. Wiff, *J. Amer. Chem. Soc.*, 1976, **98**, 8083.
[174] H. W. Roesky and B. Kuhtz, *Chem. Ber.*, 1976, **109**, 3958.
[175] J. P. Tandon, S. R. Gupta, and R. N. Prasad, *Monatsh.*, 1976, **107**, 1379.
[176] K. E. Peterman and J. M. Shreeve, *Inorg. Chem.*, 1976, **15**, 743.
[177] J.-C. Remy and Y. Pauleau, *Inorg. Chem.*, 1976, **15**, 2358.
[178] N. I. Pirtskhalava and K. P. Giorgadze, *Russ. J. Inorg. Chem.*, 1976, **20**, 1796.
[179] J. E. Drake, R. T. Hemmings, and E. Henderson, *J.C.S. Dalton*, 1976, 366.

have been proposed[180, 181] for the pyrazole derivatives of the Group IV elements (20), but the interesting mixed valence cyclodistannazane (21) was reported[182] only to show 7 i.r. absorptions in the region 550—1215 cm^{-1}.

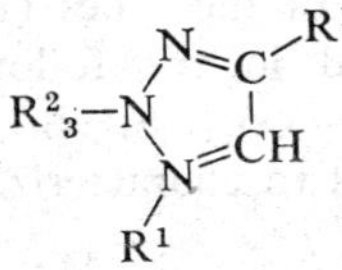

(20) R^1 = H or Me;
R^2_3M = Me_3Si, Et_3Ge, or Bu^n_3Sn

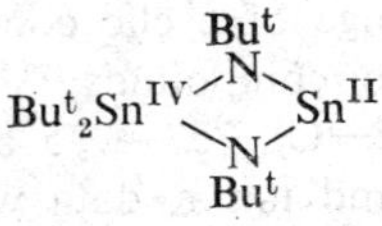

(21)

Compounds containing Si—P, Si—As, or Ge—P Bonds.—A rotational barrier of 6.32 kJ mol^{-1} was deduced from splittings in the microwave spectrum of silyl phosphine, SiH_3PH_2, and its vibrational spectra were assigned with the aid of NCA.[183] Gas-phase wavenumbers are given in Table 10. MVFF calculations

Table 10 *Gas-phase wavenumbers*/cm^{-1} *for silyl phosphine*

Symmetry	*Mode*	H_3SiPH_2	H_3SiPD_2
A'	ν_s(P—H)	2304	1587
	ν_{as}(Si—H)	2189	2198
	ν_s(Si—H)	2166	2186
	$\delta(PH_2)$	1110 [a]	810
	$\delta_{as}(SiH_3)$	955 [a]	940
	$\delta_s(SiH_3)$	940	927
	$\rho_w(PH_2)$	722	552
	$\rho_r(SiH_3)$	602	605
	ν(Si—P)	452	427
A"	ν_{as}(P—H)	2318	1597
	ν_{as}(Si—H)	2189	2198
	$\delta_{as}(SiH_3)$	—	940
	$\rho_t(PH_2)$	770 [a]	657
	$\rho_r(SiH_3)$	602	605

[a] Liquid phase value.

on Cl_3SiPMe_2 gave a Si—P force constant slightly higher than had previously been obtained for alkylsilylphosphines, but a similar effect was not found for the arsenic analogue.[184] Likewise, *f*(Si—As) in the compound F_3SiAsH_3 was not 'unusually' high.[185] The same author has also worked on the compound $F_3SiAsMe_2$ and an interesting aspect of all these papers is that MVFF force constants were transferred from molecule to molecule.[186] Values for ν(Si—P) and ν(Si—As) in $Li[E(SiH_3)_2]$ (E = P or As) are given in ref. 10. The salt-like

180 A. Marchand, M. Rivière-Baudet, J. Satgé, and M.-H. Soulard, *J. Organometallic Chem.*, 1976, **107**, 33.
181 A. Marchand, M. Rivière-Baudet, R. Gasseend, and M.-H. Soulard, *J. Organometallic Chem.*, 1976, **118**, 27.
182 D. Hänssgen, J. Kuna, and B. Ross, *Chem. Ber.*, 1976, **109**, 1797.
183 J. R. Durig, Y. S. Li, M. M. Chen, and J. D. Odom, *J. Mol. Spectroscopy*, 1976, **59**, 74.
184 R. Demuth, *Z. anorg. Chem.*, 1976, **424**, 13.
185 R. Demuth, *Z. anorg. Chem.*, 1976, **427**, 221.
186 R. Demuth, *Spectrochim. Acta*, 1976, **32A**, 1685.

constitution of the compounds $[(Me_3Si)_nPMe_{4-n}]^+[Co(CO)_4]^-$ (n = 1, 2, or 3) was confirmed by their vibrational spectra and assigned wavenumbers/cm^{-1} include the following:[187]

Vibration	$n = 1$	$n = 2$	$n = 3$
ν_s(P—C)	771	756	767
$\nu_{as}(SiC_3)$	678	695	
$\nu_s(SiCl_3)$	622	625	635
ν(Si—P)	417	379	353
$\nu_s(SiCl_3)$	240	244	
$\nu_{as}(SiC_3)$	210	198	191

Frequencies of the adducts $Me_3SiPBu^t{}_2,nAlEt_3$ (n = 0 or 1) and related adducts have been discussed.[188] Germanium(II) stabilized by triorganophosphines was formed in the compounds R_3PGeCl_2 (R = Ph or Bu^t), but i.r. absorptions in the 300 cm^{-1} region were not individually assigned.[189]

Compounds containing M—O Bonds (M = C, Si, Ge, Sn, or Pb).—Hadži *et al.* have obtained data on ^{15}N and ^2H variants of ureas $(H_2N)_2C{=}X$ (X = O, S, or Se), but were unable to find a unique force constant solution.[190]

The ν_{as}(SiOSi) vibration was given the range 1270—1275 cm^{-1} in one group of compounds[191, 192] but in other groups of quite similar compounds the range 1060—1080 cm^{-1} was quoted.[193, 194] The effect, on the character of the siloxane bond, of strongly electronegative substituents bonded to oxygen was revealed by ν(Si—O) in the compounds $Me_{4-n}Si(OC_6F_5)_n$ as follows:[195] for n = 1, 627; for n = 2, 582; for n = 3, 585; and for n = 4, 585 cm^{-1}. Bands at 522, 425, and 400 cm^{-1} were assigned to ν(Si—O) as part of the characterization of the newly prepared silicon formates.[196] Data have also been published for various other siloxanes,[197–199] and silanols.[200]

According to i.r. evidence the compound $3CaGeO_3,H_2O$ should be formulated as $Ca_3[Ge_3O_8(OH)_2]$ and i.r. bands were assigned on this basis.[201] $Pb_5Ge_3O_{11}$

[187] H. Schäfer and A. G. MacDiarmid, *Inorg. Chem.*, 1976, **15**, 848.
[188] L. Rösch and W. Schmidt-Fritsche, *Z. anorg. Chem.*, 1976, **426**, 99.
[189] W.-W. du Mont, B. Neudert, G. Rudolph, and H. Schumann, *Angew. Chem. Internat. Edn.*, 1976, **15**, 308.
[190] D. Hadži, J. Kidrič, Ž. V. Kneževič, and B. Barlič, *Spectrochim. Acta*, 1976, **32A**, 693.
[191] I. Souchek, K. A. Andrianov, L. M. Khananashvili, and V. M. Myasina, *Doklady Chem.*, 1975, **222**, 305.
[192] K. A. Andrianov, A. I. Nogaideli, N. N. Makarova, O. V. Mukbaniani, and K. N. Raspopova, *Doklady Chem.*, 1976, **224**, 575.
[193] M. G. Voronkov, S. F. Pavlov, and É. I. Dubinskaya, *Doklady Chem.*, 1976, **227**, 218.
[194] M. G. Voronkov, S. F. Pavlov, and É. I. Dubinskaya, *Doklady Chem.*, 1976, **227**, 198.
[195] N. V. Kozlova, I. F. Kovalev Yu. I. Khudobin, and M. G. Voronkov, *Doklady Phys. Chem.*, 1976, **223**, 868.
[196] R. C. Paul, T. Puri, and R. Kapoor, *Z. Natursforsch.*, 1976, **31b**, 905.
[197] K. A. Andrianov, M. A. Melomed, V. P. Roschchupkin, and V. G. Kovolev, *Doklady Phys. Chem.*, 1976, **225**, 1264.
[198] B. Becker and W. Wojnowski, *Z. anorg. Chem.*, 1976, **422**, 89.
[199] G. Becker, *Z. anorg. Chem.*, 1976, **423**, 242.
[200] K. A. Andrianov, A. I. Nogaideli, V. S. Tskhovrebashvili, L. I. Nakaidze, and D. Ya. Zhinkin, *Doklady Chem.*, 1976, 223, 416.
[201] A. Lazarev and A. K. Shirvinskaya, *Izvest. Akad. Nauk S.S.S.R., Neorg. Materialy*, 1976, **12**, 777.

undergoes a structural transition at 660 K, accompanied by changes in Raman spectra.[202]

Anhydrous tin hydroxide, $Sn(OH)_2$, has been prepared, apparently for the first time, and has bands at 575 and 340 cm^{-1} which were assigned to ν(Sn—O).[203] For some triphenyltin benzoates ν(Sn—O) was assigned to i.r. bands in the 378—293 cm^{-1} region,[204] but in some tin-containing carbene complexes, *e.g.* $(CO)_4Fe(NMe_2)OSnMe_3$ the ν(Sn—O) vibration could not be distinguished from features due to δ(FeCO) and ν(Fe—C) in the 540—660 cm^{-1} region.[205] Acetato-[206] and 2,4-dionato- tin complexes[207] have been studied.

Raman and i.r. studies of basic lead nitrates led to the suggestion that $PbO,Pb(NO_3)_2,2H_2O$ would be better formulated as $Pb_4(OH)_4(NO_3)_4,2H_2O$, but results for $3PbO,Pb(NO_3)_2,3H_2O$ were not conclusive.[208]

Silicates and Related Species.—Steger, continuing his work on sheet silicates, has studied the polarized i.r. spectra obtained from thin cleavage sheets of muscovite. His detailed comments may well be helpful for the interpretation of other silicate spectra.[209] Raman and i.r. studies of olivine-like silicates and germanates have shown that classification of vibrations into internal modes of the SiO_4 or GeO_4 tetrahedra and external modes of the lattice is a satisfactory approximation.[210] Solid solutions between silicates and germanates with a benitoite structure showed two bands in the region 650—670 cm^{-1}, which are sensitive to Si and Ge isotopic substitution, and were attributed to mixed ring structures containing $SiGe_2O_9$ or Si_2GeO_9 units.[211] Vibrations of the $[Si_2O_7]$ fragment in $3CuO,2SiO_2,nH_2O$ were characterized by i.r. spectroscopy.[212] Other silicates studied include garnets,[213] diopsides,[214] vitreous silica,[215] B_2O_3- and TiO_2-doped high SiO_2 glasses,[216, 217] aluminosilicates,[218] and glasses in the following systems: Fe_2O_3—PbO—SiO_2,[219] Na_2O—P_2O_5—GeO_2,[220] Ba, Sr, Al, Ti, and Si oxides,[221] and Li_2O—SiO_2.[222] (See also Chapter 4.)

202 K. H. Germann, W. Mueller-Lierheim, H. H. Otto, and T. Suski, *Phys. Status Solidi A*, 1976, **35**, K165.
203 W. D. Honnick and J. J. Zuckerman, *Inorg. Chem.*, 1976, **15**, 3034.
204 M. A. Mesubi, *Spectrochim. Acta*, 1976, **32A**, 1327.
205 W. Petz and A. Jonas, *J. Organometallic Chem.*, 1976, **120**, 423.
206 C. D. Garner, B. Hughes, and T. J. King, *Inorg. Nuclear Chem. Letters*, 1976, **12**, 859.
207 B. P. Bachlas and R. R. Tain, *J. Inorg. Nuclear Chem.*, 1976, **38**, 389.
208 A. G. Cram and M. B. Davies, *J. Inorg. Nuclear Chem.*, 1976, **38**, 1111.
209 E. Steger, *Spectrochim. Acta*, 1976, **32A**, 1125.
210 M. Th. Paques-Ledent, *Spectrochim. Acta*, 1976, **32A**, 383.
211 J. Choisnet, A. Deschanvres, and P. Tarte, *Spectrochim. Acta*, 1976, **32A**, 57.
212 V. V. Bogdanova, A. I. Lesnikovich, and V. V. Sviridov, *Russ. J. Inorg. Chem.*, 1976, **21**, 248.
213 T. R. N. Kutty and G. V. A. Iyer, *Indian Mineral.*, 1973 (pub. 1975), **14**, 29.
214 A. N. Lazarev, O. G. Chigareva, and N. O. Zulumyan, *Izvest. Akad. Nauk S.S.S.R., Neorg. Materialy*, 1976, **12**, 660.
215 P. H. Gaskell and D. W. Johnson, *J. Non-cryst. Solids*, 1976, **20**, 171.
216 D. Kato, *J. Appl. Phys.*, 1976, **47**, 2050.
217 T. Nakamura, Y. Suginohara, and T. Yanagase, *Yogyo Kyokai Shi*, 1976, **84**, 293.
218 G. D. Chukin and B. V. Smirnov, *Doklady Phys. Chem.*, 1976, **226**, 188.
219 H. Hagiwara and R. Oyamada, *Yogyo Kyokai Shi*, 1976, **84**, 264.
220 A. A. Margaryan, A. L. Grigoryan, M. G. Manvelyan, T. A. Sidorov, and V. P. Toroyan, *Doklady Chem.*, 1976, **237**, 313.
221 N. M. Bobkova, L. M. Sulich, L. A. Aksenovich, and L. N. Vikaruk, *Zhur. Priklad. Spektroskcopii*, 1976, **24**, 91.
222 A. M. Shevyakov, G. I. Zhuravlev, A. P. Sizonenko, and A. V. Trofimenko, *Doklady Akad. Nauk S.S.S.R.*, 1975, **224**, 160.

Compounds containing M—X Bonds (M = C, Si, Ge, or Sn; X = S, Se, or Te).— The i.r. spectra of the cyclopropenium ions (22) are dominated by a broad band at 1245 (X = S), 1220 (X = Se), or 1175 cm^{-1} (X = Te) which was attributed to a ring deformation mode coupled with ν(C—X).[223] I.r. spectra were used to

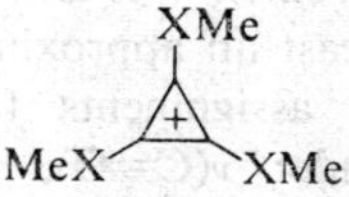

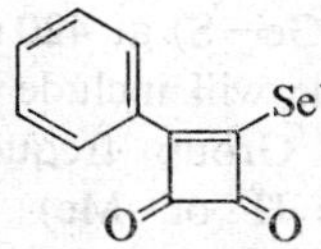

(22) X = S, Se, or Te (23)

adduce evidence for extensive delocalization in some cyclobutene selenones; the structure (23) is one of the canonical forms contributing to the resonance hybrid.[224] Largely unassigned wavenumbers have been listed for the compounds $F_3C—S_3H$,[225] $(F_3C)_2S=NH$,[226] PhN=C(Cl)SR (R = Et, Pr^n, Pr^i, or Bu^n),[227] XC(Br)=S (X = Br, Cl, or F_3CS),[228] and for various substituted thiophenes.[229]

The compound $Ag_{10}Si_3S_{11}$ was shown by *X*-ray methods to have the constitution $Ag_{10}(SiS_4)(Si_2S_7)$ and i.r. bands in the ranges 490—515 and 423—433 cm^{-1} were assigned to ν(Si—S) vibrations.[230] Vibrational spectra of the mercaptides $M(SMe)_4$ (M = Si, Ge, or Sn) revealed the presence of at least two conformers in the liquid state.[231] Raman and i.r. spectra of the compounds Ba_2MS_4 (M = Si, Ge, or Sn) and some Na, Pb, and Sr analogues have been recorded,[282] and values for ν(Si—S) and ν(Si—Se) in $Li[E(SiH_3)]$ (E = S or Se) are reported in ref. 10.

The $[Ge_4S_{10}]^{4-}$ anion has been shown to have the adamantane-like structure (24), by *X*-ray analysis of the salt $Cs_4Ge_4S_{10},3H_2O$, and correlation of the solid-state and aqueous solution Raman spectra showed that this structure persists in solution.[233] Extensive NCA calculations were performed on this anion, but

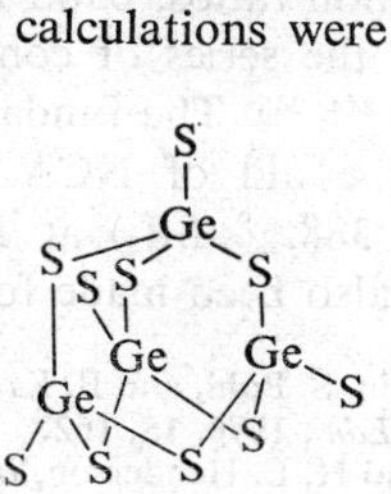

(24)

[223] R. Weiss, C. Schlierf, and K. Schloter, *J. Amer. Chem. Soc.*, 1976, **98**, 4668.
[224] A. H. Schmidt, W. Reid, and P. Pustoslemsek, *Angew. Chem. Internat. Edn.*, 1976, **15**, 704.
[225] C. A. Burton and J. M. Shreeve, *Inorg. Nuclear Chem. Letters*, 1976, **12**, 373.
[226] S. D. Morse and J. M. Shreeve, *J.C.S. Chem. Comm.*, 1976, 560.
[227] R. Appel and K. Giesen, *Chem. Ber.*, 1976, **109**, 810.
[228] G. Diderrich and A. Haas, *Chem. Ber.*, 1976, **109**, 3432.
[229] A. Haas and V. Hellwig, *Chem. Ber.*, 1976, **109**, 2475.
[230] J. Mandt and B. Krebs, *Z. anorg. Chem.*, 1976, **420**, 31.
[231] S. Shimizu and F. Watari, *Technol. Rep. Tohoku Univ.*, 1975, **40**, 239.
[232] M. Neyrand, M. Ribes, E. Philippot, and M. Maurin, *Rev. Chim. minérale*, 1975, **12**, 406.
[233] S. Pohl and B. Krebs, *Z. anorg. Chem.*, 1976, **424**, 265.

as modes are only numbered, assignments are not clear. Presumably the polarized Raman lines at 457, 346, and 193 cm^{-1} belong to ν(Ge—S)$_t$, ν(Ge—S)$_b$, and a cage breathing mode.[234] The compound $Ge_4S_6Br_4$ was shown to have a similar structure with bromide replacing the terminal sulphide atoms.[235] Preliminary assignments include ν(Ge—Br) at 192 and 188 cm^{-1} (Raman) and 300 cm^{-1} (i.r.) and ν(Ge—S) at 420 cm^{-1} (i.r.);[235] it is hoped that any eventual NCA on this molecule will include at least an approximate description of the types of vibration. Group frequency assignments for the compounds[236] R_3GeSC(O)Me (R = H or Me) include: ν(C=O), 1684—1711; ν(C—O), 1262—1286; and ν(Ge—S), 282—296 cm^{-1}. I.r. spectra were particularly useful in distinguishing the above compounds from their R_3MOC(S)Me isomers.[236] Raman spectra of the compounds GeS_2, $GeSe_2$, and $SnSe_2$ suggested the presence of S—S or Se—Se bonds.[237] Lists of i.r. absorptions were given for some Ge^{II} mercaptides,[238] and ν(Ge—S) has been assigned[239] to an i.r. band at 440 cm^{-1} in the compound $Me_2Ge(CH_2)_3P(Ph)C(S)S$.

The Raman spectrum[240] of the semiconductor tin(IV) selenide showed bands at 320 and 339 cm^{-1}. A range of triorganotin dithiocarbamates all had two strong i.r. absorptions in the ranges 417 ± 3 and 397 ± 12 cm^{-1} due to ν(Sn—S) modes.[241]

Compounds containing M—Halogen Bonds (M = C, Si, Ge, Sn, or Pb).— M—halogen vibrations are by now very well characterized and accordingly this section gives details only of those systems which are novel or interesting in the Reporter's subjective opinion. A study of tetrahaloallenes showed that ν_{as}(C=C=C) was less sensitive to change of halogen than was ν_s(C=C=C), in which the amplitude of the terminal carbon atoms' motion is greater.[242] Gas-phase i.r. data have been presented for the trifluoroacetyl pseudo-halides F_3CCOX (X = CN, NCO, NCS, or N_3).[243] Normal mode wavenumbers, absolute intensities, depolarization ratios, band widths and band contours have been extensively discussed for the series of compounds Me_nMBr_{4-n} (M = C, Si, Ge, or Sn; n = 1, 2, or 3).[244, 245] The fundamentals of methyltri-iodosilane have been assigned[246] with the aid of NCA and include ν(Si—C) at 721, ν_s(Si—I) at 252, ν_{as}(Si—I) at 388, $\delta_s(SiI_3)$ at 116, and $\delta_{as}(SiI_3)$ at 171 cm^{-1}. NCA-based assignments have also been made for various chloromethyl chloro-

234 A. Müller, B. N. Cyvin, S. J. Cyvin, S. Pohl, and B. Krebs, *Spectrochim. Acta*, 1976, **32A**, 67.
235 S. Pohl, *Angew. Chem. Internat. Edn.*, 1976, **15**, 162.
236 J. E. Drake, R. T. Hemmings, and H. E. Henderson, *Inorg. Nuclear Chem. Letters*, 1976, **12**, 563.
237 M. P. Lisitsa, V. V. Artamonov, L. I. Berezhinskii, D. I. Bletskan, and M. Ya. Valakh, *Ukrain. Fiz. Zhur.*, 1976, **21**, 218.
238 P. Jutzi and W. Steiner, *Chem. Ber.*, 1976, **109**, 1575.
239 C. Couret, J. Escudie, J. Satgé, and G. Redoules, *J. Organometallic Chem.*, 1967, **111**, 203.
240 O. P. Agnihotri, A. K. Garg, and H. K. Sehgal, *Solid State Commun.*, 1975, **17**, 1537.
241 T. N. Srivastava and V. Kumar, *J. Organometallic Chem.*, 1976, **107**, 55.
242 A. M. Snider, P. F. Krause, and F. A. Miller, *J. Phys. Chem.*, 1976, **80**, 1262.
243 W. Lidy and W. Sundermeyer, *Chem. Ber.*, 1976, **109**, 1491.
244 V. S. Dernova, I. F. Kovalev, N. V. Kozlova, R. G. Mirskov, N. P. Ivanova, and M. G. Voronkov, *Doklady Chem.*, 1976, **226**, 96.
245 I. F. Kovalev, V. S. Dernova, V. I. Anoshkin, M. D. El'kin, and M. G. Voronkov, *Doklady Phys. Chem.*, **225**, 1282.
246 A. J. F. Clark and J. E. Drake, *Spectrochim. Acta*, 1976, **32A**, 1419.

silanes.[247] Spectra of octafluorotrisilane [248] and adducts of silicon tetrafluoride [249] have been reported.

U.v. photolysis of GeF_4—F_2—O_2 mixtures at −78 °C affords the novel compound $[O_2]^+[GeF_5]^-$ which displayed ν(O—O) at 1849 cm^{-1} in freshly prepared samples; at least five ν(Ge—F) modes were observed including one at 480 cm^{-1} attributed to $\nu(Ge—F)_b$ of a NbF_5-like dimer.[250] The spectra of $(NO)_2GeF_6$ and GeF_4 were as expected in relation to their silicon analogues.[250] The salt $[NF_4]^+[GeF_5]^-$ was thought to be a *cis* fluorine-bridged polymer.[251] The base-stabilized germylenes (25) and (26) had ν(Ge—Cl) at 330 and 300,

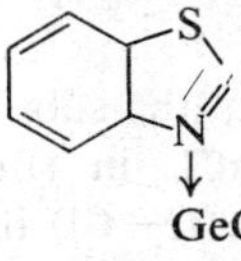

(25)

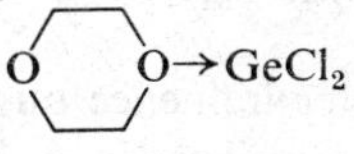

(26)

and 345 cm^{-1} respectively.[252] They react with various transition metal complexes to yield the corresponding ylide complexes with ν(Ge—Cl) in the range 385—315 cm^{-1}. An NCA of $GeCl_2$ has been performed.[253] Trichloro-germanate and -stannate complexes of various sorts gave ν(Ge—Cl) and ν(Sn—Cl) in the normal ranges.[254, 255] The phosphonium salts $[Bu^t_3PCl][MCl_3]$ (M = Ge or Sn) have unco-ordinated $GeCl_3$ and $SnCl_3$ residues.[256] The germylenes R_3PGeCl_2 show three bands attributed [257] to ν(Ge—Cl); with R = Ph, at 300, 313, and 328 cm^{-1}, but with R = Et or Bu^t at 280, 300, and 325 cm^{-1}.

There seems to be no shortage of bases which can form new bis adducts of the tin tetrahalides,[258–264] but little new information emerges from these studies, except perhaps a study of *cis–trans* isomerization in solution.[259] Aryl and alkyltin halides have also provided many unsurprising data.[265–270]

247 K. Sera, K. Suehiro, M. Hayashi, and H. Murata, *Bull. Chem. Soc. Japan*, 1976, **49**, 29.
248 F. Höfler and R. Jannach, *Monatsh.*, 1976, **107**, 731.
249 A. A. Ennan, B. M. Kats, and V. P. Petrosyan, *Russ. J. Inorg. Chem.*, 1976, **20**, 1627.
250 K. O. Christe, R. D. Wilson, and I. B. Goldberg, *Inorg. Chem.*, 1976, **15**, 1271.
251 K. O. Christe, C. J. Schack, and R. D. Wilson, *Inorg. Chem.*, 1976, **15**, 1275.
252 P. Jutzi and W. Steiner, *Chem. Ber.*, 1976, **109**, 3473.
253 J. Brunvoll and I. Hargittai, *Acta Chem. Scand.*, 1976, **A30**, 230.
254 Th. Kruck, K. Ehlert, and W. Molls, *Z. anorg. Chem.*, 1976, **422**, 59.
255 Th. Kruck and W. Molls, *Z. anorg. Chem.*, 1976, **420**, 159.
256 W.-W. du Mont, H.-J. Kroth, and H. Schumann, *Chem. Ber.*, 1976, **109**, 3017.
257 W.-W. du Mont and G. Rudolph, *Chem. Ber.*, 1976, **109**, 3419.
258 J. L. Janier-Dubry, *J. Inorg. Nuclear Chem.*, 1976, **38**, 1865.
259 S. J. Ruzicka and A. E. Merbach, *Inorg. Chim. Acta*, 1976, **20**, 221.
260 S. S. Sandhu, J. K. Sindhu, G. K. Sandhu, N. N. Greenwood, and K. G. Snowdon, *J. Inorg. Nuclear Chem.*, 1976, **38**, 989.
261 T. N. Srivastava, S. K. Tandon, and N. Bhakru, *J. Inorg. Nuclear Chem.*, 1976, **38**, 2311.
262 F.-J. Kruger and A. Schmidt, *Z. anorg. Chem.*, 1976, **427**, 61.
263 R. C. Aggarwal and S. K. Singh, *Spectrochim. Acta*, 1976, **23A**, 1587.
264 Y. Kawano, Y. Hase, and O. Sala, *J. Mol. Structure*, 1976, **30**, 45.
265 W. D. Honnick, M. C. Hughes, C. D. Schaeffer, and J. J. Zuckerman, *Inorg. Chem.*, 1976, **15**, 1391.
266 J. W. Burley, R. E. Hutton, and V. Oakes, *J.C.S. Chem. Comm.*, 1976, 803.
267 A. J. Crowe and P. J. Smith, *Inorg. Chim. Acta*, 1976, **19**, L7.
268 T. N. Srivastava, *Proc. Int. Conf. Coord. Chem.*, 16*th*, 1974, R48.
269 M. A. Mesubi, M. O. Afolabi, and K. O. Falase, *Inorg. Nuclear Chem. Letters*, 1976, **12**, 469.
270 R. C. Paul, P. Sharma, L. Subbiah, H. Singh, and S. L. Chadha, *J. Inorg. Nuclear Chem.*, 1976, **38**, 169.

Data for the compounds *cis*-$PtR(PPh_3)_2L$ (L is defined in Table 11) are interesting because they seem to indicate that the substituent R attached to

Table 11 *Wavenumbers*/cm^{-1} *for the compounds cis*-$PtR(PPh_3)_2L$

R	L	ν(Sn—Cl)
Ph	$SnPh_2Cl$	299
Ph	SnMePhCl	293
Ph	$SnMe_2Cl$	292
Me	$SnMe_2Cl$	278
m-Me-C_6H_4	$Sn(C_6H_4Me\text{-}m)_2Cl$	291

platinum has a greater influence on ν(Sn—Cl) than the substituents attached to the tin atom.[271] Solid solutions of $(Bu^n{}_4N)_2SnCl_6$ in $(Bu^n{}_4N)IrCl_6$ showed ν(Sn—Cl) increasing from 292 to 295 cm^{-1}, and ν(Ir—Cl) increasing from 307 to 310 cm^{-1} as the proportion of iridium was increased. The changes were said to be explicable in terms of increasing lattice dimensions.[272] Mössbauer and i.r. data indicated that the anion in the salt $[Me_4N][SnCl_4F]$ was probably a fluorine-bridged tetramer.[273a] In an i.r. matrix study of reactions between $SnCl_2$ or PbX_2 (X = F, Cl, Br, or I) with CO, NO, or N_2 the ligand modes increase and metal halide modes decrease (relative to free ligand or metal halide) in wavenumbers/cm^{-1}.[273b] For example:

	$\nu_{(ligand)}$	$\Delta\nu_{(ligand)}$	ν_s(MX)	$\Delta\nu_s$(MX)	ν_{a}(MX)	$\Delta\nu_{as}$(MX)
$PbF_2{}^{12}CO$	2176.4	+38.4	520.6	−10.8	496.3	−10.9
$PbF_2{}^{14}NO$	1891.4	+16.4	522.6	−8.8	498.7	−8.5
$SnCl_2{}^{12}CO$	2175.5	+37.5	343	−11	324.3	−9.9

The basic lead chloride Pb(OH)Cl, assumed to have a hydroxy-bridged folded chain structure similar to that of Pb(OH)I, displayed ν(Pb—Cl) at 340, 237 (i.r.), 331, and 276 cm^{-1} (Raman).[274] Data for clustered basic salts were also presented.[274] I.r. wavenumbers for the compounds $(C_5H_5)PbX$ (X = Cl, Br, or $MeCO_2$) were listed but not assigned.[275]

5 Group V Elements

Compounds containing N—H or P—H Bonds.—A study of association of phosphorus acid amides was conducted *via* ν(N—H) measurements.[276]

Vibrational spectra and force constants of phosphorus compounds have been reviewed.[277] K^+ and Cs^+ salts of the anion $H_2PF_4{}^-$ have been characterized. The data in Table 12, however, are the calculated wavenumbers which differ slightly from those observed.[278] Isopropyl phosphine has a low barrier to rotation of 1.09 kJ mol^{-1}, so that *gauche* and *trans* isomers are present in

271 C. Eaborn, A. Pidcock, and B. R. Steele, *J.C.S. Dalton*, 1976, 767.
272 J. E. Fergusson and Tan Wee Tee, *Inorg. Nuclear Chem. Letters*, 1976, **12**, 297.
273 (*a*) K. Dehnicke, B. Busch, and J. Pebler, *Z. anorg. Chem.*, 1976, **420**, 219; (*b*) D. Terault and K. Nakamoto, *Inorg. Chem.*, 1976, **15**, 1282.
274 P. Tsai and R. P. Cooney, *J.C.S. Dalton*, 1976, 1631.
275 A. K. Holliday, P. H. Makin, and R. J. Puddephatt, *J.C.S. Dalton*, 1976, 435.
276 B. N. Laskorin, V. V. Yakshin, and B. N. Sharapov, *Doklady Chem.*, 1976, **237**, 274.
277 J. Goubeau, *Pure Appl. Chem.*, 1975, **44**, 393.
278 K. O. Christe, C. J. Schack, and E. C. Curtis, *Inorg. Chem.*, 1976, **15**, 843.

Table 12 *Wavenumbers*/cm^{-1} *for the anion trans*-$[H_2PF_4]^-$ *assigned in* D_{4h} *symmetry*

Symmetry		Wavenumber	Symmetry		Wavenumber
A_{1g}	ν_1	2322	B_{2u}	ν_7	*inactive*
	ν_2	582	E_g	ν_8	1205
A_{2u}	ν_3	2390	E_u	ν_9	1232
	ν_4	690		ν_{10}	701
B_{1g}	ν_5	495		ν_{11}	355
B_{2g}	ν_6	394			

roughly equal proportion in fluid phases.[279] Corresponding bands are labelled *g* and *t* in Table 13. ν(P—H) was found in the range 2275—2290 cm^{-1} in a series of phosphorus-heterocyclic compounds.[280, 281]

Table 13 *Selected wavenumbers*/cm^{-1} *for isopropylphosphine* [a]

Symmetry		Me_2CHPH_2	Me_2CHPD_2
A'	ν_s(P—H)	2297 *g*	1668 *g*
		2288 *t*	1663 *t*
	$\delta(PH_2)$	1077	784
	$\rho_w(PH_2)$	813	661
	ν(C—P)	606	578
A''	ν_{as}(P—H)	2299 *g*	1673 *g*
		2288 *t*	1663 *t*
	$\rho_t(PH_2)$	809	615
	$\rho_r(PH_2)$	163	125

[a] Data are taken from the gas-phase spectra, where possible. *g* = *gauche* isomer, *t* = *trans* isomer.

Compounds containing E—C Bonds (E = P, As, or Sb).—Raman measurements on the compounds $P(CN)_3$ and $As(CN)_3$ in the gas phase supplement the previous work of F. A. Miller (1965), but no significantly new assignments were proposed.[282] Vibrational assignments of the methyl dicyano compounds $MeE(CN)_2$ (E = P or As) followed from those of the analogous methyl dichloro compounds.[283] It is noteworthy that only one ν(C—N) band was observed, at 2187 (E = P) and 2181 cm^{-1} (E = As).[283] The structure of $P(CF_3)_3$ has been reinvestigated by electron diffraction and Raman techniques, and a minor revision of the vibrational assignment was proposed.[284] Group-frequency assignments [285] for the substituted compounds $(F_3C)_3P(OMe)_2$ and $(F_3C)_3PF(R)$ (R = OMe, SMe, or NMe_2) include ν(P—F) at 690—705 and ν(POC) at *ca.* 1080 and *ca.* 815 cm^{-1}.

For the following three papers, the main objective was a study of torsional barriers, but a wealth of vibrational data is contained within them. In the first two the compounds Me_3P, Me_3As, Me_3PSe, and Me_3EX (E = P or As, X = O

[279] J. R. During and A. W. Cox, *J. Phys. Chem.*, 1976, **80**, 2493.
[280] K. Issleib, H. Winkelman, and H.-P. Abicht, *Z. anorg. Chem.*, 1976, **424**, 97.
[281] T. Saegusa, S. Kobayashi, Y. Kimura, and T. Yokoyama, *J. Amer. Chem. Soc.*, 1976, **98**, 7843.
[282] H. G. M. Edwards, J. S. Ingman, and D. A. Long, *Spectrochim. Acta*, 1976, **32A**, 731.
[283] H. G. M. Edwards, J. S. Ingman, and D. A. Long, *Spectrochim. Acta*, 1976, **32A**, 739.
[284] C. J. Marsden and L. S. Bartell, *Inorg. Chem.*, 1976, **15**, 2713.
[285] K. I. The and R. G. Cavell, *Inorg. Chem.*, 1976, **15**, 2518.

or S) were studied,[286, 287] and in the third various chlorophosphines, but $ClCH_2PCl_2$ and $EtPCl_2$ in particular, were investigated.[288] I.r. data were said to favour a twin-chain conformation of 1-phosphabicyclo[3,3,1]nonane and its P=O and P=S derivatives.[289] Spectra of compounds containing a P—C=O group[290, 291] and of other organophosphorus compounds[292, 293] have been briefly discussed.

Some vinylic diarsines such as $Ph_2AsCH{=}CHAsPh_2$ have been prepared.[294] The *cis* isomers were identified by an i.r. band at *ca.* 1570, whereas the *trans* isomers were characterized by the δ(CH) vibration at *ca.* 985 cm^{-1}. The adducts Me_2AsXL_n (n = 1 or 2; X = Cl or Br; L = thiourea) show no i.r. or Raman band which can be assigned to ν(As—X), so the ionic structures $[Me_2AsL]^+X$ and $[Me_2AsL]^+X,L$ were proposed.[295] In addition to ν_{as}(As—C) at 579—590 and ν_s(As—C) at 567—582 cm^{-1}, ν(As—S) was placed at 275 (n = 1) and 303 cm^{-1} (n = 2).[295]

Ionic structures of the type $[Me_4Sb]^+[O_2PX_2]^-$ (X = F, Cl, or H) were also proposed on the basis of i.r. and Raman evidence, with $\nu_{as}(Sb—C_4)$ assigned at 563—586 and $\nu_s(Sb—C_4)$ at 533—540 cm^{-1}. However, the compounds $Me_4SbO_2PMe_2$ and $Me_4SbOSPMe_2$ are non-ionic and show ν(Sb—O) in the 542—550 cm^{-1} region.[296]

Compounds containing N—N or P—P Bonds.—A Raman shift of 1445 cm^{-1}, assigned to (N=N), was used to support the formulation of the zwitterion (27).[297] ν(P—P) was tentatively assigned to an i.r. band at 388 cm^{-1} in the centrosymmetric phosphazene (28).[298]

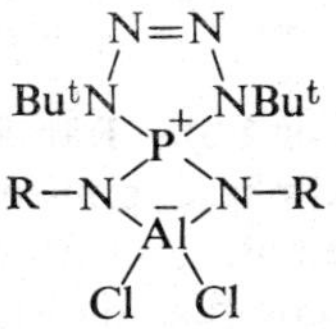

(27) R = Si(Me)$_3$

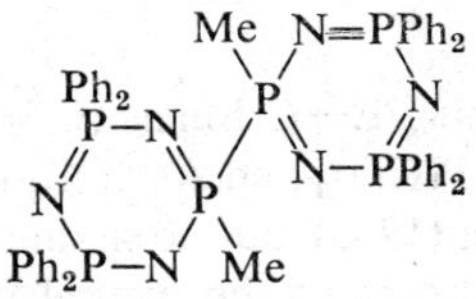

(28)

Compounds containing N—E Bonds (E = P, As, Sb, or Bi).—Volgnadt and Schmidt have courageously characterized the hexa-azidophosphate(v) ion as sodium and tetra-alkyl ammonium salts.[299] Raman data for solutions containing the $[P(N_3)_6]^-$ ion include $\nu_{as}(N_3)$ at 2146 and 2122 cm^{-1}, $\nu_s(N_3)$ at 1310

286 H. Rojhantalab, J. W. Nibler, and C. J. Wilkins, *Spectrochim. Acta*, 1976, **32A**, 519.
287 H. Rojhantalab and J. W. Nibler, *Spectrochim. Acta*, 1976, **32A**, 947.
288 A. I. Fishman, A. B. Remisov, I. Ya. Kuramshin, and I. S. Pominov, *Spectrochim. Acta*, 1976, **32A**, 651.
289 F. Krech and K. Issleib, *Z. anorg. Chem.*, 1976, **425**, 209.
290 H. W. Roesky and G. Sidiropoulos, *Angew. Chem. Internat. Edn.*, 1976, **15**, 693.
291 D. Fenske, E. Langer, M. Heymann, and H. J. Becher, *Chem. Ber.*, 1976, **109**, 359.
292 W.-W. du Mont, B. Neudert, and H. Schumann, *Angew. Chem. Internat. Edn.*, 1976, **15**, 308.
293 H.-P. Abicht and K. Issleib, *Z. anorg. Chem.*, 1976, **422**, 237.
294 Kian-Kong Chow, W. Levason, and C. A. McAuliffe, *J.C.S. Dalton*, 1976, 1429.
295 P. H. Javora, E. A. Meyers, and R. A. Zingaro, *Inorg. Chem.*, 1976, **15**, 2525.
296 B. Eberwein and J. Weidlein, *Z. anorg. Chem.*, 1976, **420**, 229.
297 E. Niecke and R. Kröher, *Angew. Chem. Internat. Edn.*, 1976, **15**, 692.
298 A. Schmidpeter and J. Högel, *Chem. Ber.*, 1976, **109**, 1911.
299 P. Volgnandt and A. Schmidt, *Z. anorg. Chem.*, 1976, **425**, 189.

and 1290 cm^{-1}, and ν(P—N) at 734 cm^{-1}. ν(P—N) in the cation $[Ph_3PN^nH_2]^+$ was observed at 987 (n = 1) or 931 cm^{-1} (n = 2) in the azide salt, but at 938 (n = 1) or 889 cm^{-1} (n = 2) in the azidopentachloroantimonate salt. The shift on changing anion was attributed to the effects of hydrogen bonding.[300] In the compound $(Me_2N)_2PHS$ ν(P—N) was associated[301] with i.r. absorptions at 692 and 723 cm^{-1}, and other assignments included ν(P—H) at 2355 and ν(P=S) at 610 cm^{-1}. The 'adduct' $[2\{(Me_2N)_2PCl\}],AlCl_3$ was shown to have the structure $[(Me_2N)_2PP(NMe_2)_2Cl]^+[AlCl_4]^-$ by n.m.r. and i.r. methods, including identification of the anion vibrations.[302] Related 'adducts' were presumed to have a similar structure.[303]

Acyclic phosphazenes[304] such as $(Me_2N)_3P{=}NSiMe_3$ exhibit ν(P=N) in the now familiar region between 1290 and 1370 cm^{-1}, but few data have appeared this year on cyclic phosphazenes.[305, 306] The novel cage-anion $[P_{12}S_{12}N_{14}]^{6-}$ is built up of P_3N_3 rings and exopolyhedral sulphur atoms.[307] Assigned vibrations include $\nu_s(NP_3)$ at 683 and 725, $\nu_{as}(NP_3)$ at 948 and 978, and ν(P=S) at 530 and 540 cm^{-1}. Routine assignments have been offered for some compounds containing the P=N—Si grouping[308] and for various compounds containing the SiSiNPN ring system.[309, 310] Otherwise i.r. data for many interesting new compounds are simply listed for characterization purposes.[311–315]

ν(As—N) was observed at *ca.* 600 cm^{-1} for many compounds of the type (29).[316] Bands due to the skeletal motions in the piperazine complexes

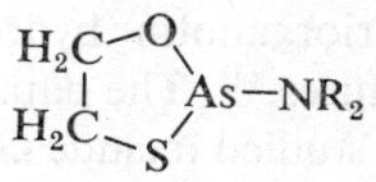

(29) R = alkyl, aryl *etc.*

MX_3,n(pipz) (M = Sb or Bi; X = Cl, Br, or I; n = 1 or 3) were difficult to assign but in all cases features at 263 and 240—250 cm^{-1} were deemed to be due to ν(M—N), at least in part.[317]

Compounds containing E—O Bonds (E = N, P, As, or Sb).—*NN*-dimethylnitrosamine, Me_2NNO, forms adducts with boron acceptors in which ν(N—O)

300 W. Buder and A. Schmidt, *Spectrochim. Acta*, 1976, **32A**, 457.
301 H. Falius and M. Babin, *Z. anorg. Chem.*, 1976, **420**, 65.
302 C. W. Schultz and R. W. Parry, *Inorg. Chem.*, 1976, **15**, 3046.
303 R. W. Kopp, A. C. Bond, and R. W. Parry, *Inorg. Chem.*, 1976, **15**, 3042.
304 O. Schlak, W. Stadelmann, O. Stelzer, and R. Schmutzler, *Z. anorg. Chem.*, 1976, **419**, 275.
305 J. Weidlein, E. Smidt, and E. Fluck, *Z. anorg. Chem.*, 1976, **420**, 280.
306 R. L. Dieck, *J. Inorg. Nuclear Chem.*, 1976, **38**, 2165.
307 E. Fluck, M. Lang, F. Horn, E. Hädicke, and G. M. Sheldrick, *Z. Naturforsch.*, 1976, **31B**, 419.
308 E. Niecke and W. Bitter, *Chem. Ber.*, 1976, **109**, 415.
309 U. Wannagat and H. Autzen, *Z. anorg. Chem.*, 1976, **420**, 132.
310 H. Autzen and U. Wannagat, *Z. anorg. Chem.*, 1976, **420**, 139.
311 U. Klingebiel, P. Werner, and A. Meller, *Monatsh.*, 1976, **107**, 939.
312 U. Wannagat and H. Autzen, *Z. anorg. Chem.*, 1976, **420**, 119.
313 G. Schöning and O. Glemser, *Chem. Ber.*, 1976, **109**, 2960.
314 H. R. Allcock and L. A. Smetlz, *J. Amer. Chem. Soc.*, 1976, **98**, 4143.
315 J. B. van den Berg, E. Klei, B. de Ruiter, and J. C. van de Grampel, *Rec. Trav. chim.*, 1976, **95**, 206.
316 O. Adler and F. Kober, *Z. Naturforsch.*, 1976, **31B**, 304.
317 G. Marcotrigiano, L. Menabue, and G. C. Pellacani, *J. Mol. Structure*, 1976, **30**, 85.

increased from 1465 cm^{-1} according to the acceptor;[318] the increase ranged from 55 cm^{-1} for BPh_3 to 100 cm^{-1} for $PhBF_2$. Detailed assignments have been given for the compound $MePOF_2$ including an i.r. band at 240 cm^{-1} due to the methyl torsional mode.[319] Spectra of the acids HPO_2F_2 and HPO_2Cl_2 have also been assigned in detail.[320] ν(P—O) was assigned in the range 987—1015 cm^{-1} in salts of the $[Me_3POH]^+$ cation.[321] The dimeric compound $[Cl_4SbO_2P(N_3)_2]_2$ showed bands at 1154, 1046, 1044, and 1068 cm^{-1}; the corresponding monomeric anion $[O_2P(N_3)_2]^-$ had bands at 1310 and 1095 cm^{-1} which were also assigned to ν(P=O) vibrations.[322] Many reports concerning ν(P=O) in phosphine oxides have appeared, but contain no surprises.[323–328]

An assignment of the spectra of tributyl phosphate has been attempted.[329] Treatment of barium phosphate with alkaline earth fluorides at 1000 °C gave rise to the anion $[O_3PF]^{2-}$, with ν(P—F) as a doublet at 780 and 755 cm^{-1} in the solid.[330] Three groups have worked on the lead phosphate $PbHPO_3$. The first group, who also studied the barium and tin analogues, made an intelligent attempt to assign the i.r. bands on the basis of a factor group analysis, although evidence from single crystals for the symmetry species involved was not available.[331] The second group examined also the diphosphates $PbH_2P_2O_5$ and $PbH_4P_2O_6$ and gave outline assignments, thus, for $PbHPO_4$ values of ν(P—H) at 2365—2500, δ(PH) at 990—1020, ν(P—O) at 970—1075, and δ(PO) at 465—595 cm^{-1} were assigned with much solid-state splitting in each range.[332] The third group examined also lead hydrogen arsenate, and the hydrogen phosphates of calcium and barium.[333] Triorganotin hydrogen phosphates gave spectra consistent with polymeric structures.[334] The data published on other phosphates were fragmentary. The systems studied include $Gd_4(P_2O_7)_3$,[335] $Zr(HPO_4)_2,H_2O$,[336] $CaBa(PO_3)_4$, $CdBa(PO_3)_4$, and $CdSr(PO_3)_4$,[337] $K_3HP_2O_7$ and K_2HPO_4,[338] $MCr(HPO_3)_2,nH_2O$ (M = Li, Na, K, or NH_4),[339] the system P_2O_5—Nb_2O_5—

[318] A. Meller, W. Maringgele, and H.-G. Köhn, *Monatsh.*, 1976, **107**, 89.
[319] J. R. Durig, K. S. Kalasinsky, and V. F. Kalasinsky, *J. Mol. Structure*, 1976, **34**, 9.
[320] E. Payen and M. Migeon, *Canad. J. Spectroscopy*, 1975, **20**, 150.
[321] F. Seel and H.-J. Bassler, *Z. anorg. Chem.*, 1970, **423**, 67.
[322] P. Volgnandt, R.-A. Laber, and A. Schmidt, *Z. anorg. Chem.*, 1976, **427**, 17.
[323] G. L. Geoffroy, D. A. Denton, and C. W. Eigenbrot, *Inorg. Chem.*, 1976, **15**, 2310.
[324] A. A. Volodin, S. N. Zelenetskii, V. V. Kireev, and V. V. Korshak, *Doklady Chem.*, 1976, **227**, 191.
[325] K. M. Abraham and J. R. Van Wazer, *Inorg. Chem.*, 1976, **15**, 857.
[326] K. G. Sharp and I. Schwager, *Inorg. Chem.*, 1976, **15**, 1697.
[327] E. A. Ishmaeva, M. Mikolaichik, A. B. Remizov, and A. N. Pudovik, *Doklady Chem.*, 1976, **223**, 410.
[328] T. Saegusa, S. Kobayashi, and Y. Kimura, *J.C.S. Chem. Comm.*, 1976, 443.
[329] E. G. Teterin and I. I. Tertishnik, *Radiokhimiya*, 1976, **18**, 17.
[330] G. P. Nikolina, *Russ. J. Inorg. Chem.*, 1976, **21**, 180.
[331] A. Muck, J. Jelinek, and B. Hajek, *Coll. Czech. Chem. Comm.*, 1976, **41**, 3354.
[332] M. Ebert and L. Kavan, *Coll. Czech. Chem. Comm.*, 1976, **41**, 3206.
[333] R. Blinc, H. Arend, and A. Kanduser, *Phys. Status Solidi B*, 1976, **74**, 425.
[334] T. Chivers, J. H. G. Van Roode, J. N. Ruddick, and J. R. Sams, *Canad. J. Chem.*, 1976, **54**, 2184.
[335] M. Kizilyalli, *J. Inorg. Nuclear Chem.*, 1976, **38**, 483.
[336] S. Yamanaka, *Inorg. Chem.*, 1976, **15**, 2811.
[337] I. A. Tokman and E. V. Poletaev, *Izvest. Akad. Nauk S.S.S.R., Neorg. Materialy*, 1976, **12**, 735.
[338] V. A. Sotnikova-Yuzhik, E. A. Prodan, M. A. Pavlyuchenko, *Russ. J. Inorg. Chem.*, 1976, **21**, 23.
[339] M. Ebert and L. Kavan, *Coll. Czech. Chem. Comm.*, 1976, **41**, 23.

$TiO_2—Fe_2O_3$,[340] calcium/cadmium hydroxyapatites,[341] lead hydroxyapatite/lead arsenate,[342] and various sodium and potassium arsenates.[343]

Quite detailed vibrational analyses have been made for the acid $MeAsO_3H$ and its anion,[344] but the i.r. data for numerous arsenous acid esters were unassigned.[345–347]

ν(Sb—O), coupled no doubt to ν(Sb—C), was placed at 394 and 536 cm^{-1} in the diacetamide derivative $Me_4SbO_2C_2Me_2N$,[90] and the range 554 ± 14 cm^{-1} was suggested for ν(Sb—O) in oxinato derivatives of antimony.[348]

Compounds containing E—S (E = N, P, or As) or E—Se (E = P or As) Bonds.— I.r. spectroscopy was used in a matrix isolation study of the products of volatization of polymeric sulphur nitride.[349] *X*-Ray diffraction and i.r. assignments for $S_{11}N_2$, $S_{16}N_2$, and $S_{17}N_2$ have been proposed.[350] Raman and i.r. spectra of the compounds $X_2BN(SCF_3)_2$ (X = Cl, Br, Me, H, or F) were 'easily' assigned, with $\nu_{as}(NS_2)$ in the 820—855 cm^{-1} region.[351] I.r. frequencies have been listed for some cyclic compounds with S_2N_2, S_3N_3, S_3N_2, and related ring systems,[352–356] and for the azulene-like cation in the salt $[S_5N_5][SnCl_5(OPCl_3)]$.[357]

I.r. spectra of P_4S_{10} show that the P_4 skeleton is similar to that of white phosphorus.[358] ν(P=S) in compound (30) was found at 690 (X = Cl) or

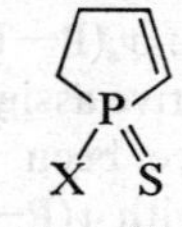

(30) X = Cl or OH

645 cm^{-1} (X = OH).[359] Both dioctylthio-phosphonic and -phosphinic acids show an intense ν(P=S) band,[360] which is evidence for the thione formulation $(C_8H_{17})_2P=S(OH)$ and $(C_8H_{17}O)_2P=S(OH)$. ν(P=Se) was found in the overall range 613—503 cm^{-1} in the compounds $(RO)_3P=Se$ (R = Me, Et, Pr^n, Pr^i, Bu^n, or Bu^i).[361]

340 N. M. Bobkova and G. E. Rachkovskaya, *Zhur. priklad. Spektroskopii*, 1976, **24**, 286.
341 N. Nayak and S. V. C. Rao, *J. Inst. Chem. Calcutta*, 1976, **48**, 31.
342 S. V. C. Rao, *J. Indian. Chem. Soc.*, 1975, **52**, 1107.
343 A. Z. Beilina, K. Zhumanova, Z. M. Muldakhmetov, S. M. Babaev, and M. I. Zhamebekov, *Zhur. fiz Khim.*, 1976, **50**, 253.
344 F. K. Vansant, B. J. Van der Veken, and M. A. Herman, *J. Mol. Structure*, 1976, **35**, 191.
345 F. Kober and W. J. Rühl, *Z. anorg. Chem.*, 1976, **420**, 74.
346 O. Adler and F. Kober, *Z. anorg. Chem.*, 1976, **420**, 109.
347 J. Kaufmann and F. Kober, *Z. anorg. Chem.*, 1976, **420**, 177.
348 Y. Kawasaki, *Bull. Chem. Soc. Japan*, 1976, **49**, 817.
349 R. A. Teichmann and E. R. Nixon, *Inorg. Chem.*, 1976, **15**, 1993.
350 H. Garcia-Fernandez, H. G. Heal, and G. Teste de Sagey, *Compt. rend. C*, 1976, **282**, 241.
351 A. Haas, M. Häberlein, and C. Krüger, *Chem. Ber.*, 1976, **109**, 1769.
352 K.-D. Schmidt, R. Mews, and O. Glemser, *Angew. Chem. Internat. Edn.*, 1976, **15**, 614.
353 Teh-Pei Lin and O. Glemser, *Chem. Ber.*, 1976, **109**, 3537.
354 D.-L. Wagner, H. Wagner, and O. Glemser, *Chem. Ber.*, 1976, **109**, 1424.
355 H. W. Roesky, G. Holtschneider, H. Wiezer, and B. Krebs, *Chem. Ber.*, 1976, **109**, 1358.
356 H. Roesky and H. Zamankhan, *Chem. Ber.*, 1976, **109**, 2107.
357 A. J. Bannister, J. A. Durrant, I. Rayment, and M. M. H. Shearer, *J.C.S. Dalton*, 1976, 928.
358 Y. Monteil and H. Vincent, *Z. Naturforsch.*, 1976, **31b**, 668.
359 K. Moedritzer, *Z. Naturforsch.*, 1976, **31b**, 709.
360 I. A. Vorsina, T. P. Eremeeva, I. S. Levin, S. N. Ryabchenko, and O. M. Frid, *Izvest. Akad. Nauk S.S.S.R., Ser. khim.*, 1975, 2718.
361 R. R. Shagidullin, I. I. Vandyukova, I. A. Nuretdinov, and Kh. Kh. Davletshina, *Doklady Phys. Chem.*, 1976, **225**, 1311.

First- and second-order RRE spectra of As_2S_3 glasses have been recorded,[362] and variations in Raman spectra following optical excitation of amorphous As_2Se_3 have been discussed.[363] Extensive i.r. spectra of triorganoarsenic sulphides include assignments of ν(As=S).[364] The structure of the complex $VO(S_2AsMe_2)_2$ in pyridine solution could not be deduced from vibrational data. The unsolvated complex shows $\nu_{as}(AsS_2)$ at 466 and $\nu_s(AsS_2)$ at 440 cm^{-1}, and a double sulphur bridge was assumed.[365] In the presence of pyridine, however, $\nu_{as}(AsS_2)$ rises to as much as 488 cm^{-1}. The corresponding vibrations in NaS_2AsMe_2 are at 466 and 427, 418 cm^{-1}.

Compounds containing E—Halogen Bonds (E = P, As, Sb, or Bi).—The i.r. and Raman spectra of the compound $PF_2{}^nH(NH_2)_2$ (n = 1 or 2) are particularly notable for showing two strong bands in the region normally associated with ν(P—H) vibrations. Because no other evidence points to the presence of conformational isomers, and because there are no binary combinations of a_1 symmetry in the correct region, it was concluded that Fermi resonance is enhancing a *ternary combination band*. Similar doubling previously observed in spectra of HPF_4, DPF_4, and $PF_2H(NHR)_2$ (R = Me, Et, or Bu^n) suggested that the enhanced ternary combination was characteristic of the PF_2HX_2 skeleton.[366] Assignments (gas phase i.r.) include: ν_s(P—F), 582; ν_s(P—F), 730; ν_s(P—N), 879; and ν_{as}(P—N), 1010 cm^{-1}. Tentative assignments of the i.r. (gas) and Raman (solid) spectra of $F_2B[NH(PF_2)]$ have been presented.[367] Bands at 965, 920, 890, and 540 cm^{-1} were associated with ν(P—F) or ν(P—N) in the compound $F_4PNMeNMePF_4$.[368]

The compound $AsCl_5$ has been prepared, amazingly, for the first time by photolysis of $AsCl_3$ in Cl_2 at −105 °C. It was characterized by Raman shifts of 437, 369(pol), 295(pol) 220, 213, and 83 cm^{-1}. These values are intermediate between those of PCl_5 and $SbCl_5$, and show that the molecule has a trigonal bipyramidal structure both in solution and in the solid state.[369] The arsenic oxychloride, $AsOCl_3$, also prepared for the first time, has Raman shifts at 987(pol) 441, 412(pol), 256, 201(pol), and 160 cm^{-1}, consistent with the expected pyramidal structure.[370]

Assignments of some of the fundamental vibrations of antimony pentafluoride-sulphur dioxide (1 : 1) are in Table 14. Shifts in the SO_2 vibrations were not regarded as diagnostic for bonding type, but it was suggested that the difference, Δ, between the two high-frequency bands of SO_2 was a useful diagnostic: if $\Delta > 190$ cm^{-1} the complex is O-bonded, but if $\Delta < 190$ cm^{-1} the complex is bonded through the sulphur atom.[371] Gas-phase spectra of monomeric SbF_5

[362] R. E. Howard, P. B. Macedo, and C. T. Moynihan, *Solid State Commun.*, 1975, **17**, 1475.
[363] J. Cernogova, F. Mollot, C. Benoit à la Guillaume, and M. Jouanne, *Solid State Commun.*, 1976, **19**, 465.
[364] R. R. Shagidullin, S. V. Izosimova, and I. L. Lamanova, *Spektrosk. Ee Primen. Geofiz. Khim.*, 1975, 227.
[365] E. D. Day and R. E. D. McClung, *Inorg. Chem.*, 1976, **15**, 571.
[366] D. E. J. Arnold and D. W. H. Rankin, *J.C.S. Dalton*, 1976, 1130.
[367] D. E. J. Arnold, E. A. V. Ebsworth, and D. W. H. Rankin, *J.C.S. Dalton*, 1976, 823.
[368] R. Goodrich-Haines and J. W. Gilje, *Inorg. Chem.*, 1976, **15**, 470.
[369] K. Seppelt, *Angew. Chem. Internat. Edn.*, 1976, **15**, 377.
[370] K. Seppelt, *Angew. Chem. Internat. Edn.*, 1976, **15**, 766.
[371] D. M. Byler and D. F. Shriver, *Inorg. Chem.*, 1976, **15**, 32.

Table 14 *Raman shifts*/cm^{-1} *for* F_5SbOSO *in solution in liquid* SO_2

Assignment	*Wavenumber*/cm^{-1}
ν(S—O′)	1325
ν(S—O)	1106
ν(SbF′)	684
$\nu_s(SbF_4)$	651
δ(O′SO)	539
$\delta(SbF_4)$	295
ν(Sb—O)	275
$\nu_{as}(SbF_4)$	605

have been studied yet again, and as usual the new work caused modifications of the old assignments to be suggested;[372] the monomer has i.r. bands at 726, 285, and 263 cm^{-1}. The compound $[SbCl_4(NCO)]_2$ is an *N*-cyanato-bridged dimer at low temperatures, but on warming it changes irreversibly to the trimer, a cyanuric acid derivative (31). The i.r. and Raman spectra of both compounds were extensively discussed.[373] The spectra of various fluoroantimonate species have been featured in six papers, with much emphasis on solid-state effects, and changes resulting from changes of cation.[374–379] The salt $NO[Sb_2F_{10}Cl]$ has ν(Sb—Cl) at 375 cm^{-1}, typical of a terminal Sb—Cl bond, so a monofluorine bridge structure was proposed for the anion,[380] with ν_{as}(Sb—F—Sb) assigned at 450 cm^{-1}, and $\nu(Sb—F)_t$ at 660, 645, 580, 570, 559, and 510 cm^{-1}. Vibrational spectra of the anions $[Ph_2SbX_4]^-$ were consistent with a *trans* structure, and the following assignments were made for ν(Sb—X): X = F, 515 (i.r.) and 496 cm^{-1} (Raman); X = Cl, 270 (i.r.) and 277 cm^{-1} (Raman); X = Br, 193 cm^{-1} (i.r.); X = N_3, 346 cm^{-1} (i.r.); and X = NCS, 302 cm^{-1} (i.r.).[381]

The remaining compounds in this section are all chloro-antimony species, and will be discussed according to the number of chlorine atoms involved. The pre-resonance Raman effect has been evaluated for $[SbCl_6]^-$ (and also $[TeCl_6]^{2-}$ and $[TeBr_6]^{2-}$, see below).[382] Other reports concerning $[SbCl_6]^-$ are of a routine nature.[383, 384]

The new species $[Cl_5SbXSbCl_5]^-$ were thought to have linear SbXSb bridges with asymmetric bridges stretching wavenumbers of 473 (X = F) and 400 cm^{-1} (X = CN).[385] Four papers describe spectra of adducts of $SbCl_5$, with ν(Sb—Cl)

[372] I. Beattie, R. Crocombe, A. German, P. Jones, C. Marsden, G. Van Schalkwyk, and A. Bukovszky, *J.C.S. Dalton*, 1976, 1380.
[373] U. Müller, *Z. anorg. Chem.*, 1976, **422**, 134, 141.
[374] B. Ducourant, R. Fourcade, E. Philippot, and G. Mascherpa, *Rev. Chim. Minérale*, 1976, **13**, 433.
[375] R. Ducourant, B. Bonnet, R. Fourcade, and G. Mascherpa, *Bull. Soc. chim. France, I*, 1976, 1089.
[376] N. Habibi, B. Ducourant, J.-F. Herzog, and R. Fourcade, *Bull. Soc. chim. France, I*, 1976, 77.
[377] P. Djahanguiri, B. Bonnet, R. Fourcade, and G. Mascherpa, *Bull. Soc. chim. France, I*, 1976, 377.
[378] B. Ducourant and R. Fourcade, *Compt. rend. C*, 1976, **282**, 741.
[379] V. V. Fomichev, K. I. Petrov, L. A. Sadokhina, G. V. Zimina, and V. E. Plyuschev, *Koord. Khim.*, 1976, **2**, 77.
[380] W. Liese and K. Dehnicke, *Z. anorg. Chem.*, 1976, **426**, 269.
[381] N. Bertazzi, *J. Organometallic Chem.*, 1976, **110**, 175.
[382] R. J. H. Clark and M. L. Duarte, *J.C.S. Dalton*, 1976, 2081.
[383] F.-J. Kruger and A. Schmidt, *Z. anorg. Chem.*, 1976, **427**, 55.
[384] G. Birke, H. P. Latscha, and A. Schmidt, *Z. anorg. Chem.*, 1976, **424**, 287.
[385] K. Dehnicke and H. G. Nadler, *Z. anorg. Chem.*, 1976, **426**, 253.

anywhere in the range 280—310 cm^{-1}.[386–389] Laber and Schmidt proposed that the 'adducts' $SbCl_4OR,ROH$ (R = Me or Et) exist as hydrogen-bonded dimers as in (32).[390] The adducts $SbCl_4(OR),L$ (R = Me, Et, or C_2H_4Cl;

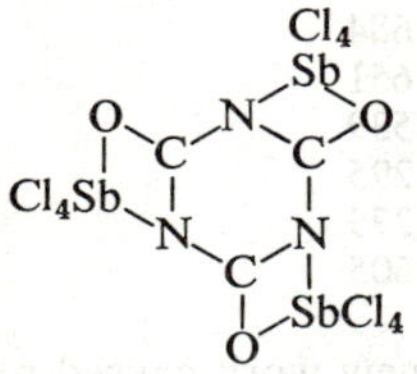

(31)

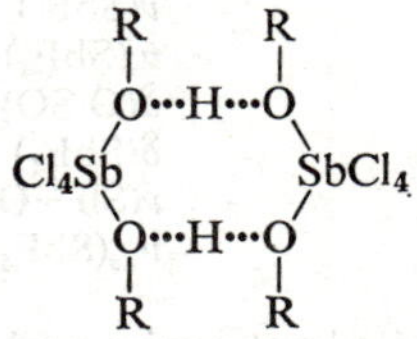

(32) R = Me or Et

L = phthalimide or dioxan) are straightforward with ν(Sb—Cl) in the narrow range 330—345 cm^{-1}.[391] Raman spectroscopy showed the formation of $SbCl_3$,tbp, $SbCl_3$,2tbp, and $SbCl_4^-$ (*sic*) when $SbCl_3$ is extracted from aqueous solution by tributyl phosphate.[392] Adducts of $SbCl_3$ and $SbCl_5$ with amine oxides have been reported,[393] but the i.r. spectra of 1 : 1 complexes of $SbCl_3$ and $SbBr_3$ with benzophenone and related compounds were almost identical with those of the ligand, indicating that the metal atom was not co-ordinated by the oxygen of the 'donor'.[394] Indeed $SbCl_3$ formed adducts with biphenyl, naphthalene, and phenanthrene,[395] which have no donor atoms. Dimethylantimonytrichloride is monomeric in solution, but forms Cl-bridged dimers with *trans* methyl groups in the solid state.[396] Raman data were listed for characterization of $SbCl_2OSeO_2Cl$,[397] and tentative assignments have been proposed for the compounds $\overline{XSb(CH_2)_nCH_2}$ (X = Me or Cl_2Me, n = 4 or 5),[398] and the compounds R_2SbClL (R = Me or Ph; L^{2-} = potentially tridentate O/N/O Schiff-base).[399]

The vibrational spectra of $SbF_5(BiF_5)_n$ (n = 1.5 or 3) have been tentatively assigned in terms of SbF_4 and BiF_4 planar units linked by fluorine bridges.[400] Polarized Raman bands in the spectra of various triphenylbismuth derivatives were assigned as follows:[401] ν_s(Bi—Ph) (so-called t mode) *ca.* 197; ν(Bi—F), 455; ν(Bi—Cl), 253; ν(Bi—NCO), 312; and ν(Bi—Br), 153 cm^{-1}.

386 K. J. Ruoff and A. Schmidt, *Spectrochim. Acta*, 1976, **32A**, 1165.
387 K.-D. Pressl and A. Schmidt, *Z. anorg. Chem.*, 1976, **422**, 266.
388 R. Ortwein and A. Schmidt, *Z. anorg. Chem.*, 1976, **420**, 240.
389 M. Burgard, J.-P. Brunette, and M. J. F. Leroy, *Inorg. Chem.*, 1976, **15**, 1225.
390 R.-A. Laber and A. Schmidt, *Z. anorg. Chem.*, 1976, **425**, 117.
391 R. C. Paul, C. M. Jaswal, and S. L. Chadha, *Chem. and Ind.*, 1976, 694.
392 E. S. Stoyanov, B. Ya. Spivakov, L. A. Gribov, and Yu. A. Zolotov, *Koord. Khim.*, 1978, **2**, 783.
393 J. R. Masaguer, M. V. Castaño, J. S. Casas, M. R. Bermejo, and J. Sordo, *Inorg. Chim. Acta*, 1976, **19**, 139.
394 I. M. Vezzosi, G. Peyronel, and A. Zanoli, *Spectrochim. Acta*, 1976, **32A**, 679.
395 H. H. Perkampus and E. Schönberger, *Chem. Phys. Letters*, 1976, **44**, 348.
396 K. Dehnicke and H. G. Nadler, *Chem. Ber.*, 1976, **109**, 3034.
397 J. Touzin, P. Bauer, and M. Jaros, *Coll. Czech. Chem. Comm.*, 1976, **41**, 2997.
398 H. Meinema, H. F. Martens, and J. G. Noltes, *J. Organometallic Chem.*, 1976, **110**, 183.
399 H. Meinema, J. G. Noltes, F. Dibianca, N. Bertazzi, E. Rivarola, and R. Barbieri, *J. Organometallic Chem.*, 1976, **107**, 249.
400 G. S. H. Chen, J. Passmore, P. Taylor, and T. K. Whidden, *Inorg. Nuclear Chem. Letters*, 1970, **12**, 943.
401 R. G. Goel and H. S. Prasad, *Spectrochim. Acta*, 1976, **32A**, 569.

6 Group VI Elements

Compounds containing Se—C, Se—Si, Se—Ge, or Te—C Bonds.—Raman spectra of the phenyl selenium derivatives $PhSeMH_3$ (M = C, Si, or Ge) showed polarized shifts at the following wavenumbers/cm^{-1}:[402]

Vibration	$PhSeCH_3$	$PhSeSiH_3$	$PhSeGeH_3$
ν(Se—M)	595	394	285
δ(CSeM)	280	250	230

The multiplicity of ν(C—N) bands observed for $Ph_3Te(NCS)$ has now been explained by an *X*-ray study, which showed the presence of NCS-bridged dimers and tetramers in the solid.[403] Outline assignments have been suggested for compound (33).[404]

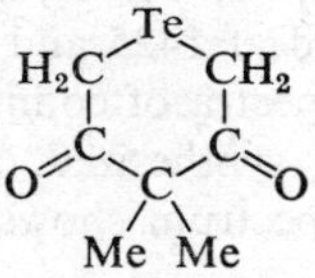

(33)

Compounds containing E—O Bonds (E = S, Se, or Te).—Raman and i.r. spectra of the compound $MeOSO_2Cl$ have been assigned with the aid of NCA, as shown in Table 15. This form of presentation makes the assignment and possible extent

Table 15 *Wavenumbers*/cm^{-1} *and approximate potential energy distribution* (%) *for* $MeOSO_2Cl$

Symmetry	*Value*/cm^{-1}	% *p.e.d.*
A′	1450	86 $\delta_{as}(CH_3)$, 16 $\rho(CH_3)_{in\ plane}$
	1443	88 $\delta_s(CH_3)$, 15 ν(C—O)
	1200	89 $\nu_s(SO_2)$
	1178	77 $\rho(CH_3)_{in\ plane}$, 13 $\delta_{as}(CH_3)$
	1607	69 ν(C—O), 12 ν(S—O), 11 $\delta_s(CH_3)$
	790	65 ν(S—O), 14 ν(C—O), 13 $\omega(SO_2)$
	600	82 $\delta(SO_2)$, 19 ν(S—Cl)
	468	58 $\omega(SO_2)$, 19 ν(S—Cl), 15 δ(OSCl)
	402	54 ν(S—Cl), 24 $\omega(SO_2)$
	252	72 δ(OSCl)
A″	1450	86 $\delta_{as}(CH_3)$, 16 $\rho(CH_3)_{out\ of\ plane}$
	1428	96 $\nu_{as}(SO_2)$
	1178	83 $\rho(CH_3)_{in\ plane}$, 13 $\delta_{as}(CH_3)$
	540	89 $\rho(SO_2)$
	310	$\tau(SO_2)$

of mixing admirably clear, and it ought to be emulated widely.[405] The vibrations $\nu_{as}(SO_2)$ and $\nu_s(SO_2)$ in the compounds Bu^t—NSN—SO_2X were observed at 1405, 1190 (X = F), 1378, 1160 (X = Cl), and 1305, 1145 cm^{-1} (X = Ph)

402 J. E. Drake and R. T. Hemmings, *J.C.S. Dalton*, 1976, 1730.
403 J.-S. Lee, D. D. Titus, and R. F. Ziolo, *J.C.S. Chem. Comm.*, 1976, 501.
404 J. C. Dewan and J. Silver, *Inorg. Nuclear Chem. Letters*, 1976, **12**, 647.
405 B. Nagel, J. Stark, J. Fruwert, and G. Geiseler, *Spectrochim. Acta*, 1976, **32A**, 1297.

respectively.[406] The same vibrations in compounds containing the $\geqslant P{=}N{-}SO_2{-}OH$ grouping were observed[407] in the ranges 1350—1360 and 1110—1120 cm^{-1}. The methoxythionyl salt $[MeOSO]^+[Sb_2F_{11}]^-$ has been studied by *X*-ray, n.m.r., and Raman methods.[408] The Raman shift of 2985 cm^{-1} was high for a methoxy group in consequence of the ionic charge, and the assignments of $\nu(S{-}O)_t$ at 1315 and $\nu(S{-}O)_b$ at 995 cm^{-1} were also proposed.[408] Addition of SO_2 to $MeF{-}SbF_5$ in SO_2ClF at −78 °C gave a white precipitate which, it was inferred, contained the cation $[H_3CO{=}S{=}O]^+$. Identification of the reaction products was not possible by Raman methods since a band at 1008 cm^{-1} was close to one of the methyl fluoride precursor at 1110 cm^{-1}, but the n.m.r. spectra showed substantial differences.[409] I.r. data have been listed for the pseudohalides $Me_3SiO{-}SO_2{-}X$ (X = NCO, NCS, $NSOF_2$, or NSO),[410] and i.r. spectra of the compounds $XOSO_2F$ (X = I, Br, ^1H, ^2H, Br_3, or O_2Cl) have been measured at 80 K.[411] $\nu(S{-}O)$ was assigned at 1425 and 1271, with $\nu(N{=}S)$ at 1182 and $\nu(S{-}F)$ at 872 cm^{-1} in the i.r. spectra of compound (34).[412] Thioformaldehyde *S*-oxide, $H_2C{=}S{=}O$, has been synthesized, and its structure determined by microwave methods;[413] its i.r. spectrum shows strong *Q*-branches at 1170 and

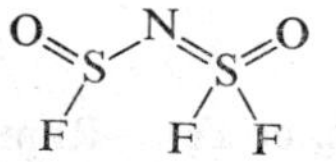

(34)

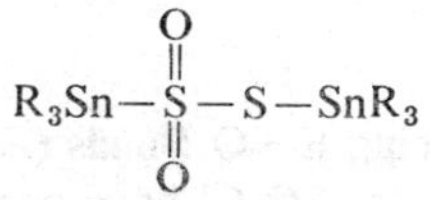

(35) R = Et or Bu^n

760 cm^{-1}. The molecule S_7O displays a single $\nu(S{-}O)$ band at 1129 cm^{-1} in CS_2 solution, but this splits into three components at 1113, 1102, and 1098 cm^{-1} in the Raman spectrum of the crystal.[414] The compound $(Bu^n{}_3Sn)_2S_2O_2$ has three i.r. absorptions in the $\nu(S{-}O)$ region, at 1127, 1103, and 950 cm^{-1}. On the basis of a comparison with spectra of sulphite compounds, the thiosulphonate structure (35) was proposed.[415]

An original study of the thiosulphate ion has been made. The isostructural salts $[M(en)_3][S_2O_3]$ (M = Zn, Cd, Fe, Ni, Co, or Mn; en = ethylenediamine) including the *N*-deuterio derivatives were prepared, and thiosulphate vibrations were identified in their spectra on the assumption, borne out by correlation diagrams, that $S_2O_3{}^{2-}$ vibrations are invariant in the series of compounds. The resulting assignment was in total agreement with that of Siebert, based on solution data, and obtained years ago.[416] A sequel to this study, of more import,

406 R. Appel and M. Montenarh, *Chem. Ber.*, 1976, **109**, 2437.
407 R. Appel, I. Ruppert, and M. Montenarh, *Chem. Ber.*, 1976, **109**, 71.
408 R. J. Gillespie, F. G. Riddell, and D. R. Slim, *J. Amer. Chem. Soc.*, 1976, **98**, 8069.
409 P. E. Peterson, R. Brockington, and D. W. Vidrine, *J. Amer. Chem. Soc.*, 1976, **98** 2660.
410 W. Buss, H.-J. Krannich, and W. Sundermeyer, *Chem. Ber.*, 1976, **109**, 1486.
411 W. W. Wilson, J. M. Winfield, and F. Aubke, *J. Fluorine Chem.*, 1976, **7**, 245.
412 H.-J. Krannich and W. Sundermeyer, *Angew. Chem. Internat. Edn.*, 1976, **15**, 311.
413 E. Block, R. E. Penn, R. J. Olsen, and P. F. Sherwin, *J. Amer. Chem. Soc.*, 1976, **98**, 1264.
414 R. Steudel and T. Sandow, *Angew. Chem. Internat. Edn.*, 1976, **15**, 772.
415 U. Kunze and R. Hengel, *Chem. Ber.*, 1976, **109**, 2793.
416 Z. Gabelica, *Spectrochim. Acta*, 1976, **32A**, 327.

was the identification of a monodentate thiosulphate group in the salt $[Pd^{II}(en)_2][Pd^{IV}(S_2O_3)_2(en)]$.[417] Vibrational spectra of the sulphate ion in $BaMg(SO_4)_2$ were assigned on the basis of C_3 symmetry.[418]

Raman and i.r. spectra of the anion SeO_2F^- were reinvestigated, and the new anion $SeF_2O_2^{2-}$ was prepared as a potassium salt.[419]

A single crystal *X*-ray study of $Te(OH)_6$,2HF indicated the presence of TeO_6 groups with near O_h symmetry.[420] The following wavenumbers were reported: $\nu(TeO_6)$, 644, 624, 609 (Raman) and 691, 672, 662 (i.r.); $\delta(TeO_6)$, 380, 370, 360, 326 (Raman); δ(TeOH), 1160, 1190 (i.r.); γ(TeOH), 935, 890, 875 (i.r.); and $\nu(OH \cdots F)$, 225 cm^{-1} (i.r.). I.r. spectra of the tellurates $Sr_{2-x}Ba_xNiTeO_6$ have been compared with those of related compounds.[421]

Compounds containing S—S, Se—Se, or Te—Te Bonds.—Internal modes of the S_8 rings in β-sulphur were found to be similar to those of α-sulphur, but the external modes showed behaviour typical of a disordered crystal.[422] Rhombic sulphur dissolved in the secondary amines R_2NH (R = Bu^n or Pr^n) to yield S_4^{2-} and S_8^{n-} ions, in a manner similar to that previously observed in primary amines.[423] The assignments proposed were disputed by Steudel [424] who claimed that the 424 cm^{-1} Raman line from S_4^{2-} was due to an impurity, but Daly and Brown maintained that this was not so.[425] A wide variety of compounds derived from $(HN)_2S_6$ and NS_7 have been characterized by i.r. spectroscopy.[426, 427]

Studies of the Raman spectra of α-selenium at 11.4 K have identified the 112 cm^{-1} band as an Se_8 ring mode.[428] The Raman spectrum [429] of an oxidation product of $Se_2(C_2F_5)_2$ showed a strong polarized band at 347 cm^{-1}, attributed to ν(Se—Se) of the cation $[\{Se(C_2F_5)\}_4]^{2+}$. The ditellurides Ph_2Te_2, $(Ph_2Te_2)CuCl$, and $(Et_2Te_2)CuCl$ displayed ν(Te—Te) at 167, 170, and 180 cm^{-1} respectively;[430] the copper compounds are discussed further in Chapter 6.

Compounds containing E—Halogen Bonds (E = S, Se, or Te).—The isocyanate O=C=N—S(O)F gave i.r. absorptions as follows:[431] ν_{as}(NCO), 2245; ν_s(NCO), 1348; ν(S—O), 1281; ν(S—F), 753; ν(S—N), 716; and δ(NCO), 610 cm^{-1}. Miscellaneous data have been reported for various other halosulphur compounds.[432–435]

417 Z. Gabelica, *Ann. Soc. Sci. Bruxelles, Sér. I*, 1976, **90**, 143.
418 G. Blasse and G. P. M. van den Heuvel, *J. Inorg. Nuclear Chem.*, 1976, **38**, 876.
419 R. J. Gillespie, P. Spekkens, J. B. Milne, and D. Moffett, *J. Fluorine Chem.*, 1976, **7**, 43.
420 R. Allmann and W. Haase, *Inorg. Chem.*, 1976, **15**, 804.
421 P. Köhl, *Z. anorg. Chem.*, 1970, **427**, 205.
422 G. Gautier and M. Debeau, *Spectrochim. Acta*, 1976, **32A**, 1007.
423 F. P. Daly and C. W. Brown, *J. Phys. Chem.*, 1976, **80**, 480.
424 R. Steudel, *J. Phys. Chem.*, 1976, **80**, 1516.
425 F. P. Daly and C. W. Brown, *J. Phys. Chem.*, 1976, **80**, 1518.
426 R. J. Ramsey, H. G. Heal, and H. Garcia-Fernandez, *J.C.S. Dalton*, 1976, 237.
427 R. J. Ramsey, H. G. Heal, and H. Garcia-Fernandez, *J.C.S. Dalton*, 1976, 234.
428 M. Gorman and S. A. Solin, *Solid State Commun.*, 1976, **18**, 1401.
429 J. Passmore, E. K. Richardson, and P. Taylor, *J.C.S. Dalton*, 1976, 1006.
430 I. Davies and W. R. McWhinnie, *Inorg. Nuclear Chem. Letters*, 1976, **12**, 763.
431 H.-J. Krannich and W. Sundermeyer, *Angew. Chem. Internat. Edn.*, 1976, **15**, 110.
432 M. Feuerhahn and R. Minkwitz, *Z. anorg. Chem.*, 1976, **426**, 247.
433 K. Seppelt, *Angew. Chem. Internat. Edn.*, 1976, **15**, 44.
434 Shin-Liang Yu and J. M. Shreeve, *Inorg. Chem.*, 1976, **15**, 14.
435 A. Haas and M. Yazdaubakhasch, *Chem. Ber.*, 1976, **109**, 1976.

The Raman spectrum of $[Se(C_2F_5)_2I]^+$ in AsF_3 showed a strong peak at 211 cm^{-1}, assigned to ν(Se—I),[436] and spectral data were given for other perfluoroalkyl selenium derivatives.[437] A T-shaped structure was proposed for the compounds Ar_3PSeBr_2 (Ar = Ph, *p*-tolyl, or *m*-tolyl). Assignments for Ph_3PSeBr_2, which are typical, include[438] ν_s(Se—Br) at 153, ν_{as}(Se—Br) at 188, and ν(Se—P) at 526 cm^{-1}.

Pre-resonance Raman spectra of $[TeCl_6]^{2-}$ and $[TeBr_6]^{2-}$ have provided accurate harmonic frequencies, and excitation-dependent intensities. For $[TeBr_6]^{2-}$ the ratio of intensities for $\nu_2 : \nu_1$ varies from 0.86 to 5.95 as the excitation wavelength varies from 647.1 to 457.9 nm.[382] Gas-phase spectra of $[TeF_4O]_2$ have been reinvestigated.[372] Other reports concerning tellurium–halogen compounds are largely superficial.[439–441]

7 Group VII Elements

Previously unknown forms of solid perchloric acid have been claimed, but it is difficult to summarize this very lengthy empirical study because the conclusions given are vague and in very bad English.[442]

Spectra of the cation $[BrO_2]^+$ showed the expected similarities with monomeric SeO_2. The three fundamentals of the bent ion were placed at 865 for ν_s(Br—O), 930 for ν_{as}(Br—O), and 375 cm^{-1} for $\delta(BrO_2)$.[443] Raman spectra of a mixture of Br_2O_5 and Br_2O_4, obtained by the action of O_3 on Br_2 at reduced temperatures, were used to probe the structure of Br_2O_5. A polymeric form similar to that of I_2O_5 was considered to be most probable.[444]

Careful analysis of the spectra of the salts $[X(SO_3F)_2]_2^+[Sn(SO_3F)_6]^{2-}$ led to the following proposals for the cations:

Vibration	*Wavenumbers*/cm^{-1} X = Br	X = I
$\nu_{as}(XO_2)$	745	712, 690
$\nu_s(XO_2)$	652, 640	663, 650
$\delta(XO_2)$	309	284

These data suggest that the authors' assignments (*Inorg. Chem.*, 1970, **9**, 2485) for the tris(fluorosulphato) halates(III) may need revision.[445]

Raman and i.r. data for the new compounds $IOSO_3F$, $IOSO_3CF_3$, and $IO_2SO_3CF_3$ as well as for I_2O_4 and $(IO)_2SO_4$ have been presented,[446] and a selection from them is given in Table 16. A polymeric structure involving $(IO)_n$ chains and bidentate bridging SO_3F groups was proposed for $IOSO_3F$ on the basis of the low ν(I—O) values and the ν(S—O) data; discrete IO- and IO_2-groups were proposed for $IOSO_3CF_3$ and $IO_2SO_3CF_3$. It was also concluded that I_2O_4

436 J. Passmore and P. Taylor, *J.C.S. Dalton*, 1976, 804.
437 C. Lau and J. Passmore, *J. Fluorine Chem.*, 1976, **7**, 261.
438 D. J. Williams and K. J. Wynne, *Inorg. Chem.*, 1976, **15**, 1449.
439 R. A. Walton, *Inorg. Nuclear Chem. Letters*, 1976, **12**, 767.
440 N. Petragnani, J. V. Comasseto, and Y. Kawano, *J. Inorg. Nuclear Chem.*, 1976, **38**, 608.
441 E. R. Clark and M. A. Al-Turachi, *J. Inorg. Nuclear Chem.*, 1976, **38**, 1443.
442 A. I. Karelin and Z. I. Grigorovich, *Spectrochim. Acta*, 1976, **32A**, 851.
443 E. Jacob, *Angew. Chem. Internat. Edn.*, 1976, **15**, 158.
444 J. L. Pascal, A. C. Pavia, J. Potier, and A. Potier, *Compt. rend. C*, 1976, **282**, 53.
445 P. A. Yeats, B. Landa, J. R. Sams, and F. Aubke, *Inorg. Chem.*, 1976, **15**, 1452.
446 J. R. Dalziel, H. A. Carter, and F. Aubke, *Inorg. Chem.*, 1976, **15**, 1247.

Table 16 *Raman shifts*/cm^{-1} *for* IO_2SO_3F *and* $IOSO_3F$

Assignment	IO_2SO_3F	$IOSO_3F$
$\nu(S{=}O)$	1335	1366, 1349
$\nu_{as}(S{-}O)_b$	1170, 1195	1112, 1148, 1160
$\nu_s(S{-}O)_b$	1070, 1030, 1010	1029, 999
$\nu(S{-}F)$	843	851, 821
$\nu_{as}(I{-}O)$	878, 900	637, 645
$\nu_s(I{-}O)$	865	590, 598
$\nu_{as}(IO)_b$	685, 660	447
$\nu_s(IO)_b$	520	428

cannot be adequately represented by formulations such as $IO_2^+IO_3^-$ or $(IO^+)_n$-$(IO_3^-)_n$.

Two new studies of the thermally and photolytically unstable compound IO_2F_3 resulted in conflicting conclusions. Beattie *et al.*[372] favour a centrosymmetric dimer in gas, liquid (melt), and solid phases, whereas Gillespie and Kraznai[447] have proposed the trimeric structure (36) for the liquid and solutions

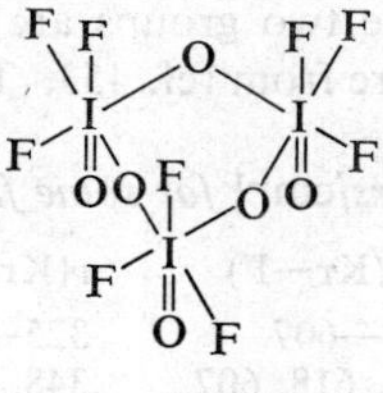

(36)

in BrF_5 and $CF_3C_6F_{11}$. This is not the first time that these two groups have produced conflicting evidence for this difficult material, and it surely will not be the last. A fairly complete assignment of the spectra of the anion $[IOF_4]^-$ in its Cs^+ and K^+ salts was made by reference to the isoelectronic $XeOF_4$ and $[TeOF_4]^{2-}$ moieties. The Raman data in Table 17 agree closely with the i.r.

Table 17 *Raman shifts*/cm^{-1} *for the compound* $CsIOF_4$

Assignment	*Wavenumber*/cm^{-1}
$\nu(I{-}O)$	888
$\nu_s(IF_4)_{in\ phase}$	533
$\delta_s(IF_4)_{umbrella}$	273
$\nu_s(IF_4)_{out\ of\ phase}$	475
$\delta_s(IF_4)_{in\ plane}$	214
$\nu_{as}(IF_4)$	485
$\delta(IF_4)_{wag}$	365
$\delta_{as}(IF_4)_{in\ plane}$	124

data obtained.[448] Wavenumbers were also reported for $HIOF_4$ and HIO_2F.[448] A polymeric structure containing I—O—I bridges and $\nu(I{-}O)$ at 430—500 cm^{-1} was proposed for the compound PhIO; for $PhIO_2$ and $PhIOF_2$ $\nu(I{-}O)$ was found in the 710—800 cm^{-1} region.[449] The compounds $IX,IF_5,3py$ (X = F,

[447] R. J. Gillespie and J. P. Krasznai, *Inorg. Chem.*, 1976, **15**, 1251.
[448] J. B. Milne and D. M. Moffett, *Inorg. Chem.*, 1976, **15**, 2165.
[449] H. Siebert and M. Handrich, *Z. anorg. Chem.*, 1976, **426**, 173.

Cl, or Br) were designated as complexes between IX,py and IF_5,py with covalent I···F···I bridges.[450] Iodine thiocyanate, stable in organic solvents, displayed ν(C≡N) at 2140 cm^{-1}, consistent with an I—SCN structure.[451]

8 Group VIII Elements

The products of reaction of KrF_2 with AsF_5, SbF_5, and PtF_6 were formulated as salts $[KrF]^+[MF_6]^-$, though the Raman spectra were assigned on the basis of a F—Kr···F···MF_5 structure.[452] Two Raman bands between 596 and 619 cm^{-1} in the spectra of the solids were assigned to ν(Kr—F); their positions were almost identical to those of the xenon analogues. Raman and n.m.r. evidence suggested that the $Kr_2F_3^+$ cation in solution had a symmetrical bridge structure similar to that of the $Xe_2F_3^+$ cation.[452] Raman spectra of the new compounds $(KrF_2)_2$-SbF_5, KrF_2,MF_5 (M = Sb or Ta), and KrF_2,$(MF_5)_2$ (M = Ta or Nb) have also been interpreted in terms of a structure containing Kr···F···M contributions to the bonding, as well as ionic formulations such as $[Kr_2F_3]^+[MF_6]^-$ and $[KrF]^+[MF_6]^-$ (M = Sb or Ta) and $[KrF]^+[M_2F_{11}]^-$ (M = Sb, Ta, or Nb).[453] The experimental data of these two groups are in good agreement where they overlap; the data in Table 18 are from ref. 453. The oxyfluoride $XeOF_2$ has been

Table 18 *Selected wavenumbers*/cm^{-1} *for some fluoro-krypton compounds*

Compound	ν(Kr—F)	ν(Kr···F)	δ(F—Kr···F)
$[Kr_2F_3][SbF_6]$	559—607	325—335	175—190
$[KrF][SbF_6]$	621, 618, 607	348, 344	174, 166, 149

synthesized in sufficient quantity to permit recording of a Raman spectrum. ν(Xe—O), at 715—750 cm^{-1} was low enough to suggest an oxygen bridged structure.[454] Data for $[XeF_3]^+$, $[XeOF_3]^+$, and $[XeO_2F]^+$ are in Chapter 4. I.r. wavenumbers have been listed without assignment for a variety of adducts of XeF_6 or XeF_2 with MnF_4, TiF_4, FeF_3, or CoF_3.[455–457]

[450] E. Lehmann, D. Naumann, and M. Schmeisser, *J. Fluorine Chem.*, 1976, **7**, 135.
[451] C. Raby, J. Claude, and J. Buxeraud, *Compt. rend.*, 1976, **282**, *C*, 345.
[452] R. J. Gillespie and G. J. Schrobilgen, *Inorg. Chem.*, 1976, **15**, 22.
[453] B. Frlec and J. H. Holloway, *Inorg. Chem.*, 1976, **15**, 1263.
[454] E. Jacob and R. Opferkuch, *Angew. Chem. Internat. Edn.*, 1976, **15**, 158.
[455] J. Slivnick, B. Žemva, M. Bohinc, D. Hanžel, J. Grannec, and P. Hagenmuller, *J. Inorg. Nuclear Chem.*, 1976, **38**, 997.
[456] B. Žemva, J. Slivnik, and M. Bohinc, *J. Inorg. Nuclear Chem.*, 1976, **38**, 73.
[457] M. Bohinc, J. Grannec, J. Slivnik, and B. Žemva, *J. Inorg. Nuclear Chem.*, 1976, **38**, 75.

6

Vibrational Spectra of Transition-element Compounds

BY M. GOLDSTEIN

1 Introduction

The modification introduced last year has been retained in this Report. The following section aims to bring together studies which are more significant from the spectroscopic view or as developing techniques, as well as collating reports which span two or more transition metal groups. There is, of course, an element of subjectivity in the selection of topics for section 2, which therefore should not be taken as a representative over-view of the subject of the whole Chapter. Some papers which are excellent in their own right have been deferred to later sections, particularly where their importance is in a restricted area or where the emphasis is on the application of established correlations *etc.* rather than on the spectroscopy *per se*. The arrangement of the material in these later sections follows the pattern of previous Reports.

2 More Significant and General Aspects

Detailed Studies.—Chapter 4 includes reports of complete studies of small, symmetric transition-metal compounds, while Chapter 7 deals with compounds in which the emphasis is on vibrations of co-ordinated ligands. In Table 1 [1–13] are collected some of the references describing the more detailed spectroscopic studies of transition-metal compounds relevant to this Chapter: the larger systems in which attention is focused on metal–ligand bonding. Most of these papers include normal co-ordinate analysis (NCA) at various levels of sophistication (some rather crude), and it is not practical to summarize the results here, but there is much to occupy the force field pundits. Of further interest in this connection is a review of some principles of the NCA of transition-metal complexes, dealing with methods of force constant calculation and including some

[1] P. L. Goggin, G. Kemeny, and J. Mink, *J.C.S. Faraday II*, 1976, **72**, 1025
[2] D. H. Finseth, C. Sourisseau, and F. A. Miller, *J. Phys. Chem.*, 1976, **80**, 1248.
[3] G. G. Barna, I. S. Butler, and K. R. Plowman, *Canad. J. Chem.*, 1976, **54**, 110.
[4] B. I. Swanson and J. J. Rafalko, *Inorg. Chem.*, 1976, **15**, 249.
[5] B. I. Swanson, *Inorg. Chem.*, 1976, **15**, 253.
[6] B. N. Cyvin, S. J. Cyvin, K. H. Schmidt, W. Wiegeler, A. Müller, and J. Brunvoll, *J. Mol. Structure*, 1976, **30**, 315.
[7] K. Ohwada, *J. Co-ordination Chem.*, 1976, **6**, 75.
[8] D. A. Brown, W. K. Glass, and M. A. Burke, *Spectrochim. Acta*, 1976, **32A**, 137.
[9] A. Müller, K. H. Schmidt, and U. Zint, *Spectrochim. Acta*, 1976, **32A**, 901.
[10] D. M. Adams and R. Appleby, *J.C.S. Chem. Comm.*, 1976, 975.
[11] D. Breitinger and K. Köhler, *Z. anorg. Chem.*, 1976, **421**, 151.
[12] R. Becker, A. Lenz , and W. Sawodny, *Z. anorg. Chem.*, 1976, **420**, 210.
[13] D. M. Adams and W. R. Trumble, *Inorg. Chem.*, 1976, **15**, 1968.

Table 1 *Compounds for which fairly complete and detailed studies have been reported*[a]

Compound		*Ref.*
CH_3HgX	X = F, Cl, Br, or I	1 [b, c]
$\{(H_2C)_3C\}Fe(CO)_3$		2 [b, c, d]
$(C_5H_5)Mn(CO)_n(CS)_{3-n}$	n = 1 or 2	3
$Cs_2Mg[Fe(CN)_6]$		4 [b, c, e]
$M_3[Co(CN)_6]_2,xH_2O$	M = Mn or Cd	5 [b, e]
$[Cu(NH_3)_4]^{2+}$		6 [b, c]
$[UO_2(NO_3)_3]^-$		7 [b]
$Cr(S_2CNR_2)_3$	R = various alkyl	8 [b, d]
$Cs_2[MS_3Se]$	M = Mo or W	9 [b]
HgX_2 [f]	X = Cl or Br	10 [e]
$(HgI_2)_2MF_6$	M = Ti, Zr, or Sn	11 [b]
$Cs_2K[MF_6]$	M = Sc, Y, La, Gd, or Yb	12 [b, e]
$CuCl_2(Me_2SO)_2$		13 [e]

[a] See also Table 2 (isotopic substitution studies) and section on matrix-isolation studies. [b] NCA carried out. [c] Deuteration or other isotopic substitution studied. [d] See also Chapter 7. See also Chapter 4. [f] High pressure phases.

calculations not otherwise in the literature.[14] An excellent review has been published too on the vibrational spectra and force constants of complexes in which NH_3 is the sole ligand.[15] There has also been some discussion [16, 17] on the extent to which the CO-factored force field interaction constants of carbonyls can be explained in terms of bonding theory, with particular reference to data in the literature on $Cr(CO)_6$.

The more important conclusions from the aforementioned studies are referred to where appropriate in later sections of this Report.

Use of Isotopic Substitution.—A marked increase has been noted in the use of stable isotopes in making vibrational assignments. Perdeuteriation of ligands is being used to considerable advantage, as emphasized in a recent review of NCA methods [14] and evident from a survey of NH_3 complexes.[15] Examples of compounds for which deuteriated ligands have been used include MeHgX (X = F, Cl, Br, or I),[1] $\{(H_2C)_3C\}Fe(CO)_3$,[2] $M_2(NMe_2)_6$ (M = Mo or W),[18, 19] $Zn(MeCN)_2X_2$ (X = Cl, Br, or I),[20] $M(C_5H_5N)_2Cl_2$ (M = Mn, Fe, Co, Ni, Cu, or Zn),[21] $[Ni\{NH(CH_2)_2NH(CH_2)_2NH(CH_2)_2\}_2](ClO_4)_2$,[22] and $[Rh(Me_2SO)_2Cl_4]^-$ and $Rh(Me_2SO)_3Cl_3$.[23]

Of greater significance is the more common use of metal isotopes, and the

[14] N. Mohan, S. J. Cyvin, and A. Müller, *Co-ordination Chem. Rev.*, 1976, **21**, 221.
[15] K. H. Schmidt and A. Müller, *Co-ordination Chem. Rev.*, 1976, **19**, 41.
[16] L. H. Jones, *Inorg. Chem.*, 1976, **15**, 1244.
[17] J. K. Burdett, R. N. Perutz, M. Poliakoff, and J. J. Turner, *Inorg. Chem.*, 1976, **15**, 1245.
[18] M. H. Chisholm, F. A. Cotton, B. A. Frenz, W. W. Reichert, L. W. Shive, and B. R. Stults, *J. Amer. Chem. Soc.*, 1976, **98**, 4469.
[19] M. H. Chisholm, F. A. Cotton, M. Extine, and B. R. Stults, *J. Amer. Chem. Soc.*, 1976, **98**, 4477.
[20] Y. Hase, C. Airoldi, Y. Gushikem, and Y. Kawano, *Spectroscopy Letters*, 1976, **9**, 177.
[21] J. E. Rüede and D. A. Thornton, *J. Mol. Structure*, 1976, **34**, 75.
[22] M. Nonoyama, *Transition Metal Chem.*, 1976, **1**, 70.
[23] Yu. V. Fadeev, Yu. N. Kukushkin, and K. A. Khokhryakov, *Russ. J. Inorg. Chem.*, 1975, **20**, 1519.

development of ^{18}O, ^{15}N, and ^{13}C studies. Table 2[4, 6, 24–44] illustrates the remarkable range and complexity of the compounds studied last year, and Table 3 gives a selection of the details.

Several of these reports are worthy of special mention. The accounts of the spectra of glycinato complexes from two different laboratories give some similar, but some different assignments.[29–31] On the one hand, the reliability of NCA is brought into question,[31] whereas the alternative (and more helpful) approach used the isotopic shift data in the calculation of potential energy distributions (PED's).[29] The latter results clearly show and quantify what must be generally obvious for chelate complexes, *viz.* considerable mixing of modes involving metal–ligand stretching and bending occurs. These papers[29–31] also pinpoint what is a remarkable paradox in vibrational spectroscopy: without the use of isotopes erroneous assignments may be made, partial isotopic shift data (*e.g.* using only metal isotopes) may help but can often be misleading, whereas the more isotopic data that are obtained the more difficult it often is to make group-frequency assignments! In several instances, it seems that virtually every significant band shifts on isotopic substitution, so that either one is reduced to using the normal empirical approach [involving a measure of subjectivity (guessing)], or one carries out sensible analyses to determine the nature of the normal modes (PED's). The problem is not confined to the glycinates[29–31] and alaninates;[32] it is more noticeable in the complexes (1) and (2) containing terdentate groups,[33–36] but is even manifest in 'simple' compounds such as Nd_2TiO_5.[39]

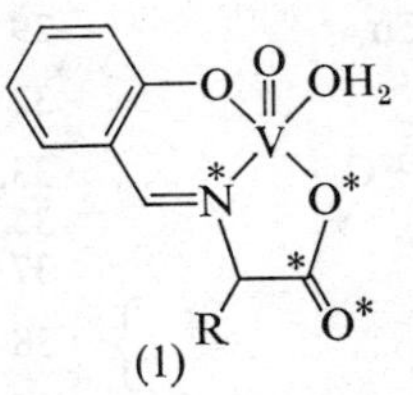

(1)

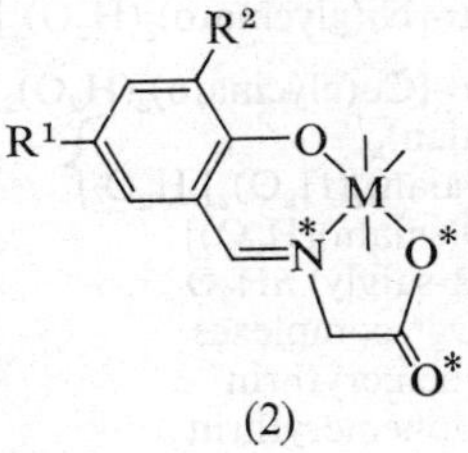

(2)

[24] H. Mahnke and R. K. Sheline, *Inorg. Chem.*, 1976, **15**, 1246.
[25] C. Engelter and D. A. Thornton, *J. Mol. Structure*, 1976, **33**, 119.
[26] J. Kincaid and K. Nakamoto, *Spectroscopy Letters*, 1976, **9**, 19.
[27] B. Hutchinson, M. Hoffbauer, and J. Takemoto, *Spectrochim. Acta*, 1976, **32A**, 1785.
[28] V. Kogan, O. A. Osipov, A. S. Egorov, and P. A. Perov, *Koord. Khim.*, 1975, **1**, 1260.
[29] J. R. Kincaid and K. Nakamoto, *Spectrochim. Acta*, 1976, **32A**, 277.
[30] G. C. Percy, *Spectrochim. Acta*, 1976, **32A**, 1287.
[31] G. C. Percy and H. S. Stenton, *J.C.S. Dalton*, 1976, 1466.
[32] G. C. Percy and H. S. Stenton, *J.C.S. Dalton*, 1976, 2429.
[33] J. B. Hodgson and G. C. Percy, *Spectroscopy Letters*, 1976, **8**, 975.
[34] J. B. Hodgson and G. C. Percy, *Spectrochim. Acta*, 1976, **32A**, 1291.
[35] G. C. Percy and H. S. Stenton, *J. Inorg. Nuclear Chem.*, 1976, **38**, 1255.
[36] G. C. Percy and H. S. Stenton, *Spectrochim. Acta*, 1976, **32A**, 1615.
[37] V. A. Vodovatov, L. G. Mashirov, and D. N. Suglobov, *Radiokhimiya*, 1975, **17**, 900.
[38] D. M. Kurtz, jun., D. F. Shriver, and I. M. Klotz, *J. Amer. Chem. Soc.*, 1976, **98**, 5033.
[39] M. Th. Paques-Ledent, *Spectrochim. Acta*, 1976, **32A**, 1339.
[40] K. H. Schmidt, A. Müller, and M. Chakravorti, *Spectrochim. Acta*, 1976, **32A**, 907.
[41] R. J. Butcher, J. R. Ferraro, and E. Sinn, *Inorg. Chem.*, 1976, **15**, 2077.
[42] Yu. G. Borodkin, *Tezisy Dokl.-Konf. Molodykh Nauchn. Rab. Inst. Neorg. Khim., Akad. Nauk Latv. SSR, 4th*, 1975, 22 (*Chem. Abs.*, 1976, **84**, 128 192r).
[43] F. Koniger, R. O. Carter, and A. Müller, *Spectrochim. Acta*, 1976, **32A**, 891.
[44] F. P. Emmenegger, C. Rohrbasser, and C. W. Schläpfer, *Inorg. Nuclear Chem. Letters*, 1976, **12**, 127.

An interesting application of ^{15}N labelling has been in the differentiation of $\nu(M{-}NCS)$ and $\nu(M{-}NH_2Ar)$ for $M(ArNH_2)_2(NCS)_2$ (Ar = Ph or *p*-tolyl; M = Co, Ni, Cu, or Zn).[25] Values for $\nu(M{-}NH_2Ar)$ are similar to those reported for chloro-analogues in 1974 (*e.g.* 376 cm^{-1} for M = Co, Ar = Ph), but $\nu(M{-}NCS)$ is now thought to be appreciably lower than previously supposed (*e.g.* 186 cm^{-1} for M = Co, Ar = Ph), and the actual description of this band

Table 2 *Compounds studied by isotopic substitution* [a]

Compound	*Isotopic substitution*	*Ref.*
$Fe(CO)_5$	^{13}C	24
$Cs_2Mg[Fe(CN)_6]$	^{13}C, ^{15}N	4
$[Cu(NH_3)_4]^{2+}$	^{15}N, $^{63,65}Cu$	6 [b]
$M(PhNH_2)_2(NCS)_2$ $M(p\text{-}MeC_6H_4NH_2(NCS)_2$	^{15}N	25 [c]
Fe(octaethylporphine)F	^{54}Fe	26 [d]
$M\{HB(pz)_3\}_2$	$^{40,44}Ca$, $^{54,57}Fe$, $^{58,62}Ni$, $^{63,65}Cu$, $^{64,68}Zn$	
$M\{B(pz)_4\}_2$ $M\{HB(3,4,5\text{-}Me_3pz)_3\}_2$	$^{54,57}Fe$, $^{58,62}Ni$, $^{63,65}Cu$, $^{64,68}Zn$	27 [e]
$Fe\{HB(3,5\text{-}Me_2pz)_3\}_2$	$^{54,57}Fe$	
$TiCl_4(benzalaniline)_2$	^{15}N	28 [f]
cis-$[Cu(glycinato)_2(H_2O)]$	^{15}N, $^{63,65}Cu$	29
	^{15}N, ^{18}O, ^{13}C	30
trans-$[Ni(glycinato)_2(H_2O)_2]$	^{15}N, $^{58,62}Ni$	29 [g]
	^{15}N, ^{18}O, ^{13}C	31
trans-$[Cu(glycinato)_2(H_2O)_2]$	^{15}N, $^{63,65}Cu$	29 [g]
$M(alan)_2$ $[Ni(alan)_2(H_2O)_2],H_2O$	^{15}N, ^{18}O	32 [h]
$[VO(salam)(H_2O)]$	^{15}N, ^{18}O, ^{13}C	33, 34 [i]
$M(R\text{-}salgly),nH_2O$	^{15}N, ^{18}O	35, 36 [j]
$AnO_2{}^+$ complexes	^{18}O	37 [k]
Oxyhemerythrin	^{18}O	38
Azidohemerythrin	^{15}N	
$NdTiO_5$	$^{46,50}Ti$	39
$(Et_4N)_2[Zn(NCS)_4]$	$^{64,68}Zn$	40
$(Et_4N)_2[Pd(SCN)_4]$	$^{104,110}Pd$	
$(EtPhNCS_2)_3Fe$	$^{54,56}Fe$	41 [l]
$Cu(8\text{-}quinolinolato)_2$ $Cu(8\text{-}quinolinthiolato)_2$	$^{63,65}Cu$	42
$TiCl_4$	^{48}Ti, ^{35}Cl	43 [m]
$CuAl_2Cl_8$	$^{63,65}Cu$	44

[a] Excluding the use of deuteriation and studies of isotopic splitting in matrix-isolated species in natural abundance. [b] Literature data used for NCA. [c] Independent labelling of the N atoms of the NCS^- and amine ligands; M = Co, Ni, Cu, or Zn. [d] Raman study in agreement with i.r. data previously reported. [e] $\{HB(pz)_3\}^-$, *etc.* are poly(1-pyrazolyl)borato ligands; M = metal isotopes indicated, Mn, or Co (spectra of complexes of the latter two elements assigned by analogy). [f] Virtually no band shifts observed in the region studied (100—400 cm^{-1}). [g] Co analogue assigned by comparison. [h] M = Cu or Ni; (alan)H = L-alanine. [i] (salam) = *N*-salicylideneglycinato dianion or alkyl derivative as in (1), which shows the isotopically labelled atoms * (various combinations studied for R = H, more restricted for R = Me or Pr^i, and no labelling for R = Bu^i, Bu^s, $PhCH_2$, or $MeSC_2H_4$). [j] (R-salgly) = substituted *N*-salicylideneglycinato dianion as in (2; R^1, R^2 = various), which shows the isotopically labelled atoms * ; M = Co or Ni. [k] An = Np, Pu, or Am; see section 14. [l] High- and low-spin states assigned; Co and Cr analogues, and related Fe complexes assigned by comparison. [m] Matrix-isolation study, using pure ^{48}Ti and ^{35}Cl isotopically substituted species.

Table 3 *Selected assignments based on isotopic shifts (data in* cm^{-1}*)* [a]

Compound	ν(NA)	$\Delta\nu(^{15}N)$	$\Delta\nu(^{18}O)$	$\Delta\nu(^{13}C)$	$\Delta\nu(^{i}M)$	*Assignment*	*Ref.*
$[Fe(CN)_6]^{4-}$	2093.6	30.5		45.7		ν(CN)	4
$Ni\{HB(pz)_3\}_2$	281.0				6.1	ν(NiN)	27
	251.2				5.4		
$Cu\{B(pz)_4\}_2$	241.2				1.5	ν(CuN)	27
	185.1				1.0		
cis-$[Cu(glycinato)_2(H_2O)]$	476 [b]	6	8	1.5		ν(CuN) + $\omega(CO_2)$	30
	478 [b]	7			<1	ν(CuN)	29
trans-$[Ni(glycinato)_2(H_2O)_2]$	289 [c]	1.0			5.0	ν(NiO)	29
	283 [c]	<1	3	<1		δ(ONiO)	31
$[Ni(alan)_2]$	462	4.5	5			ν(NiN) + ν(NiO)	32
	344	<1	7			ν(NiO)	
$[VO(salam)(H_2O)]$ [d]	536	3.5	5.5	2.5		ν(VN) + $\delta(CO_2)$	34
	483	2.5	2.5	<1		ν(VN) + ν(V—OR)	
	408	<1	6	<1		ν(V—OR)	
	379	<1	5.5	1.5		ν(V—OR) + $\delta(CO_2)$	
$(Et_4N)_2[Pd(SCN)_4]$	296.8 [e]				3.8	ν(PdS)	40
	170.0 [e]				0.5	δ(PdSC)	
$(EtPhNCS_2)_3Fe$	356 [f]				15	ν(FeS)	41
	308 [g]				4		
$Cu(8\text{-quinolinthiolato})_2$	264				0.8	ν(CuS)	42
	251				2.4	ν(CuN)	
$CuAl_2Cl_8$	444				2.1	ν(CuCl)	44

[a] Absence of an entry indicates that the substitution is not applicable or was not studied; for abbreviations of ligands, see Table 2; ν(NA) = value for species in natural abundance; $\Delta\nu(^{15}N)$, $\Delta\nu(^{18}O)$, $\Delta\nu(^{13}C)$ = shifts to lower cm^{-1} values on substitution of the isotope indicated; $\Delta\nu(^{i}M)$ = metal isotope shifts to lower cm^{-1} values, with respect to substitution in Table 2. [b, c] These bands appear to be the same. [d] Compound (1, R = H). [e] ^{104}Pd values. [f] Low-spin state, ^{54}Fe value. [g] High-spin state, ^{54}Fe value.

warrants further study to ensure it does not contain appreciable contributions from bending co-ordinates.

Two other applications of the use of isotopes, not involving establishing assignments, were the determination of the bonding mode of oxygen in oxyhemerythrin[38] and investigation of the nature of CO ligand exchange in $Fe(CO)_5$;[24] see section 8.

Matrix-isolation Studies.—Investigations of this type relevant to the present Chapter include identification of (*a*) $V(CO)_6$ and $V_2(CO)_{12}$ (see section 5),[45a, b] (*b*) $Fe(CO)_4$ (see section 8),[46] (*c*) $Ag(CO)_n$ (n = 3, 2, or 1) and $Ag_2(CO)_6$ (see section 11),[47] (*d*) $TiCl_4$ (see Table 2),[43] (*e*) $(C_2H_4)HgX_2$ (X = F, Cl, Br, or I, see section 12),[48] and (*f*) $[V(N_2)_6]_n$ (n = 1 or 2, similar to carbonyl analogues).[49]

Resonance Raman Spectra.—Compounds listed in Table 4[26, 38, 50–65] typify the range of systems now being studied by the RRE. Anharmonicity constants calculated from overtone progressions are very small.[57, 59, 60] It is gratifying to record that more caution is apparently being exercised in the assignment of metal–ligand modes in biologically relevant compounds (or models for these),[62–65] and although some tempering of enthusiasm is still called for, it is evident[64] that selective Raman intensity enhancement may have utility in such assignments for natural and synthetic macrocyclic complexes.

Mixed-valence Compounds.—It is instructive to bring together the accumulating data on this class of compound. Spectra have been obtained for $NiI_n(bqd)_2$ (see Table 4),[54] $K_3Mo_2(SO_4)_4,3\frac{1}{2}H_2O$ (see section on metal–metal vibrations),[66] compounds containing conducting linear chains of *cis*-$[Ir(CO)_2Cl_2]^{0.60-}$ units,[67] the acetate-bridged $[Rh_2(O_2CMe)_4Cl]$,[68] and $[Cu_2Au(Bu_2NCS_2)_6][Hg_2Br_6]$ which

45 (*a*) T. C. DeVore and H. F. Franzen, *Inorg. Chem.*, 1976, **15**, 1318; (*b*) T. A. Ford, H. Huber, W. Klotzbücher, M. Moskovits, and G. A. Ozin, *Inorg. Chem.*, 1976, **15**, 1666.
46 A. McNeish, M. Poliakoff, K. P. Smith, and J. J. Turner, *J.C.S. Chem. Comm.*, 1976, 859.
47 D. McIntosh and G. A. Ozin, *J. Amer. Chem. Soc.*, 1976, **98**, 3167.
48 D. Tevault and K. Nakamoto, *Spectroscopy Letters*, 1976, **9**, 185.
49 H. Huber, T. A. Ford, W. Klotzbücher, and G. A. Ozin, *J. Amer. Chem. Soc.*, 1976, **98**, 3176.
50 A. Mueller, E. Diemann, and E. Ahlborn, 'Proceedings of the 16th International Conference on Co-ordination Chemistry', University College, Dublin, Department of Chemistry, 1974, R42.
51 N. Koshizuka, Y. Yokoyama, and T. Tsushima, *Solid State Comm.*, 1976, **18**, 1333.
52 M. S. Rahaman and S. M. Korenkiewicz, *Canad. J. Chem.*, 1976, **54**, 3815.
53 T. C. Strekas and T. G. Spiro, *Inorg. Chem.*, 1976, **15**, 974.
54 T. J. Marks, D. F. Webster, S. L. Ruby, and S. Schultz, *J.C.S. Chem. Comm.*, 1976, 444.
55 R. J. H. Clark and W. R. Trumble, *Inorg. Chem.*, 1976, **15**, 1030.
56 R. J. H. Clark, M. L. Franks, and W. R. Trumble, *Chem. Phys. Letters*, 1976, **41**, 287.
57 R. J. H. Clark and M. L. Franks, *J. Amer. Chem. Soc.*, 1976, **98**, 2763.
58 J. San Filippo, jun. and H. J. Sniadoch, *Inorg. Chem.*, 1976, **15**, 2209.
59 R. J. H. Clark and M. L. Franks, *J.C.S. Dalton*, 1976, 1825.
60 R. J. H. Clark and P. C. Turtle, *J.C.S. Faraday II*, 1976, **72**, 1885.
61 G. Chottard and J. Bolard, *Inorg. Chim. Acta*, 1976, **20**, L17.
62 O. Siiman, N. M. Young, and P. R. Carey, *J. Amer. Chem. Soc.*, 1976, **98**, 744.
63 T. Kitagawa, Y. Kyogoku, T. Iizuka, and M. I. Saito, *J. Amer. Chem. Soc.*, 1976, **98**, 5169.
64 W. H. Woodruff, R. W. Pastor, and J. C. Dabrowiak, *J. Amer. Chem. Soc.*, 1976, **98**, 7999.
65 T. Kitagawa, M. Abe, Y. Kyogoku, H. Ogoshi, E. Watanabe, and Z. Yoshida, *J. Phys. Chem.*, 1976, **80**, 1181.
66 A. Loewenschuss, J. Shamir, and M. Ardon, *Inorg. Chem.*, 1976, **15**, 238.
67 A. P. Ginsberg, J. W. Koepke, J. J. Hauser, K. W. West, F. J. Di Salvo, C. R. Sprinkle, and R. L. Cohen, *Inorg. Chem.*, 1976, **15**, 514.
68 R. D. Cannon, D. B. Powell, K. Sarawek, and J. S. Stillman, *J.C.S. Chem. Comm.*, 1976, 31.

Table 4 *Resonance Raman spectral studies*

Species	*Principal assignments*	*Ref.*
$[MoOS_3]^{2-}$ $[WOSe_3]^{2-}$	ν(MO), ν(MS)	50
$CdCr_2S_4$	[a]	51
Cu^{II}-adrenaline complex	ν(CuO), ν(CuN)	52 [b]
$[(H_3N)_5Ru(pyrazine)Ru(NH_3)_5]^{n+}$	ν(RuN)	53 [c, d]
$NiI_n(bqd)_2$	$\nu(I_3^-)$	54 [e]
$Pd_2(NH_3)_4Cl_6$ $PdPt(NH_3)_4Cl_6$	ν(MCl), ν(MN), δ(MCl), δ(MN)	55 [d]
$Pt_2(EtNH_2)_8Cl_6,4H_2O$	ν(PtCl), ν(PtN)	56 [d]
$[Re_2X_6]^{2-}$	ν(ReRe)	57 [f, g]
$(Bu^n{}_3P)_2(ArCO_2)_2Mo_2X_2$ $(ArCO_2)_4Mo_2$	ν(MoMo)	58 [f, h]
$Ru_2Cl(O_2CR)_4$	ν(RuRu)	59 [f, i]
$[Bu^n{}_4N][FeBr_4]$	ν_1(FeBr)	60 [j]
$[Co\{O_2CC(NH_2)CH_2C_3H_3N_2\}_2]^{2-}$	ν(CoN)	61
'Blue' copper proteins	ν(CuS), ν(CuN)	62 [k]
Myoglobin derivatives Fe^{II} and Fe^{III} cytochrome b_5 Fe protoporphyrin IX dicyanide	ν(ligand) [l]	63
M(chel–N_4)	ν(MN)	64 [m]
Fe(octaethylporphine)F	ν(FeF)	26 [n]
M^1(oep) and M^2(oep-d_4) Fe(oep)X $[Fe(oep)(imidazole)_2]ClO_4$	[o] ν(FeX) ν(Fe–imidazole)	65 [p]
Oxyhemerythrin Azidohemerythrin	ν(OO), ν(FeO) $\nu(N_3)$	38 [q]

[a] Assignments in terms of ferromagnetic properties. [b] Structure proposed for species present at pH = 5.5. [c] n = 4, 5, or 6. [d] See section on mixed-valence compounds. [e] n = 0.50—0.70; (bqd) = bis(1,2-benzoquinone dioximate). [f] See Table 6. [g] X = Cl or Br. [h] Ar = various. [i] R = Me or Pr^n. [j] Solid and $MeNO_2$ solution. [k] Compare last year's Report. [l] Internal ligand modes. [m] M = Mn, Fe, Co, Ni, Cu, or Zn; (chel–N_4) = a π-delocalized N_4 macrocylic dianion. [n] See Table 2. [o] 488.0 nm excitation gives a rather intense, polarized line at *ca.* 670 cm^{-1}, the intensity apparently depending on the geometry of the complexes; NCA performed for M^1 = Ni, and potential energy distributions of the low-frequency modes given. [p] M^1 = Co, Cu, or M^2; M^2 = Ni or Pd; X = F, Cl, Br, or I; (oep) = octaethylporphyrin. [q] See Table 2 and section 8.

contains disordered Cu^{III} and Au^{III} atoms attached to a central Cu^{II} atom.[69] But the most interesting results coming forward relate to pyrazine-bridged $Ru^{II,\,III}$ systems,[53, 70, 71] and $Pt^{II,\,IV}$ complexes such as Wolfram's red salt [56, 72] and related compounds.[55, 72]

For $[(bipy)_2ClRu(pyrazine)RuCl(bipy)_2]^{3+}$, the appearance of a symmetric vibrational mode of pyrazine in the i.r. has been held in evidence that the two Ru sites are electronically non-equivalent on the lifetime of the measurement, *i.e.* an upper limit of 10^{12}—10^{13} s^{-1} has been placed on the rate of thermal electron transfer between the Ru atoms.[70] Similarly, resonance Raman data for aqueous solutions of $[(H_3N)_5Ru(pyrazine)Ru(NH_3)_5]^{5+}$ have been interpreted in

[69] A. W. Gal, G. Beurskens, J. A. Cras, P. T. Beurskens, and J. Willemse, *Rec. Trav. chim.*, 1976, **107**, 157.

[70] R. W. Callahan and T. J. Meyer, *Chem. Phys. Letters*, 1976, **39**, 82.

[71] J. K. Beattie, N. S. Hush, and P. R. Taylor, *Inorg. Chem.*, 1976, **15**, 993.

[72] M. Textor and H. R. Stieger, *Z. anorg. Chem.*, 1976, **423**, 185.

terms of a localized Ru^{II}(pyrazine)Ru^{III} structure.[53] In contrast, arguments (using extant i.r. and other data) have been put forward in favour of a delocalized symmetrical ground state for solid salts of the pentapositive ammine cation.[71]

Raman [56] and i.r. [56, 72] assignments of ν(PtCl) and ν(PtN) are now available (with some disagreement) for Wolfram's red mixed-valence salt [56, 72] and related complexes.[72] A very much more thorough study has been made of the spectra of $Pd_2(NH_3)_4Cl_6$ and $PdPt(NH_3)_4Cl_6$.[55] Thus resonance Raman progressions involving the three Raman-active ν(MCl) modes of the PdPt compound were observed, indicating that the resonance involves a metal–metal mixed-valence transition. A rather different situation obtained for the PdPd system, the RRE spectrum containing overtones and combinations involving several stretching and bending modes of the Pd^{IV} constituent complex. This and other effects are attributable to resonance excitation within ligand $\rightarrow$ Pd^{IV} charge-transfer bands; the mixed-valence transition is, by inference but contrary to a previous report, in the i.r. region.

Polynuclear Pseudohalide Complexes.—A significant number of compounds of general type $M^1{}_nM^2X_m$ (X = CN, NCS, NCSe, SCN, or SeCN) have now been studied by vibrational spectroscopy, and Table 5 lists those noted in the 1976 literature.[4, 5, 73–82] In these studies, attempts were made to characterize bridging

Table 5 *Polynuclear pseudohalide complexes studied by vibrational spectroscopy*

Compound	*Ref.*
$Cs_2Mg[Fe(CN)_6]$	4 [a]
$M_3[Co(CN)_6]_2,xH_2O$	5 [a, b]
$Fe[M(CN)_6]$	73 [a, c]
$M[Fe(CN)_5(NO)],nH_2O$	74 [d]
$[NiM(NCS)_6]$	75 [a, e]
$M^1M^2(NCS)_4,nL$	76 [a, f]
$CoM(NCS)_4,nL$	77 [a, g]
$M^1M^2(NCSe)_4,4L$	78 [a, h]
$CoM(NCSe)_4,nL$	79 [a, h]
$[(NC)_5Fe(CN)Cr(NH_3)_{5-n}(H_2O)_n],(1-n)H_2O$	80 [j]
$[Cu^I(NH_3)]_2[Pt(CN)_n]$ } $Cu^I{}_3[Co(CN)_6],2NH_3$ }	81 [k]
trans-$[HPt(Ph_3P)_2CN,MX_n]$	82 [l]

[a] Discussed in text. [b] Single-crystal Raman data discussed in Chapter 4; M^{II} = Mn or Cd. [c] M^{III} = Co, Rh, or Ir. [d] M^{II} = Mn, Fe, Co, Ni, Cu, Zn, or Cd; n = various. [e] M^{IV} = Ti or Sn. [f] M^1 = Ni^{II}, Fe^{II}, or Zn^{II}; M^2 = Hg^{II} or Zn^{II}; n = 2, 4, or 6; L = nicotinamide, substituted pyridine, *etc.* [g] M^{II} = Zn, Cd, or Hg; n = 2, 4, or 6; L = isothiazole or derivative, nicotinamide, *etc.* [h] M^1 = Co^{II} or Ni^{II}; M^2 = Zn^{II}, Cd^{II}, or Hg^{II}; L = py or ½(bipy). [i] M^{II} = Zn, Cd, or Hg; n = 2, 4, or 6; L = nicotinamide, substituted pyridine, *etc.* [j] n = 0 or 1. [k] n = 4 or 6. [l] MX_n = $ZnCl_2$, $CoCl_2$, $NiCl_2$, $B(OAr)_3$, AlR_3, $AlR_2(OR)$, *etc.*

73 H. Inoue, E. Fluck, and S. Yanagisawa, *Z. Naturforsch.*, 1976, **31b**, 167.
74 L. A. Gentil, E. J. Baran, and P. J. Aymonino, *Inorg. Chim. Acta*, 1976, **20**, 251.
75 P. P. Singh and S. B. Sharma, *Canad. J. Chem.*, 1976, **54**, 1563.
76 P. P. Singh and S. A. Khan, *Z. anorg. Chem.*, 1976, **423**, 173.
77 P. P. Singh, L. P. Pathak, and S. A. Khan, *J. Inorg. Nuclear Chem.*, 1976, **38**, 475.
78 P. P. Singh, A. K. Srivastava, and R. Rivest, *J. Inorg. Nuclear Chem.*, 1976, **38**, 439.
79 P. P. Singh and S. B. Sharma, *J. Co-ordination Chem.*, 1976, **6**, 65.
80 J. Casabo, J. Ribas, and S. Alvarez, *Inorg. Chim. Acta*, 1976, **16**, L15.
81 H. Siebert and D. Günauer, *Z. anorg. Chem.*, 1976, **420**, 265.
82 L. E. Manzer and G. W. Parshall, *Inorg. Chem.*, 1976, **15**, 3114.

pseudohalide ligands. However, the inherent dangers in determining the mode of co-ordination even for unidentate NCS^- and $NCSe^-$ groups from i.r. frequency data alone, together with the fearsome complexity of the spectra involving the organic ligands, make many of the structural conclusions drawn[75–79] questionable, to say the least. Moreover, the $\nu(M^{II}—NCS)$ assignments based upon ^{15}N studies,[25] if indeed correct (see discussion above), show that those proposed for the polynuclear series[75–77] must be wrong, casting further doubt on the validity of the structures put forward. Deductions for the cyano-complexes listed in Table 5 are less equivocal, but even so there is room for doubt on some of the spectral interpretations.[74, 80, 81] The appearance of a band at 250—255 cm^{-1} in the i.r. spectra of $Fe[M(CN)_6]$ complexes (M = Co, Rh, or Ir), which is not present in the $K_3[M(CN)_6]$ spectra, suggests assignment as $\nu(FeN)$, indicating MCNFe interactions.[73]

The most satisfactory work in this area last year was carried out on the $[Fe(CN)_6]^{4-}$ and $[Co(CN)_6]^{3-}$ salts listed in Table 5. For $Cs_2Mg[Fe(CN)_6]$ the study was supported by ^{13}C and ^{15}N substitution and by NCA,[4] while single-crystal Raman data were recorded for $M_3[Co(CN)_6]_2,xH_2O$ (M^{II} = Mn or Cd, see Chapter 4).[5] For all three compounds there are significant shifts in the vibrational modes on dissolution, and surprisingly the $\nu(CN)$ and $\nu(FeC)$ modes are all at higher wavenumber in the solids (*e.g.* by *ca.* 40 cm^{-1} for the ferrocyanide). A preliminary NCA on the crystalline $Cs_2Mg[Fe(CN)_6]$ showed that the observed shifts can be reproduced without changing the potential constants of the anion, except for an increase in $K(CN)$, so that the shifts arise from interionic forces.[4] The Mg—N interaction was found to be about twice as strong as the Li—N interaction in $Cs_2Li[M(CN)_6]$ (M = Fe or Co),[4] while d_π–p_π* overlap is believed to be responsible for the 'unusually strong' Mn—N forces in $Mn_3[Co(CN)_6]_2$.[5]

Metal–Metal Vibrations.—Fewer than one-third of the references cited in Table 6[18, 19, 23, 57–59, 66, 83–100] call for further comment, but they do aid our understanding of metal–metal bonded systems.

[83] J. Fuchs, S. Mahjour, and R. Palm, *Z. Naturforsch.*, 1976, **31b**, 544.
[84] G. Bor, U. K. Dietler, and K. Noack, *J.C.S. Chem. Comm.*, 1976, 914.
[85] S. Onaka and D. F. Shriver, *Inorg. Chem.*, 1976, **15**, 915.
[86] D. F. Shriver, S. Onaka, and D. Strope, *J. Organometallic Chem.*, 1976, **117**, 277.
[87] L. F. Wuyts and G. P. van der Kelen, *Spectrochim. Acta*, 1976, **32A**, 689.
[88] L. F. Wuyts and G. P. van der Kelen, *Spectrochim. Acta*, 1976, **32A**, 1705.
[89] H.-J. Haupt, W. Wolfes, and H. Preut, *Inorg. Chem.*, 1976, **15**, 2920.
[90] M. M. T. Khan, S. S. Ahamed, S. Vancheeson, and R. A. Levenson, *J. Inorg. Nuclear Chem.*, 1976, **38**, 1279.
[91] T. S. Khodashova, V. A. Varnek, É. N. Yuvchenko, and M. A. Porai-Koshits, *Doklady Chem.*, 1976, **224**, 617.
[92] R. Mattes and G. Lux, *Z. anorg. Chem.*, 1976, **424**, 173.
[93] R. Mattes, D. Altmeppen, and M. Fetzer, *Z. Naturforsch.*, 1976, **31b**, 1356.
[94] F. A. Cotton and T. R. Webb, *Inorg. Chem.*, 1976, **15**, 68.
[95] F. A. Cotton, J. G. Norman, jun., B. R. Stults, and T. R. Webb, *J. Co-ordination Chem.*, 1976, **5**, 217.
[96] R. J. Mureinik, *J. Inorg. Nuclear Chem.*, 1976, **38**, 1275.
[97] C. D. Garner, R. G. Senior, and T. J. King, *J. Amer. Chem. Soc.*, 1976, **98**, 3526.
[98] P. A. W. Dean, D. G. Ibbott, and J. B. Stothers, *Canad. J. Chem.*, 1976, **54**, 166.
[99] P. A. W. Dean and D. G. Ibbott, *Canad. J. Chem.*, 1976, **54**, 177.
[100] K. Brodersen, R. Eder, and H. Menne, *Chem. Ber.*, 1976, **109**, 2762.

Table 6 *Compounds for which metal–metal stretching modes have been assigned*

Compound	*Ref.*
$(Bu^n{}_4N)_3[HV_4O_{12}]$	83 [a]
$Co_2(CO)_8$	84, 85 [b]
$[Fe_2(CO)_8]^{2-}$	85 [b]
$[HFe_2(CO)_8]^-$	
$Co_2(CO)_8,AlBr_3$	
$(C_5H_5)_3Ni_3(CO)_2$	
$(C_5H_5)_3Ni_3(COAlEt_3)_2$	
$[(C_5H_5)Fe(CO)]_4$	86
$[(C_5H_5)Fe(COAlBr_3)]_4$	
$Ru_3(CO)_{12}$	
$Ru_3(CO)_{11}(COAlBr_3)$	
$R^1{}_nX_{3-n}SnCo(CO)_4$	
$R^2{}_mCl_{2-m}Sn\{Co(CO)_4\}_2$	87, 88 [b, c]
$R^3Sn\{Co(CO)_4\}_3$	
$[\{Mn(CO)_5\}_2In(\mu\text{-}X)]_2$	89 [d]
$RuCl(SnCl_3)(EPh_3)_3$	90 [e]
$[RuCl_2(EPh_3)_3]HgCl_2$	
$Cs_4[Rh\{SnF_2(H_2O)_2\}_2Sn_4F_{15}],4H_2O$	91 [f]
$W_2(NMe_2)_6$	19 [g]
$[Re_2X_6]^{2-}$	57 [h]
$Ru_2Cl(O_2CR)_4$	59 [i]
$Mo_2(NMe_2)_6$	18 [g]
$[Mo_2O_2F_4X_n]^{m-}$	92 [b, j]
$[Mo_2O_4Cl_4]^{2-}$	93 [b]
$K_3Mo_2(SO_4)_4,3\frac{1}{2}H_2O$	66 [b]
$Mo_2(O_2CCH_2NH_3)_4(SO_4)_2,4H_2O$	94 [b]
$Mo_2(O_2CH)_4$	95 [k]
$(Bu^n{}_3P)_2(ArCO_2)_2Mo_2X_2$	58 [l]
$(ArCO_2)_4Mo_2$	
$Mo_2(phthalate)_3$	96
$CrMo(O_2CMe)_4$	97
$Hg_2(AsF_6)_2,L$	98, 99 [m]
$[Hg_2(ArNH_2)_2]Y_2$	100 [n]

[a] This is the first known metavanadate isolated containing a discrete polyhedral unit; a band at 327 cm^{-1} is thought to be ν(VV). [b] See text. [c] R^1 = Me, Et, Bu^t, or Ph; X = Cl, Br, or I; n = 3 or 2; R^2 = Me, Pr^n, or Bu^t; m = 2 or 1; R^3 = Me, Et, Bu^t, Cl, or Br. [d] X = Cl, Br, or I; crystal structures determined; various ν(Mn—In) modes identified. [e] E = P, As, or Sb. [f] Assignments in the 150—850 cm^{-1} not distinguished between ν(SnO), ν(SnF), ν(RhSn). [g] Deuteriation shows extensive mixing. [h] X = Cl or Br. [i] R = Me or Pr^n. [j] X = H_2O (n = 2 or 0, m = 2) or F(n = 2, m = 4). [k] ν(MoMo) given as 406 cm^{-1} in Experimental section but 403 cm^{-1} in Abstract, and not otherwise mentioned. [l] Ar = Ph, *etc.*; X = Cl or Br; different values given for $(4\text{-}NCC_6H_4CO_2)_4Mo_2$ in text and Table. [m] L = benzene, naphthalene, *etc.*, or PF_3, Ph_3PS, Ph_3As, *etc.* [n] Ar = various; Y = NO_3 or ClO_4.

In Volume 8 of this series (p. 301), the description of a Raman band at 200 cm^{-1} as ν(MoMo) in $[(NH_4)MoO_2F_2,H_2O]_2$ was described as inappropriate in view of the bis(μ-oxo) bridged structure then proposed. Further studies by the same authors [92] have shown that the MoMo distance (256.6 pm) and magnetic properties are illustrative of metal–metal bonding, so that some change in view might seem required. However, the question still remains: what is the most appropriate group-frequency description for a mode of type (3)? The same 'problem' is raised by $[Mo_2O_4Cl_4]^{2-}$ (Raman band at 218 cm^{-1}).[93] The only value in such semantic arguments is that they focus attention on the nature of normal modes of vibration and might lead to sensible calculation of potential energy

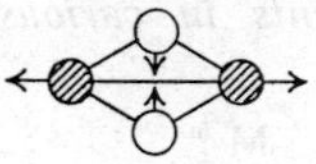

(3)

distributions. However, it must be pointed out that the influence of ligands bridging a metal–metal bonded binuclear system has been qualitatively recognized.[66, 85] Thus whereas the increased overall Mo oxidation state in $K_3Mo_2(SO_4)_4,3\frac{1}{2}H_2O$ compared with $K_4Mo_2(SO_4)_4,2H_2O$ would be expected to give rise to a lowering of 'ν(MoMo)', the effect of greater strain in the sulphato-bridges more than counteracts this so that a higher value for ν(MoMo) is found for the former compound.[66] In a similar manner, it has been found that ν(MM) for first transition series binuclear carbonyls containing bridging CO groups are above 200 cm^{-1}, whereas for single-bonded non-bridged systems ν(MM) is in the 140—190 cm^{-1} range.[85] These schematics have been used to elucidate tautomeric equilibria for several binuclear carbonyls; thus in the presence of a dipositive cation, $[Fe_2(CO)_8]^{2-}$ exists in $HCONMe_2$ as a bridged–non-bridged equilibrium mixture, $Co_2(CO)_8$ exists in three tautomeric forms, and $[HFe_2(CO)_8]^-$ contains bridging carbonyl groups.[85] Temperature dependence studies of the Raman spectrum of $Co_2(CO)_8$ indicate that the bands at 185 and 157 cm^{-1} are due to ν(CoCo) of forms II and III, respectively.[84]

Comparison of reliable data on ν(MoMo) shows there to be a fair (but not precise) inverse correlation with MoMo distance obtained crystallographically.[94]

I.r. bands assigned as ν(CoSn) in $-Co(CO)_4$ derivatives correlate well with ν(CO) values,[87] but of greater significance is the variation of ν(CoSn) in the series $Me_{3-n}X_nSnCo(CO)_4$ ($n = 0$—3; X = Cl, Br, or I).[88] Thus as n increases, ν(CoSn) increases steadily for X = Cl, but not for X = Br and with even more deviation for X = I; this is taken as evidence for increased coupling as X is varied from Cl to I.

General Series of Complexes.—Assignment of metal–ligand stretching modes in series of complexes spanning a number of periodic groups is now so commonplace that paradoxically the reliability of the assignments appears, in the opinion of this Reporter, to be on the decline. It is now so routine to obtain i.r. spectra at low wavenumbers that insufficient care is being taken to seek *evidence* for assignments. Furthermore, there is lack of attention to the nature of the modes assigned so that unacceptable descriptions are often used. Table 7 [21, 32, 41, 101—145] is remarkable not only in that it refers to several hundred ν(ML) assignments [and to even more unassigned bands in the ν(ML) regions!], but also because very few of the results are of real significance and many of the deductions do not stand up to scrutiny. In the opinion of this Reporter, refs. 105, 107—109, 115, 126, 129, 130, 134, 135, and 140 (at least) contain assignments which are too speculative or are likely to be erroneous.

A study using C_5D_5N purports to correct the literature on ν(MX) and ν(MN) of $MCl_2(py)_2$ complexes (M = Mn, Fe, Co, Ni, Cu, or Zn),[21] but the authors have totally ignored papers for 1970 and 1972 (M = Mn, Fe, Co, or Ni) and 1974 (M = Cu), reported in earlier volumes of this series (for M = Zn there was agreement with published assignments). The point group approach adopted

Table 7 *Metal–ligand assignments in various series of complexes (see also Tables* 1—6)

Series [a]	M [b]	*Assignments* [b]	*Ref.*
$[MCl_4]^{2-}$	Mn, Co, Ni, Zn, Cu	ν(MCl), δ(MCl)	101 [c]
$[(Bu^t_3P)MX_3]^-$	Co, Ni, Zn	ν(MX), ν(MP)	102
$[(HCN)_6M]^{2-}$	Mg, Fe, Co, Ni, Cu	ν(MN)	103
$[M(en)_3]^{3+}$	Cr, Co, Rh	ν(MN)	104 [d]
$MSO_4,nArNH_2$	Co, Ni, Cu, Zn, Cd	ν(MO), ν(MN)	105 [e]
$MX_2(py)_2$	Mn, Fe, Co, Ni, Cu, Zn	ν(MX, ν(MN)	21 [d]
MCl_2(2-pyNRCOMe)$_2$	Co, Ni, Cu, Zn	ν(MCl)	106 [f]
$MZ_2(4\text{-}pyCONH_2)_2$	Mn, Co, Ni, Zn, Hg	ν(MZ)	107
$MZ_2(8\text{-}NH_2C_9H_6N)_n$	Mn, Co, Ni, Cu, Zn, Cd, Pd	ν(MN)	108 [g]
$[M(4,4'\text{-bipy})(H_2O)_4]^{2-}$, $M(4,4'\text{-bipy})X_2$, *etc.*	Fe, Mn, Co, Ni, Cu, Zn, Cd	ν(MN), ν(MX)	109
$\{HB(pz)_3\}MCl_3$	Ti, V	ν(MCl)	110 [h]
$MZ_2(Ph_2oxazole)_2$, *etc.*	Co, Ni, Cu	ν(MZ), ν(MO), ν(MN)	111 [i]
$M_2X_4(diazepin)_3$, $CuX_2(diazepin)_2$	Co, Ni	ν(ML), ν(MX), ν(CuX)	112 [j]
$MX_n(benzox)_m$	Cr, Co, Ni, Cu, Zn, Cd, Hg	ν(MX), ν(MN)	113, 114 [k]
$MX_n(thiaz)_2$	Ti, Fe	ν(MX, ν(MN), ν(MO)	115 [l]
$M(bipyO)_nY_3$	Cr, Fe	ν(MO), ν(MY), ν(MN)	116 [m]
$MCl_n(phenO)_m,xH_2O$	Cr, Mn, Fe, Co, Ni, Cu, Zn	ν(MO), ν(MN), ν(MCl)	117 [n]
$MCl_n(quinoxO)_m,xH_2O$	Cr, Mn, Fe, Co, Ni, Cu, Zn	ν(MO), ν(MN), ν(MCl)	118 [o]

[101] S. V. Vokov, N. P. Evtushenko, and K. B. Yatsimirskii, *Teor. i eksp. Khim.*, 1976, **12**, 111.
[102] E. C. Alyea, G. T. Fey, and R. G. Goel, *J. Co-ordination Chem.*, 1976, **5**, 143.
[103] P. L. A. Everstein, A. P. Zuur, and W. L. Driessen, *Inorg. Nuclear Chem. Letters*, 1976, **12**, 277.
[104] J. Gouteron, *J. Inorg. Nuclear Chem.*, 1976, **38**, 63.
[105] I. S. Ahuja, R. Singh, and R. Singh, *Transition Metal Chem.*, 1976, **1**, 240.
[106] J. Bould and B. J. Brisdon, *Inorg. Chim. Acta*, 1976, **19**, 159.
[107] I. S. Ahuja and I. Prasad, *Inorg. Nuclear Chem. Letters*, 1976, **12**, 777.
[108] J. Casabo, J. Ribas, and J. Bartroli, *Rev. Chim. minérale*, 1976, **13**, 149.
[109] A. Anagnostopoulos, *Inorg. Nuclear Chem. Letters*, 1976, **12**, 225.
[110] P. Burchill and M. G. H. Wallbridge, *Inorg. Nuclear Chem. Letters*, 1976, **12**, 93.
[111] M. Biddau, G. Devoto, M. Massacesi, R. Pinna, and G. Ponticelli, *Transition Metal Chem.*, 1976, **1**, 105.
[112] C. Preti and G. Tosi, *J. Co-ordination Chem.*, 1976, **6**, 81.
[113] C. Preti and G. Tosi, *Canad. J. Chem.*, 1976, **54**, 85.
[114] C. Preti and G. Tosi, *J. Inorg. Nuclear Chem.*, 1976, **38**, 1125.
[115] P. P. Singh, O. P. Agrawal, and A. K. Gupta, *Inorg. Chim. Acta*, 1976, **18**, 19.
[116] A. N. Speca, L. L. Pytlewski, C. Owens, and N. M. Karayannis, *J. Inorg. Nuclear Chem.*, 1976, **38**, 1119.
[117] A. N. Speca, L. L. Pytlewski, and N. M. Karayannis, *Z. anorg. Chem.*, 1976, **222**, 182.
[118] D. E. Chasen, L. L. Pytlewski, C. Owens, and N. M. Karayannis, *J. Inorg. Nuclear Chem.*, 1976, **38**, 1799.

Table 7 (*cont.*)

Series [a]	M [b]	*Assignments* [b]	*Ref.*
M(pyrone)$_n$Z$_2$	Mg, Mn, Co, Ni, Cu, Zn, Cd	ν(MO), ν(MX)	119 [p]
M(thioxamide)$_n$Z$_m$	Co, Ni, Cu	ν(MN), ν(MS), ν(MX)	120 [q]
M(HOC$_2$H$_4$OH)$_2$X$_2$	Mn, Co, Ni	ν(MO), ν(MX)	121
[MCl$_2$(*o*-diphos)$_2$]$^{n+}$	Cr, Mn, Fe, Co, Zn	ν(MCl)	122 [r]
(Pdchel)MCl$_2$	Fe, Co, Ni, Cu, Zn	ν(MCl)	123 [s]
(Ptkel)MCl$_2$, (Ptkel){RhCl(CO)$_2$}$_2$	Co, Hg	ν(MCl), ν(RhCl)	124 [t]
[M(C$_2$O$_4$)$_3$]$^{3-}$, [Ir(C$_2$O$_4$)$_2$Cl$_2$]$^{3-}$, *etc.*	Cr, Fe, Co, Rh, Ir, Ru	ν(MO)	125 [d]
M(O$_2$CAr)$_2$	Co, Ni, Cu, Zn	ν(MO), ν(MN)	126 [u]
M(O$_2$CCH$_2$Cl)$_n$(py)$_m$	Ti, Zr, Fe	ν(MO)	127 [v]
M(picateO)$_n$	Cr, Mn, Fe, Co, Ni, Cu, Zn	ν(MO)	128 [w]
M(nicateO)$_2$,*x*H$_2$O	Mn, Fe, Co, Ni, Cu, Zn	ν(MO)	129 [x]
M(β-alan)$_2$(H$_2$O)$_n$	Ni, Cu	ν(MO), ν(MN)	32 [y]
M(thiazolidine)$_n$	Cu, Au, Pd	ν(MS), ν(MN)	130 [z]
M(thiaz-ine-one)$_n$	Mn, Co, Ni, Cu, Zn, Cd, Hg, Pd, Pt, Au, Ag	ν(MS), ν(MN)	131 [aa]
M(pyrazolinone)$_2$	VO, Cu	ν(MO)	132 [bb]
[M(thiaz–imid)$_3$]$^{2+}$	Mn, Co, Ni, Cu, Zn, Cd	ν(MN)	133 [cc]
M(Hchel)$_2$, M(chel), *etc.*	Mn, Fe, Co, Ni, Cu, Zn, VO	ν(MO), ν(MS), ν(MN)	134 [dd]
M(aza-ene-one)	Co, Ni, Cu	ν(ML)	135 [ee]
M(pyacac)$_2$	Be, Mn, Co, Ni, Cu, Zn, Pd	ν(MO)	136, 137 [ff]

[119] A. de Jager, J. J. de Vrieze, and J. Reedijk, *Inorg. Chim. Acta*, 1976, **20**, 59.
[120] A. C. Fabretti, G. C. Pellacani, and G. Peyronel, *J. Inorg. Nuclear Chem.*, 1976, **38**, 443.
[121] D. Knetsch and W. L. Groeneveld, *Inorg. Nuclear Chem. Letters*, 1976, **12**, 27.
[122] L. F. Warren and M. A. Bennett, *Inorg. Chem.*, 1976, **15**, 3126.
[123] B. Crociani, M. Nicolins, and R. L. Richards, *J. Organometallic Chem.*, 1976, **113**, C22.
[124] G. Minghetti, F. Bonati, and G. Banditelli, *Inorg. Chem.*, 1976, **15**, 2649.
[125] J. Gouteron, *J. Inorg. Nuclear Chem.*, 1976, **38**, 55.
[126] M. Noji and Y. Kidani, *Bunseki Kagaku*, 1975, **24**, 545.
[127] K. C. Malhotra and R. G. Sud, *J. Inorg. Nuclear Chem.*, 1976, **38**, 393.
[128] S. A. Boyd, R. E. Kohrman, and D. X. West, *J. Inorg. Nuclear Chem.*, 1976, **38**, 607.
[129] A. N. Speca, L. S. Gelfand, L. L. Pytlewski, C. Owens, and N. M. Karayannis, *Inorg. Chem.*, 1976, **15**, 1493.
[130] E. Siew, G. Klinger, and B. R. Sandheim, *J. Inorg. Nuclear Chem.*, 1976, **38**, 7.
[131] C. Preti and G. Tosi, *Canad. J. Chem.*, 1976, **54**, 1558.
[132] E. R. Menzel, D. R. Lorenz, J. R. Wasson, K. R. Radigan, and B. J. McCormick, *J. Inorg. Nuclear Chem.*, 1976, **38**, 993.
[133] R. C. Van Landschoot, J. A. M. Van Hest, and J. Reedijk, *J. Inorg. Nuclear Chem.*, 1976, **38**, 185.
[134] R. C. Aggarwal and R. B. S. Yadava, *Transition Metal Chem.*, 1976, **1**, 139.
[135] M. Vidali, U. Casellato, P. A. Vigato, and R. Graziani, *J. Inorg. Nuclear Chem.*, 1976, **38**, 1455.
[136] J. A. Faniran, K. S. Patel, and L. O. Nelson, *J. Inorg. Nuclear Chem.*, 1976, **38**, 77.
[137] K. S. Patel, J. A. Faniran, and L. O. Nelson, *J. Inorg. Nuclear Chem.*, 1976, **38**, 81.

Table 7 (*cont.*)

Series [a]	M [b]	*Assignments* [b]	*Ref.*
$[M(ttfa)_nL_{3-n}]^{(n-2)-}$ / $[M(bztfa)_nL_{3-n}]^{(n-2)-}$	Mn, Co, Ni, Cu, Zn	ν(MO)	138 [gg]
$M(oxinate)_2$	Co, Ni, Cu, Zn	ν(MO)	139 [hh]
$M(benz\text{-}ole\text{-}ate)_2$	Cu, Zn	ν(MO), ν(MN)	140 [ii]
$M(amine\text{–}CS_2)_n$	Co, Ni, Cu, Pd, Pt	ν(MS), ν(MN)	141 [jj]
$M(S_2CNR_2)_3$	Cr, Fe, Co	ν(MS)	41 [kk]
$M(S_2CNR_2)_3$	Cr, Mn, Fe, Co, Ru	ν(MS)	142 [kk]
$M(S_2CNR_2)_n$	Fe, Co, Ni, Cu, Zn, Cd, Hg	ν(MS)	143 [ll]
$M(Se_2CNR_2)_2$ / $M(Se_2CNR_2)_3$	Zn, Cd, Hg / Cr, Fe, Co	ν(MSe)	144 [mm]
$M(Se_2CNR_2)_2$	Ni, Cu, Pd, Pt	ν(MSe)	145 [mm]

[a] X = halide; Y = other anion (NCS, NO_3, ClO_4, *etc.*); Z = X or Y. [b] It is not implied that all possible combinations of M, X, *etc.* were studied, nor that all the stated assignments are given for every compound. [c] Alkali metal halide melts. [d] Discussed in text. [e] Ar = 2-naphthyl; n = 2 (M = Co, Ni or Cu) or 1 (M = Zn or Cd). [f] R = Me or Ph. [g] n = 3, 2, or 1, depending on M and Z. [h] (pz) = 1-pyrazolyl. [i] (Ph_2oxazole) = 2,5-diphenyloxazole, which is thought to be *N* or *N,O* co-ordinated depending on M, *etc.* [j] (diazepin) = 7-chloro-2-methylamino-5-phenyl-3H-1,4-benzodiazepin-4-oxide (Librium) or 1,3-dihydro-7-nitro-5-phenyl-2*H*-1,4-benzodiazepin-2-one (Mogadon); mode of co-ordination not ascertained. [k] n = 3 (M = Cr) or 2 (others); m = 3, 2, or 1 depending on M, *etc.*; (benzox) = benzoxazole-2-thione, which is thought to co-ordinate only *via* N. [l] (thia) = 2-thiazolidinethione, 2,4-thiazolidinedione, or ethylene monothiocarbonate; n = 4 (M = Ti) or 2 (M = Fe). [m] n = 1, 2, or 3. [n] n = 3 (M = Cr or Fe) or 2 (others); m = 2, 1, or $1\frac{1}{2}$; x = 2, 1, or 0. [o] (quinoxO) = quinoxaline-1-oxide; n = 3 (M = Cr or Fe) or 2 (others); m = 3, 2, 1, or $1\frac{1}{2}$; x = various. [p] (pyrone) = 2,6-dimethyl-4-pyrone; n = various. [q] (thioxamide) = *NN'*-didodecyldithio-oxamide; n = 3, 2, or 1; m = 3 (M = Co) or 2 (M = Ni or Cu). [r] (*o*-diphos) = *o*-$(Me_2P)_2C_6H_4$; n = 0 or 1. [s] Structure as in (4). [t] (Ptkel) = $L_2Pt\{C(OMe) = NAr\}_2$, where L^2 = (Ph_3P)(ArNC) or (diphos), and Ar = *p*-tolyl. [u] Ar = 2-, 3-, or 4-$H_2NC_6H_4$ or -C_5H_4N, or Ph. [v] n = 3, m = 0 (M=Fe); n = 4, m = 0 or 2 (M = Ti or Zr). [w] (picateO) = 2-picolinate *N*-oxide; n = 3 (M = Cr or Fe) or 2 (others). [x] (nicateO) = nicotinate *N*-oxide; x = various. [y] (β-alan)H = β-alanine; n = 2 (M = Ni) or 0 (M = Cu). [z] (thiazolidine) = 4-oxo-2-thione-5,5-dimethylthiazolidine anion; n = 1 (M = Au, tetrameric) or 2 (M = Cu or Pd). [aa] (thiaz-ine-one) = thiazolidine-2-thione anion; n = 1 (M = Ag) or 2 (others). [bb] (pyrazolinone) = 4-benzoyl-3-methyl-1-phenyl-2-pyrazolin-5-one anion. [cc] (thiaz-imid) = thiazolylbenzimidazole anion. [dd] (chel)H_2 = PhNHC(S)NHNHC(O)Ph. [ee] (aza-ene-one) = 6,11-dimethyl-7,10-diazahexadeca-6,10-diene-2,4,13,15-tetraone, which is an *O,O,N,N* donor; hence ν(MO), ν(MN) not distinguished. [ff] (pyacac)H = (3-pyridyl)$COCH_2COCX_3$, X = H or F. [gg] (ttfa) = α-thenoyltrifluoroacetone; (bztfa) = benzoyltrifluoroacetone; L = bipy, phen, 2py, $2NH_3$, or $2H_2O$; n = 3 or 2. (52 complexes studied.) [hh] (oxinate) = 8-quinolinolate like anion such as 2,5-dimethyl-5-hydroxyquinoxaline. [ii] (benz-ol-ate)H = 2-(*o*-hydroxyphenyl)benzoxazole or 2-(*o*-hydroxyphenyl)benzthiazole. [jj] (amine-CS_2) = 2-*N*-ethylaminocyclopentene-1-dithiocarboxylate anion; n = 1 (M = Cu), 2 (M = Co, Ni, Pd, or Pt), or 3 (M = Co). [kk] R_2N = various. [ll] R_3N = piperazino; n = 3 (M = Fe) or 2 (others). [mm] R_2N = piperidino, morpholino, or thiomorpholino.

138 G. S. Shephard and D. A. Thornton, *J. Mol. Structure*, 1976, **34**, 83.
139 K. Inagaki and Y. Kidani, *Chem. Pharm. Bull.*, 1976, **24**, 839.
140 D. R. Lorenz, T. M. Barbara, and J. R. Wasson, *Inorg. Nuclear Chem. Letters*, 1976, **12**, 65.
141 K. Nag and D. S. Joardar, *Canad. J. Chem.*, 1976, **54**, 2827.
142 G. R. Hall and D. N. Hendrickson, *Inorg. Chem.*, 1976, **15**, 607.
143 C. Preti and G. Tosi, *J. Inorg. Nuclear Chem.*, 1976, **38**, 1746.
144 D. De Filippo, P. Deplano, A. Diaz, and E. F. Trogu, *Inorg. Chim. Acta*, 1976, **17**, 139.
145 D. De Filippo, P. Deplano, F. Devillanova, and E. F. Trogu, *Inorg. Chim. Acta*, 1976, **17**, 199.

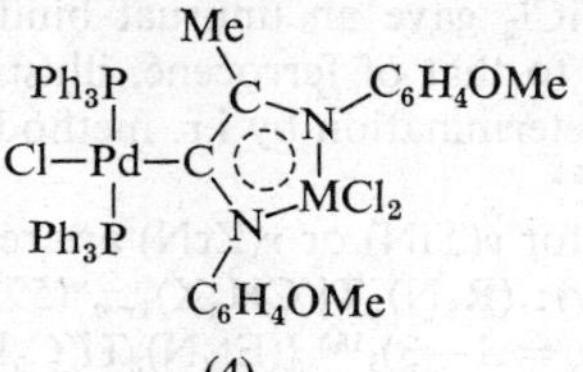

(4)

for the chain polymers is also untenable, and factor group splitting (previously shown to be important) was ignored. Thus, contrary to the implications of the paper, virtually nothing is added to knowledge or understanding in this area, and the poor refereeing is to be deplored.

On the brighter side, it must be recorded that some reliable assignments are being made, and valid structural conclusions are being drawn. Thus ν(MX) positions can help determine co-ordination geometry [111–114, 117–119, 121] or provide evidence for bridging halide ligands.[113, 118, 121] Correlations of *e.g.* ν(MO) with electronic or stability factors have again been noted.[119, 136–139] Other reports worthy of particular mention include: (*a*) distinction of *d*, *l*, and *dl* forms of $[M(en)_3]Cl_3$ and $K_3[M(C_2O_4)_3]$ salts by means of their lattice vibrational spectra;[104, 125] (*b*) recognition that ν_{sym}(MOM) is significantly lower for linear μ-oxo-bridged species (*e.g.* 225 cm^{-1} for $Cs_4Os_2OCl_{10}$) than for angular systems (*e.g.* 556 cm^{-1} for $K_2W_2O_{11}$), so may be used in structural diagnosis [*e.g.* $Cs_4Ru_2OBr_{10}$ with ν_{sym}(RuORu) at 243 cm^{-1} probably has a linear anion];[146] and (*c*) an authoritative review of the vibrational spectra of oxo-, thio-, and seleno-metallates of transition metals in the solid state ($MO_4{}^{n-}$ salts, *etc.*).[147]

Papers of significance for a single periodic group are mentioned subsequently in the appropriate section.

3 Scandium and Yttrium

Vibrational data are available for $[Y(phen)_3](NCS)_3$ [ν(YN) said to be at 420 cm^{-1}],[148] Y(thioxane oxide)$_8(PF_6)_3$,[149] and Cs_2KMF_6 (M = Sc or Y; see Chapter 4).[12]

4 Titanium, Zirconium, and Hafnium

A new complex $[C_{16}H_{16}N_2O_2Ti(BH_3)_2]_2$ has been characterized by *X*-ray diffraction, and bands due to H_2BHTi units identified.[150] A detailed reconsideration of the spectra of $Zr(BH_4)_4$ suggests that the effective overall molecular symmetry is *T* rather than T_d, but many of the 'extra' bands seen are detectable only for the solid phase.[151] A skeletal ν(ZrB) mode has been assigned [152] for $(C_5H_5)_2Zr(BH_4)_2$ at 404 cm^{-1}.

146 J. San Filippo, jun., R. L. Grayson, and H. J. Sniadoch, *Inorg. Chem.*, 1976, **15**, 269.
147 A. Mueller, E. J. Baran, and R. O. Carter, *Structure and Bonding (Berlin)*, 1976, **26**, 81.
148 Yu. G. Eremin and G. I. Bondarenko, *Russ. J. Inorg. Chem.*, 1976, **21**, 209.
149 O. A. Serva, M. Pevrier, V. K. Lakatos Osorio, and Y. Kawano, *Inorg. Chim. Acta*, 1976, **17**, 135.
150 G. Fachinetti, C. Floriani, M. Mellini, and S. Merlino, *J.C.S. Chem. Comm.*, 1976, 300.
151 B. D. James, B. E. Smith, and H. F. Shurvell, *J. Mol. Structure*, 1976, **33**, 91.
152 B. E. Smith, B. D. James, and J. A. Dilts, *J. Inorg. Nuclear Chem.*, 1976, **38**, 1973.

Reduction of $(C_5H_5)_2TiCl_2$ gave an unusual binuclear species which has an i.r. spectrum very similar to that of ferrocene, illustrating the care which needs to be taken in structure determination by i.r. methods.[153] Data are available for the new $(C_5H_5)_2Zr(CO)_2$.[154]

Assignments suggested for ν(TiN) or ν(ZrN) are reasonably consistent (largely from the same laboratory): $(R_2N)_nTi(CH_2X)_{4-n}$ (592—650 cm^{-1}; R = Me, Et, *etc.*; X = Ph or $SiMe_3$; n = 1—3),[155] $\{(Et_2N)_3Ti(C_5H_4)\}_2Ru$ (608, 615 cm^{-1}; the σ-bonded $Ti(C_5H_4)$ groups are thought to be π-bonded to Ru),[156] $M\{(NMeSiMe_2)_2Y\}_2$ (512—529 and 453—467 cm^{-1} for M = Ti,[157] *ca.* 490 and *ca.* 460 cm^{-1} for M = Zr;[158] Y = NMe, O, or CH_2), and $\{HN{=}C(Cl)C(Cl){=}NH\}TiCl_4$ (569, 535 cm^{-1}, but undergoing 30—40 cm^{-1} shifts on deuteriation).[159]

Identification of ν(MO) bands in complex systems containing these elements has usually proved to be good sport, and last year was no exception. Compounds subjected to such examination include $TiCl_3OR,L$ [R = Me or Pr^n, L = sulphur bis(t-butylamide)],[160] 'zirconium hydroxide' (formulation not given),[161] a range of titanyl(IV) porphyrins,[162] $[\{(Me_2N)_3PO\}_2TiCl_3]_2O_2$,[163] $\{(C_5H_5)_2Hf(Me)\}_2O$,[164] $ZrO(O_2CH)_2,2H_2O$,[165] a range of zirconyl(IV) complexes with carboxylic acids,[166] and a polymer $[\overline{Zr\{OC(CH{=})CHCHO}\}_2]_n$.[167] The more significant papers deal with $LnTiO_5$ (Ln = lanthanoid, including Nd; see Table 2),[39] confirmation of non-cubic structures for $CaHfO_3$ and $SrHfO_3$ from observation of first-order Raman lines (contrast $BaHfO_3$),[168] characterization of peroxotitaniumoctaethylporphyrin as the first such type of compound with the O_2 group symmetrically bonded by both atoms [*i.e.* $\mathrm{Ti}{<}\begin{array}{c}\mathrm{O}\\ | \\ \mathrm{O}\end{array}$ with ν(TiO) = 600, 645; ν(OO) = 898 cm^{-1}],[169] and the first direct spectroscopic evidence for the structure of a strongly co-ordinated species in aqueous zirconium sulphate [ν(ZrO) = 530, 700 cm^{-1}].[170]

Structures proposed on the basis of ν(MX) studies include *trans*-octahedral geometries for $TiCl_4(benzalanilide)_2$ [28] and $ZrI_4(Me_3N)_2$,[171] *cis*-octahedral

153 G. P. Pez, *J. Amer. Chem. Soc.*, 1976, **98**, 8072.
154 B. Demerseman, G. Bouquet, and M. Bigorgne, *J. Organometallic Chem.*, 1976, **107**, C19.
155 H. Bürger and C. Kluess, *J. Organometallic Chem.*, 1976, **108**, 69.
156 H. Bürger and C. Kluess, *Z. anorg. Chem.*, 1976, **423**, 112.
157 H. Bürger and K. Wiegel, *Z. anorg. Chem.*, 1976, **419**, 157.
158 K. Wiegel and H. Bürger, *Z. anorg. Chem.*, 1976, **426**, 301.
159 E. Allenstein and K. Löhmar, *Z. anorg. Chem.*, 1976, **420**, 193.
160 J. D. Wilkins, *J. Inorg. Nuclear Chem.*, 1976, **38**, 673.
161 A. I. Ryabinin, Yu. A. Afanas'ev, V. P. Eremin, V. T. Panyushkin, and N. N. Bukov, *Doklady Phys. Chem.*, 1976, **225**, 1346.
162 P. Fournari, R. Guilard, M. Fontesse, J.-M. Latour, and J.-C. Marchon, *J. Organometallic Chem.*, 1976, **110**, 205.
163 D. P. Bauer and R. S. Macomber, *Inorg. Chem.*, 1976, **15**, 1985.
164 F. R. Fronczek, E. C. Baker, P. R. Sharp, K. N. Raymond, H. G. Alt, and M. D. Rausch, *Inorg. Chem.*, 1976, **15**, 2284.
165 R. C. Paul, O. B. Baidya, and R. Kapoor, *Z. Naturforsch.*, 1976, **31b**, 300.
166 R. C. Paul, S. K. Gupta, S. S. Parmar, and S. K. Vasisht, *Z. anorg. Chem.*, 1976, **423**, 91.
167 G. A. Domrachev, É. V. Shitova, V. Yu. Vodzinskii, O. N. Suvorova, V. A. Varynkhin, and B. A. Nesterov, *Doklady Chem.*, 1976, **226**, 121.
168 Chon Il Park, R. A. Condrate, sen., and R. L. Snyder, *Appl. Spectroscopy*, 1976, **30**, 352.
169 R. Guilard, M. Fontesse, P. Fournari, C. Lecompte, and J. Protas, *J.C.S. Chem. Comm.*, 1976, 161.
170 F. G. Baglin and D. Breger, *Inorg. Nuclear Chem. Letters*, 1976, **12**, 173.
171 J. Hughes and G. R. Willey, *Inorg. Chim. Acta*, 1976, **20**, 137.

arrangements for $ZrX_4(Me_3N)_2$ (X = Cl or Br)[171] and $TiCl_4(MeCN)_2$,[172] and an edge-shared octahedral $[L_2ZrCl_{4/2}]$ arrangement for the cyclic compounds $Y(SiMe_2NMe)_2ZrCl_2$ (Y = NMe, O, or CH_2).[158] The structure of *catena*-(di-μ-chloro-bis-μ-trimethylsilylamino)di{chlorotitanium(IV)} has been established by *X*-ray studies, and this compound has $\nu(TiCl)_t$ = 427, $\nu(TiCl)_b$ = 231 cm^{-1}; the latter band disappears on dissolution, and with ν(TiCl) at 470 cm^{-1} a lower co-ordination number for Ti is indicated.[173] $[Cl_2Ti(NSiMe_3)py_2]_2$ is thought to contain no chlorine bridges [ν(TiCl) at 317 cm^{-1}].[173] Other compounds of this group for which ν(MX) assignments have been made include matrix-isolated $TiCl_4$,[43] $(C_5H_5)TiCl_3L_n$ (L = diars or bipy, n = 1 or 0),[174] MCl_4(PhNSO) and related compounds (M = Ti or Zr),[160] $TiX_4(Ph_3P)_n$ (X = Cl or Br, n = 1 or 2),[175] products from reaction of MeC(X)(Y)(Z) and $TiCl_4$ (X, Y, or Z = CH_2OMe, CH_2SMe, or CH_2NMe_2),[176] $ZrCl_4L_2$ (L = pyridine or derivative thereof),[177] as well as compounds previously mentioned in this section.[159, 160]

References to other Ti or Zr compounds are made in Table 7.[110, 115, 127]

5 Vanadium, Niobium, and Tantalum

The new species $NbH_5(Me_2PC_2H_4PMe_2)_2$ shows one i.r. ν(NbH) band,[178] at 1520 cm^{-1}.

Two papers describe formation of $[V(CO)_6]_n$ (n = 1 or 2) in low temperature matrices,[45a, b] with (thankfully!) the same overall conclusions, *viz.* the monomer is distorted (probably D_{4h}) while the dimer is $(OC)_5V(\mu\text{-}CO)_2V(CO)_5$. There is much here for the matrix-isolation pundits, particularly in the thorough studies of the more recent paper.[45b] I.r. data are available for $(C_5H_5)_2VR$ (R = various),[179] $Ta_2(CH_2SiMe_3)_4(SiMe_3)_2$,[180a] and $[Nb(CN)_8]^{4-}$ salts.[180b]

Not surprisingly there was a bumper crop of papers last year reporting ν(MO) vibrations of various kinds (especially for M = V), and even more obvious is the lack of consequence of the work. Table 8 [33, 34, 83, 132, 160, 181–207] lists for posterity

172 Y. Kawano, Y. Hase, and O. Sala, *J. Mol. Structure*, 1976, **30**, 45.
173 N. W. Alcock, M. Pierce-Butler, and G. R. Willey, *J.C.S. Dalton*, 1976, 707.
174 R. J. H. Clark, J. A. Stockwell, and J. D. Wilkins, *J.C.S. Dalton*, 1976, 120.
175 C. M. F. Rae and K. R. Seddon, *Inorg. Chim. Acta*, 1976, **17**, L35.
176 A. J. McAlees, R. McCrindle, and A. R. Woon-Fat, *Inorg. Chem.*, 1976, **15**, 1065.
177 Ts. B. Konunova, A. M. Sych, and M. F. Frunze, *Koord. Khim.*, 1976, **2**, 26.
178 R. R. Schrock, *J. Organometallic Chem.*, 1976, **121**, 373.
179 H. Bouman and J. H. Teuben, *J. Organometallic Chem.*, 1976, **110**, 327.
180 (*a*) M. H. Chisholm, F. A. Cotton, M. Extine, and B. R. Stults, *Inorg. Chem.*, 1976, **15**, 2252; (*b*) P. M. Kiernan, *Inorg. Chim. Acta*, 1976, **20**, 89.
181 H. B. Ergun, K. A. Gehring, and G. A. Gehring, *J. Phys.* (*C*), 1976, **9**, 1101.
182 E. J. Baran, *Z. anorg. Chem.*, 1976, **427**, 131.
183 E. J. Baran, I. L. Botto, and P. J. Aymonino, *Z. anorg. Chem.*, 1976, **423**, 280.
184 J. C. J. Bart and G. Petrini, *Z. anorg. Chem.*, 1976, **422**, 179.
185 I. L. Botto and E. J. Baran, *Z. anorg. Chem.*, 1976, **426**, 321.
186 L. R. Tokareva, A. P. Nakhodnova, and L. N. Kovgan, *Izvest. Akad. Nauk SSSR, Neorg. Materialy*, 1976, **12**, 870.
187 E. F. Ryannel, V. M. Kalygina, and V. I. Gaman, *Fiz. Khim. Stekla*, 1976, **2**, 109.
188 J. Dimitriev, M. Arnaudov, and V. Dimitrov, *Monatsh.*, 1976, **107**, 1335.
189 D. F. Takezhanova, D. U. Begalieva, A. K. Il'yasova, and A. B. Bekturov, *Doklady Chem.*, 1976, **226**, 85.
190 A. Marty, K. Abdmeziem, and B. Spinner, *Compt. rend.*, 1976, **283**, *C*, 285.
191 B. N. Varskoi, N. M. Spasskaya, V. M. Talanov, and A. N. Men, *Izvest. Vysshikh Uchebn. Zavedenii, Fiz.*, 1976, **19**, 16.
192 G. Colsmann, B. Reuter, and E. Riedel, *Z. anorg. Chem.*, 1976, **424**, 144.

Table 8 *Studies on compounds containing* MO *bonds* (M = V, Nb, *or* Ta)

Compound	*Ref.*
$TbVO_4$	181
$M_3(VO_4)_2$	182, 183 [a]
$Pb_5(VO_4)_nX$	183 [b]
$M_2V_2O_9$	184 [c]
$NH_4UO_2VO_4,nH_2O$	185
$[Bu^n{}_4N]_3[HV_4O_{12}]$	83
$VOWO_4,3H_2O$	186
$CaO–V_2O_5–P_2O_5$	187
$TeO_2–V_2O_5$	188
$Na_2H_2SV_2W_8O_{35}$ *etc.*	189
$[R_4N]_nH_mNb_9O_{27},6H_2O$	190 [d]
$(MV_2O_4)_x(MgFe_2O_4)_{1-x}$	191 [e]
$Zn(Zn_xFe_{1-x}V)O_4$	192 [f]
VO_2IO_3,xH_2O	193
$M^1M^2O_4$	194 [g]
$(C_5H_5)_2Nb(O)X$	195 [h]
$VOCl_n(OMe)_{3-n}$(bns)	160 [i]
$NbOCl_2(OMe)\{OP(NMe_2)_3\}_2$	196
$M_2VO_2Cl_2$	197 [j]
$[VOF_4(H_2O)]^-$	198
$(NH_4)_2[VO_2F(C_2O_4)]$	199
$[VOF(C_2O_4)_2]^{3-}$	200
VO(salam)(H_2O)	33, 34 [k]
$VO(pyrazolinone)_2$	132 [l]
$VO(S_2AsMe_2)_2,x$py	201 [m]
NiVO(chel)	202 [n]
VO(kel)	203 [o]
$VOX_2(dihyd)L_n$	204 [p]
$VO(R^1COCHCOR^2)_2$	205 [q]
$NbOCl_3\{(kel)H\}_2$	206 [r]
$NbOCl_3$ vapour	207

[a] M = Ca or Pb. [b] $n = 2$ (X = GeO_4 or SiO_4) or 3 (X = F, Cl, or Br). [c] M = Se or Te. [d] R = Me, $n = 5$, $m = 6$; R = Et, $n = 6$, $m = 3$. [e] M not given in *Chem. Abs.* [f] x = 0.1 or 0.3. [g] M^1 = Sb or Bi, M^2 = Nb or Ta. [h] X = Cl, Bu^n, α-$CH_2C_5H_4N$, or α-C_4H_3S. [i] $n = 3$ or 2, (bns) = sulphurbis(t-butylimide). [j] M = K or Rb. [k] Compound (1); see Table 2. [l] See Table 7. [m] x = 0 or *ca.* 4. [n] (chel)H_4 = a Schiff base. [o] (kel)H_2 = a Schiff base. [p] X = Cl or $\frac{1}{2}SO_4$; (dihyd) = $H_2NNHCO(CH_2)_mCONHNH_2$, m = various; $n = 0$, 1 (L = py), or 2 (L = H_2O). [q] R^1, R^2 = various. [r] (kel)H = a substituted salicylaldimine.

193 W. Meschede and R. Mattes, *Z. anorg. Chem.*, 1976, **420**, 25.
194 A. M. Sych and E. M. Verlan, *Ukrain. fiz. Zhur.*, 1976, **21**, 621.
195 D. A. Lemenovskii, T. V. Baukova, V. A. Knizhnikov, E. G. Perevalova, and A. N. Nesmeyanov, *Doklady Chem.*, 1976, **226**, 65.
196 L. G. Hubert-Pfalzgraf, R. C. Muller, M. Postel, and J. G. Riess, *Inorg. Chem.*, 1976, **15**, 40.
197 N. N. Chizhov, S. A. Kutsenko, and A. I. Morozov, *Russ. J. Inorg. Chem.*, 1976, **21**, 765.
198 H. Rieskamp and R. Mattes, *Z. Naturforsch.*, 1976, **31b**, 541.
199 H. Rieskamp and R. Mattes, *Z. anorg. Chem.*, 1976, **419**, 193.
200 H. Rieskamp and R. Mattes, *Z. Naturforsch.*, 1976, **31b**, 537.
201 E. D. Day and R. E. D. McClung, *Inorg. Chem.*, 1976, **15**, 571.
202 R. L. Lintvedt, M. D. Glick, B. K. Tomlonovic, and D. P. Gavel, *Inorg. Chem.*, 1976, **15**, 1646.
203 A. Syamal, *Transition Metal Chem.*, 1976, **1**, 103.
204 R. C. Aggarwal and B. Singh, *Transition Metal Chem.*, 1976, **1**, 161.
205 K. Dey, R. K. Maiti, and S. K. Sen, *Inorg. Chim. Acta*, 1976, **20**, 197.
206 K. Yamanouchi and S. Yamada, *Inorg. Chim. Acta*, 1976, **18**, 201.
207 N. I. Zavalishin and A. A. Mal'tsev, *Vestnik Moskov. Univ.* (*Khim.*), 1975, **16**, 741.

the relevant specimens, although a few papers are probably of some spectroscopic [33, 34, 181–183, 194] or structural [83, 193, 198–200] interest.

Metal–sulphur modes have been assigned in $Cu_{12}As_3VS_{16}$,[208] $MSBr_3$ (M = Nb or Ta),[209] $(Me_2NCS_2)_4TaCl,CH_2Cl_2$ and $[(Me_2NCS_2)_4Ta][TaCl_6],CH_2Cl_2$,[210] and $Nb_2Cl_6(SMe_2)_3$.[211] For the last-mentioned compound, a metal–metal bonded sulphide-bridged structure was proposed, with $\nu(NbS)_t$ = 285, 318 cm^{-1} and $\nu(NbS)_b$ = 238 cm^{-1} (this paper contains two points of difference between a table and the text).[211] In the case of $NbCl_5$ adducts of some cyclic polythiaethers, ν(NbCl) and ν(NbS) were not distinguished.[212]

Assignments of ν(M—halogen) modes have been reported for NH_4VF_4, $N_2H_5VF_4$, and $N_2H_6VF_5$,[213] $SbCl_5$–$NbCl_5$ mixtures,[214] and compounds previously mentioned.[160, 196, 198–200, 204, 206, 207, 210–212]

See also Table 7 for further ν(VN),[134] ν(MO),[127, 132, 134] ν(VS),[134] and ν(MX) [110, 127] reports.

6 Chromium, Molybdenum, and Tungsten

Compounds previously formulated as $M_3Mo_2X_8$ salts have now been shown to be $M_3Mo_2X_8H$ [ν(MoHMo) = 1245 (M = Cs) or 1274 cm^{-1} (M = Rb), shifting to 904—916 cm^{-1} on deuteriation; X = Cl or Br].[215] For $Al_3Mo_2(Me)_5(C_5H_5)_2(C_5H_4)_2H_2$, ν(MoHAl) [216] is at 1825 cm^{-1}. Other ν(MoH) [217] and ν(WH) [217–219] assignments have been made.

Isotopic frequency shifts for $Cr(CO)_6$ continue to be a source of debate.[16, 17] Co-condensation of Cr atoms and $PhCO_2Et$ has afforded $(PhCO_2Et)_2Cr$, characterized by i.r. [ν(Cr—ring) = 452—535 cm^{-1}] and other techniques.[220] Assignments of ν(WC) have been made in acceptable regions for various adducts of $MeWO_nCl_{5-2n}$ (n = 0, 1, or 2) [221] and $(C_5H_5)W(R)(CO)_3$ (R = CD_3, Et, or σ-allyl).[218] The new complex $K_4[Mo(CN)_7]$ appears to be of C_{2v} symmetry.[222] Other data on Cr—C,[223–225] Mo—C,[224] and W—C [180a, 224] systems are available.

Substantial mixing of ν(MN) with other modes has been shown for $M_2(NMe_2)_6$ (M = Mo or W).[18, 19] Variation of ν(WN) with R in $W(CO)_5(4\text{-}RC_5H_4N)$ has been discussed in terms of π-bonding effects.[226] Data are

208 E. Riedel and W. Paterno, *Z. anorg. Chem.*, 1976, **426**, 95.
209 G. W. A. Fowles, R. J. Hobson, D. A. Rice, and K. J. Shanton, *J.C.S. Chem. Comm.*, 1976, 552.
210 D. F. Lewis and R. C. Fay, *Inorg. Chem.*, 1976, **15**, 2219.
211 A. D. Allen and S. Naito, *Canad. J. Chem.*, 1976, **54**, 2948.
212 R. E. DeSimone and T. E. Tighe, *J. Inorg. Nuclear Chem.*, 1976, **38**, 1623.
213 S. Kladnik and P. Glavič, *J. Inorg. Nuclear Chem.*, 1976, **38**, 2317.
214 W. Bues, F. Demiray, and W. Brockner, *Spectrochim. Acta*, 1976, **32A**, 1623.
215 F. A. Cotton and B. J. Kalbacher, *Inorg. Chem.*, 1976, **15**, 522.
216 M. L. H. Green, R. E. MacKenzie, and J. S. Poland, *J.C.S. Dalton*, 1976, 1993.
217 B. R. Francis, M. L. H. Green, Tuyet Luong-thi, and G. A. Moser, *J.C.S. Dalton*, 1976, 1339.
218 G. Davidson and D. A. Duce, *J. Organometallic Chem.*, 1976, **120**, 229.
219 T. E. Dorsett and R. A. Walton, *J. Organometallic Chem.*, 1976, **114**, 127.
220 G. A. Razuvaev, G. A. Domrachev, and V. D. Zinov'ev, *Doklady Chem.*, 1976, **223**, 430.
221 C. Santini-Scampucci and J. G. Riess, *J.C.S. Dalton*, 1976, 195.
222 P. M. Kiernan and W. P. Griffith, *Inorg. Nuclear Chem. Letters*, 1976, **12**, 377.
223 A. J. Hartshorn and M. F. Lappert, *J.C.S. Chem. Comm.*, 1976, 761.
224 B. V. Lokshin, E. B. Rusach, V. S. Kaganovich, V. V. Knivykh, A. N. Artemov, and N. I. Sirotkin, *Zhur. strukt. Khim.*, 1975, **16**, 592.
225 E. Lindner and W. P. Meier, *J. Organometallic Chem.*, 1976, **114**, 67.
226 M. A. M. Meester, R. C. J. Vriends, D. J. Stufkens, and K. Vrieze, *Inorg. Chim. Acta*, 1976, **19**, 95.

available for Cr^{III} complexes of NH_3,[227] $MeNH_2$,[228] (en),[229] (tren),[230] and L-amino-acids.[231] The tentative assignment has been made for ν(CrAs) at 158 cm^{-1} in $MeC(CH_2As)_3Cr(CO)_3$.[232] The isomers $(OC)_5Cr-SPHMe_2$ [ν(CrS) = 300 cm^{-1}] and $(OC)_5Cr-PMe_2SH$ [ν(CrP) = 327 cm^{-1}] have been characterized.[225] See also Table 7.[104, 114, 116, 117]

Over fifty papers dealing with ν(MO) vibrations involving these elements were noted from the 1976 literature, representing *hundreds* of spectral band assignments. There seems as little point in itemizing all these data as in recording many of the spectra in the first place, and clearly it is impracticable to comment on the reliability of the conclusions. Interested readers (and there must be some) are referred to the original papers for: (*a*) oxides of Mo;[233] (*b*) oxyacids of Mo [234, 235] or W;[235] (*c*) oxyanionic salts of Cr,[236, 237] Mo,[238–240] or W;[146, 186, 189, 241–244] (*d*) oxyhalides of W;[245] (*e*) neutral adducts of oxyhalides of Mo [246, 247] or W;[221, 247] (*f*) mononuclear oxyhalogenoanions of Cr [248, 249] or Mo;[250, 251] (*g*) binuclear Mo [92, 93, 252, 253] or trinuclear W [254] oxyhalogenoanions; (*h*) other trinuclear Cr,[255] or binuclear (MOM) Cr [256] or Mo [257–264] complexes; and

227 P. E. Hoggard, T. Schonherr, and H.-H. Schmidtke, *Spectrochim. Acta*, 1976, **32A**, 971.
228 K. Kühn and F. Wasgestian, *Inorg. Nuclear Chem. Letters*, 1976, **12**, 803.
229 J. W. Vaughn and A. M. Yeoman, *Inorg. Chem.*, 1976, **15**, 2320.
230 S. G. Zipp and S. K. Madan, *Inorg. Chem.*, 1976, **15**, 587.
231 A. G. Galinos and J. M. Tsangaris, *Inorg. Nuclear Chem. Letters*, 1976, **12**, 755.
232 J. Ellermann and H. Schössner, *J. Organometallic Chem.*, 1976, **118**, C69.
233 Z. M. Hanafi, M. A. Khilla, and A. Abu-El Saud, *Rev. Chim. minérale*, 1975, **12**, 546.
234 L. B. Petrenko, O. V. Fedulov, Yu. F. Klyuchnikov, and A. A. Kal'kov, *V. sb., Metallurgiya i Metalloredenie, Alma-ata*, 1975, 61.
235 C. Rocchiccioli-Deltcheff, R. Thourenot, and R. Frank, *Spectrochim. Acta*, 1976, **32A**, 587.
236 V. I. Grigoruk, I. I. Kondilenko, and P. Yu. Tsyashchenko, *Ukrain. fiz. Zhur.*, 1976, **21**, 684.
237 R. L. Carter and L. K. O'Hare, *Appl. Spectroscopy*, 1976, **30**, 187.
238 T. Shigenari, Y. Takagi, and Y. Wakabayashi, *Solid State Comm.*, 1976, **18**, 1271.
239 W. Kwak and M. T. Pope, *Inorg. Chem.*, 1976, **15**, 1732.
240 W. Kwak, L. M. Rajković, J. K. Stalick, M. T. Pope, and C. O. Quicksall, *Inorg. Chem.*, 1976, **15**, 2778,
241 N. I. Bashilova and N. I. Nelyapina, *Russ. J. Inorg. Chem.*, 1976, **21**, 172.
242 L. I. Kozlova, N. A. Korotchenko, and G. A. Bogdanov, *Russ. J. Inorg. Chem.*, 1976, **21**, 773.
243 H. J. Schittenhelm and S. Kemmler-Sack, *Z. anorg. Chem.*, 1976, **424**, 211.
244 S. Kemmler-Sack and I. Seemann, *Z. anorg. Chem.*, 1976, **424**, 229.
245 V. M. Kovba, V. A. Leonov, and A. A. Mal'tsev, *Russ. J. Inorg. Chem.*, 1976, **21**, 308.
246 W. Levason, C. A. McAuliffe, and B. J. Sayle, *J.C.S. Dalton*, 1976, 1177.
247 W. Levason, F. P. McCullough, and C. A. McAuliffe, *Inorg. Nuclear Chem. Letters*, 1976, **12**, 843.
248 A. K. Banerjee and N. Banerjee, *Inorg. Chem.*, 1976, **15**, 488.
249 K. R. Seddon and V. H. Thomas, *Inorg. Chim. Acta*, 1976, **20**, L37.
250 A. Beuter and W. Sawodny, *Z. anorg. Chem.*, 1976, **427**, 37.
251 M. C. Chakravorti and S. C. Pandit, *J. Co-ordination Chem.*, 1976, **5**, 85.
252 M. Lamache, *Compt. rend.*, 1976, **282**, *C*, 1081.
253 M. S. Matson and R. A. D. Wentworth, *Inorg. Chem.*, 1976, **15**, 2139.
254 K. Mennenmann and R. Mattes, *Angew. Chem. Internat. Edn.*, 1976, **15**, 118.
255 J. Ribas, J. Casabó, and J. M. Coronas, *Rev. Chim. minérale*, 1976, **13**, 303.
256 J. Casabó, J. Ribas, and J. M. Coronas, *J. Inorg. Nuclear Chem.*, 1976, **38**, 886.
257 G. J.-J. Chen, J. W. McDonald, and W. E. Newton, *Inorg. Chem.*, 1976, **15**, 2612.
258 J. A. Zubieta and G. B. Maniloff, *Inorg. Nuclear Chem. Letters*, 1976, **12**, 121.
259 H. K. Saha and T. K. R. Chaudhuri, *Inorg. Nuclear Chem. Letters*, 1976, **12**, 465,
260 G. J.-J. Chen, J. W. McDonald, and W. E. Newton, *Inorg. Nuclear Chem. Letters*, 1976, **12**, 697.
261 R. J. Butcher, H. K. J. Powell, C. J. Wilkins, and Shim Hian Yong, *J.C.S. Dalton*, 1976, 356.
262 Yu. G. Eremin, E. F. Kolpikova, and T. V. Rodionova, *Russ. J. Inorg. Chem.*, 1975, **20**, 1815.
263 A. Komara, Y. Ikeda, and H. Imanaga, *Bull. Chem. Soc. Japan*, 1976, **49**, 131.
264 H. J. Becher, G. Müller, and N. Amsoneit, *Z. anorg. Chem.*, 1976, **420**, 203.

(*i*) a host of other complexes containing CrO [97, 265] or MoO [66, 96, 97, 257, 261, 263, 266–273] bonds. Data are also available for various forms of CrOOH,[274] and relevant compounds in Table 7.[116–118, 125, 128]

Thioanions studied include $[MSSe_3]^{2-}$ (M = Mo or W) [9] and $[MoOS_3]^{2-}$ (Table 4),[50] and systems have also been investigated in which tetrathio-species are co-ordinated to other metal atoms, *viz.* $[Sn_2(WS_4)_4]^{4-}$ [six bands 422—491 cm^{-1} deriving from $\nu(WS)_t$ and $\nu(WS)_b$],[275] $\{(Ph_2MeP)Au\}_2WS_4$,[276] and $\{(Ph_3P)Ag\}_2MS_2XY$ [M = Mo or W; X or Y = O or S; $\nu(MS)_t$ = 511—523, $\nu(MS)_b$ = 424—453 cm^{-1}].[273]

In a fairly detailed UBFF study of $Cr(S_2CNR_2)_3$ (R = various), but using only a 1 : 1 model, ν(CrS) was found to contribute substantially to bands at 445 (observed only for R = Me) and 373—378 cm^{-1} (insensitivity to R reflected in force constants).[8] Values for ν(MS) have also been given for $[Cr(H_2O)_5SH]SO_4$ (*ca.* 340 cm^{-1}),[277] $[W(chel)_3]^{2-}$ and $[M(chel)_2(S_2CNEt_2)]^-$ salts [(chel) = maleonitriledithiolato dianion],[278] and the sulphide-bridged species $[Mo_2S_2(CN)_8]^{3-}$ (320, 444 cm^{-1}) and $[Mo_2S(CN)_{12}]^{5-}$ (623 cm^{-1}).[279] See also Table 7.[41, 142, 144]

Assignments have been given for ν(CrCl),[114, 117, 118, 122, 249, 280] ν(CrBr),[114, 280] ν(CrI),[114] ν(MoF),[92, 250–252] ν(MoCl),[93, 246, 247, 264, 281–284] ν(MoBr),[281, 282] ν(WF),[250, 254] ν(WCl),[146, 221, 245, 285] ν(WBr),[245, 285] and ν(WI) [285] modes. The more detailed studies are on the i.r. spectra of vapours over WCl_6, WO_2Cl_2, $WOBr_4$, and WO_2Br_2 (partial normal mode assignments given for all these),[245] and $[MOF_5]^{n-}$ salts [M = Mo or W, *n* = 1 or 2 (*cf.* less definitive studies [251])] considered in Chapter 4.[250] Data on some binuclear oxyhalogeno species [92, 93, 252] are also worthy of note, while systems of particular structural or chemical interest include $MeWO_nCl_{5-2n}$ adducts (*n* = 0, 1, or 2),[221] isomers $MoOCl_3$-$(Ph_2PCH{=}CHPPh_2)$,[246] $[CrOCl_4]^-$ and $[CrOCl_5]^{2-}$ salts,[249] $(NH_4)_5[W_3O_4F_9]$,[254] clusters formulated as $[Mo_6Cl_8(PR_3)_6][(Mo_6Cl_8)Cl_6]$ (R = various),[284] and carbyne complexes $(RC)W(CO)_4X$ (R = Me or Ph; X = Cl, Br, or I).[285]

265 G. L. Robbins and R. E. Tapscott, *Inorg. Chem.*, 1976, **15**, 154.
266 A. M. Golub, V. V. Trachevskii, N. V. Ul'ko, and N. D. Samovskaya, *Russ. J. Inorg. Chem.*, 1975, **20**, 1495.
267 V. V. Trachevskii, A. M. Golub, and N. V. Ul'ko, *Russ. J. Inorg. Chem.*, 1975, **20**, 1656.
268 Yu. G. Eremin, E. F. Kolpikova, and T. V. Rodionova, *Russ. J. Inorg. Chem.*, 1976, **21**, 59.
269 G. Holste, *Z. anorg. Chem.*, 1976, **425**, 57.
270 R. J. Butcher and B. R. Penfold, *J. Cryst. Mol. Structure*, 1976, **6**, 13.
271 P. C. H. Mitchell and D. A. Parker, *J.C.S. Dalton*, 1976, 1821.
272 W. E. Newton, G. J.-J. Chen, and J. W. McDonald, *J. Amer. Chem. Soc.*, 1976, **98**, 5387.
273 E. Königer-Ahlborn and A. Müller, *Angew. Chem. Internat. Edn.*, 1976, **15**, 680.
274 A. N. Christensen, *Acta Chem. Scand.*, 1976, **A30**, 133.
275 A. Müller, I. Paulat-Böschen, B. Krebs, and H. Dornfeld, *Angew. Chem. Internat. Edn.*, 1976, **15**, 633.
276 J. C. Huffman, R. S. Roth, and A. R. Siedle, *J. Amer. Chem. Soc.*, 1976, **98**, 4340.
277 T. Ramasami and A. G. Sykes, *Inorg. Chem.*, 1976, **15**, 1010.
278 W. P. Bosman and A. Nieuwpoort, *Inorg. Chem.*, 1976, **15**, 775.
279 M. G. B. Drew, P. C. H. Mitchell, and C. F. Pygall, *Angew. Chem. Internat. Edn.*, 1976, **15**, 784.
280 G. Preti, G. Toshi, M. Massacesi and G. Ponticelli, *Spectrochim. Acta*, 1976, **32A**, 1779.
281 J. San Filippo, jun. and M. A. S. King, *Inorg. Chem.*, 1976, **15**, 1228.
282 H. D. Glicksman, A. D. Hamer, T. J. Smith, and R. A. Walton, *Inorg. Chem.*, 1976, **15**, 2205.
283 R. G. Weissenhorn, *Z. anorg. Chem.*, 1976, **426**, 159.
284 A. D. Hamer, T. J. Smith, and R. A. Walton, *Inorg. Chem.*, 1976, **15**, 1014.
285 E. O. Fischer and G. Kreis, *Chem. Ber.*, 1976, **109**, 1673.

Compounds of these elements for which ν(MM) data have been given are in Table 6.[18, 19, 58, 66, 92–97]

7 Manganese, Technetium, and Rhenium

Observation of two ν(ReH) i.r. bands shows that $ReH_2(NO)(PPh_3)$ in solution is *cis*.[286] A Raman spectral study of the new complex $K_4[Re(CN)_7],2H_2O$ indicates a D_{5h} structure in the solid state and aqueous solution (*cf.* V analogue).[222] The spectra of $K_2[Mn(CN)_6],3(dmf)$ have been compared with those of the known unsolvated compound.[287] Assignments for $(C_5H_5)Mn(CO)_n(CS)_{3-n}$ (n = 1 or 2) were based on local symmetries of C_{5v} for the $(C_5H_5)Mn$ and C_s for the $Mn(CO)_n(CS)_{3-n}$ moieties; surprisingly, ν(CS) are weaker than ν(CO) in the Raman spectra.[3] Values quoted for ν(MC) in $[(OC)_4MnSPMe_2]_4$,[288] $ReMe_6$, and ReO_2Me_3 (non-linear OReO)[289] are 496, 500, and 512 cm^{-1} respectively. Unusually low ν(CH) values have been discussed for $(Me_3SiCH_2)_2Mn$, $[(Me_3CCH_2)_2Mn]_4$, and $Li_2[MnMe_4]$.[290]

Several ν(MnN) assignments have been made, as in Table 7[21, 108, 109, 117, 131, 133, 134] and ref. 291, and although there is a measure of consistency, some are questionable and most are inconsequential. Values noted for ν(MP) are 334 cm^{-1} for $HSMe_2PMnBr(CO)_4$,[292] 260 cm^{-1} for the cyclic $[(OC)_4MnSPMe_2]_4$,[288] and 235—287 cm^{-1} for a range of Et_3P or Et_2PPh complexes of types $ReOX_3L_2$, $ReX_2(NO)L_3$, *etc.*[293]

The i.r. spectra of $Gd(TcO_4)_3,nH_2O$ (n = 0, 1, or 3) have been interpreted in terms of (distorted) tetrahedral technetiate(VII) anions.[294] Similarly, assignments for $Ba_5(ReO_6)_2$ have been based on the (unlikely ?) assumption that discrete $[ReO_6]^{5-}$ ions are present.[295] Values of ν(Re=O) have been given (often with little comment) for ReO_2Me_3;[289] $ReOCl_3\{(Ph_2P)_2(CH_2)_n\}$ (n = 1 or 2);[296] $Re_2O_7(py)_2$, $[ReO(OH)(py)_2X_2]$, $ReO(py)_2Br_3$, and $[ReO_2py_4]^+$ (X = Br or I);[297] and $ReOX_3(Ph_3P)L$, $ReOX_2(OEt)(Ph_3P)L$, and $ReCl_2(NPh)(Ph_3P)L$ [X = Cl or Br, L = $P(OMe)_3$ or $P\{(OCH_2)_3CEt\}$].[298] Data on ν(MnO) assignments are referred to in Table 7[117–119, 121, 128, 129, 134, 136–138] or elsewhere.[291, 299–302]

286 D. Giusto, G. Ciani, and M. Manassero, *J. Organometallic Chem.*, 1976, **105**, 91.
287 G. Trageser and H. H. Eysel, *Z. anorg. Chem.*, 1976, **420**, 273.
288 E. Lindner and H. Dreher, *J. Organometallic Chem.*, 1976, **105**, 85.
289 K. Mertis and G. Wilkinson, *J.C.S. Dalton*, 1976, 1488.
290 R. A. Andersen, E. Carmona-Guzman, J. F. Gibson, and G. Wilkinson, *J.C.S. Dalton*, 1976, 2204.
291 O. Kh. Sham'lishvili, R. I. Machkhoshvili, Yu. Ya. Kharitonov, and T. G. Dzhibladze, *Russ. J. Inorg. Chem.*, 1975, **20**, 1660.
292 E. Linder and H. Dreher, *J. Organometallic Chem.*, 1976, **104**, 331.
293 J. E. Fergusson and P. F. Heveldt, *J. Inorg. Nuclear Chem.*, 1976, **38**, 2231.
294 L. L. Zaitseva, M. I. Konarev, A. V. Velichko, and A. I. Sukhikh, *Russ. J. Inorg. Chem.*, 1975, **20**, 1152.
295 E. J. Baran, *Monatsh.*, 1976, **107**, 1327.
296 R. E. Meyers and R. A. Walton, *Inorg. Chem.*, 1976, **15**, 3065.
297 M. C. Chakravorti and C. K. Das, *Inorg. Chim. Acta*, 1976, **19**, 249.
298 N. P. Johnson and M. E. L. Pickford, *J.C.S. Dalton*, 1976, 950.
299 M. V. Goloshchapov and B. V. Martynenko, *Russ. J. Inorg. Chem.*, 1976, **21**, 757.
300 D. M. Adams and W. R. Trumble, *Inorg. Chem.*, 1976, **15**, 1972.
301 L. J Boucher and C. G. Coe, *Inorg. Chem.*, 1976, **15**, 1334.
302 R. Usón, V. Riera, M. A. Ciriano, and M. Valderrama, *Transtition Metal Chem.*, 1976, **1**, 122.

Reference to compounds for which ν(MnS) has been identified has already been made.[131, 134, 142, 288, 292]

As expected, metal–halogen stretching modes have been identified in a wide range of compounds of manganese: (*a*) see Table 7 [21, 101, 107, 117, 118, 121, 122] and earlier parts of this section;[292, 302] (*b*) $MnX_2\{OC(NMeH)_2\}_3$ [X = Cl, Br, or I; *cis* geometry proposed];[303] and (*c*) $MnX_2(o\text{-}Me_2AsC_6H_4EMe_2)_2$ (E = As or Sb; X = Cl or Br).[304] A substance thought to be mainly $(NH_4)_2[Tc(NO)Cl_5]$ [305] has $\nu(Tc^{II}Cl)$ at 312 cm^{-1}, while ν(TcX) i.r. spectra have been shown for *trans*-$[TcX_4L_2]$ and *mer*-$[TcX_3L_3]$ (X = Cl or Br; L = Ph_3P, $PhMe_2P$, or $PhEt_2P$).[306] Data are also accumulating on ν(ReX) modes: (*a*) ReF_7 (Raman data for the first time);[307] (*b*) Re_3Cl_9 (new ?) and $(Bu^tPCl_3)_2[ReCl_6]$;[308] (*c*) *cis*- and *trans*-$[ReCl_4L_n]$ *etc.* [L_n = $(Ph_2PCH_2)_2$, $2(Ph_2P)_2CH_2$, *etc.*];[296] (*d*) $ReOX_3L_2$ *etc.* (see above; different values depending on nature of *trans* ligand);[293] (*e*) *mer*-$[ReX_3\{P(OR)_3\}_3]$, *mer*-$[ReX_3\{P(OCH_2)_3CMe\}_3$, *etc.*;[298] and (*f*) eleven complexes such as $(ReCl_3)_2(Ph_2PC_2H_4PPh_2)_3$, $Re_2Br_4(Ph_2PC_2H_4AsPh_2)_2$, and Re_2Cl_6-$(Ph_2AsC_2H_4AsPh_2)_2$.[309] There seems to be nothing remarkable emerging from these studies.

Rather fewer reports of metal–metal vibrations involving these elements appeared last year (Table 6 [57, 89]) than in the recent past.

8 Iron, Ruthenium, and Osmium

A usual number of ν(FeH),[310, 311] ν(RuH),[312–320] and ν(OsH) [320] assignments was made last year, but although many of the compounds may be of chemical interest, nothing spectroscopically notable was found.

Reference has been made in section 4 to the nature of the bonding deduced for $\{(Et_2N)_3Ti(C_5H_4)\}_2Ru$ [ν(RuC) = 347, 467 cm^{-1}];[156] Chapter 7 considers the detailed study of $\{(H_2C)_3C\}Fe(CO)_3$ (and its d_6 analogue).[2] Apart from one fairly routine study,[321] the other 1976 reports on carbonyls of these elements are rather interesting, each covering very different aspects as follows. Whereas there is no intermolecular exchange between $Fe(^{13}CO)_5$ and $Fe(^{12}CO)_5$ in hexane in the absence of light for up to a few days, addition of catalytic amounts of Pd—C was found soon to increase the number of ν(CO) bands; the study showed that

303 J. P. Barbier, R. P. Hugel, P. J. van der Put, and J. Reedijk, *Rec. Trav. chim.*, 1976, **95**, 213.
304 M. H. Jones, W. Levason, C. A. McAuliffe, and M. J. Parrott, *J.C.S. Dalton*, 1976, 1642.
305 R. A. Armstrong and H. Taube, *Inorg. Chem.*, 1976, **15**, 1905.
306 U. Mazzi, G. de Paoli, P. di Bernardo, and L. Magon, *J. Inorg. Nuclear Chem.*, 1976, **38**, 721.
307 J. Burgess, J. Fawcett, R. D. Peacock, and D. Pickering, *J.C.S. Dalton*, 1976, 1363.
308 J. I. Bullock, F. W. Parrett, and N. J. Taylor, *Inorg. Chem.*, 1976, **15**, 2003.
309 J. R. Ebner, D. R. Tyler, and R. A. Walton, *Inorg. Chem.*, 1976, **15**, 833.
310 T. Ikariya and A. Yamamoto, *J. Organometallic Chem.*, 1976, **118**, 65.
311 S. D. Ittel, C. A. Tolman, A. D. English, and J. P. Jesson, *J. Amer. Chem. Soc.*, 1976, **98**, 6073.
312 G. L. Geoffroy and M. G. Bradley, *J.C.S. Chem. Comm.*, 1976, 20.
313 E. H. S. Wong and M. F. Hawthorne, *J.C.S. Chem. Comm.*, 1976, 257.
314 T. V. Ashworth and E. Singleton, *J.C.S. Chem. Comm.*, 1976, 705.
315 B. R. James, A. D. Rattray, and D. K. W. Wang, *J.C.S. Chem. Comm.*, 1976, 792.
316 D. J. Cole-Hamilton, R. J. Young, and G. Wilkinson, *J.C.S. Dalton*, 1976, 1995.
317 S. Komiya, T. Ito, M. Cowie, A. Yamamoto, and J. A. Ibers, *J. Amer. Chem. Soc.*, 1976, **98**, 3874.
318 D. G. Holah, A. N. Hughes, and B. C. Hui, *Canad. J. Chem.*, 1976, **54**, 320.
319 B. R. James and D. K. W. Wang, *Inorg. Chim. Acta*, 1976, **19**, L17.
320 P. B. Critchlow and S. D. Robinson, *J. Organometallic Chem.*, 1976, **114**, C46.
321 W. E. Stanclift and D. G. Hendricker, *J. Organometallic Chem.*, 1976, **107**, 341.

the axial–equatorial CO exchange follows an intramolecular path.[24] A combination of matrix-isolation and i.r. laser photochemistry has allowed the selective reaction of molecules of ^{13}CO-enriched [$Fe(CO)_4$] which differ only in the stereochemical distribution of the isotopic CO groups.[46] Tautomeric equilibria involving binuclear iron carbonyls in solution were considered under 'Metal–Metal Vibrations' in section 2; [$(C_5H_5)Ru(CO)_2]_2$ was also included in this study.[85] The newly-prepared anions *fac*- and *mer*-[$OsX_3(CO)_3]^-$ have been characterized by the expected number of ν(OsC) and ν(OsX) i.r. and Raman modes (X = Cl, Br, or I).[322]

The very elegant work [4] on $Cs_2Mg[Fe(CN)_6]$, and the more cursory reports on $[Fe(CN)_5(NO)]^{2+}$ salts [74] and $(NC)_5Fe(CN)Cr(NH_3)_{5-n}(H_2O)_n$ systems (n = 0 or 1),[80] have been referred to in section 2.

Reports of ν(MN) abound in last year's literature. The relevant compounds are in Table 7 [21, 103, 109, 115–117, 134] and as follows: (*a*) $[(H_3N)_5Ru(pyrazine)$-$Ru(NH_3)_5]^{n+}$ (n = 4, 5, or 6; see section 2);[53] (*b*) the poly(1-pyrazolyl)borate complexes listed in Table 2;[27] (*c*) $[Fe(oep)(imidazole)_2]ClO_4$ (209 cm^{-1}; see Table 4);[65] (*d*) (tpp)FeNFe(tpp) [(tpp) = tetraphenylporphinato; ν(FeNFe) associated with bands at 885, 910 cm^{-1}];[323] (*e*) protohemin complexes (ppFeCl) with imidazole (L) [ν(Fepp) = 381—384, ν(FeL) = 264—313 cm^{-1}];[324] (*f*) $FeCl_3$-(*N*-methylimidazole)$_3$ (140—245 cm^{-1});[325] (*g*) low-spin (298 K; 283 cm^{-1}) and high-spin (108 K; 213, 216 cm^{-1}) states of $[Fe(2\text{-Mephen})_3](ClO_4)_2$;[326] (*h*) [Fe-$(naphthy)_4]^{n+}$ salts [(naphthy) = 2-methyl- or 2,7-dimethyl-1,8-naphthyridine; n = 2 or 3; 235—245 cm^{-1}];[327] (*i*) $Os(NO)(OH)_3(H_2O)_2$;[328] and (*j*) OsX_3L_3 (X = Cl or I; L = py or methylpyrazine).[329] Refs. 21 and 27 are the most worthy contributions.

Assignments given [90] for ν(OsP), ν(OsAs), and ν(OsSb) in fourteen compounds of types $OsCl_2(SnCl_3)(PPh_3)_3$, $OsCl(SnCl_3)(AsPh_3)_4$, $\{OsCl(SbPh_3)_4\}HgCl_2$, [$OsCl_2$-$\{(Ph_2P)_2CH_2\}_2]HgCl_2$, *etc.* are all in the suspiciously narrow range 455—505 cm^{-1}. Similar situations obtain for analogous Ru complexes (410—460 cm^{-1}),[90] and for $OsCl_4L_2$, OsX_2L_4, and OsX_3L_3 isomers [X = Cl or Br; L = Ph_3P, Ph_3As, Ph_3Sb, $Pr^n{}_2PhP$, $MePh_2P$, or $CH_2(AsPh_2)_2$; 455—513 cm^{-1}].[330]

Most of the ν(FeO) modes in Table 7 [115–118, 125, 127–129, 134] are unsubstantiated; some are very dubious, but ref. 125 warrants full consideration. Values for ν(RuO) in $[Ru(Me_2SO)_6]^{3+}$ salts and $Ru(Me_2SO)_4Cl_2$ (475—485 cm^{-1}) are (apparently) not very sensitive to changes in oxidation state.[331] The layer compound [332] FeOOMe and [$OsX_4(acac)]^-$ salts [X = Cl, Br, or I; ν_{sym}(OsO) = 452—470; ν_{asym}(OsO) = 605—618 cm^{-1}] [333] have been examined in the i.r.

322 F. H. Johanssen and W. Preetz, *J. Organometallic Chem.*, 1976, **104**, 79.
323 D. A. Summerville and I. A. Cohen, *J. Amer. Chem. Soc.*, 1976, **98**, 1747.
324 T. Yoshimura, T. Ozaki, Y. Shintani, and H. Watanabe, *J. Inorg. Nuclear Chem.*, 1976, **38**, 1879.
325 S. A. Cotton, P. V. H. Pisani, and R. Stubbs, *Inorg. Nuclear Chem. Letters*, 1976, **12**, 695.
326 J. Fleisch, P. Gütlich, K. M. Hasselbach, and W. Müller, *Inorg. Chem.*, 1976, **15**, 958.
327 M. A. Cavanaugh, V. M. Cappo, C. J. Alexander, and M. L. Good, *Inorg. Chem.*, 1976, **15**, 2615.
328 N. M. Synitsyn and N. I. Klinova, *Russ. J. Inorg. Chem.*, 1975, **20**, 1734.
329 A. Scheffler and W. Preetz, *Z. Naturforsch.*, 1976, **31b**, 1099.
330 M. M. T. Khan, S. S. Ahamed, and R. A. Levenson, *J. Inorg. Nuclear Chem.*, 1976, **38**, 1135.
331 T. Bova and M. M. Singh, *J. Inorg. Nuclear Chem.*, 1976, **38**, 1815.
332 S. Kikkawa, F. Kanamaru, and M. Koizumi, *Inorg. Chem.*, 1976, **15**, 2195.
333 G. Schätzel and W. Preetz, *Z. Naturforsch.*, 1976, **31b**, 749.

A few papers considered ν(MOM) in bridged systems. For $Cs_4Ru_2OBr_2$, ν_{sym}(RuORu) at 243 cm^{-1} (Raman) indicates a linear bridged system (*cf.* Os and Cl analogues; see section 2).[146] The value proposed (650 cm^{-1}) for ν(OsOOs) in OsO_2,nH_2O [terminal ν_{sym}(Os=O) = 1070, ν_{asym}(Os=O) = 875, 945 cm^{-1}] is not unreasonable in comparison with literature data.[334] Likewise, ν(FeOFe) proposals for some binuclear chelates (830—840 cm^{-1}) are within commonly quoted ranges.[202, 335] An ingenious application of isotopes in making assignments has involved the use of $^{18}O^{16}O$ in oxyhemerythrin; the study showed that

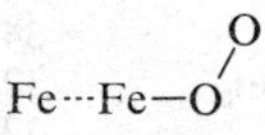

(5)

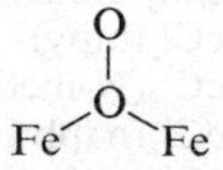

(6)

the preferred bonding of O_2 to the non-heme oxygen-carrier hemerythrin was as shown in (5) or (6), with the following assignments (calculated values):[38]

Isotopic species:	Fe—$^{16}O_2$	Fe—$^{16}O^{18}O$	Fe—$^{18}O^{16}O$	Fe—$^{18}O_2$
ν(OO)/cm^{-1}:	845	825	818	797
ν(FeO)/cm^{-1}:	504	501	485	482

The azido-analogue was also studied.[38]

A timely review has appeared on spin-transitions in hexaco-ordinate Fe^{II} complexes, which includes the i.r. spectral properties of $Fe(chel)_2(NCS)_2$ [(chel) = bipy or phen].[336] Such transitions have now been observed in $[Fe(2\text{-Mephen})_3](ClO_4)_2$ using Mössbauer and far-i.r. spectroscopy [ν(FeN) study at 298 and 108 K].[326] However, interest has converged on dithiocarbamato-complexes of Fe^{III}, and results from two laboratories have been published.[41, 142, 337] A variable-temperature study on $Fe(S_2CNR_2)_3$ and some analogues (including a Ru compound) [R = Me (high- and low-spin forms) or Et, or R_2N = pyrrolidyl][142] led to the views that (*a*) the vibrational partition function derived from the fitting of susceptibility data is *not* correct, (*b*) the energy difference between the $^6A_1$ and $^2T_2$ barycentres is probably *not* temperature dependent, and (*c*) both the spin states are very vibronic. For $Fe(S_2CNMe_2)_3$, ν(FeS) for the high-spin system was placed at 360 cm^{-1} whereas values for the low-spin analogue was given as 305, 334 cm^{-1}. These values are in the opposite relative order to that found for Fe^{II} systems,[326, 336] and possible origins of the different behaviour have been discussed.[142] However, it is noteworthy that in another i.r. investigation of $Fe(S_2CNR_2)_3$ systems, involving pressure-dependence studies and the use of $^{54,\,56}Fe$ isotopic shifts, values of ν(FeS) for high-spin–low-spin pairs (R_2N = PhEtN or $Pr^n{}_2N$) follow the 'normal' pattern of behaviour.[41] There will clearly be considerable energy expended in this area over the next year or so. The pressure-dependence work[41, 337] was able to show that the two spin states exist in rapid thermal equilibrium (rate in the 10^8—10^{12} s^{-1} range) instead of there being a single intermediate spin state as previously proposed.

[334] V. S. Khain, E. V. Fomina, and A. A. Volkov, *Russ. J. Inorg. Chem.*, 1976, **21**, 685.
[335] M. Branca, B. Pispisa, and C. Aurisicchio, *J.C.S. Dalton*, 1976, 1543.
[336] H. A. Goodwin, *Co-ordination Chem. Rev.*, 1976, **18**, 293.
[337] R. J. Butcher, J. R. Ferraro, and E. Sinn, *J.C.S. Chem. Comm.*, 1976, 911.

For some related (di)thiocarbamato compounds, assignments were not given[338] or are unreasonable.[339] In the case of $[RuCl_3(Me_2S)_3]_n$, ν(RuS) is at 315 cm^{-1} for both $n = 1$ or 2.[331] See also Table 7.[41, 134, 142, 143]

Compounds for which ν(MX) assignments have been given are in Tables 2, (ref. 26), 4 (refs. 26, 60, 65), 7 (refs. 21, 109, 115, 117, 118, 122, 123), and 9 (refs.

Table 9 *Further*[a] *references to* ν(MX) *assignments* (M = Fe, Ru, *or* Os)

Compound	*Ref.*
$FeCl_3$(dioxan)	340 [b]
$FeCl_2$(bipy)	341 [c]
$FeCl_3$(N-methylimidazole)$_3$	325
$FeCl_2$(naphthy)$_2$	327 [d]
FeCl(pp)(im)$_n$	324 [e]
$FeCl(O_2CCH_2Cl)_2$	127
RuF_8	342
$RuCl_2(Me_2SO)_4$ }	331 [f]
$[RuCl_3(Me_2S)_3]_n$ }	331 [f]
$RuCl_2(Ph_3P)_2$(CX)L	343 [g]
$RuCl_2$(alkene)(amine)$_2$	344 [h]
$[M_2OX_{10}]^{4-}$	146 [i]
$Cs[OsX_4(acac)]$	333 [j]
$(Bu^n{}_4N)[OsX_6]$	345 [k]
OsX_nL_{6-n}	330 [l]
$OsX_3(amine)_3$	329 [m]
$[OsX_3(CO)_3]^-$	322 [n]
$[OsCl_n(PMe_2Ph)_m$(thio-ate)]	346 [o]
$OsBr_nI_{3-n}(py)_2(mepyz)_2$	347 [p]
$OsCl_n(EPh_3)_m(SnCl_3)$ *etc.*	90

[a] See also Tables 2, 4, and 7. [b] Pentaco-ordinate structure proposed. [c] Orange isomer thought to be tetrahedrally co-ordinated, while rose-red isomer believed to be octahedral. [d] (naphthy) = 2-methyl- or 2,7-dimethyl-1,8-naphthyridine. [e] Protohemin and its complexes with imidazole derivatives; ν(FeCl) data. [f] $n = 1$ or 2. [g] X = O or S; L = $HCONMe_2$ or MeOH. [h] (alkene) = cyclo-octadiene or norbornadiene; (amine) = hexylamine or piperidine. [i] M = Ru or Os; X = Cl or Br. [j] X = Cl, Br or I. [k] X = Cl or Br; X-exchange observed on heating with CsY (Y = Cl or Br but ≠ X). [l] X = Cl or Br; $n = 2$, 3, or 4; L = Ph_3P, Ph_3As, *etc.* [m] X = Cl or I; (amine) = py or methylpyrazine. [n] *mer* and *fac* isomers; X = Cl, Br or I. [o] $n = 1$, 2, or 3; $m = 2$ or 3; (thiol-ate) = Ph_2PS_2, Me_2PS_2, Me_2NCS_2, or $EtOCS_2$ anions. [p] $n = 1$ or 2; (mepyz) = methylpyrazine. [q] $n = 1$ or 2; $m = 3$ or 4; E = P, As, or Sb.

90, 127, 146, 322, 324, 325, 327, 329—331, 333, 340—347. Many of the data were used for structural purposes,[115, 117, 118, 123, 322, 327, 330, 331, 340–343, 346, 347] such as overall geometry (RuF_8, cubic),[342] co-ordination number[118, 327, 340, 341] [sometimes involving identification of ν(MX) bridging modes[118, 327]], *cis* or *trans*

338 A. Malliaris and D. Niarchos, *Inorg. Chem.*, 1976, **15**, 1340.
339 H. Nakajima, T. Tanaka, H. Kobayashi, and I. Tsujikawa, *Inorg. Nuclear Chem. Letters*, 1976, **12**, 689.
340 Y. Maeda and Y. Takashima, *Inorg. Nuclear Chem. Letters*, 1976, **12**, 313.
341 H. Sato and T. Tominaga, *Bull. Chem. Soc. Japan*, 1976, **49**, 697.
342 C. Courtois and T. Kikindai, *Compt. rend.*, 1976, **283**, *C*, 679.
343 P. W. Armit, W. J. Sime, and T. A. Stephenson, *J.C.S. Dalton*, 1976, 2121.
344 C. Potvin, J. M. Manoli, G. Pannetier, R. Chevalier, and N. Platzer, *J. Organometallic Chem.*, 1976, **113**, 273.
345 W. Preetz and W. Hasenpusch, *Z. anorg. Chem.*, 1976, **420**, 97.
346 D. J. Cole-Hamilton and T. A. Stephenson, *J.C.S. Dalton*, 1976, 2396.
347 A. Scheffler and W. Preetz, *Z. Naturforsch.*, 1976, **31b**, 1347.

geometry,[330] *mer* or *fac* geometry,[322, 330, 331] or the nature of *trans* groups (*via trans* influence assertions).[343, 346]

For compounds containing metal–metal bonds, see Table 6.[59, 85, 86, 90]

9 Cobalt, Rhodium, and Iridium

About one-third of the references collected for this section relate to ν(MH) assignments of compounds of Co,[348–350] Rh,[351–358] and Ir.[356–375] There is little reason to question any of these assignments, and many were confirmed by deuteriation. The following selection should be of potential interest: (*a*) deduction of configuration from the number of ν(MH) i.r. bands in $(Ph_3P)_2IrH_2(CO)$-(σ-carb) [(σ-carb) = $(2\text{-}H\text{-}1,2\text{-}B_{10}C_2H_{10})^-$, $(7\text{-Me-}1,7\text{-}B_{10}C_2H_{10})^-$, *etc.*],[361] $[IrH_2(HPR_2)_n(CO)_{4-n}](BPh_4)$ [n = 3 or 4],[356] and $IrH_2(Ph_3E)(R_2NCS_2)$ (E = P or As; R_2N = various);[363] (*b*) determination of structure from the positions of ν(MH) modes and considering probable *trans* influences;[356, 361, 366, 370, 372] (*c*) the linear HRuNN skeleton in *trans*-$[RhH(N_2)(PhPBu^t{}_2)_2]$ (crystal structure determined);[354] and (*e*) formation of an active hydrogenation catalyst [red solution having ν(RuH) at 1944, 2260, 2269 cm^{-1}] on absorption of 2 molar equivalents of H_2 by $Rh(\eta^3\text{-}C_3H_5)\{P(OMe)_3\}_3$.[355]

348 H.-F. Klein and H. H. Karsch, *Chem. Ber.*, 1976, **109**, 1453.
349 T. Kariya and A. Yamamoto, *J. Organometallic Chem.*, 1976, **116**, 231.
350 J. Ellermann, J. F. Schindler, H. Behrens, and H. Schlenker, *J. Organometallic Chem.*, 1976, **108**, 239.
351 R. R. Schrock and J. A. Osborn, *J. Amer. Chem. Soc.*, 1976, **98**, 2134.
352 J. A. Tiethof, J. L. Peterson, and D. W. Meek, *Inorg. Chem.*, 1976, **15**, 1365.
353 Z. Nagy-Magos, B. Heil, and L. Markó, *Transition Metal Chem.*, 1976, **1**, 215.
354 P. R. Hoffman, T. Yoshida, T. Okano, S. Otsuka, and J. I. Ibers, *Inorg. Chem.*, 1976, **15**, 2462.
355 M. Bottrill and M. Green, *J. Organometallic Chem.*, 1976, **111**, C6.
356 P. Rigo and M. Bressan, *Inorg. Chem.*, 1976, **15**, 220.
357 C. W. Jung and M. F. Hawthorne, *J.C.S. Chem. Comm.*, 1976, 499.
358 D. A. Thompson and R. W. Rudolph, *J.C.S. Chem. Comm.*, 1976, 770.
359 Boon-Keng Teo, A. P. Ginsberg, and J. C. Calabrese, *J. Amer. Chem. Soc.*, 1976, **98**, 3027.
360 J. Müller, W. Holzinger, and H. Menig, *Angew. Chem. Internat. Edn.*, 1976, **15**, 702.
361 B. Longato, F. Morandini, and S. Bresadola, *Inorg. Chem.*, 1976, **15**, 650.
362 D. T. Laverty, M. A. McKervey, J. J. Rooney, and A. Stewart, *J.C.S. Chem. Comm.*, 1976, 193.
363 C. Preti and G. Tosi, *Z. anorg. Chem.*, 1976, **419**, 185.
364 L. Dahlenburg, D. Rehder, W. Reith, W. Stränz, and D. Thoennes, *Transition Metal Chem.*, 1976, **1**, 206.
365 T. L. Kelly and C. V. Senoff, *Canad. J. Chem.*, 1976, **54**, 1860.
366 L. Dahlenburg and R. Nast, *J. Organometallic Chem.*, 1976, **110**, 395.
367 J. F. van Baar, J. M. Klerks, P. Overbosch, D. J. Stufkens, and K. Vrieze, *J. Organometallic Chem.*, 1976, **112**, 95.
368 R. H. Crabtree, H. Felkin, G. E. Morris, T. J. King, and J. A. Richards, *J. Organometallic Chem.*, 1976, **113**, C7.
369 J. P. Fawcett and J. F. Harrod, *J. Organometallic Chem.*, 1976, **113**, 245.
370 M. Angoletta, L. Malatesta, P. L. Bellon, and G. Cagilo, *J. Organometallic Chem.*, 1976, **114**, 219.
371 U. Behrens and L. Dahlenburg, *J. Organometallic Chem.*, 1976, **116**, 103.
372 M. A. Bennett, R. N. Johnson, and I. B. Tomkins, *J. Organometallic Chem.*, 1976, **118**, 205.
373 S. Hietkamp, D. J. Stufkens, and K. Vrieze, *J. Organometallic Chem.*, 1976, **122**, 419.
374 C. J. Moulton and B. L. Shaw, *J.C.S. Dalton*, 1976, 1020.
375 H. D. Empsall, E. M. Hyde, E. Mentzer, B. L. Shaw, and M. F. Uttley, *J.C.S. Dalton*, 1976, 2069.

Assignments of ν(CoC) have been made for $Me_2Co(PMe_3)_3$ (470 cm^{-1}) [348] and a series of complexes $R^1{}_2Co(acac)(PR^2{}_3)_2$ (570—595 cm^{-1}; R^1, R^2 = various).[376, 377] The first NCA of $[RhX(CO)_2]_2$ (X = Cl or Br) has been claimed.[378] Data have been published on $(OC)_9Co_3C$—OCOMe (crystal structure determined) [379] and a series of $[(Me_3Si)_nPMe_{4-n}][Co(CO)_4]$.[380]

The very full study on $M_3[Co(CN)_6]_2,xH_2O$ salts (M = Mn or Cd; see Chapter 4) [5] and the less detailed work on $Fe[M(CN)_6]$ (M = Co, Rh, or Ir) [73] have been referred to in section 2.

When it comes to ν(CoN) assignments, it seems that the adage 'more means worse' is at least partly true. A high proportion of such data given in the appropriate references of Table 7 [21, 103–105, 108, 109, 111–113, 117, 118, 120, 126, 131, 133–135] and elsewhere [25, 27, 35, 61, 381–383] can be little more than guesses, and some (*e.g.* refs. 21 and 126) totally ignore the recent and substantiated literature. Nevertheless, some good science is being carried out in this area, and refs. 25, 27, and 104 may be particularly commended. Assignments of ν(RhN) modes [68, 104, 384] have also been proposed; the value of 426 cm^{-1} suggested [68] for $Rh_2(O_2CMe)_4(py)_2$ is suspiciously close to where one would expect an internal (py) mode to shift on co-ordination.

Photoreduction of $[Co(phen)_2(C_2O_4)]^+$ salts has been followed by i.r. spectroscopy.[385]

It is possible that ν(CoP) may be responsible for the i.r. band at 288—318 cm^{-1} in spectra of $CoX_2(PMe_3)_3$ (X = Me, Cl, Br, or I).[348] For $[Ir(PF_2O)_3\{PF_2(OC_3H_5)\}]^{2-}$, ν(IrP) has been quoted [386] at 380 cm^{-1}.

The comments made earlier in this section with regard to ν(CoN) assignments are generally equally applicable to ν(CoO) reports, although last year's output of papers (Table 7 [105, 111, 117–119, 121, 125, 126, 128, 129, 134–139] and other references [35, 381, 382, 387–389]) seems on balance to be broadly acceptable. Again, however, there are no obvious highlights but several apparent blunders. Values for ν(RhO) have been proposed for $Rh_2Cl_2(HCO_2)_2(phen)_2$ (397—445 cm^{-1}),[390] the single O-bonded ligand in $RhCl_3(Me_2SO)_3$ (491, shifting to 468 cm^{-1} on perdeuteriation),[23] and $Rh_2(OSCMe)_4$ (317 cm^{-1}).[391]

Metal–sulphur vibrations for these elements have been proposed for Co (Table 7 [41, 120, 131, 134, 141–143] and refs. 382 and 392), Rh,[23, 391] and Ir.[363] The most reliable would appear to be those for the $Co(S_2CNR_2)_3$ series,[41, 142] and $[Rh(Me_2SO)_2Cl_4]^-$ and $Rh(Me_2SO)_3Cl_3$ (deuteriation shifts studied).[23]

376 T. Ikariya and A. Yamamoto, *J. Organometallic Chem.*, 1976, **116**, 239.
377 T. Ikariya and A. Yamamoto, *J. Organometallic Chem.*, 1976, **120**, 257.
378 R. Prince and R. Poilblanc, *J. Chim. Phys. Phys.-Chem. Biol.*, 1975, **72**, 1087.
379 V. Bätzel and G. Schmid, *Chem. Ber.*, 1976, **109**, 3339.
380 H. Schäfer and A. G. MacDiarmid, *Inorg. Chem.*, 1976, **15**, 848.
381 H. H. Downs and D. N. Zimmerman, *J. Inorg. Nuclear Chem.*, 1976, **38**, 2177.
382 B. Anmian and R. J. Magee, *J. Inorg. Nuclear Chem.*, 1976, **38**, 2195.
383 A. Anagnostopoulos and D. Nicholls, *J. Inorg. Nuclear Chem.*, 1976, **38**, 1615.
384 J. Dehand and J. Jordanov, *Inorg. Chim. Acta*, 1976, **17**, 37.
385 H. Hennig, K. Hempel, M. Ackermann, and Ph. Thomas, *Z. anorg. Chem.*, 1976, **422**, 65.
386 J. Grosse and R. Schmutzler, *J C.S. Dalton*, 1976, 412.
387 J. Gopalakrishnan, G. Colsmann, and B. Reuter, *Z. anorg. Chem.*, 1976, **424**, 155.
388 B. Cornils, I. Förster, C. Krüger, and Y.-H. Tsay, *Transition Metal Chem.*, 1976, **1**, 151.
389 I.-P. Lorenz, W. Menzl, and W. Reuther, *Inorg. Nuclear Chem. Letters*, 1976, **12**, 141.
390 H. Pasternak and F. Pruchnik, *Inorg. Nuclear Chem. Letters*, 1976, **12**, 591.
391 I. B. Baranovskii, M. A. Golubnichaya, G. Ya. Mazo, V. I. Nefedov, Ya. V. Salyn', and R. N. Shchelokov, *Russ. J. Inorg. Chem.*, 1976, **21**, 591.

As would be anticipated, a vast array of ν(CoX),[21, 348, 381, 383, 392–396] ν(RhX),[23, 68, 372, 374, 397–406] and ν(IrX)[67, 366, 372–374, 405–410] data appeared in the literature last year (for examples, see Table 7). In many instances, there seems to have been little purpose in making the measurements, but illustrations of the uses of these ν(MX) data include: (*a*) identification of the species present in alkali metal halide melts containing $CoCl_2$;[101] (*b*) attempts to characterize pentaco-ordinate $\nu(Co^{II}X)$;[393] (*c*) determination of co-ordination geometry, *e.g.* Co(dehydrodithizone)$_4$Br$_2$,2H$_2$O is formulated as the tetrahedral cation [CoL$_2$Br$_2$],2L,2H$_2$O [ν(CoBr) = 226, 237 cm^{-1}];[392] (*d*) distinguishing between *cis* and *trans* isomers, *e.g.* [Rh{*o*-(Ph$_2$P)$_2$C$_6$H$_4$}$_2$Cl$_2$]Cl shows only one ν(CoCl) i.r. band so is probably *trans*, whereas the bromo-analogue exhibits two such bands therefore is *cis*;[397] (*e*) identification of bridging halide ligands, *e.g.* for [RhCl(CO)(PR$_3$)(CH$_3$)X]$_2$ compounds, observation of $\nu(RhCl)_t$ at *ca.* 320 and $\nu(RhCl)_b$ at *ca.* 270 cm^{-1} has led to a chlorine-bridged formulation.[403]

Table 6 includes compounds for which ν(CoM) and ν(RhM) data are available.[84–86, 91]

10 Nickel, Palladium, and Platinum

Of the large number of compounds for which ν(NiH),[411–413] ν(PdH),[374, 412, 413] and ν(PtH)[82, 374, 413–420] have been given, the following merit special mention:

392 A. C. Fabretti, G. Peyronel, and G. C. Franchini, *J. Co-ordination Chem.*, 1976, **6**, 47.
393 M. J. M. Campbell, R. Grzeskowiak, R. Thomas, and M. Goldstein, *Spectrochim. Acta*, 1976, **32A**, 553.
394 L. Y. Chia, S. A. Dias, and W. R. McWhinnie, *J. Inorg. Nuclear Chem.*, 1976, **38**, 1263.
395 J. Y. Gesland, M. Quilichini, and J. F. Scott, *Solid State Comm.*, 1976, **18**, 1236.
396 L. V. Konovalov and I. S. Maslennikova, *Russ. J. Inorg. Chem.*, 1975, **20**, 903.
397 W. Levason and C. A. McAuliffe, *Inorg. Chim. Acta*, 1976, **16**, 167.
398 M. P. Hancock, B. T. Heaton, and D. H. Vaughan, *J.C.S. Dalton*, 1976, 931.
399 H. C. Clark and K. J. Reimer, *Canad. J. Chem.*, 1976, **54**, 2077.
400 H. Pasternak, T. Glowiak, and F. Pruchnik, *Inorg. Chim. Acta*, 1976, **19**, 11.
401 H. L. M. van Gaal, F. L. A. van den Bekerom, and J. P. J. Verlaan, *J. Organometallic Chem.*, 1976, **114**, C35.
402 P. W. Lednor, W. Beck, and H. G. Fick, *J. Organometallic Chem.*, 1976, **113**, C19.
403 A. Mayanza, P. Kalck, and R. Poilblanc, *Compt. rend.*, 1976, **282**, *C*, 963.
404 C. A. McAuliffe, I. E. Niven, and R. V. Parish, *Inorg. Nuclear Chem. Letters*, 1976, **12**, 855.
405 P. W. Clark, *J. Organometallic Chem.*, 1976, **110**, C13.
406 M. A. Bennett and P. W. Clark, *J. Organometallic Chem.*, 1976, **110**, 367.
407 J. E. Ferguson and Tan Wee Tee, *Inorg. Nuclear Chem. Letters*, 1976, **12**, 297.
408 D. M. Blake and J. R. Reynolds, *J. Organometallic Chem.*, 1976, **113**, 391.
409 F. Faraone and S. Sergi, *J. Organometallic Chem.*, 1976, **112**, 201.
410 S. J. Markham, Y. L. Chung, G. D. Branum, and D. M. Blake, *J. Organometallic Chem.*, 1976, **107**, 121.
411 S. B. Miller and M. F. Hawthorne, *J.C.S. Chem. Comm.*, 1976, 786.
412 H. Imoto, H. Moriyama, T. Saito, and Y. Sasaki, *J. Organometallic Chem.*, 1976, **120**, 453.
413 R. A. Schunn, *Inorg. Chem.*, 1976, **15**, 208.
414 M. Green, J. A. K. Howard, J. Proud, J. L. Spencer, F. G. A. Stone, and C. A. Tsipis, *J.C.S. Chem. Comm.*, 1976, 671.
415 H. Kurosawa, *Inorg. Chem.*, 1976, **15**, 120.
416 B. Longato, F. Morandini, and S. Bresadola, *J. Organometallic Chem.*, 1976, **121**, 113.
417 R. Ros, R. Bataillard, and R. Roulet, *J. Organometallic Chem.*, 1976, **118**, C53.
418 R. Nast and J. Moritz, *J. Organometallic Chem.*, 1976, **117**, 81.
419 C. J. Moulton and B. L. Shaw, *J.C.S. Chem. Comm.*, 1976, 365.
420 M. Fishwick, H. Nöth, W. Petz, and M. G. H. Wallbridge, *Inorg. Chem.*, 1976, **15**, 490.
421 B. Neruda and J. Lörberth, *J. Organometallic Chem.*, 1976, **111**, 241.
422 G. C. Stocco, F. S. Gattuso, N. Bertazzi, and L. Pellerito, *J. Co-ordination Chem.*, 1976, **5**, 55.

(*a*) $H_2Pt\{o\text{-}(Bu^t{}_2PCH_2)_2C_6H_4\}$, the first *cis*-dihydride of Pt^{II};[419] (*b*) $[Et_3SiPt(\mu\text{-}H)(PR_3)]_2$ and $H(R_3P)Pt(\mu\text{-}H)_2Pt(PR_3)H$, both Pt—Pt bonded ($R = C_6H_{11}$) but containing also H-bridges [$\nu(PtH)_b$ = *ca.* 1600 and 1550 cm^{-1}, respectively];[414] (*c*) the 1,2- or 1,7-dicarba-closo-dodecaborane complexes of Ni[411] and Pt;[416] and (*d*) *trans*-$[PtH(C_nH_{2n}CN)(PPh_3)_2]$, for which $\nu(PtH)$ is significantly higher (by 77 cm^{-1}) for $n = 1$ than for $n = 3$, indicating that the cyanopropyl group has a stronger *trans*-influence.[417]

A fair measure of consistency may be recorded for $\nu(PtC)$ assignments in a diversity of studies: *cis*-$[Me_2Pt\{P(OMe)_3\}_2]$ (522—560 cm^{-1}),[421] $[Me_3PtX_3]^{2-}$ and $[Me_3Pt(bipy)(RNC)]^+$ salts (569—597 cm^{-1}; X = Cl or Br; $R = C_6H_{11}$ or *o*-tolyl; *fac* geometry proposed in each case),[422] $Pt_2Me_4(O_2CR)_2(4\text{-}MeC_5H_4N)_2$ (573—589 cm^{-1}; $R = CF_3$ or Pr^i),[423] $(Me_2PPh_2)_2\overline{PtC_3H_6CH_2}$ and I(Me)-$\overline{PtC_3H_6CH_2}$ (528—546 cm^{-1}),[424] the phenacylide complexes *cis*- and/or *trans*-$[PtX_2(PR_3)(ECHCOPh)]$ (510—544 cm^{-1}; X = Cl, Br, or I; $(PR_3) = PhPMe_2$ or PMe_3; E = $4\text{-}MeC_5H_4N^+$, $MePh_2P^+$, Ph_3As^+, or Me_2S^+).[425] Similar data are also available for some Pd analogues,[425] $Pd(CH_2CN)_2$,[426] alkene complexes of Pt,[427, 428] and cyano-complexes of Pd[429] or Pt.[81] Fragmentary data have been given for $(C_5H_5)_2(CO)_2WSbMe\{Ni(CO)_3\}$.[430]

The position of $\nu(PtC)$ varies with changing X in *trans*-$[PtX_2(PhPMe_2)$-(phenacylide)] in an order which is opposite to that expected on the basis of the *trans* influence.[425] For $K[(p\text{-}RC_6H_4CH{=}CH_2)PtCl_3]$, $\nu(PtC)$ correlates with $\sigma_p{}^+$ values of R (R = NMe_2, OEt, Me, H, NO_2, *etc.*).[427] On the basis of $\nu(PtC)$ values in *trans*-$[PtCl_2(alkene)(py)]$, vinyl acetate and vinyl methacrylate are considered to show the strongest interaction with Pt of all the alkenes studied, and this is consistent with orbital energy levels derived from photoelectron spectra.[428]

Last year saw assignment of very many spectral bands to $\nu(MN)$ modes involving Ni (Table 7 and refs. 22, 25, 27, 29, 31, 431—433), Pd (refs. 55, 108, 130, 131, 141, 384, 434—437), and Pt (refs. 55, 56, 72, 131, 141, 384, 428, 434—439). The range of N-donor ligands studied is vast, including ammonia,[55, 431]

423 J. Kuyper and K. Vrieze, *Transition Metal Chem.*, 1976, **1**, 208.
424 M. P. Brown, A. Hollings, K. J. Houston, R. J. Puddephatt, and M. Rashidi, *J.C.S. Dalton*, 1976, 786.
425 H. Koezuka, G.-E. Matsubayashi, and T. Tanaka, *Inorg. Chem.*, 1976, **15**, 417.
426 G. Oehme, K.-C. Röber, and H. Pracejus, *J. Organometallic Chem.*, 1976, **105**, 127.
427 D. G. Cooper and J. Powell, *Inorg. Chem.*, 1976, **15**, 1959.
428 M. A. M. Meester, H. van Dam, D. J. Stufkens, and A. Oskam, *Inorg. Chim. Acta*, 1976, **20**, 155.
429 A. W. Verstuyft, L. W. Cary, and J. H. Nelson, *Inorg. Chem.*, 1976, **15**, 3161.
430 P. Panster and W. Malisch, *Chem. Ber.*, 1976, **109**, 3842.
431 J. M. Janik, J. A. Janik, G. Pytasz, and J. Sokolowski, *Rap.-Inst. Fiz. Jad.* (*Krakow*), 1975, 891/PS.
432 J. R. Ferraro, L. J. Basile, L. R. Garcia-Ineguez, P. Paoletti, and L. Fabbrizzi, *Inorg. Chem.*, 1976, **15**, 2342.
433 R. H. Nuttall and D. M. Stalker, *Inorg. Nuclear Chem. Letters*, 1976, **12**, 639.
434 M. Biddau, G. Devoto, M. Massacesi, and G. Ponticelli, *Transition Metal Chem.*, 1976, **1**, 99.
435 J. Dehand, J. Jordanov, and M. Pfeffer, *J.C.S. Dalton*, 1976, 1553.
436 R. Pinna, G. Ponticelli, C. Preti, and G. Tosi, *Transition Metal Chem.*, 1976, **1**, 173.
437 K. Suzuki, J. Ooyama, and M. Sakurai, *Bull. Chem. Soc. Japan*, 1976, **49**, 464.
438 S. J. Anderson, P. L. Goggin, and R. J. Goodfellow, *J.C.S. Dalton*, 1976, 1959.
439 Ton Quang Ly, *Zhur. Priklad. Specktroskopii*, 1976, **24**, 720.

aliphatic amines,[22, 56, 72, 432] aromatic amines,[25] pyridine,[21, 428] other hetero-aromatic amino ligands,[27, 118, 384, 434–436] amino-acid anions,[29, 31, 32, 433] the nitro anion,[437, 439] the isothiocyanato anion,[25, 102, 438] and other N-donor anions.[437] Of all these studies, the most significant would seem to be as follows: (*a*) isotopic shift measurements to help identify ν(MN) [21, 25, 27, 29, 31, 32] (see section 2); (*b*) structural assignments and NCA for some $[Pt(NO_2)_nX_{6-n}]^{2-}$ salts (n = 2, 3, 5, or 6; X = Cl or Br);[439] (*c*) an investigation into the factors governing the mode of co-ordination of NCS^- in $[Pt(NCS)_2L_2]$ systems (L = NMe_3, PMe_2Et, PEt_3, $AsMe_3$, SMe_2, $SeMe_2$, *etc.*) and the use of vibrational spectroscopy in deducing the mode of co-ordination of NCS^- and the stereochemistry of complexes containing NCS^- or NCO^- ligands;[438] (*d*) a consideration of lattice as well as internal modes of several $[Ni(NH_3)_6]^{2+}$ salts (and perdeuteriates);[431] and (*e*) the effects of temperature and pressure on the spectra of $M(Et_2NC_2H_4$-$NH_2)X_2$ complexes (M = Ni or Cu; see section 11).[432]

For ν(NiP) assignments, see Table 7.[102] Values have been quoted for ν(PtP) in *trans*-$[PtX_2(PR_3)_2]$ (409—427 cm^{-1}; X = Cl or Br; PR_3 = PBu^i_3 or other bulky phosphine),[440] $(Ph_3P)_2Pt(BPh_2)Cl$ and $[(PhEt_2P)_3Pt(BPh_2)]Cl$ (420—470 cm^{-1}),[420] *cis*-$[Me_2Pt\{P(OMe)_3\}_2]$ (ν_{sym} at *ca.* 300, ν_{asym} at *ca.* 445 cm^{-1}),[421] *cis*- and *trans*-$[Pt(PEt_3)_2(R)Br]$ (410—440 cm^{-1}; R = Ph or substituted derivative),[441] and $[Pt(PF_2O)_4]^{2-}$ and *cis*-$[PtCl_2(PF_2O)_2]^{2-}$ (365—430 cm^{-1}).[386] In contrast, a detailed and convincing study of $(Bu^n_4N)[PtX_3(PY_3)]$ (X = Cl, Br, or I; Y = F, Cl, or Br) placed ν(PtP) at *ca.* 265 (Y = F), *ca.* 177 (Y = Cl), or *ca.* 148 cm^{-1} (Y = Br).[442] Some ν(PdP) data are in ref. 386.

Table 7 contains many references to ν(NiO) (refs. 32, 105, 111, 117, 119, 121, 126, 128, 129, 134—139) and ν(PdO) (refs. 125, 136, 137) assignments, and others have been mentioned earlier in this section (Pd and/or Pt).[434, 436, 437] Results have also been presented for NiO,[443] $Cs_2[Pt(OH)_6]$,[444] *cis*- and *trans*-$[PhCOCHCOMe]_2Pd$,[445] $[Ni\{2\text{-}(EtSO)C_5H_4NO\}_3](ClO_4)_2,4H_2O$,[446] Pt^{II} dipeptide complexes,[447] and $M(phenoxathiin)_2Cl_2$ (M = Pd or Pt).[448] Many of these 'results' leave much to be desired; the most illuminating data on ν(NiO), for example, pinpoint the dangers in making group-frequency assignments in chelate complexes.[29, 31, 32]

Despite the large number of ν(MS) data published for Ni,[120, 131, 134, 143, 449–451] Pd,[40, 130, 131, 141, 384, 436, 448, 451–453] and Pt,[131, 141, 438, 448, 451, 453] and the occasional

440 P. G. Leviston and M. G. H. Wallbridge, *J. Organometallic Chem.*, 1976, **110**, 271.
441 R. Romeo, D. Minniti, and M. Trozzi, *Inorg. Chem.*, 1976, **15**, 1134.
442 C. Crocker, P. L. Goggin, and R. J. Goodfellow, *J.C.S. Dalton*, 1976, 2494.
443 D. R. Smith and E. V. Loewenstein, *Appl. Optics*, 1976, **15**, 859.
444 B. N. Ivanov-Emin, L. D. Borzova, N. U. Venskovskii, A. I. Ezhov, and V. V. Kornilov, *Izvest. Vysshikh Uchebn. Zavedenii, Khim. Khim. Tekhnol.*, 1976, **19**, 552.
445 S. Okeya, H. Asai, S. Ooi, K. Matsumoto, S. Kawaguchi, and H. Kuroya, *Inorg. Nuclear Chem. Letters*, 1976, **12**, 677.
446 E. W. Benson, R. E. Kohrman, and D. X. West, *Inorg. Nuclear Chem. Letters*, 1976, **12**, 721.
447 L. E. Nance and H. G. Frye, *J. Inorg. Nuclear Chem.*, 1976, **38**, 637.
448 Whei-Lu Kwik and Sau-Fun Tan, *J.C.S. Dalton*, 1976, 1072.
449 A. Rignedoli, G. Peyronel, and W. Malavasi, *J. Inorg. Nuclear Chem.*, 1976, **38**, 1963.
450 A. Pignedoli and G. Peyronel, *Z. anorg. Chem.*, 1976, **427**, 70.
451 R. D. Bereman and D. Nalewajek, *Inorg. Chem.*, 1976, **15**, 2981.
452 S. S. Sandhu and T. Singh, *Transition Metal Chem.*, 1976, **1**, 155.
453 G. Marcotrigiano, R. Battistuzzi, and G. Peyronel, *Spectrochim. Acta*, 1976, **32A**, 291.

NCA (on *trans*-[Pt(Me_2SO)$_2$$X_2$]; X = Cl or Br),[454] little progress seems to have been made in our understanding of the spectra concerned. The quality of the reports varies enormously, from the use of metal isotopes to identify ν(PdS) and δ(PdSC) in (Et_4N)$_2$[Pd(SCN)$_4$],[40] through careful and thoughtful studies on thiocyanato complexes of type [Pt(SCN)$_2$($SbMe_3$)$_2$],[438] to quite negligent (and poorly refereed) accounts of M(thianthrene)Cl_2 and related complexes (M = Pd or Pt) where 4—5 i.r. bands in the 290—410 cm^{-1} range are said without evidence to be ν(MS).[448] Assignments of ν(MSe) modes have been made.[145]

The impracticability was commented upon in Volume 9 of listing all the compounds relevant to this section for which ν(MX) modes were assigned. The growth in this activity is now making it difficult even to ensure that all the references are collected, and it is possible that this will be the last occasion on which space will allow the references themselves even to be listed, but ν(MX) values have been published for compounds of Ni [see Table 7 (refs. 21, 101, 102, 106, 107, 111—113, 117, 118, 121—123, 126) and refs. 374 and 455], Pd (refs. 55, 123, 374, 386, 397, 402, 425, 434—436, 452, 453, 456—470), and Pt (refs. 55, 56, 72, 374, 386, 397, 402, 418—420, 422, 425, 427, 434—436, 440, 442, 447, 453, 465—470, 471—482). It is difficult with this mountain of data to detect the significant (or the erroneous), or to note trends and developments. Indeed, the general pattern seems to be essentially unchanged compared with last year, and applications may be classified as: (*a*) spectroscopic, *i.e.* developing understanding of spectra of characterized compounds (sadly, there were few such studies);[55, 56, 442] (*b*) determination of solid-state structure, *e.g.* tetrahedral or octahedral co-ordination of nickel(II);[111, 112, 117] (*c*) identification of bridging[107, 113, 118, 121, 452, 462] and/or terminal[113, 118, 464, 552] halide ligands; (*d*) determination of *cis*,[397, 459, 463, 466, 467, 473] *trans*,[397, 460, 461, 466, 468] or *fac*[422, 472] isomeric structure; (*e*) postulation of structure in terms of nature of *trans* ligand or determination of relative *trans* influence.[425, 427, 458, 465, 479]

454 M. Tranquille and M. T. Forel, *Canad. J. Spectroscopy*, 1976, **21**, 11.
455 W. Levason, C. A. McAuliffe, and S. G. Murray, *J.C.S. Dalton*, 1976, 2321.
456 A. W. Verstuyft, D. A. Redfield, L. W. Cary, and J. H. Nelson, *Inorg. Chem.*, 1976, **15**, 1128.
457 M. Nonoyama, *Inorg. Nuclear Chem. Letters*, 1976, **12**, 709.
458 J. Dehand, J. Fischer, M. Pfeffer, A. Mitschler, and M. Zinsius, *Inorg. Chem.*, 1976, **15**, 2675.
459 W. Levason, C. A. McAuliffe, and S. G. Murray, *Inorg. Nuclear Chem. Letters*, 1976, **12**, 849.
460 P. Bravo, G. Fronza, and C. Ticozzi, *J. Organometallic Chem.*, 1976, **111**, 361.
461 W. Levason, C. A. McAuliffe, and S. G. Murray, *J. Organometallic Chem.*, 1976, **110**, C25.
462 B. Crociani, M. Nicolini, and R. L. Richards, *J. Organometallic Chem.*, 1976, **104**, 259.
463 Yu. N. Kukushkin, R. A. Vlasova, E. A. Kondratov, and V. G. Pogareva, *Russ. J. Inorg. Chem.*, 1975, **20**, 1749.
464 R. Weissgerber and K. E. Schwarzhans, *Z. Naturforsch.*, 1976, **31b**, 208.
465 D. R. Coulson and L. P. Seiwell, *Inorg. Chem.*, 1976, **15**, 2563.
466 J. Reedijk and J. K. de Ridder, *Inorg. Nuclear Chem. Letters*, 1976, **12**, 585.
467 C. A. McAuliffe and S. G. Murray, *Inorg. Nuclear Chem. Letters*, 1976, **12**, 897.
468 A. Pryde, B. L. Shaw, and B. Weeks, *J.C.S. Dalton*, 1976, 322.
469 R. J. Cross, T. H. Green, and R. Keat, *J.C.S. Dalton*, 1976, 382.
470 R. Huis and C. Masters, *J.C.S. Dalton*, 1976, 1796.
471 A. Pidcock and M. Safari, *Inorg. Nuclear Chem. Letters*, 1976, **12**, 925.
472 W. Preetz and H. Kühl, *Z. anorg. Chem.*, 1976, **425**, 97.
473 A. J. Carty, *Inorg. Chem.*, 1976, **15**, 1956.
474 H. Kurosawa and G. Yoshida, *J. Organometallic Chem.*, 1976, **120**, 297.
475 S. E. Earnest and D. B. Brown, *J. Organometallic Chem.*, 1976, **120**, 461.
476 W. Levason, C. A. McAuliffe, and S. G. Murray, *J.C.S. Dalton*, 1976, 269.
477 C. Eaborn, N. Farrell, and A. Pidcock, *J.C.S. Dalton*, 1976, 289.
478 T. G. Appleton, M. A. Bennett, and I. B. Tomkins, *J.C.S. Dalton*, 1976, 439.
479 J. Chatt, R. L. Richards, and G. H. D. Royston, *J.C.S. Dalton*, 1976, 599.

11 Copper, Silver, and Gold

Compounds $ArAuCl_2$ (Ar = various) are believed to be halogen-bridged dimers, with ν(AuC) at *ca.* 450 and ν(AuCl) at *ca.* 320—380 cm^{-1} (bridged and terminal).[483] Values of ν(AuC) for the new solids $[Li(L)][Me_4Au^{III}]$ and $[Li(L)][Me_2Au^{I}]$ (490—530 cm^{-1}; L = *NNN′N″N″*-pentamethyldiethylenetriamine) are comparable with those previously found for Et_2O solutions of corresponding (uncoordinated) Li^+ salts (*cf.* last year's Report),[484] and similar to those for $[Me_2AuL_2]^+$ (L = Ph_3P, PMe_2Ph, $SbPh_3$, *etc.*).[485]

$Ag(CO)_3$, formed by co-condensation reactions, has been characterized in low temperature matrices and has been assigned a D_{3h} planar structure (supported by e.s.r. data).[47] On warming, $Ag_2(CO)_6$ is formed which is similar to $Cu_2(CO)_6$; data are also presented for (probably) Ag(CO) and $Ag(CO)_2$ (thought to be linear).[47] The formation of 'AgCN' and $[Ag(CN)_n]^{(n-1)-}$ (n = 2 or 3) in liquid NH_3 solutions of Ag^I/CN^- mixtures [486] is discussed in Chapter 4. Solution i.r. and Raman spectra for $MAu(CN)_2$ (M = Na or K) have been assigned on the basis of $D_{\infty h}$ symmetry with some breakdown in the selection rules, *e.g.* ν_1 appeared weakly in the Raman and ν_3 and ν_6 in the i.r.[486a]

		I.r.	*Raman*			*I.r.*	*Raman*
ν_1	ν(CN)	2158	2161	ν_5	δ(AuCN)	—	301
ν_2	ν(AuC)	—	445	ν_6	δ(AuCN)	410	411
ν_3	ν(CN)	2142	2141	ν_7	δ(CAuC)	125	—
ν_4	ν(AuC)	426	—		(data in cm^{-1})		

The application of $^{63,65}Cu$ and/or $^{14,15}N$ isotopic shift data in identification of ν(CuN) has been discussed in section 2 (Tables 2 and 3),[6, 25, 27, 29, 30, 32, 42] as has the use of C_5D_5N.[21] The only other notable contribution to the literature on ν(CuN) vibrations was the NCA performed on $[Cu(NH_3)_4]^{2+}$ using (*a*) C_{4h} symmetry for the whole molecule and (*b*) D_{4h} symmetry for a point mass (NH_3) model; the latter approach appeared to be a reasonable approximation for the skeletal modes.[6] Most of the assignments of ν(CuN) referred to in Table 7 cannot be regarded as confirmed (some, to this Reviewer, seem likely to be wrong), but the data serve to illustrate the range of compounds being studied. Further data are to be found in refs. 52, 81, and 487—489.

The skeletal stretching modes of $[Au(NH_3)_4]^{3+}$ (a_{1g}, 570; e_u, 556; b_{1g}, 545 cm^{-1}) [490] are at slightly higher energies than analogous modes in $[Pt(NH_3)_4]^{2+}$.

480 R. N. Haszeldine, R. V. Parish, and D. W. Robbins, *J.C.S. Dalton*, 1976, 2355.

481 A. I. Stetsenko, T. G. Abzaeva, N. D. Mitkinova, V. G. Pogoreva, L. V. Konovalov, and A. V. Barsukov, *Russ. J. Inorg. Chem.*, 1976, **21**, 97.

482 Yu. N. Kukushkin, V. F. Budanova, G. N. Sedova, V. G. Pogareva, Yu. V. Fadeev, and M. S. Soboleva, *Russ. J. Inorg. Chem.*, 1976, **21**, 689.

483 P. W. J. de Graaf, J. Boersma, and G. J. M. van der Kerk, *J. Organometallic Chem.*, 1976, **105**, 399.

484 G. W. Rice and R. S. Tobias, *Inorg. Chem.*, 1976, **15**, 489.

485 P. L. Kuch and R. S. Tobias, *J. Organometallic Chem.*, 1976, **122**, 429.

486 P. Gans, J. B. Gill, and M. Griffin, *J.C.S. Chem. Comm.*, 1976, 169.

486a B. M. Chadwick and S. G. Frankiss, *J. Mol. Structure*, 1976, **31**, 1.

487 M. Quastlerová-Hvastijová, J. Kohout, J. Gazo, and F. Valach, *J. Inorg. Nuclear Chem.*, 1976, **38**, 705.

488 T. G. Miller and L. J. Theriot, *J. Inorg. Nuclear Chem.*, 1976, **38**, 695.

489 G. W. Watt and W. J. Wells, *J. Inorg. Nuclear Chem.*, 1976, **38**, 921.

490 M. Weishaupt and J. Strähle, *Z. Naturforsch.*, 1976, **31b**, 554.

The position regarding ν(CuO) assignments is rather similar to that for ν(CuN). The most significant studies were those in which the authors, *inter alia*, set out to establish the assignments, *viz*. the single-crystal study of $CuCl_2(Me_2SO)_2$ (see Chapter 4)[13] and the investigations using isotopic shifts (Tables 2 and 3).[29, 30, 32, 42] An interesting application has been study of the effects of pressure and temperature on the electronic, mid-i.r., and far-i.r. spectra of $Cu(Et_2NCH_2$-$CH_2NH_2)_2X_2$ complexes (X = NO_3, ClO_4, BF_4, Br, or I; Ni analogues also examined) in order to gain insight into the thermochromic mechanism of these systems.[432] The results indicate that axial interaction occurs on heating but not on increasing the pressure; for example, when the nitrate is heated to 105 °C, new i.r. bands appear at 272 and 304 cm^{-1}, possibly associated with ν(CuO). Table 7 and refs. 52, 491, and 492 relate to other ν(CuO) assignments.

A sprinkling of ν(CuS) assignments appeared last year [Table 7 (refs. 120, 130, 131, 134, 141, 143) and refs. 42, 62, 69, 493, and 494], but few warrant discussion. The $^{63, 65}Cu$ shifts on the 264 cm^{-1} band in Cu(8-quinolinthiolate)$_2$ is only 0.8 cm^{-1}, yet a ν_{asym}(CuS) assignment has been proposed.[42] A study of 'blue' copper protein sites[62] is notably similar to that described in last year's Report (p. 314). Although a disordered Cu^{III},Au^{III} arrangement (about a central Cu^{II} atom) has been proposed for $[Cu_2Au(S_2CNBu^n{}_2)_6][Hg_2Br_6]$ (crystal structure determined), discrete assignments were proposed for the different ν(MS) modes (M = Cu^{III}, 398; Cu^{II}, 360; Au^{III}, 379 cm^{-1});[69] these must be heavily mixed.

I.r. data are also available for $AgAsS_2$ and $AgSbS_2$ (Ag—S bonds deduced as being present),[495] $\{(Ph_3P)Ag\}_2(MS_2XY)$ [ν(AgS) = 257—264 cm^{-1}; M = Mo or W; X = O or S],[273] and $[(R_2NCSCSNR_2)AuCl_2]^+$ and $[(R_2NCSCSNR_2)_2Au]^{3+}$ perchlorates [ν(AuS) = 354—365 and 381—388 cm^{-1}, indicating *cis* geometry; R = Me or Et].[496] See also Table 7.[130, 131]

Rather curiously, halogenocomplexes of Cu^{II} have been subject to renewed interest. A fairly comprehensive and comparative survey has appeared,[497] and a more restricted review has been published of variations in ν(CuCl) for $[CuCl_4]^{2-}$ salts with changing anion symmetry.[498] While the account of the spectra of $(R_2N{=}C{-}S{-}S{-}C{=}NR_2)^{2+}$ (the two C atoms also bridged by S) salts of $[CuBr_4]^{2-}$ (R = Me, Et, or Pr^n) or $[Cu_2Br_6]^{2-}$ anions (R = Bu^n or 1-pentyl) is brief,[499] the very detailed studies on piperizinium dication,[500] morpholinium or piperidinium,[501] and $(PhC_2H_4NH_2Me)^+$ or $(PhCHCMe_2NH_3)^+$ salts[489] of $[CuX_3]^-$, $[Cu_2Cl_6]^{2-}$, or $[CuX_nY_{4-n}]^{2-}$ anions

491 M. A. S. Goher and M. Dratovsky, *Naturwiss.*, 1976, **63**, 89.
492 H. W. Richardson, J. R. Wasson, W. E. Hatfield, E. V. Brown, and A. C. Plasz, *Inorg. Chem.*, 1976, **15**, 2916.
493 A. Pignedoli and G. Peyronel, *Spectrochim. Acta*, 1976, **32A**, 1739.
494 R. Battistuzzi, G. Peyronel, and G. Debbi, *Z. anorg. Chem.*, 1976, **427**, 274.
495 I. I. Golovach, V. S. Gevasimenko, V. Yu. Slivka, N. I. Dovgoshei, M. I. Golovei, and A. V. Bogdanova, *Izvest. Vysshikh Uchebn. Zavedenii, Fiz.*, 1976, **19**, 39.
496 A. C. Fabretti, G. C. Franchini, G. C. Pellacani, and G. Peyronel, *Inorg. Nuclear Chem. Letters*, 1976, **12**, 163.
497 D. W. Smith, *Co-ordination Chem. Rev.*, 1976, **21**, 93.
498 J. R. Ferraro, *J. Co-ordination Chem.*, 1976, **5**, 101.
499 J. Willemse, J. A. Cras, and P. J. H. A. M. van de Leemput, *Inorg. Nuclear Chem. Letters*, 1976, **12**, 255.
500 G. Marcotrigiano, L. Menabue, and G. C. Pellacani, *Inorg. Chem.*, 1976, **15**, 2333.
501 G. Marcotrigiano, L. Menabue, and G. C. Pellacani, *J. Co-ordination Chem.*, 1976, **5**, 189.

(X or Y = Cl or Br; n = 1—4) already make the reviews [497, 498] somewhat dated (see also ref. 101). It is apparent from these investigations that the cation is of major importance in determining the type and structure obtained for such salts, but it must still be emphasized that considerable restraint must be exercised when using vibrational data alone, and single-crystal *X*-ray studies [489] are desirable.

Raman spectra of superionic conductors AgI and $RbAg_4I_5$ have been measured,[502] while the β-phase of $Ag_2[HgI_4]$ has Raman bands at 80 and 205 cm^{-1} (merging to *ca.* 84 cm^{-1} in the α-phase) attributed to ν(AgI).[503]

Several papers report bands as due to copper–halogen bridging modes. Data for Cu^{II} systems are in refs. 22, 44, and 118, but more attention has apparently been given to Cu^{I} complexes.[493, 494, 504–507] A particularly interesting study on (Ph_3E) complexes of MX (M = Cu or Ag, X = Cl or Br, E = P or As) concluded that (*a*) $(Ph_3E)_3MX$ were monomeric, (*b*) $(Ph_3E)_4M_2X_2$ and $(Ph_3E)_3M_2X_2$ were X-bridged dimers, and (*c*) far-i.r. spectroscopy could be used to distinguish between cubane and chair structures (known from *X*-ray studies) for $(Ph_3E)_4M_4X_4$ clusters.[504] On the other hand, Cu(2,4-dithiobiuret)X are described as dimers,[493] whereas Cu(*N*-ethylthiourea)X are thought to be trimeric [494] (X = Cl, Br, or I). Bridging $Cu^{I}X$ groups are also believed to be present (from i.r. evidence) in Cu{poly(vinyldiphenylphosphine)}$_9Cl_4$,[505] $(R_2Te_2)CuCl$ (R = Et or Ph),[506] and in some thio-ether complexes of CuBr.[507]

AuF_5 shows a broad Raman band at 370—450 cm^{-1} which may result from AuFAu bridges.[508] Reference to $(PhAuCl_2)_2$ has already been made.[483]

Other references to $\nu(Cu^{II}X)$ vibrations are in Table 7 (refs. 21, 101, 106, 111, 112, 114, 117, 118, 120, 123, 126) or have otherwise been mentioned.[488, 492, 507] Studies of ν(AuCl) have led to *cis* formulations for $[(R_2NCSCSNR_2)AuCl_2]ClO_4$ (R = Me or Et),[496] and to deduction of a *trans* arrangement for $Cl_2(C_6F_5)Au(Ph_2PC_2H_4PPh_2)Au(C_6F_5)Cl_2$.[509]

12 Zinc, Cadmium, and Mercury

Complete assignment of the fundamentals of CH_3HgX and CD_3HgX (X = Cl, Br, or I) in solution has led to an acceptable force field; large solid-state effects were discussed.[1] The Raman spectra of a large number of other CH_3HgX compounds have been considered, but it appears that only for X = $OSeO_3^-$, SO_3^-, $OSeO_2^-$, or S^{2-} are the ν(HgC) data (537—568 cm^{-1}) new.[510] Literature data were used in force constant calculations for a different CH_3HgX series [X = CH_3, F^-, I^-, CN^-, SCN^-, Me_3As, SMe^-, $SHgMe^-$, or $S(HgMe)_2$]; K(HgC) generally correlates with the first ionization energy of X.[511] Other ν(HgC) assignments reported include those for $[MeHgL]NO_3$ (L = py, dipy,

[502] M. J. Delaney and S. Ushioda, *Solid State Comm.*, 1976, **19**, 297.
[503] D. F. Shriver, G. Joy, and D. Greig, *J. Electrochem. Soc.*, 1976, **123**, 588.
[504] B.-K. Teo and D. M. Barnes, *Inorg. Nuclear Chem. Letters*, 1976, **12**, 681.
[505] K. A. Abdulla, N. P. Allen, A. H. Badran, R. P. Burns, J. Dwyer, C. A. McAuliffe, and N. D. A. Toma, *Chem. and Ind.*, 1976, 273.
[506] I. Davies and W. R. McWhinnie, *Inorg. Nuclear Chem. Letters*, 1976, **12**, 763.
[507] E. W. Ainscough, A. M. Brodie, and K. C. Palmer, *J.C.S. Dalton*, 1976, 2375.
[508] M. J. Vasile, T. J. Richardson, F. A. Stevie, and W. E. Falconer, *J.C. S. Dalton*, 1976, 351.
[509] R. Usón, A. Laguna, J. Vicente, and J. Garcia, *J. Organometallic Chem.*, 1976, **104**, 401.
[510] D. L. Rabenstein, M. C. Tourangeau, and C. A. Evans, *Canad. J. Chem.*, 1976, **54**, 2517.
[511] N. Iwasaki, *Bull. Chem. Soc. Japan*, 1976, **49**, 2735.

phen, or derivatives thereof; 540—566 cm^{-1}),[512] $MeHg(S_2CNEt_2)$ (*ca.* 525 cm^{-1}),[513] and a series of C-bonded acetylacetone complexes of Hg, *viz.* $Hg_2(C_5H_6O_2)(C_5H_7O_2)_2$, $Hg(C_5H_7O_2)_2$, and $Hg_2(C_5H_6O_2)_2$ (408—454 cm^{-1}).[514] The vibrational spectra of $(Me_3Si)_2Hg$ show that the linear SiHgSi skeleton of the solid (known from *X*-ray studies) is retained in solution.[515]

Adducts Me_2M,2NO (M = Zn or Cd) are best formulated as MeM-{ON(NO)Me} rather than as previously supposed; ν(MC) is at 546 (M = Zn) or 480 cm^{-1} (M = Cd).[516] Some ν(M—CN) data (M = Zn or Cd) have been quoted.[517]

The quality of ν(ZnN) reports varies markedly, the best last year comprising the ^{15}N or $^{64,68}Zn$ isotopic shift studies listed in Table 2,[25, 27, 40] while the deuteriation shift investigations of $ZnCl_2(py)_2$,[21] $ZnX_2(MeCN)_2$ (X = Cl or Br),[20, 518] and $M_2[Zn(NH_2)_4]$ salts [M = K or Rb, ν(ZnN) = 370—430 cm^{-1}] [519] are considered to be reasonably plausible. Many of the ν(ZnN) assignments referred to in Table 7 [105, 108, 109, 114, 117, 118, 126, 131, 133, 134, 140] or given for other related systems,[520—522] however, are much less certain.

Data for ν(CdN) [105, 108, 109, 114, 131, 133, 520, 522] and ν(HgN) [114, 131, 522] have been referred to; Millon's base, mercury hydroxide nitride dihydrate, has also been studied.[523]

Several of the ν(ZnO) (refs. 105, 117—119, 126, 128, 129, 134, 136—140, 516, 520, 522), ν(CdO) (refs. 105, 119, 516, 522, 524), and ν(HgO) (refs. 510, 520, 522) assignments made last year seem to follow the familiar pattern of insufficient spectral evidence and over-confidence of authors. However, a number of trends of the assigned bands were noted (*e.g.* with stability constants), and such observations lend support to the assignments;[119, 134, 136—138] on the other hand, there are dangers in over-using such trends, since they could arise from second-order effects.

Evidence has been presented for S-co-ordination of the sulphite group in $MeHgSO_3$ [ν(HgS) = 206 cm^{-1}; fairly full assignments given],[510] for a linear SHgCl framework for $(F_3CS)HgCl$ [ν(HgS) = 240 cm^{-1}],[525] and for linear co-ordination of Hg in a 1 : 1 complex of Hg^{II} with $HSCH_2CH(SH)CH_2OH$ [ν_{sym}(HgS) and ν_{asym}(HgS) of $Hg(SR)_2$ given; R = Me, Et, or Bu^t].[526] Values of

512 A. J. Canty and A. Marker, *Inorg. Chem.*, 1976, **15**, 425.
513 C. Chieh and L. P. C. Leung, *Canad. J. Chem.*, 1976, **54**, 3077.
514 J. W. Macklin, *Spectrochim. Acta*, 1976, **32A**, 1459.
515 P. Bleckmann, M. Soliman, K. Reuter, and W. P. Neumann, *J. Organometallic Chem.*, 1976, **108**, C18.
516 I. S. Butler and M. L. Newbury, *J. Co-ordination Chem.*, 1976, **6**, 195.
517 I. S. Ahuja and R. Singh, *J. Co-ordination Chem.*, 1976, **5**, 167.
518 Y. Hase, C. Airoldi, Y. Gushikem, and Y. Kawano, *Spectroscopy Letters*, 1976, **9**, 105.
519 G. Lucazeau, L. Guemas, and A. Novak, *Inorg. Chim. Acta*, 1976, **20**, 11.
520 M. Massacesi, G. Ponticelli, and C. Preti, *J. Inorg. Nuclear Chem.*, 1976, **38**, 1556.
521 G. C. Pellacani, G. Peyronel, G. Pollacci, and R. Coronati, *J. Inorg. Nuclear Chem.*, 1976, **38**, 1619.
522 G. Devoto, G. Ponticelli, C. Preti, and G. Tosi, *J. Inorg. Nuclear Chem.*, 1976, **38**, 1744
523 R. Weber, *Chem.-Ztg.*, 1976, **100**, 237.
524 I. S. Ahuja, R. Singh, and R. Singh, *Spectrochim. Acta*, 1976, **32A**, 547.
525 J. Barrajo, E. L. Varetti, and P. J. Aymonino, *Spectrochim. Acta*, 1976, **32A**, 1595.
526 A. J. Canty, R. Kishimoto, G. B. Deacon, and G. J. Farquharson, *Inorg. Chim. Acta*, 1976, **20**, 161.

ν(MS) have also been given for several Zn,[131, 134, 143, 521, 526, 527] Cd,[131, 143, 526–528] and Hg[131, 143, 526, 527] complexes.

Much effort has been expended in assigning ν(MX) modes for these elements, and most of the work reported for ν(ZnX) (refs. 20, 21, 101, 102, 106, 107, 114, 117, 118, 123, 126, 518, 520—522, 527, 529, 530) seems to be reliable. Rather fewer data appeared on ν(CdX) (refs. 114, 520, 522, 527, 530), and these too are generally acceptable, except that there must be considerable doubt on the validity of assignments of cadmium–halogen bridging modes.[520, 522, 530] Indeed the same reservation must be made in respect of $\nu(ZnX)_b$ assignments,[520, 522] but it is at reports of $\nu(HgX)_b$ that most of the concern must be directed. Specifically, compounds of stoicheiometry HgX_2L (L = unidentate ligand) have generally been assumed in the literature to be halogen-bridged dimers and assignments made on that basis; generally, the evidence for such solid-state structures has rested on the assignment of one (at best two) i.r. bands, but alternative interpretations are usually possible. Last year's collection includes complexes of isoxazole and its 3,5-dimethyl derivative,[520] benzoxazole-2-thione,[114] and $(Et_2N)_2CS$.[527] A linear halogen-bridged trinuclear structure has also been suggested (on virtually no evidence) for a new phase identified $Hg_3Cl_6(py)_2$.[531] However, it has now been shown[532] that even in the related series $(R_3P)HgCl_2$ (R = Ph, Me, or Et) significant structural differences may arise in the solid state. Only for R = Ph is a simple dimeric structure obtained. Values for ν(HgCl) in the 280—300 cm^{-1} region may be correlated with the short Hg—Cl lengths found in each compound.[532] It is also now known that the extent of association of $[HgX_3]^-$ salts is markedly influenced by the cation. Thus the ammonium and alkali metal chloro-complexes usually have chain or sheet structures (see last year's Report), while with the smaller tetra-alkylammonium cations $(R_4N)^+$ the spectra may be interpreted (at least to first order) in terms of essentially monomeric cationic structures.[533, 534] With larger-sized R groups, strong halogen-bridging occurs, and the nature of the halide group is also a factor in determining the extent of association; the 'change-over' point occurs at $(Et_4N)[HgCl_3]$ for which both types of structure have been identified.[534] The surprising suggestion that piperidinium and morpholinium *tetra*-halogeno-complexes of Hg^{II} contain halide bridges[530] needs careful evaluation. There is evidence for anion–cation interactions in $Ag_2[HgI_4]$ and $Cu_2[HgI_4]$.[503] For $(HgI_2)_2MF_6$ (M = Ti, Zr, or Sn), fairly detailed analyses of the spectra have been made *via* FGA of the known $[(IHg_{2/2})_2]^{2+}$ chains; a simple force-constant calculation was performed.[11] Compounds (HgI)X (X = NO_3, BF_4, *etc.*) are thought to have a similar cationic chain structure.[11]

The high-pressure phases of $HgCl_2$(II and IV) and $HgBr_2$(II, III, and IV) have been characterized. The i.r. and Raman data are consistent with $HgCl_2$(II) being

527 G. Marcotrigiano, *Z. anorg. Chem.*, 1976, **422**, 80.
528 R. A. Haberkorn, L. Que, jun., W. O. Gillum, R. H. Holm, C. S. Liu, and R. C. Lord, *Inorg. Chem.*, 1976, **15**, 2408.
529 J. A. McCleverty and N. J. Morrison, *J.C.S. Dalton*, 1976, 2169.
530 G. Marcotrigiano, L. Menabue, and G. C. Pellacani, *J. Mol. Structure*, 1976, **33**, 191.
531 S. Ahyüz, A. B. Dempster, J. E. D. Davies, and K. T. Holmes, *J.C.S. Dalton*, 1976, 1746.
532 N. A. Bell, M. Goldstein, T. Jones, and I. W. Nowell, *J.C.S. Chem. Comm.*, 1976, 1039.
533 A. W. Herlinger, *Spectroscopy Letters*, 1975, **8**, 787.
534 R. M. Barr and M. Goldstein, *J.C.S. Dalton*, 1976, 1593.

of space-group I_h^6 ($Z = 4$) (*i.e.* isomorphous with CO_2), $HgBr_2$(III) being *Cmcm* ($Z = 4$), and $HgBr_2$(IV) having the CdI_2 structure.[10] For matrix-isolated $HgX_2(C_2H_4)$, ν(HgX) is 25 (X = F) or 12 cm^{-1} (X = Cl) lower than in the corresponding HgX_2 under comparable conditions.[48]

For many other compounds of these elements ν(MX) have been obtained, as previously referred to [1, 48, 90, 124, 525] or contained in refs. 535 and 536.

ν(HgHg) assignments may be obtained *via* Table 6.[98–100]

13 Lanthanoids

Despite the severe complexity of many of the spectra and the resultant near-arbitrary choice of bands, assignments have been made with implicit conviction for ν(YbN) in $[Yb(2\text{-}C_5H_4N{-}CH{=}NNH{-}C_5H_4N{-}2)(H_2O)_3(OH)]_2Cl_4,4H_2O$ (4 bands, 275—299 cm^{-1}),[537] and of ν(LnO) as follows (Ln = lanthanoid): $Eu_2O_3,2.8SiO_2,7.1H_2O$ (485 cm^{-1}),[538] $Eu_2(C_2O_4)_3,10H_2O$ (488—498 cm^{-1}),[539] $Eu(SCPh{=}CHCOPh)_3(pyO)_n$, *etc.* [425—455 cm^{-1}, with ν(EuS) at 300—315 cm^{-1}],[540] $[La\{(2\text{-}EtSO)C_5H_4NO\}_4](ClO_4)_3$ (332, 409 cm^{-1}),[446] $Gd_2(Pr^n_2SO)_{11}(ClO_4)_6$ (430 cm^{-1}),[541] $Ln(thioxane)_8(PF_6)_3$ (217—238 cm^{-1}; Ln = various),[149] $[Ln(chelH)_2(chel)_2]ClO_4,nH_2O$ [395—415 cm^{-1}; Ln = various; n = 0, 1, 2, or 3; (chelH) = 2-picolinic acid *N*-oxide].[542]

Values quoted for ν(LnS) are as given above [540] for ν(EuS), and at 365—372 cm^{-1} in related Nd complexes.[543]

Detailed studies, including MVFF calculations and consideration of lattice modes, on Cs_2KLnF_6 salts (Ln = La, Gd, or Yb, as well as Sc or Y) [12] are considered in Chapter 4. A systematic Raman spectral investigation has been carried out on the cubic $Cs_2NaLnCl_6$ species, including a variable temperature study.[544] In general, there was a regular dependence of the vibrational modes on the atomic number of the lanthanoid and on temperature, but for Ln = Yb the position and intensity of $\nu_2(e_g)$ showed unusual temperature effects, and this is thought to have its origin in exceptionally strong vibronic coupling. Solutions of $LnCl_3$ (Ln = various) in $(Bu^nO)_3PO$ (0.95—2.1 mol dm^{-3}) are believed, from a far-i.r. study, to contain $LnCl_3,3(Bu^nO)_3PO$ complexes and halogen-bridged polymers.[545]

535 G. Wulfsberg, *Inorg. Chem.*, 1976, **15**, 1791.
536 A. N. Nesmeyanov, V. A. Blinova, Z. S. Klemenkova, L. I. Denisovich, I. I. Kritskaya, and V. T. Aleksanyan, *Doklady Chem.*, 1976, **224**, 602.
537 E. Baraniak, R. St. L. Bruce, H. C. Freeman, N. J. Hair, and J. James, *Inorg. Chem.*, 1976, **15**, 2226.
538 S. G. Babayan, M. A. Balayan, L. Ts. Ambartsumyan, and G. T. Gazaryan, *Russ. J. Inorg. Chem.*, 1975, **20**, 1445.
539 S. S. Berdonosov, D. G. Berdonosova, M. A. Prokof'ev, and V. Ya. Lebedev, *Russ. J. Inorg. Chem.*, 1976, **21**, 644.
540 K. Nag and M. Chaudhury, *Inorg. Chem.*, 1976, **15**, 2291.
541 J. R. Behrendt and S. K. Madan, *J. Inorg. Nuclear Chem.*, 1976, **38**, 1827.
542 S. A. Boyd, R. E. Kohrman, and D. X. West, *Inorg. Nuclear Chem. Letters*, 1976, **12**, 603.
543 K. Nag and M. Chaudhury, *Inorg. Nuclear Chem. Letters*, 1976, **12**, 307.
544 H.-D. Amberger, R. D. Fischer, and G. G. Rosenbauer, *Transition Metal Chem.*, 1976, **1**, 242.
545 S. S. Korovin, O. V. Galaktionova, and E. N. Lebedeva, *Izvest. Vysshikh Uchebn. Zavedenii, Khim. Khim. Tekhnol.*, 1976, **19**, 317.

14 Actinoids

The Raman spectrum of $U(BH_4)_4$ has been assigned (again ?).[546]

I.r., Raman, and ^{13}C n.m.r. studies have established that the cubic point symmetry of the $[U(NCS)_8]^{4-}$ anion in the solid $(Et_4N)^+$ salt is maintained in $MeNO_2$ solution.[547] For complexes $AnCl_4(chelH_2)$, $An(chel)_2$, and $AnCl_2(chel)$, $\nu(AnN)$ has been placed at 576—601 cm^{-1} (An = U or Th; $chelH_2$ = Schiff base).[548, 549]

The vast array of uranyl(VI) and related complexes studied by i.r. spectroscopy continues to grow, and happily the levels of sophistication and of reliability are now much improved. Table 10 lists typical compounds for which $\nu(AnO_2)$ modes have been given; it should be noted that the list is illustrative, not exhaustive.[7, 37, 135, 185, 550–565] Little comment is warranted, save to indicate the more detailed and justified papers (from the spectroscopic view).[7, 37, 547, 550, 551]

Other compounds for which $\nu(AnO)$ has been assigned include $NpOF_4$ (976 cm^{-1}),[566] $[NpO_5]^{3-}$ salts,[567] and $Ba_2M_nU_{1-x}W_xO_6$ (M_n = Mg or $Y_{0.67}$; x = 0—1).[243, 244] Other relevant compounds for which data are available are NpO_2F_2,[566] $AnO_2(IO_3)_n$, *etc.* (An = U or Np; n = 1 or 2),[567] ζ-UO_3 and α-UO_3,$0.3H_2O$,[568] $Ba_2Ca_{1-x}Ln_{0.67x}UO_6$ (Ln = Y or Gd; x = 0—1),[569] and $U(OTeF_5)_6$.[570]

A fair number of papers deal with $\nu(AnX)$ assignments. Spectra of MUF_7 and MUF_8 (M = K, Cs, NO, NH_4, Rb, or Na) are very dependent on temperature

546 V. B. Kartha and V. N. P. Kaimal, *Indian J. Phys.*, 1976, **50**, 147.
547 G. Folcher, H. Marquet-Ellis, P. Rigny, and E. Soulié, *J. Inorg. Nuclear Chem.*, 1976, **38**, 747.
548 L. Doretti, F. Madalosso, S. Sitran, and S. Faleschini, *Inorg. Nuclear Chem. Letters*, 1976, **12**, 817.
549 L. Doretti, S. Sitran, S. Faleschini, F. Madalosso, and G. Faraglia, *J. Inorg. Nuclear Chem.*, 1976, **38**, 2123.
550 Nguyen Quy Dao and M. Knidiri, *Spectrochim. Acta*, 1976, **32A**, 481.
551 A. Perrin and P. Caillet, *Compt. rend.*, 1976, **282**, *C*, 721.
552 A. Perrin and J. Prigent, *Bull. Soc. chim. France*, I, 1976, 1070.
553 M. V. Nikonovich, A. N. Akimov, and V. V. Sikorskii, *V. sb.*, *Spektroskopiya i Lyuminestsentsiya*, 1975, 73.
554 L. Doretti, S. Faleschini, S. Sitran, and G. Faraglia, *J. Co-ordination Chem.*, 1976, **6**, 115.
555 S. Degetto, L. Baracco, G. Marangoni, and E. Celon, *J.C.S. Dalton*, 1976, 1645.
556 B. Claudel, B. Mentzen, A. Novarro, and H. Sautereau, *J. Inorg. Nuclear Chem.*, 1976, **38**, 759.
557 B. L. Khandelwal and V. P. Verma, *J. Inorg. Nuclear Chem.*, 1976, **38**, 763.
558 J. G. Ribas Bernat, J. Martin Gil, and F. J. Martin Gil, *Energia Nuc. (Madrid)*, 1976, **20**, 37.
559 B. Zarli, G. Dall'olio, and L. Sindellari, *J. Inorg. Nuclear Chem.*, 1976, **38**, 491.
560 G. Faraglia, B. Zarli, L. Sindellari, and G. Dall'olio, *Inorg. Chim. Acta*, 1976, **19**, 39.
561 R. G. Vijay and J. P. Tandon, *Monatsh.*, 1976, **107**, 95.
562 R. G. Vijay and J. P. Tandon, *Z. Naturforsch.*, 1976, **31b**, 22.
563 R. M. Costes, G. Folcher, P. Plurien, and P. Rigny, *Inorg. Nuclear Chem. Letters*, 1976, **12**, 491.
564 R. M. Costes, G. Folcher, P. Plurien, and P. Rigny, *Inorg. Nuclear Chem. Letters*, 1976, **12**, 13.
565 I. S. Ahuja, and R. Singh, *Inorg. Nuclear Chem. Letters*, 1976, **12**, 665.
566 Yu. V. Drobyshevskii, V. F. Sevik, and V. B. Sokolov, *Doklady Chem.*, 1976, **225**, 675.
567 A. Yu. Tsivadze, N. N. Krot, and B. I. Muchnik, 'Proceedings of the Moscow Symposium on Chemistry of the Transuranium Elements', 1972, ed. V. I. Spitsyn and J. J. Katz, Pergamon Press, Oxford, 1976, p. 89.
568 L. M. Kuznetsov and A. N. Tsvigunon, *Radiokhimiya*, 1976, **18**, 414.
569 S. Kemmler-Sack, I. Seemann, and H.-J. Schittenhelm, *Z. anorg. Chem.*, 1976, **422**, 115.
570 K. Seppelt, *Chem. Ber.*, 1976, **109**, 1046.

Table 10 *Examples of compounds for which* $\nu(AnO_2)$ *assignments have been made* (An = *actinoid*)

Compound	*Ref.*
$M_3[UO_2F_5]$	550 [a]
UO_2Cl_2 $[UO_2(OH)_2UO_2]Cl_2,4H_2O$	551 [b]
$K_2[UO_2(OH)_2UO_2](NO_3)_4,H_2O$	552 [c]
$M[UO_2(NO_3)_3]$	7 [d]
$UO_2(HPO_4),nH_2O$	553
$UO_2(C_9H_6NO),L$	554 [e]
UO_2(tropolonate)$_2$,L	555 [f]
UO_2(chel)X	135 [g]
$NH_4(UO_2)VO_4,nH_2O$ $(UO_2)_2V_2O_7$	185
NbO_2X,nH_2O, $CsPuO_2Cl_4$, *etc.*	37 [h]
$Na[UO_2(HCO_2)_3],H_2O$	556
$[(UO_2)_2(SeO_3)_3]^{2-}$ *etc.*	557
$(UO_2)_n(SO_4)_mM_p,xH_2O$	558 [i]
$UO_2X_2L_2$	559—565 [j]

[a] FGA carried out, but little added to extant data. [b] FGA carried out and found to be important for UO_2Cl_2 only. [c] Spectra show in-phase and out-of-phase coupling of the $\nu_{sym}(UO_2)$ and $\nu_{asym}(UO_2)$ modes, and are consistent with a model in which the two UO_2 groups are parallel to each other and perpendicular to the $U(OH)_2U$ bridge (overall D_{2h}). [d] MVFF calculations performed taking NO_3 groups as point masses. [e] L = H_2O, MeOH, py, *etc.* [f] L = Ph_3-AsO, 4-MeC_5H_4NO, *etc.* [g] (chel) H_2 = 6,11-dimethyl-7,10-diazahexadeca-6,10-diene-2,4,13,15-tetraone; X = H_2O, MeOH, py, Cu, Ni, *etc.* [h] A large number of related compounds studied; ^{18}O substitution used; X = ClO_4 or NO_3. [i] *n* or *p* = 1, 2 or 3; M = Ce, Mn, Co, Ni, Cu, Zn, Cd, or Hg; *m* = 2, 4, or 5 for electrical neutrality. [j] X = NO_3, $MeCO_2$, Et_2NCS_2, SCN, or Cl; L_2 = urea, hexamethylphosphoramide, 2(*NN*-dimethyl ethyl carbamate), Schiff base, $N(C_2H_4OC_2H_4OC_2H_4)_3N$, 18-crown-6, 1,3-di-(4-pyridyl) propane, *etc.*

and on cation size.[571] In the solids at high temperatures the disordered fluorine positions are likely to be averaged to give pseudo-D_{5h} for the $[UF_7]^-$ and pseudo-O_h for the $[UF_8]^{2-}$ anions, but at lower temperatures the spectra are indicative of symmetries no higher than C_{2v} and D_{2d}, respectively;[571] it is a pity that more specific vibrational assignments were not made. Most of the details listed for anhydrous UO_2Cl_2 and $[UO_2(OH)_2UO_2]Cl_2,4H_2O$ appear to be justified;[551] FGA was found to be more necessary for the neutral compound than for the hydroxy-species (for which assignments were aided by deuteriation). Attempts to obtain the full Raman spectra of $(Et_4N)_2[AnI_6]$ have been frustrated by fluorescence or decomposition in the laser beams;[572] only for An = Th could the three anion modes ν_1 (131), ν_3 (143), and ν_5 (apparently split: 50, 68 cm^{-1}) be observed, but in the other systems examined (An = U or Pa) ν_3 was located at 143 cm^{-1}.

One of the infuriating aspects of the chemical literature is that it is often difficult to determine how much of the work in a new paper is original; this seems to be particularly true where spectroscopic data are obtained as an adjunct to synthetic work or as a means of investigating other aspects of the systems. The 'variations on a theme' approach also seems to dominate much of the

[571] R. Bougon, P. Charpin, J. P. Desmoulin, and J. G. Malm, *Inorg. Chem.*, 1976, **15**, 2532.
[572] D. Brown, P. Lidster, B. Whittaker, and N. Edelstein, *Inor. Chem.*, 1976, **15**, 511.

spectroscopy of inorganic compounds. Relevant to this section in these respects are: (*a*) studies of the i.r. spectrum of $[UO_2F_5]^{3-}$ salts which have previously been characterized by single-crystal *X*-ray diffraction and Raman spectroscopy (11 references to previous work quoted);[550] (*b*) listing of unassigned data for the known compounds PaX_5 (X = Cl, Br, or I) and PaI_4;[573] (*c*) further studies on Cs_2UCl_6;[574] (*d*) characterization of the new compound $(H_3O)[UF_6]$ using extant data on the component ions;[575] and (*e*) identification of $M[UF_6]_n$,5MeCN [n = 2 (M = Cu or Cd) or 3 (M = Tl)].[576] A paper [546] on the Raman spectrum of UF_4 will make interesting comparison with previously confusing data (*cf.* last year's Report, p. 321).

Other compounds for which ν(AnX) data have been given include: (*a*) $NpOF_4$ (561) and NpF_5 (400—500 cm^{-1}) [bridging fluorines deduced to be present];[566] (*b*) UO_2Cl_2{1,3-di-(4-pyridyl)propane} (233 cm^{-1});[565] (*c*) $ThCl_4$,3L (244—256) and UCl_4,2L,(thf) (230—278 cm^{-1}) [L = *N*-methyl ethyl carbamate or *NN*-dimethyl ethyl carbamate];[560] and (*d*) UCl_4,2L,n(thf) (*ca.* 250 cm^{-1}; L = a substituted salicylaldimine; n = 0 or 2).[549]

15 Conclusions

It will be apparent from the increased number of references included in this Chapter over recent years that the use of vibrational spectroscopy in the study of transition-element compounds has, and still is, expanding. This, in itself, is a satisfying sign: the subject is alive and developing. Thus the range and complexity of applications are growing, and experimental limitations are disappearing in all corners of the world. But there is more to development than volume and scope; quality, significance, and reliability must not be neglected, yet there is little evidence that these important aspects are being improved. Such developments as are occurring in understanding the vibrational spectra of transition-element compounds, *e.g. via* single-crystal work and the use of isotopes to obtain more meaningful potential energy distributions, seem to be having only a limited impact on the literature covered by this Chapter. Time will tell, but it seems to this Reviewer that the energy should be expended in going forward rather than sideways. It is hoped that this Report has highlighted the needs.

573 D. Brown, G. De Paoli, and B. Whittaker, *J.C.S. Dalton*, 1976, 1336.
574 H. D. Amberger, R. D. Fischer, G. G. Rosenbauer, and A. W. Spiegl, *Ber. Bunsengesellschaft Phys. Chem.*, 1976, **80**, 495.
575 J. P. Masson, J. P. Desmoulin, P. Charpin, and R. Bougon, *Inorg. Chem.*, 1976, **15**, 2529.
576 J. A. Berry, R. T. Poole, A. Prescott, D. W. A. Sharp, and J. M. Winfield, *J.C.S. Dalton*, 1976, 272.

7

Vibrational Spectra of some Co-ordinated Ligands

BY G. DAVIDSON

1 Carbon Donors

A review of the co-ordination chemistry of allenes contains a section on i.r. spectra, in which available data on changes in ν_{as}(C=C=C) on co-ordination are listed.[1]

An adduct of Li with phenylacetylene in THF gives a band due to ν(C≡C) at 2040 cm^{-1}, compared with 2125 cm^{-1} in free phenylacetylene.[2]

ν(CH), δ(CH), and γ(CH) wavenumbers were given for $Na(C_5H_5)$-$(Me_2NCH_2CH_2NMe_2)$. Together with n.m.r. data these indicated *pentahapto*-, and mainly ionic bonding, which was confirmed by a full *X*-ray study.[3]

I.r. and Raman spectra of $MgEt_2$, $Mg(CH_2CD_3)_2$, and $Mg(CD_2CH_3)_2$ were obtained at 300 and 90 K. Differences in the spectra at the two temperatures show that the low-temperature forms possess a different chain conformation in the infinite polymeric structure.[4]

1 : 1 and 1 : 2 Adducts have been detected between Mg and butadiene. ν(C=C) bands are formed in the former at 1604 and 1575 cm^{-1}, and in the latter at 1619 and 1593 cm^{-1}. In each case the higher wavenumber was assigned to a *trans*-, and the lower to a *cis*-(C=C) vibration.[5]

Characteristic 'ionic' cyclopentadienyl bands were identified for $(C_5H_5)_2$-Ca,NH_3,THF and $(indenyl)_2Ca,NH_3$,THF.[6]

The usual bands of h^5-C_5H_5 ligands were assigned in $M(h^5\text{-}C_5H_5)_2(SPh)_2$, where M = Ti or V, and $V_2(h^5\text{-}C_5H_5)_2(\mu_2\text{-}SPh)_4$.[7]

The reaction between $CpTi(C_6H_4CH_2NMe_2\text{-}o)_2$ and CO_2 produces a compound thought to be (1), with ν(CO) at 1710 cm^{-1}.[8]

ν(C=O) for the acyl groups in $Cp_2Zr(COR)R$, 1545 cm^{-1} for R = Me, 1540 cm^{-1} for R = $PhCH_2$, have been explained by *X*-ray studies on the complex with R = Me. These revealed that the structure is (2), *i.e.* the acyl group is acting as a 3-electron donor ligand.[9]

Bis(fluorenyl)dimethylzirconium contains the unit (3). Bands due to the unique CH were detected by deuteriation of this hydrogen at 806 (shifting to

[1] F. L. Bowden and R. Giles, *Coordination Chem. Rev.*, 1976, **20**, 81.
[2] I. M. Panayotou and M. V. Bojilova, *J. Organometallic Chem.*, 1976, **105**, 281.
[3] T. Aoyagi, H. M. M. Shearer, K. Wade, and G. Whitehead, *J.C.S. Chem. Comm.*, 1976, 164.
[4] J. Kress, A. Novak, and J. Hervieu, *J. Organometallic Chem.*, 1976, **121**, 7.
[5] K. Fujita, Y. Ohnuma, H. Yasuda, and H. Tani, *J. Organometallic Chem.*, 1976, **113**, 201.
[6] M. Kirilov, G. Petrov, and C. Angelou, J. *Organometallic Chem.*, 1976, **113**, 225.
[7] E. G. Muller, S. F. Watkins, and L. F. Dahl, *J. Organometallic Chem.*, 1976, **111**, 73.
[8] L. E. Manzer, R. C. Gearhart, L. J. Guggenberger, and J. F. Whitney, *J.C.S. Chem. Comm.*, 1976, 942.
[9] G. Fachinetti, C. Floriani, F. Marchetti, and S. Murlino, *J.C.S. Chem. Comm.*, 1976 522.

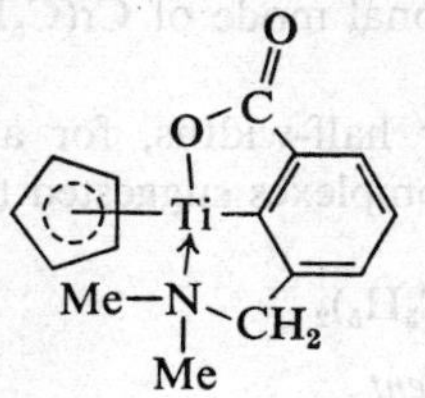
(1)

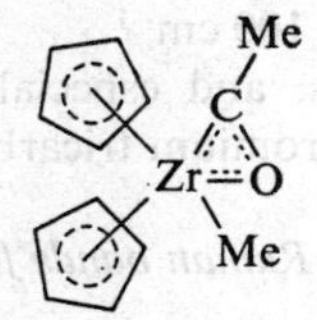
(2)

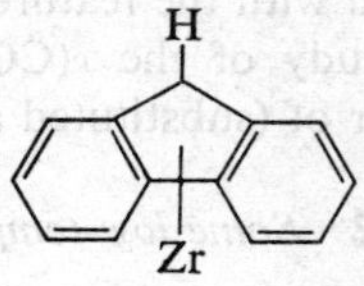
(3)

about 600) cm^{-1}, and at 767 (shifting to about 540) cm^{-1}. They were assigned as γ(CH) and δ(CH) respectively. ν(CH) was probably at about 3100 cm^{-1}, but it was difficult to distinguish from the phenyl ring CH stretches.[10]

(η^3-C_3H_5)V(CO)$_4$ and substituted allyl derivatives all give at least four ν(CO) bands in their i.r. spectra. Thus the local symmetry of C_{4v} is not adequate here. In several of the substituted allyl complexes the presence of more than 4 bands must be due to the existence of *syn*- and *anti*-isomers, *e.g.* (4a) and (4b), which would lead to a prediction of 6 bands.[11]

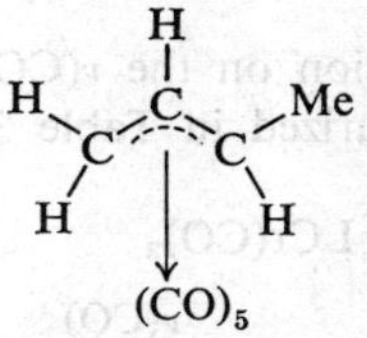
(4a)

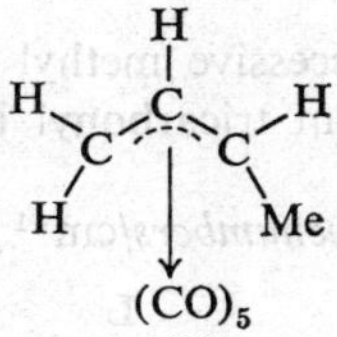
(4b)

A reasonably complete assignment has been proposed for the i.r. spectrum of K[(π-Cp)V(CO)$_3$CN], see Table 1. U.v. irradiation of a mixture of Me_4N^+-[VCp(CO$_3$)]$^-$ and Me_4N(CN) gave a species thought to be [CpV(CO)$_2$CN]$_2^{4-}$, containing bridging carbonyls, ν(CO)$_b$ at 1679, ν(CO)$_t$ at 1790 cm^{-1}, and ν(CN) at 2070 cm^{-1}.[12]

Table 1 *Vibrational assignments for* [(-πCp)V(CO)$_3$CN]$^-$

Assignment	*Wavenumbers*/cm^{-1}
ν(CN)	2063
ν(CO)	1945, 1840, 1820
ν(CC)	1425
i.p. δ(CH)	1005
o.o.p. δ(CH)	825, 815
δ(VCO)	629, 600
ν(V—CO)	546, 511, 505
δ(VCN)	478, 455
ν(V—CN)	435, 422
ν(V—Cp)	385, 360, 330

At 25 K, the Raman spectrum of chromocene develops bands not present at ambient temperatures. These are assigned to vibrational–electronic Raman

[10] E. Samuel, H. G. Alt, D. C. Hrncir, and M. D. Rausch, *J. Organometallic Chem.*, 1976, **113**, 331.

[11] M. Schneider and E. Weiss, *J. Organometallic Chem.*, 1976, **121**, 345.

[12] D. Rehder, *Z. Naturforsch.*, 1976, **31B**, 273.

transitions, see Table 2.[13] It is believed that the torsional mode of $Cr(C_5H_5)_2$ is coupled with the features below 100 cm^{-1}.

A study of the ν(CO) bands, and especially their half-widths, for a large number of (substituted arene)chromium tricarbonyl complexes suggested that in

Table 2 *Some low-temperature Raman bands for* $Cr(C_5H_5)_2$

Wavenumber/cm^{-1}	*Assignment*
35	Electronic, $E_{1g} \leftrightarrow E_{2g}$, 0—0
47	Electronic, $E_{1g} \leftrightarrow E_{2g}$, 1—2
60	Electronic, $E_{1g} \leftrightarrow E_{2g}$, 0—1
74	Electronic, $E_{1g} \leftrightarrow E_{2g}$, 0—2
85	Electronic, $E_{1g} \leftrightarrow E_{2g}$, 0—3
92	Electronic, $E_{1g} \leftrightarrow E_{2g}$, 0—4
285	ν_3, *sym* M—L stretch
359, 375	ν_{16}, *sym* ring tilt

iso-octane solution only small perturbations from C_{3v} symmetry could be detected.[14]

The effect of successive methyl substitution on the ν(CO) wavenumbers of π-pyridine chromium tricarbonyl is summarized in Table 3. The trend is as

Table 3 ν(CO) *wavenumbers*/cm^{-1} *for some* $LCr(CO)_3$

L	ν(CO)
2,6-dimethylpyridine	1986, 1925, 1913
2,4,6-trimethylpyridine	1984, 1923, 1910
2,3,5,6-tetramethylpyridine	1976, 1913, 1899
pentamethylpyridine	1972, 1908, 1891

expected from the donor power of the Me group, giving greater electron density at the Cr, and hence greater Cr—CO back donation.[15]

The complex $(OC)_3Cr[h^6\text{-}C_6H_5B(OH)_2]$ gives the following i.r. bands: ν(CO), 1920 and 2070 cm^{-1}; ν(CC) of ligand, 1530 and 1520 cm^{-1}; ν(BO), 1350 cm^{-1}; and ν(OH), 3150 cm^{-1}.[16]

The ν(CO) bands of (5a) are at 2063 and 1943 cm^{-1}; these shift to 2030, 1901, and 1850 cm^{-1} on the formation of the anion (5b), as predicted from the increase in the negative charge on the Cr.[17]

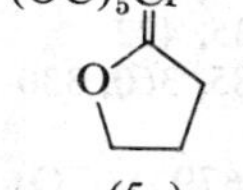
(5a)

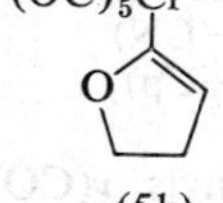
(5b)

[13] L. V. Haley, T. Parameswaran, J. A. Koningstein, and V. T. Aleksanyan, *Chem. Phys. Letters*, 1976, **42**, 13.
[14] F. van Meurs, J. M. A. Baas, and H. van Bekkum, *J. Organometallic Chem.*, 1976, **113**, 353.
[15] H.-G. Biedermann, K. Öfele, and J. Tajtelbaum, *Z. Naturforsch.*, 1976, **31B**, 321.
[16] G. A. Razuvaev, A. N. Artemov, A. A. Aladjin, and N. I. Sirotkin, *J. Organometallic Chem.*, 1976, **111**, 131.
[17] C. P. Casey and W. R. Brunsvold, *J. Organometallic Chem.*, 1976, **118**, 309.

ν(CO) in the carbene complexes (6), where R = Ph, 1-, or 2-$C_{10}H_7$, are at higher wavenumbers when X = OMe than when X = NHCHMePh or other amino groups. This is explained by the N atom forming the stronger bond to Cr by a *pπ–pπ* interaction.[18]

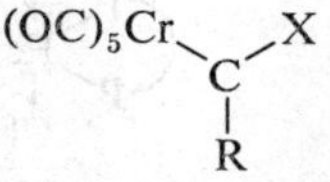
(6)

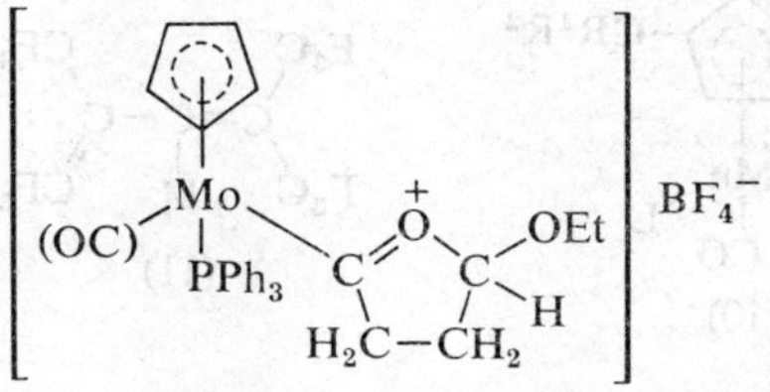
(7)

The absence of ν(CO)(acyl) from the i.r. spectrum of (7) confirms its formulation as the cyclic oxonium species.[19]

ν(CN) is at 2222 cm^{-1} in $Mo(CO)_3$(1,2-diphenylacrylonitrile)$_3$. This is practically unshifted from the value for the free ligand, and hence the co-ordination is *via* the double bond.[20]

I.r. and Raman spectra have been reported for $CpW(R)(CO)_3$, where R = H, CD_3, Et, or σ-allyl. Although local symmetry was a useful first approximation for the Cp—W (C_{5v}) and $W(CO)_3$ (C_{3v}) modes, there was definite evidence for the inadequacy of this for the Cp—W fragment, as several formally forbidden modes were observed. Quite complete assignments were made for the CD_3 and σ-allyl internal modes.[21]

The complexes (8) show bands due to $\nu(C_{carbene}—N)$ and $\nu(C_{carbene}—O)$ at 1587 and 1322 cm^{-1} respectively,[22] while (9) gives a band due to ν(C=O) at 1652 cm^{-1}.[23]

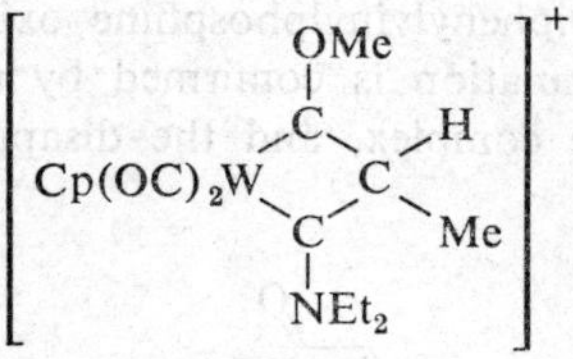
(8)

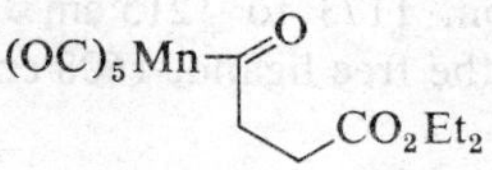
(9)

The formation of (10), where R[1], R[2] = H, Me, Et, Ph; L = PPh_3 or PPr^i_3, from the neutral carbinol complex and CF_3CO_2H is accompanied by a shift of ν(CO) by 50—100 cm^{-1}. Thus, for R[1] = H, R[2] = Me, L = PPh_3, the carbinol values are 1878 and 1940 cm^{-1}, with those for the cation being 1975 and 2022 cm^{-1}.[24]

[18] H. Brunner, J. Doppelerger, E. O. Fischer, and M. Lappus, *J. Organometallic Chem*, 1976, **112**, 65.
[19] C. M. Lukehart and J. V. Zeile, *J. Organometallic Chem.*, 1976, **105**, 231.
[20] A. H. Westlake, *Chem. and Ind.*, 1976, 372.
[21] G. Davidson and D. A. Duce, *J. Organometallic Chem.*, 1976, **120**, 229.
[22] W. Beck, H. Brix, and F. H. Köhler, *J. Organometallic Chem.*, 1975, **121**, 211.
[23] C. P. Casey, W. R. Brunsvold, and J. Koch, *Inorg. Chem.*, 1976, **15**, 1991.
[24] A. G. Ginzburg, V. N. Setkina, P. V. Petrovskii, Sh. G. Kasumov, G. A. Panosyan, and D. N. Kursanov, *J. Organometallic Chem.*, 1976, **121**, 381.

The species (11) gives bands due to ν(C=C) at 1558 and 1728 cm^{-1} when M = $CpFe(CO)_2$, and at 1570 and 1730 cm^{-1} when M = $Re(CO)_5$.[25]

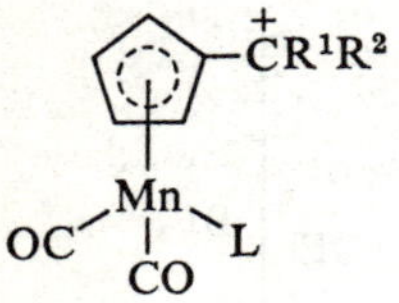

(10)

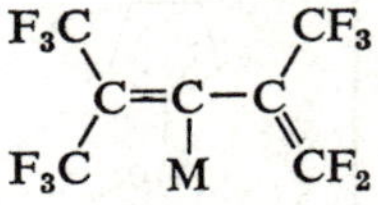

(11)

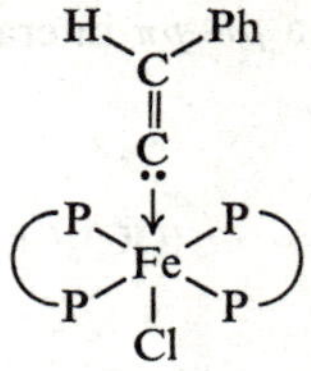

(12)

Bands involving, at least partially, ν(C=C) are seen at 1609, 1595, 1588, and 1572 cm^{-1} in the phenylvinylidene complex (12), where P—P = bis(diethylphosphino)ethane.[26]

The complex (13) gives bands due to ν(C≡C) at 2226 cm^{-1}, ν(C=C) at 1614 cm^{-1}, and three ν(CO) features at 2116, 2058, and 2037 cm^{-1}.[27]

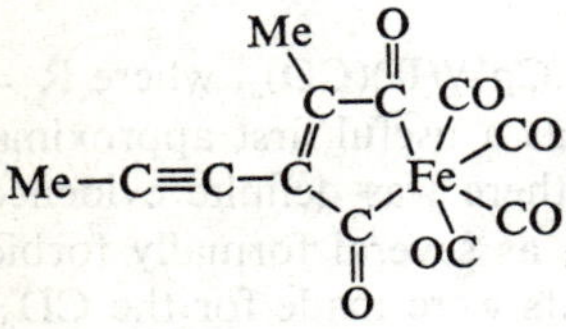

(13)

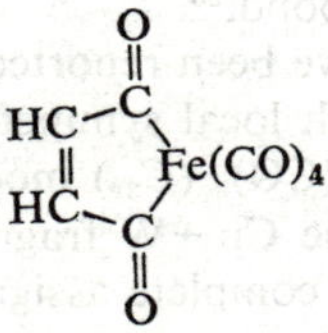

(14a)

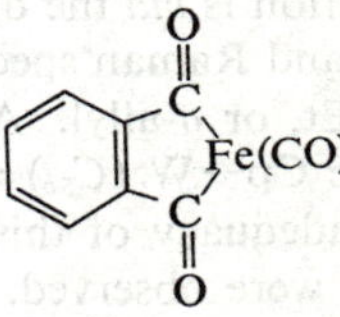

(14b)

A number of new ferracyclopentadienyl complexes have been prepared and characterized. Thus (14a) has ν(C=O) at 1675 cm^{-1}, and (14b) gives analogous features at 1706.5 and 1667.5 cm^{-1}.[28] Diphenylvinylphosphine oxide forms a complex (15) with $Fe_3(CO)_{12}$. The formulation is confirmed by the shift in ν(P=O) from 1173 to 1215 cm^{-1} in the complex, and the disappearance of ν(C=C) of the free ligand, 1608 cm^{-1}.[29]

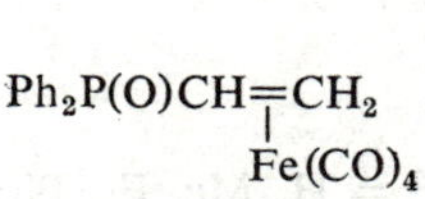

(15)

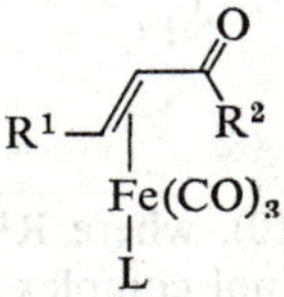

(16)

Assignments of ν(C=O) in the complexes (16) are summarized in Table 4. Its position shows that the C=O is not co-ordinated to the Fe. Also, the

[25] A. N. Nesmeyanov, N. E. Kolobova, I. B. Zlotina, B. V. Lokshin, I. F. Leshcheva, G. K. Znobina, and K. N. Anisimov, *J. Organometallic Chem.*, 1976, **110**, 339.
[26] J. M. Bellerby and M. J. Mays, *J. Organometallic Chem.*, 1976, **117**, C21.
[27] F. R. Young, D. H. O'Brien, R. G. Pettersen, R. A. Levenson, and D. L. von Minden, *J. Organometallic Chem.*, 1976, **114**, 157.
[28] F.-W. Grevels, J. Buchkremer, and E. A. Koerner Von Gustorf, *J. Organometallic Chem.*, 1976, **111**, 235.
[29] B. F. G. Johnson, J. Lewis, and S. R. Postle, *J. Organometallic Chem.*, 1976, **121**, C7.

Table 4 *ν*(C=O) *assignments for the complexes* (16) (*all wavenumbers*/cm^{-1})

R^1	R^2	L	ν(C—O)
Ph	H	PMe_2Ph	1650
Ph	H	$P(OMe)_3$	1663
Ph	Me	PMe_2Ph	1649
Ph	Me	$P(OMe)_3$	1657
Ph	Me	$P(OPh)_3$	1670
Ph	Ph	PMe_2Ph	1629
Ph	Ph	$P(OMe)_3$	1635
Ph	Ph	$P(OPh)_3$	1634

decrease with increasing donor power of L reflects the increased electron donation from the metal to the enone.[30]

The novel olefincarbamylate complexes (17), where $R^1 = R^2 =$ Me, R^1 = H, R^2 = Me, or R^1 = Me or H, $R^2_2 = (CH_2)_5$, give a band assigned to ν(N⋯C⋯O) at 1640 cm^{-1}. ν(C⋯O) is seen at 1568—1605 cm^{-1}.[31] Compound (18) gives ν(C⋯O) at 1947 cm^{-1} and ν(C=O)(acyl) at 1692 cm^{-1}.[32a]

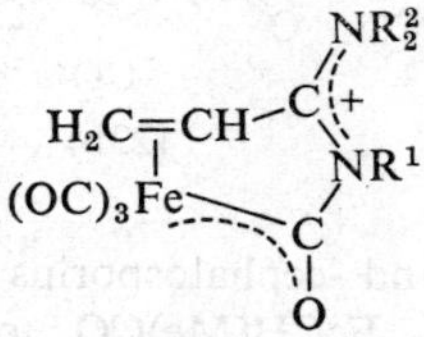

(17)

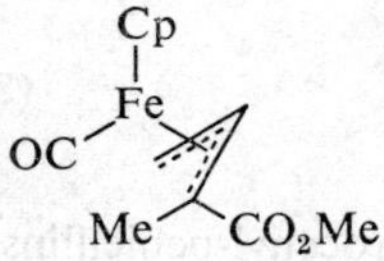

(18)

A very detailed study has been made of the vibrational spectra of $[(H_2C)_3C]Fe(CO)_3$ and its d_6-analogue. The results were analysed using a 31-parameter force field. The calculated force constants differed in several cases from those previously obtained using data for the h_6 species only.[32b]

The second known compound of the metal formyl class has been made. It is $[(PhO)_3P](CO)_3FeCHO$, for which the aldehydic CH gives a band at 2515 cm^{-1}, ν(CO) is at 1960—1872 cm^{-1}, with the formyl ν(C=O) at 1580 cm^{-1}.[32c]

The complexes (19), where L = $P(OPh)_3$ or $P(OMe)_3$, both give two ν(C≡O) bands, and no ν(C=O) in the range 1650—1700 cm^{-1}, consistent with this formulation.[33]

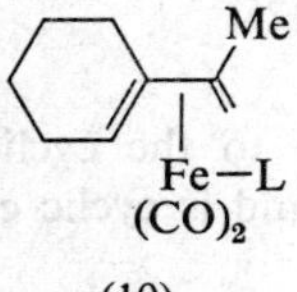

(19)

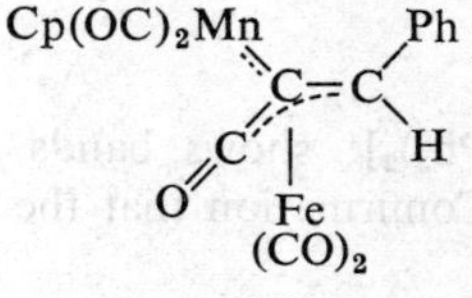

(20)

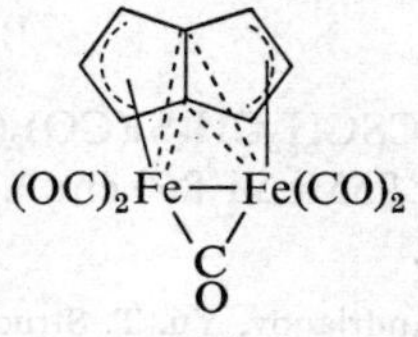

(21)

[30] A. Vessieres, D. Touchard, and P. Dixneuf, *J. Organometallic Chem.*, 1976, **118**, 93.
[31] A. N. Nesmeyanov, T. N. Sal'nikova, Yu. T. Struchkov, V. G. Andrianov, A. A. Pogrebyak, L. V. Rybin, and M. I. Rybinskaya, *J. Organometallic Chem.*, 1976, **117**, C16.
[32] (*a*) T. A. Bauch and W. P. Giering, *J. Organometallic Chem.*, 1976, **114**, 165; (*b*) D. H. Finseth, C. Sourisseau, and F. A. Miller, *J. Phys. Chem.*, 1976, **80**, 1248; (*c*) C. P. Casey and S. M. Neumann, *J. Amer. Chem. Soc.*, 1976, **98**, 5395.
[33] A. Vessieres and P. Dixneuf, *J. Organometallic Chem.*, 1976, **108**, C5.

ν(C≡O) bands are at 2086, 2055, 2029, 2000, 1970, and 1918 cm^{-1} and ν(C=O) at 1844 cm^{-1} in the trimethylenemethane-type complex (20).[34]

ν(CO) bands due to both terminal and bridging CO groups were seen in the pentalene complex of Fe, formulated as (21).[35]

New i.r. and Raman solution data have been given for $M(\eta^5-C_5H_5)_2$, where M = Fe or Ni. There do not appear to be any new assignments, but a discussion is based upon intensity data. ν_{16} (The ring tilt of e_{1g} symmetry) is assigned to the low value of *ca.* 200 cm^{-1} for M = Ni.[36]

ν(C=N) gives a band in the range 1670—1700 cm^{-1} for all of the complexes (22), where R = various alkyls.[37] Internal Cp modes, and ν(BC) and ν(BX) were assigned in the complexes $CpFe(C_5H_4R)$, where R = BX_2(F, Cl, Br, or I), BIPh, BMe_2, $B(SMe)_2$, $B(OEt)_2$, or $B(NEt_2)_2$.[38]

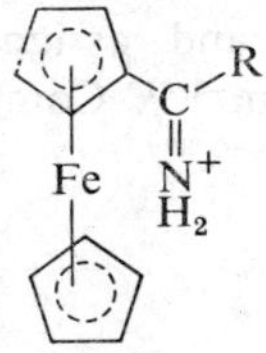

(22)

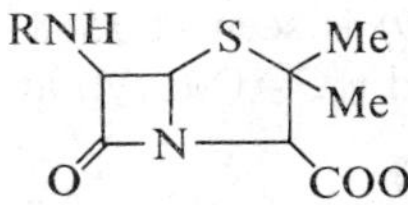

(23a)

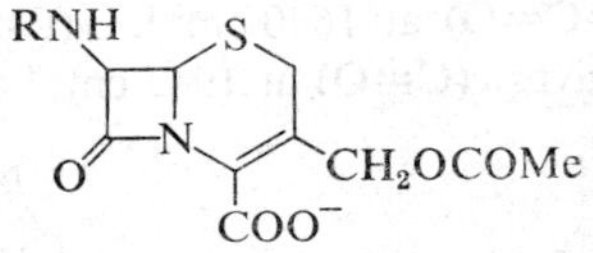

(23b)

A series of ferrocenyl-penicillins, (23a), and -cephalosporins (23b), have been prepared, where R = FcCO, $FcCH_2CO$, FcCH(Me)CO, or $FcC(Me)_2CO$; Fc = $C_{10}H_9Fe$. All show characteristic bands at about 1760 cm^{-1} due to the β-lactam carbonyl, 1640—1670 cm^{-1} (secondary amide carbonyl, and (for the cephalosporins only) at 1730 cm^{-1} (ester carbonyl).[39]

Addition of $C_7H_7^+PF_6^-$ to (cycloheptatriene)iron tricarbonyl yields polymeric species (24), where M = $Fe(CO)_3$. These give ν(CO) due to the cationic fragment at 2120 and 2080 cm^{-1}, and to the neutral fragments at 2050 and 1980 cm^{-1}.[40]

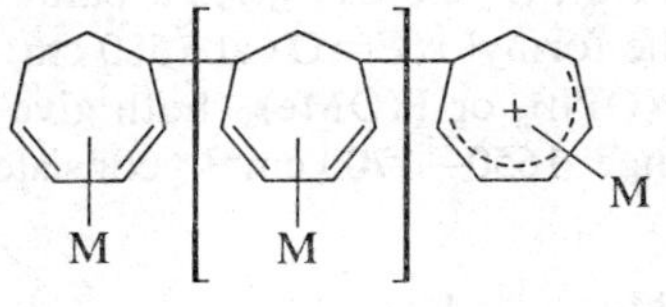

(24)

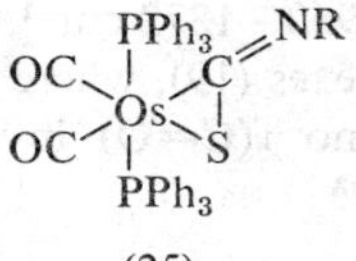

(25)

$[OsBr(CSCH_2CH_2S)(CO)_2(PPh_3)_2]^+$ shows bands due to the cyclic carbene ligand at 955 and 860 cm^{-1}. Confirmation that the ligand is cyclic came from ^1H n.m.r.[41]

[34] V. G. Andrianov, Yu. T. Struchkov, N. E. Kolobova, A. B. Antinova, and N. S. Obezyuk, *J. Organometallic Chem.*, 1976, **122**, C33.
[35] D. F. Hunt and J. W. Russell, *J. Organometallic Chem.*, 1976, **104**, 373.
[36] K. Yokoyama, S. Kobinata, and S. Maeda, *Bull. Chem. Soc. Japan*, 1976, **49**, 2182.
[37] D. Mourot and H. Patin, *J. Organometallic Chem.*, 1976, **114**, 89.
[38] T. Renk, W. Ruf, and W. Siebert, *J. Organometallic Chem.*, 1976, **120**, 1.
[39] E. I. Edwards, R. Epton, and G. Marr, *J. Organometallic Chem.*, 1976, **107**, 351.
[40] P. McArdle and H. Sherlock, *J. Organometallic Chem.*, 1976, **116**, C23.
[41] T. J. Collins, K. R. Grundy, W. R. Roper, and S. F. Wong, *J. Organometallic Chem.*, 1976, **107**, C37.

Attack of HS^- on $(OC)_2(PPh_3)_2Os(CNR)Cl^+$ produces the complex (25), where R = *p*-tolyl. Bands assigned as ν(CN) are seen at 1625, 1600, and 1580 cm^{-1}.[42]

ν(CH) of the alkyl group attached directly to Co in $CoR^1_2(acac)(PR^2_3)_2$, where R^1 = Me, Et, Pr^n, or Bu^i; R^2_3 = Et_3, Bu^n_3, $PhMe_2$, or Ph_2Me, have been assigned. Thus, when R^1 = Me, R^2 = Et they are at 2775 and 2840 cm^{-1}.[43] Similar assignments were reported for $CoMe(dpe)_2$, where dpe = 1,2-bis-(diphosphino)ethane, at 2870 and 2800 cm^{-1}. Action of CO on this complex gave (26), with ν(C≡O) 1970 and 1910 cm^{-1}, ν(C=O)(acyl) 1640 cm^{-1}, and ν(CH) (of acetyl group) 2840 cm^{-1}.[44]

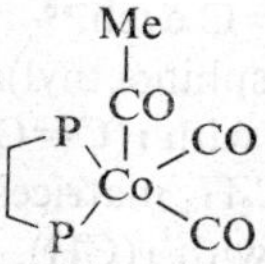

(26)

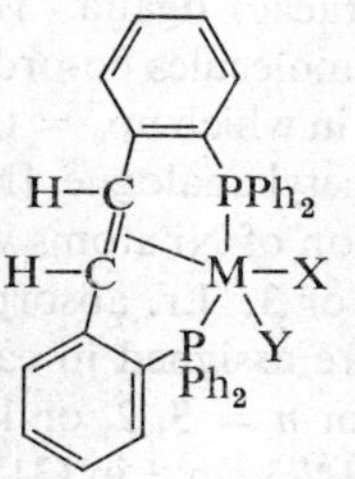

(27)

Complexes (27), with X = CO, Y = Cl, M = Rh or Ir; X = C_2H_4, Y = Cl, M = Rh or Ir; X = Y = CO, M = Rh^+ or Ir^+; X = Y = C_2H_4, M = Rh^+ or Ir^+, give an i.r. band at *ca.* 950 cm^{-1} due to the out-of-plane CH deformation of a *trans*-HC=CH, modified by co-ordination.[45]

Diagnostic bands are found in the 1400—1500 cm^{-1} region for the cobalticinium ions $[(C_5H_4R^1)Co(C_5H_4R^2)]^+$: (i) the unsubstituted salt gives one band, at 1420 cm^{-1}; (ii) 1,1′-disubstituted species give one band at 1400 cm^{-1}; (iii) monosubstituted species give two bands, at 1400 and 1420 cm^{-1}.[46]

The solid complex $RhCl(Ph_2PCH_2CH_2CH{=}CH_2)_2$ gives ν(C=C) at 1505 cm^{-1}, consistent with co-ordination of the double bond. In $CHBr_3$ solution, however, two bands are seen, at 1508 cm^{-1}, due to the five-co-ordinate species, and at 1638 cm^{-1}, due to the four-co-ordinate species. Related data were found for several similar complexes.[47]

γ-Carbon-bonded 2,4-pentanedionato complexes of Ir^{III} have been characterized by i.r., *e.g.* (28), where L = py, NH_3, *etc.*[48] The *ortho*-metallated structure, (29), L = PPh_3, has been confirmed by a diagnostic band at 810 cm^{-1}. Also the observed shift in ν(COC) was believed to suggest such a structure.[49]

42 K. R. Grundy and W. R. Roper, *J. Organometallic Chem.*, 1976, **113**, C45.
43 T. Ikariya and A. Yamamoto, *J. Organometallic Chem.*, 1976, **116**, 239.
44 T. Ikariya and A. Yamamoto, *J. Organometallic Chem.*, 1976, **116**, 231.
45 M. A. Bennett, R. N. Johnson, and I. B. Tomkins, *J. Organometallic Chem.*, 1976, **118**, 205.
46 N. El Murr, *J. Organometallic Chem.*, 1976, **112**, 189.
47 P. E. Garron, J. L. S. Curtis, and G. E. Hartwell, *Inorg. Chem.*, 1976, **15**, 3094.
48 M. A. Bennett and T. R. B. Mitchell, *Inorg. Chem.*, 1976, **15**, 2936.
49 M. Angoletta, L. Malatesta, P. L. Bellon, and G. Cagilo, *J. Organometallic Chem.*, 1976, **114**, 219.

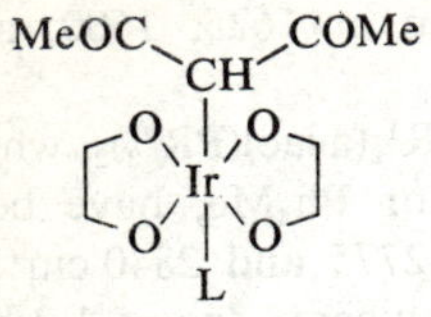

(28)

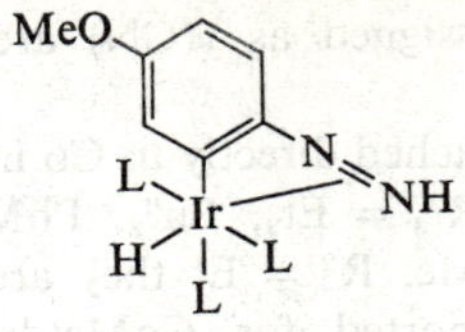

(29)

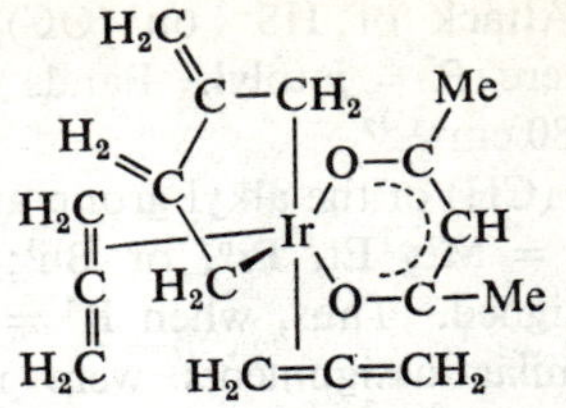

(30)

The complex (30) has i.r. bands at 1580 and 1510 cm^{-1} due to ν(C⋯O) and ν(C⋯C) of the (acac) ligand. ν(C=C) vibrations are at 1710 and 1690 cm^{-1}, from two allene molecules co-ordinated by one C=C each.[50]

$[NiMe(np_3)]^+$, in which np$_3$ = tris(2-diphenylphosphinoethyl)amine, reacts with CO to give the acetyl analogue, $[Ni(COMe)(np_3)]^+$, with ν(C=O) at 1660 cm^{-1}.[51a]

Co-condensation of Ni atoms with C_2H_4/Ar or C_2H_4 matrices gives $Ni(C_2H_4)_n$, where n = 1, 2, or 3. I.r. absorptions associated with ν(CH), ν(C=C), $\delta(CH_2)$, and $\delta_w(CH_2)$ were assigned in each case. The ν(C=C) values were 1512, 1465, and 1496 cm^{-1} for n = 3, 2, or 1, respectively.[51b]

ν(C=C) is at 1473 cm^{-1} in (31), *i.e.* about 80 cm^{-1} to lower wavenumber than in the free ligand, which agrees with the h^5-formulation given.[52]

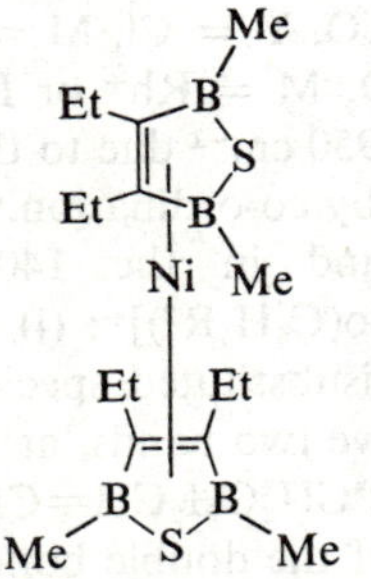

(31)

ν(C=O) of the ylide, in a number of complexes $PdCl(Bu^t_2NO)$(ylide), shift by 80—109 cm^{-1} to higher wavenumbers on complexation, as shown in Table 5.[53]

I.r. spectra of π-bonded C_2H_4 have been recorded on Pt and Pd surfaces at low temperatures.[54] ν(CC) of the allyl group in (32) is found, as expected, at 1455 cm^{-1}, whereas $\nu(CH_2)$ and $\nu(CH_3)$ are at 2955, 2912, and 2860 cm^{-1}.[55] Complex (33) produces i.r. bands assigned as ν(COC) at 1120 and 1065 cm^{-1}.[56]

[50] P. Diversi, G. Ingrosso, A. Immirizi, and M. Zocchi, *J. Organometallic Chem.*, 1976, **104**, C1.
[51] (*a*) L. Sacconi, P. Dapporto, and P. Stoppioni, *J. Organometallic Chem.*, 1976, **116**, C33; (*b*) H. Huber, G. A. Ozin, and W. J. Power, *J. Amer. Chem. Soc.*, 1976, **98**, 6508.
[52] W. Siebert, R. Full, C. Krueger, and Y.-H. Tsay, *Z. Naturforsch.*, 1976, **31B**, 203.
[53] M. Okunaka, G. Matsubayashi, and T. Tanaka, *Inorg. Nuclear Chem. Letters*, 1976, **12**, 813.
[54] Y. Soma, *J.C.S. Chem. Comm.*, 1976, 1004.
[55] P. Feldhaus, P. Ratka, H. Schmid, and M. L. Ziegler, *Z. Naturforsch.*, 1976, **31B**, 455.
[56] W. Lawrence, J. K. Ruff, and R. C. Taylor, *J.C.S. Chem. Comm.*, 1976, 9.

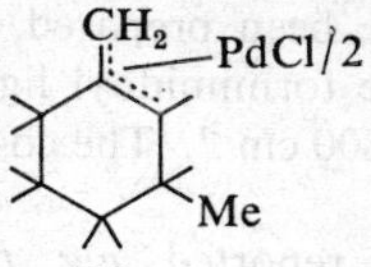

(32)

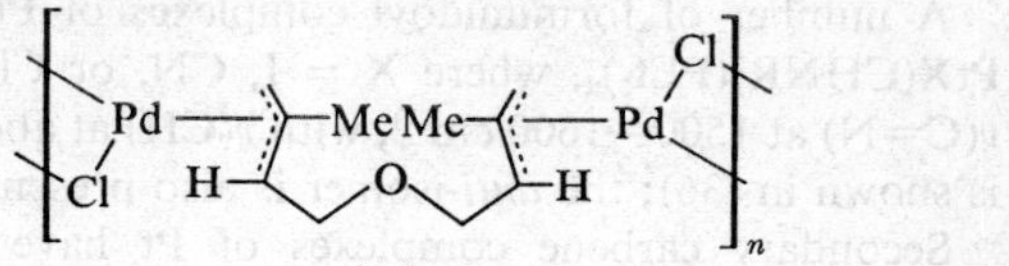

(33)

Table 5 *ν(CO) assignments for the* Pd *ylide complexes* $PdCl(Bu^t{}_2NO)$(ylide) *(all wavenumbers/cm⁻¹)*

Ylide	ν(CO)	Δν*(complex − free)*
$Ph_2MeP^+C^-HC(O)Ph$	1608	109
$Ph_3P^+C^-HC(O)Ph$	1603	80
$Ph_3As^+C^-HC(O)Ph$	1607	99
$Me_2S^+C^-HC(O)Ph$	1603	95

Carbamoyl complexes *trans*-$(Ph_3P)_2(Cl)M(CONHCHR^2CO_2R^1)$, where M = Pd or Pt; R^1, R^2 = H, Me, or Et, all show two ν(NH) bands. This is believed to be due to the presence of *cis*- and *trans*-isomers, (34a) and (34b). The

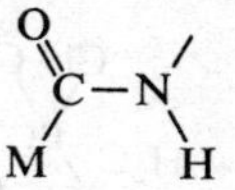

(34a)

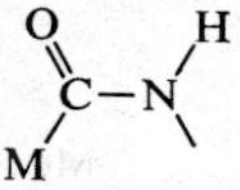

(34b)

observation was, however, only made in the solid. Bis(carbamoyl) complexes, $(R^1Ph_2P)_2Pt[-C(O)NR^2{}_2]_2$, where R^1 = Ph, R^2 = Me, Et, *etc.*, have ν(C=O) at about 1540 cm⁻¹, which is some 35 cm⁻¹ lower than in the related mono-carbamoyl derivatives.[57]

ν(C≡C) is at 1881 cm⁻¹ in $(PhC_2Ph)_2Pt$, and at 1851 cm⁻¹ in $[(Me_3P)(PhC_2Ph)Pt]_2$.[58]

Heating dicarbomethoxyacetylenebis(triphenylphosphine)platinum(0) to 130 °C in toluene solution gave a product with no bands due to σ-acetylide or π-bonded acetylene units. This product is (35), with strong i.r. bands at 1715 and 1702 cm⁻¹ due to the vinylic ester ν(C=O), and ν(C=C) of the vinyl group at 1582 cm⁻¹. In solution the i.r. spectrum contains a band at 1430 cm⁻¹, assigned as ν(C=C) of the *o*-metallated ring.[59]

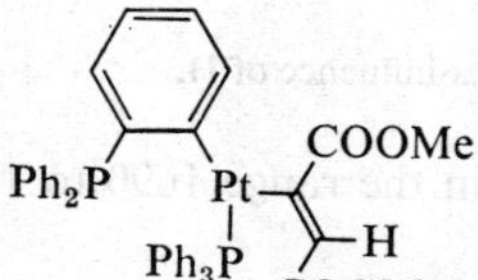

(35)

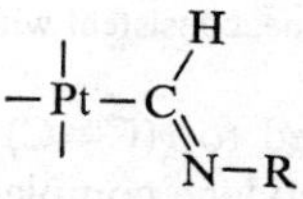

(36)

[57] W. Beck and B. Purucker, *J. Organometallic Chem.*, 1976, **112**, 361.
[58] M. Green, D. M. Grove, J. A. K. Howard, J. L. Spencer, and F. G. A Stone, *J.C.S. Chem. Comm.*, 1976, 759.
[59] H. C. Clark and K. E. Hine, *J. Organometallic Chem.*, 1976, **105**, C32.

A number of formimidoyl complexes of Pt have been prepared, *e.g. trans*-$PtX(CHNR)(PEt_3)_2$, where X = I, CN, or Cl. The formimidoyl ligand shows ν(C=N) at 1500—1600 cm^{-1}, with ν(CH) at about 2800 cm^{-1}. The co-ordination is shown in (36); the *anti*-isomer is also present.[60]

Secondary carbene complexes of Pt have been reported, *e.g. trans*-$[PtCl(CHNMeR)(PEt_3)_2]^{2+}$, with $\nu(C_{carbene}$—N) in the range 1535—1575 cm^{-1} in all cases. Some evidence was found for H-bonding interactions with the counter-ion, ClO_4^-.[61]

cis- and *trans*-Isomers of $(C_6F_5)_2PtL_2$, where L = PBu_3 or $AsPh_3$, all show bands typical of C_6F_5 at 1500, 1050, and 950 cm^{-1}; in addition, the *cis*-isomers have characteristic extra features at 790 and 780 cm^{-1}. In the Pt^{IV} complexes, $X_2(C_6F_5)_2PtL'$, where X = Cl or Br, L' = phen or bipy, bands in the 1000 cm^{-1} region were shifted by 10 to 15 cm^{-1} to higher wavenumber by comparison with the Pt^{II} values. Opposite shifts were seen in the 800 cm^{-1} region.[62]

ν(C=C) is at 1640 cm^{-1} and γ(=CH) at 965 cm^{-1} in complex (37), which are both consistent with the presence of a free C=C group.[63] ν(C≡C) is at 2110 ± 5 cm^{-1} in the series of complexes $(Ph_3P)PtL_2$, where L is one of the substituted acetylides $C{\equiv}CCMe_2(OH)$, C≡CCMe(OH)Et, or C≡CCMe=CHMe.[64]

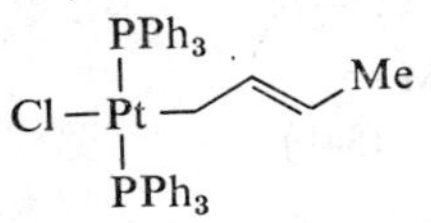

(37)

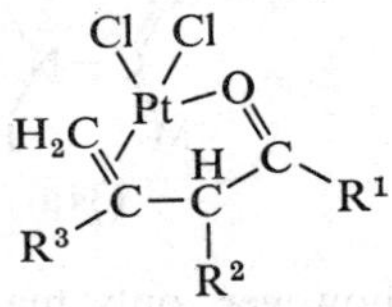

(38)

ν(C≡C) has been assigned, see Table 6, in the following σ-alkynyl complexes of Pt: *trans*-$[(Et_3P)_2(X)PtC_2C_6H_4C_2Pt(X)(PEt_3)_2]$, where X = Cl, Br, I, NCS, or H.[65]

Table 6 *Values of ν(C≡C) in some σ-alkynyl complexes of* Pt

X	ν(C≡C) *wavenumbers*/cm^{-1}
Cl	2120
Br	2120
I	2120
NCS	2100 (+νNC)
H	2095 [a]

[a] Note the low value, consistent with the large *trans*-influence of H.

A band assigned to ν(C≡C) was found in the range 1690 to 1775 cm^{-1} for a variety of Pt^0 acetylene complexes.[66]

60 D. F. Christian, H. C. Clark, and R. F. Stepaniak, *J. Organometallic Chem.*, 1976, **112**, 209.
61 D. F. Christian, H. C. Clark, and R. F. Stepaniak, *J. Organometallic Chem.*, 1976, **112**, 227.
62 R. Uson, J. Fornies, and P. Espinet, *J. Organometallic Chem.*, 1976, **116**, 353.
63 H. Kurosawa and G. Yoshida, *J. Organometallic Chem.*, 1976, **120**, 297.
64 A. Furlani, P. Carusi, and M. V. Russo, *J. Organometallic Chem.*, 1976, **116**, 113.
65 R. Nast and J. Moritz, *J. Organometallic Chem.*, 1976, **117**, 81.
66 D. Empsall, B. L. Shaw, and A. J. Stringer, *J.C.S. Dalton*, 1976, 185.

The species (38), with R^1, R^2, and R^3 = H, Me, or Et, all give ν(C=O) at about 1585 cm^{-1}, approximately 100 cm^{-1} lower than in the free ligand, which is consistent with the Pt—O=C co-ordination shown.[67]

The vibrational spectra of *trans*-$PtCl_2(C_2H_4)L$ have been reported where L = Cl^- or several O- and N-bonded ligands, *e.g.* DMF, urea, py, MeCN, aniline *etc.* The order of the *trans*-effect of the ligands is O < Cl^- < N, and this is reflected in a decrease in $\nu_s(PtC_2)$, and an increase in ν(C=C). A decrease in ν(C=C) is linearly related to an upfield shift in the ^{13}C chemical shift of the co-ordinated ethylene.[68]

I.r. data have been listed for the Zeise's salt analogue $[H_2C{=}CHCH_2NH_3{}^{+}\text{-}PtCl_3]^-$. Deuteriation was used as an aid to assignment, and the following suggestions were made: ν(Pt—C), 415 cm^{-1}; ν(Pt—Cl), 296, 336, and 340 cm^{-1}; and ν(C⋯C), 1349 or 1416 cm^{-1}.[69]

Co-condensation of $^{12}C_2H_4$ or $^{12}C_2H_4$/Ar with Cu at 10 K led to the first well-characterized Cu^0 ethylene complexes, $Cu(C_2H_4)_n$, where n = 1, 2, or 3. Proposed assignments are summarized in Table 7. These are consistent with

Table 7 *Some vibrational assignments for* $Cu(C_2H_4)_n$ (*all wavenumbers*/cm^{-1})

Vibration	$n = 1$	$n = 2$	$n = 3$
ν(CH)	3120	—	3054
ν(C—C)	1475	1508	1517
$\delta(CH_2)$	1164	1230	1252
	1155	1216	—
$\rho_w(CH_2)$	840	862	810

increased C=C and C—H bond strengths on increasing n. The C=C effects agree with the accepted notions on M—(ethylene) bonding.[70]

Similar experiments with Au gave the species $Au(C_2H_4)$, the first such Au^0 species. The vibrational assignments given, together with those for the Cu and Ag analogues, are listed in Table 8. The Au assignments were helped by

Table 8 *Some vibrational assignments for* $M(C_2H_4)$ (*all wavenumbers*/cm^{-1})

Vibration	M = Cu	M = Ag	M = Au
ν(C—C)	1475	1476	1476
$\delta(CH_2)$	1164	1152	1144
	1155	1132	1135
$\rho_w(CH_2)$	840	794	890

comparison with data on $(^{13}C_2H_4)Au$, *e.g.* ν(CC) shifted to 1452 cm^{-1}, $\delta(CH_2)$ at 1144 was shifted to 1124 cm^{-1}.[71]

The i.r. spectra of $Cu(L)(H_2O)_2ClO_4$ and $Cu_2(L)_2(H_2O)_3(ClO_4)_2$, where L = CO, C_2H_4, $CH_2{=}C{=}CH_2$, or $CH_2{=}CHCH{=}CH_2$, show shifts in ν(CO) or ν(CC) which indicate co-ordination of L to Cu^I. The data also suggested that the ClO_4 groups are weakly co-ordinated.[72]

[67] S. E. Earnest and D. B. Brown, *J. Organometallic Chem.*, 1976, **120**, 461.
[68] M. A. M. Meester, D. J. Stufkens, and K. Vrieze, *Inorg. Chim. Acta*, 1976, **16**, 191.
[69] R. Řeřicha and P. Svoboda, *Coll. Czech. Chem. Comm.*, 1976, **41**, 1014.
[70] H. Huber, D. McIntosh, and G. A. Ozin, J. *Organometallic Chem.*, 1976, **112**, C50.
[71] D. McIntosh and G. A. Ozin, *J. Organometallic Chem.*, 1976, **121**, 127.
[72] T. Ogura, *Inorg. Chem.*, 1976, **15**, 2301.

$\nu(C{=}C_{co\text{-}ord})$ has been assigned in a number of new Ag complexes containing π-bonded olefins. The values for (39a)—(39e) respectively are 1577; 1498; 1476, 1459; 1592, 1579; and 1573, 1580 cm^{-1}. These are all 50—80 cm^{-1} to lower wavenumber than in the free ligands.[73]

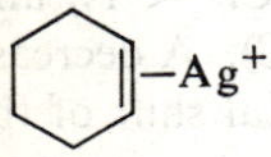

(39a)

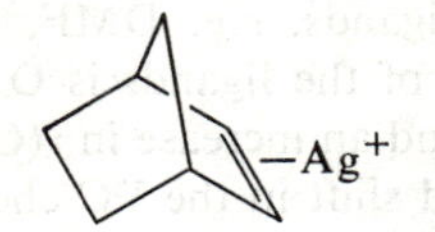

(39b)

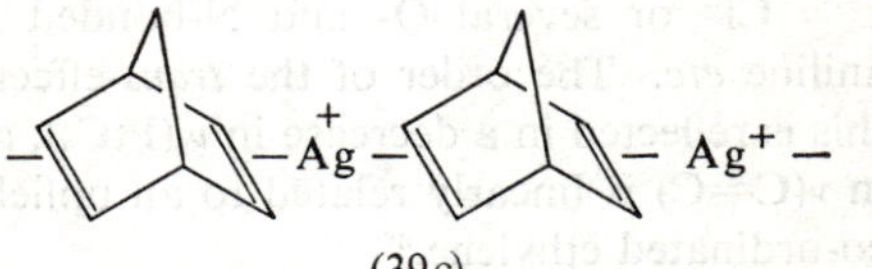

(39c)

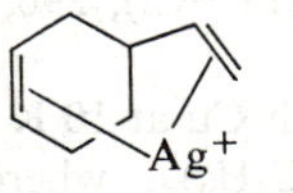

(39d)

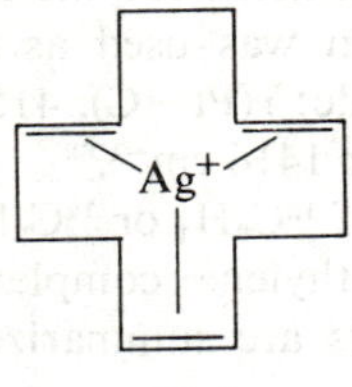

(39e)

I.r. and Raman spectra of $(C_5H_5)Ag(PPh_3)$ are consistent only with h^5-co-ordination of the cyclopentadienyl to the silver.[74]

The i.r. spectrum of $Ce(C_8H_8)_2$ contains strong bands at 692 and 902 cm^{-1}, with weaker features at 1325, 1640, and 1790 cm^{-1}. These are typical of h^8-C_8H_8, with aromatic character. $Ce_2(C_8H_8)_3$, on the other hand, gives bands due to two different types of cot ligand, and suggests that the structure is that of a triple-decked sandwich.[75]

1 : 1 Complex formation between HgX_2, where X = F or Cl, and C_2H_4 in Ar matrices at 10 K has been demonstrated by i.r. spectroscopy. Thus an increase of 25.8 cm^{-1} in $\omega(CH_2)$ compared to free C_2H_4 occurs with HgF_2; a shift of 77 cm^{-1} in this mode occurs for Zeise's salt. A normal π-type of Hg—C_2H_4 co-ordination was suggested.[76]

The position of $\nu(C{=}C)$ in the series (40, R various) and (2,3-dimethyl-2-butyl)B(CH=CHR)$_2$ is independent of the electronic nature of R, and hence is of no use in determining the extent of conjugative interactions.[77]

$Et_2Al(CH{=}CHEt)$ and $Bu^i_2Al(CH{=}CHBu^i)$ both give $\nu(C{=}C)$ at 1556 cm^{-1}. This is increased by about 30 cm^{-1} on formation of adducts with Et_2O, py, PhCN, PhOMe, or Et_3N, due to the decreased acceptor properties of the Al.[78]

I.r. and Raman spectra of the Et_2O adduct of $CpAlEt_2$ show conclusively that the Cp is h^5-bonded to the Al. For uncomplexed Et_2MCp, where M = Al, Ga, or In, the evidence suggested that h^2/h^3-co-ordination, (41), is occurring.[79]

$\nu_{as}(CO_2)$ at 1540, $\nu_s(CO_2)$ at 1395, and $\nu(C{=}C)$ at 1625 cm^{-1} are among the i.r. data reported for (42).[80]

[73] G. S. Lewandos, D. K. Gregston, and F. R. Nelson, *J. Organometallic Chem.*, 1976, **118**, 363.
[74] H. K. Hofstee, J. Boersma, and G. J. M. van der Kelen, *J. Organometallic Chem.*, 1976, **120**, 313.
[75] A. Greco, S. Cesca, and G. Bertolini, *J. Organometallic Chem.*, 1976, **113**, 321.
[76] D. Tevault and K. Nakamoto, *Spectroscopy Letters*, 1976, **9**, 185.
[77] G. Zweifel, B. M. Clark, T. Leung, and C. C. Whitney, *J. Organometallic Chem.*, 1976, **117**, 303.
[78] K. B. Starowieyski, C. Bandlow, and K. Haage, *J. Organometallic Chem.*, 1976, **117**, 215.
[89] J. Stadelhofer, J. Weidlein, P. Fischer, and A. Haaland, *J. Organometallic Chem.*, 1976, **116**, 55.
[70] S. Uemura, H. Miyoshi, H. Taro, M. Okano, and K. Ichikawa, *J.C.S. Chem. Comm.*, 1976, 218.

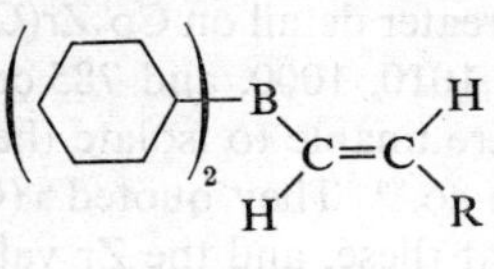
(40)

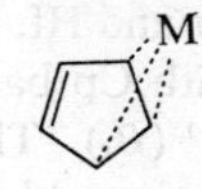
(41)

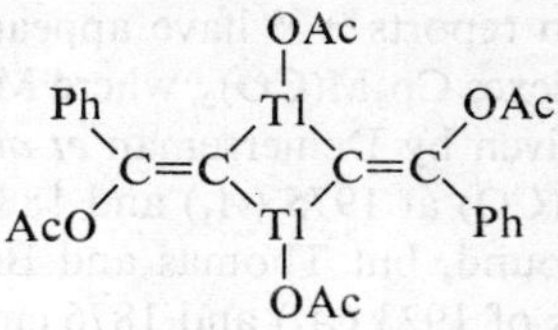
(42)

I.r. bands near 1500 and 3000 cm^{-1} indicate the presence of π-bonded species from C_2H_4 chemisorbed on SiO_2-supported Pd and Pt catalysts. These co-exist with σ-bonded M—CH_2CH_2—M species, but are readily removed by H_2.[81]

ν(C=N) has been assigned as shown in Table 9 for the *N*-(diphenylmethylene)-triorgano-element-methylamines, $Ph_2C{=}NCH_2{-}MR_3$. The value shifts to lower wavenumbers in the sequence Sn > Ge > Si > C, presumably due to conjugative interaction between M and the N=C group.[82]

Table 9 ν(C=N) *assignments in* $Ph_2C{=}NCH_2{-}MR$/ (*all wavenumbers*/cm^{-1})

MR_3	ν(C=N)
H	1629
CMe_3	1627
$SiMe_3$	1614
$SnMe_3$	1612
CPh_3	1629
$SiPh_3$	1618
$GePh_3$	1615
$SnPh_3$	1612

I.r. and Raman spectra have been obtained for $Me_3SnC_5H_5$ and $(Me_3Sn)_2C_5H_4$. These were assigned using the σ-bonded, h^1, structures deduced from electron diffraction results.[83]

2 Carbonyl and Thiocarbonyl Complexes

The first Ti^{IV} complex, (43), has been isolated. The cation gives ν(C≡N) at 2181 and 2104 cm^{-1}, and ν(C≡O) at 2055 cm^{-1}, these being differentiated by $C^{18}O$ substitution. ν(CO) is 115 cm^{-1} to higher wavenumber than in $Cp_2Ti(CO)_2$.[84]

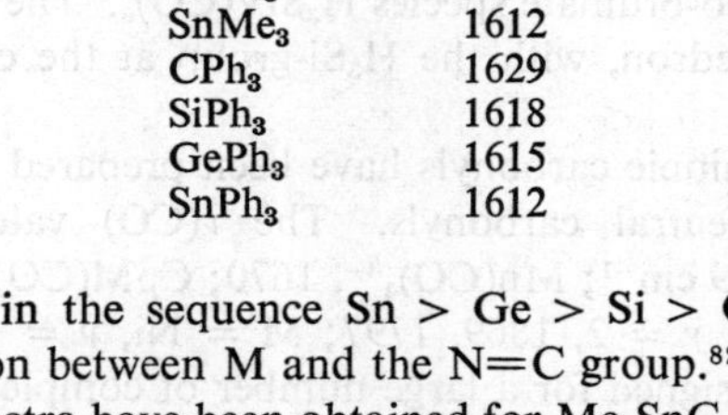
(43)

[81] J. D. Prentis, A. Lesinnas, and N. Sheppard, *J.C.S. Chem. Comm.*, 1976, 76.
[82] E. Popowski, A. Hahn, and H. Kelling, *J. Organometallic Chem.*, 1976, **110**, 295.
[83] P. G. Harrison and J. A. Richards, *J. Organometallic Chem.*, 1976, **108**, 61.
[84] B. Demerseman, M. Pankowski, G. Bouquet, and M. Bigorgne, *J. Organometallic Chem.*, 1976, **117**, C10.

Two reports [85, 86] have appeared giving details of the preparation of the new complexes $Cp_2M(CO)_2$, where M = Zr and Hf. The greater detail on $Cp_2Zr(CO)_2$ was given by Demerseman *et al.*,[85] with Cp bands at 1010, 1000, and 785 cm^{-1}, and ν(CO) at 1975 (A_1) and 1886 cm^{-1} (B_1). They were unable to isolate the Hf compound, but Thomas and Brown were able to do so.[86] They quoted ν(CO) values of 1923 (A_1) and 1876 cm^{-1} (B_1), and noted that these, and the Zr values, were markedly solvent-dependent (figures given are for hexane solutions).

$V(CO)_n$, where n = 1—5, have been characterized in matrices by i.r. spectroscopy, using $^{12,13}C$ substitution; normal co-ordinate analyses were carried out to assist the assignments, and to confirm the monomeric nature of the species, and their stoicheiometry. The following geometries were proposed: n = 5, D_{3h}; 4, T_d or D_{4h}; 3, D_{3h} or C_{3v}; 2, C_{2v} and $D_{\infty h}$. EHMO calculations were performed on V(CO) and $V(CO)_2$ as a function of VCO and CVC angles in an attempt to rationalize the observed ν(CO) values, and the remarkable observation of three different forms for n = 2.[87]

ν(CO) of A_1 and E symmetry were seen at 1975 and 1923 cm^{-1} respectively in $(C_6Ph_6)V(CO)_3$.[88]

ν(CO) bands were observed at 2088, 2026, 1990, and 1959 cm^{-1} in the i.r. spectrum of the seven-co-ordinate species $H_3SiV(CO)_6$. The most likely structure is a face-capped octahedron, with the H_3Si-group at the capping position, C_{3v} symmetry.[89]

A number of new anionic carbonyls have been prepared by the Na reduction of the appropriate neutral carbonyls. The ν(CO) values are as follows: $CpV(CO)_3^{2-}$, 1742, 1619 cm^{-1}; $Mn(CO)_4^{3-}$, 1670; $CpM(CO)_y^-$, where M = Mo, y = 3, 1786; M = Fe, y = 2, 1869, 1797; M = Ni, y = 1, 1890, 1848 cm^{-1}.[90]

ν(CS) values were assigned for a large number of complexes of CS containing Cr, Mo, or W, *e.g.* for $M(CO)_5(CS)$: 1253 (Cr), 1247 (Mo), 1258 cm^{-1} (W).[91]

Co-ordination of BF_4^- in *mer*-$(F_4B)Cr(CO)_3(PMe_3)CMe$ was confirmed by i.r. spectroscopy. B—F modes were seen at 745 cm^{-1} (ν_1 of free BF_4^-), 910/940 ($^{11}B/^{10}B$ components) and 1110 cm^{-1} (ν_3).[92]

Cotton–Kraihanzel force constants were calculated for *cis*-$(OC)_4M(L)_2$ and for $(OC)_3M(L)_3M(CO)_3$, where M = Cr, Mo, or W; L = diazabicyclo[2,2,1]-hept-2-ene. In the first type, L is co-ordinated by only one of the N lone pairs of the azo-group, whereas in the second, both N atoms are co-ordinated in the bridging L ligand. The force constant data are interpreted as supporting π-acceptor ability by the azo-group of L.[93]

I.r. spectra in the ν(CO) region of $[(OC)_4M^1(L—L)M^2(CO)_4]^{2-}$, with M^1, M^2 = Cr, Mo, or W, HL—LH = 2,2′-bi-imidazole or 5,5′-bitriazole, indicate that the two azole rings are not coplanar.[94]

[85] B. Demerseman, G. Bouquet, and M. Bigorgne, *J. Organometallic Chem.*, 1976, **107**, C19.
[86] J. L. Thomas and K. T. Brown, *J. Organometallic Chem.*, 1976, **111**, 297.
[87] L. Hanlan, H. Huber, and G. A. Ozin, *Inorg. Chem.*, 1976, **15**, 2592.
[88] M. Schneider and E. Weiss, *J. Organometallic Chem.*, 1976, **114**, C43.
[89] J. S. Allinson, B. J. Aylett, and H. M. Colquhoun, *J. Organometallic. Chem.*, 1976, **112**, C7.
[90] J. E. Ellis, R. A. Faltynek, and S. G. Hentges, *J. Organometallic Chem.*, 1976, **120**, 389.
[91] B. D. Dombek and R. J. Angelici, *Inorg. Chem.*, 1976, **15**, 1089.
[92] K. Richter, E. O. Fischer, and C. G. Kreiter, *J. Organometallic Chem.*, 1976, **122**, 137.
[93] M. N. Ackermann and L.-J. Kou, *Inorg. Chem.*, 1976, **15**, 1423.
[94] W. Beck, F. Götzfried, and M. Riederer, *Z. anorg. Chem.*, 1976, **423**, 97.

The kinetics and stereochemistry of the ^{13}CO exchange in several complexes of the type $Cr(CO)_4$(chelate), where chelate = phen, bipy, diphos *etc.*, have been examined using i.r. spectroscopy in the ν(CO) region. The frequency shifts associated with all possible degrees and positions of ^{13}CO incorporation were evaluated by using initially the Cotton–Kraihanzel approximation for the all-^{12}CO compounds.[95]

The high-resolution vapour-phase i.r. spectrum of $Cr(CS)(CO)_5$ at 312 K has been reported. P.q.r. separations were measurable for the three ν(CO) fundamentals, ν_1, ν_2, and ν_{16}. These were consistent with C_{4v}, prolate symmetrical top geometry.[96]

The presence of four ν(CO) bands in the i.r. spectra of $M(CO)_5L$, when M = Cr, Mo, or W; L = benzo[*c*]cinnoline or phenanthridine, in very similar positions to those seen in $M(CO)_5$py derivatives, suggests their formulation as (44a), rather than as the alternative, (44b), where X = N or CH.[97]

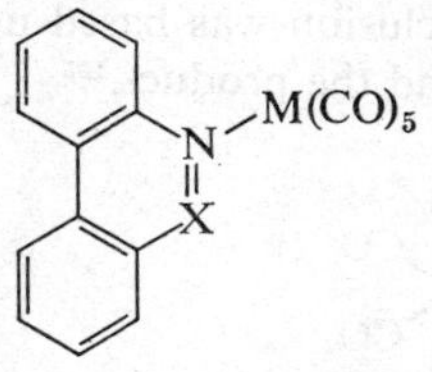

(44a)

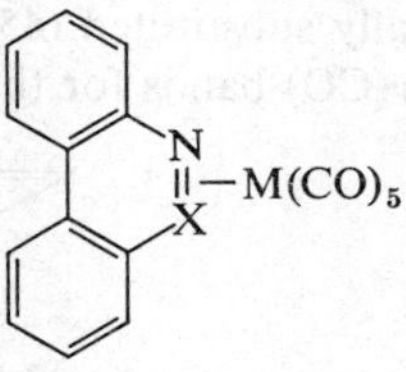

(44b)

I.r. spectroscopy in the ν(CO) and ν(CS) regions has shown that u.v. photolysis (300 ± 10 nm) of $M(CO)_5(CS)$, M = Cr or W, in Ar or CH_4 matrices at 20 K produces a mixture of two isomeric forms of $M(CO)_4(CS)$. There is no evidence for CS loss. The two isomers are thought to be both square pyramidal, one with CS axial (C_{4v}), the other with CS basal (C_s).[98]

Raman spectra of mixed crystals $M^1_xM^2_{1-x}(CO)_6$, where $0 < x < 1$; M^1, M^2 = Cr, Mo, or W, in the ν(CO) region show that bands due to molecular e_g modes are averaged. Features originating in a_{1g} vibrations, on the other hand, are superimposed. This is thought to be due to the absence of factor group splitting for the latter features. Preliminary results were also reported for $Mn_2(CO)_{10}/Re_2(CO)_{10}$ mixtures.[99]

I.r. spectroscopy in the ν(CO) region has been used in a study of surface carbonyl species formed when catalysts comprising $M(CO)_6$, M = Cr, Mo, or W, supported on Al_2O_3 or SiO_2 are activated by heating. Evidence, including the use of ^{13}CO, supports the reversible formation of surface carbonyl complexes.[100]

The ν(CO) values for $(\eta^5\text{-}C_5H_5)Mo(CO)_2(R)L$, where R = COMe, L = PPh_3, $AsPh_3$, or $SbPh_3$; R = Me, L = PPh_3 or $AsPh_3$, gave no indication of significant differences in σ-/π-interactions for the different L. The acetyl derivatives gave ν(C=O) at 1605 cm^{-1}.[101]

[95] M. A. Cohen and T. L. Brown, *Inorg. Chem.*, 1976, **15**, 1417.
[96] I. S. Butler and C. F. Shaw, *J. Mol. Struct.*, 1976, **31**, 359.
[97] M. Kooti and J. F. Nixon, *J. Organometallic Chem.*, 1976, **105**, 217.
[98] M. Poliakoff, *Inorg. Chem.*, 1976, **15**, 2022.
[99] D. Kariuki and S. F. A. Kettle, *J. Organometallic Chem.*, 1976, **105**, 209.
[100] R. F. Howe, *Inorg. Chem.*, 1976, **15**, 486.
[101] A. C. Gingell, A. Harris, A. J. Rest, and R. N. Turner, *J. Organometallic Chem.*, 1976, **121**, 205.

The presence of two equally-intense ν(CO) bands in the i.r. spectra of $(\eta^3\text{-}C_3H_4R^2)M(CO)_2(R^1N{=}CHCH{=}NR^1)X$, where M = Mo or W, R^1, R^2 = various, and X = Cl, Br, I, or NCS, shows that the two CO groups are *cis*.[102]

Ph_4As^+ salts of $CpM(CO)_3^-$, where M = Mo or W, unlike the Na^+ or K^+ salts, give simple ν(CO) spectra, entirely consistent with C_{3v} local symmetry for the $M(CO)_3$ unit. The difference was ascribed to a lack of ion-pairing in the Ph_4As^+ salts.[103]

Protonation of $M(CO)_{6-n}(PA_3)_n$, where M = Mo or W, A = Me or OMe, and n = 3 or 4, by trifluoroacetic acid leads to high-wavenumber shifts of about 90 cm^{-1} in ν(CO). Thus the proton is attached to the M in every case. Protonation also leads to structural changes, as shown by increasing numbers of ν(CO) bands, when n = 3, or by variations in their relative intensities, when n = 4. Complexes with n = 2 were not protonated under these conditions.[104]

(Bicyclo[6,1,0]nona-2,4,6-triene)$Mo(CO)_3$ reacts with ^{13}CO stereospecifically to give the axially substituted (45). This conclusion was based upon a detailed assignment of ν(CO) bands for the reactant and the product.[105]

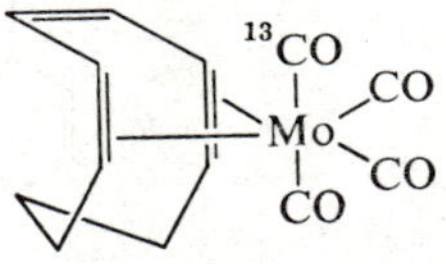

(45)

$[H_2NC_5H_{10}][(OC)_5MoP\{P(O)(OEt)_2\}_2]$ is analogous to the previously known $(OC)_5MoNH_2C_6H_{11}$ in that four i.r. ν(CO) bands are observed in CH_2Cl_2 solution. Thus, the formally-inactive B_1 mode is seen in the i.r.[106]

ν(CO) in $Cp(CO)_2W{\equiv}CR$, when R = NEt_2, are much lower than in the analogous complexes where R = Ph or substituted phenyl. This is due to the π-interaction of the W—C bond with the lone pair on the N.[107]

The mechanism of the substitution reactions of $(LL)W(CO)_4$, in which the chelating L—L is replaced by $P(OCH_2)_3CMe$, has been shown to involve ring-opening, in which the intermediate is :L—$LM(CO)_4$. ν(CO) values were quoted for this intermediate in xylene solution (L—L = 2,2,8,8-tetramethyl-3,7-dithianonane).[108]

The photolysis of $W(CO)_5(CS)$ to $W(CO)_4(CS)$ has been studied by i.r. matrix spectroscopy and other methods. *trans–cis*-Rearrangement occurs in the process, detected by ^{13}CO labelling.[109]

$Mn(CO)_3F_3$ has been formed for the first time. Two ν(CO) bands were seen, at 2037 (a_1) and 1944 cm^{-1} (e), and a weak shoulder at 1922 cm^{-1} was ascribed to crystal effects. The molecular structure is like that of $Re(CO)_3F_3$. The dimer

[102] A. T. T. Hsieh and B. O. West, *J. Organometallic Chem.*, 1976, **112**, 285.
[103] J. E. Ellis, *J. Organometallic Chem.*, 1976, **111**, 331.
[104] M. S. Arabi, R. Mathieu, and R. Poilblanc, *J. Organometallic Chem.*, 1976, **104**, 323.
[105] D. J. Darensbourg and A. Slazer, *J. Organometallic Chem.*, 1976, **117**, C90.
[106] D. Weber and E. Fluck, *Inorg. Nuclear Chem. Letters*, 1976, **12**, 515.
[107] E. O. Fischer, T. L. Lindner, and F. R. Kreissl, *J. Organometallic Chem.*, 1976, **112**, C27.
[108] L. D. Schultz and G. R. Dobson, *J. Coordination Chem.*, 1976, **5**, 163.
[109] M. Poliakoff, *Inorg. Chem.*, 1976, **15**, 2892.

$[Mn(CO)_4F]_2$ shows terminal ν(CO) bands only, and therefore the dimerization takes place solely through a Mn—Mn bond.[110]

$[Mn(CO)_3triphos]^+[Mn(CO)_5]^-$ gives ν(CO) bands due to the cation at 2030 (A_1) and 1962 cm^{-1} (E), and to the anion at 1902 (A''_2) and 1860 cm^{-1} (E'), triphos = $MeC(CH_2PPh_2)_3$.[111] $Mn_2(CO)_6(dpm)_2$, where dpm = $Ph_2PCH_2PPh_2$, is assigned structure (46), based upon the ν(CO) pattern in the i.r.[112a] $[(\eta^5-C_5H_5)$-

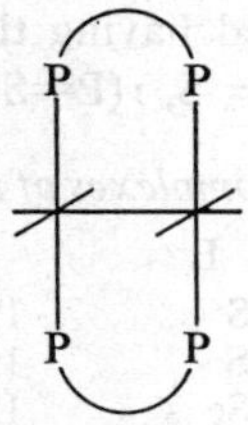

(46)

$Mn(CS)(NO)]_2$ has been claimed as the first to contain a bridging thiocarbonyl ligand. ν(NO) is at 1750 and 1718 cm^{-1}, and ν(CS) at 1150 and 1118 cm^{-1} in solution. The ν(NO) values were taken to indicate terminal NO groups only. Thus it must be doubly CS bridged, with both *cis*- and *trans*-isomers present.[112b] ν(CO) in the 1996—1966 cm^{-1} range varies according to the σ-donor strength of L in [*meso*-tetra(*p*-anisidyl)porphinato]Fe(L)(CO), where L = morpholine, pyrrolidine, piperidine, imidazole, or pyridine. The strongest donor is imidazole, with a band at 1966 cm^{-1}. There is no need to invoke π-donation from L to low-spin Fe^{II}.[113]

The presence of only one ν(CO) band in the i.r. spectra of η-$C_5H_5Fe(CO)LX$, where L = $P(OCH_2)_3CR$, R = Me, Et, or Pr; X = Cl, Br, or I, shows that all exist as single conformers, with the non-ionic form shown. All of the ν(CO) values are lower than for the corresponding dicarbonyl halides, thus L is a poorer π-acceptor than CO. Also, as expected, the order varies Cl > Br > I. Changing R, on the other hand, has little effect on ν(CO).[114]

trans-$[(\eta-C_5H_5)Fe(CS)(CO)]_2$ gives ν(CO) at 1979, and ν(CS) at 1131 cm^{-1}. The *cis*-isomer shows ν(CO) at 2011 and 1982 cm^{-1}, with ν(CS) at 1124 cm^{-1}.[115] ν(CO) and ν(CS) have also been assigned for a series of complexes containing the cation π-$(C_5H_5)Fe(CO)(CS)(L)^+$, *e.g.* L = CO, PPh_3 or $AsPh_3$.[116]

The novel ligand L = $(\eta-C_5H_5(OC)_2Fe)_3Sb$ has a high σ-donor : π-acceptor ratio, as shown by the wavenumber of the $A_1^{(2)}$ mode in $LCr(CO)_5$, 2042 cm^{-1}, and in $LW(CO)_5$, 2052 cm^{-1}.[117]

[110] M. K. Chaudhuri, M. M. Kaschani, and D. Winkler, *J. Organometallic Chem.*, 1976, **113**, 387.
[111] J. Ellermann and H. A. Lindner, *Z. Naturforsch.*, 1976, **31b**, 1350.
[112] (*a*) K. G. Caulton and P. Adair, *J. Organometallic Chem.*, 1976, **114**, C11; (*b*) A. Efrati, R. Arneri, and M. H. A. Huang, *J. Amer. Chem. Soc.*, 1976, **98**, 639.
[113] W. M. Connor and D. K. Straub, *Inorg. Chem.*, 1976, **15**, 2289.
[114] W. E. Stanclift and D. G. Hendricker, *J. Organometallic Chem.*, 1976, **107**, 341.
[115] J. W. Dunker, J. S. Finer, J. Clardy, and R. J. Angelici, *J. Organometallic Chem.*, 1976 **114**, C49.
[116] L. Busetto and A. Palazzi, *Inorg. Chim. Acta*, 1976, **19**, 233.
[117] W. Malisch and P. Panster, *Angew. Chem. Internat. Edn.*, 1976, **15**, 618.

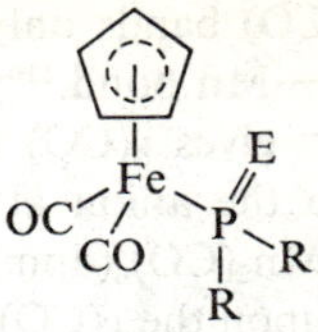

(47)

Compounds have been prepared having the structure (47), and their $\nu(CO)$ values listed, Table 10. When E = S, $\nu(P{=}S)$ is at about 620 cm^{-1}.[118]

Table 10 $\nu(CO)$ *assignments for complexes of the type* (47) (*all wavenumbers*/cm^{-1})

R	E	νCO
Me	S	1978, 2028
Et	S	1996, 2045
Ph	Se	1985, 2035
Me	SMe	2025, 2065
OEt	SMe	2023, 2065
Ph	SMe	2018, 2058
Me	SEt	2015, 2058

trans-$Fe(CO)_3(PMe_3)_2$ has two $\nu(CO)$ bands, at 1967 and 1877 cm^{-1}. Formation of the adduct with $AlMe_3$, $Fe(COAlMe_3)(CO)_2(PMe_3)_2$ leads to the appearance of bands at 1977 and 1905 cm^{-1}, together with another at 1763 cm^{-1} due to the CO—$AlMe_3$ unit.[119]

$\nu(CO)$ and oxidation potentials (*Er*) have been measured for *trans*-$Fe(CO)_3L_2$, where L = PPh_3, PPh_2Me, $P(OPh)_3$, or $P(NMe_2)_3$; there is a linear correlation between the two parameters.[120]

$\nu(CO)$ wavenumbers for *cis*- and *trans*-Fe(*trans*-σ-CF=$CFCF_3$)$(CO)_4I$, (48), are given in Table 11.[121]

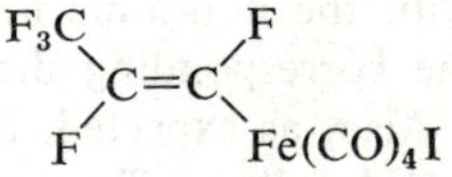

(48)

Table 11 $\nu(CO)$ *wavenumbers*/cm^{-1} *for the complexes* (48)

Isomer:	*cis*	*trans*
	2141.5 (m)	2147.2 (w)
	2093.1 (vs)	2088.2 (vs)
	2071.4 (s)	—

$\nu(CO)_b$ of $HFe_3(CO)_{11}^-$ as the Et_4N^+ salt or as the Et_3HN^+ salt in polar solvents lies between 1710 and 1740 cm^{-1}. For the Et_3HN^+ salt in non-polar solvents, however, this drops to 1650 cm^{-1}, thought to be due to the formation of a tight ion-pair under these conditions.[122]

[118] P. Piriano, F. Faraone, and M. C. Aversa, *J.C.S. Dalton*, 1976, 610.
[119] M. Pankowski, M. Bigorgne, and Y. Chauvin, *J. Organometallic Chem.*, 1976, **110**, 331.
[120] N. G. Connelly and K. R. Somers, *J. Organometallic Chem.*, 1976, **113**, C39.
[121] R. Stanley and D. W. McBride, *Canad. J. Chem.*, 1976, **54**, 1700.
[122] J. R. Wilkinson and L. J. Todd, *J. Organometallic Chem.*, 1976, **118**, 199.

The preparation of a number of arylazo-derivatives of Ru has been reported: $[Ru(p\text{-}RC_6H_4N_2)(CO)_2L_2]^+$, where L = PPh_3, R = NO_2, F, H, or OMe; L = $PPh_2(p\text{-}MeC_6H_4)$, R = OMe, as well as that for related aryl-di-imine derivatives. Two bands were assigned to ν(CO), showing that the two CO groups are *cis*. Anomalous absorption in the ν(CO) region were thought to originate from interactions with CO_2.[123]

The Ru carbonyl fluorides $[Ru(CO)_3F_2]_4$, $[Ru(CO)_3F_2.RuF_5]_2$, and $Ru(CO)_3F_3$ have been prepared; the last only is monomeric. For the tetramer, the i.r. spectrum was assignable in terms of C_{3v} local symmetry for the $Ru(CO)_3$ units. The monomer is genuinely C_{3v}; the assignments are summarized in Table 12.[124]

Table 12 *Some vibrational assignments for* Ru *carbonyl fluorides* (*wavenumbers*/cm^{-1})

$[Ru(CO)_3F_2]_4$	$Ru(CO)_3F_3$	*Assignment*	$[Ru(CO)_3F_2.RuF_5]_2$	*Description*
2140	2148	$\nu_1(a_1)$	2196	CO str
			2157	^{13}CO str
2060	2072	$\nu_3(e)$	2122	CO str
1998	1995			^{13}CO str
652	649	$\nu_2(a_1)$	654	(Ru—F) str
594	596	$\nu_4(e)$		(Ru—F) str
532	544			(Ru—C) str
502	503		533	and
484	486			CO bend

Reaction of $Ru_3(CO)_{12}$ with $Mn(CO)_5^-$ gives $Ru_6(CO)_{18}^{2-}$, which on acidification gives $HRu_6(CO)_{18}^-$. A study of the ν(CO) bands implied that the former does not possess metal atom geometry close to O_h. The *X*-ray structure of the latter, together with ν(CO) data, on the other hand, suggest that the H atom lies at the centre of an octahedron of Ru atoms, in contrast also to the Os analogues.[125]

ν(CO) i.r. wavenumbers have been reported for $Os(MR_3)(CO)_2(1,5\text{-}\eta\text{-}C_7H_9)$, where MR_3 = $SiMe_3$, $SiMe_2Ph$, or $GeMe_3$. These compounds have three strong bands, indicating that rotameric isomers are present, which was confirmed by n.m.r. results.[126]

The presence of six ν(CO) bands in the i.r. spectrum of $Os_2(CO)_6(C_8H_6)$ suggested that the structure was very unsymmetrical, (49).[127]

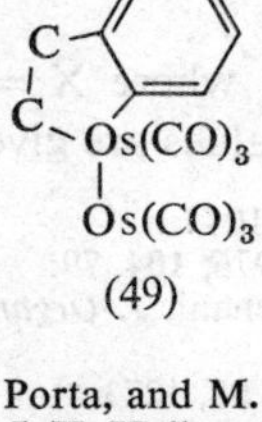

(49)

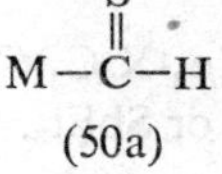

(50a)

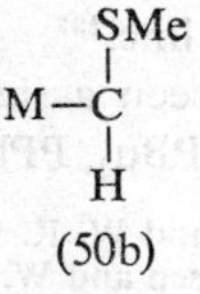

(50b)

[123] S. Cenini, F. Porta, and M. Pizzotti, *Inorg. Chim. Acta*, 1976, **20**, 119.
[124] A. J. Hewitt, J. H. Holloway, R. D. Peacock, J. B. Raynor and I. L. Wilson, *J.C.S. Dalton*, 1976, 579.
[125] C. R. Eday, B. F. G. Johnson, J. Lewis, M. C. Malatesta, P. Machin, and M. McPartlin, *J.C.S. Chem. Comm.*, 1976, 945.
[126] S. A. R. Knox, R. P. Phillips, and F. G. A. Stone, *J.C.S. Dalton*, 1976, 552.
[127] P. J. Harris, J. A. K. Howard, S. A. R. Knox, R. P. Phillips, F. G. A. Stone, and P. Woodward, *J.C.S. Dalton*, 1976, 377.

ν(CS) is at 1280 cm^{-1} in OsHCl(CS)L_3, 1295 cm^{-1} in OsHCl(CS)(CO)L_2, 1010 cm^{-1} in Os(CHS)Cl(CO)$_2L_2$, and 965 cm^{-1} in [OsCl(CHSMe)(CO)$_2L_2$]$^+$. L = PPh_3 in each case, and the values given for the last two complexes show that the co-ordination types are as shown in (50a) and (50b), respectively.[128]

Mononuclear *cis*-dicarbonyls of Os^{II} give, in suitable organic solvents at 20 °C, *mer*-[$OsX_3(CO)_3$]$^-$, and above 100 °C, the *fac*-isomers, where X = Cl, Br, or I. ν(CO) bands for all of these were assigned from solid-phase i.r. and Raman spectra, *e.g. mer*-[$OsCl_3(CO)_3$]$^-$, $\nu_1(A_1)$ 2139, $\nu_2(A_1)$ 1978, $\nu_{19}(B_2)$, 2044 cm^{-1}, and for *fac*-[$OsCl_3(CO)_3$]$^-$, $\nu_1(A_1)$ 2111, $\nu_9(E)$ 2003—2020 (split in solid) cm^{-1}.[129]

ν(CO) spectra of $H_3Os_4(CO)_{11}(HC_2XR)$ and $H_2Os_4(CO)_{11}(HC_2R)$, where R = H, Ph, or Bu^t, are very complex. Thus the clusters must have low symmetry, with different structures from the formally analogous Ru complexes.[130]

The compound (η^5-C_5H_5)Co(CO)$_3$ is exceptionally deformable; designated *fictile* on that account. In the solid phase ν(CO) are at 1833, 1775, and 1673 cm^{-1}. There is one triply bridging CO and two 'semi-bridging' groups. The wavenumbers in benzene, THF, and chloroform are quite different from the solid, and from each other. Thus the geometry of the molecule is very sensitive to subtle changes of environment.[131]

The i.r. spectrum under pressure of CO has been studied. There was no evidence for the formation of $Co_2(CO)_9$.[132]

M(CO)$_2$(HBiIm), where M = Rh or Ir, H_2BiIm = 2,2′-bi-imidazole, give two ν(CO) bands, which shows that the two CO groups are *cis*.[133]

Cotton–Kraihanzel force constant calculations were able to reproduce, within ± 0.8 cm^{-1}, all of the ν(CO) and ν(NO) wavenumbers of Co(CO)$_3$(NO), six isotopic species, providing the 'effective atomic masses' were introduced; these refer to the composite properties of the MCO and MNO groups.[134]

Relationships were found between ν_s(CO) and $E_{\frac{1}{2}}$ (determined polarographically) for [Co(CO)$_3$L]$_2$, Hg[Co(CO)$_3$L]$_2$, Hg[$Co_2(CO)_{8-n}L_n$], and Ni(CO)$_{4-n}L_n$, where L = PR_3 or P(OR)$_3$. These can be rationalized simply in terms of the expected electronic structures of these complexes.[135]

A linear relationship was found between the Cotton–Kraihanzel force constants and the Taft σ^* values of the XCH_2 groups for $R_{3-n}X_n$SnCo(CO)$_4$ compounds.[136]

ν(CO) for XCH_2SiH_2Co(CO)$_4$, where X = Cl, Br, or I, can be plotted against summed substituent electronegativities to give an almost linear correlation. Plotting against summed, modified polarity constants, however, gave a much better straight line.[137]

ν(CO) has been assigned for (μ_2-X_2C_2)$Co_2(CO)_5$L, where X = Cl, Br, or I, and L = CO, PBu_3, PPh_3, $AsEt_3$, or $SbEt_3$. Typical values are given in Table 13.

128 T. J. Collins and W. R. Roper, *J.C.S. Chem. Comm.*, 1976, 1044.
129 F. H. Johannsen and W. Preetz, *J. Organometallic Chem.*, 1976, **104**, 79.
130 B. F. G. Johnson, J. W. Kelland, J. Lewis, and S. K. Rehani, *J. Organometallic Chem.*, 1976, **113**, C42.
131 F. A. Cotton and J. D. Jameson, *J. Amer. Chem. Soc.*, 1976, **98**, 1273.
132 E. E. Ernstbrunner and M. Kilner, *J.C.S. Dalton*, 1976, 417.
133 S. W. Kaiser, R. B. Saillant, W. M. Butler, and P. G. Rasmussen, *Inorg. Chem.*, 1976, **15**, 2681.
134 G. Bor, G. Battiston, and G. Sbrignadello, *J. Organometallic Chem.*, 1976, **122**, 413.
135 D. de Montauzon and R. Poilblanc, *J. Organometallic Chem.*, 1976, **104**, 99.
136 L. F. Wuyts and G. P. van der Kelen, *Inorg. Chim. Acta*, 1976, **20**, 133.
137 J. M. Bellama and J. B. Davidson, *Inorg. Chim. Acta*, 1976, **16**, 219.

Table 13 ν(CO) *wavenumbers*/cm^{-1} *for* (μ_2-X_2C_2)$Co_2(CO)_5L$

X	L	$\nu_1(a_1)$	$\nu_4(b_1)$	$\nu_2(a_1) + \nu_6(b_2)$	$\nu_5(b_1)$
I	CO	2102.9	2071.5	2047.1	2035.8
I	PPh_3	2075.4	2033.8	2022.5	2015.1

Only four bands are seen, instead of the five expected for C_{2v} symmetry, since one band was of abnormally high intensity this was assigned to an overlap of $\nu_2(a_1)$ and $\nu_6(b_2)$.[138]

The i.r. spectra in the ν(CO) region have been reinterpreted for the following: $Co_2(CO)_6(HC{\equiv}CR)$, R = Me, Bu^n, n-C_6H_{11}, Ph, Bu^t, or 2-C_5H_{11}, in n-hexane, as well as $Co_2(CO)_6(PhC{\equiv}CPh)$. Previous work (G. Bor, *Chem. Ber.*, 1963, **96**, 2644) had only revealed five of the six possible bands. The last band, $\nu_3(a_2)$, was observed in this work. It was suggested that for some unsymmetrical acetylenes as ligands the electronic asymmetry inside the Co_2C_2 unit can be sufficient to relieve the strict C_{2v} selection rules, so as to give appreciable intensity to the ν_3 mode.[139]

The observed effects on ν(CO) bands in $(R^1C_2R^2)Co_2(CO)_6$ of alterations in the R^1, R^2 groups were used to assign individual bands to specific modes more accurately.[140]

The changes observed in the i.r. spectrum of $Co_2(CO)_8$ between -165 °C and $+210$ °C have been interpreted in terms of variations in the relative concentrations of the three isomers of this compound, assuming that the band at 2031 cm^{-1} be reassigned to the third isomer. A weak feature, observed for the first time here (but only above 120 °C), was thought to be due to $Co(CO)_4$.[141]

$AsCo_3(CO)_9$ and its 'trimer' $As_3Co_9(CO)_{24}$ are in equilibrium under CO. Their interconversion may be followed by monitoring the ν(CO) region of the spectrum.[142]

$[Rh_4(CO)_{11}(COOMe)]^-$ gives bands due to terminal CO groups at 2075, 2030, 2010, 1995, and 1965 cm^{-1}, due to bridging CO groups at 1892 and 1845 cm^{-1}, and due to ν(CO) of the carbomethoxy group at 1655 cm^{-1}. The structure is derived from that of $Rh_4(CO)_{12}$ by the replacement of a terminal CO by the carbomethoxy group.[143]

Assignments of ν(CO) in $[(OC)(C_6F_5)PdL_2]^+$ are given in Table 14. The very high values are as expected for the high formal oxidation state, +2, of the Pd.[144]

$M(CO)_n$ and $M_2(CO)_6$, where M = Cu or Ag and n = 1—3, and $Au(CO)_n$, where n = 1 or 2, have been characterized by i.r. matrix isolation spectroscopy in the ν(CO) range. $^{12,13}C$ and $^{16,18}O$ isotopic shifts were used in the assignments, as well as Cotton–Kraihanzel force constants. The $Cu_2(CO)_6$ spectrum suggests that the structure is Cu—Cu bonded, with D_3 symmetry.[145]

138 G. Varadi and G. Palyi, *Inorg. Chim. Acta*, 1976, **20**, L33.
139 G. Bor, S. F. A. Kettle, and P. L. Stanghellini, *Inorg. Chim. Acta*, 1976, **18**, L18.
140 G. Varadi, I. Vecsel, A. Vizi-Orosz, G. Palyi, and A. G. Massey, *J. Organometallic Chem.*, 1976, **114**, 213.
141 G. Bor, U. K. Dietler, and K. Noack, *J.C.S. Chem. Comm.*, 1976, 914.
142 A. Vizi-Orosz, V. Galamb, G. Palyi, L. Marko, G. Bor, and G. Natile, *J. Organometallic Chem.*, 1976, **107**, 235.
143 S. Martinengo, A. Fumagalli, P. Chini, V. G. Albano, and G. Ciani, *J. Organometallic. Chem.*, 1976, **116**, 333.
144 R. Uson, J. Forniers, and F. Martinez, *J. Organometallic Chem.*, 1976, **112**, 105.
145 G. A. Ozin, *Appl. Spectroscopy*, 1976, **30**, 573.

Table 14 ν(CO) *wavenumbers*/cm^{-1} *for* $[(OC)(C_6F_5)PdL_2]^+$

L_2	ν(CO)
$(PPh_3)_2$	2143
$(PPhEt_2)_2$	2146
$(PPh_2Me)_2$	2147
$(PEt_3)_2$	2132
bipy	2153
phen	2160
tmen	2163

The reductive carbonylation of $AuCl_3$ in $SOCl_2$ is thought to proceed *via* the mixed valence complex $Au_2(CO)Cl_4$, on the basis of an i.r. band at 2180 cm^{-1}.[146]

3 Nitrogen Donors

Molecular Nitrogen, Azido- and Related Complexes.—The crystal structure of $TiCl_3(N_3)$ shows the presence of the chain structure (51), with a distorted octahedron around the Ti. The i.r. spectrum was remeasured, and the Raman

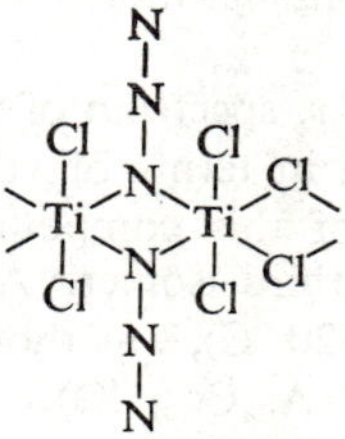

(51)

spectrum of a powdered sample recorded for the first time. Previous assignments in terms of a loose association of monomeric units were rejected.[147]

Co-condensation of Nb atoms with N_2 in Ar matrices at 14 K gives a multiplicity of bands in the ν(NN) region. No firm conclusions could be drawn as to the number or precise nature of the species formed, but all are thought to be mononuclear Nb–N_2 complexes.[148]

Similar reactions involving Cr atoms at 10 K afford $Cr(N_2)_n$, where n = 1—6. The data are summarized in Table 15. Only one ν(NN) was definitely observed

Table 15 $\nu(N_2)$ *wavenumbers*/cm^{-1} *for* $Cr(N_2)_n$

n	Ar *matrix*	N_2 *matrix*	$^{15}N_2$ *in* Ar *matrix*
1	2215		2142
2	2253		2178
3	2178		2108
4	2130	2136	2060
5	2035, 2068, 2195	2028, 2072, 2195	1966, 2000
6	2120	2112	2042

[146] D. B. Dell'Amico, F. Calderazzo, and F. Marchetti, *J.C.S. Dalton*, 1976, 1829.
[147] H.-O. Wellern and U. Müller, *Chem. Ber.*, 1976, **109**, 3039.
[148] D. W. Green, R. V. Hodges, and D. M. Gruen, *Inorg. Chem.*, 1976, **15**, 970.

for $Cr(N_2)$, therefore it is $C_{\infty v}$; a similar observation for $Cr(N_2)_2$, which must be $D_{\infty h}$. The situation for $Cr(N_2)_3$ is uncertain, but only one ν(NN) band was seen in the i.r. for $Cr(N_2)_4$, presumably D_{4h} or T_d, whereas three such features are seen for $Cr(N_2)_5$, C_{4v}. $Cr(N_2)_6$ is probably O_h. All of the above discussion assumed end-on co-ordination of the N_2 ligand.[149]

The i.r. spectra of the diazenido complexes $[MX(N_2H)(dppe)_2]$, where dppe = $Ph_2PCH_2CH_2PPh_2$, are characterized by strong, sharp bands at 1880 cm^{-1} (M = W) or 1940 cm^{-1} (M = Mo). These were assigned to ν(N=N) on the basis of ^{15}N shifts. ν(NH) was only seen at 3365 cm^{-1}, when X = F, and not for X = Cl or Br. The ligand represents the first protonated form of N_2.[150]

A number of dialkylsulphurdi-imine complexes have been prepared, (52a), with M = Cr or W, R = Et or Pr^i; (52b); with M = Cr, R = Pr^i or Bu^t;

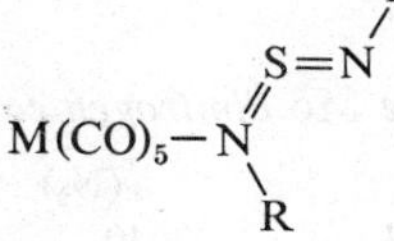

(52a)

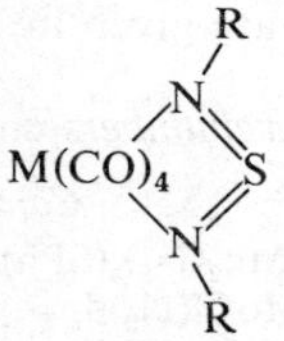

(52b)

M = Mo, or W, R = Bu^t; and $M(CO)_5(MeN{=}S{=}NMe)$, with M = Cr, Mo, or W, and the ligand is probably S-bonded to the M. ν_s(NSN) and δ(NSN) have been assigned for these complexes (Table 16). Resonance Raman spectra

Table 16 *ν_s(NSN) and δ(NSN) wavenumbers*/cm^{-1} *for complexes* (52)

R:	Me		Et		Pr^i		Bu^t	
	ν	δ	ν	δ	ν	δ	ν	δ
L	1075	807	1071	773	1056	756	1066	678
$Cr(CO)_5L$	1062	805	1055	772				
$W(CO)_5L$	1060	809	1056	770				
$Cr(CO)_4L$					1095	852	1090	755
$Mo(CO)_4L$							1089	776
$W(CO)_4L$							1070	776

show vibronic coupling of metal–L charge transfer transitions with the L vibrations for $W(CO)_4(Bu^tN{=}S{=}NBu^t)$ and $W(CO)_5(MeN{=}S{=}NMe)$.[151]

trans-$Mo(CO)(N_2)(dpe)_2$, where dpe = 1,2-bis(diphenylphosphino)ethane, gives four bands, at 2110, 2080, 1812, and 1791 cm^{-1}. ^{15}N substitution gives values of 2036, 2009, 1812, and 1789 cm^{-1}. The higher two are thus ν(NN), the lower two ν(CO) (the splitting is due to crystal field effects).[152]

$\nu(N_2)$ in $MoX(N_2CH_2COOEt)(dppe)_2$, where X = Cl, Br, or I; dppe = bis-(diphenylphosphino)ethane, is at 1508 cm^{-1}. The N_2 fragment can be protonated, which shifts $\nu(N_2)$ to an unidentified position, while ν(NH) appears at 3290 cm^{-1}.[153]

149 T. C. DeVore, *Inorg. Chem.*, 1976, **15**, 1315.
150 J. Chatt, A. J. Pearman, and R. L. Richards, *J.C.S. Dalton*, 1976, 1520.
151 R. Meij, J. Kuyper, D. J. Stufkens, and K. Vrieze, *J. Organometallic Chem.*, 1976, **110**, 219.
152 T. Tatsumi, H. Tominaga, M. Hidai, and Y. Uchida, *J. Organometallic Chem.*, 1976, **114**, C27.
153 D. C. Busby and T. A. George, *J. Organometallic Chem.*, 1976, **118**, C16.

A series of triazenido complexes, $(\eta^5\text{-}C_5H_5)(OC)_2M(R^1\text{—N}\!\cdots\!\text{N}\!\cdots\!\text{N—}R^2)$, where R^1 = aryl, R^2 = alkyl or aryl, M = Mo or W, all show bands due to the triazenido ligand at about 1600 cm^{-1}, and 1280—1300 cm^{-1}. These positions are characteristic of a chelated unit (53).[154]

(53)

$\nu(N_2)$ for several molybdenum dinitrogen complexes $[Mo_2(N_2)_3(SPh)_2dp_4]^{2+}$ and related species are given in Table 17.[155]

Table 17 *$\nu(N_2)$ wavenumbers/cm^{-1} for some Mo dinitrogen complexes*

Complex	$\nu(N_2)$
$[Mo_2(N_2)_3(SPh)_2dp_4]Cl_2,C_7H_8$	2040
$MoCl(N_2)dp_2$	1970
$MoBr(N_2)dp_2$	1965
$[Mo_2H_2(N_2)_3dp_4](ClO_4)_2,2.5C_6H_8$	2036, 1980
$Mo(N_2)dp_2$	1950

$\nu_{as}(N_3)$ in $[MoO(N_3)_5]^{2-}$ is at 2070 cm^{-1}, compared with 2049 cm^{-1} for free N_3^-, which is a value typical of a co-ordinated N_3 group.[156]

The pseudohalides $\pi\text{-}C_7H_7M(CO)_2X$, where M = Mo or W, X = N_3 or NCO; M = Mo, X = NCS, have been prepared for the first time. The N_3 complex gives $\nu(N_3)$ bands at 2100, 2075 cm^{-1} (Mo), 2105, 2080 cm^{-1} (W). The internal modes of the NCO and NCS ligands show that both are N-bonded to the metal.[157]

The bidentate dinitrogen molecule in $[\eta\text{-}C_5H_5Mn(CO)_2]_2N_2$ has $\nu(NN)$ at 1975 cm^{-1}, showing a decrease from the monodentate value (2169 cm^{-1}) which is comparable to the decrease from the free to the monodentate molecule.[158]

The compound $Mn_2(CO)_{10}CH_2N_2$, a stable derivative of diazomethane, has the structure (54). It has i.r. bands at 1644 cm^{-1} due to $\nu(C{=}O)$ and at 1575 cm^{-1} due to $\nu(N{=}CH_2)$. Eight of the nine expected $\nu(CO)$ bands are also seen. This

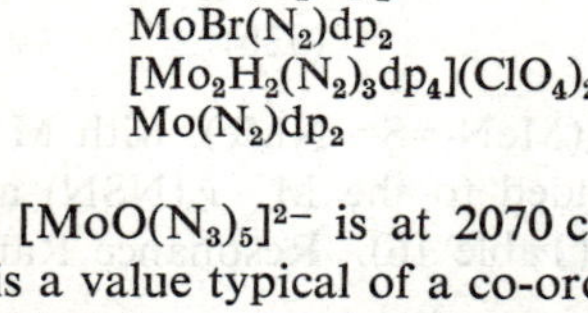

(54)

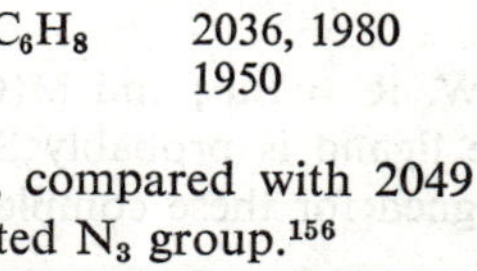

(55)

[154] E. Peiffer, J. Kuyper, and K. Vrieze, *J. Organometallic Chem.*, 1976, **105**, 371.
[155] M. Aresta and C. F. Nobile, *Inorg. Chim. Acta*, 1976, **17**, L17.
[156] A. M. Golub, V. V. Trachevskii, N. V. Ul'ko, and N. D. Samovskaya, *Russ. J. Inorg. Chem.*, 1975, **20**, 1495.
[157] G. Hoch, R. Panter, and M. L. Ziegler, *Z. Naturforsch.*, 1976, **31b**, 294.
[158] M. N. Ziegler, K. Weidenhammer, H. Zeiner, P. S. Skell, and W. A. Herrmann, *Angew. Chem. Internat. Edn.*, 1976, **15**, 695.

is the first example of CH_2N_2 co-ordinated by both N atoms.[159] The related complex (55) shows ν(NN) at 1586 cm^{-1}.[160] $\nu(N_2)$ in $Fe(N_2)$(diphos)$_2$ is at 2068 cm^{-1}.[161]

cis-$[Ru(NH_3)_4(N_2)(H_2O)]Br_2$ has ν(NN) bands at 2101 and 2105 cm^{-1}, whereas the *trans*-isomer, as the Cl salt, gives one band, at 2098 cm^{-1}.[162]

I.r. data were listed, and outline assignments proposed, for $[Co_2(N_2O_2)(NH_3)_{10}]X_4,nH_2O$, where X = Cl, n = 4; X = NO_3, n = 2. These are thought to be bridged M—N—O—O—N—M. For X = Cl, ν(O—O) is at 1020 cm^{-1}, ν_{as}(N—O) at 1481 cm^{-1}, and ν_s(N—O) at 1278 cm^{-1}.[163]

ν_{as}(NNN) in a number of triazenido complexes has been found to be diagnostic of the ligand bonding type. Thus, in $(Ph_3P)_2$(CO—DMT)XMHgY, where X = Cl or $O_2C_2F_3$, M = Ir or Rh, Y = Cl or I, the chelating ligand (56)

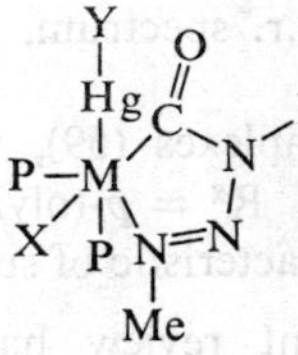

(56)

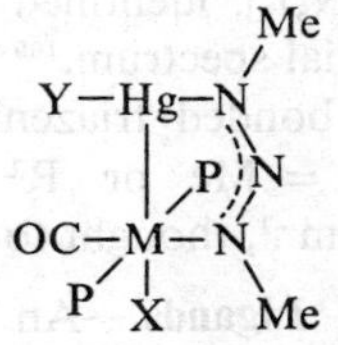

(57)

gives ν_{as}(NNN) in the range 1311—1328 cm^{-1}. $(Ph_3P)_2$(CO)XMHg(MpTT)Y, where X = Y = Cl, M = Ir, on the other hand, has structure (57), and ν_{as}(NNN) above 1350 cm^{-1}.[164]

The structure of *trans*-$[RhH(N_2)(PhPBu^t{}_2)_2]$, determined by *X*-ray diffraction, contains a linear H—Rh—N—N group. $\nu(N_2)$ was at 2155 cm^{-1}, and it was shown that for a range of dinitrogen complexes the value of $\nu(N_2)$ was a more sensitive probe for studying electronic changes than was the N—N bond distance.[165]

The aryldiazo ligand in $[IrCl(N_2Ph)(PMePh_2)_3]^+$ gives rise to two i.r. bands which are sensitive to ^{15}N substitution: 1569, 1644 (^{14}N), 1547, and 1610 cm^{-1} (^{15}N). The doublet is due to Fermi resonance with a phenyl mode, which complicates the interpretation and reduces the usefulness of the wavenumber for structural deductions.[166]

ν(NSN) features are at 850—910 cm^{-1} (*s*) and at 1190—1250 cm^{-1} (*as*) in $[(\eta^3\text{-}C_3H_5)Pd\{S\text{-}Bu^t\text{-}N,N'\text{-di(4-Cl-phenyl)NSN}\}]$ and related complexes. ν_{as}(NNN) is in the range 1350—1375 cm^{-1} in the triazenido complexes $[(\eta^3\text{-}C_3H_5)Pd(dmt)]_2$, *etc.*, where dmt = MeN_3Me; therefore it is a bridging triazenido group.[167]

159 W. A. Herrmann, M. L. Ziegler, K. Weidenhammer, H. Biersack, K. K. Mayer, and R. D. Minard, *Angew. Chem. Internat. Edn.*, 1976, **15**, 164.
160 W. A. Herrmann, M. L. Ziegler, and K. Weidenhammer, *Angew. Chem. Internat. Edn.*, 1976, **15**, 368.
161 R. A. Cable, M. Green, R. E. Mackenzie, P. L. Timms, and T. W. Turney, *J.C.S. Chem. Comm.*, 1976, 270.
162 S. S. Isied and H. Taube, *Inorg. Chem.*, 1976, **15**, 3070.
163 E. Miki and T. Ishimori, *Bull. Chem. Soc. Japan*, 1976, **49**, 987.
164 P. I. van Vliet, J. Kuyper, and K. Vrieze, *J. Organometallic. Chem.*, 1976, **122**, 99.
165 P. R. Hoffman, T. Yoshida, T. Okeno, S. Otsuka, and J. A. Ibers, *Inorg. Chem.*, 1976, **15**, 2462.
166 M. Cowie, B. L. Haymore, and J. A. Ibers, *J. Amer. Chem. Soc.*, 1976, **98**, 7608.
167 P. Hendriks, J. Kuyper, and K. Vrieze, *J. Organometallic Chem.*, 1976, **120**, 285.

I.r. spectra of Cu and Ag^{I} *SNN'*-substituted sulphurdi-imino compounds, (58), where M = Cu or Ag, R^1 = aryl, R^2 = Me or Bu^t, have ν_{as}(NSN) at 1180—1260 cm^{-1} and ν_s(NSN) at 840—910 cm^{-1}.[168]

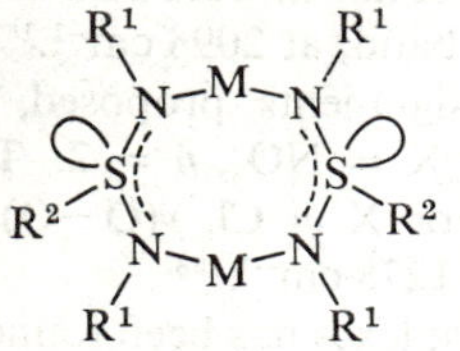
(58)

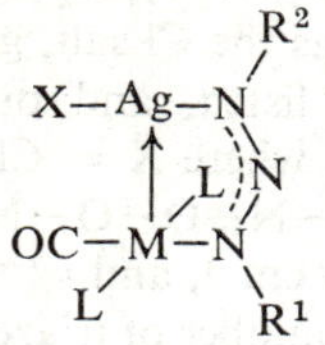
(59)

Compression of a mixture of CsN_3 and AgN_3 to 25 kbar resulted in the formation of $Cs[Ag(N_3)_2]$, identified by its i.r. spectrum, which showed bands not present in the initial spectrum.[169]

The metal—metal bonded triazenido complexes (59), where X = Cl or I, L = PPh_3, $R^1 = R^2$ = Me or R^1 = Me, R^2 = *p*-tolyl, produce ν(NNN) bands at about 1360 cm^{-1}, thought to be characteristic of such a bridging unit.[170]

Amines and Related Ligands.—An excellent review has appeared on the vibrational spectra and force constants of complexes containing only NH_3 as a ligand.[171]

$MCl_2,2N_2H_4$, where M = Mg, Sr, or Ba, give ν(NN) due to the co-ordinated hydrazine at wavenumbers 70—120 cm^{-1} higher than in free hydrazine.[172]

Some ligand vibration wavenumbers have been listed and assigned for salts containing $[Cr(NH_3)_6]^{2+}$ and $[Cr(NH_3)_5]^{2+}$, thus $\delta_r(NH_3)$ is at 675 and 700 cm^{-1} respectively.[173]

Solid-phase i.r. spectra of $[(OC)_5Cr]_2N_2H_4$ and $[(OC)_5Cr]_2N_2H_2$ have been interpreted as suggesting that (i) the N_2H_4 unit in the former has C_{2h} symmetry, and (ii) that the N_2H_2 unit in the latter is present as the *cis*-di-imine. In solution it is possible that the latter may exist as two isomers. In the N_2H_2 derivative, ν(NN) is at 1415 cm^{-1}, and in that of N_2H_4 it is at 910 cm^{-1}.[174]

Table 18 ν(NH) *wavenumbers*/cm^{-1} *for some six co-ordinate* Cr^{II} *complexes*

Complex	ν(NH)	*Complex*	ν(NH)
$Crpz_4Cl_2$	3230	$Criz_4Cl_2$	3220
$Crpz_4Br_2$	3220	$Criz_4Br_2$	3180
$Crpz_4I_2$	3230	$Cr(dmpz)_2Cl_2$	3260—3210
		$Cr(dmpz)_2Br_2$	3280—3230

I.r. bands due to ν(NH) and ν(CrN) have been assigned for six-co-ordinate Cr^{II} complexes with pyrazole (pz), imidazole (iz), or 3,5-dimethylpyrazole (dmpz) (Table 18).[175]

168 J. Kuyper, P. C. Keijzer, and K. Vrieze, *J. Organometallic Chem.*, 1976, **116**, 1.
169 B. Busch, E. Hellner, and K. Dehnicke, *Naturwiss.*, 1976, **63**, 531.
170 J. Kuyper, P. I. van Vliet, and K. Vrieze, *J. Organometallic. Chem.*, 1976, **105**, 379.
171 K. H. Schmidt and A. Müller, *Coordination Chem. Revs.*, 1976, **19**, 41.
172 Zh. G. Sakk, *Russ. J. Inorg. Chem.*, 1975, **20**, 848.
173 L. F. Larkworthy and J. M. Tabatabai, *J.C.S. Dalton*, 1976, 814.
174 D. Sellmann, A. Brandl, and R. Endell, *J. Organometallic Chem.* 1976, **111**, 303.
175 F. Mani and G. Scapacci, *Inorg. Chim. Acta*, 1976, **16**, 163.

The compound (60) displays ν(SO) at 1140 cm^{-1} in the i.r., ν(NH) is at 3238 cm^{-1}, shifting to 2410 cm^{-1} on deuteriation.[176] $\nu(NH_2)_{co\text{-}ord}$ are at 3239 and 3152 cm^{-1} in (61), with $\nu(NH)_{non\text{-}co\text{-}ord}$ at 3382 cm^{-1}, and ν(C=O) at 1520 cm^{-1}.[177]

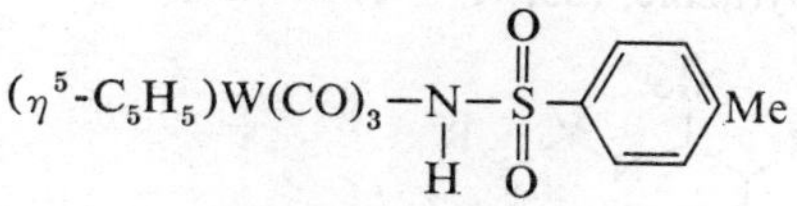

(60)

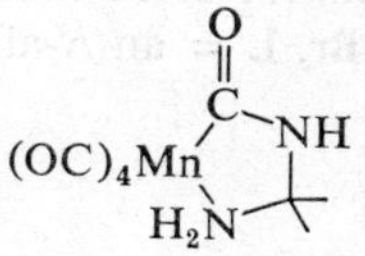

(61)

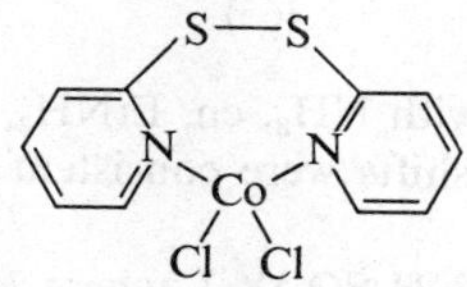

(62)

Complexes of Co and Ni with *O*-methylhydroxylamine give i.r. spectra which are very similar to that of the Pt analogue, which is known to contain M—NH_2OMe units.[178]

Co-ordination of the py groups in (62) is confirmed by internal ligand mode shifts, while ν(SS) is at 522 cm^{-1}; 25 cm^{-1} lower than in the free ligand, in accord with the observed bond length trends.[179]

Measurements of $\nu(NH_2)$ and of py internal modes in the aminopyridine complexes MCl(cod)L, where L = NH_2-2-py, NH_2-2-Me-6-py, or $(NH_2)_2$-2,6-py and M = Rh or Ir, show that no shift on complexation occurs with $\nu(NH_2)$ and slight shifts for the py modes. There is therefore no co-ordination *via* the NH_2 groups, and only weak co-ordination *via* the ring N atom. 2-Aminomethylpyridine, on the other hand, forms complexes co-ordinated by both NH_2

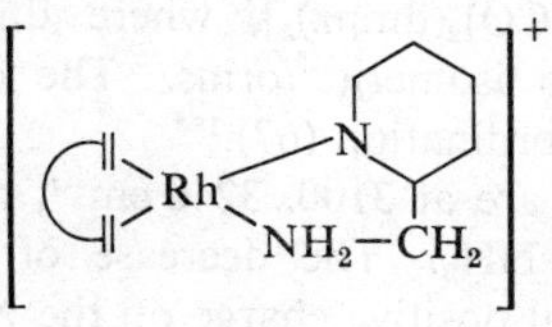

(63)

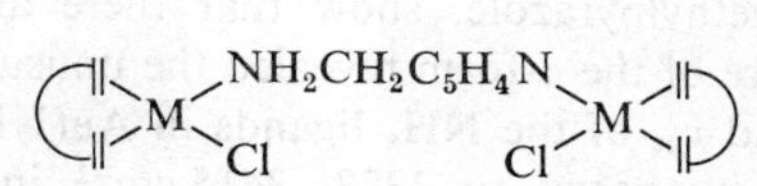

(64)

and the ring N, giving structures of the type (63). 4-Aminomethylpyridine also co-ordinates by both N atoms, to give bridging species (64).[180]

ν(NH), ν(C≡N), and ν(NiX), where X = N, S, or Cl, were assigned for a series of the new Ni^{II} complexes of the types [Ni(*pqr*-cy)$(NCS)_2$], [Ni(*pqr*-cy)Cl]$^+$,

176 W. A. Herrmann, *Chem. Ber.*, 1976, **109**, 3489.
177 W. Danzer, W. Beck, and M. Keubler, *Z. Naturforsch.*, 1976, **31b**, 1360.
178 M. A. Sarukhanov, S. S. Val'dman, and N. A. Parpiev, *Russ. J. Inorg. Chem.*, 1975, **20**, 1327.
179 M. M. Kadooka, L. G. Warner, and K. Seff, *Inorg. Chem.*, 1976, **15**, 812.
180 D. Brodzki and G. Pannetier, *J. Organometallic Chem.*, 1976, **104**, 241.

and $[Ni_2(pqr\text{-cy})_3]^{4+}$, where *pqr*-cy is a triazacycloalkane, $\overline{NH(CH_2)_pNH(CH_2)_q}NH(CH_2)_r$ and $p, q = 2$ or 3, $r = 2$—6.[181]

Ligand wavenumbers were listed and partly assigned, on the basis of standard organic chemical correlations, for MX_2L_2 and MX_2L, where M = Pd or Pt, X = Cl or Br, L = an *N*-alkylpyrazole, (65, R = H or Me).[182]

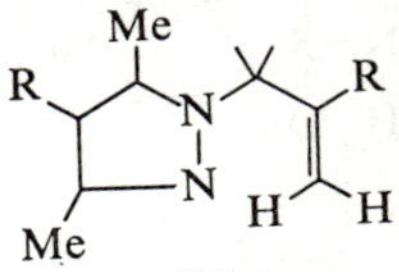

(65)

Complexes of $Cu(oxalate)_2$ with NH_3, en, $EtNH_2$, $PhNH_2$, or py were studied and the $\nu_{as}(CO_2)$ and $\nu(NH_2)$ shifts were consistent with co-ordination by both the oxalate and the amine.[183]

I.r. assignments for $M_2[Cu(NH_2SO_3)X_2]$, where M = K or NH_4, X = Cl or Br, are given in Table 19.[184]

Table 19 *Some i.r. assignments for* $M_2[Cu(NH_2SO_3)X_2]$ (*all wavenumbers*/cm^{-1})

M = K, X = Br		M = K, X = Cl		M = NH_4, X = Cl	
h_2	d_2	h_2	d_2	h_2	*Assignment*
3186	2390	3180	2391	3194	$\nu_{as}(NH_2/ND_2)$
3128	2310	3145	2316	3142	$\nu_s(NH_2/ND_2)$
1531	1133	1534	1132	1530	$\delta(NH_2/ND_2)$
1134	894	1138	899	1122	tNH_2/ND_2
734	721	761	736	757	$\nu(SN)$
707	550	694	560	693	wNH_2/ND_2

$\nu(NH)$ and ν_{8a} of the py ring were given for a series of complexes of Cu^{II} and Zn^{II} salts with (66), *n* or *m* = 2 or 3. The proposed structures varied with *n*, *m*, the metal, the stoicheiometry, and the anion. The real evidence was slight.[185]

Bands due to the NCO group in $[Cu(NCO)_2(dmpz)_2]$, where dmpz = 3,5-dimethylpyrazole, show that there are two isomeric forms. The crystal structure of the α-form revealed the unusual co-ordination (67).[186]

ν_s and ν_{as} of the NH_3 ligands in $Au(NH_3)_4{}^{3+}$ are at 3100, 3240 cm^{-1}, respectively, compared to 3337, 3415 cm^{-1} in free NH_3. The decrease of about 200 cm^{-1} is explained by the presence of a partial positive charge on the N, as a result of electron donation to the metal.[187]

$HgCl_2$[bis(2-pyridyl)disulphide] [where the ligand = (68)] contains $(-N\cdots N-HgCl_2-N\cdots N-HgCl_2-)_n$ chains. In accordance with this, characteristic shifts

181 M. Nonoyama, *Inorg. Chim. Acta*, 1976, **20**, 53.
182 T. Miyamoto, K. Fukushima, and Y. Sasaki, *Bull. Chem. Soc. Japan*, 1976, **49**, 138.
183 P. C. Scrivastava, B. J. Ansari, and B. K. Banerjee, *Indian J. Pure Appl. Phys.*, 1976, **14**, 384.
184 A. Lautie and L. Lecuir, *Spectrochim. Acta*, 1976, **32A**, 1323.
185 E. Uhlig and B. Borek, *Z. anorg. Chem.*, 1976, **425**, 217.
186 J. Kohout, F. Valach, M. Hrastijova, M. Dunaj-Jurco, and J. Gažo, *J.C.S. Chem. Comm.*, 1976, 903.
187 M. Weishaupt and J. Strähle, *Z. Naturforsch.*, 1976, **31b**, 554.

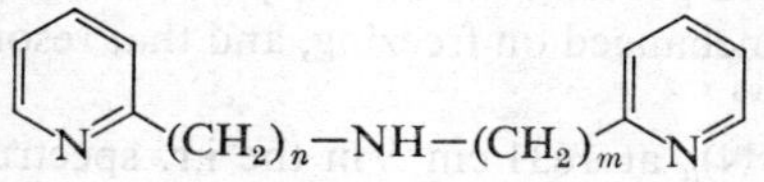

(66)

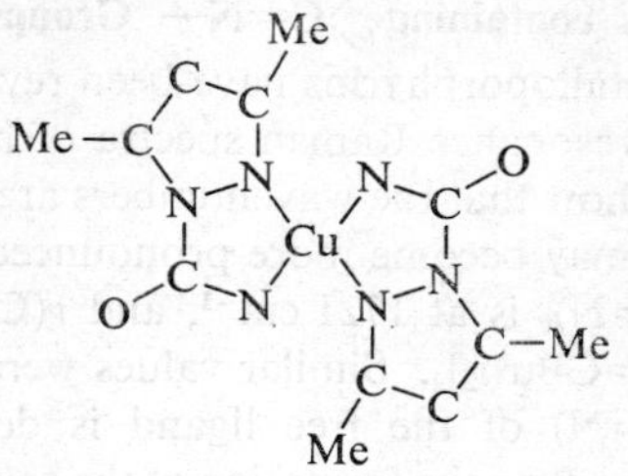

(67)

in the py modes were observed, and ν(SS) at 548 cm^{-1}, within 1 cm^{-1} of the free ligand value, *i.e.* the S is not co-ordinated.[188]

$UO_2L_2X_2$, where L = di-2-pyridylketone or di-2-pyridylsulphide and X = Cl, NCS, or ClO_4, and $UO_2L(NO_3)_2$ give patterns of ligand mode shifts consistent with *NN*-co-ordination to the UO_2 unit.[189]

The adduct of Ph_3B with 2-aminopyridine shows two sharp ν(NH) bands, 3451 and 3325 cm^{-1}. This suggests that co-ordination to the B occurs through the ring N atom, (69).[190]

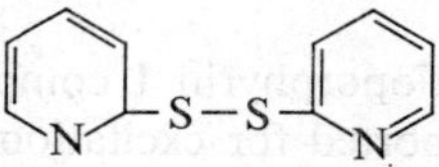

(68)

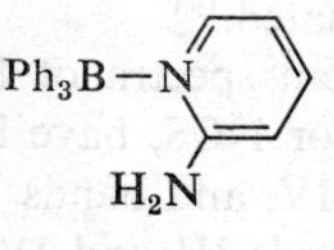

(69)

The presence of a band at 875 cm^{-1} in the i.r. spectrum of SiF_4,en, due to CH_2 rocking vibrations of *gauche* en, is thought to be evidence for a monomeric, chelate structure. Such a band is usually absent from polymeric en complexes.[191]

Sb^{III} and Bi^{III} halides form very insoluble adducts with piperazine (pipz) : MX_3,3pipz, and MX_3,pipz, where X = Cl, Br, or I. In the former the observed i.r. spectra are consistent with the pipz acting as a chelate, while in the latter series it seems to be bridging.[192]

Oximes.—A number of ligand modes have been assigned in *trans*-dioximato complexes $Co(DH)_2LX$, where L = NH_3 or thiourea, X = Cl, Br, I, SCN, NO_2, CN, or thiourea. Attempts were made to obtain relative *trans* and *cis* effects for these various ligands. Such effects could only be discerned if the Co—X and Co—L bonds were primarily σ in form, with little or no π-bonding.[193]

[CoX(dif)Sd],$n$$H_2O$, where X = Cl, Br, I, NCO, NCS, NCSe, NO_2; Hdif = diphenylglyoxime; Sd = p-$H_2NC_6H_4SO_2NHR$ where R = 2-pyridyl, give i.r. spectra which indicate that the Sd co-ordinates *via* the N of the pyridyl group, and that it is *trans* to the X.[194]

188 M. M. Kadooka, E. Hilti, L. G. Warner, and K. Seff, *Inorg. Chem.*, 1976, **15**, 1186
189 G. C. Pappalardo and A. Seminara, *J. Inorg. Nuclear Chem.*, 1976, **38**, 1993.
190 B. R. Gragg and K. Niedenzu, *J. Organometal. Chem.*, 1976, **117**, 1.
191 A. A. Ennan, L. A. Gavrilova, and V. I. Nikitin, *Russ. J. Inorg. Chem.*, 1975, **20**, 1159.
192 G. Marcotrigiano, L. Menabue, and G. C. Pellacani, *J. Mol. Struct.*, 1976, **30**, 85.
193 B. A. Bovykin and E. N. Yurchenko, *Russ. J. Inorg. Chem.*, 1975, **20**, 1574.
194 V. N. Shafranskii and D. D. Buburuz, *Zhur. obshchei Khim.*, 1976, **46**, 1133.

Ligands containing $\rangle$C=N— Groups.—The vibrational spectra of porphyrins and metalloporphyrins have been reviewed.[195]

The resonance Raman spectra of haem proteins at low temperature, down to 68 K, show that the wavenumbers are unchanged on freezing, and that resonance effects may become more pronounced.[196]

$\nu(C{=}N)_t$ is at 1721 cm^{-1}, and $\nu(C{=}N)_\mu$ at 1631 cm^{-1} in the i.r. spectrum of $[Be(N{=}CBu^t)_2]_2$. Similar values were found for the Mg and Zn analogues.[197]

ν(C=N) of the free ligand is decreased by about 45 cm^{-1}, from 1625 to 1580 cm^{-1}, on the formation of the complexes (70), where M = Mo, R = $CHMe_2$,

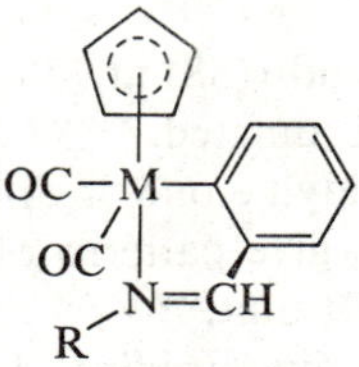

(70)

CH_2Ph, or CHMePh; M = W, R = $CHMe_2$ or CHMePh, thus, the azomethine N atom is co-ordinated.[198]

Resonance Raman spectra of Mn^{III} etioporphyrin I complexes, Mn(etio)X, where X = Cl, I, or NCS, have been reported for excitation in the regions of the bands III and IV, and bands V and V*a* of the visible absorption spectra of the complexes. Bands III and IV are 0-0 and 0-1 components of the *Q*-band. Band V contains π-π^* character and is coupled non-adiabatically to band V*a*.[199]

Another group has assigned electronic transitions of Mn^{III} etioporphyrin I using resonance Raman spectroscopy. Strong bands between 460 and 490 nm were assigned to charge transfer transitions. The excitation profiles of band V were accounted for by the presence of a second transition underlying band V, rather than by non-adiabatic coupling, see above.[200]

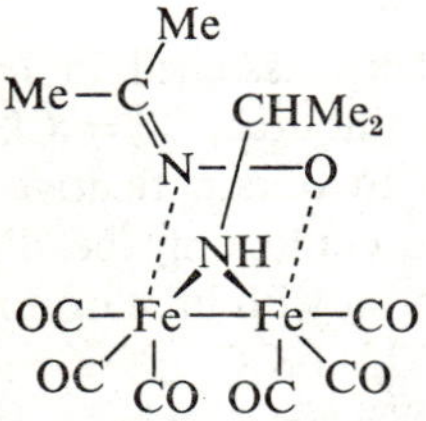

(71)

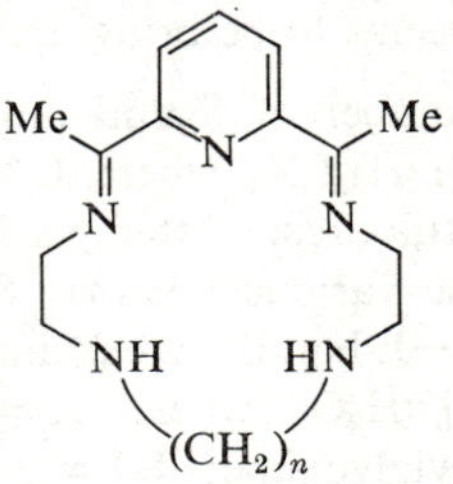

(72)

195 H. Buerger, *Porphyrins and Metalloporphyrins*, 1975, 525.
196 P. M. Champion, D. W. Collins, and D. B. Fitchen, *J. Amer. Chem. Soc.*, 1976, **98**, 7114.
197 J. B. Farmer, H. M. M. Shearer, J. D. Sowerby, and K. Wade, *J.C.S. Chem. Comm.*, 1976, 160.
198 H. Brunner and J. Wachter, *J. Organometallic Chem.*, 1976, **107**, 307.
199 J. A. Shelnutt, D. C. O'Shea, N.-T. Yu, L. D. Cheung, and R. H. Fulton, *J. Chem. Phys.*, 1976, **64**, 1156.
200 S. Asher and K. Sauer, *J. Chem. Phys.*, 1976, **64**, 4115.

The unsymmetrical nitroso binuclear compound (71) shows ν(NH) at 3262 cm^{-1}, ν(NO) at 1737 cm^{-1}, and ν(C=N) at 1601 cm^{-1}.[201]

The first macrocyclic, pentagonal bipyramidal complexes of Fe^{II} have been prepared of the type $FeLX_2$, where L = (72) (n = 2 or 3), X = halide or pseudohalide, by the dithionite reduction of the corresponding Fe^{III} compounds. The py ring vibrations at 1640 and 1590 cm^{-1} are slightly lower (8—15 cm^{-1}) than in the Fe^{III} species. When X = NCS, the positions of the NCS vibrations show that it is N-bonded.[202]

The free ligand pbim [2-(2′-pyridylbenzimidazole)], (73), has an i.r. band at 1314 cm^{-1}. This splits (1324 and 1302 cm^{-1}) in the complex $[Fe(pbim)_3]^{2+}$, indicating, it is said, that the imidazole N is bonded to the Fe.[203]

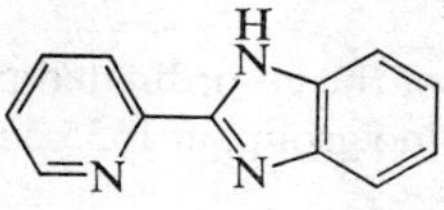
(73)

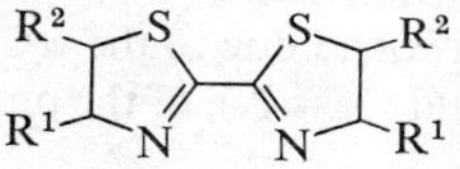
(74a)

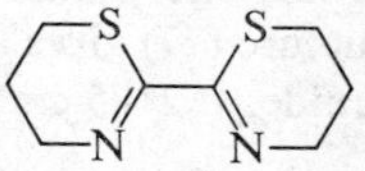
(74b)

A series of complexes of the ligands [(74a) $R^1 = R^2$ = H; R^1 = Me, R^2 = H; R^1 = H, R^2 = Me)] and (74b) has been made. They can be divided into two groups: (i) high spin complexes of Co, Zn, Hg, *etc.*, which have ν(CN) as an i.r. band 25—35 cm^{-1} to lower wavenumber than in the free ligands, and (ii) the low spin Fe^{II} and Mo^0 complexes, which show no observable ν(CN), because of M—N π-bonding.[204]

The delocalized chelate ring in (75), where M = Fe, Co, Ni, Cu, or Zn, shows only one ν(CN) band, at about 1580 cm^{-1}, whereas the free ligand has such features at 1550 and 1616 cm^{-1}.[205]

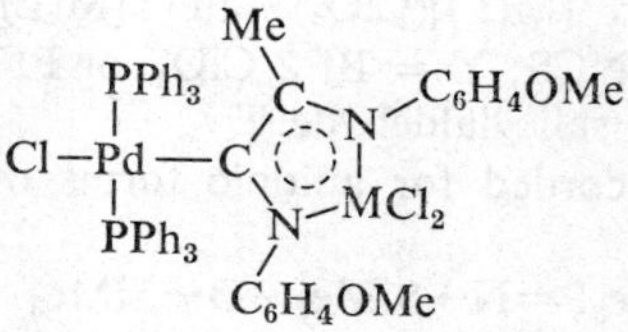
(75)

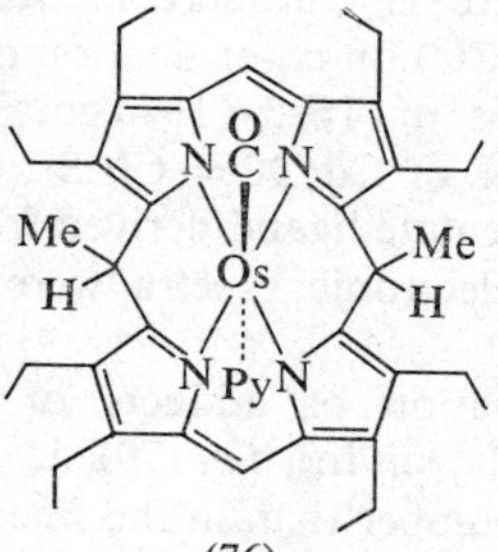
(76)

ν(CO) in the newly-prepared complex, $Os(OEPMe_2)(CO)py$, (76), is at 1863 cm^{-1}, compared to 1902 cm^{-1} in the tetrapyrrole complex Os(OEP)(CO)py. Thus, the porphodimethene system is a weaker π-acceptor than the more highly

201 S. Aime, G. Gervasio, L. Milone, R. Rosetti, and P. L. Stanghellini, *J.C.S. Chem. Comm.*, 1976, 370.
202 M. G. B. Drew, J. Grimshaw, P. P. A. McIlroy, and S. M. Nelson, *J.C.S. Dalton*, 1976, 1388.
203 J. R. Sams and T. B. Tsin, *J.C.S. Dalton*, 1976, 488.
204 J. Nelson, S. M. Nelson, and W. D. Perry, *J.C.S. Dalton*, 1976, 1282.
205 B. Crociani, M. Nicolini, and R. L. Richards, *J. Organometallic Chem.*, 1976, **113**, C22.

delocalized tetrapyrrole system. Some confirmation of this conclusion may come from the rather low wavenumber of the 'methene band' at 1592 cm^{-1}.[206]

trans-$[C_6Cl_5Ni(P)_2NH{=}C(R^1)OR^2]^+$, where R^1, R^2 = Me or $PhCH_2$, P = $PhPMe_2$ or Ph_2PMe, show ν(NH) at 3270—3350 cm^{-1} and ν(N=C) at 1623—1661 cm^{-1}.[207]

The resonance Raman spectra of some synthetic metal^{II} porphyrin analogues have been examined. Ni^{II} shows selective enhancement of low-wavenumber bands in resonance with the charge-transfer band at 585 nm, but resonant behaviour near the π–π^* transition enhances high-wavenumber ligand modes. Co^{II} shows a strong antiresonance effect near 460 nm within a weak charge-transfer band on the shoulder of the π–π^* transition. This is the first observation of Raman antiresonance involving only allowed electronic transitions. A theoretical interpretation was given.[208]

Structure (77) gives an i.r. band due to the ν(C≡N), of the co-ordinated phenyl isocyanide, at 2185 cm^{-1}, and ν(C=N), of the phenylimino group, at 1555 cm^{-1}.[209]

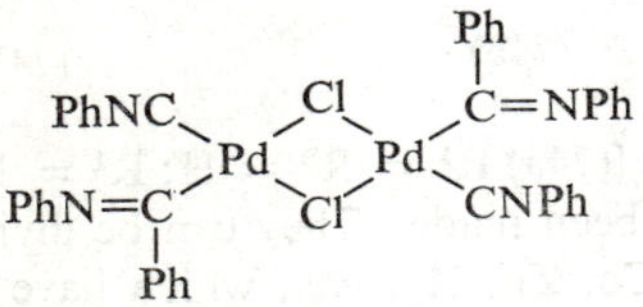

(77)

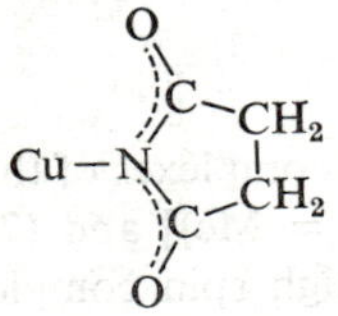

(78)

The resonance Raman spectra of copper protein haemocyanins identify the bound oxygen as O_2^{2-}, giving bands at 744 cm^{-1} (in *Cancer magister*, arthropod) and 749 cm^{-1} (in *Busycon caniliculetum*, mollusc).[210]

Shifts in ν_s and ν_{as}(CO) on co-ordination have been rationalized in terms of N-co-ordination in $Cu(NH_3)_2$(succinimidato), (78), as shown by *X*-ray diffraction results. There is considerable delocalization in the OCNCO system, which influences ν(CO) on co-ordination of the N atom.[211]

ν(C=N) is at 1625—1640 cm^{-1} in $[M(L)](ClO_4)_2$ and [M(L)X]Y, where M = Cu, Zn, or Cd; X = Cl, Br, or NCS; Y = BF_4, ClO_4, or PF_6, and L is a cyclic tetradentate ligand derived from salicylaldehyde.[212]

I.r. and electronic spectra were recorded for anionic forms of zinc aetioporphyrin.[213]

The formation of adducts of $Me_3P{=}N{-}SiMe_2{-}O{-}SiMe_3$ with MMe_3 (M = Al, Ga, or In), *i.e.* (79), leads to shifts in ν(P=N) of 134—175 cm^{-1} to lower wavenumber than in the free species.[214]

206 J. W. Buchler, K. L. Lay, P. D. Smith, W. R. Scheidt, G. A. Rupprecht, and J. E. Kenny, *J. Organometallic Chem.*, 1976, **110**, 109.
207 M. Wada and T. Shimohigashi, *Inorg. Chem.*, 1976, **15**, 954.
208 L. A. Nafie, R. W. Pastor, J. C. Dabrowiak, and W. H. Woodruff, *J. Amer. Chem. Soc.*, 1976, **98**, 8007.
209 B. Crociani, M. Nicolini, and R. L. Richards, *J. Organometallic Chem.*, 1976, **104**, 259.
210 T. B. Freedman, J. S. Loehr, and T. M. Loehr, *J. Amer. Chem. Soc.*, 1976, **98**, 2809.
211 M. D. Mazus, V. N. Biyushkin, A. V. Ablov, T. I. Malinovskii, M. A. Ablova, and T. I. Savushkina, *Doklady Chem.*, 1976, **225**, 642.
212 L. F. Lindoy, H. C. Lip, L. F. Power, and J. H. Rea, *Inorg. Chem.*, 1976, **15**, 1724.
213 N. V. Ivashin, I. F. Gurinovich, and G. P. Gurinovich, *Zhur. priklad. Spektroskopii*, 1975, **23**, 1026.
214 W. Wolfsberger and H. Schmidbaur, *J. Organometallic Chem.*, 1976, **122**, 5.

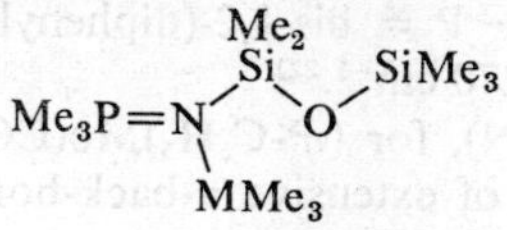

(79)

Cyanides and Isocyanides.—A review has appeared on the interpretation of vibrational force constants, and their application to the study of binding forces in metal cyano and carbonyl complexes.[215]

HCN co-ordinates *via* N in the compounds $M(HCN)_6(InCl_4)_2$, when M = Mg, Mn, Fe, Co, Ni, Cu, or Zn. ν(CH) shifts 150 cm^{-1} to lower wavenumber, ν(C≡N) 27—55 cm^{-1} to higher wavenumber, and δ(HCN) 60 cm^{-1} to lower wavenumber, compared to the values for HCN(g).[216]

I.r. and Raman spectra were recorded for 2 : 1 adducts of MeCN and MeCN-d_3 with $TiCl_4$ and $SnCl_4$ in the solid state. The data were consistent with *cis*-octahedral geometry, and a normal co-ordinate analysis was performed, using C_{2v} symmetry. The C≡N stretching force constants were about 18.0 mdyn Å^{-1} compared with 17.5 in free MeCN, consistent with the observed increases in wavenumber, *e.g.* 2304 and 2300 cm^{-1} in $TiCl_4,2CD_3CN$, and 2278 cm^{-1} in CD_3CN.[217]

A progressive shift to higher wavenumber for ν(CN) is noted on the oxidation, by $AgPF_6$, of $Cr(CNR)_6$, through $[Cr(CNR)_6]PF_6$, to $[Cr(CNR)_6](PF_6)_2$, where R = Ph, *o*-tolyl, *p*-tolyl, *p*-$MeOC_6H_4$, or *p*-ClC_6H_4.[218]

The i.r. spectra of $M(CNPh)_6$, where M = Cr, Mo, or W, $[Mn(CNPh)_6]Cl$ and $[Mn(CNPh)_6](PF_6)_2$ in the region 1800—2200 cm^{-1}, show that two bands are found, 2005 and 1950 cm^{-1}, consistent with a low site symmetry for this compound. ν(CN) in $[Mn(CNPh)_6]^+$ is at 2080 cm^{-1}, whereas for $[Mn(CNPh)_6]^{2+}$ it is at 2176 cm^{-1}.[219]

I.r. ν(CN) and ν(CO) wavenumbers were listed for complexes of the types $Cr(CO)_5L$, $Mo(CO)_5L$, *cis*-$Mo(CO)_4L_2$, and *fac*-$Mo(CO)_3L_3$, where L = Me_3MCH_2NC (M = Si, Ge, Sn, or Pb). The shifts in these bands were discussed in terms of the inductive effect of the MMe_3 group.[220]

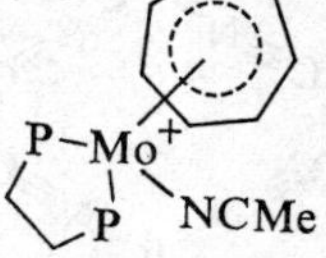

(80)

[215] L. H. Jones and B. I. Swanson, *Accounts Chem. Res.*, 1976, **9**, 128.
[216] P. L. A. Everstein, A. P. Zuur, and W. L. Driessen, *Inorg. Nuclear Chem. Letters*, 1976, **12**, 277.
[217] Y. Kawano, Y. Hase, and O. Sala, *J. Mol. Struct.*, 1976, **30**, 45.
[218] P. M. Treichel and G. J. Essenmacher, *Inorg. Chem.*, 1976, **15**, 146.
[219] K. R. Mann, M. Cimolino, G. L. Geoffroy, G. S. Hammond, A. A. Orio, G. Albertin, and H. B. Gray, *Inorg. Chim. Acta*, 1976, **16**, 97.
[220] P. M. Treichel and D. B. Shaw, *Inorg. Chim. Acta*, 1976, **16**, 199.

Structure (80), where P—P = bis-1,2-(diphenylphosphino)ethane, gives an i.r. band due to ν(CN) at 2270 cm^{-1}.[221]

I.r. data, ν(CO) and ν(CN), for $(\eta^5\text{-}C_5H_5)Mo(CO)_{3-n}(CNMe)_n$, where $n = 1$ or 2, indicate the existence of extensive π-back-bonding between the Mo atom and the isocyanide ligand.[222]

cis- and *trans*-$[Mn(CO)_2(CNMe)_4]^+$ have been prepared pure for the first time. They can be differentiated by their ν(CN) and ν(CO) value: *cis*, ν(CN) 2230, 2190, 2170, and 2150 cm^{-1}; ν(CO) 2000 and 1975 cm^{-1}; *trans*, ν(CN) 2177 cm^{-1}; ν(CO) 1985 cm^{-1}.[223] ν(CO) and ν(CN) values for $[Mn(CO)_{5-n}(CNMe)_n]^-$, where n = 1, 2, or 3, indicate an increased amount of M—CNMe π-back-bonding.[224]

I.r. spectra of $Na_3[Fe(CN)_5py],3H_2O$ and $K_2[Co(CN)_5py],H_2O$ were reported. The frequencies and intensities of the ν(CN) bands, and the proton n.m.r. shifts of the py ligand were interpreted in terms of greater metal to ligand π-back-bonding in the Fe^{II} complex than in that of Co^{III}.[225]

ν(CN) values are summarized in Table 20 for the newly-prepared complexes $LFe(CO)_2(CNR)$.[226]

Table 20 *Vibrational assignments for* $LFe(CO)_2(CNR)$ (*all wavenumbers*/cm^{-1})

L	R	*Assignments*	
butadiene	H	ν(CN) 2150	ν(NH) 3430
	D	ν(CN) 2130	ν(ND) 2377
	Et	ν(CN) 2158	
cyclohexa-1,3-diene	H	ν(CN) 2155	ν(NH) 3400
	D	ν(CN) 2125	ν(ND) 2380
	Et	ν(CN) 2150	
cyclo-octatetraene	H	ν(CN) 2148	ν(NH) 3400
	D	ν(CN) 2123	ν(ND) 2375
	Et	ν(CN) 2160	

$[CpFe(CO)(CN)_2]^-$ reacts with BX_3, where X = F, Br, or Ph, to give the 12-membered ring compounds (81). This type of co-ordination is supported by the high-wavenumber shifts observed for both ν(CN) (30—130 cm^{-1}, and ν(CO) (20—30 cm^{-1}) upon their formation.[227]

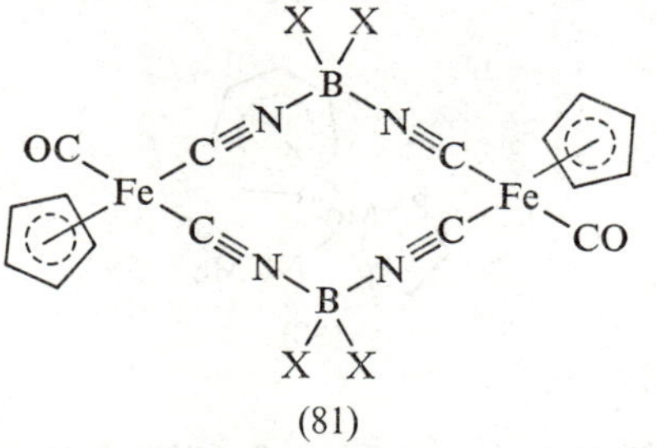

(81)

221 M. L. H. Green and R. B. A. Pardy, *J. Organometallic Chem.*, 1976, **117**, C13.
222 R. D. Adams, *Inorg. Chem.*, 1976, **15**, 169.
223 P. M. Treichel, H. J. Mueh, and B. E. Busten, *J. Organometallic Chem.*, 1976, **110**, C49.
224 R. D. Adams, *Inorg. Chem.*, 1976, **15**, 174.
225 J. A. Olabe and P. J. Aymonino, *J. Inorg. Nuclear Chem.*, 1976, **38**, 225.
226 H. Behrens, M. Moll, and P. Wüstl, *Z. Naturforsch.*, 1976, **31b**, 1017.
227 J. Emri, B. Gyöori, A. Bakos, and G. Czira, *J. Organometallic Chem.*, 1976, **112**, 325.

A study of the ν(CN) bands of $(Ph_3P)_4Ru(MeCN)_2$ reveals that one of the MeCN molecules is π-bonded to the metal, while the other is present as MeCN of solvation.[228]

The complexes $[Co(CNR^1)_3\{P(OR^2)_3\}_2]_2X$, where R^1 = Ph, *p*-FC_6H_4, *p*-ClC_6H_4, *p*-BrC_6H_4, *p*-IC_6H_4, *p*-MeC_6H_4, 2,6-$Me_2C_6H_3$; R^2 = Me, Et, or Pr^i; X = ClO_4 or BF_4, all give one strong band in solution in the i.r. due to ν(NC). Thus, regular trigonal bipyramidal co-ordination is indicated. In the solid phase, however, much more complex spectra are seen, and a distorted structure was postulated in this case. The value of the strongest ν(NC) band in these complexes is at a lower wavenumber than the corresponding feature in $[Co(CNR)_5]^+$, *i.e.* 2065—2092 instead of 2110—2120 cm^{-1}. This observation suggests increased π^* electron acceptance, and/or decreased σ-electron donation by CNR when two are replaced by trialkylphosphites.[229]

Partial i.r. assignments ν(NC) have been given for the π-olefinic complexes $Rh(dtc)(RNC)_2$(olefin), where dtc = $Me_2NCS_2^-$; R = *p*-MeC_6H_4 or 2,4,6-$Me_3C_6H_2$; olefin = tetracyanoethylene, fumaronitrile, maleonitrile, or maleic anhydride.[230]

The oxidation of Rh^I in the complex (82) is accompanied by the expected increase in ν(CN), thus when R = Me, 2174 goes to 2215 cm^{-1}; Bu^n, 2160 to 2201 cm^{-1}, and R = C_6H_{11}, 2141 to 2180 cm^{-1}.[231]

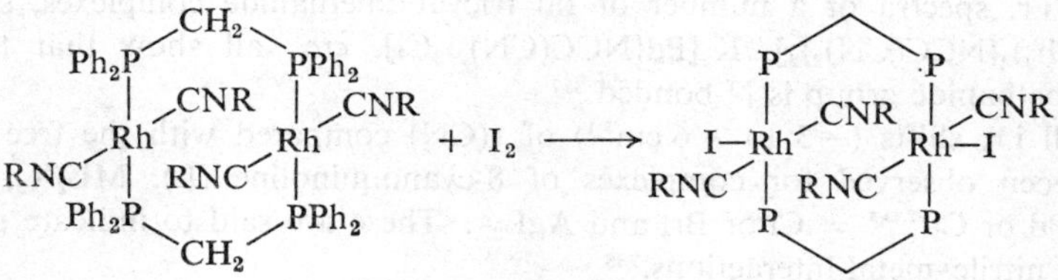

(82)

Association of $[(C_6H_{11}NC)_4Rh]^+$ with *trans*-$[(C_6H_{11}NC)_4RhI_2]^+$ gives rise to a new i.r. absorption at 2214 cm^{-1}. Isolation of the isocyanide associate as the BPh_4^- salt shows ν(CN) to be 2208 cm^{-1}. The observation of a single band supports the hypothesis of a $[X(L)_4Rh]_2^{2+}$ dimer with a Rh—Rh bond, and a probable symmetry of D_{4d}.[232]

The i.r. spectra of the dicarbene complexes (83), where R^1 = Me, R^2 = Me, Pr^n, or Bu^n, X = I; R^1 = Et, R^2 = Pr^n or Bu^n, X = I; R^1 = R^2 = Me, X = Bu^tNC, show two equally intense bands in the range 2289—2185 cm^{-1}, due to the *cis* isonitriles. A shift to lower wavenumber is consistent with increased Rh—isonitrile π-bonding, and weaker Rh—carbene π-bonding, relative to the parent isocyanide.[233]

ν(CN) in $(Ph_3P)_3Ni(CNPh)$ is at 2154 cm^{-1}, compared with 2230 cm^{-1} in free

228 E. O. Sherman and P. R. Schreiner, *J.C.S. Chem. Comm.*, 1976, 3.
229 C. A. L. Becker, *J. Organometallic Chem.*, 1976, **104**, 89.
230 T. Kaneshima, Y. Yumoto, K. Kawakami, and T. Tanaka, *Inorg. Chim. Acta*, 1976, **18**, 29.
231 A. L. Balch, *J. Amer. Chem. Soc.*, 1976, **98**, 8049.
232 A. L. Balch and M. M. Olmstead, *J. Amer. Chem. Soc.*, 1976, **98**, 2354.
233 P. R. Branson, R. A. Cable, M. Green, and M. K. Lloyd, *J.C.S. Dalton*, 1976, 12.

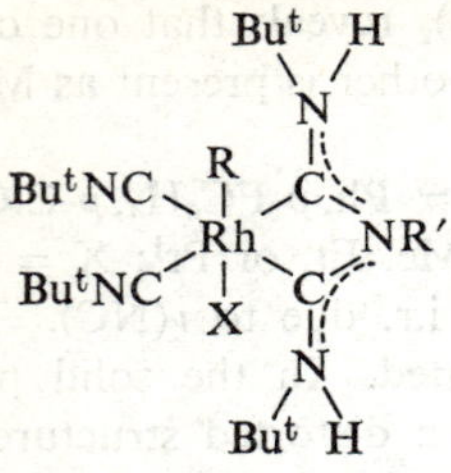

(83)

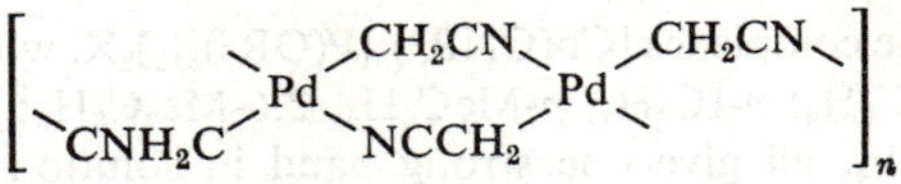

(84)

PhCN. The large shift is indicative of strong co-ordination, associated with the unusually strong π-back-bonding ability of the $Ni(PPh_3)_3$ unit.[234]

$Pd(CH_2CN)_2$ and its deuterio-derivative have been prepared for the first time. ν(CN) in the solid appears at 2239 cm^{-1} (h_4-), 2244 cm^{-1} (d_4-). The position of this band, and its breadth, suggests that the structure is polymeric, probably (84). In complexes $L_2Pd(CH_2CN)_2$, where L = PPh_3, bipy *etc.*, the value is decreased to about 2200 cm^{-1}.[235]

The presence of ν(CN) bands in the 'terminal' region only for $[Pd_2(Bu^tNC)_6]^{2+}$, 2235, 2220, and 2203 cm^{-1}, is consistent with a structure containing a Pd—Pd bond.[236]

The i.r. spectra of a number of Pd tricyanomethanide complexes, such as $[Pd(PPh_3)_2\{NCC(CN)_2\}_2]$, $K_2[Pd\{NCC(CN)_2\}_3Cl]$, *etc.*, all show that the tricyanomethanide group is N-bonded.[237]

Small i.r. shifts (-3 to -6 cm^{-1}) of ν(CN) compared with the free ligand have been observed for complexes of 8-cyanoquinoline (L), ML_2X_2, where M = Pd or Cu; X = Cl or Br, and AgL_2^+. These are said to indicate perpendicular nitrile–metal interactions.[238]

ν(C≡NR) has been assigned, to two bands at 2250 and 2210—2220 cm^{-1}, in $Pt(CNR)_4^{2+}$, where R = 4-dimethylaminophenyl, 4-azobenzene, 2-methylphenyl, or 3,4-dimethylphenyl. The splitting was due to solid-state effects.[239]

ν(C=C) and ν(C≡N) have been assigned, and compared with free ligand values, for $L^1CuSnCl_3$, $L^1_2CuSnCl_3$, $L^2CuSnCl_3$, and L^2CuCl, where L^1 = acrylonitrile, L^2 = methacrylonitrile. Co-ordination appears to be *via* N only in the latter two, and *via* N and C=C in the former two.[240]

The vibrational spectra of $Au(CN)_2^-$, as Na^+ and K^+ salts, are largely assignable in terms of $D_{\infty h}$ symmetry. Polycrystalline $MAu(CN)_2$, where M = K, Na, NH_4, or Li, can be similarly assigned, but when M = Tl, Ag the effective symmetries of the anion are C_{2v}, $C_{\infty v}$ respectively. $TlAu(CN)_2$ seems to involve a weak Tl—Au covalent bond, while $AgAu(CN)_2$ contains bridging CN units.[241]

234 I. W. Bassi, C. Benedicenti, M. Calcaterra, and G. Rucci, *J. Organometallic Chem.*, 1976, **117**, 285.
235 G. Oehme, K.-C. Röber, and H. Pracejus, *J. Organometallic Chem.*, 1976, **105**, 127.
236 M. F. Rettig, E. A. Kirk, and P. M. Maitlis, *J. Organometallic Chem.*, 1976, **111**, 113.
237 T. Likowski and J. L. Burmeister, *Inorg. Chim. Acta*, 1976, **17**, 117.
238 J. E. Sutton and J. I. Zink, *Inorg. Chem.*, 1976, **15**, 675.
239 H. J. Keller and R. Lorentz, *Z. Naturforsch.*, 1976, **31b**, 565.
240 J. A. Sampedro and T. Ogura, *Transition Metal Chem.*, 1976, **1**, 127.
241 B. M. Chadwick and S. G. Frankiss, *J. Mol. Structure*, 1976, **31**, 1.

ν(CN) is at 2169 and ν(RNC) is at 2290 cm^{-1} in Au(CN)(CNMe). The latter is 120 cm^{-1} higher than in free MeNC, and so the NC bond order is higher in the complex, as is found for metals in intermediate or high oxidation states.[242]

The new compounds $(C_5H_4Me)_3U[C(CN)_3]$ and $(C_5H_5)_3U[C(CN)_3]$ have been characterized the i.r. data indicate the presence of bridges (85).[243]

(85)

ν(C=N) has been assigned (1615—1640 cm^{-1}) in $(R^1{}_3Al{-}N{=}CR^2H)^-$, where R^1 = Et, R^2 = Ph, *p*-MeC_6H_4, *cyclo*-C_6H_{11}, or Et; R^1 = Bu|, R^2 = *cyclo*-C_6H_{11}.[244]

I.r. and Mössbauer data have been recorded for a number of complexes of TCNE and TCNQ with Sn. TCNE complexes gave one band due to ν(CN), 2210—2220 cm^{-1}, whereas the TCNQ compounds showed two or three bands.[245]

Nitrosyls.—$V(NO)_2phen_2$ gives two ν(NO) bands, at 1695 and 1547 cm^{-1}, and it is therefore *cis*. The values depend only slightly upon the nature of the counter-ion.[246]

The reaction of (arene)$Cr(CO)_3$ with NOCl at room temperature led to completely decarbonylated products, showing ν(NO) at 1852 and 1710 cm^{-1}, which could not be isolated. At -30 °C, solvated $CrCl(CO)_2(NO)$ is formed, with ν(CO) at 2024, 1942 cm^{-1}, and ν(NO) at 1681 cm^{-1}. The reaction of this with indenyl or fluorenyl lithium yields the new compounds (86), with ν(NO) at 1699 (86a), 1697 (86b) cm^{-1}.[247]

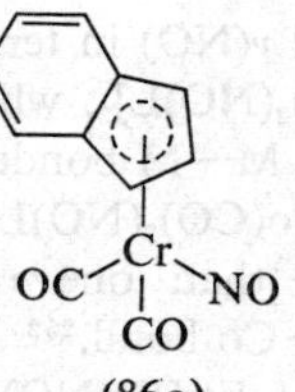

(86a)

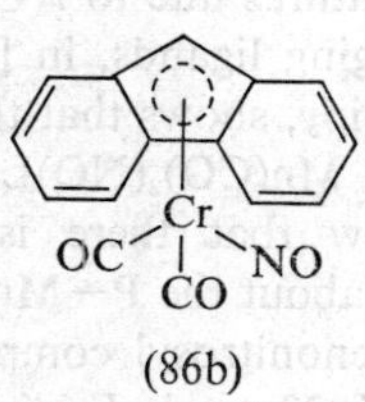

(86b)

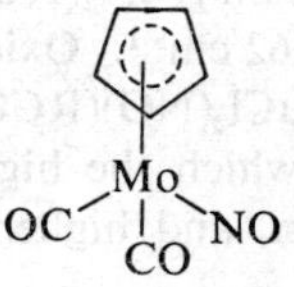

(87a)

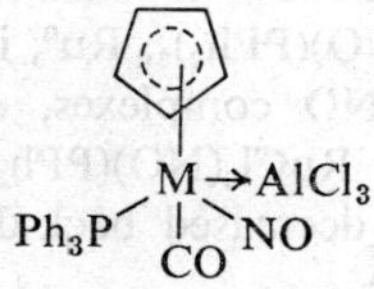

(87b)

[242] S. Esperås, *Acta Chem. Scand.*, 1976, **A30**, 527.
[243] R. D. Fischer and G. R. Sienel, *Z. anorg. Chem.*, 1976, **419**, 126.
[244] H. Hoberg and J. Korff, *J. Organometallic Chem.*, 1976, **118**, C1.
[245] A. B. Cornwell, C.-A. Cornwell, and P. G. Harrison, *J.C.S. Dalton*, 1976, 1612.
[246] S. Sarkar, R. Maurya, S. C. Chaurasia, and R. Srivastava, *Transition Metal Chem.*, 1976, **1**, 49.
[247] M. Herberhold and W. Bernhagen, *Angew. Chem. Internat. Edn.*, 1976, **15**, 617.

Compound (87a) gives ν(NO) at 1670, and ν(CO) at 1950 and 2025 cm^{-1}. Addition of $AlCl_3$ gives the formation of adducts: $-NO-AlCl_3$ [ν(NO) 1450; ν(CO) 2025, 2087 cm^{-1}] and $-CO-AlCl_3$ [ν(NO) 1710, ν(CO) 1735, 2065 cm^{-1}]. Note the increase in wavenumber for the ligand not affected by the adduct formation. Similar types of adduct were formed by a large range of other carbonyls and nitrosyls, whereas $M-AlCl_3$ adducts are formed by CpM(CO)-(NO)PPh_3, where M = Mo or W, (87b), ν(NO) > 1700, ν(CO) ~ 2070 cm^{-1}.[248]

ν(NO) wavenumbers for the tris(pyrazolyl)borate Mo complexes {MoXY(NO)-[HB(3,5-$R_2C_3HN_2$)$]_3$}$_n$ are summarized in Table 21. When R = H, ν(NO) is insensitive to changes in X and Y, suggesting that there are compensating electronic effects operating.[249]

Table 21 ν(NO) *assignments*/cm^{-1} *for some tris(pyrazolyl)borate* Mo *complexes*

X	Y	R	ν(NO)
Cl	Cl	H	1727
Cl	Cl	Me	1720
Br	Br	H	1724
Br	Br	Me	1720
I	I	H	1722
I	I	Me	1710
Cl	OEt	Me	1676
Br	OEt	Me	1673
I	OEt	Me	1678
OEt	OEt	Me	1636

Reactions of, for example, $MoCl_2(NO)_2(HMPT)_2$, $MoCl_3(NO)(OPPh_3)_2$, *etc.* with $EtAlCl_2$, Et_3Al, $AlCl_3$, or CO were monitored by following the i.r. absorption bands due to ν(NO).[250]

The presence of features due to ν(CO) and ν(NO) in terminal positions only, and not due to bridging ligands, in $[M(CO)_2(NO)L]_2$, where M = Mo or W; L = *o*-phen or 2,2′-bipy, shows that they are M—M bonded species.[251]

The i.r. spectra of $Mn(CO)_3(NO)L$ and $Co(CO)_2(NO)L$, where L = [FeCp-$(CO)_2\{P(CF_3)_2\}$], show that there is more than one ν(NO), attributed to rotational isomerism about the P—Mn or P—Co bond.[252]

The anionic halogenonitrosyl complexes of Fe:$[Fe(NO)X_3]^-$, where X = Cl or Br, have ν(NO) at 1820 cm^{-1}. Earlier data on $[FeX_2(NO)_2]^-$ were confirmed.[253]

ν(NO) in $RuCl(NO)(PPh_3)_2$, Ru^0, is at 1762 cm^{-1}. Oxidative addition gives a number of other NO complexes, *e.g.* $RuCl_2(NO)(RCO)(PPh_3)_2$, RuCl(NO)-$(PPh_3)_2$(olefin), and $RuCl_3(NO)(PPh_3)_2$, in which the higher oxidation state of the metal leads to decreased back-donation, and higher values of ν(NO), *i.e.* 1790—1875 cm^{-1}.[254]

[248] B. V. Lokshin, E. B. Rusach, N. E. Kolobova, Yu. V. Makarov, N. A. Ustynyuk, V. I. Zdanovich, A. Zh. Zhakaeva, and V. N. Setkina, *J. Organometallic Chem.*, 1976, **108**, 353.
[249] J. A. McCleverty, D. Seddon, N. A. Bailey, and N. W. Walker, *J.C.S. Dalton*, 1976, 898.
[250] R. Taube, K. Seyferth, L. Bencze, and L. Marko, *J. Organometallic Chem.*, 1976, **111**, 215.
[251] S. C. Tripathi, S. C. Srivastava, and A. K. Shrimal, *J. Organometallic Chem.*, 1976, **118**, 325.
[252] R. C. Dobbie and P. R. Mason, *J.C.S. Dalton*, 1976, 189.
[253] N. G. Connelly and C. Gardiner, *J.C.S. Dalton*, 1976, 1525.
[254] O. Gandolfi, B. Giovannitti, M. Ghedini, and G. Dolcetti, *J. Organometallic Chem.*, 1976, **104**, C41.

The complex $[RuBr_3(NO)(Et_2SO)]_2$ has ν(NO) at 1847 cm^{-1}, suggesting that the ligand is formally NO^+, with linear MNO, and ν(SO) at 920 cm^{-1}, indicative of M—O bonding. Both of these were confirmed by the crystal structure, (88).[255]

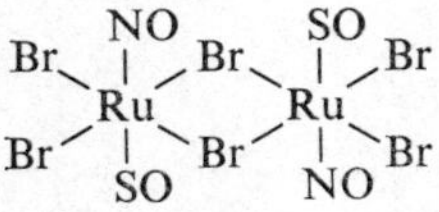

(88)

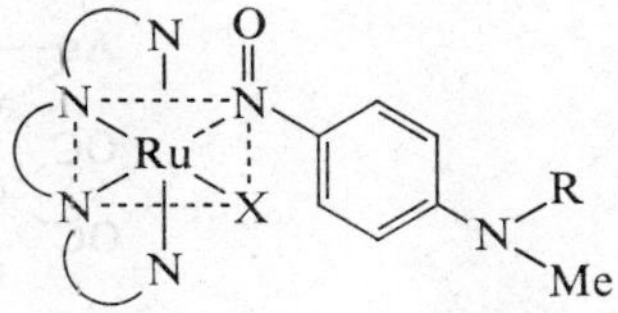

(89)

p-Nitroso-arene complexes of the type $[Ru(bipy)_2LX]^+$, (89), in which R = H or Me, X = Cl or NO_2, all give a characteristic i.r. band at 1330—1380 cm^{-1}. Assignment of ν(NO) was not clear, although ^{15}N substitution points strongly to its being at 1284 cm^{-1}.[256]

The new compound dinitrosyl(octaethylporphinato)osmium shows ν(NO) at 1798 and 1500 cm^{-1}. As the geometry of the macrocycle imposes a *trans* configuration on the NO groups, this observation was taken as evidence for the simultaneous existence of NO^+ and NO^- groups.[257]

The compound $Co(NO)D_2$, where HD = dimethylglyoxime, forms adducts with solvents, leading to large shifts in ν(NO). The MeOH adduct was studied in a number of solvents, showing that in non-co-ordinating solvents the MeOH is lost, forming a five-co-ordinate species.[258]

ν(NO) has been listed for two new series of anionic penta-co-ordinated nitrosyl complexes of Ir and Rh, $[M(NO)(chel)(PPh_3)_2]^{2+}$, where chel = phen or bipy, and $[M(NO)(MeCN)_2(PPh_3)_2]$, where M = Rh or Ir.[259]

ν(NO) for $[Ni(NO)L]BF_4$ complexes generally correlates with $^1J_{PH}$ in HL^+, and ν(NO) may therefore be correlated with the basicity of L, where L = a phosphite such as $(MeO)_3P$, $(EtO)_3P$, $(Pr^nO)_3P$, $(PhO)_3P$, *etc.*[260]

ν(NO) was recorded for a number of new nitrosyl complexes $PtA_2Cl_2(NO)X$, where A = NH_3, $MeNH_2$; X = Cl, NO_3, or ClO_4, and $[Pt(MeNH_2)_2Cl_2(NO)(NO_2)]$. The last was found as *cis*/*trans* isomers, ν(NO) being 1765 (*cis*) and 1755 cm^{-1} (*trans*).[261]

In *trans*-$[Tc(NH_3)_4(H_2O)(NO)]^{2+}$, ν(NO) is at 1680 cm^{-1}, the low wavenumber indicating great stabilization by Tc^I—NO back-bonding.[262]

4 Phosphorus and Arsenic Ligands

A review on P, As, and Sb complexes of the main group elements, Al—Tl, Si—Pb, and Group V itself, contains several references to i.r. and Raman studies.[263]

255 J. E. Fergusson, C. T. Page, and W. T. Robinson, *Inorg. Chem.*, 1976, **15**, 2270
256 W. L. Bowden, W. F. Little, and T. J. Meyer, *J. Amer. Chem. Soc.*, 1976, **98**, 44.
257 J. W. Buchler and P. D. Smith, *Chem. Ber.*, 1976, **109**, 1465.
258 C. B. Ungermann and K. G. Caulton, *J. Amer. Chem. Soc.*, 1976, **98**, 3862.
259 M. Ghedini, G. Dolcetti, O. Gandolfi, and B. Giovannitti, *Inorg. Chem.*, 1976, **15**, 2385.
260 J. O. Albright, F. L. Tanzella, and J. G. Verkade, *J. Coordination Chem.*, 1976, **5**, 225.
261 A. I. Stetsenko, O. I. Arkhangel'skaya, and N. A. Sukhanova, *Russ. J. Inorg. Chem.*, 1975, **20**, 1663.
262 R. A. Armstrong and H. Taube, *Inorg. Chem.*, 1976, **15**, 1905.
263 W. Levason and C. A. McAuliffe, *Coordination Chem. Rev.*, 1976, **19**, 173.

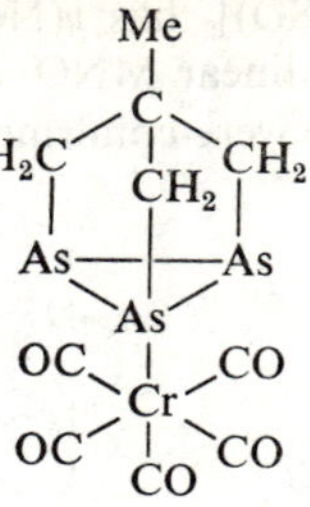

(90)

ν(AsAs) gives features at 310, 301, and 273 cm^{-1} in (90), together with the usual $Cr(CO)_5$ modes. The exact mode of co-ordination of the As_3 unit to the Cr was not definitely established, however.[264]

Cotton–Kraihanzel force constants were calculated for $Mo(CO)_5[P(CH_2)_6N_3]$ and compared with those for other $Mo(CO)_5L$ complexes. The $P(CH_2)_6N_3$ was thereby shown to be a rather good σ-donor, comparable with PPh_3.[265]

Absolute i.r. intensities were measured for ν(CO) in $LW(CO)_5$, where L = PCl_3, $P(OMe)_3$, $P(OEt)_3$, $P(OBu^n)_3$, PPh_3, or $PBu^n{}_3$. Approximately linear correlations were found between these and J(W—P), and ligand ionization energies.[266]

Reactions of bidentate L—L, $Ph_2P(CH_2)_nPPh_2$, n = 1, 2, or 3, $MeC(CH_2PPh_2)_3$, $Ph_2As(CH_2)_2AsPh_2$, o-$C_6H_4(AsMe_2)_2$, $Ph_2PCHMePPh_2$, $Ph_2PCMe_2PPh_2$, with $MeMn(CO)_4$ were monitored by ν(CO). The band patterns show that the thermodynamically favoured products $MeCOMn(CO)_3$(L—L) are *fac*-, except when L—L = $Ph_2PCMe_2PPh_2$, when the *mer*-isomer is formed.[267]

The new *ortho*-metallated complex (91) shows bands at 1555 cm^{-1}, due to stretching of the metallated ring, and 730 cm^{-1}, due to the CH o.o.p. deformation of the *ortho*-disubstituted phenyl ring.[268]

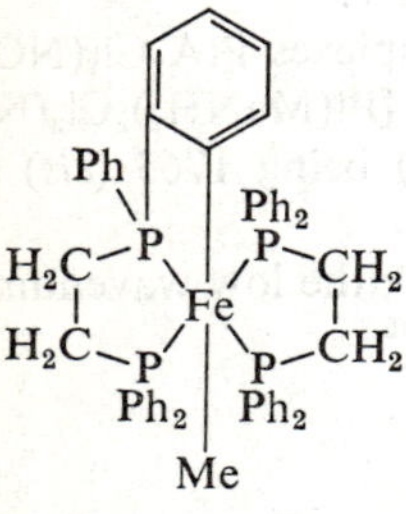

(91)

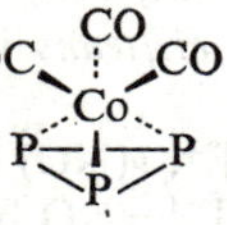

(92)

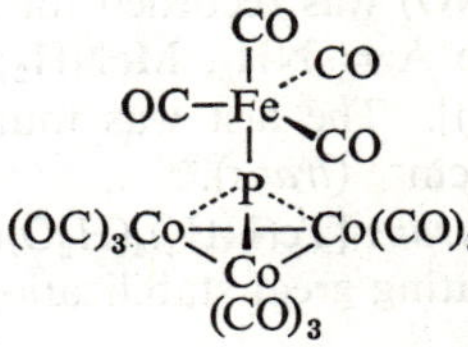

(93)

The i.r. spectrum of (92), of C_{3v} symmetry, gives ν(CO) at 2091.4 (A_1) and 2042.7 (E) cm^{-1}. The species $P[Co(CO)_3]_3$ is very unstable, but it forms an adduct with $Fe(CO)_4$, whose i.r. spectrum is consistent with the structure (93).[269]

[264] J. Ellermann and H. Schössner, *J. Organometallic Chem.*, 1976, **118**, C69.
[265] J. R. DeLerno, L. M. Trefonas, M. Y. Darensbourg, and R. J. Majeste, *Inorg. Chem*, 1976, **15**, 816.
[266] M. S. A. Abd El-Mottaleb, *J. Mol. Struct.*, 1976, **32**, 203.
[267] C. S. Kraihanzel and P. K. Maples, *J. Organometallic Chem.*, 1976, **117**, 159.
[268] T. Ikariya and A. Yamamoto, *J. Organometallic Chem.*, 1976, **118**, 65.
[269] A. Vizi-Orosz, *J. Organometallic Chem.*, 1976, **111**, 61.

$(\mu_2\text{-L})_2Co_2(CO)_5PR_3$ and $(\mu_2\text{-L})_2Co_2(CO)_5P(OR)_3$, where R = alkyl or aryl group, $L_2 = P_2$ or As_2; R = acetylene, L = CO or but-2-en-4-olid-4-ylidene, have been prepared. I.r. spectra in the ν(CO) region for all of them give the 5-line spectrum expected for either C_s or C_1 symmetry. The PR_3 or $P(OR)_3$ in every case is situated axially.[270]

Compound (94) gives i.r. bands at 980 and 892 cm^{-1} which may be characteristic of benzylic CH.[271]

trans-Difluorophosphonate complexes $MClL_2(PF_2O)$, where M = Pd or Pt, L = $PEt_{3-n}Ph_n$, n = 0—3, give ν(P=O) at 1230 cm^{-1}, suggesting bonding through P rather than O. This was confirmed by *X*-ray diffraction.[272]

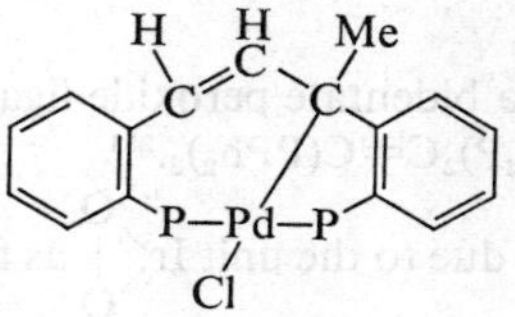

(94)

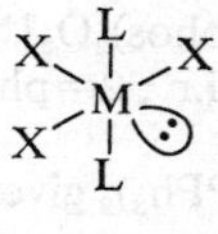

(95)

Several internal modes of $P(OMe)_3$ were assigned in the newly-prepared complex *cis*-$Me_2Pt[P(OMe)_3]_2$. Thus $\nu_{as}(PO_3)$ is at 800 (i.r.), 815 (Raman) cm^{-1}; $\nu_s(PO_3)$ 745 (i.r.), 750 (Raman) cm^{-1}; $\delta(PO_3)$ 520 (i.r.), 515 (Raman) cm^{-1}.[273]

When L = $PPh_2(2\text{-}OHC_6H_4)$ [ν(OH) = 3220 cm^{-1}] or $PBu^t_2(2\text{-}OHC_6H_4)$ (3240 cm^{-1}) from the complexes *cis*-$PdCl_2L_2$ and *trans*-$PtCl_2L_2$, ν(OH) shifts to 3220; 3495 + 3200 cm^{-1} respectively.[274]

1 : 2 Adducts of MX_3 (M = P, As; X = Cl or Br) with PMe_3-h_9 or -d_9 all give vibrational spectra consistent with their being *trans*, with geometry C_{2v}, (95). This was confirmed by comparison with the adducts MX_3,bdpe, where the ligand bdpe = 1,2-bisdimethylphosphinoethane forces a *cis* geometry for the molecular skeleton.[275]

5 Oxygen Donors

Molecular Oxygen, Peroxo- and Hydroxy-complexes.—ν(OH) bands in the i.r. spectra of $[M(H_2O)_n(NH_3)_{6-n}][Co(CN)_6]$, where M = Cr or Co, n = 1 or 2, show that the anion and the cation are strongly H-bonded together.[276]

$Cs_2WO_8O,2H_2O$ gives typical vibrations of the ring W / O—O, as follows: ν(O—O), 830 cm^{-1}, and bands with contributions from $\nu(WO_2)$ and ring deformations, 600, 580, 550, 520, and 485 cm^{-1}.[277]

270 G. Varadi, A. Vizi-Orosz, S. Vastag, and G. Palyi, *J. Organometallic Chem.*, 1976, **108**, 225.
271 M. A. Bennett, R. N. Johnson, G. B. Robertson, I. B. Tomkins, and P. O. Whimp, *J. Amer. Chem. Soc.*, 1976, **98**, 3514.
272 J. Grosse and R. Schmutzler, *J.C.S. Dalton*, 1976, 405.
273 B. Neruda and J. Lorberth, *J. Organometallic Chem.*, 1976, **111**, 241.
274 H. D. Empsall, B. L. Shaw, and B. L. Turtle, *J.C.S. Dalton*, 1976, 1500.
275 D. K. Friesen and G. A. Ozin, *J. Mol. Struct.*, 1976, **31**, 77.
276 J. Ribas, J. Casabo, and J. M. Coronas, *Rev. Chim. minérale*, 1976, **13**, 485.
277 L. I. Kozlova, N. A. Korotchenko, and G. A. Bogdanov, *Russ. J. Inorg. Chem.*, 1976, **21**, 773.

Deformation modes of Os—OH and H_2O are seen at 960 and 1630 cm^{-1} respectively in $Os(NO)(OH)_3(H_2O)_2$. ν(OH) of the OH ligand and of H_2O appears as a broad feature at 3200—3400 cm^{-1}; ν(NO) is at 1800 cm^{-1}.[278]

Raman spectral evidence has been presented for the existence of a $Co-O_2-Co$, peroxo, bridge in a complex formulated as *trans*-[(H_2O)Co(Me_2[14]diene)—O—O—Co(Me_2[14]diene)(H_2O)]$(ClO_4)_4$,H_2O, where Me_2[14]diene is an unsaturated macrocycle 5,12-dimethyl-1,4,8,11-tetra-azacyclotetradeca-4,11-diene. Thus, ν(O—O) is at 795 cm^{-1}, not at 1100 cm^{-1}, as in superoxo-bridged complexes, and compared with 800 cm^{-1} for $[(NH_3)_5Co-O_2-Co(NH_3)_5]^{4+}$.[279]

These conclusions are confirmed by further works on μ-superoxo- and μ-peroxo-dicobalt(II), with the addition that ν(O—O) is at about 1070 cm^{-1} in μ-amido-μ-superoxo complexes.[280]

The ion $[Co(2{=}phos)_2O_2]^+$ contains a bidentate peroxide ligand, with ν(O—O) at 909 cm^{-1} in the i.r.; 2=phos = $(Ph_2P)_2C{=}C(PPh_2)_2$.[281]

$Ir(Ar)(O_2)(CO)(PPh_3)_2$ gives ν(O—O) due to the unit $\mathrm{Ir}\langle^{O}_{O}|$ as follows: Ar = Ph, 825, 4-MeC_6H_4, 824, and 3-MeC_6H_4, 823. $Ir(Ar)(SO_2)(CO)(PPh_3)_2$ gives $\nu_{as}(SO_2)$ at 1174 and $\nu_s(SO_2)$ at 1033 cm^{-1}.[282]

δ(HOH) is found at very low wavenumbers in $ZnSO_4$,H_2O and $CuSO_4$,H_2O, *i.e.* 1510 cm^{-1}. No good reason was proposed, except that some covalent M—O bond character may be present.[283]

$Y(OH)Cl_2$ shows bands due to ν(OH) at 3545 and 3525 cm^{-1}, and δ(Y—O—H) at 705, 720, and 785 cm^{-1}.[284] $[M(OH)_4(H_2O)]^-$ gives bands characteristic of hydroxo-complexes at 995 cm^{-1} (M = Yb) and 992 cm^{-1} (Lu).[285]

Acetylacetonates and Related Complexes.—Trischelate complexes containing metallo-β-diketonate ligands have been characterized, *viz.*: (96), where M^1 = Mn or Re, R^1 and R^2 = Me, $PhCH_2$ *etc.*, and M^2 = Al or Ga. ν(C⋯O) is at 1525 cm^{-1}, apparently the same for all of the complexes.[286]

Characteristic bands due to the chelate (acac) are seen in the i.r. spectra of $[OsX_4(acac)]^-$, where X = Cl, Br, or I (Table 22).[287]

Table 22 *Characteristic ligand modes in* $[OsX_4(acac)]^-$ (*all wavenumbers*/cm^{-1})

Vibration X:	Cl	Br	I
ν_s(CO)	1560	1560	1560
ν_{as}(CC)[+ δ(CH)]	1540	1540	1540
ν_{as}(CO)[+ δ(CH)]	1312	1316	1326
ν_s(CC)[+ ν_s(C—Me)]	1289	1285	1285

[278] N. M. Synitsyn and N. I. Klinova, *Russ. J. Inorg. Chem.*, 1975, **20**, 1734.
[279] C. G. Barraclough and G. A. Lawrance, *Inorg. Nuclear Chem. Letters*, 1976, **12**, 133.
[280] T. Shibahara and M. Mori, *Proc. Internat. Conf. Coordination Chem., 16th*, 1974, R51.
[281] V. M. Miskowski, J. L. Robbins, G. S. Hammond, and H. B. Gray, *J. Amer. Chem. Soc.*, 1976, **98**, 2477.
[282] L. Dahlenburg and R. Nast, *J. Organometallic Chem.*, 1976, **110**, 395.
[283] V. A. Kolesova and A. N. Lazarev, *Russ. J. Inorg. Chem.*, 1975, **20**, 1576.
[284] N. V. Baryshnikov, A. V. Polonskii, and A. I. Khokhlov, *Russ. J. Inorg. Chem.*, 1975, **20**, 1757.
[285] B. N. Ivanov-Emin, S. G. Malyugina, L. D. Borzova, V. P. Dolganev, B. E. Zaitsev, and A. M. Egorov, *Russ. J. Inorg. Chem.*, 1976, **21**, 657.
[286] C. M. Lukehart, G. P. Torrence, and J. V. Zeile, *Inorg. Chem.*, 1976, **15**, 2393.
[287] G. Schätzel and W. Preetz, *Z. Naturforsch.*, 1976, **31b**, 749.

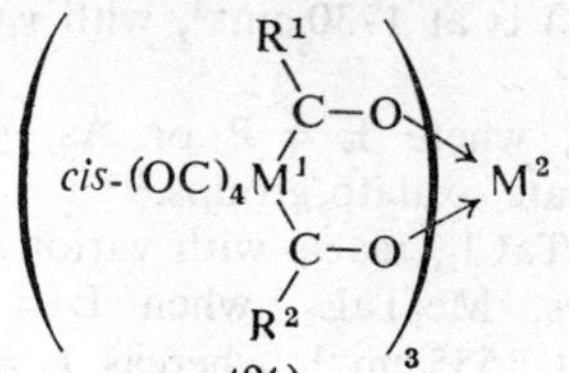
(96)

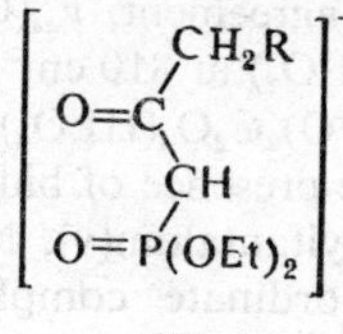
(97)

Five-co-ordinate complexes $MX(L)_2$, where M = Co, Zn; X = Br, I, NCS; L = (97), R = NEt_2, morpholine, piperidine have been prepared. The strong band at 1385 cm^{-1} for the last was assigned as OPCCO bending, suggesting that the ligand is bidentate.[288]

Two isostructural series, both monoclinic, exist for the complexes $M(AA)_4$, where M = Zr, Ce, Hf, Th, U, or Np; AA = acetylacetonate. Vibrational spectra were compared for the α- and β-modification of each complex. The main changes occur in the Me deformation and CH i.p. and o.o.p. modes. Neither the metal-sensitive band around 530 cm^{-1}, nor the ν(CO) and ν(CC) bands, 1500—1600 cm^{-1}, change significantly, suggesting that the M—O bonding in the two series is essentially identical.[289]

Monomeric keto-enolato derivatives and some related compounds of Ge^{II} have been characterized by lowering of the ν(CO) modes (1590—1510 cm^{-1}) and other data. The compounds included $Ge(acac)_2$, Ge(acac)Cl, Ge(acac)I, and $Ge(tropolonato)_2$.[290]

Carboxylato-complexes.—I.r. and Raman spectra were recorded for the isocitrate complexes Mg(iCi), Mn(iCi), Co(iCi), and $Co(iCi)_2$, where H_3iCi = CH_2COOH—CHCOOH—CH(OH)COOH. It was not possible to make an unambiguous decision about the structure, the OH and either 1 or 2 carboxylate oxygens are attached to the metal, (98a) or (98b).[291]

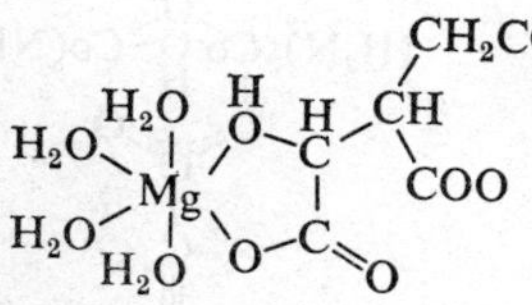
(98a)

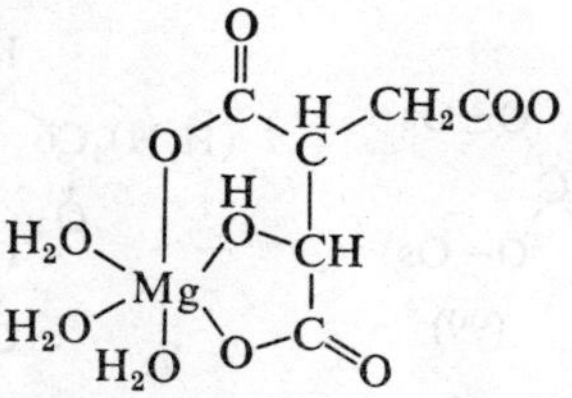
(98b)

ν_{as} and ν_s(COO) are at 1555 and 1360 cm^{-1} in $[Zr(OOCH)_4]$ or 1525 and 1345 cm^{-1} in $[ZrO(OOCH)_2,2H_2O]$. These values are not very useful for structural diagnosis, being close to the value for Na(OOCH), but presumably bridging and/or bidentate formate groups are likely.[292]

Bidentate oxalato ligands are shown by *X*-ray diffraction to be present in

[288] M. H. Youinou and J. E. Guerchais, *J.C.S. Dalton*, 1976, 293.
[289] B. Allard, *J. Inorg. Nuclear Chem.*, 1976, **38**, 2109.
[290] A. Rodgers and S. R. Stobart, *J.C.S. Chem. Comm.*, 1976, 52.
[291] G. Chottard and J. Bolard, *Bull. Soc. chim. France*, 1976, 742.
[292] R. C. Paul, O. B. Baidya, and R. Kapoor, *Z. Naturforsch.*, 1976, **31B**, 300.

$VOF(C_2O_4)_2^{3-}$, in agreement, $\nu_{as}(CO_2)$ is at 1730 cm^{-1}, with $\nu_s(CO_2)$ at 1295—1273 cm^{-1}, and $\delta(CO_2)$ at 810 cm^{-1}.[293]

The anions $[(VO)_2C_2O_4(HEO_4)]^{2-}$, where E = P or As, give i.r. spectra consistent with the presence of bidentate oxalato groups.[294]

Dichlorotrimethyltantalum(v), Me_3TaCl_2, reacts with various anionic ligands to give seven-co-ordinate complexes, Me_3TaL_2, when L = acac, bidentate groups are indicated by the band at 1555 cm^{-1}, whereas L = squarate gives bands at 1670, 1590, and 1505 cm^{-1}. These are consistent with a dimeric complex, containing bridging squarate. When L = ClO_4, features at 1185, 1100, 1015, and 910 cm^{-1} agree with bidentate co-ordination.[295]

Both terminal and bridging CO groups were detected in the ν(CO) spectrum of $[Mo(CO)_3L_2]_2$, where L = dimethyl, diethyl, di-n-butyl esters of maleic and fumaric acids. ν(C=O)(ester) is at 1715, and ν(C=C) at about 1610 cm^{-1}, a decrease of 30—40 cm^{-1} from the free esters.[296]

$[Mo_2O_3(C_2O_4)(NCS)_4]^{2-}$ has a polymeric structure, with quadridentate $C_2O_4^{2-}$ ligands bridging giving ν(CO) bands at 1635, 1350, and 1300 cm^{-1}.[297]

One product of the reaction between $W(CO)_6$ and pyridine-2-carboxylic acid (LH) is $W(CO)_3L$, $\nu_{as}(CO_2)$ is found at 1620 cm^{-1}, compared with values of about 1690 cm^{-1} in complexes such as WL_3.[298]

The CO_2 stretches in $Ru(CO_2Me)(H_2O)(PPh_3)_3$ are at 1500 and 1467 cm^{-1}, the very small splitting, $\nu_{as}-\nu_s$ = 33 cm^{-1}, is consistent only with symmetrical bidentate bonding.[299]

A larger value for this parameter, 100 cm^{-1}, is found for $RuH(O_2CR)(PPh_3)_3$, where R = Me, Pr, or Bu. This is still within the range for bidentate carboxylates, and the spectrum is the same in $CHCl_3$ solution as in the solid. Thus the structure is maintained in solution.[300]

The first compound containing two metal clusters linked by what is formally a CO_2 group has been characterized crystallographically, $[HOs_3(CO)_{10}.O_2C.Os_6(CO)_{17}]^-$. This contains the unit (99), and has ν(CO) at 1270 cm^{-1}, *i.e.* formally a single bond.[301]

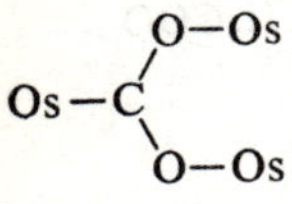

(99)

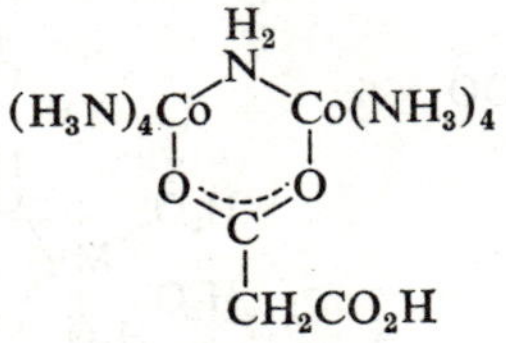

(100)

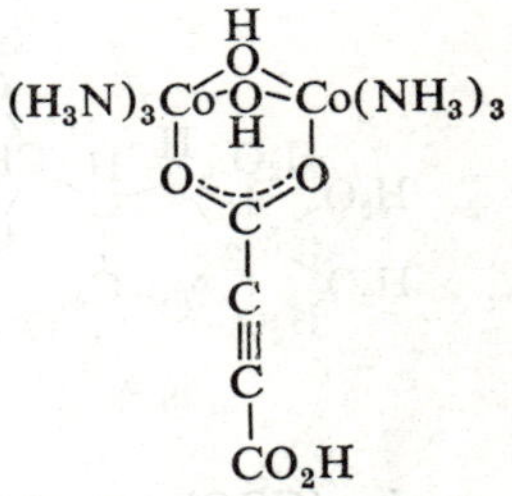

(101)

[293] H. Rieskamp and R. Mattes, *Z. Naturforsch.*, 1976, **31B**, 537.
[294] S. Pulvin, M. Ronis, and P, Courtine, *Compt. rend.*, 1976, **283C**, 585.
[295] C. Santini-Scampucci and G. Wilkinson, *J.C.S. Dalton*, 1976, 807.
[296] H. Schaper and H. Behrens, *J. Organometallic Chem.*, 1976, **113**, 361.
[297] R. Mattes, D. Altmeppen, and M. Fetzer, *Z. Naturforsch.*, 1976, **31b**, 1356.
[298] T. E. Dorsett and R. A. Walton, *J.C.S. Dalton*, 1976, 347.
[299] R. J. Young and G. Wilkinson, *J.C.S. Dalton.*, 1976, 719.
[300] A. Spencer, *Inorg. Nuclear Chem. Letters*, 1976, **12**, 661.
[301] C. R. Eady, J. J. Guy, B. F. G. Johnson, J. Lewis, M. C. Malatesta, and G. M. Sheldrick, *J.C.S. Chem. Comm.*, 1976, 602.

Carboxy groups bridging two Co atoms give $\nu_{as}-\nu_s$ of between 144 and 160 cm^{-1} in, for example, (100), which has ν_{as} at 1575 and ν_s at 1425 cm^{-1}.[302]

The μ-carboxy-group in (101) absorbs in the i.r. at 1565 and 1390 cm^{-1}, and at 1555 and 1395 cm^{-1} in the methoxy ester. When both ends of the dicarboxylic acetylene are bonded to $(H_3N)_3Co(OH)_2Co(NH_3)_3$ units the corresponding i.r. bands are at 1578 and 1395 cm^{-1}.[303]

$(MeCO_2)(C_6F_5)Pd(AsPh_3)_2$ gives i.r. bands assigned as ν_{as} and ν_s of CO_2 at 1610 and 1320 cm^{-1}. The $\Delta\nu$ value of 290 cm^{-1} shows that it is unidentate.[304]

The bridging carboxylates in (102), give an i.r. band at 1535 cm^{-1}.[305]

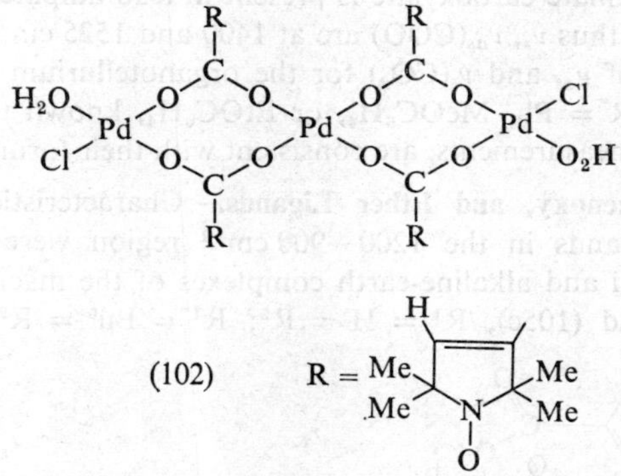

(102)

ν_{as} and $\nu_s(CO_2)$ for $(RCO_2)Cu(PPh_3)_2$, where R = Me, Et, Bu^n, or Pr^i, are at 1540—1640 cm^{-1} and 1410—1420 cm^{-1} respectively. Bidentate carboxylates are therefore present. Analogous bands in $(RCO_2)Cu(CO_2)(PPh_3)_2$ are, however, diagnostic of unidentate carboxylato co-ordination. Dithiocarboxylate analogues of the former group contain bidentate RCSS.[306]

The i.r. spectra of $L_2(CO)MAg(O_2CCF_3)_2$, where L = PPh_3 or $P(tol)_3$, M = Rh or Ir, show two $\nu_{as}(CO_2)$ bands, at 1700 and 1655 cm^{-1}. The former is due to a unidentate trifluoroacetato group, the latter to a bridging group. They are therefore formulated as (103).[307]

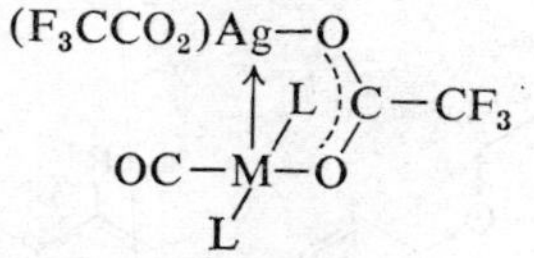

(103)

Partial i.r. spectra were reported for the complexes (SNL)Hg(OAc) and (SXZ)Hg(OAc), where SNL = spironolactone, SXZ = spiroxazone. Both are shown to contain an Hg—OCOMe group.[308]

302 B. Kipling, K. Wieghardt, M. Hery, and A. G. Sykes, *J. C. S. Dalton*, 1976, 2176
303 M. Hery and K. Wieghardt, *J.C.S. Dalton*, 1976, 1536
304 R. Uson, J. Fornies, and S. Gonzalo, *J. Organometallic Chem.*, 1976, **104**, 253.
305 R. Weissgerber and K. E. Schwarzhans, *Z. Naturforsch.*, 1976, **31b**, 208.
306 A. Miyashita and A. Yamamoto, *J. Organometallic Chem.*, 1976, **113**, 187.
307 J. Kuyper and K. Vrieze, *J. Organometallic Chem.*, 1976, **107**, 129.
308 G. Pouskouleli, P. Kouvoumakis, and T. Theophanides, *Inorg. Chem.*, 1976, **15**, 5.

$\nu_{as}(CO_2)$ (1620—1560 cm^{-1}) and $\nu_s(CO_2)$ (*ca.* 1450 cm^{-1}) values in $Ce(O_2CR)Cl_3,nH_2O$, where R = H, Me, Et, Pr^n, or Ph, n = O or 1, are consistent with symmetrical carboxylate co-ordination.[309]

Silicon(IV) formate, $Si(OOCH)_4$, appears to contain both uni- and bidentate formato ligands. The former gives $\nu(C{=}O)$ at 1715 cm^{-1}, and ν(COSi) at 1134 cm^{-1}; the latter shows ν_{as}(COO) at 1590 and ν_s(COO) at 1355 cm^{-1}.[310]

A similar study showed that $Si(O_2CMe)_4$ and $[Si(O_2CMe)_6]^{2-}$ both contain unidentate acetate, whereas $Sn(O_2CMe)_4$ and $[Sn(O_2CMe)_5]^-$ contain both uni- and bidentate.[311]

Bidentate co-ordinate carboxylate is present in lead adipate precipitated from alkaline solutions, thus ν_s, ν_{as}(COO) are at 1400 and 1525 cm^{-1} respectively.[312]

Measurements of ν_{as} and $\nu_s(CO_2)$ for the organotellurium phthalates $Ar_2Te(C_8H_4O_4)$, where R = Ph, $MeOC_6H_4$, or $EtOC_6H_4$, known to be dimers from molecular weight measurements, are consistent with their formulation as (104).[313]

Keto, Alkoxy, Phenoxy, and Ether Ligands.—Characteristic shifts to lower wavenumber of bands in the 1200—900 cm^{-1} region were observed on the formation of alkali and alkaline-earth complexes of the macrocyclic polyethers (105a), (105b), and (105c), R^1 = H = R^2; R^1 = Bu^t = R^2; R^1 = H, R^2 =

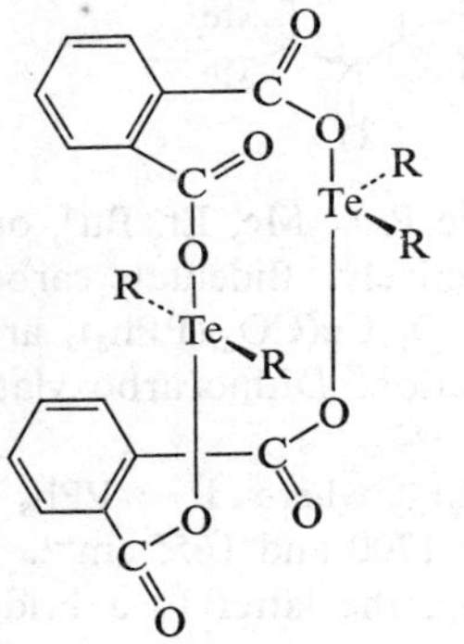

(104)

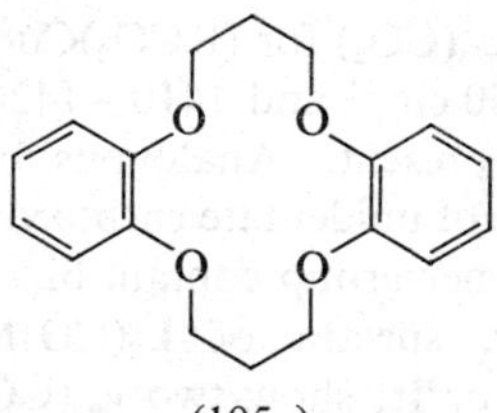

(105a)

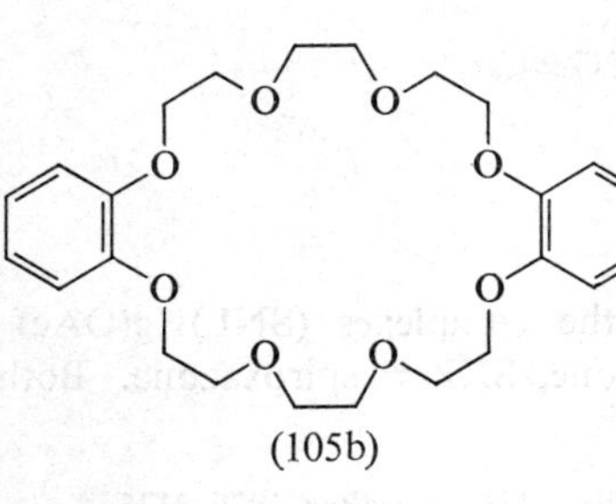

(105b)

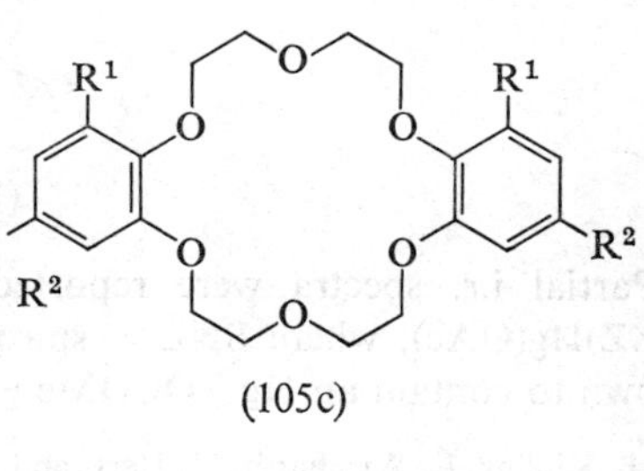

(105c)

309 B. L. Kalsotra, S. R. Gupta, and R. N. Kapoor, *Transition Metal Chem.*, 1976, **1**, 158.
310 R. C. Paul, T. Puri, and R. N. Kapoor, *Z. Naturforsch.*, 1976, **31b**, 905.
311 N. W. Alcock, V. M. Tracy, and T. C. Waddington, *J.C.S. Dalton*, 1976, 2243.
312 E. M. Urinovich, N. A. Egorushkina, N. F. Shogal, Yu. V. Oboznenko, and N. M. Dyatlova, *Russ. J. Inorg. Chem.*, 1975, **20**, 1640.
313 N. Dance and W. R. McWhinnie, *J. Organometallic Chem.*, 1976, **104**, 317.

$C_{16}H_{33}$). Typically, ν_{as}(Ar—O—CH_2) bands (1260—1220 cm^{-1}) shift by 10—30 cm^{-1}.[314]

ν(C=O) of urea in $Mg(NO_3)_2,4CO(NH_2)_2$ is at 1655 cm^{-1} (*cf.* 1680 cm^{-1} in free urea); there is, therefore, C=O—Mg co-ordination.[315]

Adsorption of acetone on MgO surfaces is accompanied by shifts of ν(C=O) to lower wavenumber, by up to 100 cm^{-1}. The shifts are smaller, however, than on adsorption on to Al^{3+}.[316]

The antopyrine (ant) in $MoO_2(NCS)_2(ant)_3$ is co-ordinated *via* the carbonyl group to the Mo^{IV}, since ν(C=O) shifts by 110 cm^{-1} to lower wavenumber.[317]

$Co(NO_3)_2,nCO(NH_2)_2,mH_2O$, where n = 2, 6, or 10; m = 2 or 4, all contain some urea molecules co-ordinated to the Co by the O atom, as ν(C=O) is lower and ν(C—N) higher than in the free urea.[318]

Some new methoxo complexes of Ni, Pd, and Pt have been reported, *i.e.* $MR(OMe)(PPh_3)_2$, where R = an alkyl or an alkenyl group. ν(CO) was seen at 1047—1070 cm^{-1}.[319]

Ligand mode assignments from the i.r. spectra of the complexes $M(DEU)_3X_2$, where M = Co^{II}, X = Cl, Br, I, or SCN; M = Ni, X = Cl, Br, or I; $Ni(DEU)_6X_2$, X = Br, I, or ClO_4; $Co(DEU)_6(ClO_4)_2$ and $Ni(DEU)_4(NCS)_2$, show that the ligand (DEU = *NN'*-diethylurea) is co-ordinated *via* carbonyl O in all cases.[320]

Decarboxylation of the new reversible CO_2 carrier $Bu^tOCO_2Cu(CNBu^t)_3$ affords $Bu^tOCuCNBu^t$, for which ν(Cu—O—Bu^t) = 938 cm^{-1}.[321]

The DMF adducts $Cu(o\text{-}XC_6H_4CO_2)_2$,DMF, where X = F, Cl, Br, or I, all give ν(C=O) of DMF at lower wavenumber than in the free ligand, *i.e.* the DMF is co-ordinated *via* the O atom. ν_s and $\nu_{as}(CO_2)$ are very similar to the values reported for the Cu^{II} alkanoates themselves.[322]

ν(C=N) and ν(C—O) are seen at 1525—1540 cm^{-1}, 1390—1415 cm^{-1} respectively in the complexes ML_2, where M = Cu, Zn, Cd, Co^{II}, or Ni^{II}, L = valeramide.[323]

Vibrational spectra for $Th(NO_3)_4.3L$, L = MeCONHEt; $M(NO_3)_4.2.5L$, M = Th, L = $MeCONPr^i_2$, and M = U, L = MeCONHEt or $MeCONPr^i_2$; $Pu(NO_3)_4.2L$, L = MeCONHEt or $MeCONPr^i_2$, and $UO_2(NO_3).2L$, L = $MeCONPr^i_2$, all show shifts by 40—70 cm^{-1} to lower wavenumbers, *i.e.* co-ordination from the O to the metal.[324]

Some *p*-benzoquinone–$AlCl_3$ complexes have been identified. Thus ν(C=O) in (106a) is at 1640 (free) and 1505 (complexed) cm^{-1}. (106b) gives a band at 1552 cm^{-1} only.[325]

314 L. Tusek, H. Meider-Gorican, and P. R. Danesi, *Z. Naturforsch.*, 1976, **31b**, 330.
315 M. N. Nabiev, I. A. Borukhov, M. T. Saibova, and L. P. Prisekina, *Russ. J. Inorg. Chem.*, 1976, **21**, 122.
316 F. Koubowetz, H. Noller, and J. Latzel, *Z. Naturforsch.*, 1976, **31b**, 922.
317 Yu. G. Eremin, E. F. Kolpikova, and T. V. Rodionova, *Russ. J. Inorg. Chem.*, 1976, **21**, 59.
318 V. G. Kuznetsov, A. P. Kinderov and L. I. Kochetkova, *Russ. J. Inorg. Chem.*, 1976, **21**, 555.
319 T. Yoshida, T. Okano, and S. Otsuka, *J.C.S. Dalton*, 1976, 993.
320 J. P. Barbier and R. Hugel, *Inorg. Chim. Acta*, 1976, **18**, 253.
321 T. Tsuda, S. I. Sano, K. Ueda, and T. Saegusa, *Inorg. Chem.*, 1976, **15**, 2329.
322 M. Melnik and J. Mrozinski. *J. Mol. Struct.*, 1976, **35**, 133.
323 Yu. Ya. Kharitonov, R. I. Machkhoshvili, and N. B. Generalova, *Russ. J. Inorg. Chem.*, 1975, **20**, 1838.
324 K. W. Bagnall, O. V. Lopez, and D. Brown, *J. Inorg. Nuclear Chem.*, 1976, **38**, 1997.
325 Z. Florjanczyk, W. Kuran, S. Pasynkiewicz, and G. Kwas, *J. Organometallic Chem.*, 1976, **112**, 21.

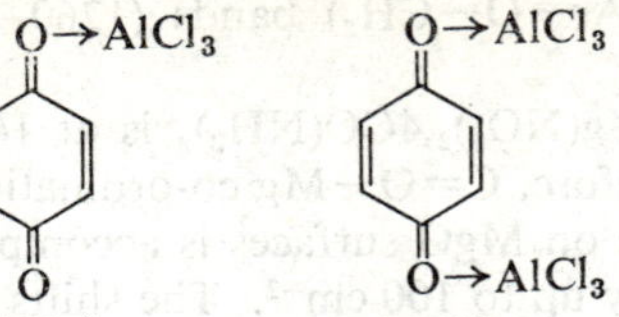

(106a) (106b)

Typical shifts in ν(C=O) to lower wavenumbers, and of ν(C—O—C) to higher wavenumbers, are seen on the formation of 1 : 1 and, except for $AlMe_3$, 1 : 2 adducts of PhCOOMe, EtCOOMe, and PhCOMe with Me_nAlCl_{3-n}.[326]

Tl(OEt), in the liquid phase, gives two bands due to ν(CO) at 1045 and 1095 cm^{-1}.[327]

ν(C=O) for Sn^{II} and Pb^{II} tropolonato and thujaplicinato complexes are summarized in Table 23. ν(CO) of the parent tropolone is reduced on co-ordination to either metal.[328]

Table 23 ν(C=O) *for some* Sn^{II} *and* Pb^{II} *complexes* (*all wavenumbers*/cm^{-1})

Complex	ν(C=O)
$Sn(trop)_2$	1587, 1565
$Sn(thuj)_2$	1587, 1569
$Pb(trop)_2$	1588, 1561
$Pb(thuj)_2$	1584, 1579

Ar_3Sb(HLO), where HLO is a terdentate ligand which can co-ordinate to Sb by a pair of *gem* diol oxygens as well as the carbonyl oxygen, give ν(C=O) at 1683 cm^{-1}, when Ar = Ph, L = tfac, 1645 cm^{-1} (Ph, tfbz), and 1682 cm^{-1} (p-ClC_6H_4, tfac).[329]

Nitrates and Nitrato-complexes.—The presence of NO_3 bands at 1490, 1260, and 970 cm^{-1} for $[Pt(Ph_3P)_2(NO_3)]BF_4$ suggests that the ligand is unsymmetrically bidentate.[330]

The i.r. spectra of various $Ln(NO_3)_5^{2-}$ anions clearly show that co-ordinated NO_3 groups are present, but do not indicate their bonding mode.[331]

Bands due to the C_{2v} co-ordinated nitrato groups were seen in $(BQ)_3Ln(NO_3)_6.HNO_3$, where BQ = *N*-benzylquinolinium and Ln = La, Ce, Pr, Nd, Sm, or Eu, ν_4–ν_1 was between 120 and 140 cm^{-1}, while the HNO_3 appears from shifts in its vibrational wavenumbers to be H-bonded to a co-ordinated NO_3 group.[332] Typical bands of co-ordinated nitrate were also reported for $Tb(NO_3)_3,4H_2O$ and $Dy(NO_3)_3,6H_2O$.[333]

326 K. B. Starowieyski, S. Pasinkiewicz, and A. Sporzinski, *J. Organometallic Chem.*, 1976, **117**, 117.
327 E. P. Turevskaya, N. Ya. Turova, and A. V. Novoselova, *Russ. J. Inorg. Chem.*, 1975, **20**, 838.
328 A. Rodgers and S. R. Stobart, *Inorg. Chim. Acta*, 1976, **16**, L13.
329 F. Ebina, T. Uehiro, T. Iwamoto, A. Ouchi, and Y. Yoshino, *J.C.S. Chem. Comm.*, 1976, 245.
330 D. A. Phillips, M. Kubota, and J. Thomas, *Inorg. Chem.*, 1976, **15**, 118.
331 G. S. Markov, B. P. Nikolskii, and M. V. Posvol'skii, *Russ. J. Inorg. Chem.*, 1975, **20**, 1155.
332 G. S. Markov, B. P. Nikolskii, and M. V. Posvol'skii, *Russ. J. Inorg. Chem.*, 1976, **21**, 518.
333 B. N. Ivanov-Emin, Z. K. Odinets, J. Del'pino, B. E, Zaitsev, and A. I. Ezhov, *Russ. J. Inorg. Chem.*, 1975, **20**, 1633.

The i.r. and Raman spectra of $K_2[UO_2(OH)_2UO_2](NO_3)_4,H_2O$ and the i.r. of its deuteriate indicate that (i) the four NO_3 groups are bidentate and (ii) that the two UO_2 groups are linked by OH bridges.[334]

Partial i.r. assignments were made for the copper complexes CuLX, where L = salicylald-4-amino antipyrine, X = NO_3 or ClO_4. The spectra were consistent with the presence of ionic perchlorate and co-ordinated nitrate.[335]

I.r. assignments for NO_3 groups in the uranyl complexes with cyclic polyethers, $UO_2L_I(NO_3)_2(H_2O)_2$ and $UO_2L_{II}(NO_3)_2(H_2O)_5$, where L_I = dibenzo-18-crown-6, L_{II} = benzo-15-crown-5, show that the NO_3 groups are co-ordinated to the U.[336]

Some assignments were proposed for the NO_3 modes in $Rb_2[Al(NO_3)_5]$: ν(N=O) of a bidentate group is at 1615 cm^{-1}, with $\nu_{as}(NO_2)$ at 1270 cm^{-1}. Unidentate nitrate groups are also present and give $\nu_{as}(NO_2)$ at 1550—1540 cm^{-1}, and $\nu_s(NO_2)$ at 1297 cm^{-1}. Only one band is seen near 1000 cm^{-1}, at 1020 cm^{-1}, which is assigned to the lowest wavenumber stretch for both the uni- and bidentate species.[337]

$K_3[Al(NO_3)_6]$ contains unidentate nitrate groups only. The i.r. spectrum was interpreted in terms of a single Al—O—NO_2 unit, of C_s symmetry, with no coupling between these units.[338]

$Ga(NO_3)_3$ has been prepared for the first time, together with $Ga(NO_3)_4^-$ as Li^+ or Na^+ salts. No features due to free NO_3^- were observed, and the large splitting ν_4–ν_1 is consistent with bidentate GaO_2NO bonding.[339]

$Ph_3Sn(OPPh_3)(NO_3)$ has been shown by *X*-ray diffraction to contain discrete five-co-ordinate monomer units. The i.r. spectrum does not, however, give unequivocal support to this, since bands are seen in the regions normally associated with bidentate NO_3 groups. This, therefore, throws some doubt on the general validity of these vibrational spectroscopic correlations.[340]

I.r. bands due to ionic and co-ordinated NO_3, and co-ordinated DMSO were listed for the seven-co-ordinate species $[SnPh_2(NO_3)(DMSO)_3]^+NO_3^-$. The splittings of ν_3 and $\nu_1 + \nu_4$ were both consistent with bidentate NO_3 co-ordination, as determined by *X*-ray diffraction. The ν(SO) bands were more in agreement with Sn—O co-ordination, since shifts of -130 cm^{-1} were seen on co-ordination.[341]

Ligands containing O—N, O—P, or O—As Bonds.—A review of co-ordination complexes of the *N*-oxides of aromatic di-imines and diazines includes many references to the literature data on i.r. results.[342]

I.r. spectra of $[Cr(PO_4)(NH_3)_5],2H_2O$ indicate that the PO_4 group acts as a unidentate ligand. In three salts of the $[Cr(H_2PO_4)(NH_3)_5]^{2+}$ ion bands due to the H_2PO_4 ligand were assigned.[343]

334 A. Perrin and J. Prigent, *Bull. Soc. chim. France*, 1976, 1070.
335 T. Radhakrishnan, P. T. Joseph, and C. P. Prabhakaran, *J. Inorg. Nuclear Chem.*, 1976, **38**, 2217.
336 A. Seminara, G. Siracusa, and A. Cassol, *Inorg. Chim. Acta*, 1976, **20**, 105
337 G. N. Shirokova, S. Ya. Zhuk, and V. Ya. Rosolovskii, *Russ. J. Inorg. Chem.*, 1975, **20**, 856.
338 G. N. Shirokova, S. Ya. Zhuk, and V. Ya. Rosolovskii, *Russ. J. Inorg. Chem.*, 1976, **21**, 527.
339 B. N. Ivanov-Emin, Z. K. Odinets, S. F. Yushchenko, B. E. Zaitsev, and A. I. Ezhov, *Russ. J. Inorg. Chem.*, 1975, **20**, 843.
340 M. Nardelli, C. Pelizzi, and G. Pelizzi, *J. Organometallic Chem.*, 1976, **112**, 263.
341 L. Coghi, C. Pelizzi, and G. Pelizzi, *J. Organometallic Chem.*, 1976, **114**, 53.
342 N. M. Karayanis, A. N. Speca, D. E. Chasan, and L. L. Pytlewski, *Coordination Chem. Rev.*, 1976, **20**, 37.
343 J. Casabo, J. M. Coronas, and M. Ferner, *Inorg. Chim. Acta*, 1976, **16**, 47.

Complexes MO_2ClL, MO_2L_2 (M = Mo or W), and $WOCl_2L_2$, where L = Ph_2POO^-, give i.r. spectra consistent with bidentate phosphinate ligands, probably bridging. Analogous species with L = Ph_2PSO^- seem to be monomeric, containing the unit (107).[344]

$M(OPR_2O)_2$, where M = Mn, Co, Ni, Cu, or Zn; R = Bu^t, Bu^n, or $MeCH_2CH_2CHMe$, have a variety of structures, depending on M and R. I.r. data due to $\nu(PO_2)$ were given, *ca.* 1130 and *ca.* 1050 cm^{-1}, and used in the assignment of structures.[345]

$\nu(P{=}O)$ shifts to lower wavenumber on going from dibutyl phosphate (DBP) to $M(DBP)_3$, where M = Fe or Al. The shifts are not so great as in the Zr^{IV} or UO_2^{2+} analogues, *i.e.* the Fe and Al give weaker co-ordination. The other DBP modes were assignable on the basis of C_{2v} local symmetry.[346]

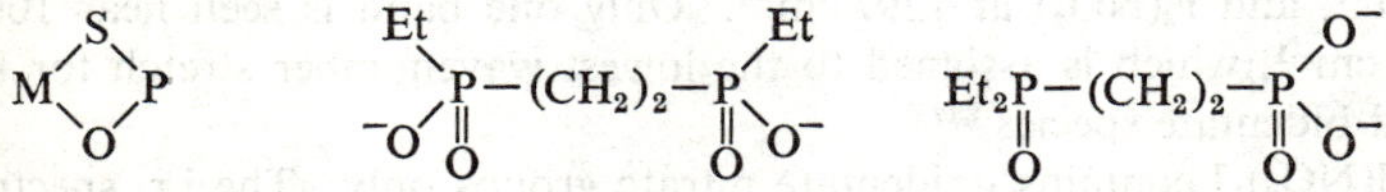

(107) (108a) (108b)

$\nu(NO)$ is at 1258 cm^{-1} (*cf.* the free ligand value of 1313 cm^{-1}) in [(pyrazine-*N*-oxide)$Ru(NH_3)_5]^{2+}$, implying that $d\pi-\pi^*$ back-bonding is antibonding for the N—O interaction.[347]

$RuCl_3(NO)(AsPh_3)(OAsPh_3)$ and $RuCl_2(CO)_2(AsPh_3)(OAsPh_3)$ both give bands due to $\nu(As{=}O)$ at about 870 cm^{-1}.[348]

Shifts in the $\nu(P{=}O)$ bands of (108a), on formation of CoL [L = (108a)], 1150 to 1138 cm^{-1}, are consistent with the co-ordination of the P=O. It is also thought that the $P{-}O^-$ is co-ordinated, giving a polymeric chain structure for CoL. Analogous conclusions were drawn for L = (108b), but here the polymeric structure is more complex.[349]

$\nu(P{=}O)$ has been assigned for a series of bis(diphenylphosphine oxide)-methane, DPMO complexes, $[M(DPMO)_3]^{2+}$, where M = Co, Ni, Cu, Zn, or Fe^{III}. $\Delta\nu(P{=}O)$ on co-ordination is 23—28 for the +2 metals, and 66 cm^{-1} for the Fe^{III} case, indicating co-ordination *via* the O atom.[350]

New Ni compounds of the type $[Ni_7(OH)_6L_6]X_2$, where X = NO_3 or ClO_4, and L = β-ketophosphonate anion ligands such as $[(EtO)_2P(O)CHCOCH_2NEt_2]^-$, show decreased values of both $\nu(C{=}O)$ and $\nu(P{=}O)$ on co-ordination. Thus both of these O atoms are engaged in bonding.[351]

Pyrophosphate ligand modes have been assigned for the i.r. spectra of complexes $Cu(P_2O_7H_2)L_n$, where n = 1 or 2; L = bipy, phen, py, py-d_5, or H_2O. The presence of four $\nu(PO)$ stretches suggests a departure from D_{3d} symmetry for the pyrophosphate ion.[352]

344 A. A. Kuznetsova, L. F. Yankina, Yu. G. Podzolko, I. A. Zakharova, and Yu. A. Buslaev, *Russ. J. Inorg. Chem.*, 1975, **20**, 891.
345 H. D. Gillman and J. L. Eichelberger, *Inorg. Chem.*, 1976, **15**, 840.
346 E. G. Teterin, A. S. Solovkin, and N. N. Shesterikov, *Russ. J. Inorg. Chem.*, 1975, **20**, 1680.
347 M. A. Blesa and H. Taube, *Inorg. Chem.*, 1976, **15**, 1455.
348 K. Natarajan, R. K. Poddar, and U. Agarwala, *Inorg. Nuclear Chem. Letters*. 1976, **12**, 749.
349 K. B. Yatsimirskii, M. I. Kabachnik, T. Ya. Medved', Z. A. Sheka, E. I. Sinyavskaya, M. A. Konstantinovskaya, and I. B. Goryunova, *Russ. J. Inorg. Chem.*, 1975, **20**, 1536.
350 C. R. White and M. D. Joesten, *J. Inorg. Nuclear Chem.*, 1976, **38**, 2173.
351 M. T. Youinou and J. E. Guerchais, *Inorg. Chim. Acta*, 1976, **19**, 257.
352 J. G. Contreras and G. V. Seguel, *J. Inorg. Nuclear Chem.*, 1976, **38**, 1371.

$ZnCl_2$ is extracted from aqueous MCl, where M = Li, Na, K, or NH_4, solutions by tributylphosphate(TBP) as TBP adducts, for which ν(P=O) is lower than for free aqueous TBP.[353]

$Ph_2P(=O)CH_2P(=O)Ph_2$, (L) co-ordinates *via* both O atoms in the complexes LMX_2, where M = Zn, Cd, or Hg; Ni, Cu, or Me_2Sn; X = Cl, Br, or I, ν(P=O) shifts by 15– 43 cm^{-1} to lower wavenumbers.[354]

The ν(PO) wavenumbers decrease, and the ν(PN) wavenumbers increase, on the formation of the complexes $HgCl_2$,2HMPA, and the Br and I analogues, where HMPA = hexamethylphosphoramide, compared to the free ligand values. These changes are as expected for Hg—O co-ordination.[355]

Ligand modes shift on co-ordination for $Ln(3\text{-MepyO})_8I_3$, where Ln = La, Tb, Nd, Sm, Dy, Yb, or Y, in the way predicted assuming co-ordination from the O atom to the metal.[356]

ν(NO) shifts only from 1290 to 1270 cm^{-1} on complexation of 3-picoline-*N*-oxide with lanthanide bromides. These shifts are much smaller than in py-*N*-oxide analogues, and so this wavenumber probably corresponds to a mixed vibration involving ν(NO) and ν(C—Me).[357]

In complexes $Ln(ClO_4)_3,3[(Me_2N)_2P(O)NMeP(O)NMe_2]_2$, co-ordination is thought to involve both O atoms of the nonamethylimidophosphoramide ligand, on the basis of ν(P=O) changes.[358]

Complex formation of HMPA with BF_3 or BCl_3 leads to increases in ν(PO). The authors claim that this indicates co-ordination *via* the O rather than the N, although this does not seem likely.[359]

$GaCl_3$,2TBP (TBP = tributylphosphate) gives ν(P—O) at 1216 cm^{-1}, in solutions of $GaCl_3$ in TBP. At higher $GaCl_3$ concentrations, $GaCl_3$,TBP is also formed, with ν(PO) at 1180 cm^{-1}.[360]

The i.r. spectra of Pb^{II}bis(diphenylphosphinate), $Pb(OPh_2O)_2$, suggest that changes in the mode of co-ordination occur on dissolution in CH_2Cl_2. In the solid, as shown by a full *X*-ray study, the structure consists of chains of Pb atoms linked by double phosphinate bridges. The detailed structure in CH_2Cl_2 could not be determined, but changes occurred in $\nu(PO_2)$ modes.[361]

Ligands containing O—S Bonds.—ν(SO) modes have been assigned as follows in $[Me_2S(=O)CH_2]_2M(CO)_4$: M = Cr, 1215, 1176 cm^{-1}; Mo, 1214, 1175 cm^{-1}; W, 1215, 1178 cm^{-1}. $[Me_2S(=O)CH_2]_3Mo(CO)_3$ gave one band at 1155 cm^{-1}, with a shoulder at 1145 cm^{-1}.[362]

The novel binuclear Fe^{III} complexes $[(Febipy)_2O](SO_4)_2,5H_2O$ and $[(Fephen)_2O](SO_4)_2,6H_2O$ have been prepared. Vibrational wavenumbers show that there

[353] E. A. Belousov and A. A. Alovyainikov, *Russ. J. Inorg. Chem.*, 1975, **20**, 1486.
[354] S. O. Grim and L. C. Satek, *J. Coordination Chem.*, 1976, **6**, 39.
[355] J. L. Vidal and G. E. Ryschkewitsch, *J. Inorg. Nuclear Chem.*, 1976, **38**, 1933.
[356] L. Ramakrishnan and S. Soundararajan, *Canad. J. Chem.*, 1976, **54**, 3169.
[357] P. V. Sivapullaiah and S. Soundararajan, *Monatsh.*, 1976, **107**, 871.
[358] C. Airoldi, Y. Gushikem, and C. R. Puschel, *J. Coordination Chem.*, 1976, **6**, 17.
[359] J. L. Vidal and G. E. Ryschkewitsch, *J. Inorg. Nuclear Chem.*, 1976, **38**, 1937.
[360] Yu. M. Globukov, S. S. Korovin, A. M. Reznik, M. R. Agevnin, and I. P. Romm, *Russ. J. Inorg. Chem.*, 1975, **20**, 1396.
[361] P. Colamarino, P. L. Orioli, W. D. Benziger, and H. D. Gilman, *Inorg. Chem.*, 1976, **15**, 800.
[362] L. Weber, *Z. Naturforsch.*, 1976, **31b**, 780.

are two types of SO_4 present, (i) one with C_{2v} symmetry, consistent with bidentate bridging between two Fe atoms, and (ii) one with T_d symmetry, *i.e.* a free ion.[363]

Ni(DBen)$_2$SO$_4$,H$_2$O shows SO_4^{2-} bands in the i.r. which are characteristic of bridging co-ordination; DBen = PhCH$_2$NHCH$_2$CH$_2$NHCH$_2$Ph.[364]

The i.r. and Raman spectra of M(etu)$_2$S$_2$O$_3$, where M = Zn or Cd, etu = ethylene thiourea, and their deuteriated derivatives, have been assigned. Splitting and shifts of the six $S_2O_3^{2-}$ modes could be used as characteristic criteria to decide between bidentate chelate and bridging $S_2O_3^{2-}$.[365]

ν(S=O) has been assigned in the following: Ln(ClO$_4$)$_3$,6Bu$_2$SO, where Ln = La—Nd, Sm—Lu, or Yb, and Bu$_2$SO = di-n-butyl sulphoxide. In each case ν(S=O) is in the range expected for O-co-ordination of the ligand.[366] The vibrational (i.r. and Raman) spectra of lanthanide(III) fluorosulphates have been assigned, and the spectra suggest that the compounds contain both bi- and tri-dentate ligands.[367]

Ligands containing O—Cl Bonds.—The i.r. spectra of a series of Cu^{II} perchlorate complexes have been examined in the light of their known solid-state structures, and spectroscopic predictions based on group theory. Complexes containing co-ordinated perchlorate have a second band in the region of 650 cm^{-1} which distinguishes them from compounds containing ionic perchlorate.[368]

Weak unidentate co-ordination ('semi-co-ordination') is shown in Cu(cyclam)-(ClO$_4$)$_2$, where cyclam = 1,4,8,11-tetra-azacyclotetradecane, for which a full crystal structure has been determined. Thus, ν_1 shows signs of splitting, and ν_4, at 625 cm^{-1} in free ClO_4^-, gives bands at 630 and 621 cm^{-1}.[369]

The ClO_4^- modes in [Cu(tmpd)OH]$_2$(ClO$_4$)$_2$ do not allow one to distinguish uni- or bidentate ligands, but they are definitely co-ordinated.[370]

6 Sulphur and Selenium Donors

MCl$_5$, where M = Nb or Ta, react with sulphur to give MS$_2$Cl$_2$. These compounds have bands due to the co-ordinated S—S units at 535 cm^{-1} (Nb) or 525—550 cm^{-1} (Ta).[371]

CS$_2$ inserts into Vcp$_2$(SR) to give the dithiocarbamate complex V(cp)$_2$(S$_2$CSR), with a strong i.r. band at about 950 cm^{-1}, probably due to the ν(CS) mode of the chelate.[372]

I.r. spectra were reported for [Cr(Ur)$_6$]$_2$(CS$_3$)$_3$ and [Cr(Ur)$_6$][Cr(CS$_3$)$_3$], where Ur = urea. The CS$_3$ groups usually have lower symmetry than D_{3h}, lifting the degeneracy of the ν_3 mode. If the CS_3^{2-} is un-co-ordinated, this splitting

363 B. Jezowska-Trzebiatowska, A. Ozarowski, H. Kozlowski, T. Cukievda, and J. Hanuza, *J. Inorg. Nuclear Chem.*, 1976, **38**, 1447.
364 K. C. Patel and L. F. Larkworthy, *J. Coordination Chem.*, 1975, **5**, 13.
365 Z. Gabelica, *Bull. Soc. Roy. Sci. Liège*, 1975, **44**, 644.
366 V. K. Lakatos Osorio and A. M. Passos Felicissimo, *Inorg. Chim. Acta*, 1976, **19**, 45.
367 W. M. Johnson and J. W. Macklin, *Inorg. Chem.*, 1976, **15**, 1216.
368 D. L. Lewis, E. D. Estes, and D. J. Hodgson, *J. Cryst. Mol. Struct.*, 1975, **5**, 67.
369 P. M. Tasker and L. Sklar, *J. Cryst. Mol. Struct.*, 1976, **5**, 329.
370 V. H. Crawford, H. W. Richardson, J. R. Wasson, D. J. Hodgson, and W. E. Hatfield, *Inorg. Chem.*, 1976, **15**, 2107.
371 S. M. Sinitsyna, V. G. Khlebodarov, and N. A. Bukhtereva, *Russ. J. Inorg. Chem.*, 1975, **20**, 1267.
372 G. Fachinetti, S. DelNero, and C. Floriani, *J.C.S. Dalton*, 1976, 203.

remains small, *ca.* 35 cm^{-1}. When it is co-ordinated, however, the splitting can exceed 100 cm^{-1}.[373]

ν(CS) in (109), M = Cr or Mo, and other trithione complexes are at *ca.* 10 cm^{-1} to lower wavenumber than in the free ligand. The ν(CO) values for these complexes suggest that the trithiones are very weak π-acceptors with respect to the zero-oxidation state metals.[374]

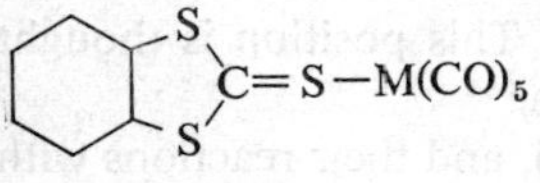
(109)

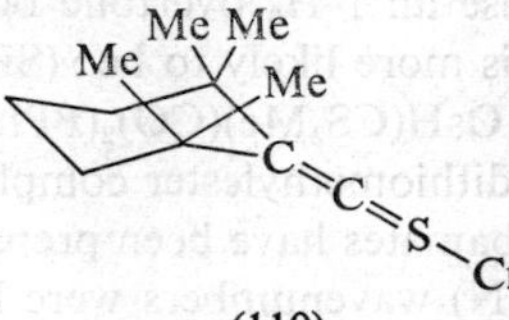
(110)

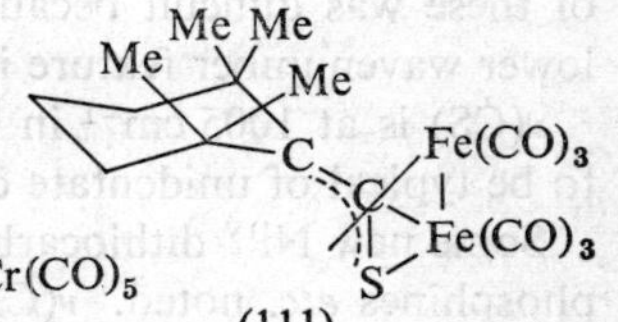
(111)

The complex (110) shows a characteristic ν(C=C=S) band at 1750 cm^{-1}, hardly shifted from the free ligand value. In the di-iron complex (111), however, this feature is absent.[375]

ν(SS) is found close to 485 cm^{-1} in $(C_5H_5)_2MS_4$, where M^{IV} = Mo or W; S_4^{2-} is a chain bidentate chelate ligand.[376]

Bands belonging to the μ-dithio groups are seen at 1025 and 870 cm^{-1} in the compounds $(dpe)_2Mo(SR)_2MoCl_4$; the values are insensitive to the identity of R.[377]

ν(CN) and ν(CS) have been reported for the compounds $Mo(R_2dtc)_4$, $[Mo(R_2dtc)_4]^+$, and $[W(R_2dtc)_4]^+$, where R = Me, Et, Pr^i, Ph, or $PhCH_2$. ν(CN) is found at higher wavenumber in the higher oxidation state.[378]

Photolysis of $W(CO)_6$ in the presence of H_2S leads to the relatively stable species $(OC)_5WSH_2$, with ν(SH) at 2560 cm^{-1}, *cf.* 2615 cm^{-1} for gaseous H_2S.[379]

Raman spectra of H_2L, MnL_2, NiL_2, and CuL_2, where H_2L = $HO_2CCH_2SCH_2$-$CH_2SCH_2CO_2H$, together with *X*-ray data on NiL_2, show that on co-ordination the ligand conformation changes from predominantly *trans* to exclusively *gauche*. ν(SS) in the *trans*-form is at 747, and in the *gauche* at 661 and 679 cm^{-1}.[380]

Symmetrical chelating xanthato-ligands, (112), in the compounds $[M(S_2COR)(CO)_4]$, where M = Mn or Re, R = Me or Et, and $[Mn(S_2COR^1)(CO)_3(PR^2_3)]$, where R^1 = Me or Et, R^2 = Et or Ph, were detected by the presence of a strong, broad i.r. band at 1250 cm^{-1}.[381]

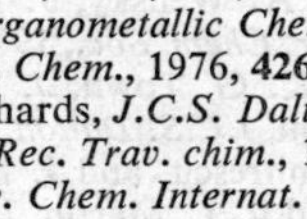
(112)

[373] A. M. Xuniguera, *Compt. rend.*, 1976, **282C**, 989.
[374] A. R. Siedle, *J. Organometallic Chem.*, 1976, **120**, 369.
[375] U. Behrens and F. Edelmann, *J, Organometallic Chem.*, 1976, **118**, C41.
[376] H. Köpf and K. S. Hazari, *Z. anorg. Chem.*, 1976, **426**, 49.
[377] J. Chatt, J. P. Lloyd, and R. L. Richards, *J.C.S. Dalton*, 1976, 565.
[378] A. Nieuwpoort and J. J. Steggerda *Rec. Trav. chim.*, 1976, **95**, 250.
[379] M. Herberhold and G. Süss, *Angew. Chem. Internat. Edn.*, 1976, **15**, 366.
[380] V. M. Miskowski, J. A. Thich, R. Solomon, and H. J. Schugar, *J. Amer. Chem. Soc.*, 1976, **98**, 8344.
[381] H. Laufen, B. Meyn, K. G. Steinhäuser, D. Vogel, and R. Kramolovsky, *J. Organometallic Chem.*, 1976, **112**, C34.

The i.r. spectra of $[M(CO)_{n-1}(ECF_3)]_2$, where E = S or Se; M = Mn, n = 4; M = Fe, n = 3, give bands due to ν(CF) in regions expected for bridging ECF_3 units, and terminal CO only.[382]

$[Ru(NH_3)_5(H_2S)](BF_4)_2$ shows no i.r. band attributable to ν(SH), as expected from the weakness of the i.r. spectrum of H_2S itself. The Raman spectrum of a fresh sample showed weak bands at 2523 and 2547 cm^{-1}, but the interpretation of these was difficult because an NH_3 overtone occurs in this region also; the lower wavenumber feature is more likely to be ν(SH).[383]

ν(CS) is at 1005 cm^{-1} in $OsH(CS_2Me)(CO)_2(PPh_3)_2$. This position is thought to be typical of unidentate dithiomethylester complexes.[384]

Some new Ni^{IV} dithiocarbamates have been prepared, and their reactions with phosphines *etc.* noted. ν(CN) wavenumbers were listed for all the compounds, *e.g.* 1528 cm^{-1} in $[Ni(S_2CNEt)_3](PF_6)$, and large differences were noted between KBr disc and CH_2Cl_2 solution values. ν(CN) decreases in the order Me > Et > Bu^n of the alkyl substituents.[385]

I.r. spectra and assignments, especially $\nu(NH_3)$ and $\nu(PS_2)$, were reported for some six-co-ordinate complexes involving $Ni[S_2P(OPh)_2]_2$ and aromatic and heterocyclic diamines. Possible structures were discussed.[386]

Bands called ν(C=C), ν(C—O), and ν(C—N) were assigned in the i.r. spectra of a series of Ni^{II} complexes of 1,3-dithio chelates, *e.g.* (113), with R^1, R^2 = OMe, OEt, NEt_2, *etc.*[387]

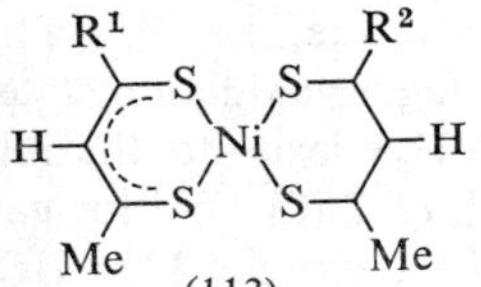

(113)

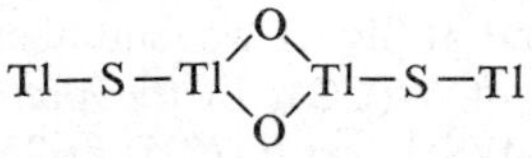

(114)

ν(PS) is lowered by about 40 cm^{-1} from the free ligand value of 628 cm^{-1} when $Ph_2P(=S)CH_2P(=S)Ph_2$ forms complexes with Cu^{II} or Ag^{I}.[388]

Oxidation of Tl_2S by O_2 produces Tl_2SO and Tl_2SO_2. The absence of ν(S—O) from the i.r. spectrum of the former indicates that it is dimeric, (114). ν(SO) in Tl_2SO_2 are at 670 and 1180 cm^{-1}.[389]

Ligand vibrational wavenumbers for $Me_2Tl(S_2COMe)$ are consistent with the structure (115). $\nu_2(CS_2)[+\ \nu_s(COC)]$ is at 935, 975 cm^{-1}, with $\nu_{as}(CS_2)$-$[+\ \rho(CH_3{-}O)]$ at 1185, 1190, and 1222 cm^{-1}.[390]

Two ν(CN) and two ν(CS) modes are detected for all of the triorganotin dithiocarbamates, $R_3Sn(dtc)$, where R = Ph, Bu^n, or $PhCH_2$, dtc = n-propyl-, n-butyl-, n-dibutyl-, benzyl-, 2-phenylmethyl-, or dibenzyl-dithiocarbamates, or pyrrolidine- or morpholine-*N*-carbadithioate. Thus they all contain unidentate

382 M. K. Chaudhuri, A. Haas, and A. Wensky, *J. Organometallic Chem.*, 1976, **116**, 323.
383 C. G. Kuehn and H. Taube, *J. Amer. Chem. Soc.*, 1976, **98**, 689.
384 T. J. Collins, W. R. Roper, and K. G. Town, *J. Organometallic Chem.*, 1976, **121**, C41.
385 J. A. McCleverty and N. J. Morrison, *J.C.S. Dalton*, 1976, 541.
386 R. Micu-Semeniuc, L. Silaghi-Dumitrescu, and I. Haiduc, *Inorg. Chim. Acta*, 1976, **17**, 5.
387 A. R. Hendrickson, J. M. Hope, and R. L. Martin, *Inorg. Chem.*, 1976, **15**, 1118.
388 E. W. Ainscough, H. A. Bergen, A. M. Brodie, and K. A. Brown, *J.C.S. Dalton*, 1976, 1649.
389 V. I. Rigin and S. S. Batsanov, *Russ. J. Inorg. Chem.*, 1975, **20**, 1283.
390 W. Schwarz G. Mann, and J. Weidlein, *J. Organometallic Chem.*, 1976, **122**, 303.

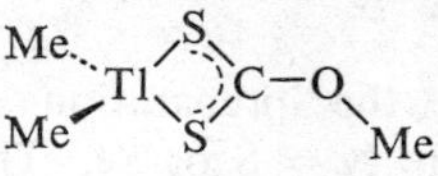

(115)

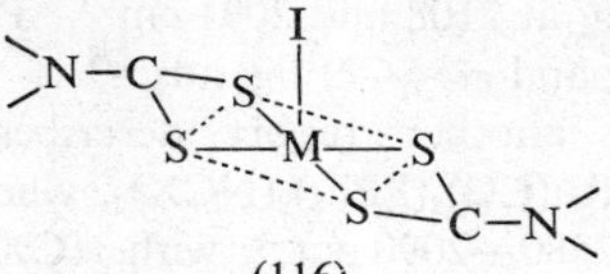

(116)

dtc ligands, Sn—S—C(=S)—NR_2, rather than the bidentate ones, which would have given one of each type of mode.[391]

The variation of ν(PS) with the nature of R in adducts of $(p\text{-}RC_6H_4)_3PS$ with $SnCl_4$ has been investigated.[392]

I.r. wavenumbers for ν(CN) and ν(CS) have been reported for As, Sb, and Bi complexes of the type $IM(R_2dtc)_2$, where R = Et, Bu^i, or $PhCH_2$, or R_2 = $(CH_2)_4$ or $(CH_2)_5$. The shifts are consistent with the dtc group acting as a bidentate ligand, giving a probable square-pyramidal arrangement about M, (116).[393]

7 Potentially Ambident Ligands

Cyanates, Thiocyanates, and their Iso-analogues.—$[(THF)_3Cl_2Mg_2OM(NCO)]$ give bands due to ν(N=C=O) at 2235 cm^{-1} (M = Ti), 2197 cm^{-1} (V), which were identified from $^{14,15}N$ isotopic shifts.[394]

ν(CN) due to Mo—N co-ordinated NCX, and to free NCX^- were identified in $[MoObipy_2X]X_2$, where X = NCO, NCS, or NCSe. The co-ordinated, and free values are 2190, 2235 (NCO); 2060, 2020 (NCS); and 2055, 2020 (NCSe) cm^{-1}.[395]

ν(CS) and δ(NCS) values for $M(NCS)(DMP)_4$, where M = Mn, Co, or Ni, and $M(NCS)(DMP)_2$, where M = Co, Zn, or Cd; DMP = 3,5-dimethylpyrazole, clearly show M—N co-ordination.[396]

The complex $ReCl(MeNCO)(dppe)_2$, containing the group (117), gives ν(CO) at 1835 cm^{-1}, *i.e.* 435 cm^{-1} lower than in the free MeNCO; dppe = bis(diphenylphosphino)ethane.[397]

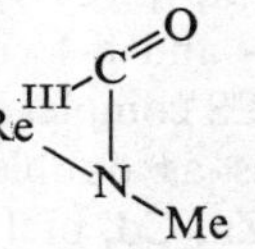

(117)

The observed splitting of ν(CN) in the complex [Fe(2,2′-bi-2-thiazoline)-$(NCS)_2$], 2083 and 2070 cm^{-1}, shows that the ligand geometry is *cis*.[398]

$Ru(CO)_3(PPh_3)_2$ undergoes oxidative addition with $(SCN)_2$, forming the complex $Ru(SCN)_2(CO)_2(PPh_3)_2$. This has ν(CS) at 836 cm^{-1}, and ν(CN), in

391 T. N. Srivastava and V. Kumar, *J. Organometallic Chem.*, 1976, **107**, 55.
392 E. G. Amanskii, A. A. Shvets, O. A. Osipov, and V. A. Likhovidov, *Koord. Khim.*, 1975, **1**, 1323
393 C. A. Tsipis and G. E. Manoussakis, *Inorg. Chim. Acta*, 1976, **18**, 35.
394 P. Sobota, B. Jezowska-Trzebiatowska, and Z. Janas, *J. Organometallic Chem.*, 1976, **118**, 253.
395 V. V. Trachevskii, A. M. Golub, and N. V. Ul'ko, *Russ, J. Inorg. Chem.*, 1975, **20**, 1656.
396 A. Anagnostopoulos, *J. Inorg. Nuclear Chem.*, 1976, **38**, 435.
397 R. Richards and G. Rouschias, *J. Amer. Chem. Soc.*, 1976, **98**, 5729.
398 V. McKee, S. M. Nelson, and J. Nelson, *J.C.S. Chem. Comm.*, 1976, 225.

CH_2Cl_2 solution, at 2108 and 2091 cm^{-1}. These observations are consistent with Ru—S bonding and *cis*-SCN ligands.[399]

Interestingly, another report describes the preparation of compounds formulated as $Ru(CO)_2(PPh_3)_2(NCX)_2$, where X = S or Se. These are said to give ν(NC) at 2080—2090 cm^{-1}, with ν(CX) at 838 (S) or 595 (Se) cm^{-1}. This would indicate N-bonding for both, whereas the observation of only one ν(NC) suggests that the two NCX groups are *trans*. Two ν(CO) bands, however, show that the CO groups are *cis*, and the structure is, therefore, (118). Similarly,

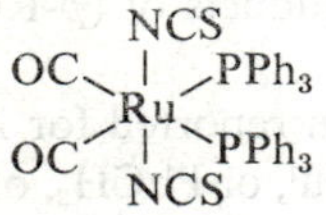

(118)

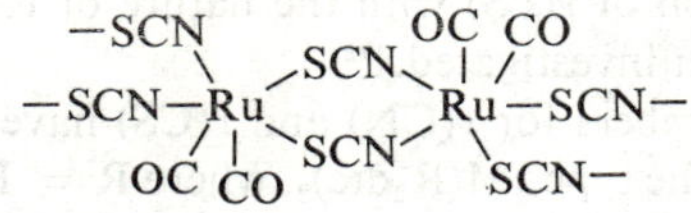

(119)

ν(NC) and ν(CO) values for $Ru(CO)_2(NCS)_2$ suggest that the NCS groups are bridging, and that the CO groups are *trans*, *i.e.* the structure is probably (119).[400]

ν(NCS) data were used in proposing that the NCS is N-bonded in *trans*-$[Co(TIM)(NCS)X]^+$, where TIM is a tetra-aza macrocyclic ligand, and X = Cl or Br.[401]

The positions of ν(CS) and ν(CN) in $[Co(phen)_2(NCS)_2]X$, where X = NCS, Br, ClO_4, or NO_3, and the bipy analogues, X = NCS or ClO_4, are consistent only with Co—N co-ordination of the NCS groups.[402]

I.r. spectra have been reported for $ML_2(NCS)_2$, where M = Co, Ni, Cu, or Zn, L = aniline or *p*-toluidine. The spectra are consistent with polymeric formulations, with octahedral co-ordination at the metal for the Co^{II} and Ni^{II} complexes, and with tetragonal co-ordination for those of Cu^{II}. The Zn^{II} complexes were assigned octahedral (L = aniline) or tetrahedral (L = toluidine) structures.[403]

Dissolving *trans*-$[Ir(pip)_4(X)H]^+$ in CH_2Cl_2 produces initially only N-bonded species, both when X = NCS or SCN. Ageing of the solutions, however, gives increasing quantities of *trans*-$[Ir(pip)_4(SCN)_2]^+$.[404]

Pt—N, Pt—S, and mixed (N- and S-) co-ordination was deduced from i.r. data for a large number of Pt–NCS complexes also containing P or As ligands.[405]

The presence of ν(CN) bands as doublets, 2156, 2140 cm^{-1}, in the nitrobenzoylhydrazine complexes of Zn, Cd, or Hg with NCS ligands was said to be evidence for the presence of both M—NCS and bridging NCS groups.[406]

I.r. and Raman spectra of $[Ln(NCO)_6]^{3-}$, where Ln = Yb or Lu, both agree with strictly octahedral co-ordination, and linear Ln—N—C—O groups. The assignments of internal ligand modes are given in Table 24.[407]

[399] H. B. Kuhnen, *J. Organometallic Chem.*, 1976, **105**, 357.
[400] F. Faraone and S. Sergi, *J. Organometallic Chem.*, 1976, **112**, 201.
[401] C. K. Poon and C. L. Wong, *Inorg. Chem.*, 1976, **15**, 1573.
[402] D. M. Palade, Yu. L. Popov, and G. V. Chudaeva, *Russ. J. Inorg. Chem.*, 1975, **20**, 1524.
[403] C. Engelter and D. A. Thornton, *J. Mol. Struct.*, 1976, **33**, 119.
[404] G. H. Weddle, J. L. Burmeister, and E. R. Birnbaum, *Inorg. Chim. Acta*, 1976, **18**, 59.
[405] R. J. Dickinson, W. Levason, C. A. McAuliffe, and R. V. Parish, *Inorg. Chem.*, 1976, **15**, 2934.
[406] P. V. Gogorishvili, S. Sh. Nagebashvili, A. E. Shvelashvili, and R. I. Machkhoshvili, *Russ. J. Inorg. Chem.*, 1975, **20**, 853.
[407] H.-D. Amberger, R. D. Fischer, and G. G. Rosenbauer, *Z. Naturforsch.*, 1976, **31b**, 1.

Table 24 *Some vibrational assignments for* $Ln(NCO)_6^{3-}$, *where* Ln = Yb *or* Lu (*all wavenumbers*/cm^{-1})

Vibration	I.R.	R	
ν(CN)	2190 t_{1u}	2235, 2183	a_{1g}, t_{2g}
ν(CO)	1333 t_{1u}	1334, 1235	a_{1g}, t_{2g}
δ(NCO)	633 t_{1u}	666, 634	a_{1g}, t_{2g}

In $Ge(NCS)_4$,4L, where L = (120), and $R^1 = R^2 = H$; $R^1 = H$, $R^2 = Me$, Br, or NMe_2; $R^1 = Me$, $R^2 = NMe_2$, ν(C=N) is shifted to values 30—40 cm^{-1} higher than in the free ligand. ν(NCS) is at 2050—2070 cm^{-1}, *i.e.* the bonding is Ge—NCS.[408]

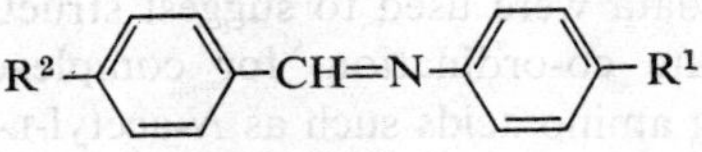

(120)

I.r. and Raman spectra have been recorded for a number of Pb^{II} thiocyanate complexes with mono- and bidentate ligands, *e.g.* $Pb(NCS)_2$,L, where L = 2py, phen, tmen, 2DMSO, *etc.* The wavenumbers were consistent with bridging and/or N-bonded NCS groups, but not S-bonded.[409]

The pseudohalide vibrations in Ph_3TeX, where X = CN, N_3, NCO, NCS, or NCSe, suggest that there is more than one type of structurally distinct pseudohalide group in each case. Thus, for X = NCO, ν_{as}(NCO) gives 2168, 2139, and 2132 cm^{-1}; ν_s(NCO) 1314 and 1304 cm^{-1}. This agrees with *X*-ray results, which show that they all have oligomeric structures.[410]

Ligands containing N or P and O Atoms.—A very thorough examination has been made of the i.r. spectra of nucleic acid–base complexes of various metal carbonyls, *e.g.* $W(CO)_5L$, and of Pd^{II} and Pt^{II} complexes. Use was made of these spectra in discussions of bonding.[411]

$Mn(pytm)(NO_2)_2$, where pytm = (121), gives $\nu_{as}(NO_2)$ at 1395 cm^{-1}, $\nu_s(NO_2)$ at 1195 cm^{-1} and $\delta(NO_2)$ at 825 cm^{-1}.[412] The NO_2 groups are therefore O-bonded.

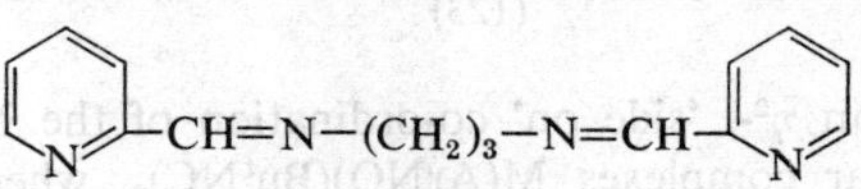

(121)

The isomers $(OC)_4BrReOP(H)Ph_2$ and $(OC)_4BrRePPh_2(OH)$ can be differentiated by the presence of ν(OH) at 3190 cm^{-1} in the latter species.[413]

The absence of ν(CO) at 1600 cm^{-1} in the amido complexes $Re(CO)_2(NHCOR)(PPh_3)_2$, R = Ph or *p*-MeC_6H_4, is indicative of bidentate bonding.[414]

408 N. I. Pirtskhalava and K. P. Giorgadze, *Russ. J. Inorg. Chem.*, 1975, **20**, 1976.
409 A. D. Baranyi, R. Makhija, and M. Onyszchuk, *Canad. J. Chem.*, 1976, **54**, 1189.
410 R. F. Ziolo and K. Pritchett, *J. Organometallic Chem.*, 1976, **116**, 211.
411 W. Beck and N. Kottmair, *Chem. Ber.*, 1976, **109**, 970.
412 W. C. Potter and L. T. Taylor, *Inorg. Chem.*, 1976, **15**, 1329.
413 E. Linder and B. Schilling, *J. Organometallic Chem.*, 1976, **114**, C23.
414 G. LaMonica, S. Ceinini, F. Porta, and M. Pizzotti, *J.C.S. Dalton*, 1976, 1777.

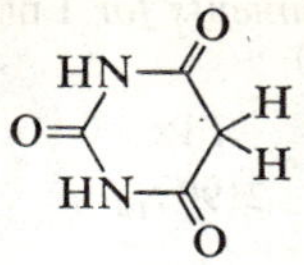

(122)

Complex barbiturates, $[Ru(NO)(B)_2Cl_3]^{4-}$ and $[Ru(NO)(B)_5]^{7-}$, have been prepared, H_2B = (122). The free ligand and both complexes give bands at 1620 and 1350—1380 cm^{-1} due to ν(C=O) and ν(C—O) respectively (from the enolate). However, bands in the complex at about 1500 cm^{-1} are not seen in the free ligands, and these are assigned as ν(C=O) for a C=O group bound to the Ru. ν(NO) is at 1840 cm^{-1} in each case.[415]

$\nu(CO_2)$ and other i.r. data were used to suggest structures, involving bridged carboxylate or unidentate co-ordination, for complexes of Co^{II} with some peptide-group-containing amino acids such as *N*-acetyl-L-alanine, *etc.*[416]

Yellow and orange isomers have been isolated for $K_6[Co_2(CN)_{10}N_2O_2]$. I.r. and Raman data suggest that they are linkage isomers, the orange being μ-(*trans*-hyponitrito-*NN'*), with ν(NN) at 1257, ν_s(NO) at 1165, and ν_{as}(NO) at 1157 cm^{-1}. The yellow form is μ-(*trans*-hyponitrito-*OO'*), with ν(NN) at 1412, ν_s(NO) at 1070, and ν_{as}(NO) at 987 cm^{-1}.[417]

I.r. data show that the azetinine-2-carboxylato-ligand, RCO_2, acts as a bidentate group, with co-ordination involving N and the carboxylato oxygen, in $Co(RCO_2)_3$ and $M(RCO_2)_2$, where M = Ni, Pt, Cu, or Zn.[418]

Complex (123) shows a strong i.r. band at 1531 cm^{-1}, assigned to ν(C$\cdots$N). It shifts to 1495 cm^{-1} on ^{15}N substitution.[419]

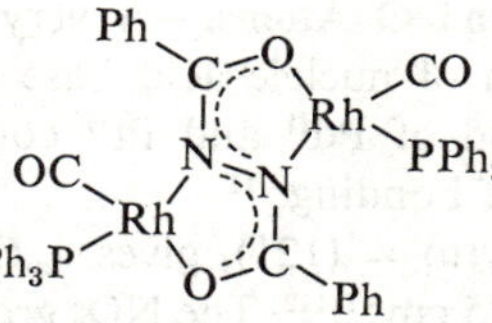

(123)

I.r. data suggest an η^2-, 'side on' co-ordination of the N=O group in the mono- and trinuclear complexes $M(ArNO)(Bu^tNC)_2$, where M = Ni, Ar = p-XC_6H_4, X = H, Cl, Br, OMe, Me, or NMe_2; $M(PhNO)(PPh_3)_2$, M = Ni, Pd, or Pt; and $Pd_3(PhNO)_3L_3$, where L = PBu^t_3, $PhPBu^t_2$. ν(NO) is generally 1025—1040 cm^{-1}, and the structural unit (124) is present.[420]

Complexes of Ni, Co^{II}, Fe^{II}, and Cu^{II} with α-naphthohydrazide are believed to contain (125). $\nu(NH_2)$ are found in the expected regions, but ν(C=O) fails to show a decrease on co-ordination, perhaps due to H-bonding in the free

[415] S. Sueur, C. Bremard, and G. Nowogrocki, *Bull. Soc. chim. France*, 1976, 1051.
[416] G. Marcotrigiano, L. Menabue, and G. C. Pellacani, *Transition Metal. Chem.*, 1976, **1**, 229.
[417] H. Okamura, E. Miki, K. Mizumachi, and T. Ishimori, *Bull. Chem. Soc. Japan*, 1976, **49**, 666.
[418] C. R. White and M. D. Joesten, *J. Coordination Chem.*, 1976, **6**, 53.
[419] A. S. Kasenally and F. M. Hussein, *J. Organometallic Chem.*, 1976, **111**, 355.
[420] S. Otsuka, Y. Aotani, Y. Tatsuno, and T. Yoshida, *Inorg. Chem* 1976, **15**, 656.

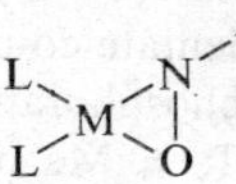

(124)

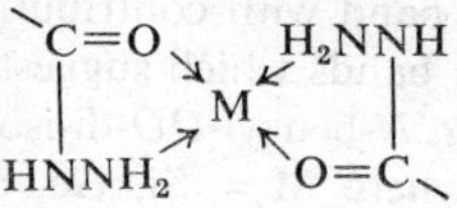

(125)

ligand.[421] Similar complexes of Ni^{II} and Cu^{II}, with *p*- and *m*-benzoylhydrazides, however, give both $\nu(NH_2)$ and $\nu(CO)$ in the required ranges.[422]

I.r. and Raman spectra of $(H_4EDTA)PdCl_2,5H_2O$ in acid aqueous solutions show absorptions due to both ionized and protonated carboxyl groups, suggesting that the EDTA is acting as a bidentate ligand.[423]

$PtMe_2(ONO)_2Lut_2$, where Lut = 3,5-lutidine, $\nu(NO)$ at 945 cm^{-1}, isomerizes to $PtMe_2(NO_2)_2Lut_2$, $\nu_{as}(NO_2)$ 1428, $\nu_s(NO_2)$ 1338 and 1328 cm^{-1}, on standing.[424]

Pt^{II} complexes have been prepared which contain alanylmethionine, $amH_2 = NH_2CHMeCONHCH(CH_2CH_2SMe)COOH$: $Pt(amH_2)Cl_2$ and Pt-(amH)Cl. Both give bands at about 1730 cm^{-1} due to the protonated carboxyl group, whereas bands from free CO_2^- of the peptide chain are found at 1550 and 1570 cm^{-1} respectively. The free amide I band of alanylmethionine is absent, and therefore the peptide group participates in co-ordination, *via* the N atom.[425]

A very thorough Raman study has been made of the interaction of the $Pt(NH_3)_2$ moiety with inosine, 1-methylinosine, and 5′-inosine monophosphate. The work provides definite evidence against the accepted view that adenine, guanosine, and inosine are bidentate.[426]

Group-frequency assignments, $\delta_{i.p.}(CH)$, $\nu(C{=}O)$, $\nu(CN)$, *etc.*, have been given for complexes formed by the reaction of (4,9-dimethyl-5,8-diazadodeca-4,8-diene-2,11-dione)copper(II), Cu(chel), with isocyanates RNCO. These complexes are of the types (126a) and (126b).[427]

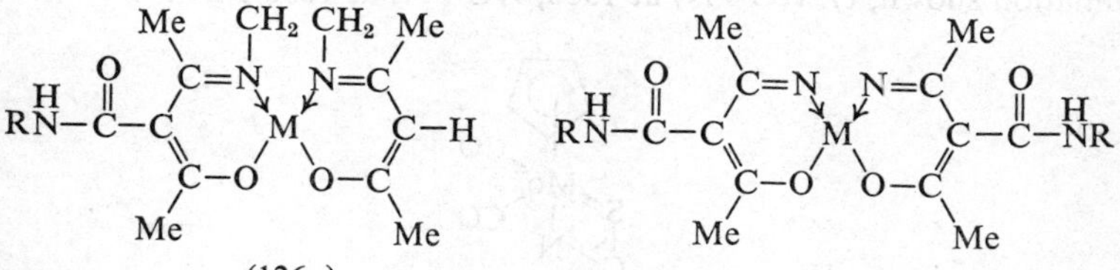

(126a) (126b)

Dioxouranium(VI) Schiff-base complexes $UO_2(SBH_2)_2(NO_3)_2$, where SBH_2 = *o*-$HOC_6H_4C(Me){=}NX^1OH$, where X^1 = CH_2CH_2, CH_2CHMe, or $CH_2CH_2CH_2$, or $MeC(OH)CHC(Me){=}NX^2OH$, where X^2 = CH_2CH_2, CH_2CHMe, CH_2-CH_2CH_2, or *o*-C_6H_4, all give $\nu(C{=}N)$ at 1640 cm^{-1} (1625 cm^{-1} in the free Schiff

421 M. V. Gerbeleu, S. Yu. Chundak, S. F. Manole, and S. S. Butsko, *Russ. J. Inorg. Chem.*, 1976, **20**, 912.
422 P. I. Sh'manko, N. V. Gerbeleu, and S. S. Butsko, *Russ. J. Inorg. Chem.*, 1975, **20**, 1334.
423 A. A. McConnell and R. H. Nuttall, *Inorg. Chim. Acta*, 1976, **19**, 253.
424 J. R. Hall and G. A. Swile, *J. Organometallic Chem.*, 1976, **122**, C19.
425 M. F. Mogilevkina, M. V. Sekacheva, I. M. Cheremisina, and E. A. Gal'tsova, *Russ. J. Inorg. Chem.*, 1976, **21**, 77.
426 G. Y. H. Chu and R. S. Tobias, *J. Amer. Chem. Soc.*, 1976, **98**, 2641.
427 P. N. Howells, J. W. Kenney, J. H. Nelson, and R. A. Henry, *Inorg. Chem.*, 1976, **15**, 124.

bases), with another band with contributions from ν(C=N) at 1601 cm^{-1}. The NO_3 ligands gave i.r. bands which suggested unidentate co-ordination.[428]

ν(P=O) values for *N*-benzyl-*OO*-di-isopropylphosphoramidates, (Me_2CHO)-PON(CH_2Ph)MR_3, where M = Si, Ge, or Sn; R = Me or Ph, are all in the range, 1220—1250 cm^{-1}, expected for N-bonded isomers.[429]

The complexes R_2SbCl(Trid), where R = Me, Ph; Trid^{2-} = potentially tridentate Schiff-base ligand with O—N—O donor atoms, have been prepared. The i.r. spectra are consistent with O—N—O co-ordination, and basically octahedral geometry, (127).[430]

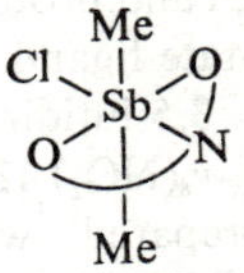

(127)

Ligands containing N or P and S Atoms.—Series of isomeric complexes $(OC)_5$MS=PR_2H and $(OC)_5MPR_2$SH, where M = Cr, Mo, or W; R = Me, Et, or Ph, have been prepared. ν(CO) bands were assigned for both series, together with ν(PS). The M—S bonded compounds gave the latter at 550—560 cm^{-1}, whereas those containing M—P bonds have ν(PS) below 510 cm^{-1}.[431]

I.r. wavenumbers due to ligand modes have been assigned for the following: Fe(ACDA)$_2$,2H_2O, Fe(ACDA)$_3$, Cr(ACDA)$_3$, 4H_2O, Co(ACDA)$_2$,3H_2O, and Co(ACDA)$_3$, where ACDA = 2-aminocyclopentene-1-dithiocarboxylic acid. The spectra suggest that for Fe^{II} and Fe^{III} bonding takes place from N and S, whereas for Co^{II}, Co^{III}, and Cr^{III} disulphur chelation occurs.[432]

A number of thioamide complexes have been reported, *e.g.* (128). This has a characteristic i.r. band at 1490 cm^{-1}, consistent with the pseudo-allylic type of co-ordination shown, *cf.* ν(C—N) at 1300, ν(C=N) at 1650 cm^{-1}.[433]

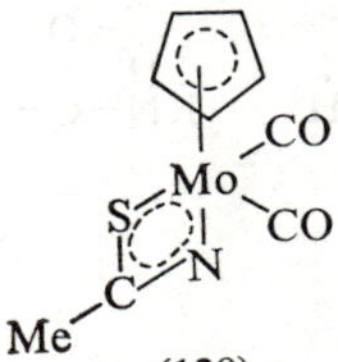

(128)

The linkage isomers $(OC)_4$XMnSPHR_2 and $(OC)_4$XMnPR_2SH, with R = Me, X = Cl or Br; R = Ph, X = Br, have been prepared. Quite detailed vibrational assignments were proposed from i.r. and Raman evidence. ν(PH) is at 2370 cm^{-1} in the S-bonded, and ν(SH) is at 2460 cm^{-1} in the P-bonded derivatives. The

428 R. G. Vijay and J. P. Tandon, *Z. Naturforsch.*, 1976, **31B**, 22.
429 C. Glidewell, *J. Organometallic Chem.*, 1976, **108**, 335.
430 H. Meinema, J. G. Noltes, F. Di Bianca, N. Bertazzi, E. Rivarola, and R. Barbieri, *J. Organometallic Chem.*, 1976, **107**, 249.
431 E. Lindner and W. P. Meier, *J. Organometallic Chem.*, 1976, **114**, 67.
432 K. Nag and D. S. Joardar, *Inorg. Chim. Acta*, 1976, **17**, 111.
433 E. W. Abel and I. D. H. Towle, *J. Organometallic Chem.*, 1976, **122**, 253.

former shows ν(P=S) at 560 cm^{-1}, while the latter have ν(P--SH) at 505 cm^{-1}. It is possible to monitor their isomerization *via* their i.r. spectra.[434]

The i.r. spectrum of the new carbonyl complex (129) has been partially assigned; four ν(CO) bands are at 2075, 2008, 1994, and 1959 cm^{-1}; $\nu_{as}(PC_2)$ at 736 cm^{-1}; $\nu_s(PC_2)$ at 702 cm^{-1} and ν(PS) at 523 cm^{-1}.[435]

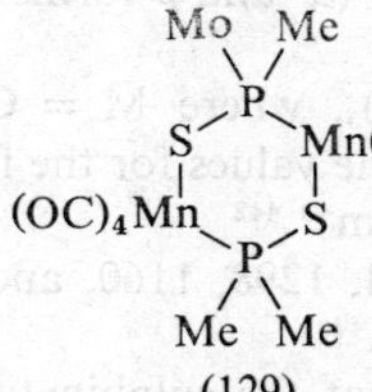

(129)

RNCS, where R = Me or Ph, react with $MHX(CO)(PPh_3)_3$, where M = Ru or Os; X = Cl or Br, $RuH_2(PPh_3)_4$ or $IrHCl_2(PPh_3)_3$ to give the new species $MX(S\cdots CH\cdots NR)(CO)(PPh_3)_2$, $Ru(S\cdots CH\cdots NR)_2(PPh_3)_2$, and $Ir(S\cdots CH\cdots NR)Cl_2(PPh_3)_2$ respectively. These show no ν(NH) or ν(SH), and weak bands at 1290—1200 and 930—890 cm^{-1}; they are formulated as containing the group (130). $\nu(C\cdots NR)$ gives a strong band at about 1500 (Ph), 1570 (Me) cm^{-1}.[436]

I.r. wavenumbers were measured for ligand modes in complexes of thiazolidine-2-thione with M^{II}, where M = Co, Ni, Cu, Zn, Cd, or Hg. For M = Zn or Cd, co-ordination is thought to be *via* the N, whereas for the others the thiocarbonyl sulphur is involved.[437]

The phosphinodithioformamates $(R_2PCS_2)^-$, where R = Ph, C_6H_{11}, or $PhCH_2$, act as *PS*-chelating agents when forming inner complexes of Ni^{II}, (131); ν(C=S) is at 1087—1100 cm^{-1}.[438]

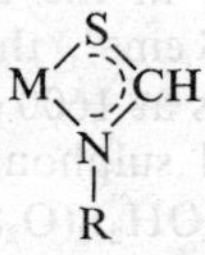

(130)

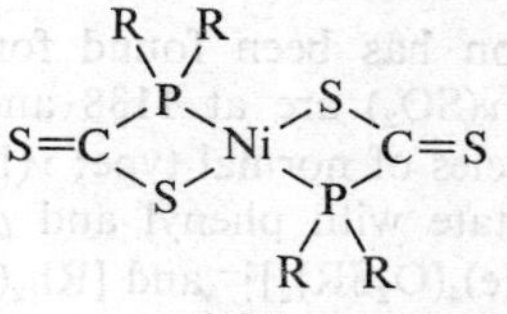

(131)

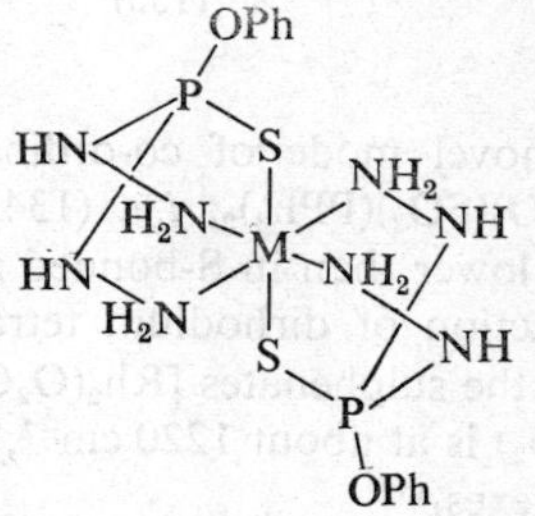

(132)

ML_2^{2+}, where M = Ni, Zn, or Cd; L = dihydrazidothiophosphoric acid phenyl ester, give i.r. and Raman spectra which show that NH_2 and PS groups are both participating in co-ordination, *i.e.* (132).[439]

434 E. Lindner and H. Dreher, *J. Organometallic Chem.*, 1976, **104**, 331.
435 E. Lindner and H. Dreher, *J. Organometallic Chem.*, 1976, **105**, 85.
436 S. D. Robinson and A. Sahajpal, *J. Organometallic Chem.*, 1976, **111**, C26.
437 K. Geetharani and D. N. Sathyanarnyana, *Indian J. Chem.*, 1976, **14**, 170.
438 J. Kopf, R. Lenck, S. N. Olafsson, and R. Kramolowsky, *Angew. Chem. Internat. Edn.*, 1976, **15**, 768.
439 U. Engelhardt and G. Scherer, *Z. Naturforsch.*, 1976, **31b**, 1553.

ν(C=S), and stretching wavenumbers involving the tertiary N in benzaldehyde thiosemicarbazone, PhCH=NNH(H_2N)C=S, complexes Hg(TC)Cl_2, $Hg(TC)_2$-Cl_2, and $Hg(TC)_4^{2+}$, are at lower wavenumber than for the free ligand. This suggests co-ordination *via* S and the N of the hydrazine residue.[440]

Ligands containing S and O Atoms.—ν(SO) data were used to show that MSO_3, where M = Ca, Sr, Ba, Pb, or Cd (α- and β-forms), and some of their hydrates contain co-ordinated S.[441]

ν(SO) in Me_2S(=O)$CH_2M(CO)_5$, where M = Cr, Mo, or W, is at about 1192 cm^{-1}, intermediate between the values for the free ylide, 1135 cm^{-1}, and the trimethysulphoxonium ion, 1228 cm^{-1}.[442]

$\nu(SO_2)$ bands are found at 1313, 1298, 1160, and 558 cm^{-1} in *cis*-$[Mo(CO)_2$-$(SO_2)(Me_2PCH_2CH_2PMe_2)_2](PF_6)_2$.[443]

I.r. bands show conclusively that the sulphinate ligands in π-$C_7H_7Mo(CO_2$-SO_2R, R = Ph or C_6H_4Me, are S-bonded to the molybdenum, *e.g.* when R = Ph, they are at 1155, 1085, 1025, 1015, and 995 cm^{-1}.[444]

Typical S-bonded sulphinato-bands were observed in the i.r. spectrum of $Cp(OC)_3WS(O)_2Me$, $\nu_{as}(SO_2)$ 1207—1196 cm^{-1}, $\nu_s(SO_2)$ 1043—1038 cm^{-1}.[445]

The complex (η^5-C_5H_5)$Mn(CO)_2SO(NEt_2)_2$ shows ν(SO) at 1127 cm^{-1}; the corresponding SOF_2 complex has this band at 1276 cm^{-1}.[446]

ν(C=O) of the dithio-oxalato complexes FeL_3^{3-} generally fall in the range 1510—1520 cm^{-1}.[447]

CpFe(CO)(PPh_3)CH_2SO_3Me gives ν(S=O) at 1350 and 1150 cm^{-1}, due to the unit (133). CpFe(PPh_3)$S(O)_2Me$ gives similar bands at 1150 and 1030 cm^{-1}.[448]

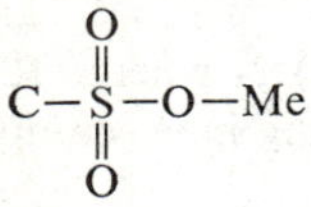

(133)

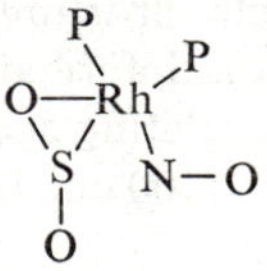

(134)

A novel mode of co-ordination has been found for SO_2 in the complex $Rh(NO)(SO_2)(PPh_3)_2$, *i.e.* (134). $\nu(SO_2)$ are at 1138 and 948 cm^{-1}, the latter much lower than in S-bonded species of normal type; ν(NO) is at 1600 cm^{-1}.[449]

Reaction of dirhodium tetracetate with phenyl and *p*-tolyl sulphonic acids yields the sulphenates $[Rh_2(O_2CMe)_4(O_2SR)_2]^{2-}$ and $[Rh_2(OH)(OH_2)_2(O_2SR)_6]^{2-}$. $\nu_{as}(SO_2)$ is at about 1220 cm^{-1}, and this was taken to indicate S-bonding in both complexes.[450]

ν(S=O) in $RhL_2Cl_4^-$ is at 1120 cm^{-1} for L = DMSO and DMSO-d_6. The position of the band is consistent with S-co-ordination, while the presence of

440 T. N. Chernova, S. V. Lugovoi, and V. D. Chistota, *Russ. J. Inorg. Chem.*, 1975, **20**, 850.
441 H. D. Lutz and S. E. Suradi, *Z. anorg. Chem.*, 1976, **425**, 134.
442 L. Weber, *J. Organometallic Chem.*, 1976, **105**, C9.
443 J. A. Connor and P. I. Riley, *J.C.S. Chem. Comm.*, 1976, 149.
444 D. Mohr, H. Wienand, and M. L. Ziegler, *Z. Naturforsch.*, 1976, **31b**, 66.
445 I.-P. Lorenz and R. Greisshammer, *Z. Naturforsch.*, 1976, **31b**, 888.
446 M. Höffler and A. Baitz, *Chem. Ber.*, 1976, **109**, 3147.
447 T. Birchall and K. M. Tun, *Inorg. Chem.*, 1976, **15**, 376.
448 T. C. Flood, F. J. DiSanti, and D. L. Miles, *Inorg. Chem.*, 1976, **15**, 1910.
449 D. C. Moody and R. R. Ryan, *J.C.S. Chem. Comm.*, 1976, 503.
450 J. G. Norman and E. O. Fey, *J.C.S. Dalton*, 1976, 765.

only one suggests *trans*-bonding. For RhL_3Cl_3 three $\nu(SO)$ bands are seen, again with no significant shift on deuteriation. The features at 1150 and 1135 cm^{-1} are due to S- and that at 935 cm^{-1} to an O-bonded ligand.[451]

The presence of ν_{as}, $\nu_s(SO_2)$ respectively at 1190, 1050—1090 cm^{-1} for (135), where L = $PMePh_2$, $x = 0$ or 1, $y = 2$, Z = Cl; L = $PMePh_2$, $x = 1$, $y = 2$, Z = Br, show that the sulphinato-group is S-bonded. For those complexes

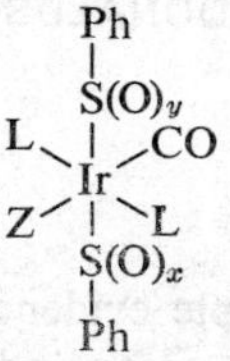
(135)

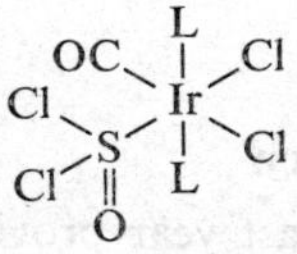
(136)

with $x = 1$, ν(S=O) of the sulphenato group is at 1042 cm^{-1}. Complexes with thionyl chloride as ligand, (136), where L = PPh_3 or $PMePh_2$, have ν(S=O) of the $SOCl_2$ at about 1100 cm^{-1}, compared with 1210 cm^{-1} for the free ligand.[452]

{Ni[$(Ph_2PCH_2)_2CMe$](SO_2)} shows $\nu_{as}(SO_2)$ at 1190 cm^{-1}, $\nu_s(SO_2)$ at 1050 and 1045 cm^{-1}, and $\delta(SO_2)$ at 540 cm^{-1}. These values agree with the idea of back-bonding to the SO_2.[453]

(η^5-C_5H_5)(PPh_3)Pd—SO_2—R gives bands due to $\nu(SO_2)$ of the Pd—S-bonded sulphinate at 1199 and 1056 cm^{-1} (R = Me), 1181 and 1029 (R = Ph).[454]

The RSO_2^- groups are co-ordinated *via* the S in $[(RS(O)_2)_2PdClL]^-$, where R = Ph or *p*-MeC_6H_4; L = MeCN, Me_2CO, or THF, $\nu_{as}(SO_2)$ is about 1215 cm^{-1}, $\nu_s(SO_2)$ 1062 cm^{-1} in each case. Each band is single except for R = Ph, L = THF where the splittings of both bands suggest that the RSO_2 groups are *cis*.[455]

ν_{as} and ν_s of SO_2 in $RTi(O_2SR)X$, where X = $MeCO_2$ or Cl, R = mesityl, are as expected for unidentate, *O*-sulphinate co-ordination. The structure of the $[RTl(O_2SR)]^+$ is probably (137), although a weakly X-bridged structure cannot be ruled out. For $R_2Tl(O_2SR)$, $\nu_{as}(SO_2)$ is at 979, $\nu_s(SO_2)$ at 941 cm^{-1}, and the structure is probably bridged, *i.e.* (138).[456]

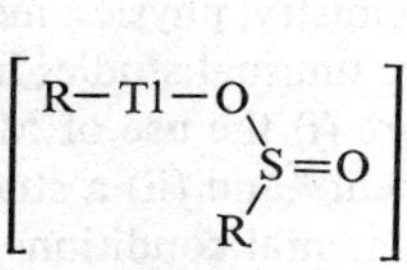
(137)

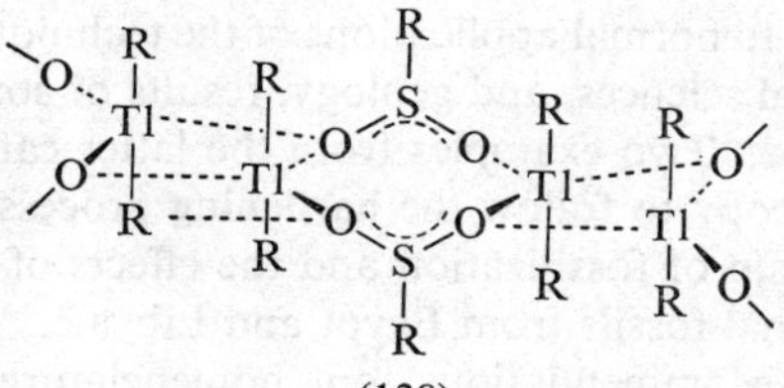
(138)

[451] Yu. V. Fadeev, Yu. N. Kukushkin, and K. A. Khokhryakov, *Russ. J. Inorg. Chem.*, 1975, **20**, 1519.
[452] S. J. Markham, Y. L. Chung, G. D. Branum, and D. M. Blake, *J. Organometallic Chem.*, 1976, **107**, 121.
[453] D. Dapporto, S. Midollini, A. Orlandini, and L. Sacconi, *Inorg. Chem.*, 1976, **15**, 2768.
[454] G. K. Turner and H. Felkin, *J. Organometallic Chem.*, 1976, **121**, C29.
[455] I.-P. Lorenz, E. Lindner, and W. Reuther, *Z. Naturforsch.*, 1976, **31b**, 1425.
[456] G. B. Deacon and I. K. Johnson, *J. Organometallic Chem.*, 1976, **112**, 123.

8
Mössbauer Spectroscopy

BY J. D. DONALDSON AND M. J. TRICKER

1 Introduction

The literature of the past year provides ample evidence of the ways in which Mössbauer spectroscopy has developed both as an independent technique and as a tool to be used in ever widening applications. The papers reviewed show the continuing advancement of a number of aspects of the subject including theoretical treatments, instrumentation, data handling, and applications in which assessments of Mössbauer spectroscopic results form the central theme of the research. Developments in the use of Mössbauer spectroscopy as applied to biological materials, in studies of compounds with unusual solid-state properties and systems of technological interest such as catalysts, and in the use of less well known isotopes perhaps warrant special mention. Papers describing sources for the rarer isotopes ^{129}Xe, ^{149}Sm, ^{157}Gd, ^{187}Os, ^{199}Hg, and ^{201}Hg were presented at the Corfu International Meeting.[1] There is an encouraging increase in reports of work combining Mössbauer spectroscopy with other techniques. Such studies are not only advisable in terms of maximizing the data available on the compounds concerned but are also of considerable value in assessing the value of Mössbauer spectroscopic results in an overall context, and, in certain cases, can also indicate directions of further research on uses of the Mössbauer effect. Another increasing trend is towards the publication of Mössbauer data, particularly for iron- or tin-containing materials, as standard spectroscopic results in publications with titles that give no indication of the presence of these data. This observation confirms that there is steady development of the technique from a new method to a standard spectroscopic tool in appropriate cases. In addition to normal applications of the technique in chemistry, physics, metallurgy, biological sciences, and geology, results of some more unusual studies have been published. Two examples from the latter category are (i) the use of Mössbauer spectroscopy to follow the hardening process of cement[2] and (ii) a study of the mechanism of fossilization and the effects of environmental conditions on plant and animal fossils from Egypt and Libya.[3]

The recommendations on nomenclature and conventions for reporting Mössbauer spectroscopic data have been published.[4] These recommendations stem from the physical chemistry division commission on molecular structure

1 Abstracts of the International Conference on the Applications of the Mössbauer Effect, Corfu (Greece), Sept., 13–19, 1976. Proceedings to be published in *J. Phys.* (*Paris*) *Colloq.*
2 N. V. Rudenko, N. S. Strashnikov, and B. L. Goncharov, *Priklad. Teor. Fiz.*, 1974, 6, 160.
3 N. A. Eissa, H. A. Sallam, B. A. Ashi, M. Y. Hassan, and S. A. Saleh, *J. Phys.* (*D*), 1976, **9**, 1391.
4 *Pure Appl. Chem.*, 1976, **45**, 211.

and spectroscopy of IUPAC and cover all aspects of data presentation including information in the text, numerical and tabulated data, and figures. A manual of terminology is presented along with a list of symbols and units for Mössbauer spectroscopy and notes on specific points such as data computation, isomer shift standards, quadrupole splittings, and recoil-free fraction. It is to be hoped that all authors of papers containing Mössbauer results will now follow these recommended conventions in presenting their results.

The layout of this chapter remains basically the same as that adopted by the previous Reviewer. The main section headings are the same but there are some changes in the layout and content of subsections. This introductory section covers books and review articles, and is followed by sections that deal with theoretical developments and with advances in instrumentation and methodology. Sections 4, 5, and 6 contain detailed reviews of the literature on iron-57, tin-119, and other isotopes respectively. The final section is a bibliography which lists papers not discussed earlier in the text.

During the year the following 39 resonances have received attention: ^{57}Fe (14.412 keV), ^{61}Ni (67.40 keV), ^{67}Zn (93.26 keV), ^{83}Kr (9.3 keV), ^{99}Ru (90 keV), ^{119}Sn (23.875 keV), ^{121}Sb (37.15 keV), ^{125}Te (35.48 keV), ^{127}I (57.60 keV), ^{129}I (27.72 keV), ^{129}Xe (39.58 keV), ^{133}Cs (81 keV), ^{149}Sm (22.5 keV), ^{151}Eu (21.6 keV), ^{153}Eu (97 and 103.2 keV), ^{154}Gd (123.1 keV), ^{155}Gd (86.54 and 105.2 keV), ^{157}Ge (64 keV), ^{160}Dy (86.79 keV), ^{161}Dy (25.65 keV), ^{166}Er (91.5 keV), ^{169}Tm (8.40 keV), ^{170}Yb (84.26 keV), ^{171}Yb (66.74 keV), ^{181}Ta (6.2 keV), ^{182}W (100.10 keV), ^{184}W (111.2 keV), ^{186}W (122.6 keV), ^{187}Os, ^{187}Re (134.2 keV), ^{193}Ir (73.1 keV), ^{195}Pt (98.8 keV), ^{197}Au (77.34 keV), ^{199}Hg (158 keV), ^{201}Hg (32.2 keV), ^{237}Np (59.54 keV), and ^{241}Am (84 keV).

Books and Reviews.—A textbook by T. C. Gibb [5] entitled 'Principles of Mössbauer Spectroscopy' has been published as part of the series 'Studies in Chemical Physics'. Two books on specialist applications have also appeared: 'The Mössbauer Effect in Geochemistry and Cosmochemistry' by Malysheva [6] and 'Mössbauer Spectroscopy as a Research Method in Applied Organic Chemistry' by Khristov, Bonchev, and Busova.[7] Volume 1 of a new series entitled 'Applications of Mössbauer Spectroscopy' was published [8] in 1976. This book contains review articles on: elements of Mössbauer spectroscopy, ferrous alloy phase transformations, corrosion and interfacial reactions, experimental techniques for biochemical studies, biological iron transport, enzyme systems, monitoring solid state reactions, analysis of archaeological artifacts, analysis of works of art, and applications in polymer science. A further volume of the Mössbauer Effect Data Index (MEDI) appeared during the year.[9] A general test [10] by Kaipov on nuclear γ-resonance and the processes accompanying it has also been published.

[5] T. C. Gibb, 'Principles of Mössbauer Spectroscopy', Chapman Hall, London, 1976.
[6] T. V. Malysheva, 'Mossbauer Effect in Geochemistry and Cosmochemistry', Nauka, Moscow, 1975.
[7] D. Khristov, T. Bonchev, and Z. Busova, 'Mössbauer Spectroscopy as a Research Method in Applied Organic Chemistry', Bulgarian Academy of Sciences, Sofia, 1975.
[8] Applications of Mössbauer Spectroscopy', Volume 1, ed. R. L. Cohen, Academic Press, New York, 1976.
[9] J. G. Stevens and V. E. Stevens, 'Mössbauer Effect Data Index Covering the 1975 literature', IFI/Plenum, New York, 1976.
[10] D. K. Kaipov, 'Nuclear Gamma Resonance and Processes Accompanying it', Nauka: Alma-Ata. Kaz. SSR., 1976.

A number of general reviews on Mössbauer spectroscopy have appeared [11–14] including the sixth review [15] in the Analytical Chemistry series which covers the literature for 1974—5 and a review [16] of the development of Mössbauer spectroscopy in the Soviet Union. More specific articles have covered: γ-resonance spectroscopy of solids under high frequency excited conditions,[17] Mössbauer emission and the Auger effect induced by electron capture degradation,[18] hyperfine interactions,[19] solvent structure studies,[20] and diffusion.[21] Uses of the Mössbauer effect in studies of materials science,[22] surface science and polymer technology [23] and ceramics [24] have been reviewed. The application of Mössbauer spectroscopy in corrosion studies, particularly of Fe, Sn, and Co alloys, has also been discussed.[25] Theoretical topics reviewed during the year considered include the interpretation of Mössbauer parameters on the basis of molecular orbital theory [26] and relaxation theories for radioactive impurities in solids.[27] A review [28] of the theoretical and experimental aspects of applications of elastic and inelastic scattering at reciprocal lattice nodes to structural phase transitions and disordered and amorphous solids also appeared. The uses of *X*-ray scattering and Mössbauer spectroscopy in the determination of crystal lattice sites have been reviewed.[29]

Discussions on Mössbauer spectroscopic results have appeared in reviews of the chemistry of various elements or compounds including, antimony,[31] tin,[32] transition-metal organometallic compounds,[33] ferrocene,[34–35] and methyltin halides and their molecular complexes.[36] Two general review articles [37–38] on the uses of the Mössbauer effect in biological sciences have been published. Reviews of more specific biological applications are concerned with porphyrins

11 A. Ito, *Oyo Butsuri*, 1976, **45**, 367.
12 J. M. Friedt, Report CRN–CNPA–76–3, 1976.
13 P. Guetlich, *Ber. Kernforschungsanlage Jeulich*, 1976, *Jeul-Conf*–22.
14 T. Zemcik, *Vesmir.*, 1975, **54**, 295.
15 J. G. Stevens, *Anal. Chem.*, 1976, **48**, 232R.
16 I. P. Suzdalev, *At. Energy* (*USSR*) (English transl.), 1976, **40**, 234.
17 E. F. Makarov and A. V. Mitin, *Usp. Fiz Nauk*, 1976, **120**, 55.
18 M. Harada and H. Sano, *Bunseki Kiki*, 1976, **14**, 455.
19 T. Zemcik, *Cesk. Cas. Fyz.*, 1975, **25**, 464.
20 A. Vertes, *Kem. Ujabb. Eredmenyei*, 1975, **28**, 263.
21 C. Janot, *J. Phys.* (*Paris*), 1976, **37**, 253.
22 S. Nasu and U. Gonser, *Zairy.*, 1975, **24**, 1083.
23 L. H. Lee, Abstracts of the Centennial ACS Meeting, New York, USA, April 4—9, 1976.
24 M. Shinada, *Seramikkusu*, 1976, **11**, 22.
25 G. W. Simmons and H. Leidheiser, Abstracts Centennial ACS Meeting, New York, USA, April 4—9, 1976.
26 A. Trautwein, *Kyoto Daigaku Genshiro Jikkensho*, 1976, *Kurri*–TR–141.
27 F. Hartmann-Boutron, *Ann. Phys.*, 1975, **9**, 285.
28 C. Ghezzi, *Comments Solid State Phys.*, 1976, **7**, 59.
29 J. F. Duncan, A. G. Freeman, and J. H. Johnstone, in Proceedings of the International Congress, Conference on Anomalous Scattering', 1974 (published 1975), p.163.
30 H. Schmidbauer, *Angew. Chem. Internat. Ed.*, 1976, **15**, 728.
31 L. D. Freedman and G. O. Doak, *J. Organometallic Chem.*, 1976, **106**, 217.
32 P. G. Harrison, *J. Organometallic Chem.*, 1976, **109**, 241.
33 P. S. Braterman, *J. Organometallic Chem.*, 1976, **123**, 75.
34 G. Marr and B. W. Rockett, *J. Organometallic Chem.*, 1976, **106**, 259.
35 B. W. Rockett and G. Marr, *J. Organometallic Chem.*, 1976, **123**, 205.
36 V. S. Petrosyan, N. S. Yashina, and O. A. Rentov, *Organometallic Chem.*, 1976, **14**, 63.
37 C. E. Johnson, *Topics in Appl. Phys.*, 1975, **5**, 139.
38 R. Cammack, *New Tech. Biophys. Cell Biol.*, 1975, **2**, 341.

and metalloporphyrins[39] and with structural investigations of peptides and proteins.[40]

2 Theoretical

This section deals with topics of a general theoretical nature. Developments relating to specific isotopes will be discussed in later sections.

The problem of the construction of a γ-ray laser has received further consideration and it has been suggested that a first step towards a solution should be the study of super-narrow Mössbauer lines of width less than 10^{-10} eV.[41] It has been predicted that during the coulombic excitation of the 14.4 keV transition of ^{57}Fe one γ-photon per 10^{12}—10^{13} electrons should be coherently emitted in a discrete crystallographic direction which depends on the crystal structure and the energy and speed of charged particles.[42] The anomalous emission of atoms located deep within solids has also been discussed in relationship to the construction of γ-lasers.[43]

In recent years there have been a number of attempts to apply delayed Mössbauer coincidence methods to the study of transient after-effects associated with the occurrence of non-equilibrium conditions during and following the decay of Mössbauer excited nuclear spin-states. General formulae have been given for Mössbauer emission spectra measured within finite time intervals following the decay event under conditions where the electronic state of the emitting atom is not in equilibrium with particular reference to changing isomer shift. It has been shown that even for short relaxation times noticeable asymmetries appear in the spectra.[44] Formulae have also been given for Mössbauer spectral lineshape for transmission, emission, and scattering experiments with and without the presence of relaxation effects. It was concluded that scattering experiments reveal little about relaxation effects, but comparison between emission and transmission spectra should reveal useful information on electronic rearrangements occurring during emission.[45, 46] Simplified derivations of Hartmann–Boutron's expressions for the shape of emission spectra in the presence of relaxation have been given.[47] A general expression has been derived for the emission probability of Mössbauer radiation in the presence of relaxation and it was shown that the recoilless fraction is independent of the mode of relaxation.[48] It has been demonstrated that application of acoustic frequencies to a γ-source gives rise to oscillations in the spectrum and that if the acoustic frequencies are less than the natural linewidth the signal broadens into a doublet.[49]

[39] P. Hambright and A. J. Bearden, *Porphyrins Metalloporphyrins*, 1975, 539.
[40] C. E. Johnson, *Amino-acids, Peptides and Proteins*, 1976, **7**, 221.
[41] Yu. A. Il'Inskii, R. V. Khokhlov, R. N. Kuz'min, and O. P. Revokatov, Dig. Tech. Pap.–Int. Quantum Electron Conf., 8th, 1974, IEEE, N.Y.
[42] V. A. Belyakov and V. P. Orlov, *Tr. Vses. Soveshch. Fiz. Vzaimodeistviya Zarazhennykh Chastits Monokrist*., 5th, 1973 (published 1974), Moscow, USSR, p.131.
[43] J. P. Hannon and G. Trammel, *Opt. Commun.*, 1975, **15**, 330.
[44] A. M. Afansev and V. D. Gorobchenko, *Phys. Status Solidi (B)*, 1976, **76**, 465.
[45] F. Hartmann-Boutron, *J. Phys. (Paris)*, 1976, **37**, 537.
[46] F. Hartmann-Boutron, *J. Phys. (Paris)*, 1976, **37**, 549.
[47] A. M. Afansev and V. D. Gorobchenko, *Phys. Status Solidi (B)*, 1976, **73**, 73.
[48] A. Gupta and K. R. Reddy, *J. Phys. Soc. (Japan)*, 1976, **41**, 1611.
[49] A. R. Bvev, *Soviet Phys. Solid State*, 1976, **18**, 1046.

A semiphenomenological model for the recoilless fraction has been described for uniaxial crystals based on the Debye frequency tensor whose principal elements are described in terms of the elastic stiffness constants and densities. The method has been applied to Sb and Te single crystals.[50] The thermal dependence of the isomer shift and log recoilless fraction has received theoretical attention and it has been shown possible to define an effective temperature, which relates directly to the lattice energy and correlates linearly with the isomer shift and log recoilless fraction.[51] Methods of measuring recoilless fractions by time filtering of resonant emission spectra[52] and by the use of both conventional and resonance detectors have been described.[53,54] The effect of dislocations on the recoilless fraction have been analysed[55] and a method for the determination of optimum absorber thickness given.[56] An iterative method for absorber thickness corrections has proved to be of particular use in the analyses of multiphase systems.[57]

The dependence of the Mössbauer resonance intensities on vibrational lattice anisotropy for the case of an axial field gradient has been discussed and polynomial expressions given to describe the functional dependence of the relative intensities on the magnitude of the anisotropy. It was shown that these expressions may be used to extract the relevant parameters from experimental data without requiring the numerical integrations implicit in the Goldanskii–Karyagin description.[58] A phenomenological model has been developed which allows the interpretation of Mössbauer spectra of paramagnetic powders in an applied field for the case of weak, intermediate, and strong magnetization when the internal fields are anisotropic for the case where the γ-ray direction is parallel to H_{ext} and for trigonal, tetragonal, or orthorhombic local site symmetries. In the low magnetization limit significant improvements in computed spectral fits were obtained; compared to fits assuming an isotropic H_{eff}. An intermediate and high magnetization, empirical parameters were introduced which allow for the saturation of the internal hyperfine fields in easy crystallographic directions. The model was used in the interpretation of the spectra of iron fluorosilicate and iron doped into zinc fluorosilicate.[59] It has been emphasized that Mössbauer spectra of ^{57}Fe subjected to combined magnetic dipole and quadrupole interaction cannot give a unique solution for all parameters, even if intensities are taken into account, but expressions were derived which give the full range of possible solutions.[60]

Using Stokes parameters on a Poincaré sphere the polarization of Mössbauer radiation for $\frac{1}{2} \rightarrow \frac{3}{2}$ nuclear transition has been shown to be sensitive to the hyperfine field parameters.[61] Mössbauer studies of relaxation effects using a

[50] Ya. A. Isoilevskii, *J. Phys. Chem. Solids*, 1976, **37**, 897.
[51] R. L. Collins and J. G. Cosgrove, *J. Inorg. Nuclear Chem.*, 1976, **38**, 507.
[52] N. Hayash, T. Kinoshito, I. Sakamoto, and B. Furubayash, *Nuclear Instr. and Methods*, 1976, **134**, 317.
[53] B. Manuchev, Ts. Bonchev, and D. Ivanov, *Nuclear Instr. and Methods*, 1976, **136**, 261.
[54] B. Manuchev, Ts. Bonchev, and D. Ivanov, *Nuclear Instr, and Methods*, 1976, **136**, 267.
[55] B. Bhattacharya, *Phys. Status Solidi (B)*, 1976, **76**, K81.
[56] J. Piekoszenski, K. Kisywska, and L. Dabrowski, *Nucleonika*, 1975, **20**, 947.
[57] D. Gunter, B. Seidel, and W. Meisal, *Nuclear Instr. and Methods*, 1976, **133**, 381.
[58] G. K. Shenoy and J. M. Freidt, *Nuclear Instr. and Methods*, 1976, **136**, 569.
[59] F. Varret, *J. Phys. Chem. Solids*, 1976, **37**, 265.
[60] J. Van. Donegan-Torman, R. Jagannathen, J. M. Troster, *Hyperfine Interactions*, 1975, **1**, 135.
[61] A. Lal, Alimuddin, and K. R. Reddy, *Indian J. Pure Appl. Phys.*, 1976, **14**, 167.

magnetically split and polarized source have been discussed and it has been demonstrated that in certain special cases where the hyperfine fields in source and absorber are perpendicular to the γ-direction, the individual components of the spectrum can be separated. The method was applied to $Fe(ClO_4)_3,9H_2O$, $(NH_4)_3FeF_6$, ferric alum, $Fe(acac)_3$, and frozen aqueous solutions.[62]

Most theoretical treatments of the calculation of Mössbauer transmission line-shapes in the presence of electronic relaxation have assumed the presence of a 'white noise energy bath' which drives fluctuations of the ionic system at frequencies much larger than the unperturbed hyperfine frequencies and as a result the relaxation matrix is independent of the hyperfine frequencies and the computational procedure is greatly simplified. It has been pointed out that although in many cases this approximation is fully justified, the assumption is questionable for a rigid lattice with spin–spin coupling (as is the case for concentrated rare-earth and actinide salts where the r.m.s. dipolar energy becomes comparable with the hyperfine energy). A discussion of the Mössbauer line-shape in the absence of the white noise approximation (WNA) has been given and the relaxation matrix shown to be dependent on the hyperfine parameters, linewidth, and bath parameters. The method was applied to the reanalysis of the relaxed ^{170}Yb Mössbauer spectrum of $Cs_2NaYbCl_6$ and significantly improved fits and values of the dipolar field and electronic spin correlation time were obtained. It was also suggested that the WNA may not be valid for iron systems in biomolecules where the dipolar fields are weak.[63] The use of Mössbauer spectroscopy for studying paramagnetic relaxation phenomena has in certain circumstances been limited due to the inadequate development of theory, for example, it has often been wrongly assumed that the electronic state is a highly anisotropic Kramer's doublet, because of the inherent simplicity of computation. The perturbation approach of Hirst previously applied to $0^+ \rightarrow 2^+$ Mössbauer transitions (see Vol. 8) has been extended to $\frac{1}{2} \rightarrow \frac{3}{2}$ nuclear transitions for axial symmetry with proper inclusion of off-diagonal terms in the hyperfine Hamiltonian. The method has been used to analyse the relaxed spectra of ^{171}Yb in $Cs_2NaYbCl_6$.[64] The influence of off-diagonal hyperfine coupling in relaxed spectra of ^{57}Fe, ^{169}Tm, and ^{166}Er has been examined for a number of pseudo-electronic states in various point symmetries in terms of the stochastic relaxation model of Blume and Clauser and the influence of external magnetic fields discussed.[65] A rapid method for the evaluation of relaxed spectra has been described which makes use of a little known method of Clauser. It was shown that the super-matrix (consisting of the matrix elements of the Louiville operator describing the relaxation process) need only be inverted once rather than at every point of the spectrum, thus saving considerable computing time. Applications to previously intractable problems such as haemoglobin cyanide, acid met-haemoglobin, and NpF_6 were made.[66] An alternative computational technique for calculation of relaxed spectra involving the approximate diagonalization of

[62] S. Moerup, *Hyperfine Interactions*, 1976, **1**, 533.
[63] G. K. Shenoy, B. D. Dunlap, S. Dattagupta, and L. Asch, *Phys. Rev. Letters*, 1976, **37**, 539.
[64] B. D. Dunlap, G. K. Shenoy, and L. Asch, *Phys. Rev. (B)*, 1976, **13**, 18.
[65] J. Sivardière, M. Blume, and M. J. Clauser, *Hyperfine Interactions*, 1975, **1**, 227.
[66] G. K. Shenoy and B. D. Dunlap, *Phys. Rev. (B)*, 1976, **13**, 1353.

the super-matrix has also been presented.[67] A computer program for the simulation of relaxed Mössbauer spectra for $\frac{3}{2} \rightarrow \frac{1}{2}$ nuclear transition and $S = \frac{5}{2}$ for powders and crystals has been written.[68]

The effect of magnetic crystallographic anisotropy on the line-shape of Mössbauer spectra in super-paramagnetic compounds has been analysed using a stochastic approach.[69]

A detailed theoretical analysis has been made which allows the interpretation of the Mössbauer spectra of molecular systems where the e.f.g. tensor is undergoing anisotropic relaxation. The method was applied to the study of a thiourea-ferrocene clathrate (see p. 399).[70]

A simple rate equation has been used to calculate Mössbauer diffusion line broadening in highly anisotropic materials with NiAs or CdI_2 structures, where diffusing atoms may occupy more than one site. Large departures from Lorentzian line-shapes were calculated and it was predicted that the use of simple crystals should yield detailed information because of the large angular dependence of the diffusional broadening.[71]

It has been shown theoretically that the exchange narrowing of Mössbauer lines can be induced by magnetic phase transitions in crystals and the nature of discontinuities in isomer shift, intensities, and the spin correlation functions, $\langle S_i . S_j \rangle$ near the critical point have been examined.[72] The Doppler line-shape of γ-lines has been calculated on the basis of Lindherd–Blangrunt theory.[73]

A derivation of an expression for the line-shape of resonantly scattered Mössbauer γ-radiation has been reported.[74] Electron diffusion theory (rather than the more commonly used exponential electron attenuation law) has been used to appraise the depth selective ability of ^{57}Fe and ^{119}Sn conversion electron Mössbauer spectroscopy and the probability of electrons emerging from surfaces with and without inert overlayers calculated. It was demonstrated that diffusion theory is more effective in accounting for the energy distribution of forward scattered electrons from a thin foil illuminated on its reverse side rather than of the back-scattered electrons.[75] The quantitative interpretation of conversion electron Mössbauer spectra has received further consideration on the assumption that the electrons obey an exponential attenuation law.[76]

The quadrupolar antishielding factors for a number of $3d^5$, $4d^5$, and $5d^5$ ions have been calculated using restricted Hartree–Fock–Slater type wave functions combined with the perturbation approach of Sternheimer [77] and recent developments in the calculation of quadrupole field gradients in non-cubic metals discussed.[78] A simple model has been devised to explain the diminution of hyperfine fields in microcrystals and it was shown that below the super-

[67] D. C. Price, *J. Phys. (E)*, 1976, **9**, 337.
[68] D. Barb and L. Diamandescu, *Inst. Fiz. At. (Rome)*, Report 1975, SR–13.
[69] G. N. Belozerskii, Yu. T. Pavlyukhin, and V. N. Gittsovich, *Zhur. eksp. teor. Fiz.*, 1976, **70**, 717.
[70] T. C. Gibb, *J. Phys. (C)*, 1976, **9**, 2627.
[71] A. T. Howe and G. J. Morgan, *J. Phys. (C)*, 1976, **9**, 4463.
[72] G. Ya. Selyutin, *Phys. Status Solidi (B)*, 1976, **75**, 233.
[73] R. B. Begzhanov and F. S. Akilov, *Dokl. Akad. Nauk Uz. SSR*, 1976, 21.
[74] K. Ruebenbauer and W. Zurek, *Phys. Status Solidi (B)*, 1976, **74**, K165.
[75] A. Minkova, B. Slavov, and Ts. Bonchev, *Bulg. J. Phys.*, 1975, **2**, 592.
[76] G. P. Huffman, *Nuclear Instr. and Methods*, 1976, **137**, 267.
[77] K. D. Sen and P. T. Narasimhan, *Phys. Rev. (A)* 1976, **14**, 539.
[78] R. S. Raghanan, *Hyperfine Interactions*, 1976, **2**, 29.

paramagnetic blocking temperature thermally excited oscillations of the magnetization around an energy minimum cause a reduction in average magnetization. The method was applied to microcrystals of α-Fe_2O_3.[79] A method for the calculation of the distribution of both magnetic and hyperfine fields in amorphous materials has been given.[80] Activation energies of solid-state reactions have also been obtained from Mössbauer studies.[81]

Small angle scattering of Mössbauer radiation from liquids near the critical point[82] and elastic scattering from molecular solids have also been theoretically analysed.[83]

3 Instrumentation and Methodology

A cyclotron has been used to produce ^{129}Xe, ^{121}Sb, and ^{83}Kr sources[84] and it has been suggested that K_2CoO_3 is a suitable matrix for ^{57}Co sources combining the advantages of present sources.[85] An improved Mössbauer spectrometer[86] and a simple circuit for a highly stable spectrometer have been described.[87] Details of an automatic constant-velocity Mössbauer spectrometer have also been given.[88]

An improved laser-driven Michelson interferometer for calibration and checking linearity of Mössbauer drive systems has been developed. The design allows the avoidance of errors down to the lowest velocities normally used and it was shown that the recoil of the drive vibrator is a major problem in obtaining a linear velocity scale.[89] Further discussion of spectrometer calibration with lasers has also been presented.[90]

Recently there has been increasing experimental interest in the Mössbauer spectroscopy of thin films of iron and it has been noted that the films can be divided into two classes (a) films thinner than about 100 nm and (b) films 100 nm to 1 μm thick and that the main difficulty in measurements is the small size of the Mössbauer effect. For films thinner than about 100 nm back-scatter methods are best employed but for thicknesses >100 nm it is better to employ transmission methods. In the past, transmission experiments of this kind have involved stacking absorber films till a total thickness of about 8 μm is reached. (In order to avoid this a 'high counting rate' Mössbauer spectrometer has been developed which has allowed a good spectrum from a single layer, 0.5 μm thick, of nickel ferrite (containing natural iron) to be obtained in 4 days with a 50 mCi source.[91]

[79] S. Moerup and H. Topsoe, *Appl. Phys.*, 1976, **11**, 63.
[80] P. Mangin, G. Marchal, M. Piecoch, and C. Janot, *J. Phys. (E)*, 1976, **9**, 1101.
[81] N. A. Eissa, A. A. Baghat, A. H. Mohamed, and S. A. Selah, *J. Amer. Ceram. Soc.*, 1976, **59**, 327.
[82] E. Iolins and V. V. Kozlov, *Latvijus PSR Zinatnu Akad. Vestis, Fiz., Teh. Zinat, Ser.*, 1976, 32.
[83] S. P. Tewari, *Phys. Letters (A)*, 1976, **56A**, 221.
[84] S. Bukshpan, F. T. ten Broek, W. Hillbrants, S. R. Reintsema, and H. De. Waard, *Int. J. Anal. Radiat. Isot.*, 1976, **27**, 15.
[85] M. I. Afanasov, L. A. Kulikov, Yu. D. Perfil'ev, and A. M. Babeshkin, *Pribory i Tekhn. Eskperim.*, 1976, 53.
[86] P. J. Ouseph, V. K. Vaidyan, U. A. Raphael, and T. A. K. Pillai, *Indian J. Pure Appl. Phys.*, 1976, **14**, 779.
[87] V. N. Belogurov and V. F. Vasil'ev, *Pribory i Tekhn. Eksperim.*, 1976, 67.
[88] M. Bhattacharya and R. Bhattacharaya, *Indian J. Appl. Physics*, 1976, **14**, 404.
[89] M. A. Player and F. W. D. Woodhams, *J. Phys. (E)*, 1976, **9**, 1148.
[90] R. Koch, S. Reiman, and H. Raudsepp, *Fuss. Uliopilastoode Kogumik, Tartu Riiklik Ulik. Uliopilaste Tead, Uhing*, 1974, 37.
[91] G. A. Fatseas, *J. Phys. (E)*, 1976, **9**, 1066.

There have been further developments in equipment associated with back-scattering experiments. A conversion electron Mössbauer spectroscopy (CEMS) study of thin films of Fe and Fe–Si 25—250 nm thick containing natural abundance of iron has been made by detection of the conversion electrons *in vacuo* using a windowless channel electron multiplier. A comparison between transmission and back-scatter methods was made and it was concluded that thin films are best studied by CEMS.[92] A He–CH_4 proportional counter for CEMS work at liquid nitrogen temperatures has been described[93] and a He–CH_4 proportional counter which allows the incident γ-photon to sample angle to be continuously varied has been constructed. Thickness effects were detected in highly enriched iron foils, and by obtaining spectra at glancing incident angles for enriched samples or by using either the high- or low-energy region of the electron pulse height spectra it was shown possible to gain some information about the depth distribution of surface phases.[94] A CEMS method for surface stress methods in the outermost few hundreds of nanometres of ferrous metals has been developed. In order to measure the small shifts in the Mössbauer lines produced by surface strain a thermal scanning technique was developed based on the second-order Doppler shift which has the advantage that the modulated emitted radiation is isotropic (no cosine effect).[95] An efficient 2π sealed proportional counter for detecting back-scattered low-energy *X*- and γ-photons has been constructed.[96]

An apparatus has been described for determining the Compton scattering profile of magnetic electrons and the first measurement of this kind carried out on an iron foil. The method makes use of the fact that the Compton scattering of circularly polarized 122 keV ^{57}Co γ-photons can be used for studying the spin density in momentum space in ferromagnetic materials because the relativistic Compton scattering cross-section depends on the electron spin. The difference in scattering from an iron foil magnetized in different directions was determined as a function of the projection of the electron momentum vector and the scattering vector (Figure 1). The results indicate that the magnetization of some electrons with low moments is opposite to the majority of electrons.[97]

The direct measurement of the Rayleigh scattering in iron metal has been made by selective-excitation-double Mössbauer technique (SEDM). An absorption was selectively excited in ^{57}Fe in iron metal which resulted in the emission of photons with different energies and since Rayleigh scattering is an elastic process its only contribution is at the energy of the emission transition which corresponds to the original absorption transition. By measurement of the line-shape of the SEDM lines and comparison with theory it was shown possible to determine the Rayleigh differential cross-sections.[98]

A liquid helium crystal coupled to an electromagnetic separator has been described for the *in situ* examination of implanted sources[99] and designs for a

[92] R. Oswald and M. Ohring, *J. Vacuum Sci. Technol.*, 1976, **13**, 40.
[93] J. A. Sawicki, B. D. Sawicka, and J. Stanek, *Nuclear Instr. and Methods*, 1976, **138**, 565.
[94] M. J. Tricker, A. G. Freeman, A. P. Winterbottom, and J. M. Thomas, *Nuclear Instr. and Methods*, 1976, **135**, 117.
[95] C. M. Yagnik, R. L. Collins, R. A. Mazak, and W. H. Baer, in Proceedings of the 10th Symposium of Nondestructive Evaluation, 1975, 194.
[96] R. A. Levy, P. A. Flinn, and R. A. Hartzell, *Nuclear Instr. and Methods*, 1975, **131**, 559.
[97] N. Sakai and K. Ono, *Phys. Rev. Letters*, 1976, **37**, 351.
[98] B. Balko and G. R. Hoy, *Phys. Rev. (B)*, 1976, **13**, 2729.
[99] S. A. Drentje and S. R. Reintsema, *Nuclear Instr. and Methods*, 1976, **133**, 421.

low-temperature chamber for Mössbauer diffraction measurement [100] and for a cryostat for magnetic studies at 4.2 K reported.[101] Details have been given for a high-field Mössbauer spectrometer using water-cooled Bitter coils capable of producing fields up to 155 kG and the problems of stray fields considered.[102] A high-pressure cell, suitable for Mössbauer and *X*-ray studies, with polycrystalline

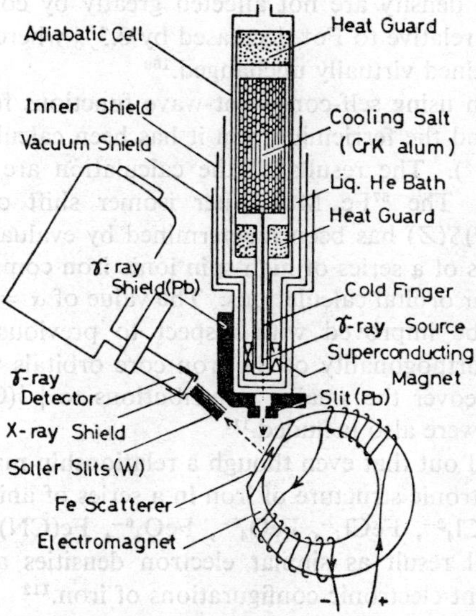

Figure 1 *Apparatus for measuring the Compton profiles by circularly polarized γ-rays* (Reproduced by permission from *Phys. Rev. Letters*, 1976, **37**, 351)

cubic boron nitride anvils producing pressures of up to 170 kbar [103] and a pressure cell capable of operating at 4.2 K and at 20 kbar especially suitable for small crystals have been described.[104]

A two-piston quenching process has been successfully used to produce amorphous Fe-P-C and F-P-B alloys in a form suitable for Mössbauer studies.[105]

The convolation of line-shapes which involve the Lorentz distribution [106] and a Mössbauer fitting program have been described.[107] A procedure has been developed which allows the comparison of Mössbauer spectra without the need for detailed analysis and the method has been shown to be particularly useful, for example, in the examination of ancient ware.[108]

[100] V. S. Zasimov and R. N. Kuz'min, *Vestn. Mosk. Univ., Fiz. Astron.*, 1976, **17**, 105.
[101] U. Helgeson, L. C. Moberg, and N. Ruus, University of Stockholm, Sweden, Report, 1975, USIP–75–18.
[102] J. Chappert and J. R. Regnard, *Colloq. Intern. Centre Nat. Rech. Sci.* (*Paris*), 1975, **242**, 257.
[103] E. V. Kapitanov, V. N. Panyushkin, L. F. Vereshchagin, E. N. Yakovlev, V. I. Novokshonov, and V. V. Evdokimova, *High Temp.–High Pressures*, 1976, **7**, 555.
[104] G. N. Stepanov and V. N. Panyushkin, *Prib. Tekh. Eksp.*, 1976, 235.
[105] H. Ino, H. Oya, K. Kawano, N. Susumi, T. Namiki, and S. Yoshioka, *Seisan Kenkyu*, 1976, **28**, 58.
[106] P. P. Schmidt, *J. Phys.* (*B*), 1976, **13**, 2331.
[107] D. P. Lazar and M. Morariu, *Inst. Fiz. At.* (*Rome*), Report 1976,–SR 17.
[108] F. Arumu, V. Maxia, and C. Muntoni, *Nuclear Instr. and Methods*, 1976, **137**, 503.

4 Iron-57

General Topics.—*Nuclear Parameters and Isomer Shift Calibrations.* Numerical multi-configuration Hartree–Fock results have been reported for the configurations $3d^6(^5D)4s^n$ (n = 0, 1, 2) of Fe^{2+}, Fe^+, and Fe with proper inclusion of degeneracy and correlation effects. It was shown that although the absolute values of the charge density are not affected greatly by correlation effects the charge density of Fe relative to Fe^+ decreased by 8.3% whereas the same density relative to Fe^{2+} remained virtually unchanged.[109]

From a calculation using self-consistent wave functions for several low-lying states of ferrocene and the ferricinium ion it has been calculated that $\Delta R/R$ for $^{57}Fe = -(4.4 \times 10^{-4})$. The results of the calculation are described in more detail on p. 398.[110] The ^{57}Fe Mössbauer isomer shift calibration constant $\alpha = \frac{2}{5}\pi Ze^2(R^2 - R_0^2)S(Z)$ has been redetermined by evaluation of the electron density at the nucleus of a series of high-spin ionic iron complexes using CNDO (open shell) molecular orbital calculations. The value of $\alpha = -0.417\ a_0^3$ mm s^{-1} was considered to be improved with respect to previous estimates because corrections for non-orthogonality of the iron core orbitals with ligand orbitals were made and moreover the smaller contributions to $|\psi(0)|^2$ from cross and distant ligand terms were also included.[111]

It has been pointed out that even though a relationship may exist between the isomer shift and electronic structure of iron in a series of anions such as FeF_6^{4-}, FeF_6^{3-}, $FeCl_6^{3-}$, $FeCl_4^{2-}$, $FeCl_4^-$, FeO_4^{5-}, FeO_4^{6-}, $Fe(CN)_6^{4-}$, and $Fe(CN)_6^{3-}$ ambiguities may still result as similar electron densities at the nucleus may correspond to different electronic configurations of iron.[112]

Alloy-type Systems. All references concerning metals and intermetallic alloys are not discussed explicitly in this section but are collected together in a bibliography at the end of the chapter. This section deals with hydrides, borides, carbides and carbon–steels, nitrides, silicides, germanides, phosphides, arsenides and antimonides, and some tellurides.

A Mössbauer study has been made of hydrides of Ni–Fe ($0 < Fe < 10\%$) prepared by electrocathodic methods and it was shown that the hydrides are ferromagnetic at low temperature with T_c increasing sharply with increasing iron content. The results were taken to suggest that the iron atoms have no H neighbours.[113] The metallurgical and electronic structure of $TiFeH_x$ ($0.1 < x < 2$) alloys has been probed by ^{57}Fe Mössbauer spectroscopy in relationship to the ability of the system to act as a vehicle for hydrogen storage and volume effects and charge flow were shown to be important factors in determining the isomer shift.[114]

Mössbauer spectra have been obtained for Fe_3C and Fe_5C_2 from 4 K to the Curie temperature of each carbide and shown to consist of superpositions of two and three components respectively. The results were discussed in terms of the

[109] C. Froese-Fischer and J. S. Carley, *J. Phys. (B)*, 1976, **9**, 29.
[110] P. S. Bagus, U. I. Walgreen, and J. Almlof, *J. Chem. Phys.*, 1976, **64**, 2324.
[111] S. Vijaya, P. K. Meurotra, and P. T. Manoharan, *Chem. Phys. Letters*, 1976, **44**, 261.
[112] V. S. Nefedov, *Koord-Khim.*, 1976, **2**, 991.
[113] T. Mizutani, T. Shinjo, and T. Takada, *J. Phys. Soc. Japan*, 1976, **41**, 794.
[114] L. J. Swartzendruber, L. H. Bennett, and R. E. Watson, *J. Phys. (F)*, 1976, **6**, L331.

iron environments and orientation of the spins.[115] A Mössbauer spectroscopic study has been made of ^{57}Fe and ^{119}Sn in perovoskite carbides of formula $M_3M'C$ (where M = Fe or Mn and M′ = Al, Ga, Ge, Zn, or Sn). All the compounds were characterized by a large quadrupole coupling of 1.3 mm s^{-1} and by low internal magnetic fields of $\simeq$ 70 kG at the M atom. The effect of carburation and the different contributions to the internal field were discussed and a number of magnetic anomolies observed which could only be accounted for in terms of itinerant magnetism.[116] An *X*-ray and Mössbauer investigation has been made of the structure and electronic properties of the cementite-like carbides $(Fe_xMn_{1-x})_3C$ ($x = 0.9$, 0.55, 0.75, or 0.64) prepared in the form of thin films by co-condensation of the components *in vacuo*. In the mixed carbides the decrease in iron isomer shift with increasing Mn concentration was interpreted on the basis of a transfer of electrons from the d-like band of Fe to the d-like band of Mn.[117] A study of cementite isolated from tempered silicon steels has also been made.[118]

A theoretical study of the effect upon the Mössbauer spectrum of the diffusion of interstitial carbon or nitrogen atoms in austenite has shown that the low-temperature spectra should consist of a superposition of doublets and an unsplit line due to iron with and without next nearest-neighbour interstitial atoms, but at higher temperatures the doublet is expected to collapse because of motional narrowing.[119]

A Mössbauer study of the low-temperature martensitic phase of a low manganese carbon steel has demonstrated the presence of a new spectral component due to ^{57}Fe atoms adjacent to carbon atoms in tetrahedral interstices. At low temperatures the populations of the tetrahedral and octahedral interstices were found to be virtually identical, but during tempering carbon atoms in tetrahedral sites jump to octahedral sites.[120] Studies have also been made of carbon distribution in the bulk and the surface of martensitic steels by use of scattering techniques[121, 122] and of the redistribution of carbon in invar upon ageing.[123]

The magnetic properties of $(Fe_{1-x}Co_x)Si$ have been studied by low temperature specific heat ^{59}Co NMR and ^{57}Fe Mössbauer effect measurements. It was shown that an Fe atom substituted for a Co atom in a CoSi gives rise to a paramagnetic moment on six nearest Co atoms suggesting that the magnetic electrons are itinerant.[124] A Mössbauer source experiment on the distribution of dilute ^{57}Co in Fe_3Si has indicated that all Co impurity atoms selectively replace Fe_{II} site atoms in agreement with previous n.m.r. studies.[125]

115 C. Le Caer, J. M. Dubois, and J. P. Senateur, *J. Solid State Chem.*, 1976, **19**, 19.

116 F. Grandjean and A. Gerard, *J. Phys. (F)*, 1976, **6**, 451.

117 B. I. Kotlyar, O. Kh. Tadeush, L. G. Dobrovol'skaya, and E. V. Kuz'mina, *Metallofiziki*, 1976, **64**, 29.

118 M. P. Arbiezov, B. I. Kotlyar, and M. M. Yatsura, *Izvest. Vyssikh Uchebn. Zavedenii, Fiz.*, 1976, **19**, 147.

119 S. Dattagupta, *Phil. Mag.*, 1976, **33**, 59.

120 C. Shiga, F. Fujita, and M. Kimura, *Nippon Kinsokie Gakkaishi*, 1975, **39**, 1205.

121 V. N. Gridnev, V. V. Nemoshkalenko, V. G. Gavrilyuk, Yu. A. Poloshkin, and O. N. Razumov, *Dopovidi Akad. Nauk Ukrain. R.S.R. Ser. A. Fiz.–Mat. Tech. Nauki*, 1976, 829.

122 L. H. Schwartz and K. J. Kim, *Metall. Trans. (A)*, 1976, **74**, 1567.

123 A. I. Zakharov, B. V. Molotilov, V. A. Makarov, and L. V. Pastukhova, *Fiz. Metal. i Metalloved.*, 1975, **40**, 668.

124 S. Kawarazaki, H. Yasuoka, Y. Nakamura, and J. H. Wernick, *J. Phys. Soc., Japan*, 1976, **41**, 1171.

125 C. Blaauw, G. R. Mackay, and W. Leiper, *Solid State Comm.*, 1976, **18**, 729.

The hexagonal and cubic phases of Fe_3Ge have been investigated by Mössbauer and magnetization techniques and it was discovered that both phases of the materials are ferromagnetic with T_c = 640 K and 740 K and $\mu_{Fe} = 2.0\mu_B$ and $2.2\mu_B$, for the hexagonal and cubic phases respectively. In the hexagonal phase a second-order spin flip transition was observed at 380 K and it was found that in the cubic phase the preferred spin direction is ⟨100⟩ and that the anisotropic part of the hyperfine field is large.[126]

Mössbauer spectra have been recorded over a range of temperatures for the antiferromagnetic intermetallic compound Fe_2As revealing the presence of two sites which were discussed in relationship to the known magnetic properties of the compound. At temperatures close to T_N it was determined that the spins lie in the basal planes but as the temperature is lowered the spins gradually rotate from the basal plane and make an angle of 16° at 77 K.[127]

Solid solutions of the type $Fe_{1-t}T_tAs$ (T = V, Cr, Mn, or Co) with the MnP structure have been studied in order to relate the variations in Mössbauer spectra to changes in composition.[128]

Mössbauer *X*-ray and conductivity studies of $FeTe_x$ (x = 0.95, 1.2, or 2.0) which crystallize as tetragonal, hexagonal, and orthorhombic phases respectively have been made and spectra obtained in the temperature range 77—700 K. At room temperature and below, all phases exhibited a quadrupole split spectra characteristic of high-spin Fe^{2+} covalently bonded to Te atoms. At 420 K the spectrum at $FeTe_{0.95}$ collapses into a single line and a large change in isomer shift occurs. It was possible to correlate these changes with a marked variation in the conductivity temperature relationship which can be explained by the conversion of Fe^{2+} to Fe^{3+} iron. Similar changes were observed at 500 K and 520 K for $FeTe_{1.2}$ and $FeTe_2$.[129]

Frozen Solutions and Diffusion Phenomena. Frozen aqueous solutions of Fe^{3+} salts have been investigated in order to study glass formation which was shown to be critically dependent on cooling rate. Changes in spin–spin relaxation times were found to correlate with average iron–iron separations. In the $FeCl_3$–H_2O systems homogeneous glasses formed for salt concentrations larger than 3.5 moles per 100 moles of H_2O but in more dilute solutions ice crystallized. The effects of glass-forming agents such as NO_3^- and ClO_4^- were also investigated.[130] Mössbauer studies have been made on thermally quenched iron(II) chloride aqueous solutions in the temperature range 10—255 K. Three phases were found connected by irreversible transitions and from the temperature dependence of the Mössbauer parameters it was concluded that in two of these phases $Fe^{2+}(H_2O)_6$ complexes exist and that the third phase contains $Fe^{2+}(H_2O)_4Cl_2,2H_2O$.[131] It has been demonstrated that spin–spin relaxation times are enhanced in frozen aqueous solutions when two different iron complexes are present.[132] A study has been made of colloidal solutions or iron ferrocyanides and stannates in glycerine.

[126] J. W. Drijver, S. G. Sinnema, and F. Van Der Woude, *J. Phys. (F)*, 1976, **6**, 2165.
[127] P. Raj and S. K. Kulshreshtha, *Phys. Scr.*, 1976, **14**, 125.
[128] K. Selte, A. Kjekshus, and T. Rakke, *Acta Chem. Scand.*, 1976, **30A**, 340.
[129] K. V. Reddy and S. C. Chetty, *Phys. Status Solidi (A)*, 1976, **37**, 687.
[130] S. Moerup, J. E. Knudsen, M. K. Nielsen, and G. Trumpy, *J. Chem. Phys.*, 1976, **65**, 536.
[131] W. Neuwirth and H. J. Schroeder, *Z. Phys. (B)*, 1976, **23**, 71.
[132] S. Moerup and J. E. Knudsen, *Chem. Phys. Letters*, 1976, **40**, 292.

The form of the adsorption line yielded information on homogeneity of particle size and indicated that the logarithm of the linewidth depends linearly on its diffusion broadening.[133]

^{57}Fe Impurity Studies. Resisitivity and Mössbauer measurements have been made of iron in solid xenon matrices containing up to 30% atomic iron, and a non-conducting to conducting transition was shown to be accompanied by a change from a paramagnetic phase of iron to a magnetically ordered phase.[134]

The advantages of conversion electron Mössbauer spectroscopy outlined in Vol. 9, p. 414, in the study of ion-implanted materials have been exploited in the past year.[93, 135–138] Studies have been made on ^{57}Fe ion implanted into a wide range of targets. A great variety of spectra were obtained showing that the state of the ^{57}Fe is strongly host-dependent.[135] In further studies of magnetic hosts (e.g. Fe, Co, and Ni) magnetic hyperfine patterns have been observed but in para- or dia-magnets (V, Cr, Mo, Pd, Re, Rh, Ir, Pt) single lines or asymmetric broad structured lines (Mn, Ti, Nb, W, Mo, Cu, Au) or doublets (Zn, Ge, Si, Ag, In, Sr, Pb) were observed. The systematics of the isomer shifts of dilute iron in various transition metal targets were found to be very similar to the behaviour observed for dilute alloys. It was concluded that the lattice damage has little effect on the isomer shift but can give rise to quadrupole interaction and that the overall pattern of the results are similar to those obtained for rapidly quenched alloys. Quadrupole splitting was observed in Fe implanted in Au and this was explained in terms of a model which attributes the origin of the e.f.g. at the iron to random defects but in order to explain the additional structure observed in the spectrum of the Fe–Cu system clustering of the iron atoms was invoked.[136] A study of ^{57}Fe implanted (*ca.* 10^{15} ions cm^{-2}) into silica has also been made. Initially a symmetric doublet was observed which became asymmetric above 600 °C where migration of iron atoms begins to occur and α-$FeSi_2$ is precipitated; at 1200 °C β-$FeSi_2$ was also precipitated. Above 1200 °C the signal disappeared as the iron atoms diffused into the bulk.[137] It has been demonstrated that magnetic coupling takes place in a aluminium foil containing 2×10^{17} ^{57}Fe implants cm^{-2} at 77 K.[93] A CEMS study of iron implanted to a depth of 100 nm in graphite consists of a broad doublet which sharpens upon annealing at 500 °C. It was suggested that one main type of iron site is occupied.[138]

Mössbauer effect studies and internal conversion experiments of ^{57}Fe impurity atoms in Cr, Ni, Cu, Pt, and Au have been made. The M- and N-shell conversion electron spectra were measured using a high resolution conversion electron β-ray spectrometer. The results indicated that the 4s electron density at the ^{57}Fe iron nucleus is nearly equal to the host metals. Examination of the relationship between $\rho_{4s}(0)$ and the isomer shift indicated that $\rho_{3s}(0)$ plays an important role in the determination of the isomer shift.[139]

133 S. Micheva and Z. Chobanova, *Kolloid. Zhur.*, 1976, **38**, 692.
134 Z. Shanfield, P. A. Montano, and P. H. Barret, *Phys. Rev. Letters*, 1975, **35**, 1789.
135 B. Sawicka, J. Sawicki, and J. Stanek, *Nucleonika*, 1976, **21**, 949.
136 B. D. Sawicki, J. Sawicka, and J. Stanek, *Phys. Letters*, 1976, **59A**, 59.
137 J. Sawicki, B. Sawicka, J. Stanek, and J. Kowalski, *Phys. Status Solidi (B)*, 1976, **77**, K1.
138 M. J. Tricker, R. K. Thorpe, J. H. Freeman, and G. A. Gard, *Phys. Status Solidi (A)*, 1976, 33, K97,
139 T. Shinohara, M. Fujioka, H. Onodera, K. Hisatakie, H. Yamamoto, and H. Watanabe, *Hyperfine Interactions*, 1976, **1**, 345.

Zeemen-field-induced quadrupole and hyperfine interactions in Jahn–Teller active ions have been theoretically considered with particular reference to Fe^{2+} ions in octahedral symmetry. An effective electron nuclear Hamiltonian was derived whose parameters were calculated by a perturbation procedure incorporating the vibronic effects in terms of the Ham reduction-factors between spin-orbit states. Relevant expressions were given so that these parameters may be determined from Mössbauer spectra and the method applied to the study of ^{57}Fe doped into $KMgF_3$ and CaO. The induced quadrupole splitting and hyperfine fields for $KMgF_3Fe^{2+}$ are linear in applied field and are, for example, -0.30 mm s^{-1} and -113.0 kG and 0.52 mm s^{-1} and -114.7 kG for a 50 kG external field in the [100] and [111] directions respectively. For CaO,Fe^{2+} the behaviour is non-linear and the induced values are -0.75 mm s^{-1}, -224.0 kG and $+0.36$ mm s^{-1} and -200.4 kG under the same conditions described for $KMgF_3$,Fe^{2+}.[140, 141]

A study of monoclinic $K_3Fe(CN)_6$ diluted in diamagnetic $K_3Co(CN)_6$ has been reported and the line-shape of the relaxed spectra analysed using the rapid Clauser technique (see p. 365). In order to describe the experimental data it was necessary to invoke different relaxation rates between iron atoms and their first and second nearest neighbours.[142]

A study of tris-(acetylacetonato)iron(III) diluted in the analogous Al(III) and Ga(III) complexes has shown that at 4.2 K paramagnetic relaxation phenomena take place. Irradiation of the samples with γ-rays from a ^{60}Co source was shown to lead to a reduction of these effects consistent with the operation of a motional narrowing mechanism.[143] Mössbauer studies of Fe^{3+} ions doped into $LiSiO_3$ and mullite in the temperature range 4—295 K reveal that the $\pm\frac{1}{2}$ and $\pm\frac{5}{2}$ doublet relaxation times are similar in mullite, and moreover the high E/D ratio for tetrahedral iron in mullite makes possible the observation of the $\pm\frac{1}{2}$ doublet in mullite even in zero-applied fields.[144] The orientation dependence of the Mössbauer spectra of Fe^{3+} doped into $Al(NO_3)_3,9H_2O$ single crystals in small magnetic fields <300 G has been studied.[145]

A study of ^{57}Fe doped into $KMnF_3$ in the temperature range 13—570 K has shown that structural transformations occur at 190 K and 90 K and that V_{ZZ} is >0 near T_N ($=$ 86 K) but <0 at temperatures less than T_N.[146] An investigation of iron doped into $Fe_3B_7O_{13}I$ has revealed information on the magnetic structure of the material.[147] The large observed quadrupole splitting of ^{57}Fe doped into Si_3N_4 has been explained in terms of a nitrogen vacancy associated with the impurity atoms.[148] Mössbauer impurity and source studies have been made of ^{57}Fe and ^{57}Co doped into pure and oxidized tin dichalcogenides. It was

140 T. Ray and J. R. Regnard, *Phys. Rev. (B)*, 1976, **14**, 1796.
141 T. Ray and J. R. Regnard, *Phys. Rev. (B)*, 1976, **14**, 1805.
142 G. K. Shenoy and B. D. Dunlap, *Phys. Rev. (B)*, 1976, **13**, 3709.
143 G. M. Bancroft and T. K. Sham, *J. Chem. Soc. Faraday II*, 1976, 1706.
144 B. L. Dickson and K. K. P. Srivastava, *J. Solid State Chem.*, 1976, **19**, 117.
145 A. M. Afanasev, V. N. Zarubin, and S. S. Yakimov, *Zhur. eksp. teor. Fiz.*, 1976, **70**, 1957.
146 S. I. Reiman and K. Mitrofanov, *Soviet Physics Solid State*, 1976, **18**, 759.
147 D. Barb, D. Tarina, S. Constantinescu, T. A. Himichi, I. S. Zheludar, and I. N. Yarmuhamedov, *Rev. Roumaine Phys.*, 1976, **21**, 105.
148 H. Yamamura, K. Kijima, S. Shirasaki, Y. Inomata, and H. Suzuki, *J. Materials Sci.*, 1976, **11**, 1754.

found that the impurity iron is present in interstitial positions as high-spin ferric.[149] A Mössbauer impurity and source study has also been made on ^{57}Fe and ^{57}Co doped into rutile.[150]

A Mössbauer investigation of iron doped into the ferroelectric material triglycine sulphate (tgs) has shown that changes occur at the iron nucleus as the properties of the tgs change with temperature. Variation of the Mössbauer spectra at temperatures above the normal Curie point of the compound (T_c = 49 °C) correlate with the polarization hysteresis loop of the undoped material. However, above 60 °C an anomalous increase in the quadrupole splitting is observed and on cooling the specimen to T_c the hysteresis loops reappear but show no bias. Similarly it was found that the quadrupole splitting did not decrease immediately. These results were discussed on the basis of an internal bias field.[151]

^{57}Co Source Experiments and Decay After-effect Phenomena. The yields of the stabilized forms of iron atoms produced by *K*-capture of ^{57}Co in glassy frozen solutions of H_2SO_4, HNO_3, and $HClO_4$ and their salts have been estimated using ^{57}Co emission spectra. The data were analysed on the basis of competition between ferric ions and other components for Auger electrons. The effectiveness of scavengers was estimated and was shown to be similar to their effectiveness for electron capture during radiolysis of aqueous solutions.[152] In a study of a number of strong-field ^{57}Co-labelled chelate complexes of cobalt(III) with ligands of varying degrees of conjugation *viz.* [Co(III)(phen)$_3$](ClO$_4$)$_3$,2H$_2$O; [Co(III)-(phen)$_2$(en)]Cl$_3$,H$_2$O; [Co(III)(phen)(en)$_2$](NO$_3$)$_3$,H$_2$O and [Co(III)(en)$_3$]Cl$_3$ by emission spectroscopy it has been concluded that high degrees of conjugation in the ligand molecule inhibit fragmentation of the complex subsequent to the decay of ^{57}Co.[153] Studies of the stabilization of daughter species following the Auger event in labelled cobalt(II) chelates, [Co(III)(phen)$_3$](ClO$_4$)$_3$,2H$_2$O and [Co(III)-(bipy)$_3$](ClO)$_3$,3H$_2$O in sucrose, polyethylene oxide has revealed that the Fe(II)–Fe(III) ratio varies from matrix to matrix. These observations were rationalized on the basis of electron acceptor and donor centres present within the solvent matrices.[154] A Mössbauer emission study has been made on ^{57}Co-labelled $CoCl_2$ dissolved in various non-aqueous media but in spite of the known electron scavenger properties of the solvents, no correlation was found between the yields of Fe^{2+} and the dielectric constants or the donacities of the solvents.[155] ^{57}Co-labelled cobalt(II) acetylacetonates have also been studied.[156]

Emission spectra of ^{57}Co-doped [Co(phen)$_2$(NCS)$_2$] and [Co(bipy)$_2$(NCS)$_2$] have been recorded in the temperature range 4.2 K to room temperature. It was

149 V. Fano and I. Ortalli, *J. Phys. Chem. Solids*, 1976, **37**, 729.
150 T. R. Sandin, D. Schrooer, and C. D. Spencer, *Phys. Rev.* (*B*), 1976, **13**, 4784.
151 I. E. G. Morrison, L. V. C. Rees, J. Silver, and E. A. D. White, *J.C.S. Dalton*, 1976, 1103.
152 Y. D. Perfil'ev, L. A. Kulkinov, L. T. Bugaenke, A. M. Babeshkin, and M. I. Afanasov, *J. Inorg. Nuclear, Chem.*, 1976, **38**, 2145.
153 M. B. Misroch, C. J. Schramm, and A. Nath, *J. Chem. Phys.*, 1976, **65**, 1982.
154 T. S. Srivastava and A. Nath, *J. Phys. Chem.*, 1976, **80**, 529.
155 M. J. Afanosov, S. Nagy, Yu. D. Perfil'ev, A. Vertes, and A. M. Babeshkin, *Radiochem. Radioanal. Letters*, 1975, **23**, 181.
156 M. I. Afanasov, Yu. D. Perfil'ev, and A. M. Babeshkin, *Vestnik Moskor. Univ.* (*Khim.*), 1975, **16**, 749.

found that the ^{57}Fe complexes formed by the decay are exclusively high-spin whereas the corresponding synthetic iron complexes exhibit temperature-dependent high-spin ($^5T_{2g},O_h$) low-spin ($^1A_{1g},O_h$) transitions. An analysis of the variations of the quadrupole splitting with temperature indicated that the ground state of iron is $^5A_2$ in C_{2v} symmetry.[157] Similar behaviour has also been observed for the analogous methylated phenathroline cobalt complexes.[158]

A detailed study of a Mössbauer study of ^{57}Co doped into NiO and CoO in the temperature range 22—1150 °C has been made. At low temperatures two lines attributed to Fe^{2+} and Fe^{3+} iron were observed but as the source temperature was raised the two lines merged into a singlet whose position was found to be dependent on the oxygen partial pressure around the source. This was interpreted on the basis of a charge relaxation valence averaging between Fe^{2+} and Fe^{3+}. Discontinuities in the recoilless fraction were also observed indicative of phase transitions at $\simeq 500$ and 300 K. The results were discussed in terms of a charge and vacancy ordering model which may be responsible for the phase transitions.[159]

A magnetic susceptibility, *X*-ray diffraction and a Mössbauer emission study of labelled $EuCoO_3$ has been made and the presence of a high-spin (Co^{3+}) low-spin (Co^{III}) cobalt spin-state equilibrium detected. The results indicated that up to 350 K both Co^{3+} and Co^{III} exist with Co^{III} as the major species. Above 350 K discrepancies between the Co^{3+}/Co^{III} ratios obtained by analysis of the magnetic susceptibility or Mössbauer data were taken to indicate an electron transfer from Co^{3+} to Co^{III} to yield Co^{II} and Co^{IV}. Finally, at temperatures above 1180 K the singlet due to Co^{3+} was found to disappear due to a localized electron-itinerant electron transition and the population of a σ-band state. The overall pattern of the results was considered to be consistent with Goodenough's description of the electronic properties of the material.[160] A study of the Co^{3+}–Co^{III} spin-state equilibrium in $LuCoO_3$ has indicated that there is a partial transformation of low-spin Co^{III} to high-spin Co^{3+} at 450 K.[161] Hopping conduction in $Ln_{1-x}Sr_xCoO_3$ (Ln = La or Nd) was studied by emission spectroscopy and the hopping times deduced from conductivity and Mössbauer measurements were found to differ widely.[162]

A Mössbauer emission effect study has been made of impurity ^{57}Co in alkali metal halides and MgF_2, both pure and doped with small concentrations of La^{3+} ions. The monovalent halides were all shown to give similar spectra consisting of broad single lines or broad overlapping doublets. Previous Mössbauer and e.s.r. spectra were reviewed briefly and discussed in relationship to vacancy formation.[163] The angular dependences of the polarized γ-rays from a reduced ^{57}Co-doped $LiNbO_3$ single crystal have shown that the principal axis of the e.f.g.

157 J. Ensling, P. Gütlich, H. M. Hasselbach, and B. W. Fitzimmons, *Chem. Phys. Letters*, 1976, **42**, 232.
158 J. Fleisch and P. Gütlich, *Chem. Phys. Letters*, 1976, **42**, 237.
159 C. Song and J. G. Mullen, *Phys. Rev. (B)*, 1976, **14**, 2761.
160 V. G. Jadhao, R. M. Singru, G. N. Rao, D. Bahadur, and C. N. R. Rao, *J. Phys. Chem., Solids*, 1976, **37**, 113.
161 D. Bahadur, *Indian J. Chem. (A)*, 1976, **14A**, 204.
162 C. N. R. Rao, V. G. Bhide, and N. F. Mott, *Phil. Mag.*, 1975, **32**, 1277.
163 A. G. Maddock, A. F. Williams, K. E. Siekierska, and J. Fenger, *Phys. Status Solid. (B)*, 1976, **74**, 183.

tensor at the Fe^{2+} site is colinear with the *c*-axis of the crystal thus excluding the formation of charge-compensating oxygen vacancies.[164]

The ability of ^{57}Co emission spectra to reveal surface information has again been demonstrated in a study of ferromagnetic metal surfaces. The emission spectra of carrier-free ^{57}Co ions electroplated onto an iron surface have revealed that all iron atoms at the surface are ferromagnetic but the observed hyperfine fields suggest that a reduction in the local iron moments occurs. Similar results were obtained for transmission studies of ^{57}Fe-enriched iron surfaces.[165] An *in situ* study of the passivation of cobalt surfaces in borate buffer at pH 8 has been made by emission spectroscopy. The method was shown to be capable of providing information about surface layers 20—200 Å thick. At low cathodic polarization 2+ and 3+ charges were detected on the ^{57}Fe daughter probe element but at higher passivating potentials only 3+ oxidation states were observed and at higher potentials still 4+ oxidation states were observed. However, interpretation of the spectra was complicated by the possible occurrence of effects leading to the stabilization of defect charge states in the daughter ion.[166]

Miscellaneous Studies. Mössbauer transmission spectra of sandwiches of eight 1000 Å thick iron films deposited on NaCl have been recorded after irradiation of the films with He^+ or N^+ ions in order to investigate possible induced structural changes. It was found that the irradiation produces peaks corresponding to new hyperfine magnetic fields in iron and the amplitude and structure of these peaks were governed by the total dose. The data were interpreted in terms of impurity-stabilized iron valency complexes which promote nucleation of a metastable h.c.p. phase of iron in the original b.c.c. lattice.[167]

The electric field gradients produced in grain-orientated specimens of silicon–iron subject to trigonal and tetragonal distortions have been measured and it was found that the e.f.g. produced by unit extension is the same in both cases in contrast to the expectations of the point charge model which predicts a change of sign.[168] Mössbauer effect measurements made in a back-scatter geometry of single crystals of iron, subjected to cyclic stress, have revealed information on fatigue crack initiation and on the eventual failure of the specimens.[169]

Mössbauer sidebands from a single parent line have been induced by radio-frequency irradiation of a sample of non-magnetic stainless steel coated with nickel. The ferromagnetic magnetostrictive nickel was shown to provide the driving force for the production of ultra-sound and hence sidebands in the sample. No sidebands were observed for stainless steel alone or stainless steel plated with copper, further supporting the view that magnetostriction effects rather than eddy currents are responsible for the appearance of the sidebands.[170] It has been shown that application of a strong external radio-frequency field to perma-alloy foils

164 U. Gonser, H. Sakai, and W. Keune, *J. Electrochem. Soc.*, 1976, **123**, 1915.
165 T. Shinjo, *IEEE Trans. Magn.*, 1976, *Mag*-12, 86.
166 G. W. Simmons, E. Kellerman, and H. Leidheiser, *J. Electrochem. Soc.*, 1976, **123**, 1276.
167 V. N. Bykov, G. G. Zdorovtseva, V. A. Troyan, and E. M. Ivanyushkin, *Soviet Physics Solid State*, 1976, **18**, 118.
168 T. E. Cranshaw and R. C. Mercader, *J. Phys. (F)*, 1976, **6**, 2129.
169 J. Jag. Singh, *Nuclear Instr. and Methods*, 1976, **134**, 363.
170 C. L. Chien and J. C. Walker, *Phys. Rev. (B)*, 1976, **13**, 1876.

results in a complete collapse of the magnetically split Mössbauer spectra to single lines.[171]

A Mössbauer investigation has been made of the mechanism of the production of colour centres in iron-doped photochromic sodalites. Bromide sodalites doped with iron were studied before and after heat treatment in H_2 atmospheres and after coloration by ultra-violet irradiation or by electron bombardment. It was discovered that the irradiation converted the original ferrous ions to ferric ions in quantities which agree with the expected number of F-centres from the results of other experiments and that the thermal recovery time at room temperature of Fe^{3+} back to Fe^{2+} is several months.[172]

Tellurium(IV) oxide and iron(III) oxide glasses [173] and the hardening of cement have also been studied by Mössbauer effect measurements.[2]

A Mössbauer study has established a correlation between the activity of unpromoted iron acting as a catalyst in ammonia synthesis and the nitriding capacity of iron. High catalytic activities were found to be accompanied by high nitride formation especially when the catalyst contains a large concentration of defects.[174] In related studies of V_2O_5 and Al_2O_3 promoted iron catalysts nitrided in ammonia, it has been shown that at 200 °C disordered 0.4 wt % solid solutions of nitrogen in α-Fe formed, but above 230 °C, ε, ε', and γ-iron nitrides containing 1—10 wt % nitrogen formed.[175]

Structural modifications in oxide catalysts in the synthesis of organic compounds from carbon monoxide and water have been investigated by Mössbauer spectroscopy.[176] Iron molybdate catalysts have received attention [177] and it has been shown that a reversible oxidation–reduction process occurs throughout the bulk of such catalysts which have been subjected to cyclic treatment with $MeOH/N_2$ streams and oxygen.[178] Mössbauer studies of the activating effect of scandium in the catalytic synthesis of ammonia over iron catalysts [179] and of the use of magnetite in this process have also been made.[180]

Compounds of Iron.—*High-spin Iron*(II) *Compounds*. The temperature dependence of the quadrupole splitting of Mössbauer spectra of divalent iron compounds in rhombic fields has been considered theoretically and the results of the analysis applied to the interpretation of iron environments in hydrated ferrous halides.[181] The relationship between the isomer shift and quadrupole coupling in high-spin ferrous compounds has also been discussed.[182]

[171] M. Kopcewicz, *Solid State Comm.* 1976, **19**, 719.
[172] I. Shiolovsky and I. Nowick, *Solid State Comm.*, 1976, **18**, 155.
[173] V. S. Kozhukharov, M. R. Marinov, and S. M. Nikolov, *Fiz. Khim. Stekla*, 1975, **1**, 499.
[174] V. K. Yatsimirski, N. I. Girenkova, and Yu. V. Maksimov, *Teor. i eksp. Khim.*, 1976, **12**, 263.
[175] Yu. V. Maksimov, I. P. Suzdalev, R. A. Arents, and M. Yu. Kushnerev, *Teor. i eksp. Khim.*, 1975, **11**, 695.
[176] Yu. V. Maksimov, I. P. Suzdalev, A. I. Mateev, I. F. Makarov, and A. V. Kravtsov, *Izvest Akad. Nauk S.S.S.R., Ser. Khim.*, 1975, 2065.
[177] Yu. V. Maksimov, I. P. Suzdalev, V. I. Goldanski, A. I. Mateev, E. F. Makarov, and L. Margolis, *Doklady Akad. Nauk S.S.S.R.*, 1975, **225**, 135.
[178] M. Carbucicchio and F. Trifiro, *J. Catalysis*, 1976, **45**, 77.
[179] L. A. Rudnitskii, P. D. Rabina, A. A. Melikyan, V. P. Korneev, L. D. Kuznetzov, T. V. Saprykina, and A. M. Alexseev, *Kinetica i Kataliz*, 1976, **17**, 989.
[180] T. Peev, *Montash.*, 1976, **44**, 345.
[181] Sh. Sh. Bashkirov, R. A. Manapov, and E. K. Sadykov, *Paramagn. Rezon.*, 1974, **8**, 118.
[182] R. Garg and V. K. Garg, *Phys. Status Solidi (B)*, 1976, **77**, K165.

A n.m.r. and Mössbauer study has been carried out on the mixed valence compound $Fe^{III}Fe^{II}F_5,2H_2O$ [183] and in a further magnetic study of the same compound by Mössbauer spectroscopy it was shown that magnetic ordering appears at T_C or T_N = 48.6 K and that a magnetic transition occurs at T_t = 26.5 K. This transition was shown to correspond to the rotation of the ferric spins whilst the ferrous spins remained fixed along the local anisotropy axis which is roughly the principle axis of the e.f.g. tensor.[184]

A Mössbauer investigation has been on the cubic elpasolites of general formulae A_2NaFeF_6 (A = K, Rb, or Cs), and it has shown that the Mössbauer spectra are broad singlets in the temperature range 4.2—600 K. The broadening was attributed to a statistical distribution of electric field gradients at the iron site of the distorted FeF_6^{3-} octahedra. It proved possible to estimate the order of magnitude of the local distortions and for Rb_2NaFeF_6, an anion displacement of q = 0.11 Å from the regular octahedral position was calculated which agrees with the r.m.s. displacement of $\bar{u}$ = 0.13 Å which results in this direction from the strongly anisotropic *F* temperature factors.[185] An *X*-ray and Mössbauer spectroscopic study of ^{57}Fe in $K(FeM)F_3$ (M = Mn, Co, Ni, or Zn) perovskites has indicated that random replacement of Fe^{2+} by M occurs but that the parent lattice is preserved. Some small induced e.f.g. were observed but these are not attributable to pressure effects.[186]

Single-crystal Mössbauer spectra of $FeCl_2,4H_2O$ subjected to cyclic pressure treatment of up to 7 kbars have been reported with the observation direction parallel to the crystal *b*-axis when the two iron sites in the monoclinic cell appear identical. It was observed that at the 7 kbar and at 4.5 K in an external field of 40 kG that the internal hyperfine field is reduced from 229.5 ± 1 to 223 ± 1 kG and that the ligand field splitting is also decreased.[187] Mössbauer data have been reported for $FeCl_2,4H_2O$.[188] A Mössbauer resonance study of $FeCl_2$ monolayers deposited on orientated based planes of graphite (Grafoil) has been made. Samples with fractional monolayer coverages of 0.2 and 0.9 were studied between 300 K and 80 K. The spectra were shown to consist of two distinct quadrupole doublets, one with a room temperature splitting of $\simeq 2$ mm s^{-1} (designated L) and another $\simeq 0.8$ mm s^{-1} (B) whose value is identical to the bulk value of $FeCl_2$. A marked difference between the intensities of the two lines of each doublet which depends on the γ-ray direction relative to the Grafoil sheets was observed and the total intensity of the B doublet relative to the L doublet is reduced with decreasing coverage. It was noted that the parameters of L are remarkable in that they differ from those of bulk $FeCl_2$ and other known Fe^{2+} compounds and also show an unusual temperature dependence. The results were taken to suggest the existence of at least two distinct phases at $FeCl_2$ in the film samples and that $FeCl_2$–graphite samples involve monolayer coverage rather than bulk $FeCl_2$ aggregates. A calculation of

183 T. Sakai and T. Tominga, *Radiochem. Radioanal. Letters*, 1975, **23**, 329.
184 P. Imbert, G. Jehanno, Y. Macheteau, and F. Varret, *J. Phys.* (*Paris*), 1976, **37**, 969.
185 J. Peblar, K. Schmidt, R. Haegle, and D. Babel, *Z. Naturforsch. B: Anorg. Chem. Org. Chem.*, 1976, **31B**, 1039.
186 J. Silver and J. D. Donaldson, *Inorg. Nuclear Chem. Letters*, 1976, **12**, 795.
187 H. Vogel and H. Spiering, *Chem. Phys. Letters*, 1976, **44**, 266.
188 J. Tatarkiewicz, *Postepy Fiz.*, 1976, **26**, 641.

the absorption binding energy from the temperature dependence of the Debye–Waller factor yields values of about 80 kcal $mole^{-1}$ and 60 kcal $mole^{-1}$ for the B and L. Possible models for the surface arrangement of ferrous chloride molecules were discussed.[189]

A new polymorph of anhydrous $FeSO_4$ has been identified during a Mössbauer study of α- and β-anhydrous $FeSO_4$ in the temperature range 4.2–30 K. A weak signal was detected in the spectrum of β-$FeSO_4$ attributed to a new γ-phase which

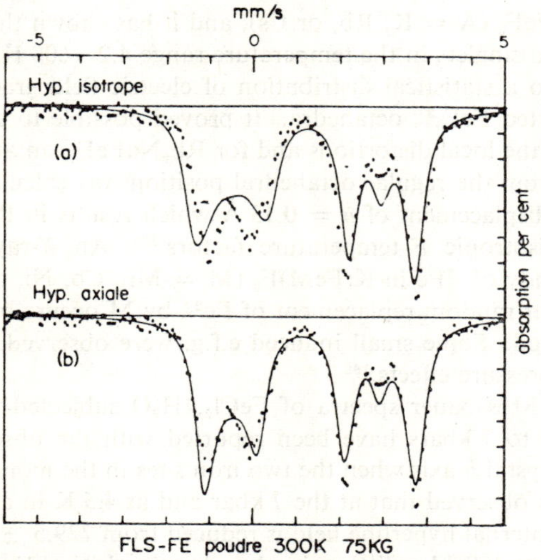

Figure 2 *Fe-fluorsilicate*, 300 K, 75 *k*Oe, (a) *fit using a Gabriel–Collins line-shape*, (b) *fit using an axial magnetization hyperfine tensor*
(Reproduced by permission from *J. Phys. and Chem. of Solids*, 1976, 37, 265)

was shown to be isomorphous with γ-$CoSO_4$ by *X*-ray studies and to be a uniaxial ferromagnet. The magnetic ordering temperatures were determined to be 24 ± 0.5 K, 17.0 ± 1.5 K, and 19.5 ± 1 K for the α-, β-, and γ-phases respectively and the Mössbauer parameters given.[190]

The importance of anisotropic hyperfine effects in the interpretation of data from polycrystalline paramagnets has been experimentally demonstrated in a detailed study of powdered ferrous fluorosilicate and ^{57}Fe-doped zinc fluorosilicate. The spectra of $FeSiF_6$ at 300 K in an applied field of 75 kG parallel to γ-ray direction showed no enhancement of the external field and the material is therefore effectively a diamagnet at 300 K, but it was demonstrated that the spectrum could only be computed satisfactorily by inclusion of anisotropic hyperfine effects (Figure 2). Similar effects were observed for Fe^{2+} iron impurities in $ZnSiF_6$. At 4.2 K in the presence of an external field of 60 kG the spectrum of powdered $ZnSiF_6$Fe was shown to resemble that of a single-crystal sample as for most directions of the external field and magnetization created in the easy plane is saturated. In the intermediate magnetization case a model was

[189] H. Shecter, J. G. Dash, M. Mor, R. Ingalls, and S. Bukshpan, *Phys. Rev.* (*B*), 1976, **14**, 1876.
[190] H. L. Wehner, H. Spiering, and G. Ritter, *Solid State Comm.*, 1976, **20**, 831.

developed which allowed for partial saturation effects by use of an empirical tanh-type relationship between the effective and external fields, and further improvements in the quality of fits to experimental data were obtained. It was concluded that by use of this model for powdered samples the hyperfine fields created in various directions and the magnitude and sign of V_{ZZ} can be obtained but accurate values of η require the onset of sizeable magnetization.[59]

The Mössbauer spectrum of ferrous zirconium double sulphate consists of a doublet ($\delta = 1.30$ mm s^{-1}, $\Delta = 3.22$ mm s^{-1}) at 77 K consistent with an octahedral Fe^{2+} environment.[191]

A Mössbauer study of the structure of $FeSnF_6$ and $FeSnF_6,6H_2O$ and the interaction of Fe^{II} hexafluorostannates with liquid HF has been made. ^{119}Sn and ^{57}Fe spectra revealed that Fe occupies two nonequivalent crystallographic positions in $FeSnF_6$. Strong anion–cation interactions are observed in $FeSnF_6$ but in $FeSnF_6,6H_2O$ the anion–cation bond is weak.[192]

The ligand field at Fe^{II} site of $FeCO_3$ has been studied by use of orientated powders at 4.2 K in the presence of applied fields of up to 80 kG. The field was applied normal to the spin direction and under these conditions the spins make an angle of 7.1° with the three-fold axis. The data were analysed assuming ^5D Fe^{2+} ground state with the inclusion of an anisotropic exchange term proportional to $\langle S \rangle$ and it was concluded that covalency effects in $FeCO_3$ are very small.[193]

The Mössbauer spectra of iron and cobalt fluoroborate hexahydrates have been shown to display a change, by a factor of two, in the quadrupole splitting at 280 K and 185 K respectively, together with a sudden drop in resonance fraction. This behaviour was explained as a first-order phase transition in which the axial parameters of the e.f.g. changes in sign and magnitude.[194]

$Fe(TCNQ)_2,3H_2O$ (TCNQ = 7,7,8,8-tetracyanoquinodimethanidetrihydrate) has been investigated by Mössbauer technique in the temperature interval 4.2—295 K and the spectra shown to consist of two superimposed quadrupole doublets. The major component ($\simeq 94\%$) was attributed to $Fe(TCNQ)_2,3H_2O$ and has parameters, at 295 K, $\delta = 1.139 \pm 003$ mm s^{-1} and $\Delta = 3.32 \pm 0.01$ mm s^{-1} indicative of the presence of high-spin Fe^{2+} with a configuration $3d^6 4s^{0\cdot5\rightarrow0\cdot9}$. The second doublet has $\delta = 0.35 \pm 0.01$ mm s^{-1} and $\Delta = 0.71 \pm 0.01$ mm s^{-1} and was attributed to Fe^{3+} iron in $Fe(TCNQ)_3,xH_2O$. The temperature dependence of the isomer shift and recoil-free fraction suggested that phonon hardening or a phase transition occurs below 50 K.[195]

The electronic properties and relaxation processes have been studied for iron in bisacetylacetonatoiron(II) dihydrate. The spectrum was measured from 4.2 to 300 K and in the presence of external fields of up to 60 kG. Strong texture effects due to the shape of the crystals were observed. Application of a 35 kG field at 300 K, where the paramagnetic susceptibility is negligible,

[191] R. P. Singh and B. N. Srivastava, *Indian J. Pure Appl. Phys.*, 1976, **14**, 71.
[192] V. K. Sokolava, V. A. Varnek, N. F. Yudanov, and I. I. Tychinskaya, *Izvest. sibirsk. Otdel. Akad. Nauk, Ser. khim. Nauk*, 1975, 13.
[193] H. Spiering, D. L. Nagy, and R. Zimmerman, *Chem. Phys. Letters*, 1976, **18**, 243.
[194] L. Asch, I. Dezsi, T. Lohner, and B. Molnar, *Chem. Phys. Letters*, 1976, **39**, 177.
[195] E. E. Chain, D. N. Kervill, C. W. Kimball, and L. W. Weber, *J. Phys. Chem. Solids*, 1976, **37**, 817.

demonstrated that $V_{ZZ} > 0$ and that the principal axes of the e.f.g. tensor lies along the [100] direction. An analysis of spectra recorded at 4.2 K in the external field revealed that $\eta \simeq 0.4$ and that the hyperfine field is colinear with V_{ZZ} and has a saturation value of -175 kG. It was concluded that under the action of spin–orbit coupling and external magnetic fields the ground state of Fe^{2+} is a spin doublet split by $\simeq 1\ cm^{-1}$.[196]

A study has been made of a range of iron(II) chloride complexes of phenylhydrazine and its derivatives, *viz.* $FeCl_2(N_2H_4)_2$, $FeCl_2(p\text{-}O_2NC_6H_4NHNH_2)_4$, $FeCl_2(o\text{-}O_2NC_6H_4NHNH_2)_2$, $FeCl_2(PhNHNH_2)_2$, $FeCl_2(m\text{-}ClC_6H_4NHNH_2)_2$, and $FeCl_2(p\text{-}ClC_6H_4NHNH_2)$, and the observed spectra interpreted in relationship to their structures.[197]

The dangers inherent in the common practice of correlating the magnitude and temperature dependence of Mössbauer quadrupole splitting values with the orbital ground states of Fe^{2+} iron have been discussed in relationship to an experimental study of the pseudo-octahedral iron complex *trans*-Fe(pyridine)$_4$-(NCS)$_2$. The Mössbauer spectrum at 4.2 K in an applied field of 60 kG was shown to consist of a triplet–doublet pattern expected for a rapidly relaxing paramagnet having $V_{ZZ} > 0$ and an internal induced field of 42 kG which suggests some degree of anisotropy in the magnetic susceptibility and a negative internal field. The crystal structure of Fe(pyridine)$_4$(NCS)$_2$ shows a D_{4h} local symmetry about iron and hence produces a tetragonal distortion incompatible with $V_{ZZ} > 0$ if the orbital ground term is ^5E as previously suggested on the basis of the magnitude ($\Delta = 1.93$ mm s^{-1}) of the quadrupole splitting. It was suggested that a more reasonable explanation is that the ground state is $^5B_{2g}$ with π-bonding effects accounting for the small value of Δ. Related systems were also discussed.[198]

An infra-red and Mössbauer study has been made of complexes of iron(II) dioximes with amino- and cyano-pyridines. The compounds FeL_2Q_2 (HL = dimethylglyoxime or diphenylglyoxime; Q = 3- or 4-amino- or 3- and 4-cyano-pyridine) were all found to have *trans*-configurations with pyridine co-ordinated to iron through the heterocyclic nitrogen atoms. The effects of various ligand substituents on the Mössbauer parameters were also discussed.[199] The thermal decomposition of [Fe(bipy)$_3$]Cl$_2$,5H$_2$O in a nitrogen atmosphere has been investigated by Mössbauer techniques and two isomers of [Fe(bipy)]Cl$_2$ have been found to occur in the deomposition sequence, one of which appears to be identical with the rose red form of Fe(bipy)Cl$_2$ prepared from solution. The structures of the two isomers were determined on the basis of their Mössbauer spectra, infra-red, and Raman and magnetic susceptibility measurements. It was concluded that the decomposition (and structure) sequence is [Fe(bipy)$_3$]-Cl$_2$,5H$_2$O $\rightarrow$ Fe(bipy)$_2$Cl$_2$ $\rightarrow$ orange Fe(bipy)Cl$_2$(monomeric, tetrahedral) $\rightarrow$ rose-red Fe(bipy)Cl$_2$(polymeric, octahedral) $\rightarrow$ $FeCl_2$.[200]

An extensive Mössbauer and thermogravimetric study has been made of the electronic and magnetic structure of the polymeric systems of di-imidazolato-

[196] J. P. Mathieu and J. Chappert, *Phys. Status Solidi* (*B*), 1976, **75**, 163.
[197] V. A. Strarodub, M. S. Novakovskii, V. G. Kirichenko, V. V. Chekin, and A. J. Velikodnyi, *Koord. Khim.*, 1975, **1**, 1706.
[198] W. M. Reiff, G. J. Long, and B. F. Little, *Inorg. Nuclear Chem. Letters*, 1976, **12**, 405.
[199] V. N. Shafranskii, V. N. Zubarev, R. A. Stukan, K. I. Turta, and D. D. Buburuz, *Izvest. Akad. Nauk, S.S.S.R., Ser. khim.*, 1976, 260.
[200] H. Sato and T. Tominaga, *Bull. Chem. Soc. Japan*, 1976, **49**, 697.

iron(II) synthesized by reaction of either ferrocene or cyclopentadienyl-irondicarbonyl with imidazole. Two different iron sites (designated A and B) were observed with A : B = 1.79 and below 13.9 ± 0.10 K the compound was found to order magnetically and had $V_{ZZ}(A) < 0$ and $V_{ZZ}(B) > 0$. The temperature dependence of the Mössbauer parameters were interpreted using a simple ligand-field picture for orbital splittings and quasi-tetrahedral and quasi-octahedral co-ordinated high-spin Fe^{2+}.[201] In a related study the Mössbauer spectra of the polymeric compounds obtained in the reaction of 2-methyl-, 2-ethyl-, or 2-phenyl-imidazole with ferrocene or cyclopentadienyl-iron-dicarbonyl have been obtained over a wide temperature range and interpreted in detail. The substituents have a pronounced effect on the electronic structure and magnetic ordering, $V_{ZZ} > 0$ and $\eta = 0$ for all compounds, and iron is tetrahedrally co-ordinated by imidazole nitrogen in a three-dimensional polymeric network. The orbital ground states and first excited states and their energy separations were determined.[202, 203]

An independent study of the Mössbauer spectra of $[Fe(2\text{-}Cl\text{-}phen)_3](ClO_4)_2,H_2O$ in the temperature range 293—5 K has reaffirmed that the iron atom possesses a spin quartet ground state throughout this temperature range in contradiction to the previously suggested existence of high-spin–low-spin transition made on the basis of magnetic susceptibility data. A theoretical analysis of the temperature dependence of the quadrupole splitting with trigonal distortion, spin-orbital coupling, and effects of covalency taken into account, resulted in a value for ground term of the iron(II) of $^5A_1(^5T_{2g})$, separated from the $^5E(^5T_{2g})$ term by *ca.* 165 cm^{-1}, and $V_{ZZ} < 0$ in agreement with previous magnetically perturbed spectra.[204] The thermal decomposition of $[Fe(phen)_3]Cl_2,7H_2O$ in a nitrogen atmosphere has been shown, by Mössbauer spectroscopy, to occur in the following sequence: $[Fe(phen)_3]Cl_2,7H_2O \rightarrow [Fe(phen)_3]Cl_2 \rightarrow Fe(phen)_2Cl_2 \rightarrow Fe(phen)Cl_2 \rightarrow Fe(phen)Cl_2$,[205] finally yielding two structural isomers of $Fe(phen)Cl_2$.

A magnetic susceptibility and Mössbauer spectroscopic investigation of quinquedentate macrocyclic nitrogen ligand complexes of iron(II) has been made. The complexes $FeLX_2,xH_2O$ and $FeClL(ClO_4),2H_2O$ [X = halogen or pseudo-halogen and L is a 15- or 16-membered N_5 macrocyclic ligand (1)] were shown to contain high-spin ferrous iron and are 7-co-ordinate. With one exception the spectra were simple quadrupole doublets with parameters which are the first reported for D_5 symmetry but fall into the range expected for high-spin iron(II) in 6-co-ordinate environments despite the 7-co-ordination. The complex $FeI_2L^1,2H_2O$ was shown to display two quadrupole doublets possibly indicative of the presence of both $[FeIL^1(OH_2)]I,H_2O$ and amounts of $[FeI_2L^1],2H_2O$ in the material.[206]

[201] H. E. Marcolin, A. Trautwein, Y. Maeda, F. Seel, and P. Wende, *Theor. Chim. Acta*, 1976, **41**, 223.
[202] A. Trautwein, Y. Maeda, F. Seel and P. Wende, *Theor. Chim. Acta*, 1976, **41**, 233.
[203] F. Seel, P. Wende, H. E. Macolin, A. Trautwein, and Y. Maeda, *Z. Anorg. Allgem. Chem.*, 1976, **426**, 198.
[204] J. Fleisch, P. Gütlich, and K. M. Hasselbach, *Inorg. Chim. Acta*, 1976, **17**, 51.
[205] H. Sato and T. Tominaga, *Radiochem. Radioanal. Letters*, 1976, **26**, 185.
[206] M. G. N. Drew, J. Grimshaw, P. D. A. McIlroy, and S. M. Nelson, *J.C.S. Dalton*, 1976, 1388.

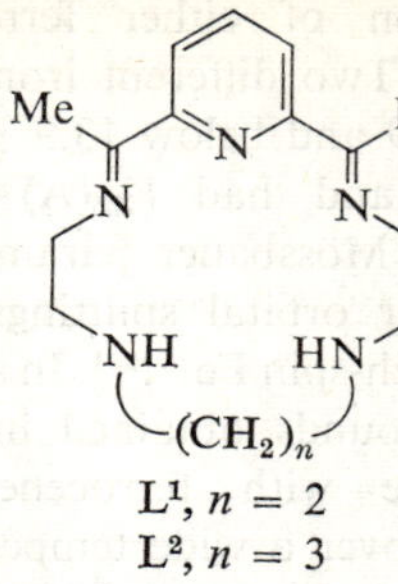

(1)

The compounds $[Fe(2\text{-mnapy})_4](ClO_4)_2,2H_2O$; $[Fe(2,7\text{-dimnapy})_3](ClO_4)_2,2H_2O$; $[Fe(2\text{-mnapy})_4](ClO_4)_3,H_2O$; $[Fe(2,7\text{-dimnapy})_4](ClO_4)_3,H_2O$; $[Fe(2\text{-mnapy})_2Cl_2],H_2O$, and $[Fe(2,7\text{-dimnapy})_2Cl_2],3H_2O$ (2-mnapy = 2-methyl-1,8-naphthyridine and 2,7-dimnapy = 2,7-dimethyl-1,8-naphthyridine) have been prepared and characterized by Mössbauer magnetic and other spectral techniques. The quadrupole splittings were indicative of the 6-co-ordination in the ferrous salts and are significantly smaller than those reported for 8-co-ordinated Fe^{II} complexes of unsubstituted 1,8-naphthyridines, suggesting that the steric restrictions prevent 8-co-ordination of Fe^{II} by the methyl-substituted bidentate ligands. The ferric complexes were thought probably to contain both monodentate and bidentate ligands.[207]

Mössbauer and magnetic studies of the sulphur-bridged iron(II) complexes, with N_2S_2 donor ligands $(FeL)_2$ and $(FeL')_2$ (2) have shown that the complexes

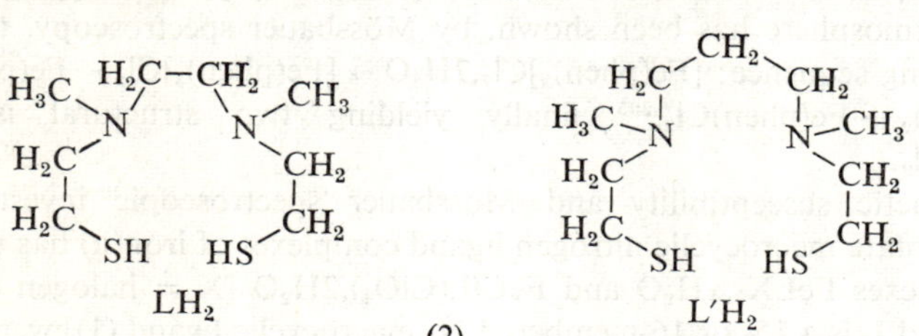

(2)

contain high-spin iron(II) ($S = 2$) atoms which are antiferromagnetically coupled leading to a overall $S = 0$ ground state.[208]

A Mössbauer investigation of the tetraethylammonium salts of bis-(cyclopentadienyldithiocarboxalato)ferrate(II) and tris-(cyclopentadienyldithiocarboxalato)ferrate(III) has shown the complexes to contain high-spin iron(II) and iron(III) respectively. The spectrum of the ferrous complex was shown to be a doublet with $\delta = 0.69$ mm s^{-1} and $\Delta = 4.52$ mm s^{-1} at 77 K, indicative of 5-co-ordinated iron. The ferric material displays a broad singlet ($\delta = 0.41$ mm s^{-1}, $\Delta = 1.71$ mm s^{-1}) at 77 K, indicative of ferric iron in a slightly distorted octahedral environment.[209]

207 M. A. Cavanagh, V. M. Cappo, C. J. Alexander, and M. L. Good, *Inorg. Chem.*, 1976, **15**, 2615.
208 K. D. Karlin and S. J. Lippard, *J. Amer. Chem. Soc.*, 1976, **98**, 6951.
209 R. D. Bereman, M. L. Good, B. J. Kalbacher, and J. Buttone, *Inorg. Chem.*, 1976, **15**, 618.

The iron complexes Fe(dpe)X_2 (X = Cl^- or Br^-) and Fe(dpe)$_2$(NCS)$_2$ (dpe = 1,2-diphenylphosphine-ethane) have been studied by Mössbauer and magnetic techniques and the halide complexes shown to display quadrupole doublets with parameters $\delta = 0.64$ mm s^{-1} and $\Delta = 2.53$ mm s^{-1} at 80 K, consistent with a distorted tetrahedral iron environment. Fe(dpe)$_2$(NCS)$_2$ was found to be diamagnetic and to contain 6-co-ordinate iron.[210]

High-spin Iron(III) *Compounds*. The calculation of the electric field gradient at the nucleus of Fe^{3+} atoms in weakly covalent compounds, by the conventional practice of using screening coefficients appropriate to the 3*d* shell in Fe^{2+}, can lead to large errors. Values for the screening coefficients of the 2*p*, 3*p*, and 3*d* shells of Fe^{3+} iron have been calculated which more correctly account for screening effects.[211]

$NaFeF_4$ has been studied by Mössbauer and magnetic susceptibility techniques and a new antiferromagnetic structure revealed. At 293 K the Mössbauer spectrum is a doublet ($\delta = 0.457 \pm 0.005$ mm s^{-1}, $\Delta = 0.685 \pm 0.005$ mm s^{-1}) but at 4.2 K a well-resolved hyperfine pattern appears with $\delta = 0.47 \pm 0.01$ mm s^{-1}, $\Delta = 0.69 \pm 0.05$ mm s^{-1}, $B_{int} = 562.3$ kG, and $\eta = 0.1$.[212] Mössbauer single-crystal measurements in the temperature range 1.4—300 K have established that $T_N = 133.8$ K in $RbFeF_4$ and that the hyperfine field is at an angle of 16.5° to the principal axis of the e.f.g. tensor. The hyperfine field was shown to lie along the crystallographic [100] axis which shows that the direction of V_{ZZ} is coincident with the shortest Fe—F bond in the FeF_6^{3-} octahedra.[213] The Mössbauer data of γ-$Na_5Fe_3F_{14}$ have been obtained and discussed in relationship to the crystal structure and environment of the iron atoms[214] and the pyrochlore $RbNiFeF_6$ has been shown to be magnetically ordered at 6 K.[215]

The room temperature Mössbauer spectrum of $XeFeF_9$ has been found to consist of an asymmetric quadrupole doublet ($\delta = 0.476$ mm s^{-1} and $\Delta = 0.531$ mm s^{-1}) consistent with the formulation of $(XeF_5^+)_n(FeF_4^-)_n$. The material has a layered structure and the asymmetrical nature of the doublet is associated with texture effects in the absorber. Below 112 K the compound was found to order antiferromagnetically ($H_{int} = 430$ kG) but some samples displayed superparamagnetism.[216, 217]

Mössbauer spectra of graphite $FeCl_3$ and graphite–$AlCl_3$–$FeCl_3$ intercalates have been reported and at 298 K the parameters of graphite–$FeCl_3$ are the same as those of anhydrous $FeCl_3$ but for graphite–$FeCl_3$–$AlCl_3$ systems with $FeCl_3 : AlCl_3$ ratios <1 two sets of quadrupole doublets appear. The results were

210 W. A. Baker and P. M. Lutz, *Inorg. Chim. Acta*, 1976, **16**, 5.
211 Sh. Sh. Bashkirov and V. N. Lebedev, *Fiz. Tverd. Tela*, 1976, **18**, 2679.
212 J. M. Dance, R. Sabatier, F. Ménil, W. Winterberger, J. C. Cousseins, G. Le Flem, and A. Tressaud, *Solid State Comm.*, 1976, **19**, 1059.
213 J. D. Rush, A. Simopoulos, M. F. Thomas, and B. M. Wanklyn, *Solid State Comm.*, 1976, **18**, 1039.
214 M. Vlasse, F. Menil, C. Moriliere, J. M. Dance, A. Tressaud, and J. Portier, *J. Solid State Chem.*, 1976, **17**, 291.
215 V. A. Bobkov, A. S. Kamzin, and P. P. Syrnikov, *Fiz. Tverd. Tela*, 1976, **18**, 2796.
216 J. Slivnik, B. Zemva, M. Bohinc, D. Hanzel, J. Grannec, and P. Hagenmuller, *J. Inorg. Nuclear Chem.*, 1976, **38**, 997.
217 D. Hanzel, *Inorg. Nuclear Chem. Letters*, 1976, **12**, 539.

accounted for in terms of transfer of graphite π-electron density to intercalated $FeCl_3$ moieties.[218]

A Mössbauer study of a range of pyrylium salts of tetrachloroferrate(III) has been made and it was shown that some anomalies in the Mössbauer spectra exist in relationship to the co-ordination of the iron atoms.[219]

The parameters and orientation of the e.f.g. tensors of the iron atoms situated at the 4*c* and 8*d* sites in Fe_3BO_6 have been determined by Mössbauer methods.[220] Spin reorientation effects have also been studied in this compound.[221]

The layered compound $FeOOCH_3$ has been prepared and its Mössbauer parameters are reported to be $\delta = 0.37$ mm s^{-1} and $\Delta = 0.60$ mm s^{-1} at room temperature.[222] The Mössbauer spectrum of tris-(tri-isopropoxy)iron(III) has been shown to be complex and indicative of two inequivalent iron sites within the compound.[223]

The chemical bonding in the Prussian blue analogues of the type $Fe[M(CN)_6]$ (M = Co, Rh, or Ir) has been examined and the trends in isomer shift related to Fe—N σ-bonding interactions.[224]

Mössbauer and other studies of mixed iron(III) and molybdenum(IV) complexes of Schiff bases, $(FeL)_2(MO_aX_b)$ [H_2L = *NN'*-ethylenebis(salicyclideniemine), X = 0, $a = b = 1$; X = S, $a = 2$, $b = 7$; or H_2L = *NN'-o*-phenylenebis-(salicyclideneimine), X = 0, $a = 2$, $b = 7$; or H_2L = 3,3'-iminodiethylenebis-(salicyclideneimine), X = 0, $a = b = 1$] have shown them to be binuclear high-spin ferric complexes with bridging molybdate or thiomolybdate anions but with reduced spin coupling compared to the analogous oxo- and thio-bridged compounds.[225] Two oxo-bridged complexes of iron(III), *cis*-$[\{Fe(bmen)\}_2O(H_2O)_2](SO_4)_2,H_2O$ (bmen = *NN'*-bis(2-methylpyridyl)ethylenediamine) and *trans*-$[\{Fe(tetpy)\}_2O](SO_4)_2,7H_2O$ (tetpy = 2,2':6,2'':6'',2'''-tetrapyridyl), have been shown to contain pairs of antiferromagnetically coupled ferric ions with $S = \frac{5}{2}$ by Mössbauer and other investigations.[226] A study of high-spin $[Fe(pmen)Cl_2]Cl$ and low-spin $[Fe(pmen)(OH_2)OH](ClO_4)_2$ (pmen = *NN'*-bis(2-pyridylmethyl)-ethylenediamine) has been made by Mössbauer and a variety of other techniques and the implications of the stereochemical demands of the complexes on the variation of spin state discussed.[227]

The formation of a highly covalent 6-co-ordinate complex when bis-(*NN'*-dithiocarbamato)iron(III) halides are dissolved in tetrahydrofuran (THF) has been proved by Mössbauer studies in the temperature range 4.2—300 K. Both infra-red and Mössbauer measurements indicated that the THF is only weakly bonded to the paramagnetic centre of the originally 5-co-ordinate complex.

218 T. Takeshi, S. Takuhiko, and T. Kimura, *Bull. Chem. Soc. Japan*, 1975, **49**, 2755.
219 N. G. Bokii, R. V. Verdrinskii, V. V. Kitaev, N. A. Lopatina, and Yu. T. Struchkov, *Koord. Khim.*, 1976, **2**, 103.
220 O. A. Bayukov, V. M. Buznik, V. P. Ikonnikov, and M. I. Petrov, *Fiz. Tverd. Tela*, 1976, **18**, 2319.
221 A. S. Kamzin, V. A. Bokov, and M. K. Chizhov, *Fiz. Tverd. Tela*, 1976, **18**, 2793.
222 S. Kikkawa, F. Kanamura, and M. Koizumi, *Inorg. Chem.*, 1976, **15**, 2195.
223 D. Benlian, R. Panossian, C. More, and Y. Richards, *Ann. Chim. (Paris)*, 1976, **1**, 9.
224 H. Inoue, E. Fluck, and S. Yanagisawa, *Z. Naturforsch., B. Anorg. Chem. Org. Chem.*, 1976, **31b**, 167.
225 P. C. H. Mitchell and D. A. Parker, *J.C.S. Dalton*, 1976, 1821.
226 M. Branca, C. B. Pispisa, and C. Aurisicchio, *J.C.S. Dalton*, 1976, 1543.
227 M. Braca, P. Checconi, and B. Pispisa, *J.C.S. Dalton*, 1976, 481.

An analysis of the quadrupole splitting, isomer shift, and hyperfine interactions showed that in the solvated complex Fe^{3+} iron has $S = \frac{5}{2}$ and that D is negative with the $\frac{5}{2}$ Kramers doublet lying lowest with $|D| \approx 1.4\ cm^{-1}$.[228]

A Mössbauer study has been made of a range of iron(III) complexes of the facultative tetradendate sulphur ligands $MeS(CH_2)_nS(CH_2)_mS(CH_2)_nSMe$ [$n = m = 2(2,2,2)$; $n = m = 3(3,3,3)$; $n = 2$, $m = 3(2,3,2)$; $n = 3$, $m = 2(3,2,3)$] which show an unexpected preference to co-ordinate to Fe^{3+} rather than Fe^{2+}. The Mössbauer spectra of $Fe_2(2,2,2)Cl_6$ and $Fe(2,3,2)Cl_6$ both consist of an overlapping doublet ($\delta = 0.65\ mm\ s^{-1}$ and $\Delta = 0.72\ mm\ s^{-1}$ and $\delta = 0.40\ mm\ s^{-1}$ and $\Delta = 0.38\ mm\ s^{-1}$) and a singlet ($\delta = 0.06\ mm\ s^{-1}$ and $\Delta = 0.19\ mm\ s^{-1}$). These data together with the magnetic susceptibility data were taken to indicate the presence of high-spin Fe^{3+} in octahedral co-ordination giving rise to the doublet and $(FeCl_4)^-$ anions the singlet.[229] Mössbauer studies have also been carried out on relaxation phenomena in dioxane complexes and $FeCl_3$,[230] ferric acetate and chloro-substituted acetates,[231] polyvinyl ferrocene complexes containing high-spin Fe^{3+} [232] and $Fe_4N_3F_3$.[233]

Spin-crossover Systems and Unusual Electric States. A dynamic low-spin–high-spin transition has been observed in $1T$-$Fe_xTa_{1-x}S_2$ ($O < x < \frac{1}{3}$). The anomalous temperature dependence of the magnetic susceptibility and ^{57}Fe isomer shift were attributed to a $^1A_{1g} \rightleftharpoons {}^5T_{2g}$ spin equilibrium occurring on a time scale $< 10^7\ s^{-1}$. The transition moved to higher temperatures with increasing iron content.[234]

Several new octahedral ferrous complexes of the bidentate ligand 2-(2′-pyridyl)-benzimidiazole (pyben) have been prepared and characterized by Mössbauer and other methods. $Fe(pyben)_3A_2,xH_2O$ ($A = L^-$, NCS^-, BF_4^-, BPh_4^-, $Cr(NH_3)_2(NCS)_4^-$, Br^-, NO_3^-, or ClO_4^-, and $x = 1, 2$) were shown to exhibit $^5T_{2g} \rightleftharpoons {}^1A_{1g}$ spin crossovers with slow relaxation indicated by superimposed doublet spectra. The detailed nature of the spin crossover was found to be sensitive to x and the anion. Further studies were made of a fully high-spin and two fully low-spin complexes in the presence of external magnetic fields and it was demonstrated that for the low-spin complexes the signs of V_{ZZ} were indeterminate due to large η values (Figure 3). These results are consistent with a *mer*-octahedral structure as for the $^1A_{1g}$ ground state the point charge model predicts $\eta = 1$ for this stereochemistry. For the high-spin case it was found that $V_{ZZ} > 0$ and $\simeq 0.5$ again indicative of a *mer*-octahedral configuration. It was also demonstrated that the complex $Fe(pyben)_2(NCS)_2$ is fully high-spin down to 4.2 K and that the molecule is tetragonally distorted and has a $|xy\rangle$ orbital singlet ground state.[235, 236]

[228] A. Milliaris and D. Niarchos, *Inorg. Chem.*, 1976, **15**, 1340.
[229] W. Levason, C. A. McAuliffe, S. G. Murray, and S. M. Nelson, *Inorg. Chim. Acta*, 1976, **19**, L15.
[230] Y. Maeda and Y. Takashima, *Inorg. Nuclear Chem. Letters*, 1976, **12**, 313.
[231] K. Mahesh, N. D. Sharama, and D. M. Puri, *Indian J. Chem. (A)*, 1976, **14A**, 338.
[232] M. H. George and G. F. Hays, *J. Polymer Sci., Part A-1, Polymer Chem.*, 1976, **14**, 475.
[233] F. Menil, M. Pezat, and B. Tanguy, *C.R. Hebd. Séances Acad. Sci. Sér. C*, 1975, **281**, 849.
[234] M. Eibschütz and F. J. Disalvo, *Phys. Rev. Letters*, 1976, **36**, 104.
[235] J. R. Sams and T. B. Tsin, *Inorg. Chem.*, 1976, **15**, 1544.
[236] J. R. Sams and T. B. Tsin, *J.C.S. Dalton*, 1976, 488.

A detailed study by Mössbauer, infra-red spectroscopy, and magnetic susceptibility measurements of tris(2-methyl-1,10-phenanthroline)iron(II) perchlorate from room temperature down to 2.6 K has revealed the presence of a high-spin–low-spin equilibrium. Two high-spin doublets were observed in the spectrum of $Fe(2\text{-}CH_3phen)_3(ClO_4)_2$ arising from two inequivalent lattice sites in the compound but both of these convert to one and the same low-spin state.

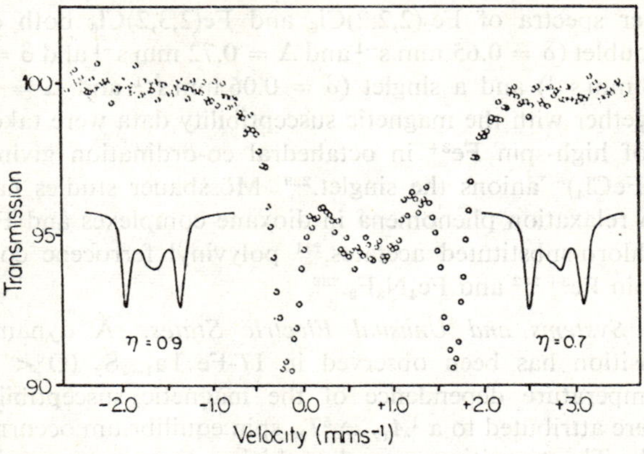

Figure 3 *Mössbauer spectrum of* $Fe(pyben)_3(BF_4)_2,2H_2O$ *at* 80 K *in a longitudinal magnetic field of* 50 kG. *Computed spectra for* $V_{ZZ} > 0$ *and* $\eta = 0.7$ *and* 0.9 *are shown in the insets for comparison*
(Reproduced by permission from *Inorg. Chem.*, 1976, **15**, 1544)

From a theoretical analysis of the temperature dependence of the quadrupole splitting it was shown that the two high-spin iron(II) doublets can be attributed to ^5E and $^5A_1$ ground states in D_3 symmetry with opposite signs of V_{ZZ}. Anomalous magnetic behaviour and discontinuities in several of the Mössbauer parameters at 70—100 K were observed and are believed to be associated with a phase transition.[237]

New examples of incomplete $^5T_2 \rightleftharpoons {}^1A_1$ spin equilibria have been discovered for several iron(III) complexes with Schiff basis, *viz.* $Fe(ppa)_2(NSC)_2$, $Fe(pa)_2(NCS)_2$, $Fe(pot)_2(NCS)_2,\frac{1}{2}CHCl_3$, $Fe(ppt)_2(NCS)_2$, $Fe(ppc)_2(NCS)_2$, and $Fe(ppa)_2(NCS)_2$, where (ppa = *N*-(2-pyridylmethylene)isopropylanene, pa = *N*-(2-pyridylmethylene)aniline, pot = *N*-(2-pyridylmethylene)-*o*-toluidene, ppt = *N*-(2-pyridylmethylene)-*p*-toluidene, ppca = *N*-(2-pyridylmethylene)-*p*-chloroaniline, ppa = *N*-(2-pyridylmethylene)-*p*-anisidine) and the nature of the magnetic change-over has been shown to be sensitive to modifications to the crystal lattice.[238]

New magnetically anomalous iron(III) complexes containing hexadentate ligands derived from triethylenetetramine and various X-substituted salicylaldehydes, *viz.* [Fe(X-Sal_2 trien)]Y (X = H, NO_2, or OCH_3; Y = PF_6^-, NO_3^-, BPh_4^-, I^-, or Cl^-), were found to display $^2T \rightleftharpoons {}^6A$ crossover with a rate constant $>10^{-7}$ s for $T < 200$ K.[239] The room temperature Mössbauer spectrum of

[237] J. Fleisch, P. Gütlich, K. M. Hasselbach, and W. Müller, *Inorg. Chem.*, 1976, **15**, 958.
[238] Y. Maeda, Y. Takashima, and Y. Nishida, *Bull. Chem. Soc. Japan*, 1976, **49**, 2427.
[239] M. F. Tweedle and L. J. Wilson, *J. Amer. Chem. Soc.*, 1976, **98**, 4824.

$[Fe(acac)_2trien]BPh_4$ has been shown to consist of two doublets at room temperature consistent with the occurrence of a $^6A \rightleftharpoons {}^2T$ spin crossover occurring on a timescale $>10^{-7}$ s.[240] A Mössbauer effect study has been made of tris-$[Fe_xZn_{1-x}(2\text{-pic})_3]Cl_2,C_2H_5OH$ (2-pic = 2-picolylamine, x = 1.0, 0.8, 0.6, or 0.2) in the temperature range 10—300 K and it was observed that all the complexes display two quadrupole doublets indicative of $^1A_1 \rightleftharpoons {}^5T_2$ spin crossover. The coexistence of two quadrupole doublets at T_c was demonstrated and this indicates the first order nature of the spin transition. The quadrupole splitting and isomer shifts were all found to be independent of x at all temperatures but T_c varies linearly with iron concentration. It was pointed out that the concentration dependence of T_c is different from that observed in normal magnetic systems showing a phase transition due to spin ordering.[241] The temperature dependence of the Mössbauer spectra of $M(FeL_2),H_2O$ ($M = H^+, Li^+, Ni^+$, K, NH_4^+, and $Zn^{2+}/2$; L = the thiosemicarbazone of 5-Cl, 5-Br, or 5-NO_2 salicycaldehyde) complexes has been reported. It was found that the ground state is typically $^2B_1$ but complexes containing 5-Cl substituted ligand molecules exhibit a $^6A_1 \rightleftharpoons {}^2B_1$ crossover with $^6A_1$ being the ground state at 273 K. For the 5-Br derivatives both ground states were observed but term crossing did not occur.[242] The $^5T_2 \rightleftharpoons {}^1A_1$ spin equilibrium has also been studied in $[Fe(papt(H)_2]X_2,nH_2O$ (papt H = 2-(2-pyridylamine)-4-(2-pyridyl)thiazole, X = halogen).[243] A Mössbauer investigation of a series of tris-(dithio-oxalate)ferrate(III) complexes has revealed that both high-spin and low-spin configurations exist but there is no evidence for a thermal spin equilibrium in any of the compounds.[244]

A ^{57}Fe and ^{129}I Mössbauer spectroscopic study of transferred hyperfine interaction and magnetic ordering has been made of a number of dithiocarbamate derivatives of iron(III), *viz.*

$Fe(S_2CN(C_2H_5)_2)_2I$, $Fe(S_2CN$  $)_2I$, and $Fe(S_2CN$ 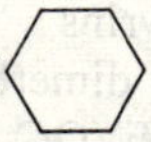 $O)_2I$.

The magnetic ordering temperatures were determined to be 2.00, 2.24, and 3.60 K respectively. The expectation values of the spin and molecular exchange field were found to be perpendicular to the iron–iodine bond but the temperature dependences of the magnetic hyperfine fields differ from the predictions of the molecular field model.[245] Details of the intermolecular magnetic exchange interactions in structural polytypes of iodobis(morpholydyldithiocarbamate) complexes of iron(III) studied at 4.2 K have been reported.[246]

Biological and Related Compounds. A theoretical analysis of the charge density at the ^{57}Fe nucleus in haemin has been made using the self-consistent charge extended Hückel approach.[247]

240 E. V. Dose, K. M. M. Murphy, and L. J. Wilson, *Inorg. Chem.*, 1976, **15**, 2622.
241 M. Sorai, J. Ensling, and P. Gütlich, *Chem. Phys.*, 1976, **18**, 199.
242 K. I. Turta, A. V. Ablov, N. V. Gerbeleu, Ch. V. Dyatlova, and R. A. Stukan, *Zhur. neorg. Khim.*, 1976, **21**, 495.
243 E. Koenig, G. Ritter, and H. A. Goodwin, *Chem. Phys. Letters*, 1976, **44**, 100.
244 T. Birchall and K. Martun, *Inorg. Chem.*, 1976, **15**, 376.
245 D. Petridis, A. Simopoulos, A. Kostikas, and M. Pasternak, *J. Chem. Phys.*, 1976, **65**, 3139.
246 J. M. Grow, G. L. Robbins, and H. H. Wickman, *Chem. Phys. Letters*, 1976, **43**, 77.
247 J. C. Chang, M. Yong, T. P. Das, and K. J. Duff, *Theor. Chim. Acta*, 1976, **41**, 37.

Mössbauer measurements have been performed on three Fe(II) complexes with tetrahedral sulphur simulating the active centres in rubredoxin and it was shown that one of the complexes has virtually identical parameters with reduced rubredoxin.[248]

Mössbauer and other studies have been reported of the newly synthesized iron(II) complexes of 13-, 14-, 15-, and 16-membered, unsubstituted, fully saturated, tetradendate macrocyclic ligands having nitrogen donor atoms. The structures are formally analogous to such natural haem proteins as cytochrome *c*. The complexes were shown to exhibit *cis*, *trans*, or dimeric 6-co-ordinate structures and spin states, which include low-spin ($S = 0$), intermediate-spin ($S = 1$), and high-spin ($S = 2$) iron. The Mössbauer iron spectra were correlated with structural features using the point charge model and the effects of ring size on partial isomer shift examined.[249, 250]

The square planar complex iron(II)octaethylporphyrin [Fe(OEP)] has been prepared and studied by Mössbauer spectroscopy and other techniques. The ferrous iron was found to have $S = 1$ and the nearly temperature-dependent quadrupole splitting suggested that there is no low-lying excited orbital state. The hexaco-ordinate adducts with pyridine, ammonia, and 3- and 4-picoline are all diamagnetic and have $V_{ZZ} > 0$ and $\eta \approx 0$.[251]

Mössbauer data have also been reported for the 6-co-ordinate carbonyl complexes of bis(morphaline), bis(pyrrolidine), and bis(piperidine) haemochromes of *meso*-tetra(*p*-anisidyl)porphenatoiron(II). The spectra were found to be sharp doublets with average isomer shifts of 0.23 mm s^{-1} which is rather smaller than the average of bis(amine)haemochromes. The reduction in isomer shift in the carbonyl complexes was attributed to strong Fe $\rightarrow$ Co π-bonding and weak Fe $\rightarrow$ Co bonding compared to the bis(amine) complexes.[252]

A Mössbauer study has been made of the axial ligation modes of five-co-ordinate iron(III)porphyrins [Fe(PPIXDMe)$SC_6H_4NO_2$)] (1) (PPIXDME = protoporphyrin 1X dimethylester), [Fe(OPE)(SPh)] (2) (OPE = octaethylporphyrin), [Fe(PPIXDME)($OC_6H_4NO_2$)] (3), [Fe(PPIXDME)(OEt)] (4), [Fe(OPE)(OMe)] (5), at 4.2 K and 77 K. All the samples were characterized by single quadrupole doublets but upon raising the temperature one arm of the doublet was observed to broaden as more slowly relaxing states are occupied. The details of this behaviour were not correlated with the nature of the axial or in-plane ligands. Spectra of compounds (1), (3), and (4) obtained in external fields up to 80 kG parallel to γ-ray direction indicate that V_{ZZ} is perpendicular to the heme plane and $\langle S \rangle$ in the plane and that V_{ZZ} is >0 for compounds (1) and (3) but <0 for (4) (Figure 4). It was concluded that the magnitude of the internal hyperfine magnetic field is sensitive to the nature of the axial ligation as oxygen ligation results in $B_{int} \simeq 500$ kG whereas sulphur ligation produces much smaller fields. Moreover, sulphur ligation produces a larger measurable rhombic distortion than oxygen ligation.[253]

248 A. Kostikas, V. Petrouleas, and A. Simpoulos, *Chem. Phys. Letters*, 1976, **38**, 582.
249 D. D. Watkins, D. P. Riley, J. A. Stone, and D. H. Busch, *Inorg. Chem.*, 1976, **15**, 387.
250 D. P. Riley, J. A. Stone, and D. H. Busch, *J. Amer. Chem. Soc.*, 1976, **98**, 1752.
251 D. Dolphin, J. R. Sams, T. B. Tsin, and K. L. Wong, *J. Amer. Chem. Soc.*, 1976, **98**, 6970.
252 W. M. Connor and D. K. Straub, *Inorg. Chem.*, 1976, **15**, 2289.
253 S. C. Tang, S. Koch, G. C. Papefthymiou, S. Foner, R. B. Frankel, J. A. Ibers, and R. H. Holm, *J. Amer. Chem. Soc.*, 1976, **98**, 2414.

A Mössbauer spectroscopic investigation has been made of the product formed by the high-temperature solution decomposition of tetraphenylporphinatoiron(III)azide (TPPFe(III)N_3). It was found that in the absence of base and oxygen a nitrogen-bridged haemin dimer TPPFeNFeTPP is formed

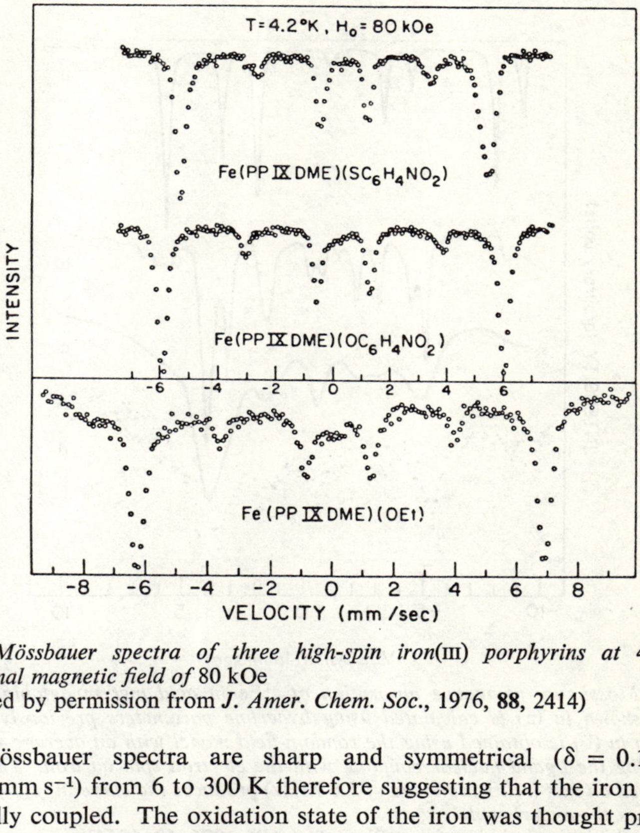

Figure 4 *Mössbauer spectra of three high-spin iron*(III) *porphyrins at* 4.2 K *in a longitudinal magnetic field of* 80 kOe
(Reproduced by permission from *J. Amer. Chem. Soc.*, 1976, **88**, 2414)

whose Mössbauer spectra are sharp and symmetrical ($\delta = 0.10$ mm s^{-1}, $\Delta = 1.08$ mm s^{-1}) from 6 to 300 K therefore suggesting that the iron atoms are magnetically coupled. The oxidation state of the iron was thought probably to be higher than III but two differing oxidation states with rapid valence oscillation could not be ruled out.[254] The anti-ferromagnetic β-phase of tetraphenylporphinatoiron(III)fluoride has been studied and the room temperature Mössbauer spectrum shown to consist of a sharp doublet. The Fe–Fe interactions in the β-phase were suggested to arise as a consequence of molecular stacking rather than the formation of a F-bridged dimer.[255] A study of the hyperfine interactions of the ferric iron in octaethylyhaemin has been made by the magnetic perturbation method and from a spin Hamiltonian treatment it was found that $B_{eff} = 495$ kG, $D = 80$ cm^{-1}, and $\Delta = 0.93$ mm s^{-1}.[256]

A preliminary Mössbauer study of Fe(III)pheophorbides has been made. A large body of data is available on porphyrins but only a little on the closely

[254] D. A. Summerville and I. A. Cohen, *J. Amer. Chem. Soc.*, 1976, **98**, 1747.
[255] I. A. Cohen, D. A. Summerville, and S. Ru Su, *J. Amer. Chem. Soc.*, 1976, **98**, 5813.
[256] B. W. Fitzimmons, J. R. Sams, and T. Tsin, *Chem. Phys. Letters*, 1976, **38**, 588.

related chlorins in which a peripheral double bond of one of the pyrrole rings has been hydrogenated. Mössbauer spectra of (Pheo)FeCl and (EtPpb)FeCl (H_2Pheo = macrocyclic ligand chlorophyll *a* and H_2EtPpb = ethylpheophorbide *a*) at 4.2 K were shown to consist of quadrupole doublets with parameters similar to analogous porphyrins.[257]

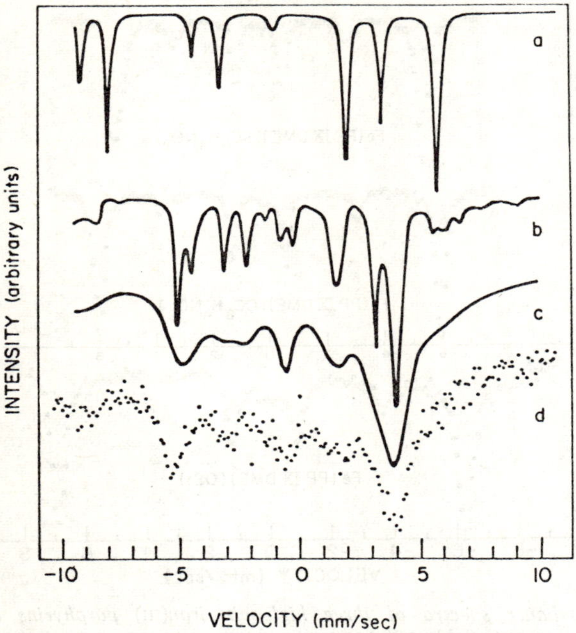

Figure 5 *Mössbauer resonance absorption of* ^{57}Fe *in acid met myoglobin. The static pattern shown in* (a) *is calculated using hyperfine parameters previously given. The spectrum in* (b) *is obtained using the random-field model with an average field of* 10 G *to describe the ligand nuclear coupling with the electron spin on iron. The relaxation superposed on this* (*relaxation rate of* 5.8 MHz) *produces the curve in* (c), *which exactly reproduces the experiment*
[Reproduced with permission from *Phys. Rev.* (*B*), 1976, **13**, 1354]

The low-temperature Mössbauer spectra of haemoglobin cyanide and acid met myoglobin have been re-analysed using the rapid method of Clauser. The spectrum of haemoglobin cyanide was satisfactorily fitted on the basis of the previously determined hyperfine parameters for a low-spin ferric state with $S = \frac{1}{2}$ and a spherical relaxation rate of 4.6 MHz. As a second application of the method the low temperature spectrum of acid met myoglobin was also re-analysed and it was found that although the Hamiltonian is axial and the $S = \pm\frac{1}{2}$ state along is occupied there are additional complications due to the coupling of the iron spin with ligand nuclear moments. In order to reproduce the experimental spectrum it was necessary to introduce this influence by an interaction $H.g.S$

[257] J. C. Fanning, J. J. Jen, and A. J. Mouchet, *J. Inorg. Nuclear Chem.*, 1976, **38**, 645.

where H is the average magnetic field set up by the ligands as well as relaxation phenomena [66] (Figure 5).

The effect of low-intensity millimetre radiation on powdered samples of oxyhaemoglobin has been investigated. Before irradiation a symmetrical doublet spectrum was obtained indicative of low-spin ferrous iron but during irradiation a second asymmetrical doublet appeared characteristic of a high-spin ($S = 2$) ferrous complex. This behaviour was taken to indicate a weakening of the $Fe-O_2$ bond and a strengthening of the iron bond to nitrogen on a neighbouring histidyl-residue. The Mössbauer spectrum was seen to return to its original form after a few hours.[258]

A Mössbauer study has been made of aggregated and anhydrated sickle-cell haemoglobin and comparison of the Mössbauer spectra of deoxygenated sickle-cell haemoglobin (Hb-S) and normal haemoglobin (Hb-A) made in the temperature range 80—250 K. Over this range the spectra of anhydrated Hb-A and anhydrated Hb-S are identical within statistical error with approximately 50% of the high-spin irons in Hb-A and Hb-S being converted to haemochrome upon anhydration. This behaviour was indicative that similar conformational changes occur at the iron sites in both the normal and sickle cell haemoglobin. At 82 K the spectra consist of single quadrupole doublets ($\delta = 0.8$ mm s^{-1}, $\Delta = 2.0$ mm s^{-1}) corresponding to high-spin ferrous but there was also evidence for a second doublet ($\simeq$10%) ($\delta = 0.35$ mm s^{-1}, $\Delta = 1.1$ mm s^{-1}) in the Hb-S sample indicative of the presence of a low-spin ferrous porphyrin complex (haemochrome). The possibility that this is related to partial anhydration induced by hydrophobic aggregation of sickle cell haemoglobin molecule was discussed.[259]

A Mössbauer and e.s.r. study has been made of the 'E–S compound' of cytochrome *c* peroxidase in the temperature range 4.2—195 K and in external applied fields. At 4.2 mK in the absence of an external field, the spectrum was shown to be a quadrupole doublet ($\delta = 0.05$ mm s^{-1}, $\Delta = 1.55$ mm s^{-1}) with no evidence for an internal hyperfine magnetic field. The spectrum remained almost unchanged up to 195 K although a small change in isomer shift which was attributed to the second-order Doppler effect occurred. Application of an external field induced a large temperature-dependent internal magnetic field. The data were interpreted in terms of a model in which the compound retains a unit unpaired-spin which experiences a large crystal field splitting of axial or near to axial symmetry and a d^4 configuration of the iron.[260]

Mössbauer spectra of haem iron in various states of cytochrome P-450$_{cam}$ from the camphor-hydroxylating system of the bacterium *Pseudomonas putida* have been obtained. Native camphor-free P-450$_{cam}$ was found to contain low-spin Fe^{III}, 50—70% of which is converted to high-spin Fe^{III} on addition of camphor. It was demonstrated that the native low-spin state forms a complex with 2-phenylimidazole and that small changes in the Mössbauer spectrum resulted. Addition of *putidaroxin* complex caused further changes. The reduced form of P-450$_{cam}$ was shown to contain high-spin Fe^{II} both in the presence and

[258] N. D. Devyatkov, V. V. Khrapov, G. F. Bakaushina, A. M. Khrapkov, A. A. Levina, and A. P. Andreeva, *Doklady Akad. Nauk S.S.S.R.*, 1975, **225**, 962.

[259] C. S. Yen, B. H. Huynh, G. C. Papaethymiou, P. H. Swerdlow, J. L. Groves, C. S. Wu, and J. F. Bertles, *Biochim. Biophys. Acta*, 1976, **453**, 233.

[260] G. Lang, K. Spartalian, and T. Yonetani, *Biochim. Biophys. Acta*, 1976, **451**, 250.

absence of camphor. Complexes of reduced P-450$_{cam}$ with oxygen were found to be diamagnetic with unusual parameters for high-spin Fe^{II} suggesting the presence of a bound superoxide $Fe^{3+}O_2$. It was concluded that the systems are analogous to haemoglobin complexes.[261]

Further studies of iron in human transferrin have been made and the Mössbauer spectrum of human serum transferrin observed in high magnetic fields at temperatures as low as 175 mK. The spectra were interpreted on the basis of a rhombic spin Hamiltonian and high-spin ferric iron and yielded $D = 0.3\ cm^{-1}$ and $\lambda = 0.26$. Higher order effects in the spin Hamiltonian were investigated and used to explain the differences in Mössbauer and e.p.r. data.[262]

A Mössbauer and e.p.r. study has been made on protocatechnate 3,4-dioxygenase from *Pseudomonas aeruginosa*. Low-temperature Mössbauer studies of the native enzyme indicated that iron is present as Fe^{III} with $S = \frac{5}{2}$ and that $B_{int} = -525$ kG consistent with a rubredoxin-like ligand environment for iron. However, the dithionate-reduced enzyme has Mössbauer parameters unlike those of reduced rubredoxin. Studies of ternary complexes of the enzyme with 3,4-dihydroxyphenylpropionate and O_2 established that the iron atoms are high-spin Fe^{III} which have a negative zero field splitting; $D \simeq -2\ cm^{-1}$.[263]

Preliminary results have been reported on the characterization of Fe^{II} bovine carbonic and human carbonic anhydrate. It was discovered that the iron is low-spin and is bound to an enzymatically active site.[264]

A Mössbauer investigation of the endogenous iron content of human lungs has demonstrated the ability of the technique to distinguish inorganic iron from storage iron in various occupational groups.[265]

Low-spin and Covalent Complexes. An approximate method for the calculation of the Mössbauer spectra of low-spin Fe^{II} complexes has been given. The main features of the method involved evaluation of the entire e.f.g. tensor from the point charges in the lattice and the 3*d* crystal field orbital populations and subsequent diagonalization of this tensor. Applications of the method were made to two classes of low-spin ferrous compounds having either pseudo-D_3 or pseudo-D_{2h} symmetry and they indicated that the approach correctly predicts the sign of the quadrupole splitting for complexes of this type. Illustrative calculations were performed on model co-ordination polyhedra for compounds where geometrical and other changes in parameters are expected to be important in determining the sign of the quadrupole splitting.[266]

A number of addition complexes of hydro ferro- and ferri-cyanides have been prepared with organic bases and characterized by Mössbauer and other methods.[267–269]

261 M. Sharrock, P. G. Debrunner, C. Schulz, J. D. Lipscomb, V. Marshall, and I. C. Gunsalus, *Biochim. Biophys. Acta*, 1976, **420**, 8.
262 C. P. Tsang, L. Bogner, and A. J. F. Boyle, *J. Chem. Phys.*, 1976, **65**, 4584.
263 L. J. Que, I. D. Lipscomb, R. Zimmerman, E. Meunch, N. R. Orme-Johnson, and W. H. Orme-Johnson, *Biochem. Biophys. Acta*, 1976, **452**, 320.
264 A. M. Friedman, S. L. Ruby, J. C. Sullivan, B. J. Zabransky, D. Lawless, D. Basile, V. Ploplis, and C. A. Hoenich, *J. Inorg. Nuclear Chem.*, 1976, **38**, 1413.
265 L. Guest, *Ann. Occup. Hyg.*, 1976, **19**, 49.
266 S. C. Jackels, E. R. Davidson, and N. J. Rose, *J. Amer. Chem. Soc.*, 1976, **98**, 5826.
267 Y. P. Singh, W. U. Malik, B. K. Jain, and K. Chandra, *J. Inorg. Nuclear Chem.*, 1976, **38**, 1451.
268 W. U. Malik, R. Bembi, and P. P. Bhargava, *Indian J. Chem. (A)*. 1976, **14A**, 523.
269 Y. P. Singh, M. P. Singh, and W. U. Malik, *Current Sci.*, 1976, **45**, 241.

Frozen solutions of alkali metal and phosphonium cyanoferrates(III) $M_3Fe(CN)_6$ (M = Na^+, K^+, Bu_4P^+, or H^+) have been studied by Mössbauer n.m.r. infra-red spectroscopic techniques. In the frozen state it was shown that the alkali metal ions were associated with the cyanoferrate ions and that these interactions could be suppressed by prior addition of quaternary phosphonium salts or glycerol to the solutions.[270]

The thermal decomposition of $Na_2[Fe(CN)_5NO],2H_2O$ has been monitored[271] and the bonding in a series of substituted pentacyanoferrate(II) complexes $Na_2[Fe(CN)_5L]$ and $Na_3[Fe(CN)_6L]$ (L = monoprotonated diamine or a nitrogen heterocycle) studied by Mössbauer spectroscopy.[272]

The influence of alkyl and aryl groups attached to the di-imine ring of the ligand (3) on the properties of several low-spin tris-di-imineiron(II) complexes

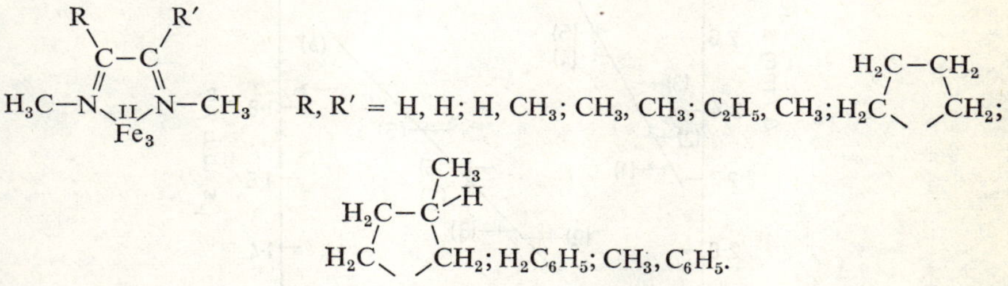

(3)

has been investigated by Mössbauer spectroscopy. The isomer shifts remain almost constant within the series at both 4.2 and 295 K but marked variations in the quadrupole splitting occur. The introduction of phenyl groups into the ligand causes an 80% increase in Δ. Measurements in applied fields at 4.2 K indicated that $V_{ZZ} < 0$ and $\eta > 0.4$ for the series. The results were discussed in relation to the electronic spectra of the compounds and it was concluded that the marked increase in quadrupole splitting is due to steric rather than π-back donation effects.[273] A number of complexes of the type $[FeL_3]X_2$ (L = 2,2′-bi-2-thiazole, or its 4,4′-dimethyl or 5,5′-dimethyl derivative; X = ClO_4^- or $FeCl_4^-$) have been studied by a variety of techniques including Mössbauer spectroscopy. The results indicate that the ligand is bonded to low-spin iron(II) through heterocyclic nitrogen rather than through sulphur atoms.[274]

Magnetic susceptibilities, electronic, infra-red, and Mössbauer spectra have been obtained for a series of complexes $Fe^{III}(qo)_3$, $Fe^{III}Cl(qo)_2$, and $M[Fe^{II}(qo)_3]$ (qoH = 1-nitroso-2-naphthol, 2-nitroso-1-naphthol, phenanthrene-9,10-quinone-mono-oxime, 5-methoxy-2-nitrosophenol, and 4,5-dimethyl-2-nitrosophenol) and it was shown that the iron atoms are low spin in all complexes.[275]

270 S. Papp, P. Kvintovics, F. Nagy, and E. Vertes, *Acta Chim. Acad. Sci. Hung.*, 1975, **87**, 343.
271 C. A. Broughton, G. R. Bradley, D. E. Matejczyk, G. R. Reed, B. J. McKinney, and J. C. Lombardi, *J. Inorg. Nucl. Chem.*, 1976, **38**, 427.
272 P. J. Aymonino, A. M. Blesa, J. A. Olabe, and E. Frank, *Z. Naturforsch., B, Anorg. Chem. Org. Chem.*, 1976, **31B**, 1532.
273 E. Baggio-Saitovich and M. A. Depaoli, *Inorg. Chim. Acta*, 1976, **17**, 59.
274 J. Nelson, S. M. Nelson, and W. D. Perry, *J.C.S. Dalton*, 1976, 1282.
275 J. Charalambous, M. J. Frazer, and R. Sims, *Inorg. Chim. Acta*, 1976, **18**, 247.

A magnetic susceptibility and Mössbauer study of tris-(*NN'*-dialkydithiocarbamate)iron(IV) complex $[Fe(S_2CNR_2)_3]^+(BF)_4^-$ has indicated it to be a d^4 complex. The magnetically perturbed spectra at 4.2 K were found to be consistent with an orbital singlet ground-state for iron.[276] A series of cationic six co-ordinate, low-spin bis[1,2-bis(dialkylphosphine)ethane]iron(II) complexes

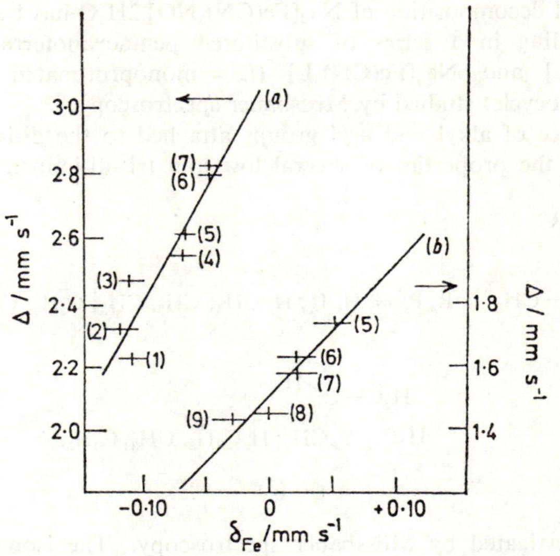

Figure 6 *Plots of quadrupole splitting against isomer shift for* (a) $[Fe(CO)_4L]$ *and* (b) $Hg[Fe(CO)_2(NO)L]_2$ *for various ligands* (L)
(Reproduced with permission from *J.C.S. Dalton*, 1976, 3220)

have been prepared and characterized by Mössbauer spectroscopy and other methods in order to elucidate their stereochemistry. By use of the partial quadrupole splitting approach and formulae appropriate to $FeABC_4$ complexes, it was possible to show that $[FeCl(NCPh)(depe)_2]^+$ (depe = $Et_2PCH_2CH_2PEt_2$) and $[FeCl(NO)(depe)_2]^{2+}$ are both *trans* and that $[Fe(CO)Cl(dmpe)_2]^+$ (dmpe = $Me_2PCH_2CH_2PMe_2$) exists as both *cis*- and *trans*- isomers.[277] A preliminary study of derivatives of the novel cations complex *trans*-[FeCl-(C=CHPh)(depe)$_2$]$^+$ has also been reported.[278]

Neutral diarsine complexes containing low-spin iron(II) of formula FeX_2-(diars)$_2$ (X = Cl, Br, I, or NCS) and FeClL(diars)$_2$Y (L = CO, P(OMe)$_3$, or P(OPh)$_3$, Y = $SnCl_3$ or BPh_4) have been assigned to *trans*- structures on the basis of this Mössbauer spectra which were analysed in terms of the point charge model. The partial isomer shift data for the series reveal that diarsine is an appreciably poorer σ-donor than analogous chelating phosphines, but a very similar σ-donor and π-acceptor to an *o*-phenanthroline.[279]

[276] C. A. Yong and B. W. Fitzimmons, *Chem. Phys Letters*, 1976, **39**, 476.
[277] J. M. Bellerby, M. J. Mays, and P. L. Sears, *J.C.S. Dalton*, 1976, 1232.
[278] J. M. Bellerby and M. J. Mays, *J. Organometallic Chem.*, 1976, **117**, C21.
[279] G. M. Bancroft and G. Cerantola, *Canad. J. Chem.*, 1976, **54**, 1285.

A series of compounds in which $Fe(CO)_2(NO)L$ (L = CO, tertiary phosphine, or arsine) are bonded to mercury, lead, or gold have been prepared and their Mössbauer spectra recorded. Large quadrupole splittings were observed which were consistent with a d^8 configuration of iron present in the mercury compounds $Hg[Fe(CO)_2(NO)L]_2$ and are indicative of a linear L—Fe—Hg—Fe—L system with two carbonyl groups and a nitrosyl group completing the trigonal pyramid about iron. It was found that the splittings change systematically with variation in the ligand and that there is also a smaller but systematic trend in isomer shift (Figure 6). Both of these trends are similar to those previously found for $Fe(CO)_4L$ complexes. If it is supposed that V_{ZZ} lies along the Hg—Fe—L direction and is positive, the quadrupole splitting is expected to increase with increasing π-acidity or decreasing σ-basicity of the substituent, whereas the isomer shift should increase with either property. The observed trends therefore suggested that σ-bonding effects predominate in this particular series.[280] It has been found that $Fe(CO)_5$, though a weak base, forms adducts with strong acid salts to form $[HFe(CO)_5]PF_6$, $[Cd\{Fe(CO)_5\}_2](AsF_6)_2$, $[Fe^{2+}\{Fe(CO)_5\}_2](SbF_6)_2$, and $[Sn\{Fe(CO)_5\}](SbF_6)_2$. It was shown that these compounds contain low-spin iron(II) in 6-co-ordination. The tin-containing compound is discussed further on p. 422.[281]

Treatment of thietes, which are four-membered, unsaturated, sulphur-containing [281] heterocyclics with metal carbonyls has been shown to lead to ring opening and the production of metal thioacroleins. A wide range of cobalt and iron containing compounds were prepared and, for example, compound (4) has Mössbauer parameters $\delta = 0.0806$ mm s^{-1} and $\Delta = 1.49$ mm s^{-1} comparable

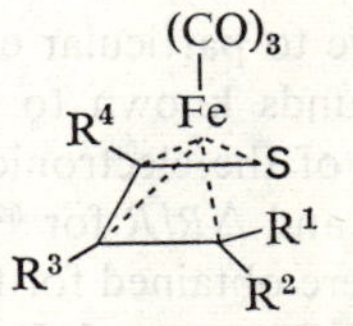

$R^1 = R^2 = R^3 = R^4 = H$

(4)

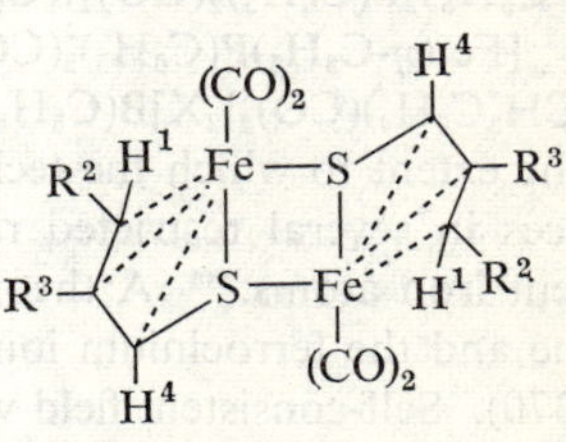

$R^2 = R^3 = H$

$R^2 = CH_3$; $R^3 = C_2H_5$

(5)

with analogous iron carbonyl–olefin complexes consistent with the presence of Fe^0. The equivalence of the iron atoms in the dimeric complex (5) was also established.[282]

An investigation has been carried out on the mechanistic aspects and on the products obtained from the halogenation of di-μ-alkylthio- and di-μ-arylthio-bis(tricarbonyliron) complexes and related substituted derivatives. Mössbauer spectra of bis-substituted bridging halogen derivatives typified by $[\{Fe(CO)_2[P(OMe)_3]SPh_2\}I](BPh_4)$ were shown to have simple quadrupole doublets consistent with the structure (6) where the iron atoms are electronically and

280 C. A. McAulliffe, I. E. Niven, and R. V. Parish, *J.C.S. Dalton*, 1976, 2477.
281 P. A. W. Dean, D. G. Ibbot, and G. M. Bancroft, *J.C.S. Chem. Comm.*, 1976, 2286.
282 D. C. Dittmer, K. Takahashi, M. Iwanami, A. J. Tsai, P. L. Chang, B. B. Bilder, and I. K. Stamos, *J. Amer. Chem. Soc.*, 1976, **98**, 2795.

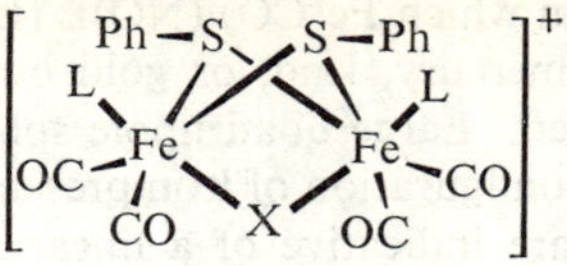

(6)

magnetically equivalent suggesting that the unpaired electron density is distributed equally over the iron centres. The Mössbauer parameters were found not only to be sensitive to the nature of the bridging SR unit, the ligand L, and the halogen X, but also to the nature of the counter-ion (e.g. BPh_4^- or PF_4^-) even though this ion lies beyond the immediate co-ordination sphere of the iron atoms. It was concluded that caution is therefore needed in comparing trends in data from a series of compounds which do not have the same counter-ion. The Mössbauer spectra of $[\{Fe(CO)_2I(SR)\}_2(Ph_2PCH_2PPh_2)]$ (R = Me or Ph) were broad singlets consistent with a dimeric structure where the R groups of the SR bridges are *syn* disposed.[283] In a related study a Mössbauer investigation has been made of a range of products obtained from the reactions of $Fe(p\text{-}C_5H_5)\text{-}(CO)_2$ and $Fe(\rho\text{-}C_5H_5)P(C_6H_5)_2(CO)_5$, *viz.* cis- and trans-$[Fe(\eta\text{-}C_5H_5)(CO)_2]_2$, $[Fe_2(\eta\text{-}C_5H_5)_2(CO)_3L]$ (L = $P(OCH_3)_3P(OC_2H_5)_3$, $P(O{-}Pr^i)_3$, $P(OC_6H_5)_3$, $P(C_2H_5)_3$, or $P(C_3H_7)_3$); $[\{Fe(\eta\text{-}C_5H_5)(CO)\}_2(C_6H_5)_2PRP(C_6H_5)_2]$ (R = CH_2, C_2H_4, or $N(C_2H_5)$); $[Fe_2(\eta\text{-}C_5H_5)P(C_6H_5)_2(CO)_6]$, $[Fe_2(\eta\text{-}C_5H_5)P(C_6H_5)_2\text{-}(CO)_5P(C_2H_5)_3]$, $[Fe_2(\eta\text{-}C_5H_5)P(C_6H_5)_2(CO)_4L_2]$ (L = $P(OC_6H_5)_3$, or $P(C_2H_5)_3$); $[\{Fe_2(\eta\text{-}C_5H_5)_2P(C_6H_5)_2(CO)_4\}(C_6H_5)_2PCH_2P(C_6H_5)_2]$ $[Fe_2(\eta\text{-}C_5H_5)P(C_6H_5)_2\text{-}(CO)_5]$, $[Fe_2(\eta\text{-}C_5H_5)P(C_6H_5)_2(CO)_4L]$ (L = $P(OCH_3)_3$ or $P(C_6H_5)_3$; and $[\{Fe(\eta\text{-}CH_3C_5H_4)(CO)_2\}_2X]B(C_6H_5)_4$ (X = Cl, Br, or I), in order to help define the extent to which the technique is sensitive to particular environmental differences in several restricted ranges of compounds known to contain non-equivalent iron atoms.[284] A theoretical treatment of the electronic structure of ferrocene and the ferrocinium ion has been made and $\Delta R/R$ for ^{57}Fe estimated (see p. 370). Self-consistent field wave functions were obtained for ferrocene and several low-lying states of the ferricinium ion using extended basis sets of Gaussian functions. The electronic structures were found to change very considerably when ferricinium is formed by removal of an Fe *d* molecular orbital of ferrocene. The charge on an Fe atom as determined by a Mulliken gross population average was calculated to be +1.47 for the ion and +1.39 for ferrocene and it was noted that this difference is less than 0.1 electrons but a value of close to 1 might be expected. The major contribution to the change comes from a greatly increased covalency of ligands of e_{1g} symmetry in the ion and this change is also responsible for the small difference found between the isomer shift of ferrocene and ferricinium salts. Direct contributions from the iron 4*s* electron were found to account for about one-half of the computed difference in electron density at the nucleus.[110] A theoretical study by the iterative extended Hückel molecular orbital method has been made, in order to examine the electromagnetic properties and to obtain the wave functions of

[283] J. Haines, J. A. de Beer, and R. Greatrex, *J.C.S. Dalton*, 1976, 1749.
[284] R. Greatrex and R. J. Haines, *J. Organometallic Chem.*, 1976, **114**, 199.

several low-lying states of bis(fulvalene)di-iron and its mono- and di-oxidized cations and of related biferrocene derivatives. These compounds are of particular interest because even though bis(fulvalene)di-iron and its ions are structurally closely similar to biferrocene and oxidized biferrocene they show no trapped valence effects resulting in inequivalent iron atoms but rather exact equivalence of the iron atoms. Comparing calculated quadrupole splitting estimated from a *d*-orbital population analysis with experiment, it was concluded that the cations of bis(fulvalene) di-iron must result from the ionization of an e_{2g} type orbital delocalized over two iron centres thus preserving the equivalence of the ions upon oxidation.[285]

A Mössbauer study of monoclinic single crystals of ferrocene has been made using polarized γ-radiation from a magnetized ^{57}Co–Fe source in order to throw more light on the behaviour of the molecule in the solid state. Analysis (see Vol. 9, p. 424, for further details) of Mössbauer spectra obtained of a matrix of single crystals of ferrocene with the polarized source magnetized parallel and perpendicular to the crystallographic *b* axis and observed along the *c* axis revealed that the e.f.g. tensor is axially symmetric with its principal axis coincident with the molecular axis. A low-temperature study demonstrated that the molecules gradually rotate by *ca.* 20° in the vicinity of a λ-transition at 164 K. A possible packing arrangement of the low-temperature phase was devised which accounts for many of the observed properties.[286] The ^{57}Fe Mössbauer spectra of a 3 : 1 thiourea–ferrocene clathrate has been shown to exhibit an unusual temperature-dependent relaxation of the e.f.g. tensor. This is caused by anisotropic re-orientation of the ferrocene molecules in the channels of the clathrate lattice. The molecular axis can lie either parallel or perpendicular to the channel axis. All molecules are effectively frozen in the lattice below 141 K. An inhomogeneous phase transition occurs between 141 and 153 K, above which there is a rapid reorientation about the channel axis of those molecules which lie perpendicular to it. However, those molecules lying along the channels remain frozen, and two distinct quadrupole doublets are seen. Above 200 K there is a progressive inward collapse of the spectrum as relaxation between the two molecular positions takes place with an activation energy of 15 kJ mol^{-1}, and the spectrum is almost fully narrowed at 295 K. The theoretical equations appropriate to anisotropic relaxation have been developed and allow a full interpretation of the data.[70] The molecular geometry of a series of trimethylene-bridged ferrocene derivatives (Figure 7) has been investigated by Mössbauer techniques. The spectra of all the compounds studied were quadrupole doublets and except for TKTMF significant decreases in isomer shift and quadrupole splittings occur as the number of bridges increases. For isomeric pairs the decreases are largest for non-adjacent bridges 13BTMF and 124TTMF. For TKTMF δ and Δ are higher than the values for three bridged isomers but intermediate between the values of the two bridged isomers. It was concluded that these results are best interpreted on the basis of non-planarity of the rings rather than tilting effects since loss of planarity would reduce metal ring bonding interactions and hence lower the

[285] R. F. Kirchner, G. H. Loew, and U. T. Meuller-Westerhoff, *Inorg. Chem.*, 1976, **15**, 2665.

[286] T. C. Gibb, *J.C.S. Dalton*, 1976, 1237.

metal *d*-orbital electron density leading to a reduction in isomer shift and quadrupole splitting.[287]

A Mössbauer investigation of the metallocenylboranes has shown that the quadrupole splitting decreases with increasing Lewis acidity of the dihaloborysyl group, whereas the isomer shift remains constant. The increase in quadrupole

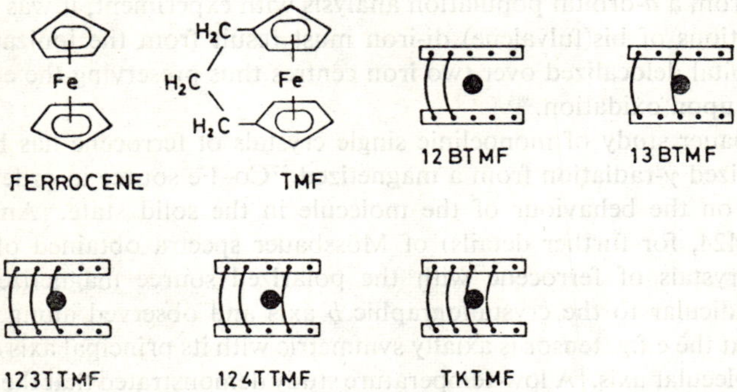

Figure 7 *Ferrocene and its poly-bridged trimethylene derivatives. In each case the rectangles represent the cyclopentadienyl rings, the small solid circles the carbon atoms of the ring, the large solid circles the iron atoms and the arcs the carbons bridges.* TMF, 1,1′-trimethyleneferrocene; 12BTMF, 1,1′,2,2′-bis(trimethylene)ferrocene; 13BTMF, 1,1′,3,3′-bis(trimethylene)ferrocene: 123TTMF, 1,1′,2,2′,3,3′-tris(trimethylene)ferrocene; 124TTMF, 1,1′,2,2′,4,4′,tris(trimethylene)ferrocene; TKTMF, 1,1′,2,2′,3,3′,4,4′-tetrakis(trimethylene)ferrocene

(Reproduced with permission from *J. Organometallic Chem.*, 1976, **117**, 55)

splitting was attributed to withdrawal of electron density from metal e_g-orbitals but this is compensated for by synergic bonding thus maintaining constancy of isomer shift.[288]

Oxide and Chalcogenide Systems containing Iron.—*Binary Oxides and Hydroxides.* From general considerations one may expect an increase of the e.f.g. in the vicinity of a surface. The existence of such an effect has been explored by an investigation of thin films of γ-FeOOH by detection of back-scattered conversion iron electrons in a β-ray spectrometer. Three samples of γ-FeOOH (enriched to *ca.* 25% ^{57}Fe) were prepared by electrodeposition, in the form of thin films of thickness 10, 20, and 50 Å on platinum substrates and were shown to have quadrupole splitting of 0.94 mm s^{-1}, 0.80 mm s^{-1}, and 0.73 mm s^{-1} compared to the bulk value of 0.55 mm s^{-1}. It was concluded that these data provide experimental evidence of the expected increase in e.f.g. in the vicinity of a surface.[289] A Mössbauer study has been made of the thermal decomposition of δ-FeOOH which was shown to lead eventually to α-Fe_2O_3 in the temperature range 80—600 K. It was further demonstrated that δ-Fe_2O_3 and a compound having a collinear ferromagnetic structure formed in the early stages of

[287] A. G. Nagy, I. Dezsi, and M. Hillman, *J. Organometallic Chem.*, 1976, **117**, 55.
[288] J. Pebler, W. Ruf, and W. Siebert, *Z. Anorg. Allgem. Chem.*, 1976, **422**, 39.
[289] A. Minkova and J. P. Schunk, *Doklady Bolgar. Akad. Nauk*, 1975, **28**, 1171.

decomposition.[290] Hydrogeothite has also been studied by Mössbauer spectroscopic techniques.[291]

Super-paramagnetism in small particles of α-Fe_2O_3 has been studied by use of Mössbauer spectroscopy and the spectra interpreted using the theory described on p. 366.[79] The magnetic structure of haematite containing small quantities of chromium atoms, $(Fe_{1-x}Cr_x)_2O_3$ ($0.005 < x < 0.02$) has been investigated by Mössbauer spectroscopy from 4.2 to 300 K. The Morin temperature (T_m) was found to decrease and broaden with increasing chromium content and, moreover, the iron atoms located near chromium impurities did not exhibit a Morin transition down to 4.2 K.[292]

Mössbauer evidence for the formation of superparamagnetic clusters of α-Fe_2O_3 within the lattice of $(1 - x)Al_2O_3xFe_2O_3$ ($x \simeq 29$ mol %) has been obtained,[293] the behaviour of $(V_2O_5)_{1-x}(Fe_2O_3)_x$ solid solutions investigated,[294] and the solubility of α-Fe_2O_3 in beryllium oxide aluminium oxide and synthetic chrysoberyl confirmed by Mössbauer spectroscopy.[295]

It has long been known that magnetite (Fe_3O_4) undergoes a metal insulator transition at the Verway point at 119 K. Traditionally it has been held that conduction occurs by electron hopping between neighbouring Fe^{2+} and Fe^{3+} on octahedral *B*-sites and a strong argument used in favour of this picture has been the broadening of the *B*-site lines in the Mössbauer spectrum of Fe_3O_4 when compared to the tetrahedral *A*-site. However, more recently, on the basis of a wider range of experimental evidence, it has been suggested that a band model of the conduction process in Fe_3O_4 might be more appropriate (see Vol. 9, p. 451). In an attempt to throw fresh light on this conflict a study of the linewidth of the *B*-site Mössbauer spectrum of an orientated Fe_3O_4 single crystal in the presence of external magnetic fields has been made at room temperature. The linewidth of the *B*-site spectrum was shown to be dependent on the orientation of the crystal with respect to the external magnetic field of 8 kG which was of sufficient magnitude to orientate the internal hyperfine field in any desired direction. For the magnetic fields along the crystallographic [001] direction the line broadening of the *B*-site sub-spectrum was completely suppressed (Figure 8). The broadening was therefore considered to arise from the combined effects of electric quadrupolar and magnetic dipolar effects as for H_{ext} along [100] and V_{ZZ} along [111], which is the local *B*-site symmetry axis of trigonal symmetry, these effects are expected to vanish because $3\cos^2\theta - 1 = 0$ (θ = angle between H and V_{ZZ}). It was concluded that these observations are more appropriate to a band model of conduction and are incompatible with a hopping model, and that any conclusions regarding the frequency of hopping derived from the temperature dependence of the linewidths of the *B*-site spectrum of Fe_3O_4 are unrealistic.[296] A Mössbauer

[290] V. A. Povitskii, E. F. Marakov, N. V. Murashko, and A. N. Salugin, *Phys. Status Solidi* (*A*), 1976, **33**, 783.
[291] T. S. Gendler, R. N. Kuz'min, and T. K. Urazaeva, *Kristallografiya*, 1976, **21**, 774.
[292] V. A. Povitskii, A. N. Salugin, E. F. Makarov, and Yu. V. Baldokhin, *Soviet Phys. Solid State*, 1976, **18**, 958.
[293] R. N. Tyagi and B. N. Srivastava, *Acta Cienc. Indica*, 1975, **1**, 331.
[294] E. Burzo and L. Staesco, *Solid State Comm.*, 1976, **20**, 653.
[295] A. P. Vinnikov, L. V. Zubenko, D. V. Balashov, and E. G. Semin, *Izvest. Akad. Nauk S.S.S.R., Neorg. Materialy*, 1975, **11**, 1860.
[296] A. M. Van Diepen, *Phys. Letters* (*A*), 1976, **57A**, 354.

study of slightly doped magnetite(Fe)($M_{0.05}Fe_{1.95}$)O_4 (M = Li^+, Ni^{2+}, Al^{3+}, or V^{3+}) has been made. Variations in the line-shape of the spectrum of the *B*-lattice were observed due to the response of the conduction electrons to the local perturbation caused by the coulomb potential of the impurity atoms. This causes both short-range spin and charge oscillation on the *B*-sublattice which

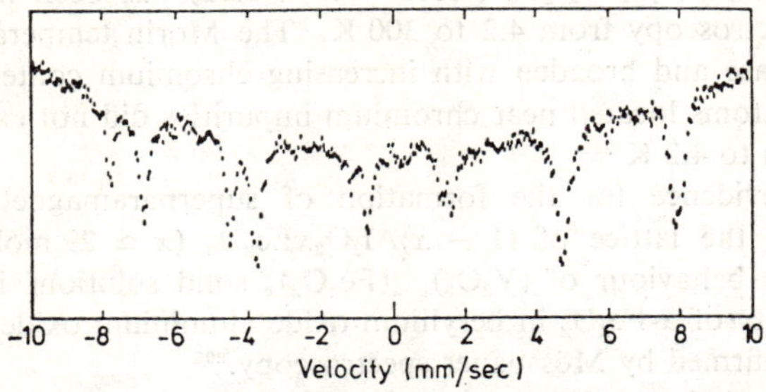

Figure 8 *Room temperature Mössbauer spectrum of* Fe_3O_4 *with the applied field in the* [001] *direction, showing that under well-chosen conditions the* A *and* B *sites give narrow lines of practically the same width*
(Reproduced with permission from *Phys. Letters* (*A*), 1976, **57A**, 354)

are interpretable in terms of a narrow band model for conduction in magnetite and may relate to the reason for the existence of the Verway transition in Fe_3O_4.[297] Non-stoicheiometry in magnetite has also been studied by Mössbauer and *X*-ray diffraction techniques.[298]

Mössbauer spectroscopy has been used to investigate the action of organic inhibitors in corrosion prevention. A hyperfine sextet due to δ-FeOOH was observed on a steel following immersion in dilute H_2SO_4 but not after pre-immersion of the steel in solutions of the organic inhibitors dibenzyl sulphoxide.[299] The oxidation of iron has been further studied by conversion electron Mössbauer spectroscopic and by γ-ray scattering techniques. Several oxide phases were identified, such as Fe_2O_3, Fe_3O_4, and FeO, and the complementary nature of the electron and γ-scattering techniques for the examination of oxide layers in the thickness ranges up to 300 nm and 10—20 μm respectively was emphasized.[300] Oxide films stripped from low carbon steels have also received attention.[301]

Spinel Oxides. The Mössbauer spectra of $Zn_xFe_{3-x}O_4$ (x = 0, 0.2, 0.4, 0.6, or 0.8) have been studied at 77 and 300 K and relaxation effects observed at room temperature far from the Néel point. It was considered that these effects could not be accounted for in terms of critical point fluctuations of the spin systems so far from T_N but simultaneous investigations of the magnetic relaxation phenomena showed that excited domain-wall oscillations occur in the series of compound with frequencies of comparable magnitude to the nuclear Zeeman frequencies thus providing an explanation for the relaxation phenomena observed in the Mössbauer spectra.[302]

297 C. Boekema, F. Van der Woude, and G. A. Sawatsky, *J. Phys.* (*C*), 1976, **18**, 2439.
298 K. Volenik, M. Seberini, and J. Neid, *Czech. J. Phys.*, 1975, **B25**, 1063.
299 I. Bibicu, G. Moisil, D. Barb, and M. Romanescu, *Rev. Roumaine Phys.*, 1975, **20**, 531.
300 A. S. Camera and W. Keune, *Corrosion Sci.*, 1975, **15**, 441.
301 K. Volenik, J. Girak, and M. Seberini, *Brit. Corros. J.*, 1975, **10**, 196.
302 C. M. Srivastava, S. N. Shringi, and R. G. Srivastava, *Phys. Rev.* (*B*), 1976, **14**, 2041.

It has been shown that ^{57}Fe impurity atoms doped into synthetic $MgAl_2O_4$ are equally distributed over octahedral and tetrahedral sites and display relaxed spectra with the tetrahedral sites having the longer relaxation time. Low-temperature spectra were analysed and yielded the following electronic zero-field splitting parameters ($D = 0.50 \pm 0.02\ cm^{-1}$ and $E = 0.130 \pm 0.003\ cm^{-1}$ for the *A*-site, and $D = -0.08 \pm 0.02\ cm^{-1}$ and $E = -0.012 \pm 0.002\ cm^{-1}$ for the *B*-site, indicating that both sites depart significantly from axial symmetry.[303] The quenching temperature dependence of the cation distribution and magnetic exchange interactions of $Fe_xMg_{1-x}(Fe_{2-x}Mg_x)O_4$ has been studied by Mössbauer spectroscopy and the experimental results satisfactorily accounted for by use of a molecular field model and an equilibrium distribution with proper inclusion of covalency effects.[304] The kinetics of the reaction between MgO and Fe_2O_3 between 573 and 1673 K have been studied by Mössbauer spectroscopy.[305]

In a Mössbauer study of the atomic distribution and magnetic structure of the spinel series $Fe_{1-x}V_{2-x}O_4$ ($0 \leqslant x \leqslant 1$) in the temperature range 80—500 K it has been demonstrated that both the Fe^{2+} and Fe^{3+} ion are present at *A*- and *B*-sites and that charge hopping occurs on the *B*-sites. The observed cation distribution was shown to predict the lattice constant, the oxygen parameter, the saturation magnetic moment, and the Curie constant.[306] In a related study ^{57}Fe impurity and ^{57}Co source experiments have been made of $Co_{1+x}V_{2-x}O_4$ ($0 \leqslant x \leqslant 1$) in the temperature range 80—500 K in order to investigate further the proposition that the daughter ion's valence is determined primarily by the chemical bonding properties of the iron in the environment into which it is created. It was found that in CoV_2O_4 ($x = 0$) doped iron replaced cobalt only at *A*-sites and that both source and absorber samples showed Fe^{3+} iron; in the remainder of the series except for Co_2VO_4 ($x = 1$) ^{57}Co at the *A*-site always decays to Fe^{3+} and is replaced by Fe^{3+} at both sites; and for Co_2VO_4 sources, ^{57}Co at *B*-sites decayed to 25% Fe^{2+} and 75% Fe^{3+} and doped iron predominantly occupied the *B*-site as Fe^{3+} with minor amounts of Fe^{2+} located on *A*-sites. It was concluded that in general the ^{57}Co at a given lattice site decays to the same charge state of iron as that taken on by the impurity atom but that detailed aspects of electronic behaviour following decay of ^{57}Co are not completely explanable in a simple way on the basis of the bonding properties of iron. Other features of the magnetic and electron structure of the cobalt–vanadium spinel were also discussed.[307]

Mössbauer spectra have been recorded for the spinel system $Co_xFe_{3-x}O_4$ ($0 \leqslant x \leqslant 2.76$) and were seen to fall into three classes depending on composition. Spinels for which ($0 \leqslant x \leqslant 1$) showed two sets of magnetically split spectra corresponding to exchanging Fe^{2+}–Fe^{3+} and to Fe^{3+} iron, for the range ($1 \leqslant x \leqslant 2$) magnetically split spectra corresponding to Fe^{3+} iron were observed and spinals ($x = 2.3$ and 2.5) contained only Fe^{3+} iron and were not magnetically ordered.[308] A selection of these spinels with $x = 1, 1.5, 2.0$, or 2.5 were further

[303] B. L. Dickson and K. K. P. Srivastava, *J. Phys. Chem. Solids*, 1976, **37**, 447.
[304] E. Degrave, C. Dauwe, A. Govaert, and J. de Sitter, *Phys. Status Solidi (B)*, 1976, **73**, 527.
[305] Yu. B. Voitkovskii, M. N. Shatalov, and Yu. S. Yusfin, *Izvest. Vysshikh Uchebn. Zavedenii, Chernaya Met.*, 1975, 27.
[306] J. Duk Lee and D. Schroeer, *J. Phys. Chem. Solids*, 1976, **37**, 739.
[307] J. Duk Lee and D. Schroeer, *J. Phys. Chem. Solids*, 1976, **37**, 747.
[308] P. J. Murray and J. W. Linnett, *J. Phys. Chem. Solids*, 1976, **37**, 619.

studied at 4.2 K and it was possible to distinguish between octahedral and tetrahedral ferric iron by application of a 60 kG transverse magnetic field to the absorber. The results were examined in terms of either site competition between Fe^{3+} and Co^{2+} ions for both sites and the relative merits of the two extreme models discussed.[309]

Magnetic and Mössbauer measurements have been made in the temperature range 77—300 K on polycrystalline YFe_2O_4. The room temperature Mössbauer spectrum was found to be a quadrupole doublet but at 90 K the spectrum of magnetically split (B_{int} = 460 kG). The experimental evidence was interpreted on the basis of the compound displaying 'induced parasitic ferromagnetism'.[310]

Mössbauer effect and other measurements have also been used in studying cation distribution and relaxation effects in: $Zn_{0.8}Co_{0.2}Fe_2O_4$,[311] $Cd_xFe_{1-x}Ni_x$-$Fe_{1-x}O_4$ ($0 \leqslant x \leqslant 1.0$),[312] $Mn_{1-x}Zn_xFe_2O_4$,[313] and $Zn_{0.3}{}^{2+}Mn_{0.7}{}^{2+}Mn_y{}^{3+}Fe_{2-y}{}^{3+}$-$O_4$.[314]

Other Oxides. A Mössbauer study has been made of fifteen silicate garnets between 4.5 and 681 K. Well resolved quadrupole doublets were observed and assigned to Fe^{2+} on dodecahedral sites and Fe^{3+} in octahedral and tetrahedral sites depending on the cation substitution at other sites. An analysis of the temperature dependence of the quadrupole splittings of the ferrous site revealed a large splitting of the lowest 3*d* state and magnetically perturbed spectra indicate $V_{ZZ} < 0$ for dodecahedral Fe^{2+} sites and andradite and $V_{ZZ} > 0$ for almadine. It was shown that the recoil free fraction of each site differs and this fact must be included in the determination of Fe^{2+}–Fe^{3+} ratios.[315] The temperature dependences of the *f*-fractions of the Mössbauer spectra of $Y_3Fe_5O_{12}$ and $Tb_3Fe_5O_{12}$ has been shown to possess an anomaly associated with a magnetic phase transition due to the exchange perturbation of the phonon spectrum.[316] The garnet $Ca_3Fe_{3.5}V_{1.5}O_2$ has also been studied by Mössbauer spectroscopy.[317]

Mössbauer studies of the defect perovskite system lanthanum sodium ferrate, $La_{1-x}Na_xFeO_{3-x}$, have revealed that a new hyperfine pattern appears in addition to that due to the perfect structure and this behaviour is attributed to a tetrahedral ferric iron site formed by creation of oxygen vacancies upon introduction of Na^+.[318]

The Mössbauer spectrum of $BiFeO_3$ has been shown to consist of superimposed magnetically split signals brought about by different quadrupole coupling constants of opposite sign.[319]

309 P. J. Murray and J. W. Linnett, *J. Phys. Chem. Solids*, 1976, **37**, 1041.
310 T. Suigimara, M. Abe, and T. Katsura, *J. Phys. Soc. Japan*, 1976, **40**, 1211.
311 M. K. Fayey, A. A. Bahgat, and Y. M. Abass, *Atomkernenergie*, 1975, **26**, 285.
312 N. A. Eissa and A. A. Bahgat, *Atomkernenergie*, 1976, **28**, 139.
313 E. V. Korneev, A. F. Belov, T. A. Khimich, and I. M. Kolesnikov, *Zhur. fiz. Khim.*, 1975, **49**, 2640.
314 R. G. Gupta, R. G. Mendiratta, and W. T. Escue, *NASA Tech. Memo*, 1975, NASA–TM–K–64961.
315 G. Amthauer, H. Annersten, and S. S. Hafner, *Kristallgeom. Kristallophys., Kristallochem.*, 1976, **143**, 14.
316 V. I. Nikolaev and V. S. Rusakov, *Zhur. eksp. teor. Fiz.*, 1976, **70**, 1949.
317 V. Belogurov, *Radiats. Fiz.*, 1975, **8**, 188.
318 H. Yamamura, S. Shirasaki, K. Taahashi, and K. Kakegawa, *J. Amer. Ceram. Soc.*, 1976, **59**, 533.
319 J. De Sitter, C. Dauwe, E. de Grave, and A. Govaert, *Solid State Comm.*, 1976, **18**, 645.

The M_2–M_3 phase transition in $V_{1-x}Fe_xO_2$ ($x = 0.024$) has been studied by Mössbauer spectroscopy.[320]

A comprehensive study of disordered and ordered scheelite-related bismuth ferrate molybdate $Bi_3(FeO_4)(MoO_4)_2$ has been made and it was shown that the phase formed from aqueous solution has iron and molybdenum cations randomally distributed over tetrahedral cation sites. It was discovered that this phase is metastable and undergoes a phase transition on heating to an ordered scheelite-like phase. The large isomer shift for tetrahedral sites ($\delta = 0.28$ mm s^{-1}) was seen to be due to long Fe—O distances in the tetrahedral sites.[321]

A Mössbauer, crystallographic, and magnetic study of hydrothermally synthesized barium ferrite $Ba_4Fe_9O_{20}H_6$ has shown that it is formulated as $Ba_4Fe^{III}(OH)_6Fe^{II}Fe_7{}^{III}O_{14}$ and is ferrimagnetically ordered below $T_N = 515$ K.[322]

The strontium europium ferrite $SrEu_2Fe_2O_7$ has been studied from 77 to 650 K and has $T_N = 544 \pm 2$ K. It was found that the values of the quadrupole could be satisfactorily calculated using a point-charge model, that V_{ZZ} is in the (110) plane, and that the direction of the magnetization is nearly perpendicular to V_{ZZ} and in the [110] direction.[323]

The spin-order and the magnetic properties of the hexagonal ferrite $BaZn_2Fe_{16}O_{27}$ have been studied by Mössbauer and magnetic measurements. The high saturation value of the magnetization at 0 K and at room temperature were explained by assuming a local reversal of the Fe^{3+} moments due to the perturbing action of Zn^{2+} ions.[324]

Spin-flopping has been observed in the antiferromagnetically ordered lattice of $K_{1+x}Fe_{11}O_{17}$ ($x = 0.35, 0.86$).[325]

An *X*-ray diffraction, high resolution electron microscopy, and ^{57}Fe Mössbauer study of the existence of iron tungsten bronzes in the WO_3–Fe systems has been made. The results showed that for Fe_xWO_3 with $x < 0.0193$ a bronze was formed with an orthorhombic or tetragonal structure but for $x > 0.0193$ $FeWO_4$ and reduced tungsten oxides formed.[326]

The site location of Fe^{2+} iron exchanged into *A*-zeolite and its interaction with ethylene have been studied in detail by Mössbauer spectroscopy.[327]

Minerals. The effects of heating and subsolidus reduction of lunar materials have been studied by Mössbauer, optical, magnetic, and *X*-ray methods. The changes produced by high-temperature reduction of two Apollo 17 soils led to an artificial maturation of the soil as indicated by enrichment of Fe^0 at the expenses of Fe^{2+} in silicates, enlargement of metal grains, partial recrystallization, and spectral darkening.[328] A study of the distribution of iron in carbonaceous meteorites has been made.[329]

[320] J. Pebler, K. Schmidt, and H. Wasinski, *Z. Naturforsch., B. Anorg. Chem., Org. Chem.*, 1976, **313**, 1367.
[321] W. Jeitschko, A. W. Sleigt, and W. McKlellan, *Acta Cryst.* (*B*), 1976, **B32**, 1163.
[322] K. Kitahama and R. Kiriyama, *Bull. Chem. Soc. Japan*, 1976, **49**, 2748.
[323] D. Hamzel, *Appl. Phys.*, 1976, **11**, 159.
[324] G. Albanese, M. Carbucicchio, and G. Asti, *Appl. Phys.*, 1976, **11**, 81.
[325] A. T. Howe and G. J. Dudley, *J. Solid State Chem.*, 1976, **18**, 149.
[326] I. J. McColm, R. J. D. Tilley, C. P. M. Barton, and N. N. Greenwood, *J. Solid State Chem.*, 1976, **16**, 265.
[327] L. V. C. Rees, Abstracts, 172nd ACS National Meeting, San Francisco, California, 1976.
[328] A. Brecher, W. H. Menke, J. B. Adams, and M. J. Gaffey, *Geochem. Comm. Chim. Acta*, 1975, **6**, 3091.
[329] G. P. Vdovykin, V. J. Grachev, T. V. Malysheva, and L. M. Satarova, *Geokimiya*, 1975, 1872.

Mössbauer spectroscopy has been applied to the study of age-dependent variation in minerals, for dating iron deposits, and for the evaluation of their thermodynamic conditions of formation with special reference to argillites and albitites.[330, 331] The thermal transformation of rocks in the overburden of the Ekibashaz coal deposit has also been studied.[332]

A Mössbauer investigation of the mechanism of the interaction of oleic acid with iron ores during flotation has been made.[333] The quantitative analysis of goethite and haematite [334, 335] containing ores, the rapid phase analysis,[336] and behaviour of nickel oxide ores during reductive roasting have been studied by Mössbauer spectroscopy.[337]

A Mössbauer study of olivine has indicated that the relationship between the Fe^{2+} site-occupancy factors and the d_{130} interplanar spacing should be used with caution.[338] Site-occupancy studies have been carried out on the pyroxenes, augite, and diopside [339] and on the tourmalines, elbaite, schorl, and dravite.[340] Optical absorption and Mössbauer studies have been used to determine the site populations in a suite of vesuvianites and it was found that in green vesuvianites 90% of iron is Fe^{3+} on Al–Fe positions but in other samples Fe^{2+} iron is on 5- or 6-co-ordinate sites depending on the titanium content.[341] Mössbauer site population analyses have been carried out on hornblendes [342] separated from apodiabasic amphibolites and minerals of the serpentine group including chrysotiles, lizardites, and antigorites.[343] A Mössbauer investigation of potash feldspars has shown that they contain ferric iron, in tetrahedral sites with $\delta = 0.21$—0.22 mm s^{-1} in contradiction to earlier reports.[344] The distribution of vacancies in α-leboite has been studied and the spectra of samples of various composition shown to consist of asymmetric doublets which can be interpreted in terms of the number and position of the vacancies in the iron first and second co-ordination sphere.[345] The interaction between γ-irradiation and octahedral iron in riebeckite has been shown to lead to ferric iron in the M1 and M2 positions whose parameters differ from the ferric iron produced by thermal

330 Kh. I. Amirkhanov, L. K. Anokhina, L. K. Palivoda, R. J. Chalabov, and R. U. Gabitova, *Doklady Akad. Nauk, SSSR*, 1975, **223**, 1218.
331 Kh. I. Amirkhanov and L. K. Anokhima, *Doklady Akad. Nauk S.S.S.R.*, 1975, **225**, 659.
332 V. P. Shomeiko and N. P. Goryonova, *Khim. Tovd. Topl. (Moscow)*, 1976, 148.
333 P. M. Soloshenkin, Yu. B. Voitkovskii, S. Sidorov, and G. G. Sidorenko, *Doklady Akad. Nauk Tadzh, S.S.R.*, 1976, **19**, 31.
334 A. Govaert, C. Dauwi, P. Plinke, J. de Sitter, E. de Grave, and G. Robbrecho, *Bull. Soc. chim. belges*, 1976, **85**, 251.
335 V. M. Zaporozhets, A. V. Polosin, T. V. Tkacheva, and F. G. Pasova, *Nov. Dannye Geol. Boksitov*, 1975, **2**, 68.
336 P. A. Ioffe, V. V. Klement'ev, L. Sh. Aleksandrova, L. Sh. Tsemekhman, M. I. Kazakov, S. E. Vaisburd, and B. F. Verner, *Izvest. Akad. Nauk S.S.S.R., Met.*, 1976, 12.
337 T. Ruskov and T. Tomov, *Iron Steel Int.*, 1975, **48**, 405.
338 J. Shinno, *Mem. Geol. Soc. Japan*, 1974, **11**, 11.
339 A. I. Bakhtin and R. A. Manapov, *Geokhimiya*, 1976, **3**, 433.
340 N. V. Gorelkiova, Yu. D. Perfil'ev, and A. M. Babeshkin, *Zap. Vses. Mineralog. Obshchestva*, 1976, **105**, 418.
341 P. G. Manning and M. J. Tricker, *Canad. Mineralogist*, 1975, **13**, 259.
342 B. A. Batievskii, V. P. Ivanitskii, A. L. Litvin, I. V. Matyash, and E. V. Polshin, *Geol. Zhur.*, 1975, **35**, 54.
343 T. V. Malsheva, V. I. Grachev, and I. Chaschukhin, *Geokhimiya*, 1976, 621.
344 H. Annersten, *Neues. Jahrb. Mineral. Montash.*, 1976, 337.
345 F. A. Sidorenko, B. V. Ryzhenko, V. Ya Elner, and R. V. Geld, *Zhur. eksp. teor. Fiz.*, 1976 **70**, 992.

oxidation.[346] Mössbauer effect studies of ^{57}Fe and ^{119}Sn in stannite, stannoidite, and maursonite have shown them to be formulated as $Cu_2^{1+}(Fe^{2+}, Zn^{2+})Sn^{4+}S_4^{2-}$, $Cu_8^{1+}Fe_2^{3+}(Fe^{2+}, Zn^{2+})Sn^{4+}S_{12}^{2-}$, and $Cu_6^{1+}Fe^{3+}Sn^{4+}S_8^{2-}$ respectively. The tin is in a strongly covalent tetrahedral site.[347] Franckeite and cylindrite [348] and the

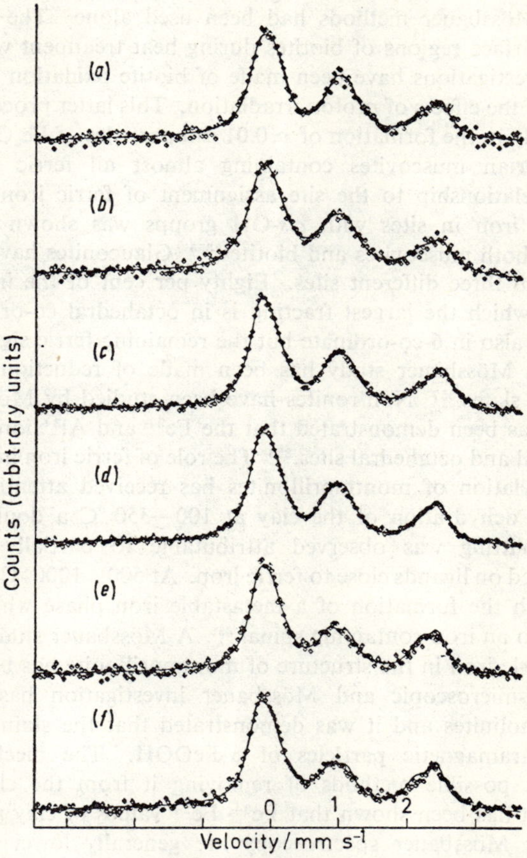

Figure 9 CEM *spectra of biotite heated at* 620 K. *The total time of heat treatment was* 114 (a), 90 (b), 78 (c), 60 (d), 43 (e), *and* 24 h (f)
(Reproduced from *J.C.S. Dalton*, 1976, 1289)

'*x*-phase' in the illmenite–haemitite series have also been studied by Mössbauer spectroscopy.[349]

A Mössbauer study has been made of sheet and boron containing silicates [350] and synthetic and natural biotites which demonstrated that the synthetic biotites

[346] V. P. Ivanitskii, I. V. Matyash, N. I. Polovko, and D. I. Shvets, *Dopovidi Akad. Nauk U.K.R., R.S.R., Ser. B*, 1976, 19.

[347] T. Yamanaka and A. Kato, *Amer. Mineralogist*, 1976, **61**, 260.

[348] R. A. Manapov and I. N. Penkov, *Doklady Akad. Nauk S.S.S.R.*, 1976, **228**, 181.

[349] J. L. Allen and P. N. Shive, *Geomagn. and Geoelectr.* 1974, **26**, 329.

[350] A. V. Astakhov, Yu. B. Voitkovskii, O. N. Generalov, and S. V. Sidorov, *Kristallografiya*, 1975, **20**, 769.

show ordering of Fe^{2+} iron on M2 octahedral sites.[351] ^{57}Fe CEMS has been shown to be a viable method in the study of the early stages of solid-state reactions which start in surface regions. Increases in the Fe^{3+} : Fe^{2+} ratio were detected in the outermost 300 nm of a biotite flake which had been subjected to mild air oxidation (Figure 9). These changes would have gone undetected if conventional transmission Mössbauer methods had been used alone. The reactions which occur in the surface regions of biotites during heat treatment were discussed.[352] Mössbauer investigations have been made of biotite oxidation and dehydroxylation [353] and of the effects of proton irradiation. This latter process was shown to result eventually in the formation of $\simeq 0.01$ μm particles of Fe_3O_4.[354] Mössbauer spectra of ferrian muscovites containing almost all ferric iron have been discussed in relationship to the site assignment of ferric iron in dioctahedral micas. Ferric iron in sites with *cis*-OH groups was shown to have similar parameters in both muscovites and biotites.[355] Glauconites have been shown to contain iron on three different sites. Eighty per cent of the iron present is as ferric iron of which the largest fraction is in octahedral co-ordination. Ferric iron is present also in 6-co-ordinate but the remaining ferric signal could not be assigned.[356] A Mössbauer study has been made of reduction spots in Welsh purple roofing slates.[357] Nontronites have been studied by Mössbauer spectroscopy and it has been demonstrated that the Fe^{3+} and Al^{3+} ions are distributed over tetrahedral and octahedral sites.[358] The role of ferric iron in the dehydration and dehydroxylation of montmorillonites has received attention. During the final stages of dehydration of the clay at 100—350 °C a doublet with a high quadrupole splitting was observed attributable to partially uncompensated charges localized on ligands close to ferric iron. At 500—1000 °C dehydroxylation took place with the formation of a metastable iron phase which subsequently recrystallized to an iron-containing spinal.[359] A Mössbauer study of the position and role of ferric iron in the structure of montmorillonite has been made.[360]

An electron-microscopic and Mössbauer investigation has been made of iron-stained kaolinites and it was demonstrated that the stain is composed of small super-paramagnetic particles of α-FeOOH. The mechanism of stain formation and possible methods of removing it from the clay surface were discussed.[361] It has been shown that Fe^{3+}–Fe^{2+} values of clay mineral fractions determined by Mössbauer spectroscopy are generally lower than the values determined by classical wet analyses.[362]

[351] H. Annersten, *Fortschr. Mineral*, 1975, 52.
[352] M. J. Tricker, A. P. Winterbottom, and A. G. Freeman, *J. Chem. Soc. Dalton*, 1976, 1289.
[353] V. P. Ivanitskii, A. M. Kalinichenko, I. V. Matyash, and T. P. Khomyak, *Geokhimiya*, 1975, 1864.
[354] A. Kotlicki, N. B. Olsen and J. S. Olsen, *Radiat. Eff.*, 1976, **28**, 1.
[355] B. A. Goodman, *Mineral. Mag.*, 1976, **40**, 513.
[356] A. Cimbalnikova, K. Raclausky, and J. Lipka in 'Proceedings of the 6th Conference on Clay Mineral Petroleum', 1973, ed. J. Karta, 1975.
[357] P. G. Manning, *Canad. Mineralogist*, 1975, **13**, 358.
[358] B. A. Goodman, J. D. Russel, A. R. Fraser, and F. W. D. Woodhams, *Clay and Clay Minerals*, 1976, **24**, 53.
[359] M. V. Eirish and A. A. Dvorecheskaya, *Geokhimiya*, 1976, 597.
[360] M. V. Eirish and A. A. Dvorecheskaya, *Geokhimiya*, 1976, 748.
[361] D. A. Jefferson, M. J. Tricker, and A. P. Winterbottom, *Clays and Clay Minerals*, 1975, **23**, 355.
[362] K. Punakivi, T. Ericsson, and R. Wappling, *Bull. Geol. Soc. Finl.*, 1975, **47**, 171.

A Mössbauer investigation has been made of transformations induced by firing of a clay fraction of Attic soil at temperatures of up to 1100 °C. It was shown that the unfired clay contained Fe_2O_3, β-FeOOH, and substitutional iron in the clay. At 200—400 °C β-FeOOH transformed to Fe_2O_3 and at 500—800 °C the clay mineral dehydroxylated. Finally, an iron-rich phase recrystallized. These changes were accompanied by an increase in the particle size of the original oxides during firing.[363] The mechanism of plant and animal fossilization has been investigated by Mössbauer spectroscopy. The samples originated from four sites near to the Mediterannean Sea in Egypt and Libya. The results indicate a two-stage process involving formation of an iron mineral which is subsequently transformed to α-Fe_2O_3 under oxidizing conditions.[3]

Chalcogenides. A Mössbauer effect study of nearly stoicheiometric FeS and $Fe_{0.93}S$ has been made between 80 and 600 K. The Mössbauer spectra of nearly stoicheiometric FeS indicate that the low-temperature 2*C* structure with spins parallel to the *c*-axis preserves the spin-orientation during the structural transformation to the 1*C* form at $T = 430$ K. However because of a miscibility gap, a few per cent of the high temperature 1*C* structure with spins perpendicular to the *c*-axis are present even at 300 K and the concentration of this component increases with temperature. At 445 K the spins in the 1*C* structure that are parallel to the *c*-axis rotate to an orientation perpendicular to the *c*-axis. In $Fe_{0.93}S$ an examination of the hyperfine splittings indicates that the spin rotation which occurs at 170 K from parallel to perpendicular to the *c*-axis is a first-order process.[364] The oxidation states of iron in pyrrhotite, troilite, and mackinwite have been elucidated by Mössbauer spectroscopy[365] and an investigation into phase transformations in highly ferruginous pyrrhotites carried out.[366]

Measurements of the quadrupole splitting of ^{57}Fe in the series of compounds FeS_2, $Fe_{0.01}Co_{0.99}S_2$, and $Fe_{0.01}Ni_{0.99}S_2$ which have either the pyrite and/or marcasite structures have been reported and the values correlated with those calculated on the basis of a point charge model. Deviations between experiment and theory were found and accounted for by inclusion of covalency effects.[367] Mössbauer data has been reported for iron pyrites.[368]

A theoretical study has been made of the bonding in metal–iron–sulphide compounds. By use of an empirical valence–bond length relationship, it was shown possible to relate the Mössbauer isomer shift to the electrostatic valence (V) by an equation of the form $\delta = 1.4—0.4V$ and hence further correlate δ with such properties as effective magnetic moments and electrical conductivity.[369]

A Mössbauer spectroscopic investigation of solid solutions of $RbFeS_2$ and $KFeS_2$ has been made and the magnetic transition temperature shown to decrease

[363] A. Simopoulos, A. Kostikas, J. Sigalas, N. H. Ganges, and A. Moukarika, *Clays and Clay Minerals*, 1975, **23**, 393.
[364] J. R. Gosselin, M. G. Townsend, R. J. Tremblay, and A. N. Webster, *J. Solid State Chem.*, 1976, **17**, 43.
[365] P. A. Ioffe, L. Sh. Tsemekhman, and S. E. Vaisbura, *Doklady Akad. Nauk S.S.S.R.*, 1976, **228**, 92.
[366] G. V. Novikov, V. K. Egurov, V. I. Popov, L. V. Sipavina, and G. Yu. Odinets, *Geokhimiya*, 1975, **12**, 1776.
[367] J. B. Ward and D. G. Howard, *J. Anal. Phys.*, 1976, **47**, 388.
[368] P. A. Montano and M. S. Seewra, *Solid State Comm.* 1976, **20**, 897.
[369] J. T. Hoggins and H. Steinfink, *Inorg. Chem.*, 1976, **15**, 1682.

linearly as the Rb mole percentage decrease. This was taken to suggest that the Fe—Fe distances between the FeS_2 chains is a linear function of rubidium content.[370] The preparation and characterization of copper iron germanium sulphide solid solution series $CuFe_xGe_{1-x}S_2$ $(0.5 < x < 1.0)$ having the chalcopyrite structure have been described and the materials shown to contain Fe^{II} and Fe^{III} on tetrahedral sites.[371]

The effect of small variations of the iron : chromium ratio on the Mössbauer spectra of $FeCr_2S_4$ has been studied and it was discovered that small deficits of iron had little effect but slight excesses of iron have a profound effect on the nature of the spectra. The Mössbauer spectrum of $Fe_{1.06}Cr_{1.94}S_4$ consists of a superposition of a singlet and doublet which was interpreted to be indicative of Fe^{2+} on *A*-sites some of which sites have a neighbouring Cr^{3+} on a *B*-site replaced by Fe^{3+}.[372] An extensive Mössbauer, *X*-ray, neutron, and magnetic study has revealed that the metal atoms in the cubic π-phases $(Fe, Co, Ni, M)_9S_8$ occupy 4-co-ordinated majority and 6-co-ordinated minority sites and that the natural π-(Fe, Ni, S) phases pentlandite display Pauli-paramagnetism down to 4.2 K with no resultant magnetic moment on the metal atoms and hence are broad band metals. The linear temperature dependences of the isomer shifts at the two types of site were interpreted in terms of the existence of a composite *s–d* conduction band and an analysis of the temperature variation of the quadrupole splitting at the majority site taken to indicate that only a few levels at the bottom of this band are occupied.[373] The structure, magnetic and electrical properties of $Ba_{15}Fe_7S_{25}$ and Ba_3FeS_5 have been studied by Mössbauer and other techniques and it was found that in both compounds the iron is tetrahedrally co-ordinated by sulphur. In Ba_3FeS_5 isolated tetrahedral were identified but in $Ba_{15}Fe_7S_{29}$ eight isolated trinuclear units, formed by a central tetrahedron sharing one edge and one corner with two terminal tetrahedra, and four isolated single tetrahedra were shown to be present. The effective magnetic moment of $Ba_{15}Fe_7S_{25}$ is 5.5 μB (at room temperature), which is the expected value for six Fe^{3+} and one Fe^{2+} ions but the Mössbauer spectrum showed signals with isomer shifts of 0.2 mm s^{-1}. This was taken to be indicative of the occurrence of electron delocalization across the edge-shared tetrahedra to give an average charge distribution of five Fe^{3+} and two $Fe^{2.5+}$ ions. The magnetic moment of Ba_3FeS_5 is 5.1 μB (at room temperature) demonstrating the presence of either Fe^{2+} or Fe^{4+}, but the Mössbauer spectrum consists of a quadrupole doublet indicative of Fe^{3+}. This observation was again reconciled by postulating that delocalization of electrons over the iron tetrahedra occurs.[374] A Mössbauer and *X*-ray diffraction study of $Fe_xCo_{1-x}Rh_2S_4$ $(0 \leqslant x \leqslant 1)$ has been carried out and the Mössbauer spectra shown to be complex.[375]

[370] H. P. Nissen and K. Nagorny, *Z. Phys. Chem.*, 1976, **99**, 151.
[371] J. Ackemann, S. Soled, A. Wold, and E. Kostiner, *J. Solid State Chem.*, 1976, **19**, 75.
[372] A. M. Van Diepen, F. K. Lotgering, and J. F. Olijhoek, *J. Magnetism Magn. Materials*, 1976, **3**, 117.
[373] O. Knop, C. Huang, K. I. G. Reid, J. S. Carlow, and F. W. D. Woodhams, *J. Solid State Chem.*, 1976, **16**, 97.
[374] J. T. Lemley, J. M. Jenks, J. T. Hoggins, Z. Eliezer, and H. Steinfink, *J. Solid State Chem.*, 1976, **16**, 117.
[375] H. Kondo, *J. Phys. Soc. Japan*, 1976, **41**, 1247.

Powdered samples of $(FeS)_{1-x}(FeSe)_x$ have been studied by Mössbauer magnetic and *X*-ray diffraction techniques, and although the samples apparently crystallize with the NiAs structure they contain increasing quantities of Fe_7S_8 together with compensating quantities of metallic iron as x increases.[376]

$Fe_{1.04}Se$ has been studied by Mössbauer spectroscopy and electrical conductivity measurements over a wide range of temperature. A structural phase transition was seen to occur at 731 K with the structure of $Fe_{1.04}Si$ changing from the anti-PbO to the NiAs structure type. The room temperature isomer shift and quadrupole splitting ($\delta = 0.47$ mm s^{-1}, $\Delta = 0.26$ mm s^{-1}) conform to known trends for related sulphides and tellurides. At 4.2 K the spectrum is not magnetically split and an application of a 50 kG field was shown to give no appreciable alignment of iron atom spins thus indicating that $Fe_{1.04}Se$ is a Pauli paramagnet.[377] The selenoferrates $MFeSi_2$ (M = K, Rb, or Cs) have been studied and shown to be linear antiferromagnets with weak interchain interaction similar to the corresponding thioferrate.[378]

A ^{57}Fe Mössbauer diffusional line broadening and neutron diffraction study has been made on $Fe_{1.33}Te$ which has the CdI_2 layer structure type. The broadened Mössbauer spectra were interpreted on the basis of rapid jumping of iron atoms both between and within the partly filled layers and D was determined to be *ca.* 10^{-12} m^2 s^{-1} at 903 K.[379]

5 Tin-119

General Topics.—A scintillation resonance detector has been described [380] for ^{119}Sn Mössbauer studies. The device is prepared by introducing lead dioxide into the molten scintillator.

The half-life of the $\frac{3}{2}$ level of ^{119}Sn has been measured [381] by delayed coincidence techniques using magnetic spectrometry associated with scintillation methods and scintillation spectrometry. The results obtained from these two measurements are 17.87 ± 0.11 and 18.48 ± 0.3 ns respectively. A weighted average of these values is taken and the conclusion is that the period of the $\frac{3}{2}$ level of ^{119}Sn is $T_{\frac{1}{2}} = 17.94 \pm 0.10$ ns. This compares with previous values of 17.75—22.4 by direct measurement and 23.2 from a Mössbauer experiment. The authors also compare the measured reduced $M1$ transition probability with single-particle and Kisslinger and Sorensen model predictions.

Some of the problems associated with the absolute determination of electron density changes from Mössbauer isomer shifts have been discussed by Kundig.[382] He makes use of data from three experimental techniques, conversion electron spectroscopy, lifetime measurements, and muonic isomer-shift studies. The influence of chemical environment on converted δ-decays was illustrated by lifetime and Mössbauer isomer-shift studies of the ^{119}Sn sources: $Sn(SO_4)_2$, SnS_2^-, β-Sn, $SnCl_2$, SnI_4, and SnS. Decay constants plotted against isomer shifts

376 M. Rea and C. A. Domenicali, *Solid State Comm.*, 1976, **20**, 325.
377 T. Tsuji, A. T. Howe, and N. N. Greenwood, *J. Solid State Chem.*, 1976, **17**, 163.
378 H. P. Nissen and K. Nagorny, *Z. Phys. Chem.*, 1976, **99**, 209.
379 A. T. Howe, P. Coffin, and B. E. F. Fender, *J. Phys.* (*C*), 1976, **9**, L61.
380 S. I. Bondarevskii and V. V. Eremin, *Pribory i Tekhn. Eksperim.*, 1975, 57.
381 M. C. Abreu and F. B. Gil, *J. Phys.* (*G*), 1976, **2**, 1976.
382 W. Kündig, *Hyperfine Interactions*, 1976, **2**, 113.

give straight-line relationships. The CNDO II MO method has been used to calculate the e.f.g. and $\Delta\psi^2(0)$, the charge density at the tin nucleus for SnX_4 (X = F, Cl, Br, and I) and SnX_6^{2-} (X = F, Cl, and I) compounds and the results are found to show good correlation with the experimental Mössbauer parameters.[383] Isomer shift and quadrupole splitting results were found to vary linearly with the calculated values for valence *s*-electron population and e.f.g. respectively. The correlation leads to a value of $\Delta R/R = 1.44 \times 10^{-4}$ for ^{119}Sn. A value of $(1.68 \pm 0.14) \times 10^{-4}$ for $\Delta R/R$ has been obtained from an isomer-shift calibration scale resulting from a comparison[384] of δ-resonance and *X*-ray electron spectroscopic data for some tin(II) compounds. Additive e.f.g. models, such as those used to predict trends in Mössbauer quadrupole splittings for Sn(IV), Sb(V), Ir(III), and low-spin Fe(I), have been extended[385] to n.q.r. data for ^{115}Ir(III) compounds and the results related to those for Sn(IV) compounds. Mössbauer isomer shifts have been used[386] in conjunction with shifts of $K_{\alpha 1}$ *X*-ray lines to study the nature of the chemical bonding in a variety of inorganic compounds including some tin-containing materials.

The Mössbauer effect has been measured[387] in amorphous and crystalline tin and in crystalline $Sn_{1-x}Cu_x$ ($x = 0.10$—0.18) over the temperature range 2.6—108 K. For thin absorbers the area under the spectrum is proportional to the Debye–Waller factor which increases when the amorphous phase changes to a crystalline one. The enhanced superconducting transition temperature in the amorphous state can be explained in terms of a change in electron–phonon interactions which in turn can be related to the measured Debye–Waller factors. The compositions and structures of highly dispersed tin prepared by electro-condensation in C_6H_6, $C_6H_6.CH_3$, $CH_3.CO$, and C_3H_7OH have been studied.[388] Isomer-shift data showed that the particles, which had diameters of about 75 Å, consisted of β-Sn and SnO_2 and that the composition was independent of the organic dispersing medium. The increase in isomer shift in going from α-Sn to β-Sn has been rationalized[389] in terms of a combination of three effects: (1) the increase in minimum bond length from 2.18 Å in α-Sn to 3.02 Å in β-Sn which would increase the *s*-electron density at the Sn nucleus in β-Sn, (2) the increased co-ordination from tetrahedral in α-Sn to distorted 6-co-ordination in β-Sn which would also result in an increased *s*-electron density and (3) the lower density (5.76 g cm^{-3}) of α-Sn compared with β-Sn (7.28 g cm^{-3}) would mean that solid-state band formation, and therefore loss of tin *s*-electron density to the band structure, would be more favourable in the β-modification. A variable temperature (4.2—295 K) Mössbauer study of the layer intercalation compound TaS_2,Sn has also been reported.[390] Isomer shift data ($\delta = 3.13$ mm s^{-1} at 78 K

[383] V. Kothekar, *Internat. J. Quantum Chem.*, 1976, **10**, 993.
[384] V. N. Gerasimov, V. M. Kuklakov, F. K. Chibirova, V. S. Shpinel, and V. I. Burlakov, *Zhur. eksp. teor. Fiz.*, 1976, **71**, 919.
[385] T. B. Brill, *Inorg. Chem.*, 1976, **15**, 2558.
[386] L. L. Markov, Y. M. Zaitsev, and Y. F. Batrakov, *Izvest. Akad. Nauk S.S.S.R., Ser. fiz.*, 1976, **40**, 399.
[387] J. Bolz and F. Pobell, in 'Proceedings of the 14th International Conference on Low Temperature Physics,' ed. M. Krusius and M. Vuorio, North Holland, Amsterdam, 1975, **2**, 425.
[388] A. S. Plachinda, O. I. Pisarenko, E. F. Makarov, and M. A. Lunina, *Zhur. fiz. Khim.*, 1976, **50**, 1832.
[389] J. Silver, C. A. Mackay, and J. D. Donaldson, *J. Material Sci.*, 1976, **11**, 836.
[390] R. H. Herber and R. F. Davis, *J. Chem. Phys.*, 1976, **65**, 3773.

and 3.06 mm s^{-1} at 295 K respectively relative to $Ba^{119}SnO_3$) and quadrupole splittings (Δ = 1.09 mm s^{-1} at 78 K and 3.06 mm s^{-1} at 295 K) are similar to the values for SnS (δ = 3.29, Δ = 0.91 mm s^{-1} at 78 K and δ = 3.23, Δ = 0.86 mm s^{-1} at 295 K) and show that Sn in TaS_2,Sn is formally Sn^{2+}. The Sn must therefore be acting as an electron donor to the TaS_2 lattice. It can be concluded from the angular dependency of the intensity ratio of the two components of the doublet spectrum at room temperature and from the insensitivity of this ratio to temperature that there is a single lattice site for Sn in the layer lattice. Figure 10 shows the difference in the spectra obtained from a

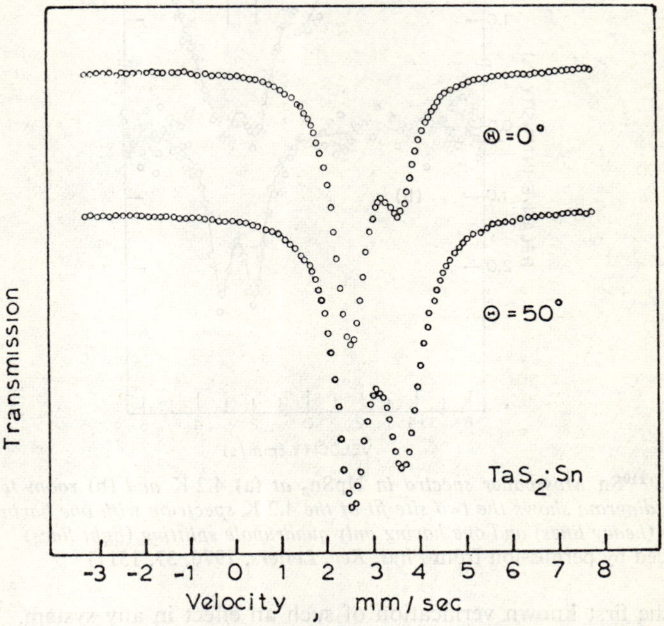

Figure 10 *Baseline corrected Mössbauer spectra of a single crystal of* TaS_2,Sn *at* 294 K. *The upper curve corresponds to parallel alignment of the optical and crystallographic* c *axes while the lower curve corresponds to an orientation angle of* 50° (Reproduced by permission from *J. Chem. Phys.*, 1976, **65**, 3773)

single crystal of TaS_2,Sn at two observational angles. From the temperature dependence of δ, it is inferred that the tin atom motion is that of a bare Sn atom in contrast to comparable data for SnS in which appreciable participation of the nearest-neighbour atoms in the motional behaviour of the Sn atom can be deduced. Using a simple Debye theory of solids model, the high-temperature region dependence of the ^{119}Sn recoil-free fraction gives effective lattice temperatures of θ_m = 176 ± 5 and 190 ± 5 K for TaS_2,Sn and SnS respectively.

A Mössbauer effect study has been carried out on strongly supersaturated solutions of Sn in germanium. Two phases are present in rapidly quenched samples of Sn(1—75 atom %) in Ge, *viz.* β-Sn containing 1% Ge and strongly supersaturated Sn in Ge. The Mössbauer spectra[391] consist of two lines, one corresponding to β-Sn and the other, with a shift close to α-Sn, due to the solid solution of Sn in the diamond-type Ge structure. Measurements of specific heat, Mössbauer effect, and electrical resistivity of $NpSn_3$ identify[392] its weak anti-ferromagnetism (T_N = 9.5 K) due to itinerant 5*f*-electrons. These results

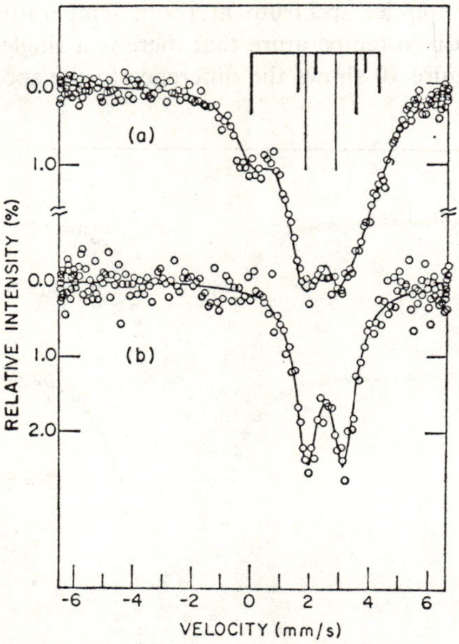

Figure 11 ^{119}Sn *Mössbauer spectra in* $NpSn_3$ *at* (a) 4.2 K *and* (b) *room temperature. The bar diagram shows the two site fit of the* 4.2 K *spectrum with one having magnetic splitting* (*heavy lines*) *and one having only quadrupole splitting* (*light lines*)
(Reproduced by permission from *Phys. Rev. Letters*, 1976, **37**, 1511)

provide the first known verification of such an effect in any system. The ^{119}Sn spectrum at room temperature is a simple quadrupole doublet but at 4.2 K the spectrum (Figure 11) shows the presence of two non-equivalent Sn sites, one with magnetic splitting and one with no magnetic splitting. This is typical of what happens in anti-ferromagnets where some sites may be surrounded by magnetic moments all oriented in the same direction whereas other sites are surrounded by equal numbers of oppositely directed spins. The electronic and magnetic properties of other ASn_3 compounds (A = La, Ce, Pr, Nd, Sm, Eu, Gd, and Yb) have also been investigated by ^{119}Sn Mössbauer studies.[393] The values of δ and Δ are nearly the same for all compounds and band structure calculations are consistent with this observation. Transferred magnetic fields

[391] V. I. Lisichenko, N. N. Petrichenko, and A. A. Yakunin, *Soviet Phys. Solids State*, 1976, **18**, 183.
[392] R. J. Trainor, M. B. Brodsky, B. D. Dunlap, and G. K. Shenoy, *Phys. Rev. Letters*, 1976, **37**, 1511.
[393] J. P. Sanchez, J. M. Friedt, G. K. Shenoy, A. Percheron, and J. C. Achard, *J. Phys.* (*C*), 1976, **9**, 2207.

and their orientations with respect to the e.f.g. axis at various Sn sites in the magnetically ordered compounds have been used to deduce simple magnetic structures (for A = Pr, Nd, and Sm), and to provide approximate descriptions of more complex magnetic structures (for A = Eu and Gd). The mechanism of spin polarization *via* the conducting electrons was also estimated from the magnitude of the transferred magnetic field. In particular the spin densities at Sn in $PrSn_3$ and $NdSn_3$ were found to be nearly equal. The hyperfine fields at Sn sites in the Heusler-type alloys Co_2VSn and Co_2NbSn have been estimated from Mössbauer effect measurements[394] at 4.2 K as +9.9 and +15.0 kOe respectively. These anomalously low values are discussed in terms of a change in sign of the polarization of the conduction bands. A temperature dependence for the isomer shift and relative recoilless fraction of the body-centred cubic compounds $V_3Ga_{1-x}Sn_x$ (x = 0.9, 0.5, and 0.1) has been observed;[395] δ is found to decrease with temperature. The point symmetry for the Sn atoms, which occupy cell corners and centres, is cubic with all nearest-neighbours being V atoms, but the resonance line is broadened because of (i) nuclear electric-quadrupole interactions resulting from composition fluctuation on the Sn–Ga sites and (ii) the presence of defects. A systematic investigation of the Mössbauer parameters of some intermetallic binary compounds and electroplated alloys of tin has been carried out.[389] Data for the binary alloys are interpreted on the basis of band theory and a tentative correlation between the observed parameters and the known properties of Ni–Sn, Cu–Sn, and Co–Sn electroplated alloys is made.

Two studies of tin-containing oxide glasses appeared in 1976. The Sn(IV) : Sn(II) ratio in silicate glasses was varied and determined by Mössbauer spectroscopy.[396] Isomer shift values were used to determine the co-ordination of Sn(II) and Sn(IV) sites. The authors also stated that spectral linewidth measurement can be used to distinguish between Sn in the vitreous state and Sn in compounds that have crystallized out from the glass. Results from a tin-containing glaze were used to illustrate this behaviour. The asymmetry of Sn(II) and Sn(IV) in alkali-free aluminoborosilicate glasses has been studied in the temperature range 80—295 K using the anisotropy of the Mössbauer effect probability.[397] The anisotropy of the Sn(IV) spectra is said to result from the non-equivalence of the bonds within a distorted octahedral environment and that of the Sn(II) spectra from the presence of a flattened SnO_4 grouping. It is difficult to imagine, however, how such groups could be aligned in a random glass structure to produce asymmetric spectra. The absolute values of the probability of resonance absorption for the Sn atoms in the glass were found to be 0.45 for Sn(IV) and 0.17 for Sn(II). Mössbauer spectroscopy has also been used to study the glass–crystal transition in the As_2Se_3–Sn system.[398] Introduction of 4—5 atom % of Sn into glassy As_2Se_3 alters the dependence of the electrical

394 I. Vincze, W. Koch, E. Baggio-Saitovich, and K. Endo, *Solid State Comm.*, 1976, **19**, 997.
395 C. W. Kimball, L. W. Weber, and F. Y. Fradin, *Phys. Rev. (B)*, 1976, **14**, 2769.
396 H. Dannheim, H. J. Oel, and G. Tomandl, *Glastech. Ber.*, 1976, **49**, 170.
397 G. M. Bartenev, G. M. Magomedev, A. D. Tsyganov, and Yu. V. Zalenev, *Fiz. Khim. Stekla.*, 1976, **2**, 114.
398 G. N. Belozerskii, V. N. Gittsovich, E. V. Shkolnikov, Z. V. Borisova, E. Y. Bessonova, and S. Z. Bashaikin, *Soviet Physics Solid State*, 1976, **18**, 184.

conductivity on the composition, probably due to the appearance of Sn(II) in the glass. At the beginning of the devitrification process the activation energy for electrical conduction falls, the room temperature conductivity increases ten-fold, and the conduction changes from p-type to n-type. Mössbauer spectra of $As_2Se_3Sn_x$ (x = 0.5, 0.26, and 0.41) were recorded for glasses prepared from specially pure components. The Sn(II) and Sn(IV) spectral lines observed have parameters similar to those of SnSe and $SnSe_2$ respectively. During isothermal recrystallization $SnSe_2$ is the predominant phase produced above 533 K and SnSe is the main phase formed below this temperature.

In another study,[399] the Mössbauer spectra of the As_2Se_3Sn phases were measured at 80 and 295 K. In the vitrification region of As_2Se_3 the spectra obtained were of a tin(IV) material with an isomer shift 0.2 mm s^{-1} different from the value for $SnSe_2$. Only at the vitrification boundary did a spectral line due to Sn(II) appear. Tin(II) was found to have a limited solubility (33 mole %) in vitreous As_2Se_3 because of the high ionicity of SnSe compared with $SnSe_2$. A decreased vitrification region results in substitution of $SnSe_2$ with SnSe.

The vibrational motion of Sn in the high critical-field superconductor $SnMoS_8$ has been studied[400] by ^{119}Sn Mössbauer spectra obtained as a function of temperature (4.2—200 K) and in an external magnetic field. The sample contained about 5% of β-Sn but its effect was stripped from the spectrum. Graphs of $K^2\langle z^2\rangle$ and $K^2\langle x\rangle^2$ against δ both show discontinuities at about 80 K. The results show (i) anisotropic mean square lattice displacements and a strong deviation from harmonic behaviour, (ii) a temperature-dependent line shift indicative of large anharmonicity and soft modes, and (iii) a possibly large contribution of soft vibrational modes to the mass enhancement and hence to the high observed superconducting transition temperature.

The ^{119}Sn Mössbauer parameters have been compared with the unit cell dimensions and minor element contents of some cassiterites[401] and data for the minerals stannite, stannoidite, and mawsonite have also been reported.[402]

There have been a number of publications describing Mössbauer measurements on various lattices doped with tin. In a study[403] of the antiferromagnetic layer phase Ca_2MnO_4 doped with $^{119}Sn^{4+}$, a hyperfine magnetic field (saturation value ≈ 38 kOe) was found at the Sn nucleus below the Néel temperature. A study of the temperature dependence of the field gave a value of β = 0.44 ± 0.10 assuming that the field obeys the power law $H(T) = H(0)D[1 - (T/T_N)]^{\beta}$ that is normally found for sub-lattice magnetization. The value of β is in reasonable agreement with values calculated from three-dimensional Heisenberg and Ising models (β = 0.31—0.38) and derived from neutron measurements (β = 0.38 ± 0.06). The Mössbauer spectra of the $^{119}Sn^{4+}$-doped antiferromagnetic, Cr_2O_3, have also been studied.[404] In the concentration range 1—8 atom % of doped Sn

[399] L. N. Vasilev and A. S. Bakhtiyarov, *Izvest. Akad. Nauk S.S.S.R., Neorg. Mater.*, 1975, **11**, 2074.
[400] C. W. Kimball, L. Weber, G. Van Landuyt, F. Y. Fradin, B. D. Dunlap, and G. K. Shenoy, *Phys. Rev. Letters*, 1976, **36**, 412.
[401] A. M. Clark, E. J. Fejer, J. D. Donaldson, and J. Silver, *Mineral. Mag.*, 1976, **40**, 895.
[402] T. Yamanaka and A. Kato, *Amer. Mineral.*, 1976, **61**, 260.
[403] F. Menil, R. Colmet, G. Le Flem, J. Portier, and P. Hagenmuller, *Solid State Comm.*, 1976, **18**, 1013.
[404] K. P. Mitrofanov, P. B. Fabrichnyi, E. V. Lamykin, A. M. Babeshkin, and L. P. Fefilatev, *Vestnik Moskov. Univ.* (*Fiz. Asronomiya*), 1975, **16**, 742.

impurity in In_2Te_3, the Sn atoms are found to be neutral.[405] This observation may explain the absence of impurity conduction in doped samples of In_2Te_3. The paper also contains an isomer-shift scale for tin compounds with tetrahedral co-ordination as a function of the effective charge of the valence *s*-electrons. Other tin-doped systems studied by their Mössbauer parameters are: the origins of hyperfine fields at ^{119}Sn impurities in $LaFeO_3$,[406] europium monochalcogenides doped with Sn(II),[407] and metal doping of zeolites.[408] There have also been reports of studies of tin-doped metal lattices.[409, 410] The temperature dependence of the electron density and magnetic hyperfine interactions at Sn impurity atoms in the Ni lattice have been studied near the critical temperature. The reasons for the sharp changes in δ with small changes of temperature in the vicinity of T_c are discussed.[409] A temperature-dependent resonance-line broadening was observed above T_c and ascribed to fluctuations in the matrix magnetic moment. For Ni containing 0.95 atom % of Sn, the magnetic hyperfine field in the temperature range $0.9678 \leqslant T/T_c \leqslant 0.9978$ was found to be proportional to $(1 - T/T_c)^\beta$ with $\beta = 0.625 \pm 0.04$. No peculiarities in the temperature dependence of the Mössbauer effect probability were observed in the vicinity of T_c. The effects of impurity ^{119}Sn atoms (<0.5 atom %) in Ge, Si, and A(III)B(V) semiconductors have been studied by Mössbauer spectroscopy.[410] Single-line spectra are found for Sn in Ge and Si and a quadrupole split doublet for Sn in A(III)B(V) phases such as InSb.

^{119}Sn Mössbauer emission studies have been carried out on ^{119}Sb atoms in SnTe [411, 412] and in SnSb.[412] The samples were irradiated with 16 MeV protons and the Mössbauer spectra obtained at 78 K. SnSb had an emission line with $\delta = 2.4$ mm s^{-1} corresponding to ^{119}Sn atoms replacing the Sb site while SnTe had two lines ($\delta = 2.3$ and 3.5 mm s^{-1}) assignable to ^{119}Sn in Te and Sn sites respectively. The energetic recoil of ^{119}Sb atoms produced by the proton reactions in SnTe was found to come to rest at both the Sn and Te sites of the matrix.

Measurement of the energy loss of electrons emitted during the absorption of γ-quanta has been used as the layer-depth criterion in a study of the structure of surface zones.[413] A thin layer of ^{119}Sn was deposited on the surface of Cu platelets 40 μ thick. Spectra of the annealed platelets show lines at 0.0, 1.8, 2.6, and 4.5 mm s^{-1}. The first three lines are due to SnO_2, Cu_3Sn, and β-Sn respectively and the fourth is said to be due to compounds arising from reaction with the oxidation products of diffusion pump oil. The intensity of the SnO_2 line decreases rapidly with depth while that due to Cu_3Sn increases with annealing time.

[405] V. Koshkin, E. E. Ovechkina, and V. P. Romanov, *Zhur. eksp. teor. Fiz.*, 1975, **69**, 2218.
[406] N. S. Ovanesyan and V. A. Trukhtanov, *Soviet Phys. Solid State*, 1976, **18**, 925.
[407] N. Bykovetz, *Diss. Abs. (B)*, 1976, **37**, 1763.
[408] V. N. Bogomolov, A. I. Zadorozhnii, and N. A. Klushin, *Fiz. Tverd. Tela*, 1975, **17**, 2452.
[409] N. N. Delyagin, Yu. D. Zonnenberg, and V. I. Nesterov, *Zhur. eksp. teor. Fiz.*, 1975, **69**, 1372.
[410] A. K. Kamalov, M. S. Yunusov, P. P. Seregin, M. A. Sagatov and F. S. Nasredinov, *Izvest. Akad. Nauk Uz. S.S.R., Ser. Fiz. Mat. Nauk*. 1976, 60.
[411] F. Ambe and S. Ambe, *Chem. Phys. Letters*, 1976, **39**, 294.
[412] F. Ambe and S. Ambe, *IPCR Cyclotron Prog. Rep.*, 1974, **8**, 108.
[413] A. Minkova, Le Val Shang, and Ts. Bonchev, *Fiz. Metal. i Metalloved.*, 1976, **41**, 983.

Tin(II) Compounds.—Mössbauer effect studies on a variety of tin(II) compounds including some organotin(II) compounds have been described.

Tin(II) hydroxide has been synthesized [414] by an anhydrous organometallic method involving exchange of groups between a triorganotin hydroxide and tin(II) chloride.

$$2R_3SnOH + SnCl_2 \longrightarrow Sn(OH)_2 + 2R_3SnCl$$

The amorphous product, which analysed as $Sn(OH)_2$, was characterized by its vibrational and Mössbauer spectra. The values of δ and Δ for $Sn(OH)_2$ at 77 K, 2.83, and 2.20 mm s^{-1} respectively, are within experimental error the same as those obtained previously for hydrous tin(II) oxide samples precipitated from aqueous solution with varying amounts of water, *viz.* $5SnO,2H_2O$ and $3SnO,H_2O$. There seems little doubt that the Sn environments in the hydrous oxide samples and in $Sn(OH)_2$ are similar. A combination of Mössbauer spectroscopy and *X*-ray electron spectroscopy has been used to obtain information [384] on the effective valence shell electron configurations of Sn in SnO, $SnSO_4$, SnC_2O_4, $SnCl_2,2H_2O$, and $SnC_4H_4O_6$. The *s*- and *p*-electron densities were determined by combining Mössbauer results and the differences in binding energy of the $3d\ (\frac{5}{2})$ and $4d$ Sn electrons determined by electron spectroscopy with the Hartree–Fock–Slater method of calculation.

The Mössbauer spectra for the phases found in the Eu_xWO_3–Sn_yWO_3 system in the isothermal section ($x = 0.05$—0.27, $y = 0.07$—0.2) show the presence [415] of both Sn^{II} and Eu^{II}. This is in marked contrast to the parent Eu bronze in which Eu^{III} was found. The change in oxidation state of the Eu must be due either to the presence of Sn^{II} or to the 12×3 tetragonal structure. The ^{119}Sn isomer shifts are in the range 3.53—3.73 mm s^{-1} and the quadrupole splittings have values of 1.05—1.39 mm s^{-1}. The quadrupole splittings observed for solid solutions of SnTe in SnS and SnSe have been found to decrease monotonically with increasing SnTe content,[416] while introduction of SnS or SnSe into cubic SnTe leads to a slight distortion of the structure and to a broadening of the singlet resonance line. The isomer shifts for the SnS–SnTe and SnSe–SnTe solid solutions also decrease monotonically from the SnS or SnSe values towards that for SnTe. In the binary Sn compounds with *p*-block elements including SnSe and SnTe there is an increase in isomer shift [389] with the increasing number of *p*-electrons available and the relevance of this observation to band population has been discussed.

Among the publications concerned with tin(II) halide derivatives was the report of an investigation [417] into the stannous chloride acceleration of the resin vulcanization of butyl rubber. The process depends upon the liberation of HCl on heating $SnCl_2$ with the rubber. The Mössbauer spectra of the products obtained when $SnCl_2$ is heated with carbon black show that the tin can be completely converted to SnO_2 by heating for a sufficiently long time or at a

414 W. D. Honnick and J. J. Zuckerman, *Inorg. Chem.*, 1976, **15**, 3034.
415 I. J. McColm, R. Steadman, C. S. Dimbylow, N. N. Greenwood, and G. E. Turner, *J. Solid State Chem.*, 1976, **18**, 161.
416 L. N. Vasilev, Y. V. Krylnikov, B. T. Melekh, T. V. Sinko, and P. P. Seregin, *Izvest. Akad. Nauk S.S.S.R., Neorg. Materialy*, 1976, **12**, 119.
417 D. Khristov and Ts. Bonchev, *Kautschuk Gummi. Kunstst.*, 1976, **29**, 116.

sufficiently high temperature. The accelerating effect of $SnCl_2$ on the vulcanization reaction was also found to become increasingly poor with increasing conversion of $SnCl_2$ to SnO_2. The products obtained from the molten and aqueous systems $SnF_2 : MI$ (M = Na, K, Rb, or NH_4) have been studied and Mössbauer data for the phases with 1 : 1, 2 : 1, and 3 : 1 $SnF_2 : MI$ ratios presented.[418] The spectra of the $MSnIF_2$ phases (M = K, Rb, or NH_4) show evidence of two Sn(II) sites. One of the sites has a doublet spectrum with parameters consistent with the presence of SnF_2I^- ions. The spectra for the other sites are single lines with larger isomer shifts indicative of Sn(II) in a higher symmetry environment. The relative intensities of the spectra from the two sites vary with temperature, suggesting a temperature-dependent reversible solid-state equilibrium between the two tin environments. The Mössbauer spectra for the MSn_2IF_4 compounds show the presence of only one Sn site and the data are consistent with the presence of $Sn_2F_4I^-$ ions similar to $Sn_2F_5^-$ but with an I bridging atom. All of the phases from the M : Sn(II) : I : F system are coloured and the reasons for these colours are discussed with reference to their Mössbauer data. Tri-tert-butylphosphine reacts with $SnCl_2$ and $SnBr_2$ to give the complexes $Bu^t{}_3P,SnX_2$ (X = Cl or Br). The Mössbauer spectra of the complexes[419] show the expected lower isomer shifts and larger quadrupole splittings compared with those for the parent tin(II) halides (*e.g.* for $Bu^t{}_3P,SnCl_2$ $\delta = 3.20$, $\Delta = 1.43$ mm s^{-1}). Reaction of $SnBr_2$ with diethylacetylene dicarboxylate (deadc) gives[420] a 2 : 1 complex $(SnBr_2)_2$(deadc) with parameters ($\delta = 3.65$ and $\Delta = 1.44$ mm s^{-1}) consistent with complex formation by Sn(II)—O bonding. Many reactions similar to those, however, result in the formation of tin(IV) compounds and are discussed in the following subsection.

Three papers on crystal structures of tin(II) chelate complexes published during the year also contained the Mössbauer data (Table 1) for the complexes.

Table 1 ^{119}Sn *Mössbauer data for tin*(II) *chelate complexes of known structure*

	δ* mm s^{-1}	Δ mm s^{-1}	*T*K	*Source*	Ref.
Sn^{II}(edda)	2.76	1.65	77	$BaSnO_3$	421
$Sn^{II}(S_2COMe)_2$	3.60	0.56	77	$BaSnO_3$	422
$Sn^{II}(dedtc)_2$	3.26	1.20	77	$BaSnO_3$	422
$Sn^{II}(dedtc)_2$	3.38	1.05	83	Pd/Sn	423
$Sn^{II}(dedtc)_2$	3.38	1.05	4.2	Pd/Sn	423

* Relative to Ba $^{119}SnO_3$.

All three structures have tin in distorted 4-co-ordinated environments with sterically active lone-pairs. The monoclinic crystals of tin(II) *NN'*-ethylenebis-(acetylideneimine), Sn(edaa), have Sn in distorted square-pyramidal sites[421] with two bonds to O (2.08—2.20 Å) and two to N (2.23—2.29 Å) and the fifth positions occupied by lone-pairs of electrons. The Sn atoms lie about 0.94 Å above the basal plane. The Sn—O bonds are very short compared with those

[418] J. D. Donaldson and J. Silver, *J. Solid State Chem.*, 1976, **18**, 117.
[419] W.-W.Dumont, B. Neudert, G. Rudolf, and H. Schumann, *Angew. Chem. Internat. Edn.*, 1976, **15**, 308.
[420] P. F. R. Ewings and P. G. Harrison, *Inorg. Chim. Acta*, 1976, **18**, 165.
[421] P. F. R. Ewings, P. G. Harrison, and A. Mangia, *J. Organometallic Chem.*, 1976, **114**, 35.

in most other compounds of known structure and this is reflected in the relatively low isomer shift. Sn(edda) is air-sensitive and undergoes oxidative hydrolysis which is shown by the appearance of a singlet resonance line at 0.12 mm s^{-1} in the Mössbauer spectrum. Monoclinic tin(II) bis(*o*-methyl dithiocarbonate), $Sn(S_2COMe)_2$, has Sn chelated by two ligand groups resulting in a pseudo-trigonal bipyramidal arrangement of electron-pairs.[422] The Sn—S (axial) bond distances [2.817 (4) Å] are longer than the Sn—S (equatorial) distances [2.623 (3) Å] and the remaining equatorial position is occupied by a sterically active lone-pair. The S—Sn—S angles involving axial and equatorial Sn—S bonds are 140.8 (1) and 99.3 (1)° respectively. The Mössbauer data for the complex are compared in Table 1 with those of the related compound bis-(diethyldithiocarbamato)tin(II), $Sn(dedtc)_2$, which is also of known structure. The higher δ and lower Δ for $Sn(S_2COMe)_2$ are said to reflect, in valence-bond terminology, a greater contribution of the canonical form $[Sn^{2+}]2[S_2COMe]^-$ or, in molecular orbital terms, a greater localization of the lone-pair in the 5*s* orbital for $Sn(S_2COMe)_2$. These two compounds do, however, emphasize a point made by the authors and by the Reviewer on many occasions that for tin(II) chemistry differing Mössbauer parameters, particularly differing quadrupole splittings, do not necessarily exclude isostructurality and *vice versa.* The synthesis, crystal structure and Mössbauer spectrum of $Sn(dedtc)_2$ has also been reported by Hoskins *et al.*[423] who also find that the Sn atoms are in a pseudo-trigonal bipyramidal environment of electron-pairs with two long (axial) Sn—S distances of 2.78 and two short (equatorial) Sn—S distances of 2.57 and 2.58 Å. Although the Mössbauer parameters reported by the two sets of workers differ (Table 1) the trends discussed by Ewings, Harrison, and King[422] remain correct. The spectra of three other sulphur-chelated tin(II) compounds have also been obtained[423] at 83 K relative to a Pd–^{119}Sn source, *viz.* bis(ethylxanthate)tin(II) for which δ = 1.93 and Δ = 0.75 mm s^{-1}; tetramethylammonium triethyl-xanthatostannate(II) for which δ = 2.10 and Δ = 0.80 mm s^{-1}; and bis-(diethyldithiophate)tin(II) for which δ = 2.29 and Δ = 0.94 mm s^{-1}. Hoskins *et al.*[423] have also produced a graph of the isomer shifts, relative to Pd–Sn for selected tin(II) compounds against quadrupole splitting values but there seems little value in attempting, as they have done, to divide a set of *selected* tin compounds into two categories according to the value of their quadrupole splitting parameter. They do, however, make the valid point that while the quadrupole splitting for tin(II) compounds is, in general, consistent with a sterically active lone-pair of electrons, its magnitude appears to be more sensitive to the nature of the ligand than to tin stereochemistry. In a study of complex formation between tin(II) compounds and tetracyanoethylene, 7,7,8,8-tetracyano-quinodimethane (tcnq), 2,3,5,6-tetrachlorobenzoquinone (tcbq), and tetracyano-ethylene (tcne), it was found[424] that [Sn(tcne)$_2$,thf] and the tcnq complexes of tin(II) bis(β-ketoenolates) exhibited resonances due to both Sn^{II} and Sn^{IV}. The shifts and splittings in the Sn^{II} region of the spectra were in the range 3.15—3.39 and 1.78—2.10 mm s^{-1} respectively. The single-line Sn^{IV} spectra for the

[422] P. F. R. Ewings, P. G. Harrison, and T. J. King, *J.C.S. Dalton*, 1976, 1399.
[423] B. F. Hoskins, R. L. Martin, and N. M. Rhode, *Austral. J. Chem.*, 1976, **29**, 213.
[424] A. B. Cornwell, C. A. Cornwell, and P. G. Harrison, *J.C.S. Dalton*, 1976, 1612.

β-ketoenolates had shifts of 0.05—0.19 mm s^{-1} and the parameters found for the Sn^{IV} site in [Sn(tcne)$_2$,thf] were δ = 0.42 mm s^{-1} and Δ = 1.09 mm s^{-1}. It is suggested that the Mössbauer and i.r. data for these complexes are consistent with linear chain structures in which the organic molecule bridges adjacent Sn atoms. The observed decrease in the recoilless fraction of the complexes as the bulk of the β-ketoenolate group increases can also be rationalized in terms of these structures, since the Sn^{II} site with its sterically active lone-pair would be very susceptible to perturbations arising from the bulky ligands. The Mössbauer isomer shifts of the complexes [Co(CO)$_4$(SnX)] where X is a β-ketoenolate residue have shifts (3.35—3.68) in the tin(II) range above those of the corresponding SnX_2 compounds.[425] Since the formulation of the complexes cannot, therefore, involve direct Sn—Co bonding, which would lead to a decrease in shift of >1.5 mm s^{-1}, structures (7) involving $Co(CO)_4^-$ anions co-ordinated to

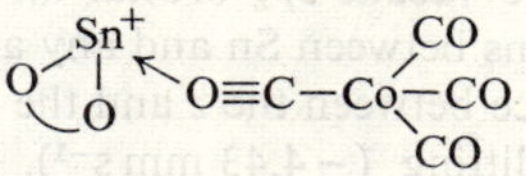

(7)

$[SnX]^+$ cations *via* carbonyl oxygen are proposed. A variable-temperature (77—295 K) Mössbauer study has been reported[426] along with a crystal and molecular structure determination of the mixed valence compound, octakis-μ-(*o*-nitrobenzato)-di-μ_3-oxo-bis(tetrahydrofuran) ditin(II)-ditin(IV). The different nature of the bonding of the Sn^{II} and Sn^{IV} atoms in the structure is reflected in the temperature-dependent recoilless fractions of the tin atoms. For the Sn^{II} site, the isomer shift varies from 3.60 mm s^{-1} at 77 K to 3.91 mm s^{-1} at 295 K, the quadrupole splitting from 1.82 to 2.13 mm s^{-1} and the resonance peak area from 15.46 to 1.07 over the same temperature range. The changes for the Sn^{IV} site are δ, 0.07—0.02 mm s^{-1}; Γ of singlet 1.18—1.05 mm s^{-1} and area 22.29—2.32. The recoilless fractions are unequal implying significantly different lattice vibrational spectra for the sites. The Debye temperatures determined from the data for the Sn^{II} and Sn^{IV} atoms, 104.1 and 119.6 K respectively, are in accordance with the crystal structure which shows that the Sn^{IV} atoms are more strongly held in the lattice.

The Mössbauer parameters for a number of organometallic tin(II) derivates have been reported. The data for Sn[CH(SiMe$_3$)$_2$]$_2$ [427] are δ = 2.16 mm s^{-1} and Δ = −2.31 mm s^{-1}. The isomer shift lies only just above the value for α-tin (2.05 mm s^{-1}) which should be taken as the borderline between Sn^{II} and Sn^{IV} and not the β-Sn value used by some workers. A magnetic Mössbauer measurement on the compound showed that V_{zz} was positive. The dimeric structure (9) in which there are weak Sn–Sn interactions between the lone-pair orbitals on the Sn atoms with empty orbitals on adjacent Sn atoms is consistent with the Mössbauer data (V_{zz} lying along the direction of the weak Sn–Sn interaction) and with its known crystal structure and magnetic susceptibility data. The crystal

[425] A. B. Cornwell and P. G. Harrison, *J.C.S. Dalton*, 1976, 1608.
[426] P. F. R. Ewings, P. G. Harrison, A. Morris, and T. J. King, *J.C.S. Dalton*, 1976, 1602.
[427] J. D. Cotton, P. J. Davidson, M. F. Lappert, J. D. Donaldson, and J. Silver, *J.C.S. Dalton*, 1976, 2286.

structure and Mössbauer parameters also distinguish $Sn[CH(SiMe_3)_2]_2$ from formally similar compounds of the type $Sn(C_5H_5)_2$, $Sn(C_5H_5)Cl$, and $(SnBu^n{}_2)_n$. The cyclopentadienyl compounds have no Sn—Sn bonding and have isomer shifts characteristic of tin(II). $(SnBu^n{}_2)_n$ and similar compounds, on the other hand, have strong Sn—Sn bonds and contain quadrivalent tin. The zero quadrupole splitting for $(SnBu^n{}_2)_n$ is characteristic of tin(IV) compounds containing both Sn—C and Sn—Sn bonds and is in sharp contrast to the splitting of 2.31 mm s^{-1} found for $Sn[CH(SiMe_3)_2]_2$ which has only weak Sn—Sn interactions. The quadrupole splitting values recorded for 3-co-ordinate $Sn[CH(SiMe_3)_2]_2$-transition metal complexes are much larger [427] than that for the parent dialkyl derivative and in fact are among the largest recorded for organotin systems. A magnetic Mössbauer study of $[Cr(CO)_5\{Sn[CH(SiMe_3)_2]\}]$ has been carried out and like $Sn[CH(SiMe_3)_2]_2$ has V_{zz} positive. As before the direction of V_{zz} can be identified with that of the vacant $5p_z$ orbital on tin. In this case, however, there are no weak interactions between Sn and any atom in the *z* direction. This produces a larger *p*-imbalance between the *z* and the *x* and *y* directions and leads to a larger quadrupole splitting (-4.43 mm s^{-1}). The isomer shifts for the

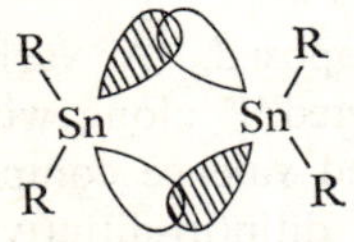

(8)

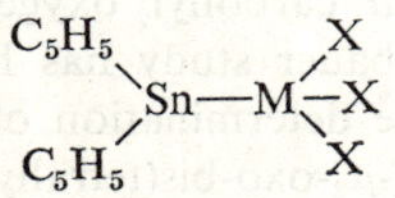

(9)

stannylene–Cr compound and for four similar complexes do not differ markedly from that for $Sn[CH(SiMe_3)_2]_2$ showing that the *s*-densities are similar in every case. Moreover, no significant change in δ value is found in going from mono- to di-substituted products. The Mössbauer spectrum of $[PtCl(PEt_3)Sn$-$[CH(SiMe_3)_2]Sn[CH(SiMe_3)_2]_2Cl$ proved useful in characterizing this complex. Two sets of parameters were identified: (i) a set with $\delta = 2.05$ and $\Delta = 4.23$ characteristic of a co-ordinated $Sn[CH(SiMe_3)_2]_2$ group and (ii) a set with $\delta = 1.73$ and $\Delta = 2.66$ mm s^{-1} characteristic of a Sn^{IV} environment containing an Sn-transition metal bond. On the basis of ^{57}Fe data and a ^{119}Sn isomer shift ($\delta = 3.76$ and $\Delta = 1.40$) in the Sn^{II} region, it has been stated that $[Sn\{Fe(CO)_5\}]$-$(SbF_6)_2$ is best regarded as a complex of Sn^{IV} in which the $Fe(CO)_5$ group is behaving as a weak Fe-donor ligand.[428]

The reactions of dicyclopentadienyl tin and bis(methylcyclopentadienyl) tin with some B and A1 trihalides give complexes of composition R_2Sn,MX_3 ($R = C_5H_5$, $MX_3 = BF_3$, BBr_3, $AlCl_3$, or $AlBr_3$ and $R = MeC_5H_4$, $MX_3 = AlCl_3$). The ^{119}Sn Mössbauer spectra of the complexes consist [429] of broad single resonances with shifts in the range 3.71—3.83 mm s^{-1}. These values are very similar to those of the free stannocenes confirming the divalent nature of the tin in the complexes and excluding the possibility of insertion of the stannylene into the group(III) metal–halogen bond to form Sn^{IV}. The rather large linewidths

[428] P. A. W. Dean, D. G. Ibbott, and G. M. Bancroft, *J.C.S. Chem. Comm.*, 1976, 901.
[429] P. G. Harrison and J. A. Richards, *J. Organometallic Chem.*, 1976, **108**, 35.

(Γ = 1.24—2.08 mm s^{-1}) suggest the presence of unresolved quadrupole splittings. Both the Mössbauer and i.r. data for the complexes are consistent with the formation of Lewis base–Lewis acid complexes with structures (9) retaining the geometry of the $(C_5H_5)_2Sn$ residue. The Mössbauer data for the 1 : 1 complex of cyclopentadienyl chloride with $AlCl_3$ are also given and discussed. With BCl_3, $SnBr_4$, and $FeCl_3$ exchange of C_5H_5 groups for halogen takes place and the Mössbauer spectra of the products show mixtures of both Sn^{II} and Sn^{IV} resonances. The complexes of $[Sn(C_5H_5)_2]$ with tcne and tcnq have[424] isomer shifts of 3.03—3.84 mm s^{-1} and quadrupole splittings of 1.30—2.37 mm s^{-1}, showing that the Sn is present as Sn^{II}. The data suggest that the compounds should be formulated as charge-transfer complexes involving the π-electron system of the C_5H_5 ring. With tcne this interaction causes some electron drift away from tin as shown by a decrease in the isomer shift from 3.84 in $[Sn(C_5H_5)_2]$ to 3.03 and 3.12 mm s^{-1} in the complexes. At the same time the quadrupole splitting increases from 0.84 to 2.37 and 2.26 mm s^{-1}. The interaction with tcnq is much weaker and although the quadrupole splitting increases the isomer shift is unchanged. The product of the reaction between $Sn(C_5H_6)_2$ and $(C_5H_5)W(CO_3)H$ was formulated as tin(II) derivative $Sn[W(CO)_3(C_5H_5)]_2$ on the basis of its Mössbauer data.[430]

Papers dealing with the following compounds $SnCl_2$,[382] SnS,[382] Ta_2S,Sn,[390] SnTe,[411, 412] As_2Se_3Sn,[398, 399] silicate glasses,[396] borosilicate glasses,[397] and Sn-doped Eu chalcogenides[497] that have resonances in the tin(II) region of the isomer shift range have been discussed earlier in this chapter.

Inorganic Tin(IV) Compounds.—The Mössbauer spectra of ferrocyanides and of stannates in glycerin at 200—280 K have been reported and the nature of the resonance lines said to provide information on the homogeneity of the sample and on particle size.[133] The effect has also been used to study the In_2O_3–SnO_2 system.[431] ^{119}Sn spectra have been recorded for a number of ternary tin(IV) sulphides.[432] Five compounds are isolated from the Na_2S–SnS_2 system, *viz.* Na_4SnS_4, $Na_6Sn_2S_7$, $Na_4Sn_3S_8$, and high- (HT) and low- (LT) temperature forms of Na_2SnS_3. The crystal structures of all of these compounds, except Na_2SnS_3 (HT), are known. Na_4SnS_4 and $Na_6Sn_2S_7$ contain tetrahedrally co-ordinated tin, the latter with bridging S atoms. Na_2SnS_3 (LT) has Sn in octahedral sites while $Na_4Sn_3S_8$ has tin in both tetrahedral and trigonal bipyramidal co-ordination. The Mössbauer spectrum of $Na_4Sn_3S_8$ can be fitted to two sites consistent with the structure, taking into account the fact that there are twice as many 5- as 4-co-ordinated tin atoms. The Mössbauer parameters for these sites are δ = 1.19, Δ = 0.54 mm s^{-1} and δ = 1.30, Δ = 1.40 mm s^{-1} respectively. The spectra (Figure 12) for the ternary sulphides clearly reflect the differences in tin environments. The data (δ = 1.31, Δ = 1.14 mm s^{-1}) for Na_2SnS_3 (HT) suggest the presence of Sn in a grossly distorted tetrahedral environment. The spectra for the compounds Ba_2SnS_4 (HT and LT) and $Ba_3Sn_2S_7$ are consistent with their known structures containing highly distorted $SnS_4{}^{4-}$ tetrahedra. The

[430] A. B. Cornwell, P. G. Harrison, and J. A. Richards, *J. Organometallic Chem.*, 1976, **108**, 47.
[431] M. B. Varfolomeev, A. S. Mironova, F. Kh. Chibirova, and V. E. Plyushchev, *Izvest Akad. Nauk S.S.S.R., Neorg. Materialy*, 1975, **11**, 2242.
[432] R. Greatrex, N. N. Greenwood, and M. Ribes, *J.C.S. Dalton*, 1976, 500.

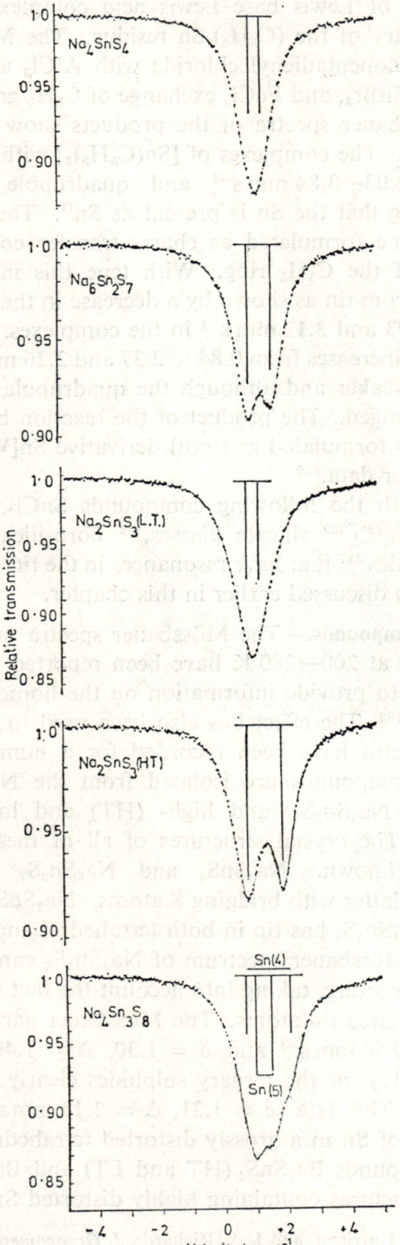

Figure 12 ^{119}Sn *Mössbauer spectra at 77 K for compounds in the system* Na_2S–SnS_2 (Reproduced from *J.C.S. Dalton*, 1976, 500)

spectrum of $BaSnS_3$, which is of unknown structure, has been fitted to two tetrahedral Sn sites ($\delta = 1.31$, $\Delta = 0.73$ and $\delta = 1.37$, $\Delta = 1.61$ mm s^{-1}). The only compound isolated from the PbS–SnS_2 system, $Pb^{II}Sn^{IV}S_3$, has tin surrounded by six S atoms and has the Mössbauer parameters, $\delta = 1.13$, $\Delta = 0.45$ mm s^{-1}. It is concluded that the following tentative correlation exists between the isomer shifts and tin co-ordination in these sulphides: δ (6-co-ordinate Sn) 1.02—1.13 mm s^{-1}; δ (5-co-ordinate Sn) 1.19 mm s^{-1}; and δ (4-co-ordinate Sn) 1.19—1.35 mm s^{-1}. The trend implying an increase in *s*-density at the tin nucleus with decreasing co-ordination number is found in other Sn^{IV} systems and is probably best interpreted in terms of a decrease in the occupation of tin *5d*-orbitals as the basis set for bonding changes from sp^3d^2 through sp^3d to sp^3. The Mössbauer effect in $Sn_{1-z}Eu_zMo_6S_3$ ($z = 0, \frac{1}{2}, \frac{2}{3}$) has also been studied.[433] The spectra consist of quadrupole doublets with unequal intensities at all temperatures (4.9—293 K). The Debye–Waller factor, isomer shift, and asymmetry of the quadrupole components show irregularities between 50 and 110 K which are said to be due to a phase transition. No quantitative explanation is given for the large values of the asymmetry but evidence is presented to suggest that this behaviour is connected with the properties of the librational modes of the Mo_6O_8 units in the crystal. The existence of the compounds Tl_2Se_3Sn and Tl_4SnSe_4 in the Tl_2Se–$SnSe_2$ systems has been proved by Mössbauer spectroscopy.[434] A temperature dependence of the Mössbauer effect probability in the Sn^{IV} derivatives Cu_2SnZ_3 and Ag_2SnZ_3 (Z = S, Se, and Te) has been reported[435] and their Debye–Waller factors and mean-square Sn atom amplitudes obtained.

The tin-containing compounds Fe_3SnC and Mn_3SnC were included in a ^{57}Fe study of perovskite carbides.[116] The ^{119}Sn isomer shifts for the relatively wide single-line spectra were 1.60 mm s^{-1} for the Fe compound and 1.70 for Mn_3SnC. The latter was found to have a very much smaller percentage resonance effect.

^{119}Sn Mössbauer and i.r. data have been used to characterize the tin(IV) products formed as a result of some oxidative addition reactions of tin(II) halides.[420] $SnOCl_2$,1.5THF, which is precipitated when oxygen is passed through a solution of $SnCl_2$ in THF, has a singlet resonance line with $\delta = 0.03$ mm s^{-1}. The tin(IV) halide carboxylates $SnX_2(O_2CR)_2$ (X = Cl or Br, R = CH_3 and C_6H_5), prepared by the reaction of tin(II) halides with lead tetra-acetate or dibenzyl peroxide, have isomer shifts in the range 0.21—0.47 mm s^{-1} and zero quadrupole splitting. These data suggest octahedral arrangements of ligand atoms around the Sn^{IV} with, on the evidence of broad resonance lines, some distortion occurring in the case of the acetates. The reactions of $SnCl_2$, $SnBr_2$, and SnI_2 with dimethyldiacetylene dicarboxylate (dmadc) and of $SnCl_2$ with diethylacetylene dicarboxylate (deadc), in contrast to the reaction of $SnBr_2$ with deadc, result[419] in 1 : 1 Sn^{IV} adducts. This suggests that carbenoid interactions of the tin(II) halide with the C≡C have occurred. The reaction of $SnBr_2$ with dimethylsulphide in thf also results in the formation of a solvated tin(IV) compound which

433 J. Bolz, T. Hauck, and F. Pobell, *Z. Phys. (B)*, 1976. **25**, 351.
434 L. N. Vasilev, A. Sh. Bakhtiyarov, Yu. P. Kostikov, and P. P. Seregin, *Izvest. Akad. Nauk S.S.S.R., Neorg. Materialy*, 1976, **12**, 117.
435 M. Balciuniene, A. Dragunas, and K. Makariunas, *Lietuvos Fiz. Rinkinys.*, 1975, **15**, 1009.

gives a single line spectrum with $\delta = 0.44\ mm\ s^{-1}$. The Mössbauer effect was used to study the dynamics of the molecules $SnCl_4$, SnI_4, and $Sn(CH_3)_4$ absorbed on graphite.[436] The mean square displacement normal to the basal plane was obtained from the temperature dependence of the recoilless fraction and the results used to establish the bonding energy of the absorbed molecule. The isomer shift, linewidth, and resonance probability parameters at 95 and 295 K have been studied for a series of thirteen hexafluorostannates(IV).[437] A correlation between isomer shift and the polarizing capacity of the cation present was found. The ^{119}Sn Mössbauer effect has also been used[438] in conjunction with Raman spectroscopy to follow the structural phase transition in K_2SnCl_6. Me_4NF and Me_4NSCN react with $SnCl_4$ in liquid SO_2 to give $Me_4N[SnCl_4F]$, $Me_4N[SnCl_4(NCS)]$, and $[Me_4N]_2[SnCl_4(NCS)_2]$. The vibrational and Mössbauer spectra of these complexes suggest[439] that $Ne_4N[SnCl_4F]$ ($\delta = 0.29$, $\Delta = 0.7$, $\Gamma = 1.01\ mm\ s^{-1}$) has a fluorine-bridged tetrameric structure similar to that of the isoelectronic $SbCl_4F$. A dimer structure with S-bonding was proposed for $Me_4N[SnCl_4(NCS)]$ ($\delta = 0.59$, $\Delta = 0.4$, $\Gamma = 0.97\ mm\ s^{-1}$) in contrast to an octahedral tin environment with N-bonded isothiocyanate groups for $[(Me_4N)_4]_2[SnCl_4(NCS)_2]$ ($\delta = 0.38$, $\Delta = 0$, $\Gamma = 1.02\ mm\ s^{-1}$). The data ($\delta = 0.41$, $\Delta = 0.36$, $\Gamma = 0.94\ mm\ s^{-1}$) for $[NMe_4]_2[SnCl_4(N_3)_2SnCl_4]$ are also discussed.

There have been a number of reports concerned with Mössbauer studies of the complexes of the tin tetrahalides. Equilibrium constants have been estimated[440] from frozen solution data obtained from mixtures of 1 mole of SnX_4 (X = Cl, Br, or I) with 2 moles of the donor molecules acetic anhydride, acetonitrile, acetone, and tetrahydrofuran. The constants found for the SnI_4–THF system were $k_1 = 1.1 \pm 0.3$ and $k_2 = 0.06 \pm 0.02\ min^{-1}$. The equilibrium constants for the solvation of SnX_4 by thf decreased with increasing anion size.[441] U.v., i.r., and Mössbauer spectra have been used to show that the molecules $(EtO)_2PNEt_2$ and $(EtO)_2PNHPh$ act as P-donor ligands to form 2 : 1 complexes with $SnCl_4$ although under some conditions reduction occurs to give $SnCl_2$.[442] The Mössbauer spectra have also been obtained[443] for the complexes of SnX_4 (X = Cl and Br) with $[PR(Z)(NMe)_2]_2$ (R = OMe, CH_2Cl, NMe_2, Ph, Me, and Et; Z = O, S, and Se) and for $SnCl_4$ with $[PR(NMe_2)_2]_2$ (R = Et, Ph, and Me_2N). The isomer shifts for the first series of complexes with $SnCl_4$ are in the ranges 0.25—0.36, 0.65—0.67, and 0.73—0.82 $mm\ s^{-1}$ for the O-, S-, and Se-containing ligands respectively. The isomer-shift range for the second series is 0.74—0.85 $mm\ s^{-1}$. The $SnBr_4$ complexes all have larger isomer shifts than the corresponding $SnCl_4$ compounds. It is suggested that all of the complexes have

436 S. Bukshpan, T. Sonnino, and J. G. Dash, *Surface Sci,*, 1975, **52**, 466.
437 V. A. Varnek, I. T. Tychinskaya, N. F. Yudanov, and L. N. Mazalov, *Zhur. strukt. Khim.*, 1976, **17**, 579.
438 J. Winter, K. Rossler, J. Bolz, and J. Pelzl, *Phys. Status Solidi (B)*, 1976, **74**, 193.
439 K. Dehnicke, B. Busch, and J. Pebler, *Z. anorg. Chem.*, 1976, **420**, 219.
440 A. Vertes. I. Nagy-Czako, and K. Burger, *J. Phys. Chem.*, 1976, **80**, 1314.
441 A. Vertes and I. Nagy-Czako, *Magyar Kém. Folyóirat*, 1976, **82**, 200.
442 A. A. Muratova, V. P. Plekhov, E. S. Zinurova, and A. N. Pudovik, *Zhur. obshchei Khim.*, 1976, **46**, 1729.
443 Sh. Sh. Bashkirov, F. G. Vagizov, I. Ya. Kuranshin, A. A. Muratova, A. N. Pudovik, and A. S. Khramov, *Zhur. strukt. khim.*, 1976, **17**, 177.

the ligands in *trans*-positions. The Mössbauer data for the 1,2-ethylenebis-diphenylarsine (eda) and 1,4-butylenebisdiphenylarsine (bda) complexes, $SnCl_4eda$, $SnCl_4bda$, $SnBr_4eda$, $(SnBr_4)_2(bda)_3$, and $SnCl_4,eda$, have been reported.[444] The lowering of the isomer shift from the $SnCl_4$ value on complex formation is found as expected but the effect is less than that with N- or O-donor ligands and about the same as found for P- and S-ligands. The isomer shift is unusually low for a bromine complex in $(SnBr_4)_2(bda)_3$ and it is possible that one of the biarsine molecules is unco-ordinated.

The Mössbauer spectra of the halogenotin complexes [424] $[SnX_2(tcne),THF]$ (X = Cl, Br, and I) and $[NEt_3H][SnCl_3(tcne)]$ consist of sharp singlet resonances ($\Gamma < 0.7$ mm s^{-1}) in the range 0.38—0.60 mm s^{-1} indicative of almost cubic disposition of electron density around tin. These data, taken with the intense colours of the complexes, suggest that they are tetrahedral tin-substituted radical species (10).

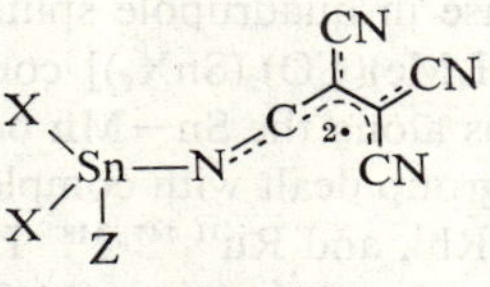

X = Cl; Z = Cl
X = Cl, Br or I; Z = thf

(10)

Mössbauer studies of a wide variety of compounds containing tin-transition metal bonds have been reported. The majority of these compounds have spectra with isomer shift values in the Sn^{IV} region.

Mössbauer quadrupole splittings along with i.r. stretching frequencies have been used to assign structures to nine new 6-co-ordinate tin(IV) compounds,[449] *viz*. $[M(SnCl_3L)]$ [M = $Mn(CO)_5$, $Mn(CO)_4(PPh_3)$, or $Fe(cp)(CO)_2$; L = phen or bipy] and $[M(SnXL'_2)]$ (X = Cl or Ph; L = quin or pd), and $[M\{SnPhCl_2(phen)\}]$. The ^{119}Sn isomer shifts and quadrupole splittings for the compounds are smaller than those for the 4-co-ordinate species $[M(SnCl_3)]$. Partial quadrupole splittings $(p.q.s.)_L^{oct}$ have been calculated from the experimental data and a good linear relationship, $(p.q.s.)_L^{oct} = 0.73(p.q.s.)_L^{tet} — 0.01$, found with previously calculated values for the corresponding tetrahedral compounds. The gradient of the quotient of $(p.q.s.)_L^{oct}$ and $(p.q.s.)_L^{tet}$, 0.73, is said to compare well with a value of 0.67 calculated by simple molecular orbital arguments. The values for the isomer shifts of the compounds are considerably larger than those of their Ph–Sn analogues because of the larger *s*-character in the Sn—M bonds. The order of increasing *s*-character in the L—Sn bond is Cl, Br $\ll$ Ph < Me <

[444] S. S. Sandhu, J. K. Sindhu, G. K. Sandhu, N. N. Greenwood, and K. G. Snowdon, *J. Inorg. Nuclear Chem.*, 1976, **38**, 998.
[445] G. M. Bancroft and T. K. Sham, *J.C.S. Dalton*, 1976, 467.
[446] A. B. Cornwell and P, G. Harrison, *J.C.S. Dalton*, 1976, 1054.
[447] V. A. Varnek, E. N. Yurchenko, G. L. Elizarova, A. I. Shan'ko. L. G. Matvienko, P. G. Antonov, and Yu. N. Kukushkin, *Koord. Khim.*, 1975, **1**, 161.
[448] E. N. Yurchenko, V. A. Varnek, G. L. Elizarova, L. G. Matvienko, and M. S. Ioffe, *Koord. Khim.*, 1975, **1**, 1406.
[449] G. M. Bancroft, V. G. Kumar Das, and T. K. Sham, *J.C.S. Dalton*, 1976, 643.

$Mn(CO)_5 \leqslant Mo(cp)(CO)_3 < Mn(CO)_4(PPh_3) \leqslant Fe(cp)(CO)_2$. The reaction of bis($\beta$-diketoenolate)tin(II) compounds, SnX_2, with octacarbonyldicobalt in benzene at room temperature yields products of composition $[\{Co(CO)_4\}_2SnX_2]$, $[Co_2(CO)_7(SnX_2)]$, or $[Co(CO)_4(SnX)]$ depending upon the nature of the β-ketoenolato-residue, X. Tin-119m Mössbauer data for $[\{Co(CO)_4\}_2SnX_2]$ are very similar [425] to those of $[Co_2(CO)_7(SnX_2)]$ which have the same structure as that determined for $[\{Co(CO)_4\}_2Sn(pd)_2]$ showing that the $Co(CO)_4$ groups occupy *cis*-positions at octahedrally co-ordinated tin(IV). The compounds $Co(CO)_4(SnX)]$ have resonance lines in the Sn(II) region and have been discussed earlier. The Mössbauer spectra of the complexes of composition $[Mn(C_5H_4Me)(CO)_2(SnX_2)]$ where X is a β-diketonate ligand have also been reported. The spectra consisted of doublets with isomer shifts in the narrow range 1.88—1.99 mm s^{-1} and quadrupole splittings varying from 2.26 to 2.77 mm s^{-1}. The data for the complexes where X = Cl or Br were also similar. The reduction in isomer shifts and the increase in quadrupole splittings found when the β-keto-enolates form the $[Mn(C_5H_4Me)(CO)_2(SnX_2)]$ complexes can be attributed to synergic ($\sigma + \pi$) interactions along the Sn—Mn bond.[446]

Two papers by the same group dealt with complexes containing Sn—M bonds where M = Pt^{II}, Pd^{II}, Ir^{III}, Rh^{I}, and Ru^{II}.[447, 448] The spectra of 22 complexes [447] of these platinum group metals of the type $[M(SnX_n)_5]^{m-}$ (X = Cl, F, C_2O_4, or OH and n = 2 or 3) were collected at 78 K. The shifts and splittings of the Pt-containing complexes were similar and the high energy side of the doublet spectra was broadened in comparison with the lower energy peak, the order of broadening being F > C_2O_4 > Cl. It was suggested that this effect was due to the existence of several crystal modifications. The broadening was, however, much less for the complexes of Pd, Ir, Rh, and Ru and did not occur in frozen solution spectra. A combination of Mössbauer spectroscopy, *X*-ray electron spectroscopy, i.r., and MO calculations was used [448] to characterize the σ and π properties of Sn ligands containing Cl, F, OH, and C_2H_4 bonded to Pt, Ir, and Rh.

Papers dealing with a number of compounds with ^{119}Sn isomer shifts in the tin(IV) region have been discussed earlier in this chapter. These contain data for the following tin(IV) species SnF_4,[383] $SnCl_4$,[383] $SnBr_4$,[383] SnI_4,[382, 303] $Sn(SO_4)_2$,[382] SnS_2^-,[382] SnX_6^{2-} (X = F, Cl, or I),[383] silicate [396] and borosilicate [397] glasses, minerals,[401, 402] $As_2Se_3Sn_x$,[398, 399] Sn-doped oxides,[403, 404, 406] Sn deposits on Cu,[413] and a mixed valence compound.[426]

Organotin(IV) Compounds.—There has been a considerable addition to the literature on organotin compounds over the past year.

Bancroft and his co-workers [449] have extended their additive model for ^{119}Sn Mössbauer quadrupole splitting to 5-co-ordinate organotin(IV) compounds. They report the parameters of ten complexes of the type $[R_3SnL_2][BPh_4]$ (R = alkyl or phenyl and L = electronegative ligand). Isomer shifts for the compounds are in the range 1.18—1.37 mm s^{-1} and quadrupole splittings are 3.29—4.10 mm s^{-1}. Mössbauer–Zeeman experiments show that the sign of eQV_{zz} is negative for both $[Me_3Sn(bipyo)][BPh_4]$ and $[Ph_3Sn(bzbz)]$. The authors give details of a regression method which they use to distinguish structural isomers (*e.g.* of R_3SnL_2 and R_2SnL_3) of trigonal bipyramidal species by their

^{119}Sn quadrupole splittings. Data for 65 compounds were used in the regression analysis and the calculation of p.q.s. parameters. Comparison of the results with experiment indicates that the additive model gives a consistent account of the relationship between Δ and the stereochemistry in trigonal bipyramidal organotin(IV) compounds. A table of estimated p.q.s. values for 41 ligands in this geometry has been given and the observed and calculated values of Δ listed for 46 compounds. The failure of the point-charge model to predict quadrupole splitting values is considered in a study of the tin–ligand bond character in methyltin iodides by ^{119}Sn and ^{129}I Mössbauer spectroscopy.[450] ^{119}Sn spectra for Me_3SnI, $MeSnI_3$, Ph_3SnCl, and $PhSnCl_3$ and for their frozen solutions in inert solvents are described. The data show that there are no major structural differences between the neat solids and their frozen solutions indicating that intermolecular forces even in the solids must be small. Use of these data along with ^{129}I Mössbauer parameters and n.m.r. tin–proton coupling constants provides, for the first time, a consistent interpretation of the data. The conclusion reached is that the ^{119}Sn Mössbauer spectra of alkyl tin iodides (and perhaps other alkyl tin halides as well) can be explained by the imbalance in the *p*-orbital distribution around tin without involving π-bonding interactions between Sn and halogen.

Tin-119 Mössbauer data have been reported[451] for a number of diorganotin bis(arylsulphonates) and triorganotin monoarylsulphonates. Isomer shift values are high compared with other R_3SnX and R_2SnX_2 compounds being in the ranges 1.38—1.70 and 1.41—1.52 mm s^{-1} for $R_3Sn(O_3SAr)$ and $R_2Sn(O_3SAr)_2$ respectively. These results are indicative of the ionic nature of the Sn—O bonds. The quadrupole splittings of the bis(arylsulphonates) (Δ = 4.64—4.91 mm s^{-1}) are somewhat less than those for the stronger F-, Cl-, F_3C-, and alkyl-sulphonic acids (Δ = 4.91—5.54 mm s^{-1}) but serve to suggest that they have structures similar to that for $Me_2Sn(O_3SF)_2$ in which the sulphonate groups bridge linear moieties. The $R_3Sn(O_3SAr)$ compounds also have large quadrupole splittings (Δ = 3.90—4.31 mm s^{-1}) suggesting an infinite chain structure with sulphonate groups bridging planar R_3Sn moieties. The crystal structure of trimethyltin benzenesulphonate is reported in the paper along with its Mössbauer parameters, δ = 1.53, Δ = 4.20 mm s^{-1}. The Sn atoms are in trigonal bipyramidal co-ordination with Me groups in the equatorial plane and oxygen atoms, one from the sulphonate group (Sn—O = 2.37 Å) and one from a water molecule (Sn—O = 2.30 Å) in axial positions. The origins of the possible small electric field gradients in several trimethyl tin amines, phosphines, arsines, stibines, and sulphides are said[452] to arise from polarizing effects and not *pπ–dπ* interactions. The isomer shifts of the series SnR_3Cl, SnR_3Br, and $SnMeR_2I$ [R = $CH(SiMe_3)_2$] are similar to those of other trialkyltin halides.[427] The quadrupole splitting values for SnR_3Cl (2.18 mm s^{-1}) and SnR_3Br (2.05 mm s^{-1}) are however smaller than normal. This suggests either that $CH(SiMe_3)_2$, when attached to Sn, is less electron withdrawing than Me, Et, or Ph or, more probably, that the bulky

450 A. P. Marks, R. S. Drago, R. H. Herber, and M. J. Potasek, *Inorg. Chem.*, 1976, **15**, 259.
451 P. G. Harrison, R. C. Phillips, and J. A. Richards, *J. Organometallic Chem.*, 1976, **114**, 47.
452 H. Schumann, W. W. Dumont, H. J. Kroth, N. Neumann, P. J. Corvan, and J. J. Zuckerman, *J. Organometallic Chem.*, 1976, **121**, 321.

halides are monomeric in the solid state possibly with significant ionic character. Mössbauer spectra of Me_2SnCrO_4, $(Me_2Sn)_2O(CO_3)$, and $(Ph_2Sn)_2O(CO_3)$ have been obtained at 110 and 295 K. The data indicate[453] substantial differences between the Me—Sn—Me and Ph—Sn—Ph angles in the two basic carbonates. The Me compound has a relatively narrow doublet showing that the two Sn resonances coincidently overlap as in $[Me_2Sn(NCS)]_2$ because of similar Me—Sn—Me angles. In contrast the two doublets found in the spectrum of the Ph compound suggest that one Ph—Sn—Ph angle is close to 90° while the other is about 180°. Polymeric structures with bridging CrO_4^{2-} or CO_3^{2-} groups have been proposed. The Mössbauer spectra of $(Me_3Sn)_2PO_3H$, $(Ph_3Sn)_2PO_3H$, and $(Me_3Sn)_2PO_3H,2H_2O$ suggest[454] that the two Sn atoms are in similar, probably 5-co-ordinated environments, implying at least one 3-co-ordinate oxygen per formula unit. The vibrational and Mössbauer spectra for $(Me_3Sn)_3PO_4$ are consistent with a weakly associated structure in which all three Sn atoms have similar tetrahedral environments that are distorted towards trigonal bipyramidal. Mössbauer and other measurements have been used[455] to show that *N*-trimethyltin aziridine ($\delta = 1.21$, $\Delta = 2.24$ mm s^{-1}) is a one-dimensional polymer. Some related compounds, for example the 2-Me and 2,2′-dimethylaziridine complexes and homologues where Sn is attached to cyclic amines with 3- to 7-membered rings, however have smaller Δ values (≈ 1 mm s^{-1}) and are monomers. Reaction of trimethyltin aziridine with BF_3 in ether produces a 1 : 1 adduct which is monomeric and has $\Delta = 3.03$ mm s^{-1}. A general correlation has been found between the ESCA Sn-3*d* linewidths and Mössbauer quadrupole splittings and between ESCA bonding energies and isomer shifts for solid organotin(IV) compounds.[456] Eleven (triphenylstannyl)cyanamide–organic isocyanate adducts were converted to anionic complexes by reaction with Et_3NHCl. The i.r. spectra of the products indicate Sn—O not Sn—N bonding and the Mössbauer parameters of two of the complexes ($\delta = 1.31$, $\Delta = 3.11$ mm s^{-1}) are said to have quadrupole splittings consistent with trigonal bipyramidal tin co-ordination with Ph groups in the equatorial positions.[457] Mössbauer parameters for Me_4Sn,[436] $C_6H_5C{\equiv}CSnR_3$ (R = Me, Et),[458] 1 : 1 and 1 : 2 complexes of Bu_2SnH_2 and Bu_3SnH with donor solvents[459] and the diorganotin(IV) adenine compounds, R_2SnAd_2 (R = Bu^n, Oct^n, or Ph),[460] have been reported.

Organotin(IV) complexes with chelating ligands have been investigated by a number of workers. The parameters found[421] for $Me_3Sn(Hedaa)$ and $Me_2Sn(edaa)$ are $\delta = 1.40$, $\Delta = 3.49$ mm s^{-1} and $\delta = 1.12$, $\Delta = 3.07$ mm s^{-1} respectively. The structure of the latter probably resembles that of $Me_2Sn(salen)$

[453] R. G. Goel, H. S. Prasad, G. M. Bancroft, and T. K. Sham, *Canad. J. Chem.*, 1976, **54**, 711.
[454] T. Chivers, J. H. G. Van Roode, J. N. Ruddick, and J. R. Sams, *Canad. J. Chem.*, 1976, **54**, 2184.
[455] M. E. Bishop and J. J. Zuckerman, *Inorg. Chim. Acta*, 1976, **19**, L1.
[456] G. M. Bancroft, I. Adams, H. Lampe, and T. K. Sham, *J. Electron Spectrosc. Related Phenom.*, 1976, **9**, 191
[457] E. J. Kupchik and D. K. Parikh, *J. Organometallic Chem.*, 1976, **121**, 327.
[458] V. S. Zavgorodnii, E. T. Bogoradovskii, V. L. Maksimov, V. B. Lebedev, B. I. Rogozev, and A. A. Petrov, *Zhur. obshchei Khim.*, 1975, **45**, 2466.
[459] V. O. Reikhsfeld, V. A. Ivanov, and I. E. Saratov, *Zhur. obshchei Khim.*, 1975, **45**, 2243.
[460] L. Pellerito, G. Ruisi, N. Bertazzi, M. T. Logiudice, and R. Barbieri, *Inorg. Chim. Acta*, 1976, **17**, L9.

(δ = 1.13, Δ = 3.46 mm s^{-1}) in which the central $[SnO_2N_2]$ unit is essentially planar although the smaller quadrupole splitting for the edaa complex is likely to correspond to a much smaller (*ca.* 130°) C—Sn—C bond angle. The synthesis and Mössbauer spectra of the R_2Sn\{glycylglycinato-*ONN*(2-)\} compounds with R = Me, Bu^n, and Oct^n have been reported.[461] The ligand is the simplest dipeptide which could behave as a tetradentate ligand but the data (δ in the range 1.26—1.37 mm s^{-1} and Δ = 3.19—3.43 mm s^{-1}) are interpreted in terms of trigonal bipyramidal tin co-ordination with the ligand acting as a tridentate (2N, 1O) donor. The structures of the oximes formed from (4-oxopentyl) $(Ph)_3Sn$ and (4-oxopentyl)$(Ph)_2SnCl$ have been determined[462] by a combination of ^{119}Sn Mössbauer, p.m.r., and i.r. data. The triphenyl derivative, which has a singlet Mössbauer spectrum with an isomer shift of 1.29, is tetrahedral. The Ph_2SnCl derivative, on the other hand, has a quadrupole split spectrum (Δ = 3.34 mm s^{-1}) with a shift of 1.47 mm s^{-1} suggesting, by comparison with the parent compound, a trigonal bipyramidal structure with Cl and O atoms in the axial positions. Tin-119m Mössbauer quadrupole splitting data (Δ = 3.8—4.1 mm s^{-1}) specify *trans*-$(Me)_2Sn$ configurations for ten substituted 1,10-phenanthroline and two substituted 2,2′-bipyridine 1 : 1 complexes with Me_2SnCl_2.[463] The isomer shifts are, within experimental error, the same with a value of about 1.43 mm s^{-1}. The suggested structure from the Mössbauer parameters is confirmed by i.r., n.m.r., equilibrium, and thermodynamic measurements on the complexes.

The addition of R_2Sn (R = C_5H_5 and $CH_3C_5H_4$) to THF solutions of $M(CO)_5$ (M = Cr, Mo, and W) yields[430] the complexes $R_2SnM(CO)_5$. Complex formation is accompanied by a change in isomer shift value from that of the tin(II) stannylene to a tin(IV) value of between 1.86 and 2.00 mm s^{-1} and by an increase in quadrupole splitting from about 0.8 to 2.51—3.13 mm s^{-1}. The Mössbauer data for the dimeric complex $[R_2SnFe(CO)_4]_2$ and for the products of its reaction with strong bases (B), formulated as B,$R_2SnFe(CO)_4$ containing 5-co-ordinated Fe, were also given. The product of the reaction of SnR_2 with $Co_2(CO)_8$ was the tin(IV) compound $Sn[Co(CO)_4]_4$. Mössbauer isomer shift values suggest[464] high *s*-character in the Sn—Cd and Sn—Zn bonds in Ph_3SnMCl,tmed (M = Cd and Zn) and $(Ph_3Sn)_2CdL_2$ (L = tmed, 2,2′-bipyridine, and *o*-phenanthroline). The experimental values of Δ and the calculated partial quadrupole splittings suggest reduced donor activities towards the Sn of the Ph_3Sn moieties by Cd and Zn. Data[427] have also been collected for four new complexes containing Sn^{IV}–transition metal bonds, *viz.* $[M(\eta\text{-}C_5H_5)(CO)_2(SnR_2X)]$ [R = $CH(SiMe_3)_2$ and for M = Fe, X = Cl or Me and for M = Mo, X = H, or Me]. The isomer shifts agree well with those reported for analogous molecules. The small quadrupole splitting (0.67 mm s^{-1}) for $[Mo(\eta\text{-}C_5H_5)(CO)_3(SnR_2H)]$ is interesting and suggests that there is little difference in the character of the Sn—C, Sn—H,

[461] L. Pellerito, M. T. Logiudice, G. Ruisi, N. Bertazzi, R. Barbieri, and F. Huber, *Inorg. Chim. Acta*, 1976, **17**, L21.
[462] S. Z. Abbas and R. C. Poller, *J. Organometallic Chem.*, 1976, **104**, 187.
[463] W. D. Honnick, M. C. Hughes, C. D. Schaeffer, and J. J. Zuckerman, *Inorg. Chem.*, 1976 **15**, 1391.
[464] R. Barbieri, L. Pellerito, N. Bertazzi, G. Alonzo, and J. G. Noltes, *Inorg. Chim. Acta*, 1975, **15**, 201.

and Sn—Mo bonds. The larger splitting (2.37 mm s^{-1}) for [Fe(η-C_5H_5)$(CO)_2$-(SnR_2Cl)] reflects the *p*-orbital imbalance between the Sn—Cl and Sn—C bonds. The ^{119}Sn data for several trimethyltin amines, phosphines, and arsines (quadrupole splittings zero or very small) are compared[452] with the larger values (0.86—1.32 mm s^{-1}) found for the complexes of phosphine derivatives with transition metal carbonyls. The values of Δ are seen not to be sensitive to the nature of the transition metal carbonyl moiety [$Ni(CO)_3$, $Fe(CO)_4$, $Mo(CO)_5$, $Mn(CO)_4Br$, $Re(CO)_4Br$, and $Re(CO)_4$] but only to co-ordination by the P lone-pair. The higher effective electro-negativity of the bound P-atoms thus perturbs the electron density around the attached Sn giving rise to quadrupole splittings.

6 Other Elements

This section reviews the data for elements other than iron and tin. In each of the three main subsections, main group elements, transition-metal elements, and lanthanide and actinide elements, the isotopes are treated in order of increasing atomic number.

Main-group Elements.—*Krypton* (^{83}Kr). The Mössbauer parameters for KrF_2, KrF^+SbF_6, and the α- and β-forms of $KrF^+AsF_6^-$ have been reported.[465] Changes in quadrupole coupling were interpreted in terms of Kr—F bond lengths. A value of $Q_{ex}/Q_{gr} = 1.9580 \pm 0.0024$ was obtained for the ratio of the quadrupole moments of the 9.3 Kev and ground states of ^{83}Kr.

Antimony (^{121}Sb). The electronic parameters in ^{121}Sb Mössbauer spectroscopy, the e.f.g., and the change in *s*-electron density at the Sb nucleus, $\Delta\psi^2(0)$ have been calculated[383] by the CNDO/II MO method for the compounds SbX_3 (X = F, Cl, Br, and I), SbX_5 (X = FCl and) and SbX_6^- (X = F, Cl, and I). The correlation between these calculated parameters and the experimental data gave values of 1.266×10^{-3} for $\Delta R/R$ for ^{121}Sb and of 0.211 for Q. The additive model for quadrupole splittings in 5-co-ordinate organotin(IV) compounds has been extended[449] to Sb^{V}. Studies of isostructural Sn^{IV} and Sb^{V} compounds show that ^{121}Sb coupling constants are +6.76 times the corresponding ^{119}Sn quadrupole splittings. P.q.s. values for 11 ligands for ^{121}Sb Mössbauer spectroscopy have been calculated and the calculated quadrupole coupling constants for 15 organoantimony compounds compared with the experimental values. Good agreement is found for Ph_3SbX_2 compounds but the values calculated for Ph_4SbX compounds tend to be overestimates. A method[466] has also been presented which uses both Mössbauer isomer shifts and quadrupole coupling constants to obtain the electron populations in the hybrid atomic orbitals used by antimony(V) to form bonds. Although the method was applied to a series of compounds in which Sb was bonded to Ph and/or Cl it should have general applicability. Known Mössbauer data were used along with new data for seven compounds obtained with a $Ni_{21}{}^{121}Sn_2B_6$ source (δ to InSb = -1.65 ± 0.02 mm s^{-1}). Relationships were given which apply to most common geometries for Sb^{V} and the results for some of the compounds studied are in Table 2. Measurements of δ and e^2qQ were not sufficient to determine all of the σ values

465 J. H. Holloway, G. J. Schobilgen, S. Bukshpan, W. Hilbrants, and H. de Waard, see ref. 1.
466 L. H. Bowen and G. C. Long, *Inorg. Chem.*, 1976, **15**, 1039.

Table 2 ^{121}Sb *Mössbauer parameters and orbital populations in some* Sb^V *compounds* [466]

Compound	δ*	e^2qQ	n_s	n_p	σ_R (R = Ph or Me)	σ_X (X = F, Cl, or Br)
$pyH^+(PhSbCl_5^-)$	4.23	12.3	0.68	2.04	1.37	0.54
$pyH^+(Ph_2SbCl_4^-)$	2.73	24.9	0.79	2.37	1.35	0.51
Ph_3SbF_2	4.13	−22.0	0.70	2.26	1.00	0.26
Ph_3SbCl_2	2.80	−20.5	0.79	2.52	1.07	0.38
Ph_4SbF	4.26	−7.2	0.68	2.10	0.78	0.29
Ph_4SbCl	3.61	−6.6	0.73	2.27	0.83	0.38
Ph_4SbBr	3.30	−6.8	0.75	2.29	0.84	0.38

*Relative to InSb.

for a given compound but the authors, by considering the relationships found between Sb and Sn Mössbauer parameters and halogen n.q.r. results on some of the Sb compounds, obtained σ values that were consistent with a large body of experimental data. The electron population along the Sb—R bond, σ_R (R = Ph or Me), tends to increase as the number of electronegative groups attached to Sb increases. The 5*s*-character of the apical bonds in R_3SbX_2 and R_4SbX compounds was found to be appreciable but to vary only slightly among the compounds. Evidence was presented that bridging chlorine atoms produce a somewhat larger σ_{Cl} and a smaller $\langle r^{-3}\rangle p$ than non-bridging Cl atoms. In their conclusion the authors state that the orbital population method seems to be generally useful and they also claim to have applied their method to enough cases to show its limitations as well as its usefulness.

The recoilless fraction for ^{121}Sb in antimony single crystals has been calculated [50] over the temperature range 0—320 K for directions parallel and normal to the principal crystallographic axis. The calculated value at 77 K was found to be in good agreement with the experimental result. The crystallographic and magnetic transitions in MnSb have been followed [467] by the Mössbauer effect. Measurements made at temperatures lower than the magnetic ordering point made it possible to establish the nature of the changes in the effective magnetic field with changes in Mn concentration. Antimony-121 Mössbauer spectra have been used to study hyperfine magnetic interactions in the ^{121}Sb-doped antiferromagnetic, FeF_3. The spectra [468] show the presence of a magnetic transfer region around Sb. The size of H_{eff} at 77 K (110 ± 15 kOe) in comparison with that for $^{119}Sn^{4+}$ doping suggests a different type of transference in the magnetic hyperfine region.

Antimony Mössbauer studies have shown that compositions approaching $VSbO_4$ in the V—Sb—O system contain Sb only in the pentavalent state and that they are not mixed valence compounds of Sb^{III} and Sb^{IV}.[469] The Mössbauer spectra of USb_3O_{10}, $USbO_5$, and a catalyst containing 60 wt % $USb_{4.6}O_x$ and 40 wt % of SiO_2 were obtained at 77 K.[470] The spectra show that there is only one Sb site in $USbO_5$ but that USb_3O_{10} has more than one Sb environment. The

[467] K. K. Valiev, K. A. Dyuldina, M. A. Atef, and R. N. Kuzmin, *Tezisy Dokl. Vses. Konf. Kristallokhim. Intermet. Soedin.*, ed. R. M. Rykhal and R. M. Lvov, 2nd, 1974, 36.
[468] P. B. Fabrichnyi, M. Danot, V. N. Onuchak, A. M. Babeshkin, J. M. Dance, J. Portier, and P. Hagenmuller, *Solid State Comm.*, 1976, **18**, 1017.
[469] T. Birchall and A. W. Sleight, *Inorg. Chem.*, 1976, **15**, 868.
[470] B. J. Evans, *J. Catalysis*, 1976, **41**, 271.

Sb^V shifts in all of these compounds are larger than that of Sb_2O_5. The Mössbauer parameters for some Sb-containing sulphide and oxide minerals have been determined.[471]

Antimony-121 Mössbauer spectra have been recorded at 4.6 K for a number of complex anions derived from antimony(III) fluoride.[472] Data for the compounds $M[SbF_5]$ (M = Na, K, and N_2H_6); $M[SnClF_3]$ (M = K and NH_4); $Na[SbClF_3],H_2O$; $M[Sb_3F_{10}]$ (M = Na, Tl, and NH_4) and $M[Sb_4F_{13}]$ (M = K and Cs) are compared with the known stereochemistry of the Sb in the complexes of known crystal structure. Comparisons of isomer shifts for structurally related compounds show that there is a general increase to more positive values as the Sb : F ratio approaches 1 : 1. In general, the Mössbauer parameters were consistent with the known stereochemistry and in particular the generally large positive values of the quadrupole coupling constant were said to be in agreement with the presence of a stereochemically active lone-pair orbital on the antimony. The data for the $[SbClF_3]^-$ salts (*e.g.* $\delta = -5.78$, $eQV_{zz} = 19.2$ for $K[SbClF_3]$) are very similar to those for SbF_3, suggesting a similar Sb environment. This suggestion is borne out in practice because the environment found for Sb in $K[SbClF_3]$ is similar to that in SbF_3 in that it has three F atoms at 1.94, 1.95, and 1.96 Å with F—Sb—F bond angles of 82.2, 85.6, and 87.9°. The Sb—Cl distances are much longer being 3.09, 3.11, and 3.16 Å. The Mössbauer data for the complexes $SbCl_3$,RCN and $2SbCl_3$,RCN (R = Me and Ph) and for $SbCl_3$,PhCONHCOR (R = Me, Pr, Bu^i, and Ph) have been reported. The isomer shifts of the first two series of complexes were found to be less than that for $SbCl_3$ and this was interpreted [473] in terms of interactions between Sb *p*-orbitals and the π-electrons of the C≡N group. The isomer shifts for $SbCl_3$,PhCONHCOR are, however, larger than that for $SbCl_3$ and this was presumed to result from Sb—O bond formation. Longworth and Day have reported [474] a variable temperature study of the intervalence compound Cs_2SbCl_6 (Table 3). As previous

Table 3 *Mössbauer data* [474] *for* $Cs_4Sb^{III}Sb^VCl_{12}$

Temperature (K)	Sb *site in* $Cs_4Sb^{III}Sb^VCl_{12}$	δ* mm s^{-1}	Γ mm s^{-1}
4.2	V+	−3.00	2.61
	III+	−19.02	3.78
77	V+	−3.09	3.49
	III+	−18.88	3.37
120	V+	−2.40	4.71
	III+	−18.63	3.95

* Relative to $Ba^{121}SnO_3$.

authors have done, they have identified both Sb^{III} and Sb^V Mössbauer resonances. Their comments on a paper dealing with the difficulties in assigning different vibrational spectra to the two antimony sites and which they have taken to

471 J. G. Stevens, see ref 1.
472 J. G. Ballard, T. Birchall, R. Fourcade, and G Mascherpa, *J.C.S. Dalton*, 1976, 2409.
473 T. N. Sumarokova, E. F. Makarov, D. Kh. Kamysbaev, A. Yu. Aleksandrov, I. I. Amelin, and M. I. Usanovich, *Izvest. Akad. Naul Kaz. S.S.R., Ser. khim.*, 1975, **25**, 9.
474 G. Longworth and P. Day, *Inorg. Nuclear Chem. Letters*, 1976, **12**, 451.

interpret as a suggestion that the materials do not contain both Sb^{III} and Sb^{V} are, however, invalid. Moreover, they offer no explanation for the unexpected decrease in ionic character for the Sb^{III}—Cl environment with temperature. This decrease would in fact be consistent with transfer of Sb lone-pair electron density to the band structure of the material. The isomer shift for the Sb^{V} at 120 K does have the smallest negative value showing that the electrostatic character of the Sb^{V}—Cl bond is greatest at this temperature. Antimony-121 Mössbauer evidence for the Sb oxidation states present in intercalation compounds of SbF_5 and $SbCl_5$ in graphite was reported.[475] The pentahalides are found to be intercalated rapidly into the graphite lattice to give free-flowing powders suitable for use as Mössbauer absorbers. The spectra showed, in addition to the expected strong resonances in the Sb^{V} region, a weaker absorption in the Sb^{III} region with parameters close to those for SbF_3 and $SbCl_3$. Clearly the pentahalides do not, contrary to some previous reports, enter the graphite lattice without some reduction occurring. The two halides when intercalated had slightly higher Sb^{V} isomer shifts than when sublimed directly on to the Mössbauer probe. $SbCl_5$, whether intercalated or sublimed, consistently gave a small negative eQV_{zz} regardless of the sign of the initial estimate of this parameter for computer fitting. The data are thus consistent with a trigonal bipyramidal structure for $SbCl_5$ and not with a composite of $SbCl_4^+$ and $SnCl_6^-$. A Mössbauer investigation of the hydrates of antimony pentachloride has also been carried out.[476]

The Mössbauer data for seven organometallic binuclear antimony complexes have been reported.[477] The compounds $[NMe_4][Cl_3Me_2SbXSbMe_2Cl_3]$ (X = CN or F) have isomer shifts of about 1.3 mm s^{-1} and large quadrupole coupling constants (*ca.* 30 mm s^{-1}) consistent with a structure containing *trans*-Me groups in complexes with X—Sb—X bridging in contrast to the $[NMe_4][Cl_5SbXSbCl_5]$ compounds which have shifts of about 5.9 mm s^{-1} and negligible quadrupole couplings. The complexes $[Cl_2Me_2SbX]_2$ (X = N_3, NCO, and Cl) also have large eQV_{zz} values indicative of *trans*-Me groups on the Sb atoms. The Mössbauer spectra have been obtained[478] for a series of mono- and di-organotimonate(V) salts containing the octahedral species $[SbRX_5]^-$ and $[SbR_2X_4]^-$ (R = Me or Ph; X = F, Cl, Br, N_3, or CNS). The data were analysed using a transmission integral analysis and the importance of using this approach is emphasized. All of the compounds studied showed a positive sign for e^2qQ indicating an excess of electron density along the principal axis of the e.f.g. tensor. A ratio of 1 : 2 was found for the e^2qQ values for corresponding mono- and di-organo-species as predicted by a simple point-charge model for the octahedral $[SbRX_5]^-$ and $[trans\text{-}SbR_2X_4]^-$ geometries. V_{zz} in both cases would lie along the Sb—R bond axis and this is consistent with all of the data. The authors showed that a more detailed interpretation of the ^{121}Sb Mössbauer parameters could be achieved by comparison with the data for the corresponding organotin(IV) compounds. A close parallel of shift data for the Sb^{V} and Sn^{IV} compounds suggests that

[475] J. G. Ballard and T. Birchall, *J.C.S. Dalton*, 1976, 1859.
[476] J. G. Stevens and J. L. Resce, see ref. 1.
[477] J. Pebler, K. Schmidt, H. G. Nadler, and K. Dehnicke, *Z. anorg. Chem.*, 1976, **427**, 166.
[478] N. Bertazzi, T. C. Gibb, and N. N. Greenwood, *J.C.S. Dalton*, 1976, 1153.

similar redistributions of *s*- and *p*-electron densities occur in the M—X and M—C bonds in the series SbR_nX_{5-n} and SnR_nX_{6-n}, *etc.* Striking confirmation of the existence of a relationship between the e^2qQ parameters for corresponding organotin and organoantimony compounds is also presented. The Mössbauer parameters for $SbMe_2Cl_3$ are consistent [478] with a 6-co-ordinated Sb environment with bridging Cl atoms and *trans*-organic groups.

Tellurium (^{125}Te). Some tellurium compounds were included in a compilation [386] of *X*-ray chemical shifts and Mössbauer isomer shifts. The recoilless fraction for ^{125}Te in tellurium single crystals has been calculated [50] over the range 0—320 K for directions parallel and perpendicular to the principal crystallographic axis. The calculated value at 77 K was found to be in good agreement with the experimental value. Te-125 data have been mentioned in a paper considering the influence of chemical environment on conversion electron spectra.[382] The effects of nuclear resonance absorption on Mössbauer diffraction have been considered [479] using a kinematic theory.

A comparison [480] of the spectra for tellurium metal obtained with Zn, ^{125m}Te, and ^{125}Sb–Cu sources gave an experimental linewidth of 5.93 (8) mm s^{-1} for the ^{125}Sb–Cu source which was smaller than that for Zn ^{125m}Te source [6.50 (8) mm s^{-1}].

The electronic wave functions for the systems Te-chain and Te-atom present as substitutional impurities in Se chains and in Se- and S-rings have been obtained by the extended Hückel procedure. By using these wave functions, the field gradients at the ^{125}Te nuclei in all of these systems were investigated.[481] Very good agreement was found between theoretical and experimental values of the magnitude and sign of the quadrupole coupling constant and the asymmetry parameter for the Te-chain system. For ^{125}Te in the rings, the calculated quadrupole couplings are about 82% of the observed values. The signs of eQV_{zz} are not known for the ring systems, but the calculations indicated that the sign for the Se-ring should be the same as in the Te-chain but opposite to that for S-rings. Experimental asymmetry parameters are not available but the calculations give large η (0.87 and 0.98 for Se- and S-rings respectively) values. For the Se-chain system, the calculated sign of the quadrupole coupling agrees with the measured sign but its magnitude is only one-half of the experimental value. The calculated value of η for the Se-chain system is 0.13. The magnetic hyperfine interactions of $^{125}Te^{VI+}$ impurities in α-Fe_2O_3 have also been examined [482] by the Mössbauer effect and the results compared with those for $^{121}Sb^{V+}$ in the same lattice.

The tellurium-125 Mössbauer effect has been used to study Te in a number of alloy phases. The isomer shift data for ^{125}Te in the diamond-structure semiconductors ZnTe, CdTe, and HgTe showed [483] an increase in *s*-electron density at the Te nucleus in the order ZnTe < CdTe < HgTe. The electron distribution, e.f.g., and the local Te environment for phases in the Cu–Te systems were studied

[479] M. A. Andreeva, R. N. Kuzmin, and O. Y. Tikhomirov, *Kristallografiya*, 1976, **21**, 1026.
[480] M. Takeda and N. N. Greenwood, *J.C.S. Dalton*, 1976, 631.
[481] A. Coker, T. Lee, and T. P. Das, *Phys. Rev. B*, 1976, **13**, 55.
[482] P. B. Fabritchnyi, V. N. Onoutchak, A. M. Babechkin, J. Fava, and M. Danot, *Solid State Comm.*, 1976, **20**, 497.
[483] A. K. Dragunas and K. V. Makaryunas, *Soviet Phys. Solid State*, 1976, **18**, 807.

by Mössbauer spectroscopy.[484] Samples were prepared with an antimony-125 dopant as the ^{125}Te precursor. The electron density at the Te nucleus was found to increase with Cu content and the change in quadrupole coupling during the transition from Te to $Cu_{4-x}Te_2$ or $Cu_{2-x}Te$ showed a decrease in the anisotropy of the electron distribution and a simultaneous increase in metallic bonding. The local structure of As_2Te_3, $As_4Te_5I_2$, and $TlAsTe_2$ in both the vitreous and crystalline states has been the subject of a Mössbauer study.[485] The similarity in the ^{125}Te Mössbauer parameters for $As_4Te_5I_2$ and As_2Te_3 suggests the presence of similar Te environments in these compounds. Two Te resonances were found in the spectrum of crystalline $TlAsTe_2$ – one corresponded to a Te site similar to that in As_2Te_3 and the other to that of Te in Te–Tl compounds. The transition of $TlAsTe_2$ to the vitreous state can be followed by the development of a single-site Te spectrum indicative of Te in a low symmetry environment. The Mössbauer effect has also been used [486] for structural investigations of amorphous thin films and bulk glasses of composition As_xTe_{1-x} ($0.10 \leqslant x \leqslant 0.57$). The amorphous samples showed quadrupole doublets characterized by well resolved splittings which increased in value smoothly as x decreased. The amorphous to crystalline transition in the phase with $x = 0.3$ was followed by the appearance of the resonance line for two Te sites corresponding to Te metal and monoclinic As_2Te_3. The spectra used to follow the crystallization process in the phase with $x = 0.5$ revealed the presence of intermediate phases. The ^{125}Te Mössbauer effect was used to examine the nature of the spontaneous magnetic moment in samples of EuTe below the Néel temperature. At 80 K the Mössbauer spectrum consisted of a singlet. At 4.2 K the broad line was said [487] to be equivalent to a single magnetic hyperfine field at Te of about 73 kOe.

The ^{125}Te Mössbauer spectra of eleven Te^{IV} oxides and three oxyfluorides have been reported.[488] Their isomer shifts, relative to an I–Cu source, and quadrupole splittings span only very small ranges of values, 0.34—0.78 and 5.9—9.5 mm s^{-1} respectively. All of the oxides considered have lone-pair distorted Te^{IV} environments with either distorted square-pyramidal or distorted trigonal pyramidal oxygen co-ordination. The general features of the Mössbauer parameters for these oxides and oxyfluorides have been explained in terms of the use of about 10% of the Te 5*s*-electron density in bonding with the observed quadrupole couplings arising from the stereochemically active lone-pair. The Te Mössbauer spectra at 4.2 K for the crystalline tellurium subhalides Te_3Cl_2, Te_2X (X = Br or I), $Te_2Br_{0.75}I_{0.21}$, and α- and β-TeI show [480] the presence of at least two different Te sites in each subhalide lattice. This observation is consistent with the known structures of the halides which are shown in schematic form in Figure 13. The isomer shifts, relative to Zn ^{125m}Te at 4.2 K, in four different types of Te co-ordination were found, *viz.* (1) 0.6 mm s^{-1} for sites

[484] E. N. Efremov and R. V. Baranova, *Izvest. Akad. Nauk S.S.S.R., Neorg. Materialy*, 1976, **12**, 23.
[485] L. N. Vasilev, Yu. V. Krylnikov, A. K. Kamolov, and P. P. Seregin, *Izvest. Akad. Nauk S.S.S.R., Neorg. Materialy* 1976, **12**, 350
[486] P. Boolchand, M. Tenhover, and R. E. Flasch, *AIP Conf. Proc.*, 1976, **31**, 102.
[487] N. A. Blum and R. B. Frankel, *AIP Conf. Proc.* (Magnetism and magnetic Materials Annual Conference, 21st, 1975), 1976, **29**, 416.
[488] P. Dobud and C. Jones, *J. Solid State Chem.*, 1976, **16**, 201.

consisting of Te surrounded by a trigonal bipyramidal arrangement of electron pairs consisting of two bonds to Te, two bonds to Cl, and one lone-pair, (2) 0.8—1.0 mm s^{-1} for Te surrounded in a distorted octahedral arrangement of two bonds to Te, two bonds to X (X = Br or I), and two lone-pairs, (3) 0.3—0.8 mm s^{-1} for the pseudo-tetrahedral distribution of two bonds to Te, one to X

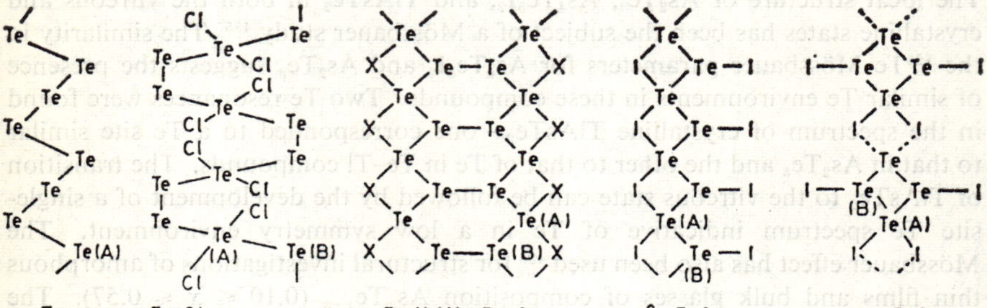

Figure 13 ***Schematic structures of tellurium and its subhalides***
(Reproduced from *J.C.S. Dalton*, 1976, 631)

(X = Te or I), and one lone-pair, and (4) 0.8 mm s^{-1} for the pseudo-tetrahedral electron pair distribution of two bonds to Te and two lone-pairs. The quadrupole splittings for the four types of environment were 11.0, 12.1—13.7, 3.4—6.1, and 8.1 mm s^{-1} respectively. In Figure 13 the four types of co-ordination are shown by (1) Te(A) in Te_2Cl, (2) Te(A) in Te_2X, α-TeI, and β-TeI, (3) Te(B) in Te_2X, α-TeI, and β-TeI, and (4) Te(A) in Te and Te(B) in $TeCl_2$ respectively. The spectra for the vitreous phases $Te_2Br_{0.75}I_{0.25}$ and Te_2Br were different from those of the corresponding crystalline compounds but again showed the presence of at least two Te sites in each compound.

Three papers on the Mössbauer spectra of organotellurium compounds have appeared. The parameters for the single Te site found for cyclo-1,4-dimethylene-tetratellurium [δ = 0.27 (^{125}Sb–Cu), Δ = 9.33 mm s^{-1}] provide evidence [489] for its formulation as the tetratellurium compound and not as CH_2Te_2 as previously reported. The mass spectrum of the compound was consistent with the formulation suggested by the Mössbauer results. The ^{125}Te data for a wide range of organotellurium compounds have been reported [490] including R_2Te, Ar_2Te, Ar_2Te_2, R_2TeX_2, Ar_2TeX_2, and $ArTeX_3$ where R = alkyl, Ar = aryl, and X = Cl, Br, and I. The isomer shifts were found to lie in the order tellurides $\leqslant$ ditellurides < dihalides < trihalides corresponding to an increase in *s*-electron density at the nucleus in that order. The shifts suggest the presence of a small amount of 5*s*-character in the Te—C bonds. The quadrupole splittings have been rationalized assuming ideal geometries for the molecules and that the bonding occurs predominantly through Te 5*p*-orbitals. The Townes and Dailey theory then predicts that the organodihalides and trihalides would have the same value for Δ, as is observed. The authors also state that the Δ values for (*p*-$MeOC_6H_4$)$TeCl_3$(py), (*p*-$MeOC_6H_4$)$TeCl_3$(tmtu), and $TeCl_4$(py) were

[489] F. J. Berry and B. C. Smith, *J. Organometallic Chem.*, 1976, **110**, 201.
[490] C. H. W. Jones, R. Schultz, W. R. McWinnie, and N. S. Dance, *Canad. J. Chem.*, 1976, **54**, 3234.

consistent with the structures that have been proposed for these compounds and that δ for $(p\text{-EtOC}_6H_4)Te(O)Cl$ was more consistent with 4- rather than 5-co-ordination. The ^{125}Te Mössbauer data measured at 4.2 K have also been reported[491] for the compounds Ar_2TeF_2, $ArTeF_3$, $ArTe(OCOCH_3)_2$, $ArTe(OCOC_6H_5)_2$ (where Ar = Ph, $p\text{-EtOC}_6H_4$, $p\text{-MeOC}_6H_4$, and $p\text{-C}_7H_7$), and for $Ph_2(CH_3)TeI$ and $Ph(CH_3)_2TeI$. Evidence was presented to suggest that the difluorides have similar structures to the other dihalides but that the co-ordination around Te in the trifluorides was not the same as in the other trihalides. The dicarboxylates have parameters similar to the dihalides, although the small Δ values for diphenyl tellurium dibenzoate and diacetate appear to suggest subtle changes in the co-ordination about Te with changes in the aryl ligand. The Mössbauer data for the monoiodides support their formulation as $R_2R'Te^+I^-$.

Iodine (^{127}I *and* ^{129}I). Some iodine compounds have been included in a compilation[386] of *X*-ray and Mössbauer chemical shifts and a method for the simultaneous analysis of iodine *X*-ray and Mössbauer shifts proposed.[492] The influence of chemical environment on electron capture decay of ^{125}I has been compared with ^{127}I and ^{129}I Mössbauer results.[382]

Iodine-129 Mössbauer parameters at 4.2 K are used along with ^{119}Sn data for Me_3SnI and $MeSnI_3$ to study the Sn—I bond character and the failure of the point-charge model to predict quadrupole splitting magnitudes.[450] The isomer shift for Me_3SnI relative to the Zn ^{129m}Te source ($-0.15\ mm\ s^{-1}$) was smaller that that for $MeSnI_3$ ($+0.26\ mm\ s^{-1}$) indicating a greater *s*-electron density at the I nucleus in $MeSnI_3$. The quadrupole splittings for the two compounds showed that the I environment was more symmetrical in Me_3SnI than in $MeSnI_3$. This observation taken with the ^{119}Sn data showed that $p\pi$–$d\pi$ bonding was not important in these compounds (see p. 429).

A series of ^{129}I Mössbauer measurements have been completed on 4-n-hexoxybenzylidine-4-^{129}iodoaniline dissolved in the liquid crystal 4-n-hexoxybenzylidine.[493] Ordered solute monocrystals were produced by ordering the samples in an external magnetic field. The sample showed three liquid-crystalline phases:

$$\text{smectic H} \underset{}{\overset{66\,^\circ\text{C}}{\rightleftharpoons}} \text{smectic A} \overset{68\,^\circ\text{C}}{\rightleftharpoons} \text{nematic}$$

The spectra of all three phases show resolved quadrupole splittings with e^2qQ negative and $\eta \approx 0$. Analysis of the ^{129}I Mössbauer data for the smectic H phase, which is a three-dimensional ordered structure, gave a value of $50 \pm 4°$ for the tilt angle in the liquid crystal.

The presence of a segregated AgI phase in I-doped AgBr crystals in molar concentrations as low as 1.5% can be detected by ^{129}I Mössbauer spectroscopy,[494] and the results were used with data from the photoluminescence spectrum of the material to provide information on the electronic configuration of the doping I atoms. The Mössbauer effect was also used to study ^{129}I dissolved in bromoiodide

491 F. Berry and C. H. W. Jones, *Canad. J. Chem.*, 1976, **54**, 3737.
492 L. I. Malkanov, Y. S. Grushko, J. M. Band, G. A. Shadrina, and M. B. Trizhaskovskaya, *Zhur. eksp. teor. Fiz.*, 1976, **70**, 2218.
493 M. J. Potasek, P. G. Debrunner, and G. DePasquali, *Phys. Rev. (A)*, 1976, **13**, 1605.
494 P. Junod, H. Hediger, and W. Kuendig, *Phot. Sci. Eng.*, 1976, **20**, 47.

photographic emulsions and a model was proposed to explain the trapping properties of iodide centres in AgBr. A simple perturbation approach interpretation of the data obtained from isomer shift studies of substitutionally implanted ^{119}Sn and ^{129}I in group IV semiconductors has been proposed.[495] The isomer shifts showed that the *s*-electron densities at the impurity nuclei decreased from Ge to Si to $C_{diamond}$. There have also been reports of ^{129}I Mössbauer experiments on matrix-isolated alkali iodide molecules[496] and of iodine in tobacco mosaic virus.[497]

The combination of resonance Raman and ^{129}I Mössbauer spectroscopy has been shown to be an attractive tool for studying electron distribution in iodine-oxidized mixed valence materials where the iodine can be present as I_3^-, I^-, or I_2.[498] For example, the iodine oxidation of bis(1,2-benzoquinone dioximato) (bdq) complexes of Ni or Pd yields crystalline materials of composition $(bdq)_2MI_n$ (n = 0.5—0.7). The ^{129}I data for the Ni complex are in good agreement with the published data for I_3^- and with the Raman spectra. The *cis*- and *trans*-influences in the square planar Pd^{II} iodide complexes, $Pt(L)I_3$ and *cis*- and *trans*-$Pt(L)_2I_2$ (L = PH_3, C_2H_4, CO, H_2S, NH_3, H_2O, H, SiH_3, CN, CH_3, I, Cl, OH, and NO_2), have been studied.[499] Comparison of the results between the series shows that the *cis*-influence is similar in magnitude to the *trans*-influence although the two ligand series are not parallel. Variations in the I orbital populations were compared with the experimental values for U_p (the *p*-electron imbalance) and h_s (the I *s*-orbital population) obtained from Mössbauer data using the formula $h_s = 9.1(-\delta + 1.5h_p - 0.54)$ where δ is the isomer shift (mm s^{-1}) relative to ZnTe. The experimental trends in U_p are reflected in the calculation but significant differences have been found between the observed and calculated values of h_s. The differences probably arise from the assumption, implicit in the above expression, that the I atoms in IO_3^-, IO_4^-, and IO_6^{5-} are in limiting tetrahedral or octahedral forms. Details of a Mössbauer study[245] of transferred hyperfine interactions and magnetic ordering in Fe^{III} bis(dithiocarbamato) iodides have been given earlier in this chapter (see p. 389). The ^{129}I Mössbauer isomer shift, the principal components of the e.f.g. tensor, and the transferred hyperfine magnetic field have been measured in ferromagnetic CrI_3.[500] The hyperfine parameters were used to determine the electron and spin transfer coefficients for CrI_3 and the results compared with those for other Cr compounds. The hyperfine interactions at I atoms in VI_2 have also been reported.[501]

Xenon (^{129}Xe). The preparation of a Mössbauer source to study the ^{119}Xe resonance has been reported[502] and Mössbauer effect measurements have been made on the ^{129}Xe resonance following ^{133}Xe implantation in Fe.[503]

495 E. Antonick, *Hyperfine Interactions*, 1976, **1**, 329.
496 M. Pasternak, S. Shamai, and T. Sonnino, see ref. 1.
497 H. Haffner, A. Andle, H. Appel, and G. Buche, see ref. 1.
498 T. J. Marks, D. F. Webster, S. L. Ruby, and S. Schultz, *J.C.S. Chem. Comm.*, 1976, 444.
499 D. R. Armstrong, R. Fortune, R. G. Perkins, R. J. Dickinson, and R. V. Parish, *Inorg. Chim. Acta*, 1976, **17**, 73.
500 J. P. Sanchez, B. Djermouni, J. M. Friedt, and G. K. Shenoy, *Hyperfine Interactions*, 1976, **1**, 313.
501 J. M. Freidt and J. P. Sanchez, see ref. 1.
502 S. Jha, D. C. Liu, and R. Leonard, see ref 1.
503 J. Odeurs, J. de Bruyn, H. Pattyn, M. Van Rossum, and R. Coussement, see ref. 1.

Caesium (^{133}Cs). The ^{133}Cs resonance has been studied after ^{133}Xe implantation in Fe.[503, 504]

Transition-metal Elements.—*Nickel* (^{61}Ni). The 67.5 keV transition in ^{61}Ni has been used to study[505] the hyperfine interactions at Ni in $Ni[P(OEt)_3]_4$, $Ni(NO_3)_2,6H_2O$, $NiCl_2$, and $NiCl_2,6H_2O$. Recoilless fractions, isomer shifts from a ^{61}NiV source, and magnetic hyperfine fields were obtained from theoretical curves computer fitted to the experimental data. The chlorides were found to have f_A of 3.4—3.5 and isomer shifts of 40 mm s^{-1} with H_{eff} being 33 ± 1 kOe for $NiCl_2$ and 46 ± 1 for $NiCl_2,6H_2O$. A direct correlation between isomer shifts and ligand electronegativity was found for the tetrahedral $Ni[P(OEt)_3]_4$ ($\delta = -37$ mm s^{-1}) and for six pseudo-octahedral compounds including the chlorides, $Ni(NO_3)_2,6H_2O$ ($\delta = -23$ mm s^{-1}) and three compounds, $KNiF_3$, NiF_2, and NiO, for which Mössbauer data had been previously reported.

The Mössbauer charge density for $^{61}Ni^{2+}$ ($3d^8$) in NiO was calculated using Hartree–Fock wave functions.[506] All the core *s*-functions ϕ_{ns} ($n = 1$, 2, and 3) of Ni^{2+} and their overlaps with six O^{2-} ions arranged in an octahedron corresponding to Ni have been calculated as a function of the interatomic distance R and the results tabulated. It was found that as a result of the variation of overlap integrals between the core *s*-functions and the nearest-neighbour ligand functions, the isomer shift varies with interatomic separation. For small variations the shift is proportional to the inverse of the seventh power of interatomic distance. This result can be used to predict the effect of thermal expansion of the host lattice as well as the variation of line shift for the same nuclei in going from one material to another.

Zinc (^{67}Zn). The Mössbauer charge density for $^{67}Zn^{2+}(d^{10})$ in a tetrahedral environment of O^{2-} ions, corresponding to ZnO, has also been calculated.[506] In this case the isomer shift again appears to depend upon the interatomic separation, R, and a power law, R^{-7}, was predicted. This was the first reported calculation of Mössbauer charge density for ^{67}Zn.

The natural linewidth of ^{67}Zn has been achieved in a measurement of the spectrum of a ZnO absorber with a ZnO(^{67}Ga) source.[507] The width of the inner three lines of the spectrum corrected for absorber thickness was 0.33 ± 0.04 μm s^{-1} compared with a minimum observable value of 0.31 μm s^{-1}.

Ruthenium (^{99}Ru). The isomer shifts of the ^{99}Ru resonances in Na_3RuO_4 and $Na_4Ru_2O_7$ measured at 4.2 K confirm that the Ru atoms are present in the 5+ oxidation state.[508] The simple two-line spectrum of $NaRu_2O_4$, on the other hand, is consistent with the presence of both Ru^{III} and Ru^{IV} and shows the absence of an averaging process by charge hopping on the nanosecond time scale. The magnetic hyperfine pattern for Na_3RuO_4 is the largest yet observed for any Ru system. The flux density at the nucleus is said to arise from antiferromagnetic coupling of Ru ions in spite of an apparently unfavourable

504 S. R. Reintsema, S. A. Drentje, J. L. Stavast, and H. de Waard., see ref. 1.
505 F. E. Obenshain, J. C. Williams, and L. W. Houk, *J. Inorg. Nuclear Chem.*, 1976, **38**, 19.
506 K. N. Shrivastava, *Phys. Rev. (B)*, 1976, **13**, 2782.
507 W. Potzel, A. Forster, and G. M. Kalvius, see ref. 1.
508 N. N. Greenwood, F. M. DaCosta, and R. Greatrex, *Rev. Chim. minéralé*, 1976, **13**, 133.

geometric distribution of $Ru_4O_{16}^{2-}$ units. A similar flux density was observed for $Na_4Ru_2O_7$ although the spectrum was more complicated.

Other ^{99}Ru Mössbauer studies reported during the year were: the combination of ESCA and Mössbauer data to provide a means of monitoring the surface and bulk properties of metallic and supported heterogeneous catalysts;[509] hyperfine interactions in $RuSn_3$ [510] and the ^{99}Ru effect in perovskite phases.[511]

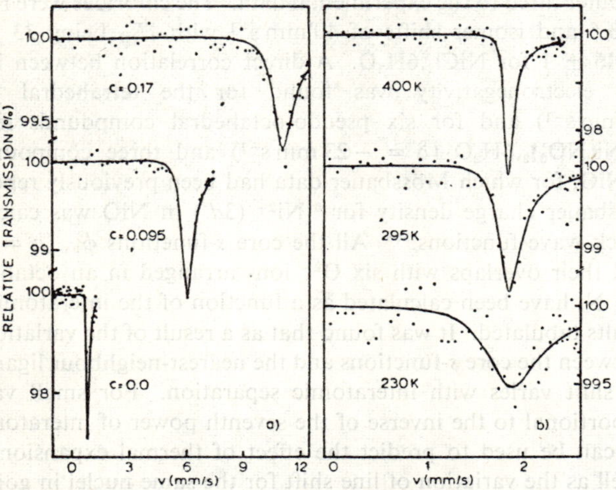

Figure 14 ^{181}Ta *Mössbauer absorption spectra of* H-*loaded* Ta *foils*: (a) *at* 295 K *for various* H *concentrations* c *and* (b) *at* c = 0.018 *for different temperatures*
(Reproduced by permission from *Phys. Rev. Letters*, 1976, **36**, 214)

In order to study the covalency effects in the Mössbauer hyperfine interactions of ruthenium, MO calculations using the SCCC–MO method have been performed [512] for six Ru complexes, *viz.* $Ru(CN)_6^{4-}$, $Ru(NH_3)_6^{2+}$, $Ru(NH_3)_6^{3+}$, $Ru(CN)_5NO^{2-}$, $RuCl_5NO^{2-}$, and $Ru(NH_3)_5NO^{3+}$.

Tantalum (^{181}Ta). The effects of H in Ta metal on the isomer shift and linewidth of the 6.2 keV Mössbauer resonance of ^{181}Ta have been studied as a function of H concentration and temperature.[514] The experiments were performed on H-loaded Ta foil (99.996% pure) as absorbers with a single line source of ^{181}W diffused into a single crystal of W metal. Some of the spectra obtained are shown in Figure 14. The data were fitted to a single Lorentzian line modified by a constant dispersion term. Two prominent effects can be seen from Figure 14: (1) with increasing H concentration, the resonance line is broadened and strongly shifted to higher energies (the increase in δ corresponds to a decrease in *s*-density

[509] C. A. Clausen and M. L. Good, Abstracts of the Centennial ACS Meetihgs, New York, USA, April 4-9, 1976.
[510] M. M. Abdelmegid and G. Kaindi, see ref. 1.
[511] F. M. DaCosta, R. Greatrex, and N. N. Greenwood, see ref. 1.
[512] D. Guenzburger and J. Danon, see ref. 1.
[513] E. Bodenstedt, *Hyperfine Interactions*, 1976, **2**, 1.
[514] A. Heidemann, G. Kaind, D. Salomon, H. Wiff, and G. Wortmann, *Phys. Rev. Letters*, 1976, **36**, 213.

at Ta with increasing H-concentration) and (2) the H-induced line broadening exhibits a strong temperature dependence but the H-induced line shift does not change with temperature. The results have been used to derive values for the activation energy (0.14 ± 0.01 eV) of the H diffusion and for the mean residence time of the H atoms in Ta. The correlation time for the motional narrowing of the ^{181}Ta Mössbauer line by H diffusion has also been calculated.[515] By solving the rate equation in a finite crystal and extrapolating with respect to time and crystal size a value of the mean resonance time of the H atoms in Ta was obtained.

Tungsten (^{182}W, ^{184}W, and ^{186}W). Mössbauer isomer shifts, along with observations of shifts in muonic *X*-rays, show that $\Delta R/R$ values are much smaller than those predicted by the simple centrifugal stretching model. ^{182}W, ^{184}W, and ^{186}W experimental results for $\delta\langle r_p^2\rangle + 1$ are compared with theoretical values.[513]

Osmium (^{187}Os). A source suitable for Mössbauer effect studies of ^{187}Os has been described.[502]

Rhenium (^{187}Re). The Mössbauer effect of the 134 keV transition in ^{187}Re has been used to determine the Debye–Waller factors of Re in different environments.[516] The Debye temperatures found for Re(Re) (291 ± 10 K) and Re(W) (298 ± 5 K) were compared with results obtained by other methods.

Iridium (^{193}Ir). Most of the Ir compounds investigated during the year were carbonyl complexes with metal–metal bonds.

The stoicheiometry, magnetic susceptibility, and Mössbauer spectrum of a one-dimensional iridium carbonyl chloride conductor has been described.[517] The Mössbauer data for the material $Ir(CO)_3Cl_{1+x}$ ($x = 0.10 \pm 0.03$) show that the charges arising from the partial oxidation are not localized on individual Ir atoms. The ^{193}Ir Mössbauer isomer shifts have also been measured [518] for a set of closely related iridium carbonyl chlorides and were found to decrease with increasing oxidation state 1+, 1.1+, 1.4+, and 2+. This trend is the reverse of that normally observed when *s*-electron density is deshielded on removal of *d*-electrons and has instead been attributed to the combined effects of strong π-bonding removal of electrons with some 6*s* character in a changing oxidation state. Complexes of the type $Ir(CO)_2LL$ or $Ir(CO)_2LCl$, where LL is a singly charged bidentate chelating ligand and L is a monodentate neutral organic base, were the subject of an Ir-193 Mössbauer study [519] of the formation of metal–metal bonds in planar complexes. The Mössbauer data were consistent with the existence of three categories of compound: (1) $Ir(CO)_2LL$ complexes with derivatives of the phenyl-imino-naphthalato and phenylazo-naphthalato ligands in which Ir is bonded to one O and one N of the ligands. This group includes four pairs of compounds with identical ligands but which crystallize in different modifications – one with Ir—Ir bonds and one containing monomer units. (2) $Ir(CO)_2LL$ complexes of the acetylacetonato and tropolonato derivative

[515] R. H. Swendsen and K. W. Kehr, *Solid State Comm.*, 1976, **18**, 541.
[516] E. Radeschutz and H. Langhoff, *Z. Phys.* (*B*), 1976, **25**, 37.
[517] A. P. Ginsberg, R. L. Cohen, and F. J. DiSalvo, in 'Proceedings of the 16th International Conference on Coordination Chemistry', 1974, 2. 19a.
[518] A. P. Ginsberg, J. W. Koepke, R. L. Cohen, and K. W. West, *Chem. Phys. Letters*, 1976, **38**, 310.
[519] R. Aderjan, H. J. Keller, H. H. Rupp, F. E. Wagner, and U. Wagner, *J. Chem. Phys.*, 1976, **64**, 3748.

ligands. Only one of the complexes in this series had two crystal forms, and (3) $Ir(CO)_2LCl$ complexes (L = analine derivative or CO), all of which contain Ir—Ir bonds. Mössbauer parameters have been used to distinguish between complexes with and without Ir—Ir bonds as illustrated for the phenylimino naphthalato derivative in Figure 15. The results showed that formation of

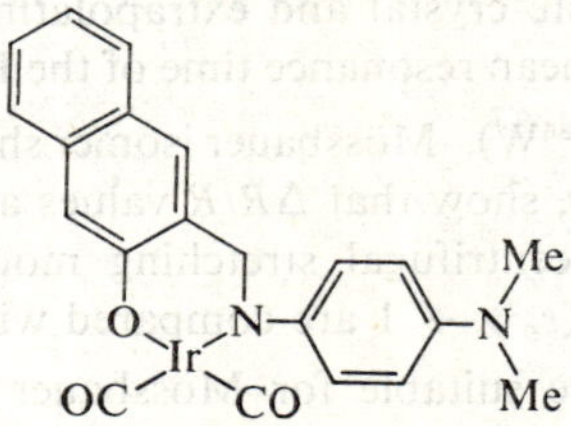

^{193}Ir *Mössbauer parameters for* (a) *modification with* Ir–Ir *bonding and* (b) *modification with no* Ir–Ir *bonding*

	δ*/mm s^{-1}	ΔE_Q/mm s^{-1}
(a)	0.15 (4)	4.73 (8)
(b)	0.33 (2)	6.12 (8)

* *Relative to* Ir *metal.*

Figure 15 ^{193}Ir *Mössbauer parameters for an* Ir phenylimino-naphthalato *derivative in modifications with and without the presence of* Ir–Ir *bonds*

columnar stacks with Ir—Ir bonds resulted in a reduction both in the electrical quadrupole interaction and in the charge density at the Ir nucleus. The pressure dependence and the sign of the electric field gradient have also been studied.[520] Another series of anionic linear chain iridium carbonyl halides, $K_{0.6}Ir(CO)_2Cl_2,0.5H_2O$, $(ttf)_{0.61}Ir(CO)_2Cl_2$ (ttf = tetrathiafulvalenium), $K_{0.57}Ir(CO)_2Br_2,0.2CH_3COCH_3$, and $Cs_{0.6}Ir(CO)_2Br_2$ has been investigated [521] by i.r., electrical conductivity, magnetic susceptibility, and Mössbauer measurements. All of the complexes were found to contain conducting linear chains of *cis*-$[Ir(CO)_2X_2]^{0.6-}$ units (X = Cl or Br). The apparently well defined compounds $K_{0.98}Ir(CO)_2Cl_{2.42},0.2CH_3COCH_3$, and $Na_{0.93}Ir(CO)_2Cl_{2.32},0.3CH_3COCH_3$, have been reformulated as $K_{0.6}Ir(CO)_2Cl_2,0.4KCl,0.2CH_3COCH_3$ and $Na_{0.61}Ir(CO)_2 0.32NaCl,0.3CH_3COCH_3$ containing interstitial KCl and NaCl respectively. The Ir Mössbauer data are reported for all of these compounds and for the square planar Ir^I complex $[Ir(CO)_2Cl_2]^-$ and for the halogen-bridged Ir^{II} dimer $[Ir_2(CO)_4Cl_6]^{2-}$. The complexes containing Ir—Ir bonds have isomer-shift values close to that of Ir metal ($\delta = -0.084$ to -0.164 mm s^{-1} relative to Ir metal) and quadrupole splittings of 3.54—4.00 mm s^{-1}. A comparison of the Mössbauer spectra of $[Ir(CO)_2Cl_2]^-$, $Ir(CO)_3Cl_{1.1}$, $[Ir(CO)_2Cl_2]^{0.6-}$, and $[Ir_2(CO)_4Cl_6]^{2-}$ led the authors to the following order of isomer shifts with oxidation states: Ir(2+), -0.228; Ir(1.4+), -0.116 to -0.084; Ir(1.1+), -0.04; and Ir(1+), $+0.063$ mm s^{-1}. The intermediate value of the isomer shift for the chain complexes between the values for Ir(1+) and Ir(2+) provides evidence

[520] A. Vasquez, F. E. Wagner, U. Klein, J. Moser, and G. Wortmann, see ref. 1.
[521] A. P. Ginsberg, J. W. Koepke, J. J. Hauser, K. W. West, F. J. DiSalvo, C. R. Sprinkle, and R. L. Cohen, *Inorg. Chem.*, 1976, **15**, 514.

for the delocalization of the charge carriers. The quadrupole splittings in some *cis*- and *trans*-Ir^{III} complexes have also been reported.[522]

Platinum (^{195}Pt). The ^{195}Pt and ^{197}Au resonances were used to study hyperfine fields at Pt and Au atoms in ferromagnetic Pt alloys.[523]

Gold (^{197}Au). Several compounds containing the novel AuF_6^- ion have been prepared and characterized. The Mössbauer parameters[524] of the complexes of the type $XeF_{11}AuF_6$ confirm that the Au is in the 5+ oxidation state and also suggest octahedral fluorine co-ordination for the Au atoms. An investigation of Au^I complexes by ^{35}Cl n.q.r. and ^{197}Au Mössbauer spectroscopy has been reported.[525] Electrical conductivity measurements, optical absorption edge data, and Mössbauer spectroscopy have been used[526] to study the effects of Au impurity atoms in vitreous AsSe, $AsSe_{1.5}$, $AsSe_{2.5}$, and $AsSe_4$. Introduction of Au into the glasses was found to be accompanied by the formation of trivalent Au compounds of As and Se. Magnetic and electric hyperfine interactions in solids were studied[527] by the ^{197}Au effect in $AuCl_2$, $AuGa_2$, $AuIn_2$, and $AuSb_2$.

The effects of pressure (up to 80 kbar) on the isomer-shift and quadrupole-splitting parameters of ^{197}Au in AuCN and $KAu(CN)_2$ measured at 4.2 K have been reported.[528] An increase in the electron density and a decrease in the absolute value of the e.f.g. at the Au nucleus with pressure was found for both compounds. This observation contrasts with the correlation that has been established within a series of Au^I compounds at ambient pressure and suggests that π-bonding increases strongly under pressure. *X*-ray diffraction data for AuCN under pressures of up to 115 kbar together with investigations of the Goldanskii–Karyagin effect on AuCN at ambient pressure support the previous assumption that the sign of the field gradient at the Au sites is negative. The largest value of the quadrupole-splitting parameter (11.30 mm s^{-1}) yet found for a gold(I) compound has been observed in $\{Au[o\text{-}C_6H_4CH_2N(CH_3)_2]_2\}Li$ for which an isomer shift of 5.26 mm s^{-1} has also been found.[529]

ESCA and ^{197}Au Mössbauer spectra of a series of ylide complexes of Au^I, Au^{II}, and Au^{III} have been recorded.[530] The Mössbauer parameters for the Au^I ylide, μ-bis(depdm)digold (depdm = diethylphosphoniodimethylene), are $\delta = 3.76$, $\Delta E_Q = 9.60$ mm s^{-1}, those for the Au^{II} ylid, μ-bis(depdm)-bis(AuBr), are $\delta = 2.67$, $\Delta E_Q = 6.71$ mm s^{-1} and those for the Au^{III} ylid, μ-bis(depdm)-bis($AuBr_2$), are $\delta = 2.02$, $\Delta E_Q = 5.52$ mm s^{-1}. The isomer shifts and quadrupole splittings show a steady decrease with increasing Au oxidation state and these data correlate well with the ESCA results.

522 V. P. Gespo and A. G. Maddock, *Radiochem. Radioanal. Letters*, 1976, **25**, 305.
523 I. Vincze, F. E. Wagner, E. Baggio-Saitovitch, and W. Koch, see ref. 1.
524 K. M. Leary, Report LBL–3746, 1975, *E.R.D.A. Res. Abs.*, 1976, **1**, 1632.
525 H. Schmidbaur, D. Koth, and F. E. Wagner, *Z. Naturforsch., B*, 1976, **31**, 1607.
526 P. P. Seregin, V. P. Sivkov, F. S. Nasredinov, N. F. Kartenko, L. N. Vasil'ev, and A. Sh. Bakhtigarov, *Fiz. Tekh. Poluprovodn.*, 1976, **10**, 1325.
527 F. E. Obenshain, J. O. Thomson, P. G. Huray, J. C. Love, G. Petitt, and M. Jaeger, *Annual* Progress Report of the Oak Ridge National Laboratory (Physical Division) 1974, ORNL–5025 1, 212-18.
528 H. Posser, G. Wortmann, K. Syassen, and W. B. Holzapfel, *Z. Phys.* (*B*), 1976, **24**, 7.
529 M. P. A. Viegers, J. M. Trooster, G. van Koten, and J. G. Noltes, see ref 1.
530 H. Schmidbaur, J. R. Mandl, F. E. Wagner, D. F. Van de Vondel, and G. P. Van der Kelen, *J.C.S. Chem. Comm.*, 1976, 170.

Mössbauer studies have been carried out on ternary gold alloys with the NaTl structure,[531] Li_2AuX (X = Ga, In, Tl, Ge, Sn, Pb, and Bi) and $Li_2Au_{2-x}In_x$ (x = 1.0—1.75). Isomer shifts from the single-line spectra were correlated with the Allred–Rochow electronegativity for the first three co-ordination spheres around the Au atoms, normalized to the number of outer electrons and corrected for the distance from the Au atom. It was suggested that substantial *5d–6s* mixing occurs with nearly matching charge compensation and additional *5d* depletion through an interaction of the *5d* band of Au with host orbitals of the correct symmetry. A net electron flow off the Au site occurs in every case. The values of the absolute ^{197}Au absorber recoilless fraction for 2% Au in Cu and 5% Au in Ag have been measured[532] and the results interpreted in terms of the Mannheim lattice dynamical model. The hyperfine field at Au nuclei in the Heusler alloy, Au_2MnAl, has also been determined.[533]

Mercury (^{199}Hg and ^{201}Hg). Sources for studying the ^{199}Hg and ^{201}Hg Mössbauer effects have been prepared. The ^{199}Hg source was a 98% ^{198}Pt-enriched foil activated by the reaction ^{198}Pt$(n,\gamma)^{199}$Pt. The source activity was about 4 Ci with a half-life of 3.15 days. The source was used to study Hg–Pt alloys.[534] Mössbauer effect measurements on the 32.2 keV state in ^{201}Hg were made with a strong activity carrier-free source of 73.5 hr ^{201}Tl.[502]

Lanthanide and Actinide Elements.—*Samarium* (^{149}Sm). A series of papers by Coey and co-workers on the ^{149}Sm Mössbauer effect with a source of ^{149}Eu in EuF_3 have been published. The isomer shifts obtained in these studies have been collected together in Table 4. The authors express surprise that the isotope

Table 4 ^{149}Sm *isomer shifts relative to* ^{149}Eu *in* EuF_3

Metal and alloys	δ mm s^{-1}	Sm^{II} *compounds*	δ mm s^{-1}	Sm^{III} *compounds*	δ mm s^{-1}
Sm	0.25 (4)	SmS	−0.72 (2)	SmN	0.07 (2)
				SmP	0.04 (2)
$SmNi_5$	0.20 (4)	SmSe	−0.71	Sm_2O_3	0.04 (2)
SmAg	0.18 (4)	SmTe	−0.72	SmH_3	0.04 (2)
				Sm_2S_3	0.03 (4)
SmZn	0.13 (3)			Sm_2Se_3	−0.02 (3)
				Sm_2Te_3	−0.10 (6)
SmAs	0.06 (4)			SmF_3	0.00 (2)
SmSb	0.00 (4)			$SmCl_3$	−0.01 (4)
SmB_6	−0.33			$Sm_2(SO_4)_3,8H_2O$	−0.01 (3)

has received so little attention in view of the fact that its resonance is easily observed at room temperature. The observed shifts are well correlated with those for the corresponding compounds of Dy and Gd – the ratio of the $\Delta\langle R^2\rangle$ values for Sm : Dy : Gd is 1 : 12.5 : −6.5. In a study[535] of Sm^{III} compounds and alloy phases the shifts for SmN and SmH_3 compared with that for Sm metal are said to indicate a large transfer of Sm 6*s* density either to the 5*d*-band or to

[531] P. Geutlich, S. Odar, and A. Weiss, *J. Phys. Chem. Solids*, 1976, **37**, 1011.
[532] J. F. Prince, L. D. Roberts, and D. J. Erickson, *Phys. Rev. (B)*, 1976, **13**, 24.
[533] W. Leiper, C. Blaauw, G. R. Mackay, and F. W. D. Woodhams, *J. Phys. (F)*, 1976, **6**, 2191.
[534] W. Wurtinger, see ref. 1
[535] J. M. D. Coey, *J. Inorg. Nuclear Chem.*, 1976, **38**, 1139.

suitable ligand orbitals. Room temperature Mössbauer studies of Y-doped SmS have been carried out.[536] The discontinuous concentration-dependent electronic phase transition in $Sm_{1-x}Y_xS$ was found to be accompanied by a sharp increase in isomer shift. The Sm electron configuration, considered as a mixture of $4f^6$ and $4f^55d^1$, was derived as a function of Y concentration and the results were analysed in terms of delocalization of the 5*d* electrons. The Mössbauer spectra of the same system under applied pressures (0—11 kbar) have also been obtained[537] for $Sm_{1-x}Y_xS$ (x = 0—0.34). There was a sharp increase in isomer shift at P = 6 kbar and x = 0.15, accompanying the transformation from the black to the gold phase. This work on SmS is said to be the first use of the Mössbauer effect to follow a metal-insulator transition as a function of temperature. The characteristic time for any electronic configuration is less than a nanosecond in the gold phases of SmS under pressure and in $Sm_{1-x}Y_xS$. The intermediate valence configurations have been determined in both systems, assuming negligible 6*s*-character, but the results neither confirm nor deny the configuration fluctuation picture. The Sm has less trivalent character in the gold phase of $Sn_{1-x}Y_xS$ than in SmS under pressure and there is some indication that the 5*d* electron is rather localized in the intermediate valence state. Calculations have been made with a Falicov Hamiltonian in which a term for 4*f*–5*d* hybridization was included and this model was found to account for the experimental variations in S electron density.

Europium (^{151}Eu and ^{153}Eu). The close relationship between the isomer shift in muonic atoms and the Mössbauer isomer shift has been demonstrated[382] for ^{153}Eu. The 97 and 103 keV resonances of ^{153}Eu have also been used in an investigation of the time dependence of Eu^{2+} formation after the electron-capture decay of ^{153}Gd in anhydrous rare-earth chlorides.[538]

In a low temperature study[539] of europium hydride, the Mössbauer spectra of $EuH_{1.90}$ were obtained at 78 and 4.2 K. At 78 K a broad unsplit line was obtained but the spectrum at 4.2 K was magnetically split with asymmetries ascribable to the presence of a small quadrupole interaction. Final parameters at 4.2 K were δ = 11.87 ± 0.012 mm s^{-1}, H_{eff} = −274 ± 0.3 kOe, and e^2qQ = 97.98 ± 2.7 MHz. The Eu^{2+} ions in $EuH_{1.90}$ do not have cubic symmetry and the dipolar contribution to the magnetic hyperfine field was estimated at −20 kOe which would result in a non-dipolar part of −254 kOe. The isomer shift at 78 K was −12.03 mm s^{-1} but the spectrum may have been broadened by an unresolved quadrupole interaction. The Mössbauer parameters of the europium chalcogenides have been rationalized in terms of a magnetic exchange mechanism and the parameters for the hydride can also be explained in this way. It was also suggested that the vacancy electrons in $EuH_{1.9}$ would be bound in deep donor levels and that they would be unlikely to take part in the exchange mechanisms which produce the long-range ordering in EuH_2. The Mössbauer parameters would not be affected by the vacancy electrons because their wave functions would be described by a combination of neighbouring *d*-orbital behaviour.

[536] M. Avignon, S. K. Ghatak, and J. M. D. Coey, *J. Magnetism Magnetic Materials*, 1976, **3**, 88.
[537] J. M. D. Coey, S. K. Ghatak, M. Avignon, and F. Holtzberg, *Phys. Rev.* (*B*), 1976, **14**, 3744.
[538] U. Wagner, F. E. Wagner, J. M. Freidt, D. Mihaila-Tarabasanu, and G. M. Kalvius, see ref 1.
[539] A. K. Zaheer, W. G. Bos, P. J. Ouseph, and S. L. Ruby, *J. Inorg. Nuclear Chem.*, 1976, **38**, 103.

^{151}Eu and ^{57}Fe Mössbauer spectra for $EuCoO_3$ have been recorded[160] in the temperature range 78—1200 K. The Eu spectra were fitted to single Lorentzian lines with isomer shifts of 0—1.0 mm s^{-1} from the $^{151}Sm_2O_3$ source which are characteristic of Eu^{III}. The broadening of the line (Γ = 3.6 mm s^{-1}, *cf.* 3.0 mm s^{-1} for Eu_2O_3) was said to be due to a non-spherical distribution of *f*-electrons. The isomer shift decreases with temperature suggesting ion displacements as well as second-order Doppler effects. The Mössbauer spectra of the phases found in the Eu_xWO_3–Sn_yWO_3 system ($x = 0.05$—0.27, $y = 0.07$—0.2) show the presence of both Eu^{II} and Sn^{II}. This is in marked contrast to the parent Eu bronze in which Eu^{III} was found. Europium-151 Mössbauer studies[540] of the series of tetragonal scheelite-type compounds, Eu_xMoO_4 ($0.67 \leqslant x \leqslant 1.00$), have removed the uncertainty concerning the oxidation state of Mo. The spectra show a complete absence of magnetic hyperfine interaction at 4.2 K and the presence of substantial amounts of Eu^{II} (see Table 5). The experimental values of the

Table 5 ^{151}Eu *Mössbauer parameters for* Eu_xMoO_4 phases[540]

	Eu^{II}			Eu^{III}		
		% Eu^{II}			% Eu^{III}	
	δ* mm s^{-1}	*Exp.*	*Theor.*	δ* mm s^{-1}	*Exp.*	*Theor.*
$Eu_{0.9}MoO_4$	−11.7	73 ± 2	77.8	+0.6	27 ± 1	22.2
$Eu_{0.7}MoO_4$	−12.1	15 ± 2	14.3	+0.6	85 ± 2	85.7
$Eu_{0.67}MoO_4$	—	0 ± 1	0	+0.7	100 ± 0.1	100
$Gd_{0.5}Eu_{0.5}MoO_4$	−12.0	55 ± 0.8		+0.9	45 ± 0.7	

* Relative to EuF_3.

Eu^{II} + Eu^{III} percentages were calculated from the computer-fitted resonance assuming similar recoilless fractions. The theoretical values were calculated according to a cation vacancy defect model with no contribution from Mo^V. The results ruled out the presence of $(Mo^VO_4)^{3-}$ in the Eu_xMoO_4 series although when Ga was present the observed Eu^{II} : Eu^{III} ratio could only be explained by the simultaneous partial reduction of Mo^{VI} or the much less likely reduction of Ga^{III}.

The specific heat, Mössbauer effect, and low-temperature magnetic properties of Eu_3S_4 have been reported.[541] The effect of pressure on the isomer shift and the magnetic hyperfine field in divalent Eu chalcogenides has also been investigated[542] at 4.2 K.

The thermal annealing of Eu^{2+} to Eu^{3+} in CaF_2 injected with oxygen was followed, and evidence from Mössbauer spectroscopy has been found for the existence of four distinct defect-stabilized Eu^{2+} species in the CaF_2. The involvement of oxygen in the stabilizing process was strongly indicated and CaF_2 doped with Eu_2O_3 and with EuF_3 gave equivalent results. The study showed that the Mössbauer effect was very useful in accounting for all ions simultaneously in this type of conversion process.[543] Enhanced optical fluorescence has also been

[540] N. N. Greenwood, F. Viegas, E. Banks, and M. Nemiroff, *Inorg. Chem.*, 1976, **15**, 2317.
[541] O. Massanet, J. M. D. Coey, and F. Holtzberg, *J. Phys. Paris Colloq.*, 1976, **3**, 297.
[542] U. F. Klein, G. Wortmann, and G. M. Kalvius, *J. Magnetism Magnetic Materials*, 1976, **3**, 50.
[543] R. L. Lambe and D. Schroeer, *Phys. Rev. Letters*, 1976, **36**, 45.

observed[544] following resonant absorption of Mössbauer γ-rays by Eu^{3+} nuclei in CaF_2. The Mössbauer spectrum of $^{151}EuCl_2$ matrix isolated in argon has been described.[545]

A number of reports on studies of Eu containing metallic phases have been published. A strong pressure-induced shift of the resonance lines in $Eu_{0.25}$-$La_{0.75}Rh_{2.00}$ has been described in terms of an increase of the energy separation between Eu^{2+} and Eu^{3+} valence states.[546] The decrease in isomer shift observed on hydrogenation of a dilute Eu–Pd alloy (2.5 atoms % Eu) has been discussed[547] and the dependence of interconfiguration excitation energies on the local environment, composition, and temperature in the phases $EuA_{2-x}B_x$ (A = Rh or Ir, B = Pt or Al) has been followed by ^{151}Eu Mössbauer spectroscopy.[548] The Mössbauer resonance has also been studied in Eu metal at pressures of up to 16 kbar. The pressure coefficient of the isomer shift $d\delta/dp = 4.8\,(5) \times 10^{-2}$ mm s^{-1} was interpreted as a congruent compression of the 6*s* conducting electrons. The magnetic hyperfine splitting also decreased with pressure and the relationship $dB/dp = +0.133\,(8)$ Tesla kbar^{-1} was said to arise predominantly from an increase in the conduction electron polarization.[549]

Gadolinium (^{154}Gd, ^{155}Gd, and ^{157}Gd). The isomer shifts of the 86.5 keV ^{155}Gd resonance in the binary compounds of Gd with Ag, In, Cu, and Ga were measured.[550] The Goldanskii–Karyagin effect in $Gd_2M_2O_7$ and $Eu_2M_2O_7$ using the ^{154}Gd and ^{155}Gd resonances was reported[551] and a source for studying the ^{157}Gd effect has been developed.[502]

Dysprosium (^{160}Dy and ^{161}Dy). The 86.79 keV gamma transition of ^{160}Dy has been employed to study the paramagnetic hyperfine structure of Dy impurities in Zr metal.[552]

When hyperfine interactions are non-diagonal with respect to the electronic states the effective-field approximation does not hold but the influence of relaxation on paramagnetic Mössbauer spectra can be described by stochastic theory[553] and calculations have been given for ^{161}Dy.

A Mössbauer study (1.8—300 K) of single crystal and polycrystalline samples of the antiferromagnetic material $DyPO_4$ (Néel temperature 3.39 K) has been reported by Forester and Ferrando.[554, 555] The spectrum at 3 K is shown in Figure 16. For the range 0—10 K the spectra can be analysed in terms of a static $S = \frac{1}{2}$ spin Hamiltonian and the measured components of the magnetic hyperfine tensor parallel and perpendicular to the C_4 axis of the crystal are -831 ± 4 and -4 ± 5 MHz. This was the first report of the measurement of a perpendicular

[544] C. P. Lichtenwalner, H. J. Guggenheim, and L. H. Pfeiffer, *Phys. Letters* (*A*), 1976, **56**, 117.
[545] E. Baggio-Saitovitch, F. J. Litterst, and H. Micklitz, See ref. 1.
[546] G. Wortmann, J. Moser, and U. F. Klein, *Phys. Letters* (*A*), 1976, **55**, 486.
[547] M. Meyer, J. M. Friedt, L. Iannarella, and J. Danon, Report, 1975, crn–CNPA–75–15, *Nucl. Sci. Abs.*, 1976, **33**, 20694.
[548] E. R. Bauminger, I. Felner, D. Levron, I. Nowik, and S. Ofer, *Solid State Comm.*, 1976, **18**, 1073.
[549] U. F. Klein, G. Wortmann, and G. M. Kalvius, *Solid State Comm.*, 1976, **18**, 291.
[550] J. N. Ross and J. Sigalas, *J. Phys.* (*F*), 1976, **5**, 1973.
[551] E. R. Bauminger, A. Diamant, I. Nowik, and S. Ofer, see ref. 1.
[552] J. Stoehr, J. D. Cashion, W. Wagner, and L. Asch, *Solid State Comm.*, 1976, **18**, 35.
[553] J. Sivardiere and M. Blume, *Hyperfine Interactions*, 1975, **1**, 283.
[554] D. W. Forester and W. A. Ferrando, *Phys. Rev.* (*B*), 1976, **13**, 3991.
[555] D. W. Forester and W. A. Ferrando, *Rep. N.R.L. Prog.* 1975, 30.

component in Mössbauer studies of Dy compounds. The ratio of the quadrupole parameters for the excited and ground states of ^{161}Dy was given as 0.986 ± 0.003. For temperatures above 10 K the spectra of $DyPO_4$ are relaxation broadened and were described in terms of a time-dependent Hamiltonian. The spectra were analysed using a random fluctuation stochastic model. The analysis then depends upon a single relaxation parameter $\Omega(T)$. Between $T = 16$ and $T = 36$ K the relaxation is said to be dominated by an Orbach spin-relaxation process of the form $\Omega = \Omega_0 \exp(-\Delta G T^{-1})$ where $\Delta G \approx 70$ cm^{-1}. In $DyFeO_3$ the ^{161}Dy

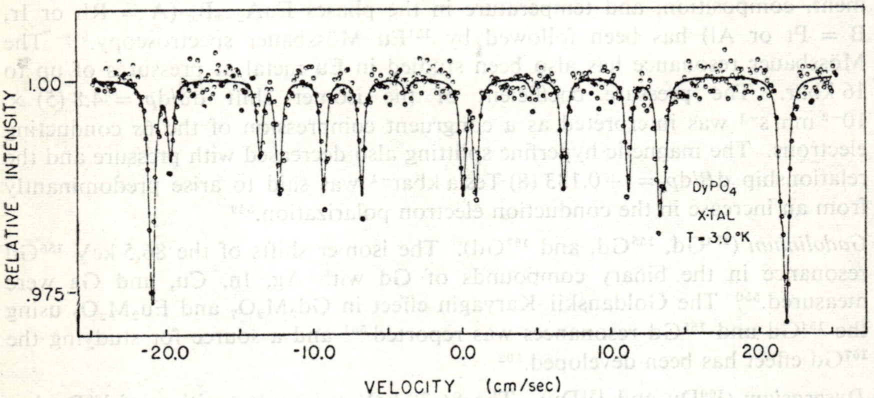

Figure 16 *Mössbauer spectrum of* ^{161}Dy *in a single crystal of* $DyPO_4$ *at* 3 K. *The direction of the gamma-ray propagation was perpendicular to the tetragonal* c*-axis* (Reproduced with permission from *Phys. Rev.* (*B*), 1976, 13, 3991)

resonance has been recorded [556] at 300, 4.2, and 1.6 K on single-crystal platelets perpendicular to the (100), (010), and (001) directions. A significant difference was observed between the spectra corresponding to (100) and (010) platelets so that a helicoidal structure is unlikely. The hyperfine interactions at ^{161}Dy in Gd_2O_3 and Gd metal have been observed.[527]

In a study of rare-earth impurities implanted in iron and nickel, data for ^{161}Dy, ^{166}Er, and ^{169}Tm are presented.[557] In almost all cases about one-half of the impurities end up in substitutional sites and the rest are probably associated with vacancies. The Mössbauer spectrum of ^{161}Dy in Ni, for example, consists of two well-defined components with relative abundance 40 : 60. The less abundant component has the smaller hyperfine field and quadrupole interaction and this is said to result from Dy in substitutional sites where these parameters would be reduced by cubic crystal-field effects. The fact that the spectrum of this component decreases in intensity on annealing is also consistent with it being due to substitutional Dy atoms. ^{161}Dy spectra have also been used [558] to show that the Dy moments in amorphous $DyCo_{3.4}$ are strongly coupled to the local crystal field axis. The magnetic structure is one in which the Co sublattice

556 M. Belakhovsky, M. Borge, J. Chappert, and J. Sivardiere, *Solid State Comm.*, 1976, **20**, 473.
557 L. Niesen, *Hyperfine Interactions*, 1976, **2**, 15.
558 J. M. D. Coey, J. Chappert, J. P. Rebouillat, and T. S. Wang, *Phys. Rev. Letters*, 1976, **36**, 1061.

is strongly ferromagnetic and the Dy moments are almost randomly distributed in directions between antiparallel and perpendicular to the cobalt.

Erbium (^{166}Er). Data for ^{166}Er implanted in iron and nickel [557] (see the dysprosium subsection) have been reported. The influence of relaxation on paramagnetic Mössbauer spectra has been described by the stochastic theory and calculations have been given for ^{166}Er in a cubic environment.[553]

Thulium (^{169}Tm). Accurate values for hyperfine interaction parameters were obtained for Tm metal by using an implanted ^{169}Er in Al source and Moiré velocity calibration.[559] The magnetic moment ratio was found to be 2.223 (13), the ground-state magnetic coupling was 33.26 (9) cm s^{-1} and an excited-state quadrupole coupling strength, e^2qQ/h = 1933 (12) MHz was found. In discussing the assumptions made in considering the influence of chemical environment on conversion electron spectra, results for ^{169}Tm have been mentioned.[382] Data for ^{169}Tm implanted in iron and nickel have been reported [557] (see the dysprosium subsection).

Ytterbium (^{170}Yb and ^{171}Yb). The method [63] of calculating Mössbauer paramagnetic relaxation spectra without the 'white noise approximation' (see p. 365) has been applied to ^{170}Yb in $CsNaYbCl_6$ which contains Yb in a strictly O_h site. The Hirst theory for the line shape of Mössbauer spectra (see p. 365) has been applied [64] to the $\frac{1}{2} \rightarrow \frac{3}{2}$ transition of ^{171}Yb also in $CsNaYbCl_6$.

Neptunium (^{237}Np). Literature data for 47 Np compounds have been used to construct a correlation diagram between isomer-shift values and the oxidation state of Np.[560]

The Mössbauer spectra of ^{237}Np in NpH_{2+x} (x = 0.1, 0.3, and 0.5) consist [561] of several absorption lines which have been attributed to the presence of non-equivalent Np sites with different numbers of hydrogen neighbours. The relative intensities of the lines do not, however, correspond to random distribution of hydrogen. Isomer-shift values increased with x showing that electrons were donated to hydrogen and that the compound was hydridic.

Possible structures have been suggested for the compounds $M_3[NpO_4(OH)_2]$ (M = Li and Na) as a result of Mössbauer studies.[562] Mössbauer effect measurements in ^{237}Np have been carried out using the β-decay of ^{237}U.[563] The isotopic purity of the UO_2 used for the source was such that the material could be irradiated and run directly. It was found that this β-decay source did not provide a significantly narrower line than had been achieved with α-decay sources. Below the UO_2 antiferromagnetic transition a small line broadening was observed indicating only a very small moment on the Np ions in the ordered state. Calculations of the s-electron densities at Np in NpF_6^{3-}, NpF_6^{2-}, NpF_6^{-}, NpF_6, NpO_2^{+}, NpO_2^{2+}, and NpO_2^{3+} and based on $5f$, $6d$, and $7s$ occupations have been carried out.[564] The spectrum of NpF_6 has also been used to illustrate a method

559 H. P. Wit and L. Niesen, *Hyperfine Interactions*, 1976, **1**, 501.
560 A. A. Chaikhorskii, *Radiokhimiya*, 1975, **17**, 910.
561 M. H. Mintz, J. Gal, Z. Hadari, and M. Bixon, *J. Phys. Chem. Solids*, 1976, **37**, 1.
562 K. V. Ilyatov, M. Yu. Matuzenko, L. M. Krizhanskii, and A. A. Chaikhorskii, *Radiokhimiya*, 1975, **17**, 905.
563 R. D. Meeker, B. D. Dunlap, and D. Cohen, *J. Phys. Chem. Solids*, 1976, **37**, 551.
564 V. S. Nefedov, *Koord. Khim.*, 1976, **2**, 1120.

of calculation of line-shape for relaxed Mössbauer spectra.[66] The magnetic properties of $NpBr_4$ were studied from 4.2 to 20 K by Mössbauer and magnetic susceptibility measurements.[565] Within experimental error both methods gave the same antiferromagnetic transition temperature (12.5 ± 0.5 K). The Mössbauer spectra (Figure 17) showed magnetic hyperfine splitting over the

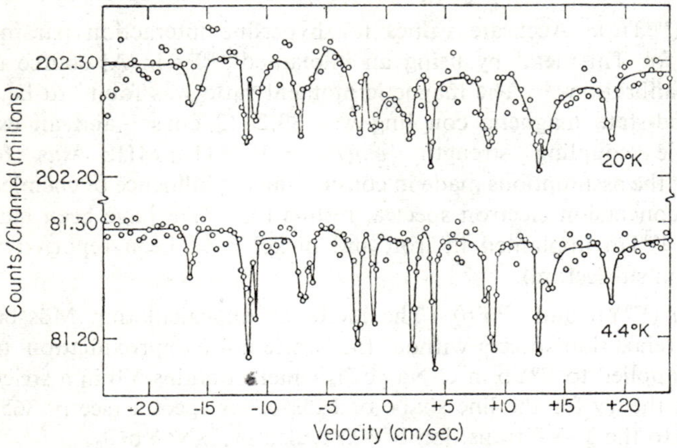

Figure 17 *Mössbauer spectra of* $NpBr_4$ *above and below the Néel temperature. The hyperfine splitting is nearly identical due to paramagnetic relaxation effects. Broadening of the lines is apparent at the higher temperature*
(Reproduced with permission from *J. Chem. Phys.*, 1976, **64**, 1492)

temperature range with $g_0\mu_n H_{eff}$ = 10.4 cm s^{-1} at 4.2, a small drop in this value was found for temperatures above 12 K.

$(C_5H_5)_3NpC_4H_9$ was synthesized by the reaction of n-butyl-Li with $(C_5H_5)_3NpCl$. The Mössbauer spectrum of the product showed the presence of $(C_5H_5)_4Np$, presumably a by-product in the preparation, and two lines broadened by paramagnetic relaxation effects.[566] These lines were interpreted as the magnetically split ($g_0\mu_n H_{eff} \approx$ 5.8 cm s^{-1}) spectrum of $(C_5H_5)_3Np^nC_4H_9$. The isomer shift of 0.27 cm s^{-1} indicated a smaller covalency than in $(C_5H_5)_4Np$ (δ = 0.72 cm s^{-1}) consistent with the σ-character of the $^nC_4H_9$—Np bond. Attempts to prepare $(C_5H_5)_3NpC_6H_5$ were unsuccessful, the products isolated and identified by their Np spectra were $(C_5H_5)_4Np$ and $(C_5H_5)_3Np$,THF.

Phases of a number of alloys have been studied using the ^{237}Np Mössbauer effect. These included $NpSn_3$ [392] and cubic $NpPd_{3-x}Rh_x$ (x = 0—3) in which an extremely large hyperfine field (4.8 MOe) was observed at 4.2 K.[567] The spectra of the compounds $NpMnSi_2$ and of $NpME_2$ (M = Cr, Fe, Co, Ni, and Cu; E = Si and Ge) have been measured.[568, 569] The Si compounds [568] show a well-defined single magnetic hyperfine field that decreases with temperature, but only

565 J. L. Stevens, E. R. Jones, and D. G. Karraker, *J. Chem. Phys.*, 1976, **64**, 1492.
566 D. G. Karraker and J. A. Stone, in 'Abstracts 172nd ACS National Meeting,' San Francisco, Aug. 29–Sept., 3, 1976.
567 J. Gal, M. Kroupp, and Z. Hadari, *Solid State Comm.*, 1976, **19**, 73.
568 J. Gal, M. Kroupp, and Z. Hadari, *Solid State Comm.*, 1976, **20**, 421.
569 J. Gal, M. Kroupp, Z. Hadari, and I. Nowik, *Solid State Comm.*, 1976, **20**, 515.

$NpCuSi_2$ has a first-order magnetic phase transition (at $T_N = 34$ K). The Ge compounds[569] all order ferromagnetically and have Néel temperatures in the range 28—62 K.

Mössbauer emission spectra have been observed[570] for ^{241}Am sources in ThO_2 with a NpO_2 absorber at temperatures between 77 and 223 K. The emission process resulted in the formation of both Np^{IV} and Np^{V}.

Americium (^{243}Am). Isomer shifts, linewidths, and resonance intensities of several Am compounds have been measured with various ^{243}Pu sources.[571] The pentavalent Am compounds, $KAmO_2CO_3$ and $KAmO_2F_2$, gave five line absorption spectra attributable to electric hyperfine interaction. A complex spectrum interpreted in terms of magnetic hyperfine interaction was also observed for the $PuFe_2$–AmO_2 system. The isomer shifts obtained for Am compounds (relative to a standard PuO_2–AmO_2 system) were $AmF_3 = +52$, $Am_2(C_2O_4)_3 = +49$, AmEDTA = +48, $AmO_2 = 0$, $KAmO_2F_2 = -2$, and $KAmO_2CO_3 = -8$ mm s^{-1}. The differences in shift with oxidation were attributed to increases in apparent *s*-electron density caused by removal of 5*f*-electrons.

7 Bibliography

The following is a list of references on Mössbauer spectroscopy that have not been mentioned in the main text The references are ordered alphabetically with respect to the first author.

N. E. Alekseevskii and A. P. Dodokin. Study of the Mössbauer effect on iron-57 and tin-119 in Nb–Sn–Fe, V–Sn–Fe, and V–Si–Fe compounds, *Doklady Akad. Nauk SSSR*, 1976, **226**, 609.

C. Bansal and G. Chandra. Mössbauer studies in disordered $(NiFe)_{1-x}Cr_x$, *Solid State Comm.*, 1976, **19**, 107.

C. Bansal, J. Ray, and C. Chandra. Effect of environment on hyperfine and exchange fields of iron-57 atoms in disordered $Fe_{0.48}Ni_{0.52}$ alloy studied by the Mössbauer effect, *Proc. Nucl. Phys. Solid State Symp.*, 1974, **17C**, 369.

Sh. Sh. Bashkirov, G. D. Kurbatov, E. S. Makhnev, and V. A. Christyakov. Use of the Mössbauer effect for measuring the parameters of short range interactions in solid solution, *Doklady Akad. Nauk SSSR*, 1975, **223**, 622.

P. Brill and G. Wortmann. Hyperfine fields and Mössbauer isomer shifts in PdAgFe alloys, *Phys. Letters A*, 1976, **56**, 477.

S. J. Campbell, P. E. Clark, and T. J. Hicks. Mössbauer study of the magnetic properties of CuFe alloys, *J. Phys. (F)*, 1976, **6**, 249.

G. Chandra, J. Ray and C. Bansal. A Mössbauer study of very dilute iron impurities in Ni–Mn systems, 'Proceedings of the 14th International Conference on Low Temperature Physics, ed. M. Krusius and M. Vuorio, North Holland, Amsterdam, 1975, **3**, 358.

A. J. Dekhtyar and P. S. Nizin. Mössbauer effect in iron-57 nuclei in deformed Mo single crystals, *Soviet Phys. Solid State*, 1976, **18**, 854.

S. G. Denner and R. D. Jones. The use of transmission iron-57 Mössbauer spectroscopy to study kinetics of the hot dip aluminizing of iron, *J. Material Sci.*, 1976, **11**, 1777.

S. M. Dubiel. Mössbauer effect study of the magnetic hyperfine fields at iron-57 nuclei in Fe–Cr (1–15 atoms %) alloys, *Acta Phys. Pol.*, 1976, **49A**, 619.

J. Dudas and T. Zemcik. Mössbauer study of the magnetic transformation of the Ni_3Al–Co-57+Fe-57 surface layer, *Cesk. Casopis Fyz.*, 1975, **25**, 476.

G. A. Fatseas. Some remarks on a recent Mössbauer effect study of the *β*-phase iron germanide, *Canad. J. Phys.*, 1976, **54**, 1850.

W. Fischer and R. N. Kuz'min. Magnetic hyperfine fields of iron-57 in inter-metallic compounds of heavy rare-earth (RE)–iron alloys of the type $REFe_3$, *Magn. Eigenschaften Festkoerpern (Vortr. Metalltag. DDR.)*, 10th, 1974, 143.

[570] V. M. Filin, V. I. Goldanskii, and V. S. Nefedov, *Radiokhimiya*, 1976, **18**, 166.
[571] D. D. Bohde, J. F. Wild, and E. K. Hulet, *J. Inorg. Nuclear Chem.*, 1976, **38**, 1291.

D. W. Forester. Mössbauer studies of amorphous rare-earth (RE)–iron alloys $REFe_2$, *AIP Conf. Proc.*, 1976, **31**, 384.

V. V. Gorbachev, V. A. Linskii, and I. M. Putilin. Mössbauer effect and band structure of some semiconductor compounds, *V. sb. Khim. Svyaz. u. Kris. i. ikh. Fiz Svoistva.*, 1976, 117.

K. Hanisch and M. Drosg. On the width of the central peak in the critical scattering of γ-rays by $KMnF_3$, *Phys. Letters* (*A*), 1976, **58**, 415.

I. R. Harris and G. Longworth. *X*-ray and Mössbauer studies of the pseudo-binary system Ce–Fe–Ni, *J. Less-Common Metals*, 1976, **45**, 63.

R. Hasegawa and C. L. Chien. Mössbauer effects in iron-rich glassy alloys, *Solid State Comm.*, 1976, **18**, 913.

A. S. Ilyushin and W. E. Wallace. Magnetic and structural studies of rare-earth–iron–manganese ternary phases (part I), *J. Solid State Chem.*, 1976, **17**, 131.

A. S. Ilyushin and W. E. Wallace. Magnetic and structural studies of rare-earth–iron–manganese ternary phases (part II), *J. Solid State Chem.*, 1976, **17**, 373.

V. G. Jadhao, G. N. Rao, R. M. Singru, G. Bahadur, and C. N. R. Rao. Mössbauer spectroscopic studies of $Eu^{III}Co^{III}$ oxide using Fe-57 and Eu-151 gamma rays, 'Proceedings in Nuclear Physics and Solid State Physics Symposium, 1974, **17C**, 361.

C. Janot and P. Delcroix. Mössbauer study of metallic alloys under mechanical stress, *Acta Metallurgica*, 1976, **24**, 197.

M. Kanashiro and N. Kunitomi. Iron-57 Mössbauer spectrum of partially ordered Ni_3Fe, *Solid State Comm.*, 1976, **19**, 1127.

V. Kumeishin and O. A. Ivanov. Studies of relaxation processes in nickel near the Curie point using nuclear gamma resonance, *Fiz. Metal. Metalloved.*, 1975, **40**, 1295.

M. M. L. Kwan and R. W. Hoffman. Ferromagnetism in thin amorphous cobalt films, Technical Report No. 80, Case West Reserve University Reports 1974, C00–623–200 from ERDA Research Abstracts, 1976, 1416.

R. A. Levy and J. A. Rayne. Spin behaviour in the Pd–Ag–Fe system, 'Proceedings of the 14th International Conference on Low Temperature Physics, ed. M. Krusius and M. Vuorio, North Holland, Amsterdam, 1975, **3**, 202.

W. Lin, H. Spalt, and B. W. Batterman. Study of the ω-phase in Zr–Nb alloys by Mössbauer spectroscopy and *X*-ray diffuse scattering, *Phys. Rev.* (*B*), 1976, **13**, 5158.

R. Lindsey. Measurement of Mössbauer diffusion broadening in iron stabilized with vanadium, *Phys. Status Solidi* (*B*), 1976, **75**, 583.

J. P. Mathieu. Chemical effects related to the (*n*,γ) reaction and electron capture dehydration of $BaCl_2$, H_2O, Report FRNC–TH–592, 1975.

N. M. Matveeva, A. A. Kolyada, R. N. Kuzmin, and I. M. Sharshakov. Structural transformation in Fe (30 atom %)–Ni alloy, *Metalloved. i Term. Obrabotka Metal.*, 1975, 15.

R. I. Mints and V. A. Semenkin. Effect of plastic deformation on the probability of the Mössbauer effect in multiphase Fe–Mn alloys, *Izvest. Vysshikh Uchebn. Zavedenii Chem. Metal.*, 1976, 107.

I. N. Nikolaev, L. S. Pavlyukov, V. P. Potapov, and V. P. Marin. Influence of pressure on isomer shift and effective magnetic fields at iron-57 in the $Pt_{0.72}Fe_{0.28}$ alloy, *Soviet Phys. Solid State*, 1976, **18**, 692.

K. R. P. M. Rao and P. K. Iyengar. Mössbauer spectroscopic study of short range ordering in iron-57 doped Co–Ga intermetallic compounds, Proceedings in Nuclear Physics and Solid State Physics Symposium, 1974, **17C**, 361.

J. Ray and C. Chandra. Iron-57 Mössbauer studies in a single crystal of $Ni_{0.83}Fe_{0.17}$, *Solid State Comm.*, 1976, **19**, 101.

J. Ray, C. Chandra, and C. Bansal. Distribution of iron-57 hyperfine fields in Ni–Mn alloys–a Mössbauer effect study, *J. Phys.* (*F*), 1976, **6**, 265.

J. D. Rush, C. E. Johnson, and M. F. Thomas. A Mössbauer study of magnetic order in iron–ruthenium alloys, *J. Phys.* (*F*), 1976, **6**, 2017.

M. Shiga and Y. Nakamura. Mössbauer study of Fe–Al alloys near the critical concentration, *J. Phys. Soc. Japan*, 1976, **40**, 1295.

E. W. Skluzacek. Analysis of the Mössbauer spectra of $NdCo_5$, United States NTIS AD Report 1976, AD–AO27288.

M. Stupel, M. Ron, and B. Z. Weiss. Phase identification in Ti-rich Ti–Fe systems by Mössbauer spectroscopy, *J. Appl. Phys.*, 1976, **47**, 6.

K. Sumiyama, M. Shiga, K. Tachi, and Y. Nakamura. Iron-57 Mössbauer and platinum-195 studies of the ordered and disordered states of $Fe_{72}Pt_{28}$ Invar alloy, *J. Phys. Soc. Japan*, 1976, **40**, 1002.

S. C. Tsai, K. S. V. L. Narasimhan, C. J. Kunish, and R. A. Butera. Mössbauer and crystallographic studies of $DyFe_{3-x}Ni_x$ compounds, 'Proceedings of the 10th Rare Earth Research Conference', ed. C. J. Kevane, 1973, **2** (*Conf*–730402–*P2*), p. 1107.

M. Vano, J. Sitek, and J. Cirak. A contribution to the investigation of the properties of the Au–Fe system, *Acta Phys. Slovaca.*, 1976, **26**, 201.

P. O. Voznyuk, I. M. V'yunnik, and V. N. Dubinin. Nuclear gamma resonance study steel irradiated with gamma quanta and fast neutrons, *Tezisy Dokl. Resp. Soveshch.*, 'Radiats. Pourezhdeniva Tverd. Telakh.', 1974, **1**, 125.

I. M. V'yunnik. Nuclear gamma resonance study of carbide phases formed as a result of tempering of steel under gamma radiation, *Fiz. Metal. Metalloved.*, 1975, **39**, 1284.

J. B. Ward. Iron-57 and Cobalt-57 Mössbauer study of vacancy impurity bonding in aluminium, *Scr. Metall.*, 1975, **9**,1211.

J. M. Williams, J. B. Dunlap, and B. Leontic. Iron-57 Mössbauer measurements on Fe doped Zn alloys, *Mater. Sci. Eng.*, 1976, **23**, 307.

B. Window. Properties of magnetically dilute alloys, *J. Magnetism and Magnetic Materials*, 1975, **1**, 167.

G. Wortmann and D. L. Williamson. Electric field gradient at Fe in h.c.p. transition metals from Mössbauer measurements, *Hyperfine Interactions*, 1975, **1**, 167.

Author Index